Nanotube Superfiber Materials
Changing Engineering Design

Nanotube Superfiber Materials
Changing Engineering Design

Edited by

Mark J. Schulz

Vesselin N. Shanov

Zhangzhang Yin

AMSTERDAM • BOSTON • HEIDELBERG • LONDON
NEW YORK • OXFORD • PARIS • SAN DIEGO
SAN FRANCISCO • SINGAPORE • SYDNEY • TOKYO

William Andrew is an imprint of Elsevier

William Andrew is an imprint of Elsevier
The Boulevard, Langford Lane, Kidlington, Oxford OX5 1GB, UK
225 Wyman Street, Waltham, MA 02451, USA

First edition 2014

Copyright © 2014 Elsevier Inc. All rights reserved.

No part of this publication may be reproduced, stored in a retrieval system or transmitted in any form or by any means electronic, mechanical, photocopying, recording or otherwise without the prior written permission of the publisher

Permissions may be sought directly from Elsevier's Science & Technology Rights Department in Oxford, UK: phone (+44) (0) 1865 843830; fax (+44) (0) 1865 853333; email: permissions@elsevier.com. Alternatively you can submit your request online by visiting the Elsevier web site at http://elsevier.com/locate/permissions, and selecting *Obtaining permission to use Elsevier material*

Notice

No responsibility is assumed by the publisher for any injury and/or damage to persons or property as a matter of products liability, negligence or otherwise, or from any use or operation of any methods, products, instructions or ideas contained in the material herein. Because of rapid advances in the medical sciences, in particular, independent verification of diagnoses and drug dosages should be made

British Library Cataloguing in Publication Data
A catalogue record for this book is available from the British Library

Library of Congress Cataloging-in-Publication Data
A catalog record for this book is availabe from the Library of Congress

ISBN–13: 978-1-4557-7863-8

For information on all William Andrew publications
visit our web site at store.elsevier.com

Printed and bound in the US

13 14 15 16 17   10 9 8 7 6 5 4 3 2

Contents

Preface xvii
Acknowledgment xxiii
Editor Biographies xxv

CHAPTER 1 Introduction to Fiber Materials 1
    1.1 Fibers and Nanofibers 1
        1.1.1 The nature of fiber 1
        1.1.2 The nature of nanofiber 2
        1.1.3 Carbon nanotubes, the archetypal nanofiber
             and *bête noire* of early industrial furnaces 3
        1.1.4 Carbon nanotube structure at first approximation—the
             nanoscopic begets the macroscopic 4
        1.1.5 CNT structure in more detail 8
        1.1.6 A note on carbon nanofibers 10
        1.1.7 Perturbations of the CNT structure 11
        1.1.8 Carbon nanotube synthesis 12
    1.2 The Challenge of CNT Yarn Fiber Fabrication 14
        1.2.1 A brief diversion into conventional yarn fiber
             fabrication 16
        1.2.2 CNT yarn fiber production 17
        1.2.3 On the physical properties of CNT yarn 20
    1.3 Conclusion 25
        References 26

**CHAPTER 2 New Applications and Techniques for Nanotube
Superfiber Development 33**
    2.1 New Applications for Nanotube Superfiber Development 33
        2.1.1 Sustainable nanotube materials for electromagnetics
             applications 35
        2.1.2 Nanoelectric motor design 37
        2.1.3 Nanovivo robots to change interventional medicine 39
        2.1.4 Nanodevices made from one long CNT 42
        2.1.5 Biomedical applications of nanotube thread 43
    2.2 New Techniques for Nanotube Superfiber Development 43
        2.2.1 Reactor design for growing carbon nanotubes 44
        2.2.2 Thermal annealing of carbon nanotubes 45
        2.2.3 Substrate patterning for nanotube growth control 47
        2.2.4 Spinning carbon nanotube thread 53
    2.3 Conclusions 56
        Acknowledgments 57
        References 57

CHAPTER 3 Tailoring the Mechanical Properties of Carbon Nanotube Fibers 61

- 3.1 Introduction 61
- 3.2 Irradiation Cross-Linking: Strong and Stiff CNTs and CNT Bundles 64
- 3.3 Reformable Bonding: Strong and Tough CNT Bundles and Fibers 68
- 3.4 Materials Design: Optimized Geometry and Structure 72
- 3.5 Summary 79
- Acknowledgments 80
- References 80

CHAPTER 4 Synthesis and Properties of Ultralong Carbon Nanotubes 87

- 4.1 Introduction 87
  - 4.1.1 Structure and growth mechanism of ultralong CNTs 88
- 4.2 Synthesis of Ultralong CNTs by CVD 101
  - 4.2.1 Catalysts and CVD process 101
  - 4.2.2 Carbon sources 102
  - 4.2.3 Substrates 102
- 4.3 Tuning the Structure of Ultralong CNTs 102
  - 4.3.1 Tuning the number of walls and diameters of ultralong CNTs 104
  - 4.3.2 Tuning the chiral consistency of ultralong CNTs 106
  - 4.3.3 Tuning the electrical properties of ultralong CNTs 108
  - 4.3.4 Tuning the morphology of ultralong CNTs 113
  - 4.3.5 Mechanical properties of ultralong CNTs 122
  - 4.3.6 Potential applications of ultralong CNTs 127
- 4.4 Conclusions 129
- References 130

CHAPTER 5 Alloy Hybrid Carbon Nanotube Yarn for Multifunctionality 137

- 5.1 Introduction 137
  - 5.1.1 Interface between metal nanoparticles and CNTs 138
- 5.2 Electrical Conductivity of CNT Yarns 140
- 5.3 Metal Deposition on CNT Macrostructures 142
  - 5.3.1 Metal nanoparticle deposition on CNT yarns 143
- 5.4 Gas Sensing Applications 150
  - 5.4.1 Sensing mechanisms 152
  - 5.4.2 Gas sensing using CNT macrostructures 154
- 5.5 Summary 161
- References 161

CHAPTER 6 Wet Spinning of CNT-based Fibers 167
6.1 Introduction to Wet Spinning 167
6.2 Fibers Obtained from the Coagulation of Carbon Nanotubes 170
    6.2.1 Direct coagulation in bad solvents 170
    6.2.2 Polymer-induced coagulation 173
6.3 Fibers Obtained from the Coagulation of CNT−Polymer Mixtures 185
    6.3.1 High-performance synthetic polymers 186
    6.3.2 CNT−conductive polymer composite fibers 195
    6.3.3 CNT−natural polymer composite fibers 198
6.4 Conclusions 201
    References 202

CHAPTER 7 Dry Spinning Carbon Nanotubes into Continuous Yarn: Progress, Processing and Applications 211
7.1 Introduction 212
    7.1.1 Wet spinning 212
    7.1.2 Direct spinning of CNT aerogel 213
    7.1.3 Dry spinning 213
7.2 Basis of CNT Assembly in Macroscopic Structures 213
    7.2.1 Interrelations of CNTs in macroscopic structures 214
    7.2.2 Potential energy function for the interaction between CNTs 215
7.3 From Textile Spinning Technology to Dry CNT Spinning 216
    7.3.1 CNT forest/matrix 217
    7.3.2 CNT web: formation and model 217
    7.3.3 Aligning and tensioning system 219
    7.3.4 Packing fibers in CNT yarn 219
7.4 Multistep Spinning Process Using a Drafting System 222
    7.4.1 Separating a CNT web using an electrostatic approach 222
    7.4.2 Multistep CNT dry spinning process and results 224
7.5 Several Treatments for CNT Yarn Improvement 225
    7.5.1 Compaction of a CNT yarn 225
    7.5.2 Heat treatment of a web 226
    7.5.3 Improving the strength of CNT yarn using functionalization 226
    7.5.4 Multiply twisted yarns 227
7.6 CNT-Based Composite Yarns 227
    7.6.1 CNT−polymer composite yarns 227
    7.6.2 Biscrolling CNT web and particles into yarn 230
    7.6.3 CNT−metal composite yarns 230

7.7 Applications of CNT Yarns 231
    7.7.1 Textile electrodes and supercapacitors 232
    7.7.2 CNT yarns for actuators...232
    7.7.3 CNT yarn sensors..232
    7.7.4 CNT yarns in bioengineering...233
    7.7.5 CNT-based high-performance yarns..............................234
7.8 Conclusion...234
Acknowledgments..234
References...235

CHAPTER 8 Synthesis and Properties of Boron Nitride Nanotubes 243

8.1 Introduction..243
8.2 Nanotubes: Basic Structure 246
8.3 Synthesis of BNNTs 249
    8.3.1 Arc discharge 249
    8.3.2 Ball milling 249
    8.3.3 Carbothermal synthesis..250
    8.3.4 Chemical vapor deposition 251
    8.3.5 Laser heating and ablation...258
8.4 Properties of Boron Nitride Nanotubes 259
    8.4.1 Electrical properties 259
    8.4.2 Mechanical properties..259
    8.4.3 Optical properties...260
    8.4.4 Thermal properties, thermal stability, and wetting behavior...260
8.5 Comparison of BNNTs and CNTs 261
8.6 Summary 263
Acknowledgments..263
References...263

CHAPTER 9 Boron Nitride Nanotubes, Silicon Carbide Nanotubes, and Carbon Nanotubes—A Comparison of Properties and Applications 267

9.1 Introduction..267
9.2 BNNT and SiCNT Structure and Synthesis 268
    9.2.1 Structure 268
    9.2.2 Synthesis..270
    9.2.3 Property comparison of SiCNT and BNNT 274
9.3 Composites Reinforced with High-Temperature Nanotubes...276
    9.3.1 BNNT composites..276
    9.3.2 SiCNT composites 281

9.4 Applications of High-Temperature Nanotubes 281
9.5 Concluding Remarks 285
References 285

CHAPTER 10 Carbon Nanotube Fiber Doping 289
10.1 Introduction 289
10.2 Doping 290
    10.2.1 Types of doping 291
    10.2.2 Functionalization 296
10.3 Single-Walled Carbon Nanotube Doping 296
    10.3.1 Redox potential 297
    10.3.2 Work functions 298
10.4 Multiwalled Carbon Nanotube Doping 299
10.5 Characterization of Doped CNTs 300
    10.5.1 Electrical conductivity of doped MWCNT fibers 300
    10.5.2 Raman, XPS, ^{13}C-nuclear magnetic resonance 301
10.6 Experimental Challenges in Characterization 303
    10.6.1 Electrical contact 303
    10.6.2 Diameter measurements 304
10.7 Summary 305
Acknowledgments 306
References 306

CHAPTER 11 Carbon Nanofiber Multifunctional Mat 313
11.1 Introduction 313
11.2 Development of Carbon Nanofiber Mat 315
    11.2.1 Carbon nanofiber 315
    11.2.2 Carbon nanofiber mat 317
    11.2.3 Fabrication of CNF mat-reinforced composites 321
    11.2.4 Short beam shear testing 322
11.3 Conclusion 328
Acknowledgments 328
References 328

CHAPTER 12 Direct Synthesis of Long Nanotube Yarns for Commercial Fiber Products 333
12.1 Introduction 333
12.2 Direct Synthesis of Long CNT Yarns 335
12.3 Growth of High-Quality CNTs 340
12.4 Applications of CNT Yarns/Fibers 341
12.5 Conclusions 345
Acknowledgments 345
References 346

CHAPTER 13 Carbon Nanotube Sheet: Processing, Characterization and Applications 349

- 13.1 Introduction 350
- 13.2 Two-Dimensional Films, "Buckypapers" and Sheets of Carbon Nanotubes 351
  - 13.2.1 "Buckypaper" and thin films with limited/random nanotube alignment 351
  - 13.2.2 Dry CNT sheets pulled from aligned arrays of MWNTs grown via CVD 352
  - 13.2.3 Alternative techniques for CNT sheet production 353
- 13.3 Functionalization and Characterization of CNT Sheets 354
  - 13.3.1 Techniques for functionalization of CNTs 355
- 13.4 CNT Sheet Products Manufacturing 361
  - 13.4.1 CNT sheet manufacturing 361
  - 13.4.2 Atmospheric pressure plasma functionalization 362
  - 13.4.3 Characterization 363
  - 13.4.4 Applications of CNT sheet 373
- 13.5 Conclusions and Future Work 382
- Acknowledgments 382
- References 382

CHAPTER 14 Direct Dry Spinning of Millimeter-long Carbon Nanotube Arrays for Aligned Sheet and Yarn 389

- 14.1 Introduction 390
- 14.2 Highly Spinnable MWCNT Arrays 391
  - 14.2.1 Growth of millimeter-long spinnable MWCNT arrays 391
  - 14.2.2 Characterization of MWCNT arrays 392
  - 14.2.3 Dry spinning of webs from millimeter-long CNT arrays 393
- 14.3 Unidirectionally Aligned CNT Sheet 394
  - 14.3.1 Fabrication of unidirectionally aligned CNT sheet 394
  - 14.3.2 Electrical properties 395
  - 14.3.3 Mechanical properties 396
  - 14.3.4 Thermal properties 397
  - 14.3.5 Discussions 398
- 14.4 Mechanical Properties of CNT Yarn 399
  - 14.4.1 Fabrication of CNT yarn by dry spinning 399
  - 14.4.2 Yarn evaluation methods 400
  - 14.4.3 Mechanical property of a CNT web 401
  - 14.4.4 Postspin twisting of as-spun yarns 402
  - 14.4.5 Multiply twisted CNT yarns 407
  - 14.4.6 Pressed CNT yarns 409
  - 14.4.7 Discussions 410

14.5	Conclusions	411
	Acknowledgments	412
	References	412

CHAPTER 15 Transport Mechanisms in Metallic and Semiconducting Single-walled Carbon Nanotubes: Cross-over from Weak Localization to Hopping Conduction ... 415

15.1	Introduction	415
15.2	Relationship between MS Ratio and Conductivity of SWCNT Networks	416
15.3	Summary	422
	References	423

CHAPTER 16 Thermal Conductivity of Nanotube Assemblies and Superfiber Materials ... 425

16.1	Introduction	425
16.2	Thermal Conductivity and Measurement Issues for CNT Materials	426
16.3	Individual Carbon Nanotubes	430
16.4	Carbon Nanotube Bundles	432
16.5	Carbon Nanotube Composites	435
16.6	CNT Buckypaper and Thin Films	437
16.7	CNT Superfiber Materials	441
	16.7.1 CNT arrays	441
	16.7.2 CNT sheets and yarns	445
16.8	Boron Nitride Nanotubes	449
16.9	Challenges and Opportunities	449
	Acknowledgments	451
	References	451

CHAPTER 17 Three-dimensional Nanotube Networks and a New Horizon of Applications ... 457

17.1	Introduction	458
17.2	Nanotube Network Types	459
	17.2.1 Covalent CNT junctions	460
	17.2.2 van der Waals or noncovalent CNT junctions	461
17.3	Theoretical Studies	461
	17.3.1 The simplest nanotube junction: bent nanotube by 5-7 pairs	461
	17.3.2 Schwarzites	461
	17.3.3 CNT junctions and networks: structure, properties and transport	462

17.4 Synthesis of CNT Networks 468
    17.4.1 Template approaches 469
    17.4.2 Synthesis of van der Waals or noncovalent CNT networks 470
    17.4.3 Electron beam irradiation at high temperatures 472
    17.4.4 Secondary growth approaches 473
    17.4.5 The role of sulfur during CNT synthesis 475
    17.4.6 Carbon nanotube sponges 477
17.5 Applications 479
    17.5.1 Electronic applications: memory devices, transparent electrodes, transistors, sensors 479
    17.5.2 Bioapplications: biosensors, artificial muscles, cell growth scaffolds 483
    17.5.3 Mechanical applications 484
    17.5.4 Oil absorption applications 486
17.6 Perspectives 487
Acknowledgments 487
References 487

CHAPTER 18 A Review on the Design of Superstrong Carbon Nanotube or Graphene Fibers and Composites 495

18.1 Introduction 495
18.2 Hierarchical Simulations and Size Effects 497
18.3 Brittle Fracture 499
18.4 Elastic-Plasticity, Fractal Cracks and Finite Domains 503
18.5 Fatigue 504
18.6 Elasticity 505
18.7 Atomistic Simulations 506
18.8 Nanotensile Tests 507
18.9 Thermodynamic Limit 508
18.10 Sliding Failure 513
18.11 Conclusions 515
References 516

CHAPTER 19 Transition from Tubes to Sheets—A Comparison of the Properties and Applications of Carbon Nanotubes and Graphene 519

19.1 Overview 520
19.2 Electronic Band Structures of Monolayer Graphene and Carbon Nanotubes 520
    19.2.1 Electronic band structure of monolayer graphene 520
    19.2.2 Band structure of SWNTs derived from graphenes 522

  19.2.3 Band structures of graphene nanoribbons 524
19.3 Comparison of Physical Properties and Device Applications between Graphenes and Carbon Nanotubes 526
  19.3.1 Electrical properties 527
  19.3.2 Mechanical properties 541
  19.3.3 Optical and optoelectronic properties 547
19.4 Summary 556
  References 558

CHAPTER 20 Multiscale Modeling of CNT Composites using Molecular Dynamics and the Boundary Element Method 569

20.1 Introduction 570
  20.1.1 Nanocomposites and the challenges in the modeling and simulations 570
  20.1.2 Literature review 571
  20.1.3 A hierarchical multiscale approach to modeling CNT composites 573
20.2 Nanoscale Simulations Using Molecular Dynamics 574
  20.2.1 Basics of molecular dynamics 574
  20.2.2 Developing a cohesive interface model for CNT composites using MD 575
20.3 Microscale Simulations Using the Boundary Element Method 578
  20.3.1 The boundary element method 578
  20.3.2 The fast multipole method 580
20.4 Numerical Examples 581
  20.4.1 Further study of CNT pull-out tests using MD 581
  20.4.2 Fracture analysis of CNT/polymer composites using MD 584
  20.4.3 Large-scale BEM models for CNT composites 584
  20.4.4 Effect of the cohesive interface in modeling CNT/polymer composites 586
20.5 Discussions 589
  Acknowledgments 590
  References 591

CHAPTER 21 Development of Lightweight Sustainable Electric Motors 595

21.1 Electromagnetic Devices with Nanoscale Materials 595

21.2 Electric Motor Development 598
    21.2.1 Rationale for developing electric motors
           using nanoscale materials 598
    21.2.2 Material preparation and technical design..................601
21.3 Conclusions 620
    References 620

CHAPTER 22 Multiscale Laminated Composite Materials 627

22.1 Introduction 627
22.2 Fabrication and Characterization of MWCNT
     Array-Reinforced Laminated Composites................................628
    22.2.1 CNT array synthesis 629
    22.2.2 Transfer of CNT arrays onto prepreg laminae............629
    22.2.3 Consolidation of IM7-CNT array layup 630
    22.2.4 Short beam shear test setup 631
    22.2.5 Three-point bending test setup....................................632
    22.2.6 Iosipescu interlaminar shear test setup 633
22.3 Results and Discussion 634
    22.3.1 Short beam shear test 634
    22.3.2 Three-point bending test 637
    22.3.3 Iosipescu interlaminar shear test.................................641
22.4 Conclusions 645
    References 645

CHAPTER 23 Aligned Carbon Nanotube Composite Prepregs 649

23.1 Introduction 649
23.2 Recent Advances in the Fabrication of Aligned
     Composite Prepregs 652
    23.2.1 Fabricating aligned prepregs from drawable
           superaligned CNT sheets 652
    23.2.2 Fabricating aligned prepregs directly
           from aligned CNT arrays by shear pressing................657
23.3 Mechanical and Physical Properties of CNT
     Composite Prepregs 660
    23.3.1 Effect of CNT alignment and straightness..................660
    23.3.2 Effect of CNT volume fraction....................................662
    23.3.3 Effect of CNT types 662
    23.3.4 Effect of matrix types 663
23.4 Opportunities and Challenges...663
23.5 Conclusions and Outlook..664
    References 665

CHAPTER 24 Embedded Carbon Nanotube Sensor Thread for Structural Health Monitoring and Strain Sensing of Composite Materials 671

- 24.1 Introduction 671
- 24.2 Embedded Sensing Proof of Concept 676
- 24.3 CNT Sensor Thread Performance 680
  - 24.3.1 Sensitivity (gauge factor) 682
  - 24.3.2 Hysteresis 688
  - 24.3.3 Consistency 691
  - 24.3.4 Stability 692
  - 24.3.5 Bandwidth 695
- 24.4 Carbon Nanotube Thread SHM Architectures 699
- 24.5 Areas of Strong Multifunctional Potential 704
- 24.6 Future Work 708
- Acknowledgments 709
- References 709

CHAPTER 25 Tiny Medicine 713

- 25.1 The History of Tiny Machines 714
- 25.2 Nanoscale Materials 718
  - 25.2.1 Carbon nanotube materials 719
  - 25.2.2 Magnetic nanotube and nanowire materials 721
  - 25.2.3 Magnetic nanoparticle materials 724
  - 25.2.4 Superparamagnetic core design and testing 725
- 25.3 A Pilot Microfactory for Nanomedicine Devices 727
- 25.4 Tiny Machines Concepts and Prototype Fabrication 730
  - 25.4.1 Electromagnetic devices using CNT yarn 731
  - 25.4.2 Failure mechanisms for in vivo devices 734
  - 25.4.3 Nanomaterial electric motor 735
  - 25.4.4 Thermal analysis of nanomaterial components 737
  - 25.4.5 Composite electromagnetic material 738
  - 25.4.6 Biodegradability 739
  - 25.4.7 Carbon nanotube wire 740
  - 25.4.8 Telescoping nanotubes 742
  - 25.4.9 Communication with tiny machines 743
- 25.5 Summary and Conclusions 743
- Acknowledgments 744
- References 744

CHAPTER 26 Carbon Nanotube Yarn and Sheet Antennas 749

- 26.1 Introduction 749
- 26.2 Carbon Nanotube Thread Antennas 751

26.2.1 Electromagnetic theory of carbon nanotube thread dipole antenna 752
26.2.2 Carbon nanotube thread conductivity 755
26.2.3 Electromagnetic simulation of carbon nanotube thread dipole antenna current distribution 759
26.3 Carbon Nanotube Sheet Antennas 765
26.3.1 Carbon nanotube sheet fabrication 765
26.3.2 Carbon nanotube sheet patch antenna 765
26.4 Multifunctional Carbon Nanotube Antenna/Gas Sensor 773
26.4.1 Meshed carbon nanotube thread patch antenna/gas sensor 773
26.4.2 Effect of carbon nanotube thread spacing on antenna performance 778
26.4.3 Effect of meshed carbon nanotube thread ground plane and feedline layers on antenna performance 781
26.5 Summary 785
References 785

CHAPTER 27 Energy Storage from Dispersion Forces in Nanotubes... 789
27.1 Introduction 789
27.2 Idealized Parallel-Plate System 791
27.3 Orders of Magnitude 792
27.4 Performance Simulations 797
27.5 Conclusions 800
Acknowledgments 802
References 803

Index 807

Preface

INTRODUCTION TO NANOTUBE MATERIALS

The main audience of this book has already read, or heard, about the exciting advances related to nanotube materials. Nevertheless, it is difficult to have a clear idea of the current status in the field. Researchers must understand the promise, practicality and prospects of nanotube superfiber materials. This includes asking the questions: How do the properties of nanotube fibers really compare with other materials? How do experiments compare with often too optimistic theoretical predictions? How scalable are the production processes? What are the realistic perspectives in the field? All these questions are open and the chapter authors review the main advances in the field and provide realistic perspectives and critical analyses of the results recently achieved in the field. The book offers continuity and a comprehensive coverage of the subject, not available in journal papers.

Nanotube Superfiber Materials refers to a family of forms of nanotube fibrous materials that have a unique suite of functional properties. Different forms of superfiber materials include nanotube arrays, ribbon, scrolls, yarn, braid, fabric, tape, and sheet. These materials have combinations of properties not available in existing materials. Superfiber materials are typically produced by drawing, spinning, plying, or directly depositing nanotubes to form macroscopic or bulk materials. Terminology used to describe nanotubes and their different material forms varies somewhat in the field. Nanotubes or strands of nanotubes may be considered to be a fine fiber or staple fiber, e.g. like cotton, but much smaller in diameter and possibly shorter or longer than cotton. Since individual nanotubes cannot be produced as a continuous material yet, methods are needed to form macroscale and bulk materials from short nanotubes. The process of twisting nanotubes together to form a fine thread is called spinning. Single or multi-ply thread is called yarn. Yarn is a long continuous length of interlocked nanotube strands suitable for producing textiles, weaving, and rope-making. Thread is a type of fine fibrous material intended for sewing by hand or machine. Spun yarns may contain a single type of fiber or may be a blend of various fibers that can have the same or different material or architectures. Yarns are often made up of a number of plies, each ply being a single spun thread. Small-diameter thread is produced in order to gain improved properties; multi-ply yarn is useful to allow easier handling. Plies of yarn can be twisted together (plied) in the opposite direction to make a thicker yarn. Depending on the direction of this final twist, the yarn will be known as s-twist or z-twist, determined by looking at the direction of twist from the end of the yarn. There are different diameters and lengths and numbers of walls in nanotubes produced with different process conditions. There are also different approaches for producing thread and yarn from nanotubes, and the structure and properties of the thread such as the orientation of nanotubes and strength of the thread can be different for the different types of thread and yarn produced.

Nanotubes grown on a substrate self-assemble to form an array or forest with billions of nanotubes per square centimeter. A thin ribbon or sheet of nanotubes may be drawn from the forest. Twisting the ribbon is one way of forming a thread or yarn. The nanotubes when drawn from the forest tend to stick together in parallel in bundles and the bundles overlap each other at the ends in the ribbon. The bundles may be of different diameters and are also called strands. The nantoubes in ribbon and thread are mostly aligned along the longitudinal direction of the ribbon or thread when produced from nanotube forests. Nanotube thread may also be produced directly from a furnace in a simple continuous high-rate-of-production process, but the nanotubes may not be as well aligned along the thread axis. Nanotube thread is also produced in a wet spinning process and the nanotubes are well aligned. Nanotube sheet can be formed by direct deposition of nanotubes on a rotating belt inside a furnace, by solution evaporation of nanotube powder to form buckypaper, or by polymer infiltration of nanotube powder. Another approach to form tape and sheet is by rolling over nanotube forests. The different techniques for synthesizing nanotubes and producing macroscale functional materials are comprehensively described in the book.

The main focus of the book is on carbon nanotube (CNT)-based materials, but boron nitride nanotubes (BNNT) and silicon nitride nanotubes are also discussed. Thus, it is obvious that there are many parameters that define nanotube materials and that these parameters must be optimized along with the synthesis and postprocessing methods to put the materials into commercial applications. Synthesis of nanotubes involves a large number of parameters. Nanotube reactors are usually based on the chemical vapor deposition method and types include horizontal and vertical tube, cold and hot wall, plasma assisted, field assisted, and others. There are different precursor gases, possibly water, and other gases that moderate and control the reaction. Reactions at different temperatures produce different growth rates and quality and lengths of nanotubes. Catalysts are used in a floating method or on a substrate, or a combination of the two. Buffer layers are used sometimes and a promoter metal may be used with the catalyst. Furnaces can use thermal heating, radio-frequency heating, or a combination of the two. Post processing of nanotubes may involve removing the catalyst particles, thermal annealing to heal defects, plasma or wet functionalization and cleaning, doping, and other methods. Overall, producing nanotube materials is a complicated process that is being continually improved. The continuous improvements in the properties of bulk nanotube materials gives hope that eventually fibers will have properties close to those of individual nanotubes. Bulk fiber materials with the properties of nanotubes would be a superfiber material and change engineering design, which means that large manufacturers would be willing to change their processes to incorporate nanotube materials because the improvements possible are too great to ignore.

GOALS OF SUPERFIBER RESEARCH

A goal of the book is to speed up tuning all the parameters discussed above to get nanotube materials into applications quickly. Various forms of nanotube materials are in development, they are emergent materials, and they are beginning to be put into applications. This book may be the first broad manuscript on the subject of nanotube superfiber materials. The aim of the book is to outline the paths to move forward from development of nanotubes which are molecules to development of superfiber materials (ribbon, yarns, and sheet) which are macroscale or bulk materials. The book also provides ideas for putting the materials into applications. Fully transitioning from molecules to bulk materials with good properties will be a breakthrough in materials technology and it will positively impact most industries and areas of society. A new industry of superfiber materials production is arising and a common goal of researchers around the world is to accelerate the industrialization of nanotube materials thus creating jobs and improving our economies.

Another goal of the book is to address the universal need to better understand the nucleation and growth of nanotubes. Thus, several chapters are provided in the book on the fundamentals of nanotube growth. Then the focus moves to discussion of processing thread, yarns, and sheet. There are currently a number of approaches to fabricate such materials, and to date there is not a full understanding of the process—property relationships for nanotube bulk materials. The electrical and thermal conduction mechanisms, and the mechanical failure mechanisms of the macroscale materials, are also not fully understood. The collection of the chapters of the various state-of-the-art activities will be helpful to those entering the field to develop such materials, as well as to those deciding on the value and/or maturity of such materials. The various chapter authors performed in-depth summaries of various materials/processes to ensure that the aggregate of the chapters provides a holistic assessment of the field. The first chapter provides a general background on fiber materials and then describes the lay of the land regarding nanotube fiber development. Each chapter thereafter has a survey of the broad state of the art so it can be understood within the context of the variety of technology. There are many varieties of nanoscale materials and a main strength of the book is the focus on a small subset of nanomaterials and fibrous materials and the attempt to address processing, properties, and applications holistically. Processing is composed of several different approaches all aiming toward the exploitation of nanotubes within a higher scale material. Several different synthesis routes are included for yarn, and several different material forms are included (yarn, sheet). The collection of diverse materials/approaches/processes provides a basis from which to compare and draw conclusions especially for industries looking to put the materials into applications. The book provides different viewpoints and extends the current understanding and comprehension of the field. Subtle factors for integrating nanotube materials into bulk materials are considered. An example is that CNTs and yarn are highly anisotropic material forms and must be modeled and tested in composites. The tension and compression axial properties are different, and these may be different from the transverse properties of the thread. Also, all the properties depend on the infiltration and bonding of the thread to the matrix material. The investigation of properties of nanotube thread composites leads into the area of smart nanocomposites which are materials that can monitor their own health, which are also covered in the book.

FUTURE PROSPECTS

Nanoscale fibers in general, and CNTs in particular, are a new kind of material. From a physical perspective, fiber is a particular state of matter characterized by the capacity to exert or sustain force, power, heat, signals etc. quasi-one-dimensionally. Nanoscale fibers have the additional property of vast surface area, enabling them to act as efficient conductors, interface materials, collectors, and also sensors. Their near-perfect structure and nanodimension bring them close to the theoretical maximum in physical properties and into the quantum realm where new properties emerge. The promise has been, and the challenge remains, to translate the material properties at the nanoscale into reality at the macroscale. Nanotube macroscopic yarns have not yet reached more than a fraction of the promised properties. Indeed, nanoscale fibers other than nanotubes, such as polymer nanofibers, mineral nanofibers, electrospun nanofibers, are also of interest in the field of new fibers, textiles and composites. But this book is concerned with nanotube superfiber materials, which are expected to achieve properties above existing and other new types of fibers. Thus, polymer nanofibers and electrospun fibers are not covered in this book. CNTs are the first nanoscale fibers out of the starting blocks, but others will follow in quick order. Other types of nanotubes such as BNNTs are coming on strong but do not have electrical conductivity, do not possess the suite of properties that CNTs do, and are not mass producible yet. But BNNTs have their own unique properties such as very high temperature capability, high strength, radiation shielding, and piezoelectric properties in bundles, and thus they are promising future materials covered in the book. The properties of CNTs and graphene are also compared.

MAJOR AREAS OF NANOTUBE RESEARCH

Superfiber materials development may be considered to fall within the three main areas of *processing, properties,* and *applications. Processing* involves nanotube synthesis and macroscale material formation methods. There are three principal methods of material formation currently in use to form yarn and/or sheet: the continuous flow floating catalyst process, dry spinning nanotubes from the substrate, and coagulation wet spinning. The merits of each method are discussed along with ideas for improving the methods. *Properties* relate to methods to model and characterize the mechanical, electrical, chemical and other properties of nanotubes and macroscale materials. Materials can be scaled up from the nanosize to the macrosize, but understanding how to scale the properties is the key to developing superfiber materials. Perspectives and techniques for growing high-quality long nanotubes and spinning the nanotubes into yarn are given in the book. *Applications* of nanotube superfiber cover a huge field and superfunctional materials might also be used to describe the materials. The book gives broad ideas including electronics, composites, and carbon machines. Key properties of superfiber materials are high flexibility and fatigue resistance, high energy absorption, high strength, good electrical conductivity, high maximum current density, reduced skin and proximity effects, high thermal conductivity, lightweight, good field emission, piezoresistive, magnetoresistive, thermoelectric, etc. These properties will open up the door to dozens of applications discussed in the book including space and aerospace composites, replacement of copper wire for high-frequency conduction, electromagnetic interference (EMI) shielding, coax cable, use as carbon biofiber, bulletproof vests, impact-resistant glass, wearable antennas, biomedical microdevices, biosensors, self-sensing composites, supercapacitors, superinductors, hybrid superconductors, reinforced elastomers, nerve scaffolding, energy storage, and many others.

BACKGROUND NEEDED FOR STUDYING NANOTUBE MATERIALS

The readership objective of the book is to help students, researchers, managers, and investors understand the state of the art of nanotube and superfiber materials, the commercial potential of the materials, and how to put superfiber materials into applications. This book is needed because CNT material formation is highly specialized and most industries are standing by on the sidelines waiting for the technology to be developed. Once the technology is available, they will replace their incumbent materials such as copper, aluminum, and carbon fiber with superfiber materials that are lighter, tougher, and carry more electrical current than existing materials. The bottom line is that superfiber innovations are on the verge of producing material with extraordinary properties, the manufacturing is scalable, and now we are at the point where all kinds of industries soon will be knocking at the doors of superfiber materials' producers. Nanoscience, nanotechnology, and nanomedicine researchers and developers will be interested in the book. Engineers in the aerospace and defense industries where the weight of material matters and low cost is not as critical will be especially interested in superfiber materials. Subsequent applications will be in the area of biomedical microdevices where high magnetic density devices built using nanotube cable will have better performance than conventional materials. The target audience of the book includes materials scientists and technologists, including materials and nanomaterials science later-year undergraduate and postgraduate students and their academic advisors and professors in academia, students, researchers, industry professionals, engineers, and medical doctors. Undergraduate-level science and a bachelor's degree are helpful to understand the book.

Engineering researchers interested in using nanomaterials in their research, managers who need to improve the performance of their products, and medical professionals will benefit from the book. In the race for technology and a competitive advantage, in the long term, industries will naturally adopt superfiber materials.

The current understanding of the overall class of nanotube materials must be expanded and a comprehensive picture of the various approaches and material forms within the subset of nanotube superfiber materials is needed by researchers and investors. Different authors propose various processing techniques based on structure—property relations to convert the wonderful but hard-to-use nanotubes into yarn and sheet that are readily useable. Nanotube fiber production should be viewed as an emerging area and the book covers the state of the art. A coherent book was obtained by requesting the authors to limit general introductory material that is contained in many other books and jump straight into their specific background material and topic. The publisher also conducted an extensive review of the book plan. Mr Wayne Yuhasz, senior technical editor from Elsevier, made suggestions on the book and obtained recommendations from six experts in the field whose comments tremendously improved the book. Experts in the field have contributed 27 chapters to *Nanotube Superfiber Materials*, which will serve as a current reference and teaching book. Product development managers will keep the book in their briefcase as a valuable reference to help them understand the technology and how to keep their company in the game in the future.

Acknowledgment

The most important duty of the editors is to personally acknowledge and thank the contributors to the book. The authors of the chapters in this book have contributed their knowledge to benefit the field of nanotechnology. The authors have helped to move nanotechnology research forward and toward commercialization to create jobs and provide economic benefit to the world. Another important aspect of superfiber materials is their potential as a sustainable renewable material that can improve our environment. The editors acknowledge and sincerely appreciate the work of the authors in preparing this book which we hope will heighten the enthusiasm for nanotube materials and accelerate their development and utilization. The diversity and intersection of ideas in the book—with contributions from universities, government agencies, industries, engineering, physics, and chemistry and from researchers around the world—makes the book valuable to the field. The editors also particularly acknowledge the contribution to the organization and theme of this book by Stephen C. Hawkins, Principal Research Scientist, Project Leader CSIRO Materials Science and Engineering, Australia. The managers, typesetters, and editors at Elsevier are thanked for producing this excellent book. Finally, our deepest appreciation goes to the families of the chapter authors for supporting preparation of this important book.

Editor Biographies

Mark J. Schulz is Professor of Mechanical Engineering at the University of Cincinnati (http://www.min.uc.edu/me), Co-director of the Nanoworld Laboratories at the University of Cincinnati (www.min.uc.edu/nanoworldsmart), one of two deputy directors of the National Science Foundation's Engineering Research Center for Revolutionizing Metallic Biomaterials (erc.ncat.edu), co-founder and Director of Advanced Concepts at General Nano LLC, a nanotube materials manufacturing company in Cincinnati, OH (http://generalnanollc.com), and is founder and Director of Clinical Nano Ltd, a nanotechnology and nanomedicine development company in OH (mjsclinicalnano@gmail.com). He teaches *an Introduction to Smart Structures* course.

Vesselin Shanov is Professor of Chemical and Materials Engineering, School of Energy, Environment, Biological, and Medical Engineering at the University of Cincinnati. He is co-founder and co-director of the teaching and research facility NANOWORLD Laboratories at the same university (www.min.uc.edu/nanoworldsmart) and co-founder of General Nano LLC. His recent research focuses on synthesis, characterization, processing, and application of nanostructured materials with emphasis on carbon nanotubes, graphene, as well as on biodegradable Mg for medical implants. Dr. Shanov has more than 280 scientific publications, including 15 patents, 10 provisional patents, and 6 books, cited in about 1,000 different references. He teaches a course called Nanostructured Materials Engineering.

Dr. Zhangzhang Yin is a research associate in the Department of Mechanical Engineering at the University of Cincinnati. He is also a project manager in the National Science Foundation's Engineering Research Center for Revolutionizing Metallic Biomaterials. His current research interest is biodegradable implants, smart implants based on nanotechnology, and metal corrosion. Dr. Yin received his Ph.D. in Materials Engineering from the University of Cincinnati in 2009.

CHAPTER 1

Introduction to Fiber Materials

Stephen C. Hawkins

Principal Research Scientist Materials Science and Engineering
Commonwealth Scientific and Industrial Research Organisation Australia

CHAPTER OUTLINE

1.1 Fibers and Nanofibers	1
1.1.1 The nature of fiber	1
1.1.2 The nature of nanofiber	2
1.1.3 Carbon nanotubes, the archetypal nanofiber and *bête noire* of early industrial furnaces	3
1.1.4 Carbon nanotube structure at first approximation—the nanoscopic begets the macroscopic	4
1.1.5 CNT structure in more detail	8
1.1.6 A note on carbon nanofibers	10
1.1.7 Perturbations of the CNT structure	11
1.1.8 Carbon nanotube synthesis	12
1.2 The Challenge of CNT Yarn Fiber Fabrication	14
1.2.1 A brief diversion into conventional yarn fiber fabrication	16
1.2.2 CNT yarn fiber production	17
1.2.2.1 By dispersion	17
1.2.2.2 By direct spinning	18
1.2.2.3 By direct synthesis	19
1.2.3 On the physical properties of CNT yarn	20
1.3 Conclusion	25
References	26

1.1 FIBERS AND NANOFIBERS

1.1.1 The nature of fiber

Fibers (or "fibres" outside the United States) are discrete "slender" objects able to transmit tensile but not compressive axial loads [1]. They have a relatively high aspect or length-to-diameter ratio (not universally agreed but certainly $\gg 100$), that distinguishes them from rods or bars. In singular, "a fiber" is one such object. Collectively, "fiber" refers to a type of material with extended morphology such as wool, cotton, carbon, glass, polymer fiber, and so on. The meaning of the word "fiber" is somewhat blurred by its application to different structures and concepts that would otherwise be labeled as a (singular) carbon nanotube (CNT), a fibril, a monofilament, a staple yarn, a monofilament yarn, as well as the (collective) material from which these items are made. As will be seen, it is important to distinguish between these in order to understand how the singular fiber contributes to the behavior, properties and applications of the collective fiber (that is, yarn) structure, particularly at the nanoscale. Somewhat irrationally, the term "carbon nanofiber" (CNF) has arisen to distinguish fibers of the same composition and morphology as CNTs but having distinctly different internal structures, and hence, physical properties.

Fiber is possibly the first form of material that humans ever worked, though this would be hard to prove given the ephemeral nature of the natural product. Certainly, over 30,000 years ago [2], the twisting, weaving and dyeing of flax and tur (goat) hair was already established. The capacity to gather fiber from plants or animals for use in the bindings, bowstrings and decorations; the processing and application of fiber for shelter, clothing, food gathering and communication; and the working of skins and wood—the first fiber-reinforced composites—seem rooted deep in the human psyche. It is not difficult to understand why as fiber enables the exertion of force and the shaping of structure needed by weak, defenseless, intelligent beings to survive. This is because of its quasi-one dimensionality which enables the simple linear transmission of tensile load, interweaving to create a layer, and shaping to enclose a volume.

1.1.2 The nature of nanofiber

Natural fibers, perhaps with the exception of silk, are of limited length, commonly about 20–150 mm for wool and 10–15 mm for cotton. Synthetic fibers are extruded in unlimited length, thereby being termed "monofilament". Monofilament may also be chopped into short lengths to facilitate various textile and fabrication processes such as blending, twisting, compositing, and so on. This "short" material, be it wool, cotton, chopped glass, chopped carbon, chopped nylon, etc., is referred to as "staple fiber". It should be borne in mind that these materials conventionally have diameters in the few to tens of micrometers. As such, their properties are not greatly different from the macroscopic and they experience insignificant van der Waals attraction, generally only interacting through roughness, adhesion or radial compression-induced friction. Although particles and films of nanometer dimension were well known many decades ago, the advent of sensibly long, potentially atomically perfect fibers at the nanometer-diameter scale, falling as it does between the molecular and the macroscopic, introduced an entirely new realm of possibilities and challenges, as described in the following chapters.

CNTs are the first and by-far the most advanced of the new nanofiber generation. As will be seen, non-carbon nanofibers, most notably boron nitride (BN), are rapidly achieving prominence. Despite the obvious differences in electrical and thermal conductivity, optical absorption, thermal and chemical stability, etc., the relevance of a discussion on CNTs to other nanofibers lies in several important areas. First, it maps out the path that has been created by CNT research and shows both what is possible and interesting and what opportunities there remain. For example, some of the attractions of BN are that it is *not* CNT—i.e. it is *not* black, electrically conductive or flammable! Second, just as CNTs quickly built upon the infrastructure—the processes, the reactors, the analyses, and the conceptual climate—developed for fullerene research, other nanofibers can build upon the infrastructure developed for CNTs. Third, many of the issues encountered in the production, processing and application of CNTs are ones of scale—diameter, aspect ratio, tangling, alignment, and interaction—nanometer fibers for millimeter applications. These material matters will apply in varying degree to all nanofibers. Finally, the opportunity exists to create composite structures with combinations of nanomaterials that were previously inconceivable.

1.1.3 Carbon nanotubes, the archetypal nanofiber and *bête noire* of early industrial furnaces

Interest in CNTs has been driven in turn by curiosity for this new form of matter; concern for the damage they cause in furnaces; admiration for their esthetic and geometric perfection; and most recently by their technological promise. CNTs comprise single or concentric-multiple cylinders (also termed walls or shells) of graphene or sp^2 (hexagonally)-bonded carbon of one-atom thickness. The diameter range for CNTs can be considered to be from about 0.68 nm (the diameter of a C60 fullerene, as discussed below), or more practically about 1 nm, up to about 100 nm. As the minimum energy conformation for graphene is to be flat (or gently rippled in the freestanding state [3]), the smaller the CNT diameter, the more highly curved and strained the wall, and hence, the less stable. Freestanding CNTs have been reported down to about 0.43 nm [4] but these are highly unstable even to electron microscopic analysis, and doubt existed as to the accuracy of transmission electron microscopy (TEM)-based diameter measurements at this scale [5]. Nanotubes grown or constrained within a larger tube have been observed possibly to 0.3 nm [6] but certainly down to 0.41 nm with accurate assignment of every atom position [7]. The upper limit of 100 nm is more arbitrary as no significant change in properties occurs at or approaching this value.

Although CNTs can be said to have been brought to public and scientific attention in 1991 by Iijima [8], much about the structure and most methods of synthesis was known long before [9]. An understanding of CNT structure slowly developed with the advent of TEM and diffraction. The first observation of the tubular nature of CNTs was attributed in 1955 by Hofer [10] to Radushkevich and Luk'yanovich in 1952 for observing the CNT interior and catalyst particle [11]. However, carbon, described variously as filaments [12], vermicules or threads [13], tubules [10], whiskers [14], etc., had been observed to destroy industrial furnace insulation since the late nineteenth century and considered a nuisance for the first 80 years of the twentieth century [15]. Often much was known about the process of carbon deposition and catalysis but little about the structure as the interest then was more on the cause and prevention of the impressive damage [16].

As with many "discoveries", the rise of CNTs was precipitated by the times and some luck. The 1991 report [8] came amid the excitement engendered by the "discovery" in 1985 of the icosahedral ("soccer ball shaped") C60 fullerene structure by Kroto et al. [17], predictions of natural formation around stars [18] as finally observed in 2010 [19], and large-scale laboratory synthesis in 1990 by Krëtschmer et al. [20]. (Ironically, C60 itself (and its congeners) was observed or predicted some time before 1985 but not pursued or recognized as such [18,21].) The first CNTs of the nanotechnology era were synthesized in laser ablation and carbon arc systems, set up originally to produce carbon clusters and fullerenes. Later processes utilized chemical vapor deposition (CVD) processes [22]. Yet, most of the early work utilized CVD processes and the CNT structure and growth mechanisms were very nicely summarized by Tibbetts in 1984 [23], while Bacon in 1960 used arc discharge to produce "graphite whiskers" with superlative qualities of strength, stiffness, electrical and thermal conductivity [14].

CNTs could well have been in use for over a century already if the 1889 patent by the Britons Hughes and Chambers [24,25] for the growth of strong, flexible, hair-like carbon filaments of 4–5 inches (100–125 mm) in length had managed to displace the use of bamboo and other carbonized natural fibers in incandescent lamps as intended. From the description, this material (and that of Bacon [14]) is likely to be CNTs, possibly thickened or cemented with amorphous carbon, although it may also be CNF, the distinction being outlined below. Continuous improvement of carbon globe filaments, followed by the invention and commercialization of tungsten filaments from 1904 eliminated that opportunity for CNTs. Not, perhaps, forever as this was one of the potential applications proposed in 2002 for directly spun CNT yarn [26].

Producing graphene as a continuous, defect-free sheet requires (at least currently) a continuous, defect-free substrate. The ability to synthesize graphene as a freestanding CNT cylinder rather than a sheet achieves several things. First, it eliminates the long sheet edge and the discontinuity that represents (the ends are a minor constituent). Second, it removes the one-to-one correspondence of substrate or catalyst particle and product surface area—a fiber can be of effectively unlimited length. Third, the fiber form is the one that is understood and used across a broad range of applications (although as will be seen, the CNT fiber form is not as understood as initially thought) and hence of immediate material and technological interest.

1.1.4 Carbon nanotube structure at first approximation—the nanoscopic begets the macroscopic

Consider first only a single sheet of graphene, being a single-atom-thick hexagonal array of sp^2-bonded carbon. The sheet can be rolled into a cylinder with contiguous (i.e. edge connected) hexagons falling in rows parallel to the axis, resulting in a uniform "armchair" (i.e. up-down–down-up) arrangement of atoms at the end and hence termed an "armchair CNT" (Fig. 1.1(a)). Orienting the rows so that they spiral around the tube axis results in a "chiral CNT" (Fig. 1.1(b)). Orienting the rows so that contiguous hexagons fall in circular bands around the axis gives a uniform "zigzag" (i.e. up-down—up-down) arrangement at the tube end and hence a "zigzag CNT" (Fig. 1.1(c)). (Chiral tubes have a combination of armchair and zigzag features.) The tube ends may be closed by graphene caps like fullerene hemispheres or cones depending upon the growth process and condition of the tube. The tube size

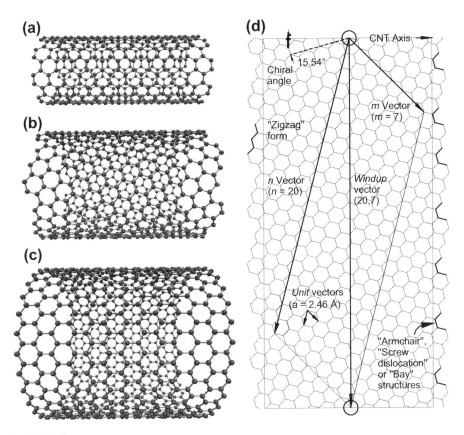

FIGURE 1.1

(a) Armchair (9,9), (b) chiral (20,7) and (c) zigzag (33,0) SWCNT structures. Note that these CNTs vary by 0.68 nm in diameter and hence could form an MWCNT. Note also that these representations give an erroneous impression in three respects. First, when illustrated with their van der Waals atom radius, it is clear that there is essentially no vacant volume between MWCNT walls or within the hexagons. Second, the physical outside diameter is 0.34 nm greater than the calculated (geometric) diameter. Third, scaled to the ~3 cm diameter of the top illustration, a 1-nm-diameter SWCNT with a length of just 1 μm would be 30 m long, and the recently reported 3-nm-diameter 55-cm CNT would be 5500 km long. (d) An illustration of a section of (20,7) CNT cut parallel to the axis and laid flat.

and character are determined by the number and orientation of the hexagons around the tube axis and can be defined using a system of tube indices (n,m) introduced by Hamada et al. in 1992 [27] and further developed and expressed by Jishi et al. [28] (see also Refs [29,30]).

The tube indices are most easily found by envisaging the tube cut parallel to the axis with the cut bisecting a carbon atom (circled) as illustrated (Fig. 1.1(d)), and laid flat. A line joining a bisected carbon atom at the top cut edge (define this as the "origin") with the matching bisected carbon atom on the bottom edge (and hence normal to the cut edges) is termed the "wrapping" or "rollup" vector. This vector can be decomposed into two vectors from the origin. One vector follows the line of contiguous hexagons closest to the wrapping vector and drawn through the row of carbon atoms including the origin and defines "n". The second vector drawn from the origin at 60° to the first follows a row of contiguous hexagons next to the nearest wrapping vector and passing through a row of carbon atoms including the origin, and defines "m". The vector diagram is completed by joining the distal end of the wrapping vector to the "n" vector by a line parallel to the "m" vector and vice versa. The intersections of the lines with the vectors fall on two carbon atoms, one along the "n" vector and the other along the "m" vector. The number of hexagons crossed from the origin to the intersection on the "n" vector and to the intersection on the "m" vector gives the indices "n" and "m", respectively. By convention, $n \geq m \geq 0$ and the CNT illustrated has indices (20,7) (Fig. 1.1(b),(d)).

The significance of the (n,m) parameter in defining the size and properties of a CNT cannot be overemphasized. For example, the angle made by the row of contiguous hexagons closest to aligning with the tube axis is the "chiral angle", φ (phi), and calculated from

$$\tan \varphi = \frac{(n-m)}{\sqrt{3}\,(n+m)} \qquad (1.1)$$

This angle can vary between 0° and 30° (and by symmetry −30° to 0°). Note that an alternative expression of chiral angle relates to the angle, θ (theta), of the "n" vector with the wrapping vector, where $\varphi + \theta = 30°$. Expressing the chiral angle in terms of the fiber axis seems both sensible and consistent with yarn/fiber practice [1].) All (n,0) tubes are zigzag, having a chiral angle, φ, of 30°, while all (n, m = n) tubes are armchair, having a chiral angle of 0°, with neither category being actually chiral as they are mirror symmetric. All other (n,m) give values of $0° < \varphi < 30°$ (and $-30° < \varphi < 0°$) and hence are actually chiral. The armchair and zigzag tubes and the chiral tubes with $n - m = 3\ell$ (ℓ = integer) are all metallic in character [28], all other tubes are semiconducting.

Similarly, the diameter, d, in Angstroms of a CNT is calculated from the (n,m) indices [31]

$$d = \frac{a}{\pi}\sqrt{(n^2 + m^2 + nm)} \qquad (1.2)$$

where $a = 2.46$ Å, the unit vector (i.e. the width of one graphene hexagon).

Thus, the chiral angle, φ, and diameter of the (20,7) CNT illustrated in Figs 1.1 (b), (d) are 15.54° and 1.900 nm, respectively, this being the chiral CNT illustrated in Fig. 1.1. Note that this is the geometric diameter, whereas the physical diameter as measured will be greater due to the van der Waals radii of the carbon atoms (2 × 1.7 Å) and blurring or imperfect resolution, depending on the measurement technique. Figure 1.1 also shows examples of the zigzag and armchair structures, illustrating the pattern of carbon atoms that would be seen at the CNT end for tubes of these respective types. The pattern of carbon atoms at the end of a chiral CNT such as the (20,7) example exhibits a series of up to m "armchairs", "screw dislocations", or "bays" as illustrated in Fig. 1.1(d). As these locations require only an additional two carbon atoms to complete the next ring, and can be stabilized as a pentagonal intermediate by a metal atom for example [32], they are favored in the growth process.

The discussion thus far relates to the formation of a single-walled carbon nanotube (SWNT or SWCNT). Addition of concentric graphene shells or walls to give a double (DWCNT), triple (TWCNT) or multiwall CNT (MWCNT) requires that each additional outer tube be about 0.68 nm greater in diameter, reflecting the approximately 0.34 nm (often quoted as 0.344 nm) wall spacing typical of turbostratic graphite rather than the 0.336 nm of crystalline graphite [33,34]. This amounts to almost nine contiguous hexagons in circumference [28]. The spacing can be somewhat larger than this as discussed below but not substantially smaller as it signifies the van der Waals atomic radii of the carbon atoms. The observed diameter of a nanotube may not directly reveal its chirality as different (n,m) indices can produce the same or very similar size. Thus, a nanotube of 1.22 nm diameter may be the chiral (15,1) or (10,8) or else the armchair (9,9) type (Fig. 1.1(a) and Table 1.1).

Taking the (9,9) armchair type (Table 1.1, bold values) and a conservative range of suitable CNT wall spacing (0.336–0.350), there are seven possible next larger tubes with $d = 1.221 + (0.672-0.700)$ nm with the "best fit" (i.e. +0.68 nm) being the two chiral tubes (15,13) and (20,7). Taking the (20,7) CNT as the next size (Table 1.1, bold values; Fig. 1.1(b),(d)), then for this tube, there would be 11 possible next larger tubes within the same limits with the best fit being (21,17) (Table 1.1, bold values) closely followed by the zigzag CNT (33,0) (Table 1.1, italic values; Fig. 1.1(c)) illustrating that a three-wall MWCNT could be composed of the armchair–chiral–zigzag combination ((9,9)(22,7)(33,0)) (Fig. 1.1). Thus, an MWCNT of a given diameter may comprise a number of different (n,m) combinations including both chiral and achiral, semiconducting and metallic varieties. The conduction between shells is substantial so that the presence of a metallic shell will impart metallic character to the whole [35]. The "closeness of fit" of successive walls in an MWNT is of relevance in influencing the physical properties of the fiber and in revealing how it formed. The tubular structure and fit is established at the moment of formation [23] as a function of the catalyst diameter and the energetics of graphene basal plane formation, interaction and curvature.

Table 1.1 $d = 1.219$ nm Armchair Type and Next Possible Diameters +(0.672–0.7 nm) (Optimal Diameters in Bold Values)

n	m	d	φ
15	1	1.216	26.80
9	**9**	**1.221**	**0.00**
10	8	1.223	3.67
14	14	1.899	0.00
22	4	1.899	21.79
15	**13**	**1.900**	**2.36**
20	**7**	**1.900**	**15.54**
16	12	1.905	4.71
17	11	1.913	7.053
24	1	1.920	27.98
19	19	2.577	0.00
26	11	2.577	13.17
20	18	2.578	1.74
21	**17**	**2.582**	**3.48**
33	0	2.584	30.00
22	16	2.588	5.21
32	2	2.588	27.00
29	7	2.589	19.43
23	15	2.596	6.93
27	10	2.596	14.86
31	4	2.598	24.01

1.1.5 CNT structure in more detail

The relatively simple picture presented above is good as a first approximation, however, the actual structure and hence behavior of CNTs is both more complex and more interesting, and of significance in understanding the potential and the problems encountered. Multiple sheets of planar graphene can be either disordered, with no indexing of one layer to another, or may be aligned in various symmetries, such as AAAA, ABAB, ABCA, and so on. Rolling the graphene sheets into a closed cylinder destroys this simple symmetry because neither the wall spacing nor the orientation fits neatly with the size of the hexagons. This leads to two types of dislocation—diameter and chirality.

The chirality of successive shells can vary from perfectly ordered to completely disordered depending upon such factors as the inside diameter and growth process and temperature [36]. Early diffraction studies [8,33] of small arc-grown MWNTs showed the same chiral angle over 3–5 layers. The precision of the angular and diameter matching or the amount of distortion that this entails is unclear since, as can be seen from Table 1.1, there are few sequences of well-fitted tubes with identical chiral angles. Remarkably, for larger MWNTs, the chiral angle was observed to increase stepwise, amounting to a chiral dislocation, for every 4 ± 1 walls from the center in either regular increments (e.g. $6°$) or irregularly (about $3°-6°$), but nevertheless in groups of shells, suggesting that the chirality and diameter of the innermost shell sets the template or memory [33,37] of successive outer shells.

Polygonization, or the flattening of the sides of a CNT (or indeed a fullerene) at some points (usually ≥ 5) with commensurate increase in curvature at others, is observed in some samples. Occurring at the end cap of a CNT [38] and maintaining a 0.34 nm wall spacing at the apices, these structures arise through insertion of heptagons and pentagons in the graphene structure to achieve curvature normal to the CNT axis. CNTs that exhibit polygonal walls were presumed to do so by a similar mechanism [33], however, this seems implausible as it would require precise alignment with the tube axis regardless of the chirality. A "diameter dislocation" resulting in a shell being slightly larger than necessary to encompass the next smaller one could be accommodated symmetrically if they both become polygonal, with the walls flattening for most of their circumference and adopting the optimum 0.34 nm spacing. The excess length of the outer shell, and perhaps a chiral mismatch or dislocation, is then accommodated through the more widely spaced and tightly curved apices, the excess strain energy being compensated for by the optimal wall spacing [39]. The observation that this can occur with the retention of stepwise chiral transitions at 4 ± 1 walls [33] emphasizes the concerted (rather than a stepwise, shell-by-shell) nature of the growth process even across chiral and diameter dislocations.

Many, if not most, authors consider the 0.34 nm MWCNT wall spacing to be practically immutable, even to using it as an internal length standard in TEM images, e.g. Ref. [6]. However, as indicated by the observation of polygonization, the wall spacing can be somewhat greater than this. Spacing may be seemingly random, such as in the intensively studied four-wall example (32,1)(26,24)(39,25)(62,2) with spacings of 0.36, 0.55 and 0.42 nm, respectively [40]. (Curiously, the outer and inner shells have identical chirality and the overall spacing is almost enough for five walls, suggesting that things started out differently.) The wall spacing may also be highly regular but >0.34. For example, Bretz et al. [41] reported a fairly uniform 0.375 nm wall spacing for a 16-wall CNT, again with a strong influence of the chiral angle which alternates between $0°$ and about $1°$ from one wall to the next. Kiang et al. [42] observed spacings that declined exponentially from around 0.39 nm at the innermost wall, asymptoting to 0.34 nm at the outer layers of large MWNTs. Smaller spacings than 0.34 nm are also occasionally observed, for example down to 0.335 nm for a DWCNT including the smallest diameter yet reliably imaged [7]. The presence of an enclosed metal catalyst particle can cause local spacing to be as low as 0.32 nm [43].

The significance of spacing is that it influences the wall-to-wall interactions with, perhaps unsurprisingly, larger values resulting in less interaction as modeled by pullout resistance [44]. This suggests that required force is up to about 10% of the nominal strength and greatest for armchair CNTs of any given spacing. Modeling also shows [45] that broken ends and other "defects" require 3—4 times higher force to pullout than do closed ends, and external pressure normal to the tube surface increases the pullout force required still further. Although there are often shells of close or identical size that could fit within a given MWCNT structure, as noted above, the interaction between the walls is influenced by the chiral angle. In general, the energy differences are relatively modest [36], with an energy difference between the very few optimum and many arbitrary arrangements equivalent to only about 30 K. This suggests that chiral disorder is greater at higher growth temperatures. The complexity and imperfect understanding mean that the modeled and measured properties of CNTs can produce some contradictory results, for example whether wall interactions within MWCNTs are so slight as to rival frictionless bearings or whether so strong as to contribute substantially to overall tensile strength [46].

1.1.6 A note on carbon nanofibers

To this point, the discussion has been restricted to CNTs with walls that are parallel to the tube axis and have clear, continuous interiors or "lumens". Cylindrical graphitic structures can also form, in which the walls are not parallel to the axis but rather are convergent with an angle (cone or tilt angle) to the axis varying from almost $0°$ (i.e. almost tubular) [47—49] to $90°$ (i.e. like stacked plates) with intermediate angles having a stacked cup, herringbone or chevron appearance [50,51]. By strict definition, any such structure with a nonzero angle is termed a CNF [50,51] and, although journals do not strictly enforce this definition, it will be adopted here for the purposes of discussion. Where the tilt angle is near zero, the CNF is essentially tubular but consists of nested cones (Fig. 1.2). Close to $90°$, the CNF is essentially solid.

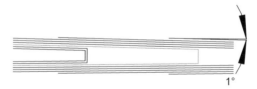

FIGURE 1.2

Scaled schematic of a section of CNF with tilt angle (α) $\sim 1°$ showing internal membrane and external edges at ~ 19 nm intervals. One wall is colored to facilitate tracking. Also shown is a double wall internal bridge to illustrate that this does not affect order. (For color version of this figure, the reader is referred to the online version of this book.)

The wider edges of the cones are exposed on the surface of the CNF, while the narrower lie in the interior and are often bridged or closed by a graphene shell, both with an interval nominally equal to $(0.34/\sin \alpha)$ nanometers, or about 19 nm for $1°$ angle. This bridge very often comprises two or more shells at commensurately greater intervals; however, this does not affect the wall order. (The pairing or grouping of shells is a recurrent phenomenon in CNT and CNF structures.) Annealed CNFs of large diameter and cone angle may also display "loop structures" where the inner edges of around 2–4 cones curve tightly around to join the edges of the next 2–4 cones [52]. (This also occurs on the outer surface but between conical and very disordered (turbostratic) cylindrical layers.) Under the influence of a crystalline Fe_3C catalytic particle stabilized by nitrogen, CNFs can exhibit an exceptional level of crystallographic registration [47,49], with around 90 walls having the same chirality. In this case, it is the catalyst particle alone sculpting the CNF structure.

Although CNFs are anticipated to have exceptional stiffness or tensile modulus, approaching that of SWCNTs (~1 TPa), when the angle is close to zero [53] and decreasing to higher angles and large diameters [54], the tensile strength of about 3–4 GPa [52] is relatively modest. This is because tensile failure primarily entails one cone sliding out of another rather than the breaking of covalent bonds [54] and illustrates both the nominally weak van der Waals interaction between successive walls and the unique character of CNTs as compared to CNFs.

1.1.7 Perturbations of the CNT structure

In addition to the polygonization, diameter and chiral mismatches, broken ends and other defects discussed above, CNTs can exhibit other perturbations from their idealized structure. Simply rolling up a graphene sheet distorts the C–C bonding symmetry and energy—there is now an inside and outside, an axial and a circumferential direction. The energy scales as the inverse of the radius squared, hence is only significant for the smallest CNTs. A consequence of this is a lower axial stiffness (Young's modulus) for smaller tubes, with zigzag tubes lower than armchair [55] and a "relaxation" with the whole structure expanding by up to several percent [56] through a slight increase in bond length, circumferential more than axial [57].

A single-CNT shell is the ultimate "thin-walled cylinder" and hence prone to buckling or flattening. Insofar as they act independently, multiple CNT walls can behave similarly, with the propensity to collapse increasing with diameter and decreasing with wall number. Thus as early as 1995 [58], MWNTs of 6–9 walls were observed to spontaneously flatten into ribbons with a "dumb-bell" cross section. The higher strain energy at the more highly curved edges is counterbalanced by the decreased overall curvature plus the increased van der Waals interaction between the flat walls, particularly the innermost one. The predicted limit for an eight-wall CNT to collapse was >17 nm but CNTs of ~15 nm are seen to flatten, with some suggestion that the flattening is initiated by local mechanical processes such as contact, bending or twisting.

The effect of external hydrostatic pressure on the collapse and rebound of SWCNTs was modeled and tested [59] to demonstrate that at atmospheric pressure, SWCNTs up to about 4.2 nm diameter are stable, assisted by polygonization. A pressure of about 2.1 GPa will collapse an array of ~1.4 nm (10,10) fibers, predicted to rebound at a pressure of about 1.5 GPa, the hysteresis being due to the van der Waals attraction within the collapsed structures. More recently an SWCNT of

4.6 nm was found to be collapsed but a DWCNT of the same diameter was still open. These were within a continuously formed yarn structure comprising CNTs mostly collapsed into aligned ribbons [60] and in agreement with the modeling. Hydrostatic pressure is also predicted to cause surface axial corrugations [61] in larger many walled CNTs.

Torsion, axial compression and bending result in buckling which, if it does not result in complete collapse of the CNT structure [58], is fully reversible. These perturbations were very recently and ably reviewed [62] and are notable for the nonlinearity or suddenness of their onset. Cylindrical structures under these types of load remain essentially unchanged until abruptly crimping or inflecting at a critical threshold.

All the perturbations discussed above relate to isolated CNTs subject to various stresses or dislocations. Another type of deformation arises when CNTs come into contact with other CNTs or surfaces under the influence of van der Waals attraction. As early as 1993 [63], a 10-wall $\sim$10.5-nm and a 12-wall $\sim$11.6-nm MWCNT were observed to be substantially flattened together. Of particular interest is that the inner wall spacings adjacent to the flat surfaces were slightly reduced, and increased elsewhere, indicating a deeper interaction than might be anticipated. Modeling [64] predicted flattening of 42% for the smaller 5.4-nm SWCNT on a flat surface, decreasing to 25%, 5% and <1% for two, four and eight walls, respectively. Very recently, a more subtle effect was reported [65] wherein long-chain polymer molecules wrapped around an MWCNT can compress the wall structure.

The underlying significance of the order and disorder, the dislocations and the perturbations is that they bear upon the interaction of one CNT wall with another, either internally across or within a single or multiwall structure, or externally in a multi-CNT assemblage. The strength or weakness of that interaction will influence the strength, stiffness, conductivity, etc. of individual CNTs and, more importantly in the context of this book, the translation of those properties from the nano- to the macroscale.

1.1.8 Carbon nanotube synthesis

There are numerous and frequent reviews of this topic, so it will only be briefly summarized here. Nevertheless, it is instructive to observe how the understanding of the mechanisms has evolved or, often, been correct from the outset (see for example Refs [23,25,50,51]). The basic requirements (which have been touched upon above) for the synthesis of CNTs are heat, a carbon source and a catalyst, and even the catalyst may be optional [25]. The syntheses have been greatly refined but not fundamentally changed since the first reports of "filamentous carbon" found in furnaces and carbon arc electrode deposits. At the simplest level, CNTs grow from a catalyst particle (or an apex of one) of similar dimensions to the CNT by accumulation of carbon from a reactive atmosphere. The carbon is transported to a point on the catalyst and assembles into a hemispherical graphene cap that breaks away and develops into a tube as more material is added to the edge. The graphene cap itself may act as a catalyst, so a separate particle may not be essential. Many different materials have been used as catalysts including metals, refractory oxides and preformed carbon structures (fullerenes and nanotube fragments). Various reactor structures including cold wall, plasma-enhanced, low pressure and high pressure, and many variations on the basic recipe to control amorphous carbon accumulation, catalyst size, dispersity, activity and longevity have been developed.

The CNT growth processes comprise two broad groups based on either CVD or carbon vaporization. The CVD group is divided into methods entailing the addition of catalyst with the gaseous or liquid carbon source (catalyst co-injection, CCI) or with catalyst predeposition (CPD) on a substrate that is subsequently exposed to the carbon source [66]. The CCI method is itself subdivided into the low-temperature (about 650–750 °C) form where the catalyst particles settle out of the reactor atmosphere onto a substrate from which the CNTs grow, as exemplified by Andrews et al. [22], and the high-temperature form (about 1000–1050 °C) wherein the catalyst particles remain afloat in the gas stream and produce nanotubes as they pass through the reactor growth zone, as exemplified by the HiPCO process of Bronikowski et al. [67]. CNTs produced in this way come into random contact and bind by van der Waals forces and amorphous carbon deposition into a loose mesh within the reactor space or into tangled masses at the collector. Some localized close-packed bundling (essentially crystallization) particularly of SWCNTs with similar diameters also occurs [60].

The CPD approach entails the deposition of a catalyst or precursor upon a compatible substrate which, if disordered and amorphous, is analogous to the early observations of CNT growth in furnace linings. Use of smooth glass substrates by, for example, Ren et al. in 1998 [68] produced highly aligned forests of fiber. The methods of catalyst deposition and the substrate types are many and varied, ranging from electron-beam evaporation of iron onto an e-beam-deposited alumina buffer layer over thermally oxidized semiconductor-grade silicon through to coprecipitation of cobalt–molybdenum with magnesium hydroxide. They have in common that at the commencement of growth, the catalyst is nanoparticulate or nanostructured and activated to carbon deposition.

Catalyst particles deposited on substrates often bind strongly, so that the CNT grows out from this "base". In addition to diameter, the growth rate and direction are a function of the particle activity, which can range widely, and orientation, which is generally away from the substrate but otherwise random. If the particles are widely spaced, the CNTs will be tangled, of different lengths, and prostrate. However, if closely packed, the CNTs will very soon come into contact near their tips and immediately and strongly bond, initially by van der Waals interaction, possibly augmented by carbon deposition. In having their tips locked together and continuing to grow from the base, the CNTs are forced to grow in parallel, at the same linear rate and normal to the substrate surface. Hence, a "forest", "mat" or "lawn" of adjacent-aligned CNTs is generated. If the particles are homogeneous in size and activity, the CNTs below the forest "canopy" will be very straight. If, however, they are polydisperse, the slowest growing CNTs may be pulled from their catalyst and terminated, the fastest growing will be held back and forced to buckle, giving a tortuous or curly appearance to the forest. As the CNTs are fixed at the tip and base, they are not topologically knotted or tangled but they become bound at each point of contact. Thus, "straight" nanotubes will have relatively few points of contact and be easily separated, whereas "curly" nanotubes will be highly interlinked and difficult to disperse.

If the deposited catalyst particle breaks up [49] or does not interact strongly with the substrate, it can be carried at the end of the nanotube distal to the substrate—i.e. the tip. This produces nanotubes that grow largely independently of their neighbors and hence in random directions and lengths. The direction can be controlled by entraining the CNT in a laminar gas stream. The careful control and free access to reactants, maintenance of catalyst activity by the addition of water [69] and absence of constraint enable the most active catalyst particles to grow faster and longer. Hence, this method has been used to produce the longest SWCNTs (18.5 cm) [70] and MWCNTs (20 cm [71,72] and 55 cm [73]) reported to-date.

As noted, evaporation of carbon is achieved with high-temperature processes such as laser ablation [32,74,75] and arc discharge [8,76—79] originally developed for carbon cluster and fullerene research. Initial reports [8,74] did not use catalysts, which were introduced later to improve yield and diameter control, particularly for the synthesis of SWCNTs. This and the observation [75,77] that even with metals present, nanotubes may grow without a catalyst particle attached, suggested the fullerene end cap alone forms a viable basis for CNT initiation and that the role of the metal in these systems is, at least in part, for isolated metal atoms to stabilize and anneal the tubular structure as it forms [32], hence minimizing defects or premature closure [80]. As with floating catalyst-grown CNTs, evaporation methods tend to produce loose, highly tangled masses except at the arc electrode where they can be densely packed. Direct conversion of polymer structures into CNTs was also recently reported [81] although even here, there is a 1971 precedent [82].

1.2 THE CHALLENGE OF CNT YARN FIBER FABRICATION

From the earliest manifestation of CNTs as a useful material [14,24] and detailed recognition of the CNT structure [8], they have been predicted to have exceptional properties of tensile strength, modulus (stiffness), thermal and electrical conductivity, stability, semiconduction, etc. as found from the earliest successful efforts to measure them [83,84]. Based upon these expectations, visions have been cast of power cables and polymer composites, of carbon computing and bulletproof paper. Perhaps, the most ambitious and inspiring such vision is of the "space elevator" or "elevator cable to earth orbit" which has been imagined for many decades and the parameters of which were summarized in 2000 [85]. Such a cable would tether an asteroid counterweight at some height, say 40,000 km or 4000 km above the 36,000 km geostationary earth orbit (GEO), to hold a station at GEO. The concept moved from the merely fanciful to the conceivable with the realization that such a cable made from CNTs would need be only 0.26 mm in diameter—indeed a yarn fiber—at its thickest and weigh around 9.2 tons, compared with 2 km and 6×10^{13} tons for one made of a high-strength polymer such as ultrahigh molecular weight polyethylene or polybenzobisoxazole.

The prospects of actually building an elevator from Earth to orbit may be small in the extreme, beset as it is with technical, environmental, social, economic and political challenges. This does not detract from the fundamental objective, and the purpose of the current work, which is to realize the extraordinary properties of nanofibrous materials at the macrofibrous or yarn scale. While the ultimate manifestation of this might be a 40,000 km CNT cable, there are eminently worthy and useful objectives at less spectacular heights, such as space tethers, aerospace structures, ultrahigh-strength pure and composite materials, high-current conductors, and so on.

Implicit in the vision for a space elevator is the conception that the cable will comprise defect-free, single or small MWCNTs stretching unbroken from ground to orbiting counterweight. The CNTs would be monodisperse, close packed in multiple strands and the strength of the yarn would equal the strength of the individual CNTs. Early observations of self-assembly of exceptionally homogeneous SWNTs into beautifully ordered "crystalline ropes" [32] and bundles [79] gave promise of the great potential of this material. The longest SWCNT reported is 18.5 cm [70] and a TWCNT reported very recently is about 55 cm [73]. A 20 cm CNT of three walls and around 3 nm diameter by the same method [72] exhibits a strength, modulus and strain to break of 200 GPa, 1.34 TPa and 17.8%, respectively, close to theoretical maximum values assuming only the outer wall bears the tensile load [86]. Note that the quoted strength is generally based upon an assumption that only the outermost wall bears the tensile load and the cross section of this is used in the calculation. Given that the inner shells and hollow space are nonfunctional from an engineering viewpoint, a more pragmatic expression of the strength would include the total CNT cross section. Thus, a 3-nm TWCNT has an outer wall comprising about 37% of the cross section total, so a "200 GPa" strength would amount to about 74 GPa over the full area. Electron beam irradiation has been shown to bind the walls together and substantially increase load to break [86] so the CNT interior may perform more than a passive function. Quoting a strength of 200 GPa for a 1-nm SWCNT would amount to a functional strength of about 152 GPa.

To put the challenge of creating macrostructures from nanomaterials into some perspective, the 3 nm $\times$ 55 cm TWCNT [73] was grown at the rate of about 5 mm or 4.2×10^{10} carbon atoms/min to a total of about 4.6×10^{12} near perfectly placed carbons using a single iron catalyst particle of a similar or slightly larger diameter and hence comprising about 2×10^3 iron atoms. This accomplishment notwithstanding, it would take over 15,000 years to grow 40,000 km using $\sim 3.4 \times 10^{20}$ (about 7 mg) carbon atoms, presumably also with only 2×10^3 Fe atoms. Apart from the challenge of scaling the production to the tons required, greatly increasing the growth rate of a CNT while maintaining quality would be unlikely, simply from kinetic considerations. Thus, any realistic process to produce yarn in sensible quantities for this or any other macroscopic materials application requires the capacity to grow CNTs in massively parallel processes and assemble them into ordered structures. Although single CNTs have been joined seamlessly end-to-end, the prospect of doing this on any significant scale seems remote. Therefore, the route to and target of yarn formation in this context is to synthesize and assemble large quantities of "short" CNTs (i.e. staple fiber as described in Section 1.1.2) such that their interaction under tension is practically equivalent to a continuous monofilament.

1.2.1 A brief diversion into conventional yarn fiber fabrication

An un- or very slightly twisted bundle of conventional fibers, be they monofilament or staple, is called a "tow" (rhymes with "toe") or else a slubbing, top, sliver, roving etc.—there is a wealth of traditional textile terminology from which to choose (see [1] Ch.2-4). A tow of staple fibers has a tensile strength approaching zero, or at most a very small fraction of the collective fiber strength. In order to achieve a tensile strength approaching that of the constituent fibers, the tow (or variants of it) must be twisted to form a "yarn". The success of this exercise depends upon two linked factors. The first is that, absent any adhesive materials or knotting, the staple fibers have almost no interaction prior to twisting. This enables them to be ginned, carded, combed, gilled or otherwise processed to align, beneficiate (remove short and tangled fibers often caused by the processing itself) and blend them, and then to be drawn or drafted down into fine yarns of perhaps only a few tens of fibers in cross section. The second factor is that upon twisting into a yarn, the fibers acquire a frictional interaction through residual torsion and radial compression induced by axial tension that is sufficiently strong that they will break before they slip.

The frictional interaction between twisted staple fibers can only occur because the twisting process causes them to "migrate" or individually thread their way repeatedly between the center and the periphery of the yarn. Migration occurs as fibers experience differential tensions due to different path lengths and depends upon minimal initial fiber interaction. In doing this, their ends are captured in the yarn structure and hence able to withstand a tensile force. Without migration, the outer fibers would simply be wound around the inner, and would unwind of their own accord. The inner fibers would be straighter than the outer and have even less interaction. Twisting a staple tow also increases the strain-to-break of the yarn compared with the constituent fibers, which are now oriented at an angle (the "twist angle") to the yarn axis.

It is not essential to twist a tow of monofilaments because the transmission of load is directed along the full length of each filament and friction between them is not required. It is also not desirable to twist monofilament tows, beyond a few turns to keep them tidy, because the outer filaments will follow a longer path than the inner, causing excessive tension in the former, and buckling in the latter. Thus, monofilaments are used in the tow form in processes such as winding/setting or braiding to create ropes, or weaving to create fabrics such as carbon fiber and Kevlar. Twisting and plying are also utilized in high-modulus monofilament rope to increase extensibility and reduce shock loading.

Conventional bulk staple fibers such as wool and cotton are produced at the scale of millions of tons per annum, yet processed into yarns at the rate of kilograms per hour per spinning machine. To be economic, processors must run nonstop mills that are hectares in extent, comprise hundreds of thousands of individual spinning elements and use very low-cost labor or intensive automation. This processing bottleneck exists because the mechanical processing reaches down to the individual fibers to clean, blend and align them. Wool and cotton form by a "base growth" process and hence are not significantly tangled or knotted. They become more so due to the handling, washing and processing necessary to harvest them, to remove the entrained dirt, vegetation and natural oils and to convert them from their natural forest-like growth (parallel adjacent alignment) to yarn (axial alignment). All this is feasible because there is no significant attraction between fibers or to machine surfaces. (Some interlocking of wool surface scales can occur, resulting in felting.)

1.2.2 CNT yarn fiber production

CNT production has increased by about six orders of magnitude between 2001 and 2010 or so, with reactors going from grams to tons per hour, though quality is at times indifferent. Until such time as highly aligned CNTs can be produced in essentially unlimited lengths at technologically significant production rates, CNT yarn will be produced from staple fiber. As-produced staple CNTs range from aligned to more or less tangled, and from almost pure to grossly contaminated with amorphous carbon and catalyst residue and hence requiring substantial purification (itself often causing severe tangling and damage) prior to sale or use. Converting this adjacent-aligned or tangled mass into an axially aligned yarn by any conventional textile process is infeasible. These fibers, particularly toward the finer diameters [63], interact so strongly with one another and other surfaces, that no physical combing process could be contemplated even were it possible to build such a device. Furthermore, 1 cm^3 of CNTs of, say, 1.5 nm diameter would contain around 10^{12} times as many fibers as 1 cm^3 of cotton fiber of the same aspect ratio (or about the same number in 1 mg of CNTs as in 1 ton of cotton) so any process such as combing that entails individualizing fibers would require proportionally more operations to achieve a significant production rate.

1.2.2.1 By dispersion

Formation of CNT yarn has taken three distinct routes that can be described as dispersion, direct spinning and direct synthesis as very recently reviewed [87]. The earliest approach was to disperse the CNTs in a liquid and extrude the mixture as a filament, with the aim of achieving a level of alignment through the extrusion and subsequent drawing processes. The first example of this from 1999 was of an SWCNT composite with isotropic pitch to produce relatively low-performance carbon fiber [88]. Melt stiffness, tensile strength, modulus and conductivity were all significantly better with up to 5% CNTs although at 8% or more, the mixture could not cohere. Similar results were reported soon after with polymethylmethacrylate and SWCNTs [89], symptomatic of fundamental problems with the dispersion approach. Initial separation or dispersion of the CNTs is difficult and the high aspect ratio of CNTs results in very high viscosities at even low concentrations [90] with rheological and electrical percolation (interaction) thresholds of <1 wt%, a tendency to clump and with marked shear-dependent properties [91].

Much of the study of CNT dispersion rheology has entailed the use of, for CNTs, relatively low (10^2-10^3) aspect ratio fibers (see, for example, Refs [90–94]) despite which network (gel) formation is seen to occur at only around 2 wt%. Extreme conditions such as concentrated acid [92], hours of high-shear mixing [93], or sonication [90,94] are often used to disperse the CNTs for this work which is mainly aimed at composite behavior. Nonetheless, a review [95] recently noted the paucity of studies in CNT rheology. For yarn applications at least, it may be that little more can be learned as the aspect ratio of CNTs for materials applications has climbed through the 10^4-10^5 into the 10^6-10^8 range and the challenge becomes one of the growing aligned CNTs rather than dispersing and organizing them, particularly if the aim is to achieve very high CNT loadings or pure yarn.

An elegant approach to high-load or pure CNT yarn formation was initiated in 2000 by Vigolo et al. [96] with the coagulation in polyvinylalcohol (PVA) of a surfactant-stabilized stream of 0.35 wt% SWCNT dispersed in water. The very low viscosity and high CNT load of the dispersion allowed some alignment of CNTs and reasonable cohesion after the surfactant and entrained PVA are washed out and the yarn dried. Although the initial yarns were fragile, great improvements have been made and the process developed for many applications [87,97,98]. Similarly, in analogy to the rigid-rod aramid polymers which form self-aligning liquid-crystal phases [99], direct solvation (albeit requiring mixing for at least 3 days) of the SWCNTs in superacids such as 102% H_2SO_4 gave a nematic 10 wt% dispersion that could be extruded [100] or (at 8 wt%) coagulation-spun to give a pure CNT yarn after the acid is washed out [101]. An approach to pure CNT yarn formation that combines the polymer dispersion and coagulation approaches was patented by Lucent in 2005 [102], whereby a dispersion of CNTs in a UV curable monomer is extruded through a tapered tube and drawn to align them, then cured to a monofilament. This can then be heated to depolymerize and evaporate the carrier (the curing step presumably providing some cohesion benefit) to produce a substantially pure yarn.

1.2.2.2 By direct spinning

A more direct method of yarn formation was reported in 2002 [26] with the discovery that certain substrate-grown MWCNT forests could be drawn from the edge into a continuous filament or, if drawn across a broad front, a wide continuous web with a large range of potential applications [103,104]. By carefully grasping the CNTs at the front face of the "directly spinnable" forest and drawing them horizontally away the contiguous ranks of CNTs, attached by just sufficient van der Waals or other bonding, would then be drawn out in turn. The phenomenon of "direct spinnability", depending as it does upon many delicately balanced variables such as the

CNT-to-CNT and CNT-to-substrate interactions, the length, diameter, dispersity and areal density of the CNTs [105,106], remains a challenge to achieve reliably.

The initial conception of "directly spinnable CNTs" was that the CNTs were connected head-to-head and tail-to-tail and hence drawn out in an essentially ordered linear fashion [107]. Subsequent studies show that yarn (and web) formation is more in the nature of an expanding mesh with multiple points of contact such that, while still being aligned in the draw direction, any one CNT may be folded many times [108]. The structure also contains a proportion of cohesive bundles which are more resistant to folding and contribute to the very low yarn density and large Poisson ratio (decrease in diameter relative to length) under tension [109].

The very poor strength of the as-spun yarns was improved somewhat (but still <1 GPa) by the introduction of twist in very explicit analogy to conventional textile processing by Zhang et al., in 2004 [110]. Twist was briefly suggested by Qian et al. in 2003 as a mechanism to increase the interaction between CNTs, however, the model discussed assumed no cohesion between the CNTs and no migration of them from the edge to the center of the yarn. Interestingly, the Lucent patent [102] lodged early in 2004 included the step of twisting the product to improve the density and closeness of the CNTs to each other. However, the device illustrated is actually a "false twister" that temporarily inserts twist for processing purposes and then releases it.

The early "poor" tensile performance (up to 460 MPa [110]) of directly spinnable yarn was attributed [111] in part to the short (<300 μm) CNT length and efforts made to grow them considerably longer. As a result, strengths of 1.91 GPa [112] and 3.3 GPa [113] were achieved with CNTs of 0.65 and 1 mm length, respectively. Of particular note are that the structure is very disordered and of low density, perhaps as the CNTs were double walled and of relatively large diameter (around 7 nm) and hence prone to collapsing into a ribbon [60]. Also that the yarns exhibit "nonbrittle" tensile failure in the form of a yield point at very low ($\ll 1\%$) strain followed by a long period of very uneven plastic behavior to 2–8% strain prior to failure at a fraction of the maximum tensile stress. This is interpreted as the yarn comprising a bundle of poorly aligned and hence unequally loaded strands which break in turn, transferring load to the remaining structure as the broken and weakly cohering CNTs slide due to the relatively low shear strength between them as modeled by Vilatela et al. [114,115]. This behavior is in distinct contrast to the earlier observations [110] which are of a brittle character (i.e. elastic strain under increasing load to the point of failure) albeit at lower loads and with shorter CNTs.

1.2.2.3 By direct synthesis

The zenith of CNT production for materials applications would be continuous, defect-free single- or small-multiwall filaments formed at some sensible rate in perfect order and alignment (provided such a structure does not hold any unwelcome surprises). The "block diagram" for such production would have reactants entering one side and yarn (or perhaps tow) exiting the other. This vision moved closer to realization with the report by Zhu et al. [116] in 2002 of SWCNTs self assembled into strands or ropes up to 20 cm in length and 300 μm in diameter. This material exhibited some beautifully ordered "crystallite" regions and a tensile strength without any twisting or densification approaching 1 GPa, again with evidence of plastic deformation but little indication of slippage or "pullout" at the break. The self-assembly occurs by aggregation of growing CNTs in the reactor atmosphere and this approach was developed into a continuous process as reported by Li et al. [117] in 2004.

Self-assembly to produce yarn uses the high-temperature (1100−1200 °C) CCI process to grow CNTs in the reactor atmosphere. As they drift down the reaction zone, the CNTs interact and bond (at least initially) by van der Waals attraction. At a sufficiently high density, this produces a cohesive aerogel or "elastic smoke" that will drift down and stick to the cooler areas of the reactor if left undisturbed. If the aerogel, which is in the shape of a diaphanous cylinder or tube, is captured and drawn continuously out of the reactor space as it forms, it collapses into a yarn (or flattens into a sheet) and can be wound directly to a bobbin [117]. The CNTs, which can be single wall or multiwall, grow randomly but acquire some degree of alignment in the gas stream and drawing operation.

As with the directly spun yarns [110], the CNTs of directly synthesized yarn [117] are initially bonded together (which is what enables them to be drawn) becoming more so as the structure is collapsed and densified. Early yarn strengths of 0.1−1 GPa were reported [117] although the MWCNTs were only around 30 μm in length and heavily contaminated with catalyst. The process has been greatly refined since then with an emphasis on SWCNTs and DWCNTs of large diameter (4−10 nm) and hence prone to collapse [60,118], which ostensibly increases the contact area between them [118]. Despite having only some tens of seconds within the growth zone, CNT lengths of about 1 mm are produced, again with some highly ordered regions. Although measured tensile strengths were highly gauge length dependent (indicative of frequent defects) with the majority around 1−2 GPa, strengths of up to about 8.8 GPa were indicated [118], with a very few short-gauge length (1 mm) samples greatly exceeding this. Also of note are that the yarns drawn most quickly were most aligned, thinnest and strongest, and showed brittle rather than ductile (plastic or shear) behavior. Although it engages a reactor to operate continuously at 1200 °C to make a single yarn, this approach to pure CNT yarn production is the first one to achieve commercial success (Nanocomp Technologies Inc.).

1.2.3 On the physical properties of CNT yarn

In the two decades since CNTs were finally appreciated, research has grown exponentially. Production quantity and quality has grown similarly. An understanding of the catalytic processes has blossomed and now much of the periodic table (and a few things not on the periodic table) has been bent to this aim. Where a few microns in length were impressive, we now look forward to the first meter-long fiber. As techniques in nanomanipulation and measurement matured, CNTs quickly and rather pleasingly lived up to the predictions of strength, modulus, electronic and thermal behavior. By any measure, these developments have marked a technological revolution.

Although the focus of this work is on nanotube yarn (or fiber), immense strides have been made in the areas of composites, sensors, electronics etc.—indeed areas too numerous to name. The aim of making a pure CNT yarn is that it exemplifies the challenge of capturing the nanoscale properties—in this context the tensile strength, modulus and conductivity—at the macroscale. Despite all the strides made in the control of diameter, chirality, length, purity, dispersity, surface chemistry, wall number etc., persuading CNTs from their random or parallel-adjacent growth habit to a parallel-axial (yarn) structure has made only incremental progress. Yarn strengths have generally remained around 1–2 GPa or one to two orders of magnitude below constituent fiber strength as summarized in Table 1.2 reproduced from Lu et al. [87].

Approaches to improving yarn strength have emphasized length, alignment and twist as exemplified by the direct synthesis process [118]. Thus, there appears to be some correlation of length and strength. However, although the length of the fibers constituting the CNT yarn was increased from as low as 30 μm up to 1 mm, this 33-fold increase has resulted in only a two- to fourfold increase in strength (albeit with some strikingly strong outliers). Alignment can be influenced by, for example, draw speed, with an approximately eightfold increase in strength from a fourfold increase in speed, limited by the system stability.

Twist has been a rather unsuccessful and perhaps misunderstood approach to improving yarn strength. As noted, it was suggested as early as 2003 [119] as a means to increase cohesion, but presented a model that assumed no contact and no fiber migration and hence was inappropriate to staple fiber. Other models similarly demonstrate winding rather than twisting [120]. The Lucent patent [102] included twisting but illustrated false twisting. The first direct yarn synthesis in 2004 [117] illustrated a mechanism to simultaneously twist and wind yarn on a single axis; however, this is not possible as two normal axes are necessary for this (or, as in cottage and mule spinning, the yarn is bent around the axis and alternately twisted and wound). The most detailed analysis of twist as a classical process applied to directly spinnable CNT yarn was also published in 2004 [110]. However, twist applied to a staple yarn depends upon two related factors ([1], Chapter 3). The first is that there is essentially no cohesion between the fibers (i.e. CNTs) in the untwisted yarn. The second is that the fibers will be free to migrate between the center and the periphery of the yarn as it is twisted, to accommodate the change in path length and hence, tension produced. Cohesion is generated by radial compression-induced friction. A corollary of this is that upon untwisting, there is essentially no residual cohesion ([1], Chapter 7, p. 279). Of course, these constraints do not apply to continuous filament yarn ([1], Chapter 4).

The capacity to "accept exceptional levels of twist", together with "modest levels of shear strength between the nanotubes" is seen as endowing the drawn CNT structure with "yarn-like qualities" [114,115]. However, it is a defining feature of CNTs

Table 1.2 Mechanical and Physical Properties of CNT as Reported in the Literature

		Wet Spinning			
				CNT Characteristics	
Spinning Technique	**Ref.**	**Comments**	**Type**	**Diameter (nm)**	**Length (mm)**
Surfactant dispersion coagulated in PVA solution	[12]	As-spun	SWCNT	–	–
	[13]				
	[33]	Stretched			
	[86]	Annealed			
	[34]	Hot-stretched			
			MWCNT		
	[87]	As-spun	MWCNT	10–15	
Surfactant dispersion coagulated in ethanol/glycerol	[30]	As-spun	SWCNT	–	–
Surfactant dispersion coagulated in diethyl ether	[26]	Heated	MWCNT	50 ± 25	0.1 ± 0.08
Surfactant dispersion coagulated in acid/base	[28]	Annealed	SWCNT	–	–
Surfactant dispersion coagulated in PEI solution	[27]	As-spun	SWCNT	–	–
Sulfuric acid dispersion coagulated in water	[14]	Annealed	SWCNT	–	–
	[29]	As spun			
		Annealed			
Chlorosulfonic acid dispersion coagulated in aqueous sulphuric acid	[31]	As-spun	SWCNT	1.95	~0.5
		Dry Spinning			
Spinning from CNT aerogel	[18]	As-growth	SWCNT	1.1–1.7	~200
	[19]	As-spun	SWCNT/MWCNT	1.6–3.5/30	~0.03
	[8,20,50]	Liquid densification	DWCNT	4–10	~1
	[52]			8–10	–
Spinning from CNT array	[7]	Twisted	DWCNT	7	1.0
	[16]		MWCNT	10	0.1
	[90]			10	0.3
	[58]	Untwisted		10	0.65
		Twisted			
	[39]	Twisted		–	0.17
	[81]			15	0.5
Spinning from CNT array	[59]	Twisted	MWCNT	9–12	4–6
	[41,85,42]	Twisted under tension		7.5–8.5	0.3–0.4

1.2 The Challenge of CNT Yarn Fiber Fabrication

	Fiber Characteristics	Mechanical Properties			Electrical Conductivity (S/cm)	Thermal Conductivity (W/m·k)
Components	Diameter (μm)	Modules (Gpa)	Strength (Gpa)	Toughness (J/g)		
			Wet Spinning			
CNT/PVA	10–100	15	0.15	–	10	–
	50	80	1.8	570	–	
	35	40	0.23	–		
CNT	–	–	–	–	200	10
CNT/PVA	30	45	1.8	40–60	–	–
		35	1.4			
	~50	–	–	–	~1	
CNT	20–30	2	Weak	–	6.7	–
						–
CNT	10–80	69 ± 41	0.15 ± 0.06	–	80	
CNT	10–50	12 Gpa/g-cc	65 MPa/g cc	–	140	–
CNT/PEI	15–50	6 Gpa/g-cc	70–100 MPa/g-cc	5–6	100–200	–
CNT	~52	12.0 ± 10	0.16 ± 0.01	–	5000	21
	60–220	–	–		<4000	~5
					<400	<19
CNT	~13	120	0.05–0.15	–	8300	–
			Dry Spinning			
CNT	5–20	~77	~0.8	–	1400	–
	20	–	0.1–1.0	–	830	
	7–20	78–397	1.3–8.8	13–21	–	
	10–200	–	0.4–1.25	–	5000	
CNT	–	100–263	1.35–3.3	110–975	–	–
	1–10	5–30	0.15–0.46	11–20	300	
	8.5	–	–	–	250	26
	4	275	0.85		170	–
	3	330	1.91		410	
	2–10	–	0.4–0.6		–	–
	10–34	–	–		600	60 ± 20
CNT	10–70	6.5	0.28	–	–	–
	16–48	–	1.2		170–370	
	20–30	37	0.6		–	

(Continued)

Table 1.2 Mechanical and Physical Properties of CNT as Reported in the Literature—Cont'd

			Wet Spinning		
				CNT Characteristics	
Spinning Technique	Ref.	Comments	Type	Diameter (nm)	Length (mm)
	[17]	Liquid densification		15	–
	[40]	Twisted/Liquid densification		10	0.235
	[43]			~10	1.0
	[45]	Twisted/Polymer infiltrated	DWCNT	4–7	0.72
	[44]	Twisted/Polymer infiltrated	MWCNT	10	0.25/0.425
	[46]	Twisted under tension/Polymer infiltrated		7.5–8.5	0.3–0.4
	[88]	Twisted under tension/Particle coated		7.5–8.5	0.3–0.4
Twisting/rolling from CNT film	[21]	Twisted/Liquid	SWCNT	1–2	–
	[54]	Twisted/Polymer infiltrated		1–2	
	[64]	Twisted/Polymer infiltrated	DWCNT	2.2	

Table from Ref. [87].

that they adhere upon intimate contact through the van der Waals interaction. Indeed, directly synthesized or spinnable CNT yarns depend upon such adhesion for them to be drawn out. In this respect, CNT yarn is more akin to a highly porous monofilament than to a staple tow. However, unlike a fully dense filament, which would strongly resist twisting through torsion, the porous structure readily accommodates twisting by compressing, much like a wrung cloth.

As a yarn is twisted, it contracts in overall length in proportion to the diameter and number of turns (hence the resultant surface twist angle) as the peripheral fibers follow a longer path length than the interior fibers ([1], p. 71). A conventional staple yarn would accommodate this by fiber migration, however, as the CNT structure is locked from the outset, contraction would buckle the yarn interior fibers as happens with continuous filament yarn when twisted. Although a staple yarn has near-zero shear strength at zero twist, twist-induced radial compression increases this to the point where a well-aligned and ordered ("worsted process") wool yarn can capture 90% of the constituent fiber strength ([1], p. 304). Even the less well-aligned and ordered ("woollen process") wool yarns can capture around 20%. Poor shear

		Wet Spinning				
	Fiber Characteristics	Mechanical Properties			Electrical Conductivity (S/cm)	Thermal Conductivity (W/m·k)
Components	Diameter (μm)	Modules (Gpa)	Strength (Gpa)	Toughness (J/g)		
	4–34	48–56	0.6-1.1		900	
	~10	8.0 ± 1.0	0.5 ± 01	30 ± 4	500	
CNT/PEI	15–20	12.0 ± 23	2.5 ± 0.31	–	–	
CNT/PVA	4–24	70–120	1.5–2.0		920	
CNT/PU	15–20	~147	1.6–2.0		–	
	13	–	1.0	500		
CNT	30 50	0 15	0.55 0.8			
CNT/PVA	20–36	20–35	0.7–1.3	50		
CNT	27–55	–	~1.6	~30		

strength between CNTs [114] as between wool fibers could be countered if twisting was effective. Note however that poor interfacial shear strength between CNT yarns and polymers is a separate issue that bears closely upon, for example, the reliable tensile testing of these materials [121,122].

Twisting does have the benefit of circularizing the yarn, increasing strain to break and enabling plying (although braiding is perhaps a more efficient approach to assembling heavier CNT yarns). Twisting also increases the CNT-to-CNT interaction, an effect that can be achieved with solvent evaporation [114] or lateral compression (e.g. passing over guides [123]). This is effectively an irreversible process that will persist even after the twisted yarn is untwisted. Thus, the untwisted or solvent densified structure has substantial strength [114,124].

1.3 CONCLUSION

In conclusion, CNT yarns are neither conventional staple structures nor monofilaments. The various approaches to making them share the common features of disorder and low packing efficiency, largely independent of CNT length and whether single wall or small multiwall. Efforts to reliably increase the yarn strength to significant fractions of the constituent fiber strength—be they SWCNTs or MWCNTs and however expressed—have made only incremental progress over a decade of effort. It is frustrating though perhaps also gratifying that CNTs have not behaved as expected. Nevertheless, spectacular progress has been made in the quality and quantity of those CNTs and they have lived up to predictions of individual strength and modulus. Advances in the capacity to grow and order CNTs will most directly affect the realization at the macro-scale of their extraordinary physical properties but also bear upon their application across a wide and ever-expanding field of uses.

References

[1] J.W.S. Hearle, P. Grosberg, S. Backer, Structural Mechanics of Fibers, Yarns, and Fabrics, vol. 1, Wiley-Interscience, 1969.

[2] E. Kvavadze, O. Bar-Yosef, A. Belfer-Cohen, E. Boaretto, N. Jakeli, Z. Matskevich, T. Meshveliani, 30,000-year-old wild flax fibers, Science 325 (2009) 1359.

[3] K.S. Novoselov, A.K. Geim, S.V. Morozov, D. Jiang, Y. Zhang, S.V. Dubonos, I.V. Grigorieva, A.A. Firsov, Electric field effect in atomically thin carbon films, Science 306 (2004) 666−669.

[4] T. Hayashi, Y.A. Kim, T. Matoba, M. Esaka, K. Nishimura, T. Tsukada, M. Endo, M.S. Dresselhaus, Smallest freestanding single-walled carbon nanotube, Nano Letters 3 (7) (2003) 887−889.

[5] C. Qin, L.M. Peng, Measurement accuracy of the diameter of a carbon nanotube from TEM images, Physical Review B 65 (2002), 155431-1−7.

[6] X. Zhao, Y. Liu, S. Inoue, T. Suzuki, R.O. Jones, Y. Ando, Smallest carbon nanotube is 3Å in diameter, Physical Review Letters 92 (12) (2004), 125502-1−3.

[7] L. Guan, K. Suenaga, S. Iijima, Smallest carbon nanotube assigned with atomic resolution accuracy, Nano Letters 8 (2) (2008) 459−462.

[8] S. Iijima, Helical microtubules of graphitic carbon, Nature 354 (1991), 56−58.

[9] M. Monthioux, V.L. Kuznetsov, Who should be given the credit for the discovery of carbon nanotubes? Carbon 44 (2006) 1621−1623.

[10] L.J.E. Hofer, E. Sterling, J.T. McCartney, Structure of the carbon deposited from carbon monoxide on iron, cobalt and nickel, Journal of Physical Chemistry 59 (1955) 1153−1155.

[11] V.M. Luk'yanovich, Structure of the carbon produced in the thermal decomposition of carbon monoxide on an iron catalyst, Zhurnal Fizicheskoi Khimii 26 (1952) 88−95. (CAS 1953:36778). 1953:36778.

[12] R. Iley, H.L. Riley, The deposition of carbon on vitreous silica, Journal of the Chemical Society (1948) 1362−1366.

[13] W.R. Davis, R.J. Slawson, G.R. Rigby, An unusual form of carbon, Nature 171 (1953) 756.

[14] R. Bacon, Growth, structure and properties of graphite whiskers, Journal of Applied Physics 31 (1960) 283−290.

[15] K.P. de Jong, J.W. Geus, Carbon nanofibers: catalytic synthesis and applications, Catalyst Reviews in Science and Engineering 42 (4) (2000) 481−510.

[16] B.M. O'Harra, W.J. Darby, The disintegration of refractory brick by carbon monoxide, Journal of the American Ceramic Society 6 (8) (1923) 904–914.

[17] H.W. Kroto, J.R. Heath, S.C. O'Brien, R.F. Curl, R.E. Smalley, C60: buckminsterfullerene, Nature 318 (1985) 162–163.

[18] H.W. Kroto, Space, stars, C60, and soot, Science 242 (1988) 1139–1145.

[19] J. Cami, J. Bernard-Salas, E. Peeters, S.E. Malek, Detection of C60 and C70 in a young planetary nebula, Science 329 (2010) 1180–1182.

[20] W. Krätschmer, L.D. Lamb, K. Fostiropoulos, D.R. Huffman, Solid C60: a new form of carbon, Nature 347 (1990) 354–358.

[21] P. Thrower, Novel carbon materials – what if? Carbon 37 (1999) 1677–1678.

[22] R. Andrews, D. Jacques, A.M. Rao, F. Derbyshire, D. Qian, X. Fan, E.C. Dickey, J. Chen, Continuous production of aligned carbon nanotubes: a step closer to commercial realization, Chemical Physics Letters 303 (1999) 467–474.

[23] G.G. Tibbetts, Why are carbon filaments tubular? Journal of Crystal Growth 66 (1984) 632–638.

[24] T.V. Hughes, C.R. Chambers, Manufacture of carbon filaments, US Patent, 1889, 18 June, 405480, Filed 30th Aug. 1886.

[25] M.H. Rümmeli, A. Bachmatiuk, F. Börrnert, F. Schäffel, I. Ibrahim, K. Cendrowski, G. Simha-Martynkova, D. Plachá, E. Borowiak-Palen, G. Cuniberti, B. Büchner, Synthesis of carbon nanotubes with and without catalyst particles, Nanoscale Research Letters 6 (2011) 303-1 9.

[26] K. Jiang, Q. Li, S. Fan, Spinning continuous carbon nanotube yarns, Nature 419 (2002) 801.

[27] N. Hamada, S. Sawada, A. Oshiyama, New one-dimensional conductors: graphitic microtubules, Physical Review Letters 68 (10) (1992) 1579–1581.

[28] R.A. Jishi, M.S. Dresselhaus, G. Dresselhaus, Symmetry properties of chiral carbon nanotubes, Physical Review B 47 (24) (1993) 16671–16674.

[29] T.W. Odom, J.L. Huang, P. Kim, C.M. Lieber, Atomic structure and electronic properties of single-walled carbon nanotubes, Nature 391 (1998) 62–64.

[30] J.W.G. Wildöer, L.C. Venema, A.G. Rinzler, R.E. Smalley, C. Dekker, Electronic structure of atomically resolved carbon nanotubes, Nature 391 (1998) 59–62.

[31] M.S. Dresselhaus, G. Dresselhaus, R. Saito, Physics of carbon nanotubes, Carbon 33 (7) (1995) 883–891.

[32] A. Thess, R. Lee, P. Nikolaev, H. Dai, P. Petit, J. Robert, C. Xu, Y.H. Lee, S.G. Kim, A.G. Rinzler, D.T. Colbert, G.E. Scuseria, D. Tománek, J.E. Fischer, R.E. Smalley, Crystalline ropes of metallic carbon nanotubes, Science 273 (1996) 483–487.

[33] M. Liu, J.M. Cowley, Structures of the helical carbon nanotubes, Carbon 32 (3) (1994) 393–403.

[34] S. Amelinckx, A. Lucas, P. Lambin, Electron diffraction and microscopy of nanotubes, Reports of Progress in Physics 62 (1999) 1471–1524.

[35] A. Stetter, J. Vancea, C.H. Back, Conductivity of multiwall carbon nanotubes: role of multiple shells and defects, Physical Review B 82 (2010) 115451-1–5.

[36] W. Guo, Y. Guo, Energy optimum chiralities of multiwalled carbon nanotubes, Journal of the American Chemical Society 129 (10) (2007) 2730–2731.

[37] M. Endo, H.W. Kroto, Formation of carbon nanofibers, Journal of Physical Chemistry 96 (1992) 6941–6944.

[38] S. Iijima, T. Ichihashi, Y. Ando, Pentagons, heptagons and negative curvature in graphite microtubule growth, Nature 356 (1992) 776–778.

[39] A.H.R. Palser, Interlayer interactions in graphite and carbon nanotubes, Physical Chemistry and Chemical Physics 1 (1999) 4459–4464.
[40] Z. Liu, Q. Zhang, L.C. Qin, Accurate determination of atomic structure of multi-walled carbon nanotubes by nondestructive nanobeam electron diffraction, Applied Physics Letters 86 (2005) 191903-2–3.
[41] M. Bretz, B.G. Demczyk, L. Zhang, Structural imaging of a thick-walled carbon microtubule, Journal of Crystal Growth 141 (1994) 304–309.
[42] C.H. Kiang, M. Endo, P.M. Ajayan, G. Dresselhaus, M.S. Dresselhaus, Size effects in carbon nanotubes, Physical Review Letters 81 (9) (1998) 1869–1872.
[43] J. Gallego, J. Barrault, C. Batiot-Dupeyrat, F. Mondragon, Intershell spacing changes in MWCNT induced by metal-CNT interactions, Micron 44 (2013) 463–467.
[44] K.I. Tserpes, Role of intertube spacing in the pullout forces of double-walled carbon nanotubes, Materials and Design 28 (2007) 2197–2201.
[45] Z. Xia, W.A. Curtin, Pullout forces and friction in multiwall carbon nanotubes, Physical Review B 69 (2004) 233408-1–4.
[46] A. Kis, A. Zettl, Nanomechanics of carbon nanotubes, Philosophical Transactions of the Royal Society A 366 (2008) 1591–1611.
[47] K.K.K. Koziol, M. Shaffer, A.H. Windle, Three dimensional internal order in multiwalled carbon nanotubes grown by chemical vapor deposition, Advanced Materials 17 (6) (2005) 760–763.
[48] C. Ducati, K.K.K. Koziol, S. Friedrichs, T.J.V. Yates, M.S. Shaffer, P.A. Midgley, A.H. Windle, Crystallographic order in multi-walled carbon nanotubes synthesized in the presence of nitrogen, Small 2 (6) (2006) 774–784.
[49] K.K.K. Koziol, C. Ducati, A.H. Windle, Carbon nanotubes with catalyst controlled chiral angle, Chemistry of Materials 22 (2010) 4904–4911.
[50] K.B.K. Teo, C. Singh, M. Chhowalla, W.I. Milne, Catalytic synthesis of carbon nanotubes and nanofibers and Nalwa, 2004 nanotubes and nanofibers, in: H.S. Nalwa (Ed.), Encyclopedia of Nanoscience and Nanotechnology, vol. 1, American Scientific Publishers, 2004, pp. 665–686.
[51] A.V. Melechko, V.I. Merkulov, T.E. McKnight, M.A. Guillorn, K.L. Klein, D.H. Lowndes, M.L. Simpson, Vertically aligned carbon nanofibers and related structures: controlled synthesis and directed assembly, Journal of Applied Physics 97 (2005) 041301-1–39.
[52] T. Ozkan, M. Naraghi, I. Chasiotis, Mechanical properties of vapor grown carbon nanofibers, Carbon 48 (2010) 239–244.
[53] C. Wei, D. Srivastava, Nanomechanics of carbon nanofibers: structural and elastic properties, Applied Physics Letters 85 (12) (2004) 2208–2210.
[54] J.G. Lawrence, L.M. Berhan, A. Nadarajah, Elastic properties and morphology of individual carbon nanofibers, ACS Nano 2 (6) (2008) 1230–1236.
[55] D.H. Robertson, D.W. Brenner, J.W. Mintmire, Energetics of nanoscale graphitic tubules, Physical Review B 45 (21) (1992) 12592–12594.
[56] R.S. Ruoff, D. Qian, W.K. Liu, Mechanical properties of carbon nanotubes: theoretical predictions and experimental measurements, Comptes Rendus Physique 4 (2003) 993–1008.
[57] M.F. Budyka, T.S. Zyubina, A.G. Ryabenko, S.H. Lin, A.M. Mebel, Bond lengths and diameters of armchair single wall carbon nanotubes, Chemical Physics Letters 407 (2005) 266–271.

[58] N.G. Chopra, L.X. Benedict, V.H. Crespi, M.L. Cohen, S.G. Louie, A. Zettl, Fully collapsed carbon nanotubes, Nature 377 (1995) 135–138.
[59] J.A. Elliot, J.K.W. Sandler, A.W. Windle, R.J. Young, M.S.P. Shaffer, Collapse of single-wall carbon nanotubes is diameter dependent, Physical Review Letters 92 (9) (2004) 95501-1–4.
[60] M. Motta, A. Moisala, I.A. Kinloch, A.H. Windle, High performance fibres from 'dog bone' carbon nanotubes, Advanced Materials 19 (2007) 3721–3726.
[61] H. Shima, M. Sato, Multiple radial corrugations in multiwalled carbon nanotubes under pressure, Nanotechnology 19 (2008) 495705-1–8.
[62] H. Shima, Buckling of carbon nanotubes: a state of the art review, Materials 5 (2012) 47–84.
[63] R.S. Ruoff, J. Tersoff, D.C. Lorents, S. Subramoney, B. Chan, Radial deformation of carbon nanotubes by van der Waals forces, Nature 364 (1993) 514–516.
[64] T. Hertel, R.E. Walkup, P. Avouris, Deformation of carbon nanotubes by surface van der Waals forces, Physical Review B 58 (20) (1998) 13870–13873.
[65] M. Giulianini, E.R. Waclawik, J.M. Bell, M. De Crescenzi, P. Castrucci, M. Scarselli, M. Diociauti, S. Casciardi, N. Motta, Evidence of multiwall carbon nanotube deformation caused by poly(3-hexylthiophene) adhesion, Journal of Physical Chemistry C 115 (2011) 6324–6330.
[66] S.C. Hawkins, J.M. Poole, C.P. Huynh, Catalyst distribution and carbon nanotube morphology in multilayer forests by mixed CVD processes, Journal of Physical Chemistry C 113 (2009) 12976–12982.
[67] M.J. Bronikowski, P.A. Willis, D.T. Colbert, K.A. Smith, R.E. Smalley, Gas-phase production of single-walled nanotubes form carbon monoxide via the HiPCO process: a parametric study, Journal of Vacuum Science and Technology A 19 (4) (2001) 1800–1805.
[68] Z.F. Ren, Z.P. Huang, J.W. Xu, J.H. Wang, P. Bush, M.P. Siegal, P.N. Provencio, Synthesis of large arrays of well-aligned carbon nanotubes on glass, Science 282 (1998) 1105–1107.
[69] K. Hata, D.N. Futaba, K. Mizuno, T. Namai, M. Yumura, S. Iijima, Water-assisted highly efficient synthesis of impurity-free single-walled carbon nanotubes, Science 306 (2004) 1362–1364.
[70] X. Wang, Q. Li, J. Xie, Z. Jin, J. Wang, Y. Li, K. Jiang, S. Fan, Fabrication of ultralong and electrically uniform single-walled carbon nanotubes on clean substrates, Nano Letters 9 (9) (2009) 3137–3141.
[71] Q. Wen, R. Zhang, W. Qian, Y. Wang, P. Tan, J. Nie, F. Wei, Growing 20 cm long DWNTs/TWNTs at a rapid growth rate of 80–90 μm/s, Chemistry of Materials 22 (2010) 1294–1296.
[72] R. Zhang, Q. Wen, W. Qian, D.S. Su, Q. Zhang, F. Wei, Superstrong ultralong carbon nanotubes for mechanical energy storage, Advanced Materials 23 (2011) 3387–3391.
[73] R. Zhang, Q. Wen, W. Qian, D.S. Su, Q. Zhang, F. Wei, Ultralong Carbon Nanotubes for the Storage of Mechanical Energy, NT12, The 13th International Conference on the Science and Technology of Carbon Nanotubes (Abstract), NT12, June, 2012.
[74] T. Guo, P. Nikolaev, A.G. Rinzler, D. Tomanek, D.T. Colbert, R.E. Smalley, Self assembly of tubular fullerenes, Journal of Physical Chemistry 99 (1995) 10694–10697.

[75] T. Guo, P. Nikolaev, A. Thess, D.T. Colbert, R.E. Smalley, Catalytic growth of single walled carbon nanotubes by laser vaporization, Chemical Physics Letters 243 (1995) 49–54.

[76] T.W. Ebbesen, P.M. Ajaayan, Large-scale synthesis of carbon nanotubes, Nature 358 (1992) 220–222.

[77] S. Iijima, T. Ichihashi, Single-shell carbon nanotubes of 1-nm diameter, Nature 363 (1993) 603–605.

[78] D.S. Bethune, C.H. Klang, M.S. de Vries, G. Gorman, R. Savoy, J. Vazquez, R. Beyers, Cobalt catalysed growth of carbon nanotubes with single-atomic-layer walls, Nature 363 (1993) 605–607.

[79] C. Journet, W.K. Maser, P. Bernier, A. Loiseau, M. Lamey de la Chapelle, S. Lefrants, P. Deniard, R. Lee, J.E. Fischer, Large-scale production of single-walled carbon nanotubes by the electric-arc technique, Nature 388 (1997) 756–758.

[80] C.D. Scott, S. Arepalli, P. Nikolaev, R.E. Smalley, Growth mechanisms for single-wall carbon nanotubes in a laser ablation process, Applied Physics A 72 (2001) 573–580.

[81] H. Qiu, G. Yang, B. Zhao, J. Yang, Catalyst-free synthesis of multi-walled carbon nanotubes from carbon spheres and its implications for the formation mechanism, Carbon 53 (2013) 137–144.

[82] M.L. Lieberman, C.R. Hills, C.J. Miglionico, Growth of graphite filaments, Carbon 9 (1971) 633–635.

[83] M.F. Yu, B.S. Files, S. Arepalli, R.S. Ruoff, Tensile loading of ropes of single wall carbon nanotubes and their mechanical properties, Physical Review Letters 84 (24) (2000) 5552–5555.

[84] M.F. Yu, O. Lourie, M.J. Dyer, K. Moloni, T.F. Kelly, R.S. Ruoff, Strength and breaking mechanism of multiwalled carbon nanotubes under tensile load, Science 287 (2000) 637–640.

[85] D.V. Smitherman Jr, Space Elevators – An Advanced Earth-Space Infrastructure for the New Millennium, NASA Conference Proceedings, NASA/CP-2000-210429 (2000) pp. 1–38.

[86] B. Peng, M. Locascio, P. Zapol, S. Li, S.L. Mielke, G.C. Schatz, H.D. Espinosa, Measurements of near-ultimate strength for multiwalled carbon nanotubes and irradiation-induced crosslinking improvements, Nature Nanotechnology 3 (2008) 626–631.

[87] W. Lu, M. Zu, J.H. Byun, B.S. Kim, T.W. Chou, State of the art of carbon nanotube fibers: opportunities and challenges, Advanced Materials 24 (2012) 1805–1833.

[88] R. Andrews, D. Jacques, A.M. Rao, T. Rantell, F. Derbyshire, Y. Chen, J. Chen, R.C. Haddon, Nanotube composite carbon fibers, Applied Physics Letters 75 (2000) 1329–1331.

[89] R. Haggenmueller, H.H. Gommans, A.G. Rinzler, J.E. Fischer, K.I. Winey, Aligned single-wall carbon nanotubes in composites by melt processing methods, Chemical Physics Letters 330 (2000) 219–225.

[90] G. Hu, C. Zhao, S. Zhang, M. Yang, Z. Wang, Low percolation thresholds of electrical conductivity and rheology in poly(ethylene terephthalate) through the networks of multi-walled carbon nanotubes, Polymer 47 (2006) 480–488.

[91] I. Alig, T. Skipa, D. Lellinger, P. Pötschke, Destruction and formation of a carbon nanotube network in polymer melts: rheology and conductivity spectroscopy, Polymer 49 (2008) 3524–3532.

[92] I.A. Kinloch, S.A. Roberts, A.H. Windle, A rheological study of concentrated aqueous nanotube dispersions, Polymer 43 (2002) 7483–7491.

[93] Y.Y. Huang, S.V. Ahir, E.M. Terentjev, Dispersion rheology of carbon nanotubes in a polymer matrix, Physical Review B 73 (2006) 125422-1–9.

[94] S. Giordani, S.D. Bergin, V. Nicolosi, S. Lebedkin, M.M. Kappes, W.J. Blau, J.N. Coleman, Debundling of singe-walled nanotubes by dilution: observation of large populations of individual nanotubes in amide solvent dispersions, Journal of Physical Chemistry B 110 (2006) 15708–15718.

[95] E.K. Hobbie, Shear rheology of carbon nanotube suspensions, Rheologica Acta 49 (2010) 323–334.

[96] B. Vigolo, A. Penicaud, C. Coulon, C. Sauder, R. Pailler, C. Journet, P. Bernier, P. Poulin, Macroscopic fibers and ribbons of oriented carbon nanotubes, Science 290 (2000) 1331–1334.

[97] A.B. Dalton, S. Collins, E. Muñoz, J.M. Razal, V.H. Ebron, J.P. Ferraris, J.N. Coleman, B.G. Kim, R.H. Baughman, Super tough carbon nanotube fibres, Nature 423 (2003) 703.

[98] A.B. Dalton, S. Collins, J. Razal, E. Muñoz, V.H. Ebron, B.G. Kim, J.N. Coleman, J.P. Ferraris, R.H. Baughman, Continuous carbon nanotube composite fibers: properties, potential application, and problems, Journal of Materials Chemistry 14 (2004) 1–3.

[99] S.L. Kwolek, P.W. Morgan, J.R. Schaefgen, L.W. Gulrich, Synthesis, anisotropic solutions and fibers of poly(1,4-benzamide), Macromolecules 10 (6) (1977) 1390–1396.

[100] V.A. Davis, L.M. Ericson, A.N.G. Parra-Vasquez, H. Fan, Y. Wang, V. Prieto, J.A. Longoria, S. Ramesh, R.K. Saini, C. Kittrell, W.E. Billups, W.W. Adams, R.H. Hauge, R.E. Smalley, M. Pasquali, Phase behavior and rheology of SWNTs in superacids, Macromolecules 37 (2004) 154–160.

[101] C. Guthy, A.N.G. Parra-Vasquez, M.J. Kim, S. Ramesh, R.K. Saini, C. Kittrell, G. Lavin, H. Schmidt, W.W. Adams, W.E. Billups, M. Pasquali, W.F. Hwang, R.H. Hauge, J.E. Fischer, R.E. Smalley, Macroscopic, neat, single-walled carbon nanotube fibers, Science 305 (2004) 1447–1450.

[102] D.S. Greywall, B. Yurke, Carbon Particle Fiber Assembly Technique, US Patent, USP, 2005, 0189671A1, Filed 27 Feb, 2004.

[103] M. Zhang, S. Fang, A.A. Zakhidov, S.B. Lee, A.E. Aliev, C.D. Williams, K.R. Atkinson, R.H. Baughman, Strong, transparent, multifunctional, carbon nanotube sheets, Science 309 (2005) 1215–1219.

[104] A.E. Aliev, C.D. Williams, R.H. Baughman, Multifunctional carbon nanotube yarns and transparent sheets: fabrication, properties and applications, Physica B 394 (2007) 339–343.

[105] C.P. Huynh, S.C. Hawkins, Understanding the synthesis of directly spinnable carbon nanotube forests, Carbon 48 (2010) 1105–1115.

[106] C.P. Huynh, S.C. Hawkins, M. Redrado, S. Barnes, D. Lau, W. Humprhies, G.P. Simon, Evolution of directly spinnable carbon nanotube growth by recycling analysis, Carbon 49 (2011) 1989–1997.

[107] K. Jiang, S. Fan, Q. Li, Method for Fabricating Carbon Nanotube Yarn, US Patent, USP, 2004, 0053780A1, Filed 31 Dec, 2002.

[108] A.A. Kuznetsov, A.F. Fonseca, R.H. Baughman, A.A. Zakhidov, Structural model for dry-drawing of sheets and yarns from carbon nanotube forests, ACS Nano 5 (2) (2011) 985–993.

[109] M. Miao, J. McDonnell, L. Vuckovic, S.C. Hawkins, Poisson's ratio and porosity of carbon nanotube dry-spun yarns, Carbon 48 (2010) 2802–2811.

[110] M. Zhang, K.R. Atkinson, R.H. Baughman, Multifunctional carbon nanotube yarns by downsizing an ancient technology, Science 306 (2004) 1358–1361.

[111] Q. Li, X. Zhang, R.F. DePaula, L. Zheng, Y. Zhao, L. Stan, T.G. Holesinger, P.N. Arendt, D.E. Peterson, Y.T. Zhu, Sustained growth of ultralong carbon nanotube arrays for fiber spinning, Advanced Materials 18 (2006) 3160–3163.

[112] X. Zhang, Q. Li, Y. Tu, Y. Li, J.Y. Coulter, L. Zheng, Y. Zhao, Q. Jia, D.E. Peterson, Y.T. Zhu, Strong carbon-nanotube fibers spun from long carbon-nanotube arrays, Small 3 (2) (2007) 244–248.

[113] X. Zhang, Q. Li, T.G. Holsinger, P.N. Arendt, J. Huang, P.D. Kirven, T.G. Clapp, R.F. DePaula, X. Liao, Y. Zhao, L. Zheng, D.E. Peterson, Y. Zhu, Ultrastrong, stiff, and lightweight carbon nanotube fibers, Advanced Materials 19 (2007) 4198–4201.

[114] J.J. Vilatela, A.H. Windle, Yarn-like carbon nanotube fibers, Advanced Materials 22 (2010) 4959–4963.

[115] J.J. Vilatela, J.A. Elliott, A.H. Windle, A model for the strength of yarn-like carbon nanotube fibers, ACS Nano 5 (3) (2011) 1921–1927.

[116] H.W. Zhu, C.L. Xu, D.H. Wu, B.Q. Wei, R. Vajtai, P.M. Ajayan, Direct synthesis of long single-walled carbon nanotube strands, Science 296 (2002) 884–886.

[117] Y.L. Li, I.A. Kinloch, A.H. Windle, Direct spinning of carbon nanotube fibers from chemical vapor deposition synthesis, Science 304 (2004) 276–278.

[118] K.K.K. Koziol, J. Vilatela, A. Moisala, M. Motta, P. Cunniff, M. Sennett, A.H. Windle, High performance carbon nanotube fiber, Science 318 (2007) 1892–1895.

[119] D. Qian, W.K. Liu, R.S. Ruoff, Load transfer mechanisms in carbon nanotube ropes, Composites Science and Technology 63 (2003) 1561–1569.

[120] I.J. Beyerleine, P.K. Porwal, Y.T. Zhu, K. Hu, X.F. Xu, Scale and twist effects on the strength of nanostructured yarns and reinforced composites, Nanotechnology 20 (2009) 485702-1–10.

[121] F. Deng, W. Lu, H. Zhao, Y. Zhu, B.S. Kim, T.W. Chou, The properties of dry-spun carbon nanotube fibers and their interfacial shear strength in an epoxy composite, Carbon 49 (2011) 1752–1757.

[122] M. Zu, Q. Li, Y.T. Zhu, M. Dey, G. Wang, W. Lu, J.M. Deitzel, J.W. Gillespie Jr., J.H. Byun, T.W. Chou, The effective interfacial shear strength of carbon nanotube fibers in an epoxy matrix characterized by a microdroplet test, Carbon 50 (2012) 1271–1279.

[123] C.D. Tran, W. Humphries, S.M. Smith, C.P. Huynh, S. Lucas, Improving the tensile strength of carbon nanotube spun yarns using a modified spinning process, Carbon 47 (2009) 2662–2670.

[124] S. Fang, M. Zhang, A.A. Zakhidov, R.H. Baughman, Structure and process-dependent properties of solid-state spun carbon nanotube yarns, Journal of Physics: Condensed Matter 22 (2010) 334221-1–6.

CHAPTER 2

New Applications and Techniques for Nanotube Superfiber Development

Mark J. Schulz[1], Brad Ruff[1], Aaron Johnson[1], Kumar Vemaganti[1], Weifeng Li[1], Murali M. Sundaram[1], Guangfeng Hou[1], Arvind Krishnaswamy[1], Ge Li[2], Svitlana Fialkova[3], Sergey Yarmolenko[3], Anli Wang[1], Yijun Liu[1], James Sullivan[1], Noe Alvarez[1], Vesselin Shanov[1], Sarah Pixley[4]

[1] College of Engineering and Applied Science, University of Cincinnati, Cincinnati, OH USA, [2] General Nano, Cincinnati, OH USA, [3] Center for Advanced Materials and Smart Structures, North Carolina A&T State University, Greensboro, NC USA, [4] College of Medicine, University of Cincinnati, Cincinnati, OH USA

CHAPTER OUTLINE

2.1 New Applications for Nanotube Superfiber Development 33
    2.1.1 Sustainable nanotube materials for electromagnetics applications 35
    2.1.2 Nanoelectric motor design 37
    2.1.3 Nanovivo robots to change interventional medicine 39
    2.1.4 Nanodevices made from one long CNT 42
    2.1.5 Biomedical applications of nanotube thread 43
2.2 New Techniques for Nanotube Superfiber Development 43
    2.2.1 Reactor design for growing carbon nanotubes 44
    2.2.2 Thermal annealing of carbon nanotubes 45
    2.2.3 Substrate patterning for nanotube growth control 47
    2.2.4 Spinning carbon nanotube thread 53
2.3 CONCLUSIONS 56
Acknowledgments 57
References 57

2.1 NEW APPLICATIONS FOR NANOTUBE SUPERFIBER DEVELOPMENT

The initial enthusiasm for nanotechnology was driven largely by the promise of what these materials could do. It has now been long enough for many to discover what can currently be achieved, to gain some insight into what the problems are, and to predict and chart a trajectory to a realistic target [1–24]. Thus, using carbon nanotubes (CNTs) as a model, there is the theory of how strong individual CNTs should be and the practicality of how strong they are as measured by various groups. Information is presented in this chapter, and in this book, in the recurring structure of what is the promise, what is the reality, and where is reality falling short of the promise and why? Thus for CNTs, the promise is of a strength of about 35 GPa and a modulus of about 1 TPa, based on perfect nanotubes and using the cross-sectional area of the nanotube to compute properties. Some researchers report higher strength based on just the area of the nanotube wall. Considering the cross-sectional area of the tube, which is used for engineering design, the properties of nanotubes still significantly exceed the properties of existing materials. The reality for single-wall nanotubes is that issues of defects, measurement, and interpretation still complicate the picture. Multiwall nanotubes properties are sometimes based on considering the outer wall only. Multiwall tubes present the challenge of promoting inner wall utilization. The problems of dispersion, alignment and matrix interaction, which have been a challenge when using the powder form of nanotubes, are mostly solved using continuous materials such as yarn and sheet. Continuous materials provide much greater improvement in properties than powder forms of nanotubes. These practical aspects are reviewed in this chapter and techniques are suggested that may allow superfiber materials to more closely approach the engineering properties of nanotubes.

In considering the utilization of superfiber materials, the promise is "the elevator cable to orbit" and the reality is superfiber yarns are currently on par with high-end existing materials. We explain where we are with pure CNT yarns and what are the problems, approaches and prospects. This chapter addresses nanotube materials as a uniquely valuable form of matter. Our perspective is that *Nanotube Superfiber Materials will engage a new generation of materials scientists and technologists working across a spectrum of disciplines from electronics to aerospace structures to biomedical microdevices. Bringing together theory and practice, the promise and the reality, the problems, and the future prospects of this most valuable of new materials will put us in a good position to improve our material world.*

The promise of any material starts with the potential applications it can fulfill. Many applications of CNT threads have already been considered including reinforcements for polymer composites with increased strength as well as electrical and thermal conductivities, structural health monitoring sensors, windings for carbon motors, and filaments for enhanced heat dissipation in rescue garments. In addition, more general uses of CNT threads include power conduction, electromagnetic interference (EMI) shielding (coaxial cables), biomedical fibers, ballistic protection garments and panels, impact-resistant glass, wearable antennas, biosensors, self-sensing composites, supercapacitors, batteries, hybrid superconductors, and reinforced elastomers and cables for high magnetic field applications. In applications ranging from aerospace to medical and beyond, it can be anticipated that fibers based on nanotechnology will play a significant role in future technological development. Some of the more intriguing uses of CNT yarns are described next. The chapter is an overview of work being performed by a large number of researchers investigating new applications and techniques for developing nanotube superfiber.

2.1.1 Sustainable nanotube materials for electromagnetics applications

Electromagnetic devices are the main energy prime movers in our society. Motors and generators that are lightweight and built without relying on diminishing supplies of permanent magnet materials and copper would be helpful for developing strong economies and a sustainable environment for the world. Traditional materials used in electric motors are bulk metals and alloys. The properties of these materials can be slightly tuned by varying the composition of the alloys or taking advantage of geometry, i.e. lamination of thin steel sheets to reduce the eddy currents. But ultimately, the properties are limited. In the future, multifunctional electromagnetic materials including CNT wire and superparamagnetic nanoparticle magnetic composite core materials will be used to replace copper and iron in conventional motors and design lightweight high-performance electric motors [23,24]. Currently, CNT ribbon that is postprocessed might achieve close to the specific conductivity of copper. This is good enough for some applications, but the properties are limited due to the length and quality of the CNTs in the ribbon. CNT ribbon in the future will be manufactured using new procedures based on long, high-quality nanotubes.

The impact of sustainable nanoscale materials will be to replace metals in electric motors with (1) CNT material, (2) magnetic nanoparticle composites, and (3) carbon fiber/nanotube composites. This will be the greatest change in electric motors design in the last century. Replacing copper with superfiber is an important advance to be made in terms of sustainability. Copper is primarily made from the ore mined through a process of open pit mining. Open pit mining is strictly regulated by the federal government. However, there are still downsides to the practice. These open pit mines can reach massive size as the dirt removed becomes less concentrated with the desired element. The Bingham Canyon Mine, located in Salt Lake County, Utah, is the largest man-made excavation site in the world and is visible from space. Assuming that copper demand in the United States does not increase much above current demand, which is a conservative estimate, the total carbon footprint of copper mining will increase from 20 to over 55 million tons of CO_2 per year. This is mainly due to the decrease in the quality of copper ore that can be mined. The highest quality copper ore is being depleted rapidly and the industry is forced to use lower quality ore which is more energy and carbon intensive per pound of copper produced. One way to reduce the carbon footprint of copper is to use more recycled copper. Currently, only 18% of copper comes from recycling. If the demand for copper was reduced by 1% annually, then there would only be a need for 36% of copper to come from recycling, which could reduce CO_2 emissions by 60% by 2050. If the replacement of copper with CNT ribbon was to curb consumption of copper by only 1% each year, the environmental effects would be tremendous. As nations like China build massive infrastructure to support its population, they consume copper (Fig. 2.1). This increase in global demand with the decrease in quality of copper ore may drive the price of copper up. With rising copper prices, the world will begin to look for alternative sustainable materials

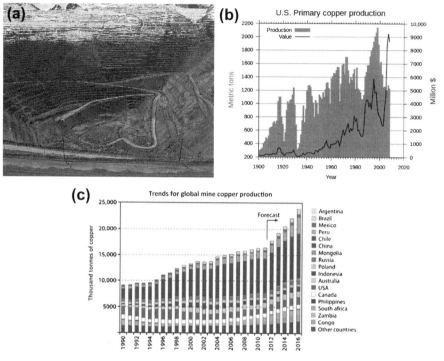

FIGURE 2.1

Copper manufacturing: (a) Bingham Canyon Copper Mine, the world's largest man-made excavation (from http://en.wikipedia.org/wiki/File:Bingham_Canyon_April_2005.jpg); (b) US copper production and value (http://en.wikipedia.org/wiki/File:US_Primary_Copper_Production_v2.svg); (c) Global copper consumption for developing nations (http://oracleminingcorp.com/copper/). (For color version of this figure, the reader is referred to the online version of this book.)

for electrical conductors. If CNT wire can be developed to support part of this demand, it would help our economy and protect our environment.

Another area to replace metal is in tubing. Carbon tubing can replace copper and steel tubing to increase stiffness and reduce weight for fluid-flow applications. Carbon tubing can also replace composite tubes for structural applications. Carbon tubing is formed by drawing nanotube ribbon from an array, infiltrating the ribbon with a polymer such as epoxy and wrapping the ribbon on a mandrel. The prospect of replacing huge copper mines and smelting operations with nanotube reactors that can be operated at any small business to produce nanotube wire and tubing is extremely attractive based on sustainability of our planet. Economically, the new technology may also spawn new industries manufacturing nanomaterial electric motors.

2.1.2 Nanoelectric motor design

CNT fiber used as an electrical conductor can lead to new and revolutionary applications. Copper wire is a great electrical conductor but it has some downsides. The largest downside is that copper wire is heavy. It is also not suited for use at elevated temperatures due to the decrease in electrical and mechanical properties. CNT superfibers have the potential to change that. Not only does the mechanical strength of the CNT fibers remain stable but also the conductivity increases with temperature. The direct current (DC) conductivity of individual CNTs matches that of copper. If these properties can be realized in a continuous fiber, then the applications are open ended. One application of the CNT superwire is in an electric motor. The properties of the CNT wire can lead to a fundamental change in electric motor design. When the weight of the electrical conductors is decreased by a factor of 5, then centripetal and gyroscopic forces are proportionally reduced. This expands the design possibilities of motors to include large diameters, lightweight, and high-RPM motors. These design parameters can be used to increase the torque and power of electric motors. CNT fibers also have better high-frequency performance than copper wire. The CNTs in the fibers can be insulated from each other to form a nano-litz wire that limits the radial and hoop direction conductivity, and therefore, the skin effect within the fiber. Skin effect is an electromotive force that increases the impedance of wire at high frequency and is increased with increasing diameter of the conductor. Allowing high-frequency current operation with no skin effect can allow electric motors to have a high number of poles and to operate at high RPMs with lower losses in the conductors [22,23].

These design features can lead to new motor designs that differ substantially from traditional motor design. Some design concepts are shown below. The first design concept is a radial flux motor that utilizes a Halbach array formation of motor poles, which is shown in Fig. 2.2(a). In this design, the Halbach array configuration

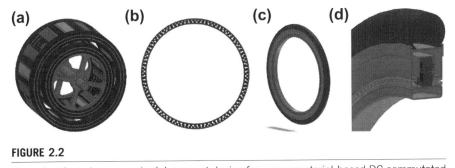

FIGURE 2.2

Nanoelectric motor concepts: (a) concept design for a nanomaterial-based DC commutated series wound Halbach array motor; (b) design of a ring motor with a very high number of poles to take advantage of the nanomaterial properties; (c) a hubless wheel motor based on the nanomaterial ring motor; (d) cross-section of the hubless wheel motor. The brown area is the motor within the wheel structure. (For interpretation of the references to color in this figure legend, the reader is referred to the online version of this book.)

allows for a better utilization of space to create a one-sided magnetic flux path. It also uses a CNT composite as both an electrical power conduit and the mechanical structure of the motor. This saves weight and space in the motor. This motor can be operated as a series wound commutated DC motor where the input voltage can be a DC or an alternating current. This configuration has advantages because the speed of the motor is not affected by the frequency of the input voltage. Then, the high frequency impedance drop of the CNT fiber can be taken advantage of without affecting the operation speed. In order to make this design operate at high frequency, the magnetic core material must also be suited for use at high frequency.

The second design of a CNT fiber-based electric motor takes advantage of the lightweight and high strength of the CNTs. This design is based on the idea that electric motor torque is increased as the diameter of the motor is increased. In addition, the torque is increased as the number of poles in the motor increases. As the number of poles and diameter increases, the amount of support material in the center of the motor also increases. This lends this motor concept toward a hubless wheel design shown in Fig. 2.2(b–d). This type of design would be impractical with a traditional brushless DC motor because of the use of permanent magnets. Permanent magnets used in motor design are heavy and expensive. Even in small motors, the cost of the magnets is often the most expensive component. Using nanomaterials, we can eliminate the need for permanent magnets.

There is well-established motivation for improving electric motors. Conventional linear and rotary electric motors, generators, and actuators are limited because they have to balance and tradeoff torque, speed, efficiency and weight to be used in developing advanced lightweight vehicles and machines. However, nanoelectric motors can overcome to a large extent the compromises made using traditional metal material. The proposed nanomaterials and design also naturally fall within a category of motors known as ring or wheel motors. Ring motors have a hollow center section that can be left as empty space or connected to a shaft via spokes. This design is perfect for many advanced concepts of hubless wheels that current motor technology does not support. Ring motors are also perfect for use with turbo fan propulsion for electric aircraft. Figure 2.3 shows different applications that could benefit from a ring nanomotor. The technology already exists for making practical electric cars like in the Chevy volt, Nisson leaf, AMP motors, etc. for everyday commuters. However, the best-selling electric car of all time is the Tesla Roadster because of its excellent performance and styling. Having electric motor technology will bring an aura to electric cars that cannot be matched by gasoline powered cars. This will increase consumer adoption of electric vehicles which is barred by range anxiety and general skepticism. Here, the motor is in the wheels and the battery is located in the frame where the gasoline engine typically is. Designing the motor based on nanoscale material allows the optimum performance of the motor to be achieved as compared to just putting nanoscale materials into an existing design of motor. The development of the nanomotor is a research project underway. It is also exciting imagining what designs can come from using lightweight thin section wheel motors with huge torque, high speed, and no gearbox.

FIGURE 2.3

Application concepts for nanoelectric motors: (a) concept electric motorcycle (https://encrypted-tbn0.gstatic.com/images?q=tbn:ANd9GcQ_FYz82G86i681NfhUejUp02Mx-b0ysuZT_22sXmGOCxPoWRV7); (b) Nike One Concept car (http://robson.m3rlin.org/cars/wp-content/uploads/2008/01/nike_one_10_1920-copy.jpg); (c) experimental vertical take-off and landing aircraft (https://encrypted-tbn0.gstatic.com/images?q=tbn:ANd9GcSlpHjEeGnA9SsrODxpHpLyPQ3fLhBGGYa4ZpALMuOB_19-p9WNCg); (d) hubless wheel chair (https://encrypted-tbn2.gstatic.com/images?q=tbn:ANd9GcTKew3zo4wFVj9Yy5epSmC4KYugOSMOCVaysqAm5wb6WkIU4XWXhA). (For color version of this figure, the reader is referred to the online version of this book.)

The main goal in the motor design is to take advantage of the lightweight and high maximum current density of CNT materials. Properties of CNT and conventional materials are listed in Table 2.1. Using long CNT will increase the electrical conductivity.

In electric motor design, many factors must be considered and weighed based on the application. The most important factor is thermal management. Other important considerations in motor designs are efficiency, weight, size, speed and torque. Electrical conductors will always generate heat equal to I^2R. This heat has to be dissipated or it will build up in the motor. Using magnetic plastics as a core and structural material, and superfiber to form electromagnets, the geometric freedom is nearly unlimited. Any shape that can be molded could be made into a core structure. This allows design of complex shapes that permit convection cooling directly to the environment. The rotation of the motor can be used to force air past the coils and cool them.

2.1.3 Nanovivo robots to change interventional medicine

Nanotube superfiber can enable building a family of micron–millimeter-scale robots to diagnose and treat cancer and other ubiquitous, incurable, and disabling diseases. These types of robots have never been built before and this is a new application of superfiber. Nanovivo robots can be built using *nano*scale materials and used in vivo to treat patients in a minimally invasive way which is transformational in terms of engineering design and revolutionary in terms of medical innovation [21]. Nanoscale materials and nanomanufacturing processes are allowing miniaturization of robotic devices that was previously impossible. And the design of a tethered robot using a micron-size nanotube cable overcomes the long-standing problems of how to power and communicate with robots in the body. These miniature robotic devices will move around the body, take measurements locally, and

Table 2.1 Electrical Properties of MWCNT Materials and Metals*

Material	Density (g/cm^3)	Resistivity ($\Omega \cdot$cm)	Skin Effect	High Frequency Impedance	High Temperature Resistivity
Copper	8.9	1.7×10^{-6}	Large	Increases	Increases
Aluminum	2.7	2.8×10^{-6}	Large	Increases	Increases
Iron	7.9	9.6×10^{-6}	Large	Increases	Increases
MWCNT (Short, no defects)	1.8	$10^{-5} - 10^{-6}$	No	Constant	Decreases
MWCNT (Short, with defects)	1.8	$10^{-4} - 10^{-5}$	No	Constant	Decreases
MWCNT Yarn (MWCNT short with defects)	1.0	$10^{-3} - 10^{-4}$	Small	Decreases	Constant or Decreases

*Comparison of the resistivities of nanotube materials in the table suggests there are three problems that must be solved to produce CNT yarn with similar DC electrical properties to metals: 1-reduce the number of defects in CNTs, 2-reduce the number of junctions in yarn, and 3-increase the density of yarn. Thermal annealing, synthesis of long CNT, and spinning straight CNT are approaches being investigated to solve the three problems. CNT materials may out-perform metals at high frequency and high temperature. The low density of CNT materials is a competitive advantage for many applications.

perform functions like taking biopsies or tumor ablation. The robots will provide new treatment options and will be of great benefit in medicine. A concept robot is shown in Fig. 2.4.

Nanorobots have long been fantasized starting from 1959 with Richard Feynman's ideas for tiny machines. But nanorobots have not been built because of the lack of interdisciplinary expertise, facilities, nanoscale materials, and nanomanufacturing technology. Also, access to medical schools and physicians to assist in the design and testing of nanodevices has been limited. In this project, engineers, scientists, and physicians have teamed together to design nanovivo robots and to test the devices in their labs. Our goal/vision is that mobile microscale robots will move through the cardiovascular system to access disease sites throughout the body. This will bring transformational changes to detection and treatment of a wide variety of disorders. Later robots will be designed to work within tissue. Our first application in mind is directed navigation of robots through the cardiovascular system to sites of possible lung cancer (e.g. lung nodules) and simultaneously measuring variables such as pressure, temperature, velocity, chemicals, cancer markers, and pH, and using these data to perform computer-aided differential diagnosis, which is the use of computer algorithms to assist in making a differential diagnosis of disease.

2.1 New Applications for Nanotube Superfiber Development

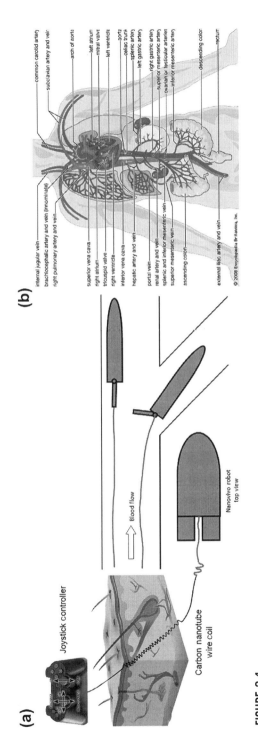

FIGURE 2.4

The nanovivo robot: (a) is shown schematically moving along the inside of blood vessels. A CNT electrical wire tether provides power and communication to the robot. Two independently actuated flaps allow steering. (b) The robots will operate in the cardiovascular system and possibly also within organs in the body.

If needed, the physician (or later robots) can then take biopsy samples, perform surgery, or deliver drugs. Nanovivo robots providing real-time in vivo sensor data to a computerized medical database to diagnose disease will be a breakthrough in advanced interventional medicine.

Conventional wisdom has been that tiny robots cannot be built to work inside the body. Robots have trouble picking up signals and communicating because radio signals do not travel far in the body. A medical microbot must also steer itself through an intricate network of fluid-filled tubes deep inside the body. And because it is so small, it must do so without a battery-powered motor. Past attempts involved building robots with metallic materials and guiding them with a magnet from outside the body. Magnet-tipped surgical catheters work this way. But a magnet poses several problems. The magnet must be positioned quite close to the robot to exert enough force to attract it. Also, propelling and powering robots wirelessly requires magnetic field gradients so strong that they might disturb the body's musculature or nervous systems. Our solution is to build the robot using a nanotube superfiber and polymer electroactive material, and it will be both lightweight and maneuverable. A CNT superfiber tether that is flexible and strong powers and controls the robot, thus countering conventional wisdom. Small robots will move in the direction of blood flow and return to their insertion point by winding in the tether, or be withdrawn downstream from the entry point. The robot is small and will not damage the vessel wall. The robot will be tracked via signals and/or by X-ray imaging. Testing has already been done in our Nanoworld Lab to prove the feasibility of the robots. Moreover, the University of Cincinnati (UC) is developing a Pilot Microfactory for Nanomedicine Devices to build prototype robots. No one has built a tethered robot that can move in the circulatory system. Nanotube superfiber is enabling the development of nanovivo robots which will be transformational to the field of interventional medicine.

2.1.4 Nanodevices made from one long CNT

Super long CNT (centimeter and longer) will be a novel building block in new electrical applications. The chirality, diameter, and number of walls are fixed along the CNT. It will make one CNT nanodevice feasible. One CNT nanodevice is a circuit made from one CNT and it can perform some functions, like sensing the environment, transforming the sensing results to an electrical signal, and transmitting the electrical signal wirelessly. The simplest oscillating circuit is the resistor, inductor, capacitor (RLC) circuit. It is possible to make many sets of RLC elements with the help of a micromanipulator. Because all the elements are from the same CNT, the electrical properties are constant. That means the tiny circuit can be mass produced and keep the same properties. Another important reason to use one CNT is to ease the process of welding RLC elements together. Jin and Iijima showed successfully connecting two same diameter CNTs together by Joule heating, but failed to connect two different diameter CNTs [19]. The tiny circuits are in demand in the biomedical field. For example, a nano- or micro-size circuit is needed to monitor the blood flow speed at one fixed point in a biomedical application. An aligned CNT sensor (100×200 μm) will

FIGURE 2.5

CNT thread for biomedical applications: (a) neurons that have attached to a CNT thread and extended neurites that follow the twist of the thread; (b) neurons growing above CNT sheet that was placed on glass coverslips. (c) CNT yarn used as a wire for crossing the skin to an implanted electrode. (For color version of this figure, the reader is referred to the online version of this book.)

Sources: Figures (a) and (b) from Chapter 7, Progress in the use of aligned carbon nanotubes to support neuronal attachment and directional neurite growth; Keith A. Crutcher, Chaminda Jayasinghe, Yeo-Heung Yun, and Vesselin N. Shanov, From Nanomedicine Design of Particles, Sensors, Motors, Implants, Robots, and Devices, edited by Mark J. Schulz, Vesselin N. Shanov, and YeoHeung Yun, October, 2009, Artech House Publishers.

be needed to attach to the wall of a blood vessel. The flow of the blood may generate a changing voltage at the millivolt level [20]. This information can be wirelessly sent out by the RLC circuit. From the receiver outside the human body, the real-time blood speed can be monitored.

2.1.5 Biomedical applications of nanotube thread

One of the more intriguing and new uses of CNT yarns is for biomedical applications. The first application is to study neural scaffolds. In this application, the geometry and electrical properties of the CNT threads offer the possibility for neurite regeneration. It has been demonstrated that neuron growth can occur longitudinally along the axis of the thread with minimal branching. The high aspect ratio, flexibility, chemical stability and electrical conductivity of CNT threads contribute to this directional growth. The growth of the neurons along the thread will follow the twist in the thread suggesting that the cells may communicate or be stimulated by the electrical conductivity of the thread. Figure 2.5(a) and (b) shows neuron growth on CNT threads. Another application is to replace metal wire for crossing the skin. Figure 2.5(c) shows a CNT thread that is used to cross the skin and is connected to a disk electrode in the body and a copper tab outside the body. This experiment was performed using a mouse model.

2.2 NEW TECHNIQUES FOR NANOTUBE SUPERFIBER DEVELOPMENT

Materials scientists want to know what it is about nanotube yarn and sheet that is both promising and challenging, whereas technologists are concerned about where and how the materials in different form factors can be applied. The nanotube properties of surface area, connectivity, conductivity, dimensional alignment, chemical stability or reactivity are important to balance the focus from strength alone. A basic understanding of the principles of fiber, fiber composite, and yarn physics and an outline of the mechanisms by which fibers interact between themselves and with their matrix is also important. For example, trying to create a CNT yarn has been driven by classical understanding. This works for micron-sized fibers where the interactions are primarily frictional and yarns depend on twist to generate cohesion. For nanotubes, the forces holding nanotubes to each other are entanglement and van der Waals attraction which both partially obviate the effect of twist and introduce new difficulties with dispersion and alignment. The vastly greater aspect ratio of nanotubes compared to traditional fibers is also an underappreciated difference when upsizing materials. Some of these challenges of developing nanotube superfiber materials are discussed in this section (also see Chapter 1 of this book).

2.2.1 Reactor design for growing carbon nanotubes

Nanotube new reactor design may be the most important factor in improving nanotube properties and producing superfiber. Advantages and limitations of various methods of synthesizing CNTs are listed below. An overall goal in new reactor design is to overcome the limitations of existing methods to develop a process that can continuously produce nanotubes that are long, pure, high quality, aligned, and low cost.

1. Chemical vapor deposition (CVD) using a substrate in a horizontal reactor
   a. Typically growth from a substrate such as a silicon wafer produces forests of aligned nanotubes usually by a base growth method.
   b. Nanotubes are of high purity.
   c. Short forests that have uniform and straight nanotubes can be drawn into sheets or drawn and twisted into thread.
   d. Typically multiwall nanotubes are grown.
   e. Long nanotubes can be grown but the quality decreases with growth time and length; postprocessing may be needed to heal defects.
   f. Recent progress in substrate engineering is allowing greater control of nanotube morphology and the morphology of the forest.
   g. Substrate preparation involves using an electron beam system to deposit a thin buffer layer and catalyst layer on the substrate.
   h. Magnetron sputtering and evaporation in the CVD chamber are the other ways to put the catalyst on the substrate.
   i. The sputtering process and cost of wafers and being a batch process may somewhat limit scale-up for mass production.
   j. Liquid catalyst methods are being developed to obviate the need for the e-beam or sputtering systems.

k. Producing spinnable arrays on large substrates becomes more difficult because the gas concentration and the temperature distribution in the horizontal reactors can vary in three directions.
l. Large wafers are being used to grow large area arrays and quartz and steel are also used as substrates.
m. Different processes use water or not, different precursor gases, low or atmospheric pressure, most use iron catalyst but some alloy it with other metals.
n. Using a load lock system or another way to prevent Oswald ripening of the catalyst and to prevent air from entering the reactor have made synthesis of spinnable arrays much more reproducible.
o. Drawability and spinnability of the forest may be affected by the weather (pressure, temperature, humidity).

2. Synthesis using a substrate in a vertical cold wall reactor
   a. The cold wall reactor allows preheating the gases and the substrate separately minimizing Oswald ripening of the catalyst which causes nonuniformity in the forest and reduces spinnability.
   b. Plasma-enhanced CVD is also available to help align the nanotubes in the forest.

3. Synthesis using a floating catalyst reactor
   a. This method injects catalyst as a liquid and catalyst particles form and float down the reactor tube along with precursor and other gases and nanotubes grow on the catalyst particles.
   b. A challenge is to nucleate nanotubes at the maximum number of catalyst particles.
   c. This may be the highest volume and most scalable method for producing fiber and sheet which come directly from the reactor.
   d. Research is focusing on preventing starving the catalyst for fuel.
   e. The nanotubes have catalyst impurities that can be removed by acid treatment.
   f. The nanotubes in the yarn and sheet produced are single-wall or double-wall high-quality because the growth process is very short and occurs at high temperature.

4. Methods under development
   a. Various improvements to existing synthesis methods are under development.
   b. Details on the methods are partly restricted because the methods may be intellectual property or trade secrets.
   c. In general, there is effort to improve the quality and spin long forests.
   d. There are many research parameters to be explored using cold wall vertical reactors.
   e. Several groups are working on horizontal synthesis of long nanotubes.

2.2.2 Thermal annealing of carbon nanotubes

High-temperature annealing is a traditional technique used in the carbon industry. Annealing is now being investigated to improve the quality of CNTs. In previous work, Joule heating was used to anneal carbon nanomaterials. Here we discuss the use of thermal annealing to heal defects in CNT sheets. Figure 2.6(a) shows the top-loading ultrahigh temperature furnace (MRF, Inc., Suncook, NH) used to carry out the thermal annealing. Specimens for annealing are taken from a single CNT sheet and loaded into the furnace. Next, the system is purged using ultrahigh purity Argon for several hours. The temperature is then ramped up to the soak temperature at a rate of about 40° C/min, and maintained for the duration of the experiment. An

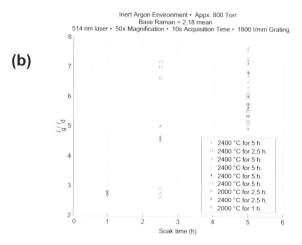

FIGURE 2.6

Thermal annealing of carbon nanotubes: (a) ultrahigh-temperature furnace; (b) I_g/I_d ratio from Raman spectroscopy of annealed CNT sheets as function of temperature and time. (For color version of this figure, the reader is referred to the online version of this book.)

inVia Raman spectroscope (Renishaw, UK) is used to characterize the specimens before and after annealing.

Figure 2.6(b) shows the results from postannealing Raman spectroscopy on various specimens from a single CNT sheet, where the I_g/I_d ratio of the absolute values of the peaks is plotted against the time spent by each specimen in the furnace. The temperature and duration for each specimen are shown in the legend, and the data are grouped according to the specific run. For example, two separate runs are carried out at 2400 °C for 2.5 h and these are represented separately in Fig. 2.6(b). The baseline I_g/I_d ratio for as-grown tubes is 2.18 (mean), and that for annealed tubes ranges from 2.63 to 7.67. The improvement in I_g/I_d values shows large variation. The largest improvement is seen at the following conditions: 2400 °C for 5 h and 2400 °C for 2.5 h. More scatter is seen at 2400 °C for 5 h. It is interesting to note that the tubes annealed at 2400 °C for 2.5 h show very little scatter within a run. The difference between the runs is believed to be due to the difference in the environmental purity, but further work is needed to confirm this.

Thermal annealing shows great promise in healing the defects in as-grown carbon nanomaterials. Our preliminary work in this area shows that thermal annealing can deliver considerable improvement in the Raman characteristics of the materials. Significant challenges and opportunities lie ahead. There is a need to optimize the process parameters to achieve reliable and repeatable results. In particular, it may be necessary to purge the water molecules before the temperature in the furnace is ramped up to the soak temperature. This could be achieved by increasing the temperature in the furnace to about 500 °C, which is below the degradation temperature of amorphous carbon, and maintaining that value for a soak time before reaching the final value. Further work is needed to study the changes in the electrical and mechanical properties due to annealing, and to understand the fundamental mechanisms that affect the various properties of the nanomaterials. Also, there are some unanswered questions. Annealing substrate-grown nanotubes increases their quality. However, annealing spun thread produces only a modest improvement in quality and the reason is not well understood. Annealing nanotubes that contain significant amount of catalyst was not successful in initial studies. The iron in the nanotubes affects the annealing.

2.2.3 Substrate patterning for nanotube growth control

Substrate patterning represents a new family of techniques to improve control over nanotube growth. CNTs have extraordinary properties which could enable a wide range of applications in sensors, transistors, biodevices, etc. One of the problems restricting these applications is the lack of CNTs with well-defined parameters such as location, orientation and geometry [25,26]. Precise control of direct CNT growth on a catalyst patterned substrate is one promising way to overcome this barrier. If successful, precise control over CNT diameter, length, growth direction and even mechanical/electrical properties may be possible. Two major strategies have been employed to obtain precise control of CNT for device applications. The first one is to grow CNTs selectively on catalyst patterned substrates. The other is to postassemble CNTs after their growth. The postassembly method requires the treatment and manipulation of CNTs, which may lead to the degradation of their intrinsic properties [27]. Using a catalyst patterned substrate to obtain the desired properties and location of CNTs is a better approach for fabricating nanodevices and for producing nanotubes with desired morphology for various applications. In order to fabricate precisely patterned catalyst substrates, a number of techniques have been explored, such as photolithography, microcontact printing (μCP), electron-beam lithography (EBL)/ion-beam lithography, dip-pen lithography, etc. These methods are briefly described in Refs [25−47].

A. Photolithography. Photolithography is widely used in microelectronics. Due to its capability to pattern over large areas, photolithography could pattern catalyst and thus the growth of CNTs on a wafer scale. Photolithography has been successful for patterning catalyst, but it also has some limitations, such as requirement of a photomask and clean room. As a multiple-step process, it is time-consuming, and the photoresist and etching solution may leave contamination. Moreover, photolithography does not have the resolution to produce particle-level nanometer-scale patterns, and patterning catalyst at the single-particle level at specific locations for growth of CNTs is a challenging task for photolithography [25,32−34].

For certain applications, there is a growing demand to synthesize aligned arrays of individual CNTs with large diameter of a few 100 nanometers. As an intermediate substitute, we are currently synthesizing arrays of microbundles of CNTs. This aligned array of CNT bundles is grown on silicon substrate, which is patterned using optical lithography technique. As shown in Fig. 2.7(a−c), patterned dots of metal catalyst are deposited and grown into aligned bundles of CNTs. The height of the CNT bundles is governed by the metal catalyst composition and is proportional to the growth time.

Every order of size and position of CNTs have a unique application. One such application of patterned arrays of CNT bundles is in the electron source for field-emission displays. Another such application lies in composites, where the CNTs of specific sizes can reinforce the matrix for specific properties. In the field of

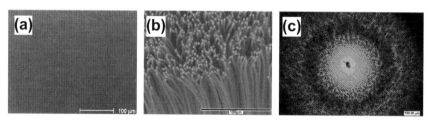

FIGURE 2.7

Photolithography results: (a) scanning electron microscope (SEM) micrograph of patterned metal catalyst dots on silicon substrate; (b) towering bundles of CNTs grown on the patterned substrate; (c) the height of posts can be tailored radially. (For color version of this figure, the reader is referred to the online version of this book.)

drug delivery and drug targeting systems, various sizes of CNT array could be used either for in vivo drug delivery as drug-carrier systems or for drug injection as an array of microneedles. Applications such as these are by no means exhaustive, but they indicate the wide range of possible applications that can be obtained even with the smallest modification in the substrate's pattern by substrate engineering.

B. Electron-beam lithography. EBL is a technique using an electron beam to scan a surface covered with resist film and patterns will be formed after the liftoff process. It utilizes the electron beam as the source of energy to pattern resists on surfaces. It is analogous to photolithography because both are energy-driven patterning of surfaces. However, since the electron wavelength is much smaller than the wavelength of light, EBL can generate nano-patterns with defined shapes and substantially smaller sizes [35].

C. Focused ion-beam lithography (FIB lithography). FIB lithography is similar to EBL but uses a high-energy ion beam instead of an electron beam to pattern surfaces [25]. These two patterning lithographic techniques are advantageous for depositing catalyst with nanoresolution at specific locations for growth of CNTs. Despite their widespread use, these methods suffer from low-throughput, high-cost, and complicated experimental conditions, making them unsuitable for large-area patterning of catalyst and wafer-scale CNT arrays [25,31].

D. Microcontact printing (μCP). μCP has been reported to be a suitable microfabrication tool for patterning small molecules and nanomaterials in the micrometer scale over a large surface area. A patterned stamp with raised and recessed features is coated with desired materials and brought in contact with a substrate to transfer materials in patterns. Advantages of this technique include low cost and easy fabrication of the stamp and easy operation of the printing process. It has been proven that catalyst nanoparticles can be directly patterned on wafers by μCP for growth of CNTs [25]. μCP is fundamentally different from traditional photolithography techniques. First, only a polymer stamp, i.e. poly(dimethylsiloxane) (PDMS), with a predesigned pattern is required. In contrast to the sophisticated equipment required in photolithography, this provides its almost universal accessibility. The stamps can be conveniently fabricated and, importantly, they can be repeatedly used. Second, the μCP technique is compatible with a variety of materials. Third, this technique is suitable for printing on planar, curved, 3D and flexible substrates, making it possible to fabricate 3D and flexible electronics [31]. However, the masks used to make the micropatterned stamps are expensive, and each pattern modification requires redesigning the mask. Also, it is not easy to generate submicron features using the conventional PDMS stamp [26,34] and it is usually difficult to control the exact position [37].

E. Scanning probe microscopy (SPM). SPM provides the ability to observe surface structures at atomic resolution. It is also a powerful tool to modify the sample surface, and various methods have been developed to pattern surfaces based on SPM. SPM-based techniques can be divided into two major categories, namely constructive and destructive, in which the probe tip is used to generate patterns by delivering materials onto the surface or to damage or scratch the surface, respectively [31]. The

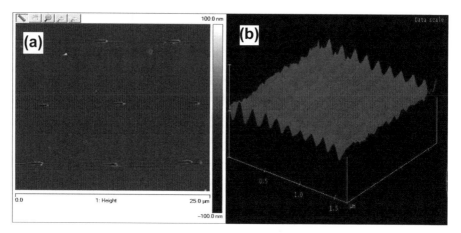

FIGURE 2.8

Surface patterns by tip-based nanofabrication on Single Crystal Silicon (Type 100): (a) 3 × 3 array of nanocavities, depth: 20–50 nm, width: 100–150 nm; (b) 10 × 10 array of nanogrooves, depth −30 nm. (For color version of this figure, the reader is referred to the online version of this book.)

destructive methods can generate predesigned nanoscale patterns and placing catalyst particles on these patterned areas could lead to fine control of CNT growth. The constructive methods directly put catalyst on the sample surface. Tip-based nanofabrication techniques have been used to produce nanoscale patterns on silicon substrate as shown in Fig. 2.8. Once these complex patterns are made, various catalyst particles can be placed at a specific point to investigate and control the CNT growth.

A nanoindentation process uses a sharp indenting probe in order to penetrate the substrate surface to depths ranging in nanometers [38]. Primarily, this technique is used to study the surface deformation of a material where the indenter, usually an atomic force microscopy (AFM) probe tip, is lowered to the workpiece surface while controlling the force and the resistance to motion is measured. The concept of the nanoindentation technique can be extended for machining nanoscale substrates by applying a high-indentation load on the probe tip and thus fabricate nanometer-deep holes [39]. An extension of the nanoindentation technique would be the nanoscratching process where the AFM tip is dragged over the substrate surface while ensuring the tip is still in contact with the surface thus forming nanochannels [40]. Nanoscratching operations can be performed in either the static mode or dynamic mode with an oscillating tip [39]. Recent advances in nanoscratching-based techniques include conducting nanoscratching in liquid medium [41] and inducing a phase transition on monocrystalline silicon using nanoscratching [42].

Another process is SPM local anodic oxidation, which is an electrochemical process. SPM anodic oxidization can fabricate nanoscale patterns by oxidizing the sample surface through anodic reaction between the tip and the surface. The SPM

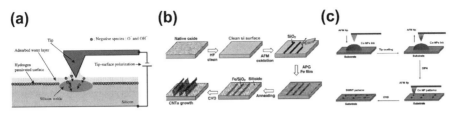

FIGURE 2.9

Schematic representation of SPM methods: (a) local oxidation mechanism. (b) example of SPM local oxidation for CNT growth. (c) DPN process used for CNT growth. (For color version of this figure, the reader is referred to the online version of this book.)

Sources: (a) Figure from Ref. [43], (b) Figure from Ref. [33], (c) Figure from Ref. [25].

tip is used as a cathode and the water meniscus between the tip and sample surface function as the electrolyte. The OH$^-$ ions from the water provide the oxidant for the electrochemical reaction. A schematic of SPM local anodic oxidation is shown in Fig. 2.9(a) [43]. SPM local oxidation can change the morphology and properties of metallic or semiconductor surfaces for fabrication of nanodevices. An example of SPM local oxidation used for CNT growth is shown in Fig. 2.9(b) [33]. One limitation of SPM local oxidation is that the tip and substrate should be conductive or somewhat semiconductive. Sometimes the tip has to be coated to become conductive in order to obtain a good oxidation result. And the oxide thickness is limited because the electric field decays and the oxide growth process self-terminates along with the oxidation process [44].

Dip-pen nanolithography (DPN) is a widely used SPM-based technique for surface modification. DPN has been one of the most popular AFM nanolithography techniques since its invention, in which materials initially on the tip are transferred to the surface while scanning in either static or dynamic mode. In this process, the SPM probe acts as a source of the "ink". The "ink" usually consists of nanoparticles suspended in liquid, or inorganic or biological molecules in a solvent. An example of DPN for CNT growth is shown in Fig. 2.9(c) [25]. Unlike the traditional lithographic methods, DPN is a maskless and single-step direct-writing method, and can be carried out under moderate operating conditions (does not require high-vacuum or high-energy ions or beams), which eliminates the possibility of cross-contamination and sample damage. More importantly, in principle, DPN is capable of delivering any kind of materials precisely to a specifically designated location, where "inks" may form any desired pattern with feature sizes down to sub-100 nm. This is crucial for nanodevice fabrication in complex integration systems [26]. DPN consequently provides a number of advantages in patterning catalytic precursors for subsequent CVD growth of CNTs. However, the reliability and reproducibility of the technique depends critically on the properties of the ink used in the process and ambient conditions such as humidity and temperature during the patterning process. In addition, the sizes of the catalyst particles that are formed

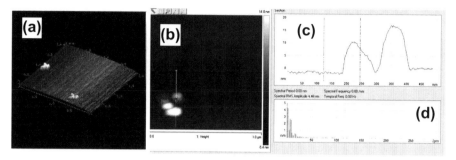

FIGURE 2.10

DPN deposition of catalyst particles: (a) AFM image of particles on a silicon surface (diameter ~100 nm and height of ~15 nm); (b) close-up image of particle; (c–d) topography of particle. (For color version of this figure, the reader is referred to the online version of this book.)

on the substrate determine the yield of CNTs after the CVD process. And the water meniscus between the tip and substrate is commonly believed to be responsible for ink transport, the dimension of which is largely influenced by the relative humidity [34,36].

DPN is a direct and nondestructive technique with good patterning ability of micro- and nanoprecision. Experiments were conducted using DPN to deposit iron solution on a silicon surface. Particles deposited are shown in Fig. 2.10. The next step is using CVD to grow CNT which is being carried out.

F. Nanoimprint lithography (NIL) for substrate preparation. Not very often in the history of science has mankind come across a single material which has remarkable mechanical, electrical and chemical properties; CNT is one such wondrous material discovered in the late twentieth century! Since then, CNTs have attracted a great deal of research attention. Since the properties of nanotubes depend on their geometry, a scalable method is sought that can control the geometry and hence properties to some degree of nanotubes. Also, the possible properties of a material drive widespread research for specific applications. To gain control over nanotube properties, stable and reliable synthesis of CNTs is of utmost importance. CNT's remarkable properties stem from geometrical factors such as small diameters, symmetrical arrangement of carbon atoms, etc. Hence, it is important to synthesize CNTs with well-defined diameters and lengths. The challenges lie in the fabrication of these tailored CNTs with controlled sizes and positions in a large area substrate. In many ways, the current capabilities of synthesis methods limit the pace of both research and commercialization of new materials and devices.

In our research, we strive to control the size, shape, spacing, growth direction and quality of grown CNTs by substrate engineering. This approach involves patterning the silicon substrate for uniform and distributed catalyst deposition, leading to improved diffusion of carbon precursor gases to the metal catalyst. The aspects of substrate engineering that are the focus of our work are (1) catalyst patterning by NIL and (2) catalyst patterning by optical lithography. Conventional CNT growth depends on self-assembly processes in nature wherein CNTs are grown from catalyst thin film. At Nanoworld, we use lithography techniques to produce nano-sized patterns of catalyst. Patterns of dots and lines with feature size ranging from a few micrometers to tens of nanometers are engineered on silicon substrates.

Lithography techniques such as E-beam lithography described above produce extremely small features with a precise size and pitch of lines and dots. However, it is a serial process, which means creating large patterned areas is tedious and expensive. These problems can be erased with the use of a technique called NIL, one in which nano-patterns can be stamped on substrates! Unlike e-beam, this technique can also cater to large area patterning by a step and repeat process. The only time-consuming and expensive part of the NIL process lies in writing nanostructures with the e-beam to define a stamp. However, this stamp can be repeatedly used to imprint nanostructures into a polymer film coated on a substrate thus turning the fabrication of nano-patterned catalyst substrates into a parallel imprint process capable of volume production. So far, we have successfully achieved patterning catalyst dots and lines of sizes ranging from a few tens of nanometers to few 100 nanometers as shown in Fig. 2.11(a) and (c). Further, these patterned catalyst substrates were used to grow short CNTs of a few microns in length as shown in Fig. 2.11(b) and (d).

NIL basically involves using silicon stamps (e.g. with a pattern of nanoscale diameter posts) to imprint a pattern (e.g. a pattern of holes which is the reverse image of the stamp) into a polymer resist and thermally cure the polymer, remove the stamp, plasma etch through the residual layer of polymer at the bottom of the holes to expose the substrate, deposit the catalyst, rinse off the remaining resist, and then synthesize nanotubes using normal techniques. Any application that can benefit from forests of nanotubes may get better when the geometry of the individual nanotubes and geometry of the forest can be custom designed for that application. The applications we have in mind that can benefit from controlling the design of the nanotube forest include supercapacitors, field-emission displays, thermal interface materials, composite materials reinforcement, improvement of thermal and electrical conduction in materials, optics including stray light absorption, potentially metamaterials, and electromagnetic nanocomposites. Various other applications such as EMI shielding, structural health monitoring, and thermoelectrics will use superfiber yarn and sheet materials produced from the arrays [48].

2.2.4 Spinning carbon nanotube thread

The long-range goal of the spinning efforts is to use patterned nanotubes and bundles of nanotubes that have uniform morphology and that are long to improve the properties of superfiber. Individual CNTs have exceptional mechanical, electrical, and thermal properties [1,2] and industry is waiting to adopt this material to produce

FIGURE 2.11

Nanoimprint lithography results: (a) SEM micrograph of a top view of 300 nm average diameter metal dots on a Si substrate, formed by imprinting into poly(methyl methacrylate) (PMMA) and a liftoff process; (b) SEM micrograph of isotropic view of CNTs grown from patterned metal dots; (c) SEM micrograph of top view of 280-nm wide lines of metal on Si substrate; (d) SEM micrograph of top view of CNTs grown from patterned lines of metal catalyst. (For color version of this figure, the reader is referred to the online version of this book.)

new exciting products that were never feasible before. The problem thwarting this "nano-age" is the material size is on the nanoscale. Although nanotubes have fantastic properties, creating a product that can be used by industry on a larger scale has proven difficult. Ultimately, the end use of CNT superfiber will be largely dependent on the properties of the nanotubes which are contained within the macrofiber structure. There are numerous factors (length, diameter, number of walls, purity, and quality) which affect the overall properties of CNTs [3], and hence superfiber derived from them. The previous section on patterning discussed controlling the geometry of nanotubes. Using the patterned arrays, various methods of spinning CNT thread, yarn and superfiber represent processes for converting individual nanotubes and bundles of nanotubes into continuous monofilaments, multifilament thread, as well as plied yarn. A number of approaches have been developed in the literature intending to translate the exceptional properties of CNTs to the macroscale [4,5]. Of these, creating CNT yarns through a spinning process has shown some of the most promising results [6,7]. Currently there are two main methods used to make CNT yarns: dry spinning and wet spinning (coagulation spinning).

A. Dry spinning. In 2002, it was shown that a CNT web can be pulled from a superaligned CNT array and then twisted into a yarn [8]. After that discovery, many research teams have focused on the synthesis of spinnable arrays, the mechanisms in which the CNT web is formed, and ways to improve the strength of CNT yarns [7,9,10]. To improve the spinnability of CNT arrays and yarn properties, the main mechanisms involved in the web formation process must be well understood. The CNT array alignment, diameter, length, spatial density, and alignment are the most important factors in determining the spinnability of the CNT arrays [11–13]. Many factors and treatments including yarn twist angle, fiber alignment, CNT quality/annealing, and plying of yarns affect the yarn properties [7,11,12,14,15]. Nanoworld is currently studying the effects of yarn geometry (twist angle and diameter) on the yarn properties, effects of annealing the yarn, and optimizing the CVD recipe in the creation of superaligned CNT arrays for spinning CNT yarn. The CNT yarns formed using the dry-spinning technique use short CNT, on the order of 500 μm. Growing and spinning longer CNT into yarns is a future goal to increase the properties of the yarn. A high-speed "forced" ring-spinning machine is shown in Fig. 2.12(a). This machine is used to quickly produce CNT yarns and study the effects of yarn geometry on yarn properties. Regarding electrical results, twist angle appeared to have little effect, and resistivity was an intrinsic property of CNT thread. The resistivity of as-spun thread is about 3×10^{-3} Ω cm. A dry-spun CNT yarn is shown in Fig. 2.12(b). The strength is about 1 GPa and increases with decreasing diameter. Various post- and pretreatments can improve the properties of the yarn.

B. Wet spinning. Since 2000, research labs from around the world have applied the traditional textile process of wet spinning to form yarns out of CNT [16]. This method of creating CNT textiles has been applied for both multi-wall carbon nanotubes (MWCNTs) and single-wall carbon nanotubes (SWCNTs) [16,17]. Wet spinning works on the principle that a fibrous material can be dispersed into a solution and extruded into a second solution, where the first solution is soluble in the second and the fiber is not, and the fiber coagulates into a yarn [18]. The wet-spinning

 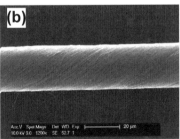

FIGURE 2.12

Dry spinning thread: (a) one of the two current dry-spinning machines at Nanoworld; (b) thread produced by dry spinning. (For color version of this figure, the reader is referred to the online version of this book.)

FIGURE 2.13

Wet spinning thread: (a) wet-spinning setup; (b) thread produced by coagulation spinning. (For color version of this figure, the reader is referred to the online version of this book.)

process has the potential to be a continuous process. Also it allows treatment of CNTs before and after spinning. Different techniques have been developed to wet spin CNT fibers. Both neat and composite CNT fibers can be spun. A wet spinning setup is shown in Fig. 2.13(a). So far, the CNTs that have been used in wet spinning process are typically micrometer long (usually <10 μm). Sonication, usually adopted in dispersing CNTs, would shorten the CNTs. Long CNTs are supposed to contribute to the strength of wet-spun fibers. Spinning centimeter-long CNTs is one of the main goals. For most polymers investigated, the presence of well-dispersed CNTs helps increase mechanical properties. With longer CNTs, the properties would be further improved.

However, there are several problems to be solved. One problem lies in dispersing long CNTs, which is much more difficult than dispersing short ones. Another problem is how to achieve a high-volume fraction of long CNTs and keep them aligned in fiber. Sonication is an approach to untangling and shortening of CNTs. Therefore, it would be helpful to determine the optimal sonication energy to maintain the length of CNTs while untangling the CNT bundles in a solvent. Super acid is found to be a good solvent but only for perfect CNTs. Still, it is interesting to explore alternative solvents for CNTs with lower quality. Moreover, annealing of CNTs seems to be another promising way to solve the above problems. Annealing will improve strength of individual CNT and straighten it, which helps maintain the length of CNTs during sonication and align CNTs in fiber. In addition, conditions of coagulation are crucial in achieving good properties of fibers, so it is important to fully understand the mechanism. Wet-spun fiber is shown in Fig. 2.13(b). The fiber presently has low strength. Further exploration and use of long nanotubes is expected to lead to novel fibers with improved properties.

2.3 CONCLUSIONS

There are many potential and new applications described waiting for a material like nanotube superfiber. If the properties of individual nanotubes can be scaled up, nanotube superfiber will change engineering design. Cooperative efforts and new approaches such as patterning will bring up the properties of superfiber most quickly.

Acknowledgments

We are grateful to the funding agencies: DURIP-ONR grant Mold Tool Set to Produce Breakthrough Carbon Nanotube Arrays, ONR STTR Project Substrate Engineering to Produce Device Quality CNT Arrays for Electronics Applications; The University of Cincinnati Education and Research Center Targeted Research Training Program (UC ERC-TRT Program) provided support for CNT materials development; NSF MRI: Development of a Pilot Microfactory for Nanomedicine Devices; NSF Engineering Research Center (ERC) for Revolutionizing Metallic Biomaterials CMMI-07272500 with North Carolina A&T State University lead and University of Pittsburgh Partner; NSF SNM GOALI, Carbon Nanotube Superfiber to Revolutionize Engineering Designs 1120382, and the National Science Foundation under the Grant Number CBET-1239779.

References

[1] M.M.J. Treacy, T.W. Ebbesen, J.M. Gibson, Exceptionally high Young's modulus observed for individual carbon nanotubes, Nature 381 (6584) (1996) 678−680.
[2] A. Thess, et al., Crystalline ropes of metallic carbon nanotubes, Science 273 (5274) (1996) 483 487.
[3] W. Hu, D. Gong, Z. Chen, L. Yuan, K. Saito, C.A. Grimes, P. Kichambare, Growth of well-aligned carbon nanotube arrays on silicon substrates using porous alumina film as a template, Applied Physics Letters 79 (19) (2001) 3083−3085.
[4] P.D. Bradford, et al., A novel approach to fabricate high volume fraction nanocomposites with long aligned carbon nanotubes, Composites Science and Technology 70 (13) (2010) 1980−1985.
[5] K. Koziol, et al., High-performance carbon nanotube fiber, Science 318 (5858) (2007) 1892−1895.
[6] A.B. Dalton, et al., Super-tough carbon-nanotube fibres, Nature 423 (6941) (2003) 703.
[7] M. Zhang, K.R. Atkinson, R.H. Baughman, Multifunctional carbon nanotube yarns by downsizing an ancient technology, Science 306 (5700) (2004) 1358−1361.
[8] K. Jiang, Q. Li, S. Fan, Nanotechnology: spinning continuous carbon nanotube yarns, Nature 419 (6909) (2002) 801.
[9] A.A. Kuznetsov, et al., Structural model for dry-drawing of sheets and yarns from carbon nanotube forests, ACS Nano 5 (2) (2011) 985−993.
[10] W. Liu, et al., Producing superior composites by winding carbon nanotubes onto a mandrel under a poly(vinyl alcohol) spray, Carbon 49 (14) (2011) 4786−4791.
[11] X. Zhang, et al., Spinning and processing continuous yarns from 4-inch wafer scale super-aligned carbon nanotube arrays, Advanced Materials 18 (12) (2006) 1505−1510.
[12] F. Shaoli, et al., Structure and process-dependent properties of solid-state spun carbon nanotube yarns, Journal of Physics: Condensed Matter 22 (33) (2010) 1−6.
[13] M. Motta, et al., Mechanical properties of continuously spun fibers of carbon nanotubes, Nano Letters 5 (8) (2005) 1529−1533.
[14] C.D. Tran, et al., Improving the tensile strength of carbon nanotube spun yarns using a modified spinning process, Carbon 47 (11) (2009) 2662−2670.

[15] L. Kai, et al., Carbon nanotube yarns with high tensile strength made by a twisting and shrinking method, Nanotechnology 21 (4) (2010) 045708.
[16] B. Vigolo, et al., Macroscopic fibers and ribbons of oriented carbon nanotubes, Science 290 (5495) (2000) 1331–1334.
[17] S. Zhang, et al., Macroscopic fibers of well-aligned carbon nanotubes by wet spinning, Small 4 (8) (2008) 1217–1222.
[18] K. Jiang, et al., Superaligned carbon nanotube arrays, films, and yarns: a road to applications, Advanced Materials 23 (9) (2011) 1154–1161.
[19] Chuanhong Jin, Kazu Suenaga, Sumio Iijima, Plumbing carbon nanotubes, Nature Nanotechnology 3 (2008) 17–21.
[20] Jianwei Liu, Liming Dai, Jeff W. Baur, Multiwalled carbon nanotubes for flow-induced voltage generation, Journal of Applied Physics 101 (2007) 064312, http://dx.doi.org/10.1063/1.2710776.
[21] Mark Schulz, John Yin, Yonghai Zhang, Sarah Pixley, David Mast, Vesselin Shanov, Yonghai Zhang, Nanovivo Robots to Change Interventional Medicine, University of Cincinnati Invention Disclosure, 3/2/2013.
[22] Rajiv Venkatasubramanian, Mark J. Schulz, Brad Ruff, Weifeng Li, Vesselin Shanov, David Mast, Noe Alvarez, John Yin, Adam Hehr, Joe Sullivan, Electromagnetic Nanocomposite Materials for Machines and Biomedical Devices, UC Invention Disclosure, July 2012.
[23] J.S. Mark, L. Weifeng, R. Brad, N.S. Vesselin, UC Invention Disclosure No. 112-016, Carbon Electric Motors, August 12, 2011.
[24] L. Weifeng, S. Mark, S. Vesselin, S. Joe, UC Invention Disclosure 111-023 Carbon Electromagnetic Materials for Electric Motors and Actuators.
[25] X. Zhou, F. Boey, H. Zhang, Controlled growth of single-walled carbon nanotubes on patterned substrates, Chemical Society Reviews 40 (11) (2011) 5221–5231.
[26] B. Li, et al., Patterning colloidal metal nanoparticles for controlled growth of carbon nanotubes, Advanced Materials 20 (24) (2008) 4873–4878.
[27] J.-M. Bonard, et al., Monodisperse multiwall carbon nanotubes obtained with ferritin as catalyst, Nano Letters 2 (6) (2002) 665–667.
[28] M. Kumar, Y. Ando, Chemical vapor deposition of carbon nanotubes: a review on growth mechanism and mass production, Journal of Nanoscience and Nanotechnology 10 (6) (2010) 3739–3758.
[29] I. Ibrahim, et al., CVD-grown horizontally aligned single-walled carbon nanotubes: synthesis routes and growth mechanisms, Small (Weinheim an der Bergstrasse, Germany) 8 (13) (2012) 1973–1992.
[30] C. Journet, M. Picher, V. Jourdain, Carbon nanotube synthesis: from large-scale production to atom-by-atom growth, Nanotechnology 23 (14) (2012) 142001.
[31] S. Esconjauregui, C.M. Whelan, K. Maex, Patterning of metallic nanoparticles for the growth of carbon nanotubes, Nanotechnology 19 (13) (2008) 135306.
[32] N.R. Franklin, et al., Patterned growth of single-walled carbon nanotubes on full 4-inch wafers, Applied Physics Letters 79 (27) (2001) 4571–4573.
[33] C.-C. Chiu, M. Yoshimura, K. Ueda, Patterned growth of carbon nanotubes through AFM nano-oxidation, Diamond and Related Materials 18 (2–3) (2009) 355–359.
[34] I. Kuljanishvili, et al., Controllable patterning and CVD growth of isolated carbon nanotubes with direct parallel writing of catalyst using dip-pen nanolithography, Small (Weinheim an der Bergstrasse, Germany) 5 (22) (2009) 2523–2527.

[35] M. Häffner, et al., E-beam lithography of catalyst patterns for carbon nanotube growth on insulating substrates, Microelectronic Engineering 85 (5–6) (2008) 768–773.

[36] X. Zhou, et al., Chemically functionalized surface patterning, Small (Weinheim an der Bergstrasse, Germany) 7 (16) (2011) 2273–2289.

[37] H. Ago, et al., Ink-jet printing of nanoparticle catalyst for site-selective carbon nanotube growth, Applied Physics Letters 82 (2003) 811.

[38] A. Gouldstone, K.J. Van Vliet, S. Suresh, Simulation of defect nucleation in a crystal, Nature 411 (6838) (2001) 656.

[39] S. Diegoli, et al., Engineering nanostructures at surfaces using nanolithography. Proceedings of the Institution of Mechanical Engineers, Part G: Journal of Aerospace Engineering 221 (4) (2007) 589–629.

[40] T. Tsui, et al., Nanoindentation and Nanoscratching of Hard Carbon Coatings for Magnetic Disks, Oak Ridge National Lab., TN, United States, 1995. Oak Ridge Associated Universities, Inc., TN (United States); Lawrence Berkeley Lab., CA (United States).

[41] J.W. Lee, et al., Nano-scratching and nano-machining in different environments on Cr2N/Cu multilayer thin films, Thin Solid Films (2011).

[42] Y. Wu, et al., Nanoscratch-induced phase transformation of monocrystalline Si, Scripta Materialia 63 (8) (2010) 847–850.

[43] D. Stiévenard, B. Legrand, Silicon surface nano-oxidation using scanning probe microscopy, Progress in Surface Science 81 (2–3) (2006) 112–140.

[44] A.A. Tseng, A. Notargiacomo, T.P. Chen, Nanofabrication by scanning probe microscope lithography: a review, Journal of Vacuum Science and Technology B: Microelectronics and Nanometer Structures 23 (2005) 877.

[45] Y.Y. Tsai, J.S. Su, C.Y. Su, A novel method to produce carbon nanotubes using EDM process, International Journal of Machine Tools and Manufacture 48 (15) (2008) 1653–1657.

[46] M. Dahmardeh, A. Nojeh, K. Takahata, Possible mechanism in dry micro-electro-discharge machining of carbon-nanotube forests: a study of the effect of oxygen, Journal of Applied Physics 109 (9) (2011) 093308.

[47] H.Y. Miao, J.T. Lue, M.S. Ouyang, Surface modification by electric discharge implemented with electrodes composed of carbon nanotubes, Journal of Nanoscience and Nanotechnology 6 (5) (2006) 1375–1380.

[48] M.J. Schulz, et al., Carbon Nanotube Superfiber Development, in: Recent Advances in Circuits, Communications and Signal Processing, WSEAS, 2013.

CHAPTER 3

Tailoring the Mechanical Properties of Carbon Nanotube Fibers

T. Filleter[1], A.M. Beese[2], M.R. Roenbeck[2], X. Wei[2], H.D. Espinosa[2]

[1] *Department of Mechanical & Industrial Engineering, University of Toronto, Toronto, ON, Canada,* [2] *Department of Mechanical Engineering, Northwestern University, Evanston, IL, USA*

CHAPTER OUTLINE

3.1 Introduction	61
3.2 Irradiation Cross-Linking: Strong and Stiff CNTs and CNT Bundles	64
3.3 Reformable Bonding: Strong and Tough CNT Bundles and Fibers	68
3.4 Materials Design: Optimized Geometry and Structure	72
3.5 Summary	79
Acknowledgments	80
References	80

3.1 INTRODUCTION

Carbon-based fibers have been widely used in the engineering of lightweight materials and components due to their high specific modulus and specific strength. Applications range from structural components of aircraft that require maximum performance and energy efficiency to automotive components, which require high cost efficiency [1]. Ever-increasing performance and efficiency demands in such industries are pushing a need for advanced materials that can outperform existing technologies. A promising new frontier for carbon fibers are a class of materials that incorporate carbon nanotubes (CNTs) to enhance specific mechanical properties such as strength, stiffness, and toughness (Fig. 3.1). These range from traditional carbon fibers grafted with CNTs [2,3] (Fig. 3.1(a)) to carbonized fibers that use a small number of internal CNTs as templates [4] (Fig. 3.1(b)) to fibers formed almost entirely from CNTs [5,6] (Fig. 3.1(c)). The last class of fibers, which contain primarily CNTs, has attracted a great deal of attention recently thanks to the discovery of an industrially scalable chemical vapor deposition (CVD) CNT approach to fiber spinning [5]. This approach has allowed for the continuous spinning of kilometer-long commercially available fibers and mats [7].

One of the major roadblocks to unlocking the full advantages of macroscopic fibers based on CNTs is controlling and optimizing the shear interactions and structural organization of the internal CNTs. A number of fundamental studies have shed light onto the nature of shear interactions that exist at different length scales within

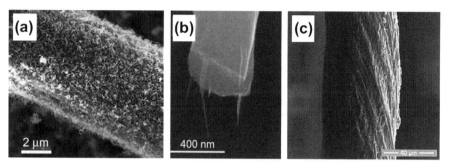

FIGURE 3.1

CNT-based carbon fibers. (a) Carbon fibers with CNTs grown on surface. (b) Carbonized polyacrylonitrile (PAN) fibers with embedded CNTs. (c) Fiber composed primarily of CNTs.

Sources: (a): Adapted with permission from Ref. [3]. Copyright 2010 Royal Society of Chemistry. (b): Adapted with permission from Ref. [4]. Copyright 2013 American Chemical Society. (c): Adapted with permission from Ref. [8]. Copyright 2010 American Chemical Society.

CNT fibers including shell—shell interactions within multi-walled carbon nanotubes (MWNTs) [9—15]; CNT—CNT interactions within CNT bundles [6,16]; and interactions between adjacent CNT bundles within CNT fibers [17]. In particular, nanomechanical experimental studies have provided insight into the shear interactions between adjacent graphitic sheets, most notably within MWNTs. Cumings and Zettl performed an experiment in situ a transmission electron microscope (TEM), which showed that once the inner walls of an MWNT were pulled out of the outer wall and then released, the inner tubes retracted back into the outer wall, pointing to a reversible and almost friction-free interaction [13]. Additional studies have been aimed at quantifying the forces required to pull internal walls of MWNTs out from the outer shells [11,14], or between adjacent pristine MWNTs [15], and have reported interfacial shear strengths between adjacent graphitic sheets to be in a wide range from ∼0.08 to ∼56 MPa. Thus, the weak interfacial shear strength afforded by van der Waals forces alone is not sufficient to efficiently transfer loads between adjacent nanotubes, and is therefore a factor that limits the attainable mechanical properties of macroscopic fibers produced from CNTs. The necessity of improving the shear interfaces is apparent in the mechanical testing of CNT-based fibers, which typically fail through shear of the weak junctions between CNTs as illustrated in Fig. 3.2 [18]. All these studies point to the need to optimize the shear interactions within fibers by engineering the chemical bonds and cross-linking at the interfaces as well as geometrical factors, such as the CNT overlap length.

Several approaches have been pursued in order to optimize the mechanical behavior of CNT fibers. The first is through covalent cross-linking of adjacent CNT shells and tubes. This approach has the effect of creating strong bonds between CNTs, which can lead to improvements in the specific strength and stiffness through

FIGURE 3.2

(a) Model of a fiber composed of discrete fibrous elements. (b) When tension is applied to the model in (a), the fiber fails by shear failure between adjacent elements. (c) SEM image of a CNT fiber that has undergone the shear failure depicted in (b). (For color version of this figure, the reader is referred to the online version of this book.)

Source: Adapted with permission from Ref. [18]. Copyright 2011 American Chemical Society.

direct load transfer to concentric shells in an MWNT. This has primarily been achieved using high-energy particle irradiation of CNTs [12,19—22]. Another approach is through reformable or rehealable bonding between CNTs, such as van der Waals interactions or hydrogen bonding, which can exist between CNTs bound together with polymer molecules [8,17]. Finally, the third approach is the application of optimized geometrical fiber designs, which are needed to complement (and take advantage of) the improvements in intrinsic properties from the first two bonding strategies. These approaches are in large part motivated by nature, which uses a variety of bond types and groupings in optimized orientations. Between 1991 and 2010, substantial efforts have been carried out to investigate the mechanisms that lead to the superior mechanical properties of natural materials [23], which can achieve both high stiffness and toughness. Through intensive experimental and theoretical studies [23—34], the community has realized that the key to the outstanding performance of biological materials such as nacre, spider silk or tendon lies in their hierarchical bonding strategies and hierarchical structures (Fig. 3.3). In this chapter, we will discuss and review these three approaches for developing high-performance CNT fibers and provide an outlook of their potential impact on advanced carbon-fiber manufacturing.

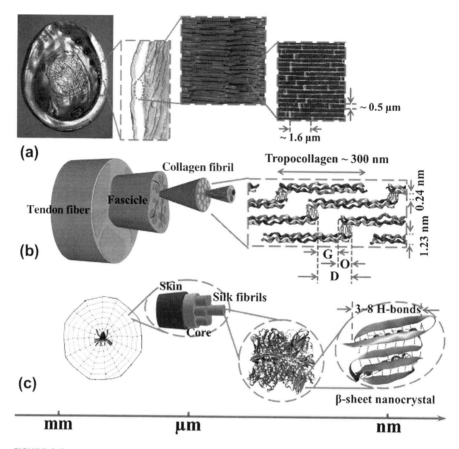

FIGURE 3.3

Hierarchical structures of three typical natural composites—(a) nacre shell, (b) tendon and (c) spider silk, scaling from micrometers down to nanometers. (For color version of this figure, the reader is referred to the online version of this book.)

Source: Adapted with permission from Ref. [15]. Copyright 2012 American Chemical Society.

3.2 IRRADIATION CROSS-LINKING: STRONG AND STIFF CNTs AND CNT BUNDLES

CNTs are by nature a layered material with strong in-plane bonding and weak out-of-plane bonding between their tubes. This leads to a significant disadvantage when they are used as load-bearing fibers, as a force applied to outer CNTs within a bundle or fiber is not adequately distributed over the entire cross-section and, therefore, leads to high stress on the outer layers of the material. This limitation can be overcome if strong bonds are created between the internal CNT layers, which effectively transfer the force to the internal CNTs. Early irradiation studies of CNTs demonstrated that this is possible by irradiating with high-energy particles, such as electrons or ions, which can "knock" out carbon atoms within the CNT layers creating reactive sites on the CNTs that can form covalent bonds between the adjacent graphitic layers within the CNT materials at either voids or carbon interstitials [35,36]. The presence of such covalently bonded sites increases the shear strength of the graphitic layer interfaces as a result of the relatively high C—C bond energies ($\sim$0.88 eV/atom) [37] as compared to the interplanar binding energy of graphite sheets ($\sim$0.02 eV/atom) [38]. Although the strength and stiffness of the interface is increased with the presence of covalent bonding, the interface also becomes more brittle in nature which can translate to lower macroscopic ductility as graphitic sheets can no longer easily slide with respect to one another.

The bridging mechanism, which was documented by Telling et al. [36], motivated many theoretical and experimental studies on how it could be applied to cross-link CNT materials [12,19,20,35,39—44]. In addition to predicting the presence of such cross-links by various particle irradiation methods [44,45], molecular dynamics (MD) simulations have also predicted that interstitial cross-linking can increase the strength of discontinuous CNT bundles to strengths of up to $\sim$60 GPa [22] which approaches the values for individual CNTs ($\sim$100 GPa). This is similar to MD predictions of carbon-ion irradiated continuous CNT bundles that have shown that irradiation only reduces the bundle tensile modulus to 50—70 GPa from that of continuous pristine bundle of CNTs while increasing shear and toughness parameters by an order of magnitude [43,44].

Cross-linking bridges can be formed both between adjacent CNTs within bundles and fibers (Fig. 3.4(a) and (b)) as well as between layers within MWNTs (Fig. 3.4(c)). This allows the beneficial effects of cross-linking to be applied at different length scales within CNT fibers. Experimental studies on electron irradiation strengthening and stiffening of CNTs and CNT bundles have confirmed the predicted beneficial effects of irradiation cross-linking (Fig. 3.5). Studies have shown that irradiation of CNTs and CNT bundles with electron energies in the range of 80—200 keV can lead to effective stiffness [47] values of $\sim$590 GPa for MWNTs [12], $\sim$750 GPa for single-walled carbon nanotube (SWNT) bundles [18], and $\sim$700 GPa for double-walled carbon nanotube (DWNT) bundles [20] and effective strengths of $\sim$35 GPa for MWNTs [12] and $\sim$17 GPa for DWNTs [20] (Fig. 3.5). Experiments predict that the optimized irradiation fluency is in the range of $\sim 5-10 \times 10^{20}$ e/cm^2 in order to achieve high strength and modulus. Higher irradiation levels have been found to lead to a high density of defects formed, which eventually leads to a weaker amorphous structure as revealed through TEM imaging of highly irradiated bundles [19,20]. As expected, this less-ordered structure will lead to significant reductions in the strength and modulus, and therefore, optimization between forming cross-links and creating defects needs to be considered. In addition to the increases in mechanical properties, the mechanism of load transfer to inner layers of CNT shells and tubes was also clearly demonstrated through in situ TEM measurements. Tensile measurements of individual MWNTs revealed that

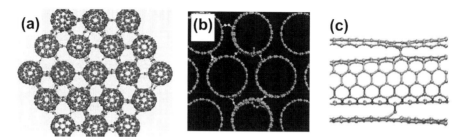

FIGURE 3.4

Simulations of cross-linked CNT materials. (a) MD simulation of SWNT bundle cross-linked with interstitial carbon atoms. (b) Simulation of ion irradiated SWNTs showing cross-linking atoms. (c) Molecular mechanics simulation of cross-linked DWNT. The simulation setup. Note that atoms in the front and back of the outer tube are not shown for the sake of clarity. Actual simulated tubes were also much longer. Tube ends were not terminated with hydrogen atoms. All atomistic visualizations were created by Visual Molecular Dynamics (VMD) [21]. (For color version of this figure, the reader is referred to the online version of this book.)

(a): Reprinted with permission from [22]. Copyright 2011, American Institute of Physics. (b): Figure adapted by permission from Macmillan Publishers Ltd: Nature Materials [46], copyright 2004. (c): Figure adapted from [14] with kind permission from Springer Science and Business Media.

the typical sword-in-sheath failure mechanism was modified from just the outer shell failing to inner shells failing, which confirmed that cross-links were transferring force to inner shells [12]. A similar observation was made for DWNT bundles in situ TEM for which uncross-linked bundles were found to fail by fracture of only the outer ring of DWNTs in the bundle, whereas optimally cross-linked bundles fractured across the entire cross-section of DWNTs [20], again confirming force transfer to the inner layer of material. The failure mechanisms observed in situ TEM for cross-linked MWNTs and DWNT bundles also demonstrate a reduction in ductility imparted by the presence of the cross-linking, as the failure is brittle in nature. In the case of DWNTs in the absence of cross-links, the bundles exhibit high ductility (accompanied by low yield stress) as inner DWNTs slide with respect to outer DWNTs [6], whereas cross-linked bundles exhibit significantly higher yield stress with reduced ductility [20].

Despite the demonstration of mechanical property improvements in smaller scale CNT materials such as bundles, particle irradiation cross-linking has not yet been successfully applied to macroscopic fibers. Such progress toward improving macroscopic CNT fibers is hampered by several roadblocks. These include achieving significant penetration depth of irradiation particle within macroscopic fibers with large diameters as well as matching the required high current densities for electron irradiation of large-scale samples [48]. Future work must focus on either overcoming these technical irradiation limitations or identifying alternative methods of internally cross-linking CNTs within CNT fibers. One possible alternative approach would be to cross-link CNTs via a chemical treatment

3.2 Irradiation Cross-Linking: Strong and Stiff CNTs and CNT Bundles

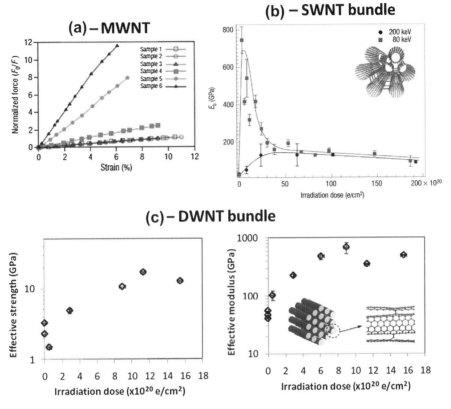

FIGURE 3.5

Electron irradiation strengthening and stiffening of CNT materials. (a) Increased tensile load carrying capacity for irradiated MWNTs. (b) Bending stiffness increases for irradiated SWNT bundles. (c) Tensile strength and modulus increases for irradiated DWNT bundles. (For color version of this figure, the reader is referred to the online version of this book.)

(a): Adapted by permission from Macmillan Publishers Ltd: Nature Nanotechnology [12], copyright 2008. (b): Adapted by permission from Macmillan Publishers Ltd: Nature Materials [19], copyright 2004. (c): Figure adapted with permission from [20], copyright ©WILEY-VCH Verlag GmbH & Co. KGaA, Weinheim.

approach. For example, stiffening of layered graphene oxide (GO) films has been demonstrated via a chemical treatment with borate to covalently crosslink GO layers within the film [49]. However, similar challenges of penetration depth of cross-linking may still need to be addressed.

As another alternative to particle irradiation of CNTs, electromagnetic irradiation has also been proposed to enhance mechanical properties through cross-linking within CNT fibers. In this case, irradiation does not lead to "knock" on damage of the CNTs but instead modifies bonding between polymer chains that bridge CNTs within a material. This approach has not been tested extensively for CNT bundle or fiber geometries but has been demonstrated with other CNT layered materials. These studies include ultraviolet irradiation covalent cross-linking of layer-by-layer polymer/SWNT [50] and dimethylformamide/CNT materials [51], and microwave irradiation bonding of polyethylene terephthalate and polycarbonate to CNTs [52]. The one approach that has been tested on CNT fibers, gamma irradiation, was found to lead to an initial improvement in strength and stiffness of the CNT fibers for a low dose of irradiation followed by a plateau in properties at higher irradiation levels [53]. This behavior was attributed to the formation of interstitial carboxyl like groups between CNTs within the fibers, which enhanced shear interactions [53]. These alternative irradiation approaches may open the door for the application of alternative or complimentary bonding types in addition to the covalent bonding achieved by particle irradiation.

3.3 REFORMABLE BONDING: STRONG AND TOUGH CNT BUNDLES AND FIBERS

As discussed in the previous sections, the inherent interactions between adjacent CNTs or within concentric walls of MWNTs are due to relatively weak van der Waals forces. This type of interaction can break and reform as adjacent CNTs or concentric walls of CNTs slide past one another. The interfacial shear properties are critical in CNT fibers as these dictate how the load is transferred between adjacent CNTs within a fiber.

In an effort to increase the shear interactions between tubes, chemical methods have been applied that modify the inherently chemically inactive surfaces of CNTs by introducing various chemical functionalizations [8,54–57]. The functionalized CNTs may then bind more strongly to each other or an external matrix material through hydrogen bonding or covalent bonds. These approaches toward CNT surface functionalization include wet treatments, for example with nitric acid to introduce oxygen-containing side groups [54–56], dry treatments, such as plasma treatment [57], and functionalization during CNT fabrication directly in the CVD reactor [8].

Nanocomp Technologies, Inc. and coworkers have developed high-strength (3–5 GPa) and high-stiffness (80–200 GPa) fibers from aerogel-spun CNT fibers that are subject to chemical treatments of acetone and a proprietary chemical [58]. In addition, they have performed chemical treatments prior to resin infusion in the development of CNT fibers to be used as torsion sensors [56]. Specifically, these fibers were treated with an acetone rinse followed by immersion in a nitric acid solution prior to drying [56]. This treatment facilitates the formation of oxygen-containing functional groups that can serve to bond the CNTs to the epoxy resin that is subsequently introduced into the fiber [59].

The wet and dry treatments applied after CNT fabrication typically introduce structural defects into the CNTs [55,60,61]. Researchers have speculated that

3.3 Reformable Bonding: Strong and Tough CNT Bundles and Fibers

structural defects in CNTs can serve as sites for covalent cross-linking between the CNTs and the matrix (e.g. [62–64]); however, covalent cross-links do not afford significant ductility to the interface, as discussed in the previous section. In addition, the introduction of structural defects in the CNTs degrades the inherent mechanical properties, thermal stability, and electrical conductivity of the CNTs themselves.

As an alternative to chemical treatments after CNT fabrication, Naraghi et al. report the functionalization of DWNT bundles in situ the CVD reactor during DWNT manufacturing by MER Corporation [8]. The polymeric coating was characterized through X-ray photoelectron spectroscopy, Fourier transform infrared spectroscopy [6], thermogravimetric analysis (TGA), mass spectroscopy [8], ^{1}H nuclear magnetic resonance (NMR) and ^{13}C NMR [65]. It was determined that the coating is largely a combination of oligomeric chains that consist of approximately 10 monomeric units of substituted acrylic acids and esters, and the inherent polymeric coating contains both hydrogen-bond donor and hydrogen-bond acceptor functional groups [6,8,65].

Notably, in contrast to the functionalization treatments applied post-CNT fabrication, the in situ CVD functionalization did not adversely affect the structure of the produced DWNT bundles, which is evident through the low-defect density observed in Raman spectroscopy [6,8]. Fibers fabricated from these DWNT bundles had high strength (up to ∼1.4 GPa) and toughness (up to ∼100 J/g), which approach values of spider silk [66,67]. Removal of the inherent polymeric coating via heat treatment resulted in a drastic loss of ductility and toughness (from an average of ∼90 J/g to an average of ∼30 J/g), illustrating that the polymeric coating was critical to obtain high toughness in these fibers.

To elucidate the mechanics of the deformation at the nanoscale, and the apparent enhancement of interfacial properties in the fibers containing DWNT bundles coated with polymer over the heat-treated samples where the intrinsic polymer had been removed, Naraghi et al. performed experimental and computational studies aimed at determining the shear properties of DWNT bundles [17]. In particular, they performed experiments to quantify the shear interfacial stress between DWNT bundles coated with the intrinsic polymer layer for varying overlap lengths. They also performed MD and coarse grained simulations to visualize the nanoscale deformation mechanisms that are present while shearing two DWNT bundles coated with the intrinsic polymer. In the simulations, the polymeric coating was represented as poly(methyl methacrylate) (PMMA) of 4–8 repeating units covalently attached to the DWNT bundles. The computational simulations showed that at the nanoscale, the PMMA that was wrapped around the CNT bundles provided an interlocking mechanism during shear deformation. Thus, as two CNT bundles were sheared over each other, the van der Waals interactions between the CNTs and between the adjacent PMMA as well as the interlocking between opposing PMMA units worked in concert to stretch the polymer coating molecules, in turn providing large shear deformations of the interface. Thus, at small deformations, the interlocking mechanism provided stiffness, while at larger deformations, the elongation of polymer chains allows for energy absorption, and thus a high-toughness material [17].

In addition to providing mechanistic understanding of the deformation, this study reports interfacial shear strengths between two bundles coated with the inherent polymer that range from 300 to 400 MPa [17]. These values are significantly higher than shear strengths measured in other CNT-based composites demonstrating the effectiveness of in situ CVD functionalization in increasing shear interactions between CNTs [11,63,64].

To identify other chemical bonds that could be introduced between the CNTs to enhance load transfer, deformation mechanisms in nature can be examined. Natural materials such as nacre, bone, tendon, and spider silk are composed of stiff building blocks that are joined by compliant links [30,68−74]. The stiff components afford mechanical stiffness and strength to the macroscopic material, while the compliant links allow the material to absorb energy during deformation, providing for more ductility than the stiff components alone could support. This is strikingly evident in nacre, in which the stiff aragonite is joined with biopolymers in a brick and mortar geometrical arrangement, and the toughness of the hierarchically structured composite material is as much as 3000 times that of monolithic aragonite alone [24,33,75].

In addition, spider silk has been widely studied due to its high toughness (reaching $\sim$165 J/g) and high specific strength ($\sim$1.3 GPa, with $\rho \sim$ 1.3 g/cm^3) [66,67]. Spider silk also has a hierarchical structure, composed of stiff beta-sheets and compliant amorphous protein regimes as well as an abundance of hydrogen bonds that have the ability to break and reform with deformation. MD simulations have revealed the nanoscale deformation mechanisms responsible for the high mechanical properties in spider silk, showing that as tension is applied to a system of stiff beta-sheets and compliant semi-amorphous proteins, the hydrogen bonds in folded semi-amorphous regions break and reform to support the applied deformation, and the straightening or uncoiling of the semi-amorphous regimes allows for the formation of additional beta-sheets, affording increased strength to the material (Fig. 3.6) [73,76,77]. Thus, the breaking and reforming of hydrogen bonds is primarily responsible for the high toughness observed experimentally in spider silk.

Inspired by the toughening mechanisms present in spider silk, Beese et al. have performed studies to investigate the effectiveness of incorporating a hydrogen-bond network between DWNT bundles inherently functionalized with a polymeric coating during CVD fabrication [65]. As previously pointed out, the inherent polymeric coating on the DWNT bundles fabricated by MER Corporation contains hydrogen-bond donating and hydrogen-bond accepting functional groups. Thus, the introduction of a polymer with hydrogen-bonding capabilities could provide for a hydrogen-bond network between adjacent DWNT bundles in macroscopic fibers fabricated from these bundles, which may act to enhance the toughness of the CNT−CNT interfaces. This approach represents one avenue for exploiting the inherent functionalization present on the DWNT bundles in order to modify the shear interactions for the development of high-performance CNT-based fibers.

In the study presented in Ref. [65], mechanical tests and computational simulations were performed to evaluate the effectiveness of introducing poly(vinyl alcohol)

3.3 Reformable Bonding: Strong and Tough CNT Bundles and Fibers

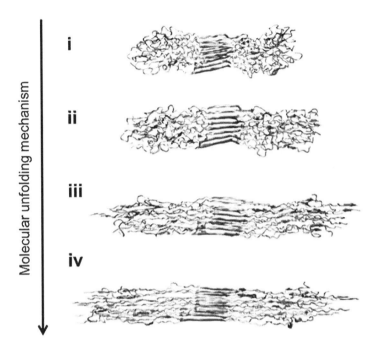

FIGURE 3.6

Molecular unfolding mechanism of the protein nanocomposite and associated nanomechanical and structural properties as a function of increasing end-to-end length. Qualification of the four-stage deformation mechanism of a single silk repeat unit. Initially (i), the protein nanocomposite consists of coiled semi-amorphous domains exhibiting hidden length embedding a stiff beta-sheet crystal. As the protein begins to unfold (ii), amorphous domains unravel, revealing their hidden length. As the amorphous domains stretch, new hydrogen bonds are created and beta-crystal units begin to form within the amorphous domains, sustaining the stress on the system as amorphous regions are extended to their limit (iii). A final regime is observed as covalent bond rupture forces many hydrogen bonds to break rapidly within the amorphous domains (iv), initiating sliding of beta-strands which ultimately leads to failure. (For color version of this figure, the reader is referred to the online version of this book.)

Source: Figure adapted from Ref. [78] with kind permission from Springer Science and Business Media.

(PVA), which contains pendant hydroxyl groups that are capable of acting as hydrogen-bond donors and hydrogen-bond acceptors, into the macroscopic fibers fabricated from inherently functionalized DWNT bundles. Tensile experiments on the fibers indicated that there is a parabolic relationship between the energy to failure and stiffness and the PVA weight percentage in the fibers. MD simulations were performed to evaluate the nanoscale deformation mechanisms as well as to quantify the hydrogen-bond density increase as a function of added PVA. These simulations showed that the stiffness of the interface between functionalized

CNTs with added PVA was parabolic with PVA content, in agreement with experimental results. In addition, the number of hydrogen bonds increased linearly with added PVA, as hydrogen bonds are formed between the inherent polymeric coating and the PVA, as well as between the PVA molecules themselves. However, the number of *productive* hydrogen bonds, or the number of hydrogen bonds existing only between the polymeric coating and the PVA molecules and excluding the hydrogen bonds between PVA molecules themselves, increased sharply with small additions of PVA, but increased much more slowly with added PVA once the optimal PVA content, defined as that which provided the highest interfacial stiffness, had been exceeded. Thus, through a combined experimental–computational approach, an optimal PVA content or a hydrogen-bond density was identified, which is sufficient to strongly link adjacent DWNT bundles while also afford ductility and therefore toughness or energy absorption to the macroscopic fiber.

Although further work is needed to optimize the structure of CNT-based fibers, as discussed in the next section, this study is encouraging in that it identifies a method through which an external material can be introduced into CNT-based fibers to take advantage of the inherent polymer coating grafted onto the CNTs during CVD fabrication, resulting in improved macroscopic fiber behavior [65]. Further advances in functionalization methods that take advantage of reformable or healable bonds should attribute high deformability and thus energy absorption to CNT fibers.

3.4 MATERIALS DESIGN: OPTIMIZED GEOMETRY AND STRUCTURE

In addition to designing bonding schemes to bridge CNTs within a fiber, various geometrical aspects play a critical role in CNT-based fiber design. While classical models of yarns' or fibers' mechanical properties can be applied to guide development of these materials, the design of fibers with nanoscale constituents presents many additional challenges beyond those in traditional fiber design. Optimizing CNT-based fiber design requires enhancing interactions across multiple length scales. At the smallest level, tube–tube interactions are largely defined by tube lengths, diameters, and various other features. Yet optimizing interactions at the nanometer-length scale alone is insufficient for creating lightweight, high-strength fibers. Global structural features such as the degree of fiber twist and porosity also play vital roles in effectively exploiting the superior mechanical properties of CNTs. A combination of analytical, experimental, and computational studies performed to-date provide fundamental insights into the role of various geometrical parameters as well as strategies for optimizing these features for use in effective CNT-based fibers. Figure 3.7 shows a summary of the geometrical and structural factors that influence CNT fiber mechanical properties at both the nano- and macroscale.

Several analytical models have been proposed to guide the development of fibers with optimized mechanical properties. Perhaps the most widely used model is that of

3.4 Materials Design: Optimized Geometry and Structure

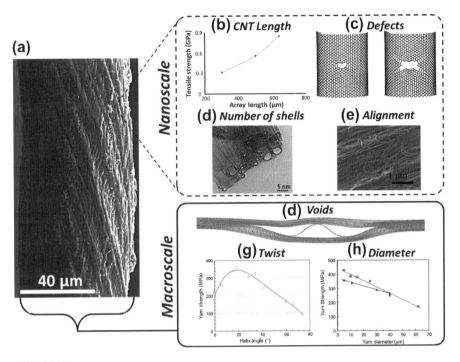

FIGURE 3.7

Overview of nanoscale and macroscale geometry considerations for CNT-based fiber design. (a): SEM image of spun DWNT fiber. (b): Fiber tensile strength increases with CNT length. (c): Crack propagation due to initial defect from atomistic simulation. (d): TEM image of collapsed DWNTs. (e): SEM image of fiber section with significant waviness. (f): Schematic of macroscale voids that separate CNTs within a fiber. (g): Dependence of fiber tensile strength on the twist (helix) angle for array-spun MWNT fibers. (h): Fiber strength increases with decreasing fiber diameter for high-twist fibers (triangles) and low-twist fibers (circles). (For color version of this figure, the reader is referred to the online version of this book.)

Sources: (a): Adapted with permission from [8]. Copyright 2010 American Chemical Society. (b): Figure adapted with permission from [79]. Copyright © 2007 WILEY-VCH Verlag GmbH & Co. KGaA, Weinheim. (c): Reprinted figure with permission from [80]. Copyright 2002 by the American Physical Society. (d): Figure adapted with permission from [80], Copyright © 2007 WILEY-VCH Verlag GmbH & Co. KGaA, Weinheim. (e): Reprinted from [82] with permission from Elsevier. (f): Reprinted from [84] with permission from Elsevier. (g): Adapted with permission from [83]. Copyright 2010 IOP Publishing Ltd. (h): Adapted with permission from [83]. Copyright 2010 IOP Publishing Ltd.

Hearle [85], who used the concept of friction between the filaments in traditional textile fibers and derived that the ratio of a yarn's tensile strength σ_y to the tensile strength of its filaments σ_f is approximately given by

$$\sigma_y/\sigma_f \approx \cos^2 \alpha \left(1 - (dQ/\mu)^{1/2} \operatorname{cosec} \alpha/3L\right) \quad (3.1)$$

where $k = (dQ/\mu)^{1/2}/(3L)$, d is the filament diameter, α is the twist angle of the filaments on the yarn or fiber surface, μ is the friction coefficient between filaments, L is the filament length, and Q is the migration length, which has the physical meaning of the distance over which a filament shifts from the yarn exterior to the yarn interior and back again. This theoretical prediction has been verified by experimental studies of CNT-based fibers that have isolated specific variables and investigated their effects on mechanical properties [79,83,86,87]. In light of the experimental evidence that supports the predictions of the Hearle model and its applicability to both nanoscale and macroscale geometry, this analytical model is broadly applicable to the development of high-strength CNT-based fibers.

At the nanoscale, there are several geometrical parameters that play key roles in defining the material properties of CNT-based fibers. Included among these are tube length, overlap length, tube diameter, tube thickness (i.e. number of CNT shells), defect density, and [15,88] alignment [88]. While previous studies have provided means of optimizing several of these parameters, other features possess inherent tradeoffs that must be carefully considered in accordance with the desired mechanical properties of the fiber.

One feature for which several studies have encountered a clear optimization trend is CNT length. Long CNTs provide several benefits over shorter tubes. First, CNT–CNT interactions are dominated by van der Waals forces that exist between the outermost shells of adjacent tubes. Consequently, the larger the contact area, the larger the adhesive forces will be between tubes. In this context, tube ends can be regarded as defects as van der Waals forces will be reduced at the ends of tubes [88]. Longer tubes lead to fewer breaks in these interactions which, in turn, should improve the integrity of the yarn, or fiber, at the nanoscale. In addition, longer tubes have a greater probability of twisting together, which further prohibits tubes from sliding past each other as compared to the case of parallel tubes [88]. These mechanical improvements afforded by longer tubes have been verified experimentally by Zhang et al. [79], who found that CNT-based yarns, or fibers, exhibit significantly improved tensile strength when composed of longer CNT arrays.

While increasing the total length of CNTs provides the opportunity for enhanced intertube interactions, the characterization and effect of optimal overlap length between tubes must also be considered. Several analytical studies have been performed to-date to understand the role of overlap length in defining the strength of hierarchical structures. Gao et al. [34] developed a tension-shear chain model to demonstrate that there is an optimal mechanical structure that ensures optimum strength and maximum tolerance of flaws. Wei et al. [15] used an analytical model (Fig. 3.8) based on Cox's classical shear lag model [89] to link the mechanical

3.4 Materials Design: Optimized Geometry and Structure

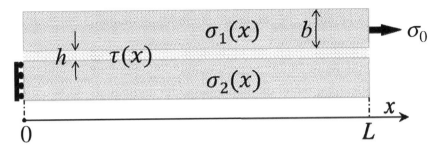

FIGURE 3.8

Schematic of 2D unit cell used in the analytical model simplified from the natural composites with brick-and-mortar structure.

properties of building blocks and their geometric arrangement with the mechanical performance of the macroscopic composites.

In the elastic regime, it can be shown that the elastic strain energy density of the unit cell (i.e. the load transfer efficiency) takes the form

$$w_{el} = \frac{\tau_f^2}{G}\frac{h}{2b+h}\left[\tanh\left(\frac{\lambda L}{2}\right)\left(\frac{1}{2}\tanh\left(\frac{\lambda L}{2}\right)+\frac{1}{\lambda L}\right)\right] = \frac{\tau_f^2}{G}\frac{h}{2b+h}f(\lambda L) \quad (3.2)$$

where

$$f(\lambda L) = \tanh\left(\frac{\lambda L}{2}\right)\left(\frac{1}{2}\tanh\left(\frac{\lambda L}{2}\right)+\frac{1}{\lambda L}\right), \text{ and } \lambda = \sqrt{\frac{2G}{Ehb}}. \quad (3.3)$$

In the above equations, E is the elastic modulus of the platelet, G is the shear modulus of the interface, and τ_f is the shear strength of the interface. b and h are the platelet thickness and interface thickness in the unit cell, respectively. Solving Eqns (3.2) and (3.3) numerically suggests that to maximize the elastic strain energy, the characteristic overlap length should be

$$L^* \sim 2.318\sqrt{\frac{Ebh}{G}}. \quad (3.4)$$

Thus, Eqn (3.4) links the material's toughness with the geometric arrangements and the mechanical properties of their constituents.

In addition to interface failure, the composite strength is also limited by the platelet rupture. Therefore, in the plastic regime, another characteristic overlap length is given by

$$L^{**} = \frac{\sigma_f}{\tau_f}b \quad (3.5)$$

where σ_f is the platelet tensile strength. The relationship given in Eqn (3.5) coincides with that reported in Ref. [34], which links the strength of the composite materials with the geometries and properties of their constituents. By regarding CNTs and their interfaces as the brick and mortar structures, respectively, this analytical model can be used to guide the selection of optimal CNT—CNT overlap length.

Another geometrical parameter of CNTs with significant impact on mechanical properties is the number of shells within the tubes. Jia et al. [82] performed tensile tests on spun fibers composed of CNTs with distinct geometries in order to isolate the role of various parameters. Their comparison of fibers fabricated with tubes of similar outer diameters, but a significant difference (approximately a factor of two) in the number of tubes showed that tubes with fewer shells enhance fiber properties. They attribute this improvement to several factors. First, tubes with fewer shells exhibited a lower defect density based on Raman spectroscopy of the two sets of tubes [82]. Lower defect density may lead to increased van der Waals interactions between the tubes, as discussed earlier in the context of tube ends as one type of CNT defect. In addition, tubes with fewer numbers of shells will flatten more under tensile loading. Motta et al. [81] demonstrated this in an experimental study in which adjacent tubes with few walls within a bundle significantly flattened against one another. The role of this flattening behavior was also investigated by Zhao et al. [90], who reported a double-peak in tensile strength and elastic modulus under increasing twist. They attributed this second peak to tube flattening, which enhances van der Waals interactions within bundles by increasing contact area between tubes within a bundle. Finally, as specific tensile properties normalized by unit mass are always of interest when considering lightweight, high-strength fibers, Jia et al. [82] noted that fibers spun from these tubes with different numbers of shells exhibited a roughly proportional difference in mass. This is consistent with the earlier experimental studies discussed in Section 3.2, such as that conducted by Peng et al. [12], which showed that the outermost shells of MWCNTs bear the majority of the applied load in the absence of cross-linking to inner shells. Consequently, reducing the number of interior shells will yield higher specific strength on the basis of mass alone, and this result is only augmented when considering the increased flattening afforded by a fewer number of inner tubes.

Tube diameter is another parameter that has been investigated experimentally and computationally for its effect on fiber strength. Vilatela et al. [18] demonstrated that from a theoretical standpoint, a large diameter tube would be especially beneficial because of its ability to collapse extensively and yield large contact areas between adjacent tubes. In practice, though, it is difficult to decouple tube diameter from the number of shells, as shown in experiments [82]. The results of Jia et al. [82] suggest that the number of shells is an important factor in fiber design, but the effect of diameter alone is less clear.

In addition to geometrical parameters pertaining to tube aspect ratios, the number of atomic-scale and nanoscale defects on tube surfaces and within tubes also plays a vital role in the nanomechanics of CNT systems. Defects exist in various forms, including vacancies, bends, and Stone—Wales 5-7-7-5 pairs [91—93]. Defects can form from various sources including high-energy irradiation [12,14,19,20,39] and chemical treatments [94]. Experimental characterization of defect density in CNTs is typically performed by measuring the ratio of the D-peak to the G-peak in Raman spectroscopy [8,82,90,95] as well as TGA [8]. At the nanoscale, the tradeoffs associated with defects are significant. Defects enable the formation of additional bonding schemes beyond van der Waals tube—tube interactions, which can significantly enhance mechanical properties of CNT bundles [14,19,20]. Yet atomistic simulations have shown that the superior mechanical properties of pristine CNTs [80] and bundles [96] significantly degrade with even a small number of defects. As such, an optimal concentration of defects must be achieved that enables effective load transfer between tubes without significantly degrading the strength of individual tubes. This area of study is still of significant interest to the scientific community in the context of the improvements afforded by chemical functionalization and covalent cross-linking.

The final nanoscale geometric aspect of interest discussed here is alignment. Degree of CNT alignment can be characterized in several ways. Transmission electron microscopy diffraction patterns can be used to characterize the alignment of tubes within individual bundles, though the technique may be complicated by coupling variations in chiral angle with misalignment [8,97]. Scanning electron microscope (SEM) imaging of the outer surfaces of fibers can be used to qualitatively determine whether or not bundles of CNTs are well-aligned with each other (e.g. [82]). X-ray scattering and polarized Raman spectroscopy can also be used to characterize alignment within fibers (e.g. [81,87].). Alignment at the fiber level is typically enhanced by the application of mechanical loads such as tension, twisting, or pressure (e.g. through an aerogel) [8,81,83,87]. The impacts of variations in alignment as well as these proposed enhancements have been investigated through several experimental studies. For example, Jia et al. [82] characterized alignment by imaging fiber surfaces in SEM and performing tensile tests on fibers composed of tubes with varying alignment. They experimentally demonstrated a tradeoff between failure strain and fiber modulus: fibers with poor alignment exhibited high strain-to-failure with low modulus, while fibers with high alignment exhibited low strain-to-failure but high modulus [82]. At the same time, they also concluded that alignment did not significantly affect tensile strength due to the fact that the fibers could be spun at high angles [82]. In addition, Koziol et al. [87] showed that specific strength and specific stiffness of CNT-based fibers improved when alignment within the fiber was enhanced. Fibers spun at faster rates exhibited a higher degree of alignment, as determined by a combination of small-angle X-ray scattering and D-peak to G-peak ratios from Raman spectroscopy [87]. Motta et al. [81] likewise found that fibers wound at the fastest possible extrusion rate from a CVD furnace exhibited better alignment, strength, and stiffness than tubes prepared with a standard extrusion rate. However, this came at the expense of energy-to-failure as failure strains were substantially reduced when fibers were spun at faster rates. The results of these studies suggest that the alignment of tubes plays a critical role in defining the mechanical properties of CNT-based fibers.

In addition to designing optimal nanoscale systems by exploring geometry at the level of individual tubes and bundles, macroscale geometric considerations are also crucial for developing high-strength CNT-based fibers. Twist, alignment, fiber diameter, and large-scale defects are among these factors that merit investigation through theoretical, experimental, and computational studies.

The macroscale helical structure from twisting plays an important role in the mechanical properties of spun fibers. The load transfer between tubes within fibers is dependent on not only the geometric arrangement of the tubes but also the normal pressure between the tubes. Increasing the twist angle causes a high transversal pressure inside fibers, thus enhancing the interaction between tubes [86]. This theoretical prediction has been verified by Fang et al. [83] through the studies of a series of MWCNT fibers with various twist angles. As shown in Fig. 3.9, the fiber strength initially increases with growing twist angle and reaches the peak tensile strength at the twist angle of approximately 20°. Then, the fiber strength decreases with continuously growing twist angle >20° due to the negative effect from the $\cos^2\alpha$ term, which arises from a tradeoff between increasing external pressure and maintaining alignment between the tensile axes of CNTs and the main axis of the fiber itself. Therefore, Hearle's model and the experimental observations suggest that there is also an optimal twist angle for improving the fiber strength. The optimal twist angle is highly sensitive to the pressure dependence of the interaction between filaments (CNTs in this study) and the microstructures of the filaments. Therefore, further experimental and computational investigations are desired to obtain the essential inputs for Hearle's model.

The final two macroscale geometry factors—fiber diameter and large-scale defects—are intrinsically linked. Experimental studies [83,98] have shown that fibers with smaller diameters tend to exhibit higher tensile strength than larger diameter fibers. This result can be attributed to two factors: a higher exerted pressure on

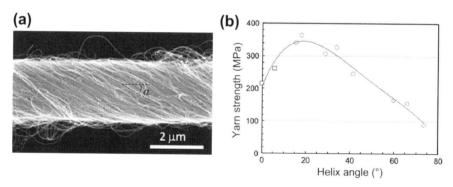

FIGURE 3.9

(a) SEM image of a twist-spun fiber from MWNT forest with the twist angle α. (b) Dependence of fiber tensile strength on the twist (helix) angle for array-spun MWNT fibers in Ref. [83]. Fibers having very low twist (squares) were liquid densified before twist insertion, and the remaining fibers (circles) were twist-spun without any prior treatment. The solid line is Hearle model prediction. The fiber diameter was kept constant at ~20 μm, and the CNT forest used for spinning had a height of 350 μm. (For color version of this figure, the reader is referred to the online version of this book.)

Sources: (a) Adapted with permission from Ref. [83]. Copyright 2010 IOP Publishing Ltd. (b) Adapted with permission from Ref. [83]. Copyright 2010 IOP Publishing Ltd.

smaller diameter fibers and fewer weak connections resulting from voids and other macroscale imperfections in the specimens [83,88,98]. Perhaps the most common type of large-scale defects in a CNT-based fiber is voids, which result from misalignments between CNT bundles. Miao et al. [84] studied the role of voids in their experimental investigation of Poisson's ratio of CNT-based fibers. They attributed the large Poisson's ratio resulting from their tensile tests to, above all, the closing of voids within the fiber that could be hundreds of nanometers across. In addition, they noted that these large gaps cause other taut CNTs to bear extra load, which can lead to localized failure that would then propagate throughout the fiber. Mechanically loading the fibers through twisting, stretching, and liquid densification during the fiber fabrication process (e.g. [8,83,84]) can significantly reduce the sizes of voids, which should enhance the integrity of the entire CNT-based fiber.

The design of next-generation CNT-based fibers relies on a complex combination of multiscale mechanical factors. Both nanoscale and macroscale geometric features must be analyzed to optimize the interactions between tubes and bundles within a fiber. Significant scientific effort remains ongoing to investigate these multiscale phenomena in the push toward macroscale fibers that effectively exploit the phenomenal mechanical properties of CNTs.

3.5 SUMMARY

Optimal designs that can tailor the mechanical properties of CNT fibers to the needs of applications in commercial markets, such as the aerospace and automotive industries, require the cooperative consideration of different internal bonding types between nanoscale constituents as well as geometrical and structural characteristics at different length scales such as characteristic overlap length, alignment, defect size and density within CNT shells, and larger scale defects such as porosity and voids. Toward the development of such designs, a great deal of experimental and theoretical work has been conducted that has yielded important insights into how CNT fibers can be advanced and improved.

Covalent cross-linking of CNTs by irradiation within fibers has been proven on smaller length scales (individual MWNTs and CNT bundles with diameters of tens of nanometers) to lead to order of magnitude improvements in strength and stiffness. Effective strengths on the order of ~17 GPa have been demonstrated for cross-linked CNT bundles [20], which provides a guideline for achievable upper limits in macroscopic fibers for which current technologies exhibit lower strengths on the order of 1−5 GPa for longer (a few millimeters to centimeters) specimens [8,58], and 9 GPa for short (~1 mm) specimens [87]. Experiments have also revealed the brittle nature of covalently cross-linked CNT materials, which can reduce ductility and toughness.

The technical challenges of scaling up irradiation cross-linking of CNT fibers, and the limitation of reduced ductility, have led to alternative methods of enhancing shear interactions between internal layers. One such example is the inclusion of polymer intermediaries between CNTs, which has allowed for bio-inspired strategies to be applied in attempts to improve mechanical performance. In this case, hydrogen-bond networks, similar to those found in spider silk, have been utilized to enhance the toughness of CNT-polymer fibers with an intrinsic polymer coating resulting from CVD synthesis [8] as well as CNT–PVA fibers [65]. This approach has led to CNT fibers that exhibit energy-to-failure (or toughness) of up to ~ 100 J/g, approaching that of spider silk [66,67]. Studies of CNT–PVA fibers have also demonstrated the great importance of utilizing an experimental–computational approach, which was used to optimize the polymer content and hydrogen-bond density in order to both strongly link adjacent DWNT bundles and impart sufficient ductility to achieve macroscopic fibers with high toughness [65].

Ultimately analytical and numerical models, in conjunction with multiscale experimental inputs, are needed to predict the optimal structure of a synthetic CNT fiber and guide design. For example, the application of the analytical model developed by Wei et al. [15] requires the identification of the inputs to Eqns (3.4) and (3.5), through fundamental experimental and computational work to obtain the material properties of the constituents. Naraghi et al. [17] addressed this with in situ shear experiment on two DWNT bundles along with multiscale modeling to probe the mechanical properties of the polymer interface between bundles (the building blocks for their DWNT fibers [8]). The experiments revealed that a transition from interface failure to bundle rupture occurs at the overlap length >850 nm which could be understood through the model developed by Wei et al., which predicts that at a sufficiently long overlap length, the critical junction force reaches saturation and the failure mode shifts to bundle failure [15]. It is the combination of studies such as this that will provide guidelines for the continued future development of high-performance CNT fibers. Future combined studies of experiments and modeling will require a detailed knowledge of both the nature of the interfacial CNT–CNT bonding, intrinsic CNT characteristics such as length, defect density, and CNT alignment within fibers, as well as macroscopic fiber characteristic such as twist, fiber diameter, and large-scale defects which all play a role in the mechanical properties.

Acknowledgments

This work was supported by ARO MURI award No. W911NF-09-1-0541. M.R.R. gratefully acknowledges support from the Department of Defense (DoD) through the National Defense Science & Engineering Graduate (NDSEG) Fellowship Program.

References

[1] S. Chand, Carbon fibers for composites, Journal of Materials Science 35 (6) (2000) 1303–1313.
[2] J.O. Zhao, et al., Growth of carbon nanotubes on the surface of carbon fibers, Carbon 46 (2) (2008) 380–383.

References

[3] H. Qian, et al., Carbon nanotube-based hierarchical composites: a review, Journal of Materials Chemistry 20 (23) (2010) 4751–4762.

[4] D. Papkov, et al., Extraordinary improvement of the graphitic structure of continuous carbon nanofibers templated with double wall carbon nanotubes, ACS Nano Vol. 7 (2013) Pg. 126–142.

[5] Y.-L. Li, I. Kinloch, A. Windle, Direct spinning of carbon nanotube fibers from chemical vapor deposition synthesis, Science 304 (2004) 276–278.

[6] T. Filleter, et al., Experimental-computational study of shear interactions within double-walled carbon nanotube bundles, Nano Letters 12 (2) (2012) 732–742.

[7] Nanocomp Technologies Inc.

[8] M. Naraghi, et al., A multiscale study of high performance double-walled nanotube-polymer fibers, ACS Nano 4 (11) (2010) 6463–6476.

[9] A. Kis, et al., Interlayer forces and ultralow sliding friction in multiwalled carbon nanotubes, Physical Review Letters 97 (2) (2006), 025501.

[10] M.F. Yu, et al., Strength and breaking mechanism of multiwalled carbon nanotubes under tensile load, Science 287 (5453) (2000) 637–640.

[11] R.S. Ruoff, M.F. Yu, B.I. Yakobson, Controlled sliding and pullout of nested shells in individual multiwalled carbon nanotubes, Journal of Physical Chemistry B 104 (37) (2000) 8764–8767.

[12] B. Peng, et al., Measurements of near-ultimate strength for multiwalled carbon nanotubes and irradiation-induced crosslinking improvements, Nature Nanotechnology 3 (10) (2008) 626–631.

[13] J. Cumings, A. Zettl, Low-friction nanoscale linear bearing realized from multiwall carbon nanotubes, Science 289 (5479) (2000) 602–604.

[14] M. Locascio, et al., Tailoring the load carrying capacity of MWCNTs through intershell atomic bridging, Experimental Mechanics 49 (2) (2009) 169–182.

[15] X.D. Wei, M. Naraghi, H.D. Espinosa, Optimal length scales emerging from shear load transfer in natural materials: application to carbon-based nanocomposite design, ACS Nano 6 (3) (2012) 2333–2344.

[16] M.F. Yu, et al., Tensile loading of ropes of single wall carbon nanotubes and their mechanical properties, Physical Review Letters 84 (24) (2000) 5552–5555.

[17] M. Naraghi, et al., Atomistic investigation of load transfer between DWNT bundles "crosslinked" by PMMA oligomers, Advanced Functional Materials (2012).

[18] J.J. Vilatela, J.A. Elliott, A.H. Windle, A model for the strength of yarn-like carbon nanotube fibers, ACS Nano 5 (3) (2011) 1921–1927.

[19] A. Kis, et al., Reinforcement of single-walled carbon nanotube bundles by intertube bridging, Nature Materials 3 (3) (2004) 153–157.

[20] T. Filleter, et al., Ultrahigh strength and stiffness in cross-linked hierarchical carbon nanotube bundles, Advanced Materials 23 (2011) 2855–2860.

[21] W. Humphrey, A. Dalke, K. Schulten, VMD - Visual Molecular Dynamics, J. Molec. Graphics. vol. 14 (1996). pp. 33-38.

[22] C.F. Cornwell, C.R. Welch, Very-high-strength (60-GPa) carbon nanotube fiber design based on molecular dynamics simulations, Journal of Chemical Physics 134 (20) (2011), 204708.

[23] P. Fratzl, et al., Structure and mechanical quality of the collagen–mineral nano-composite in bone, Journal of Materials Chemistry 14 (14) (2004) 2115–2123.

[24] J. Currey, J. Taylor, The mechanical behaviour of some molluscan hard tissues, Journal of Zoology 173 (3) (1974) 395–406.
[25] S. Kotha, Y. Li, N. Guzelsu, Micromechanical model of nacre tested in tension, Journal of Materials Science 36 (8) (2001) 2001–2007.
[26] A.Y.M. Lin, M.A. Meyers, K.S. Vecchio, Mechanical properties and structure of *Strombus gigas, Tridacna gigas*, and *Haliotis rufescens* sea shells: a comparative study, Materials Science and Engineering: C 26 (8) (2006) 1380–1389.
[27] F. Barthelat, H. Espinosa, An experimental investigation of deformation and fracture of nacre–mother of pearl, Experimental Mechanics 47 (3) (2007) 311–324.
[28] F. Barthelat, et al., On the mechanics of mother-of-pearl: a key feature in the material hierarchical structure, Journal of the Mechanics and Physics of Solids 55 (2) (2007) 306–337.
[29] E. Munch, et al., Tough, bio-inspired hybrid materials, Science 322 (5907) (2008) 1516–1520.
[30] H.D. Espinosa, et al., Merger of structure and material in nacre and bone-perspectives on de novo biomimetic materials, Progress in Materials Science 54 (8) (2009) 1059–1100.
[31] Z. Zhang, et al., Mechanical properties of unidirectional nanocomposites with non-uniformly or randomly staggered platelet distribution, Journal of the Mechanics and Physics of Solids 58 (10) (2010) 1646–1660.
[32] G. Liu, et al., Analytical solutions of the displacement and stress fields of the nanocomposite structure of biological materials, Composites Science and Technology 71 (9) (2011) 1190–1195.
[33] A.P. Jackson, J.F.V. Vincent, R.M. Turner, The mechanical design of nacre, Proceedings of the Royal Society of London 234 (1277) (1988) 415–440.
[34] H. Gao, et al., Materials become insensitive to flaws at nanoscale: lessons from nature, Proceedings of the National Academy of Sciences 100 (10) (2003) 5597–5600.
[35] F. Banhart, Irradiation effects in carbon nanostructures, Reports on Progress in Physics 62 (8) (1999) 1181–1221.
[36] R.H. Telling, et al., Wigner defects bridge the graphite gap, Nature Materials 2 (2003) 333–337.
[37] G. Leroy, et al., Refinement and extension of the table of standard energies for bonds involving hydrogen and various atoms of group-iv to group-vii of periodic-table, Journal of Molecular Structure 300 (1993) 373–383.
[38] M.C. Schabel, J.L. Martins, Energetics of interplanar binding in graphite, Physical Review B 46 (11) (1992) 7185–7188.
[39] B.W. Smith, D.E. Luzzi, Electron irradiation effects in single wall carbon nanotubes, Journal of Applied Physics 90 (7) (2001) 3509–3515.
[40] A.V. Krasheninnikov, K. Nordlund, Irradiation effects in carbon nanotubes, Nuclear Instruments and Methods in Physics Research Section B-beam Interactions with Materials and Atoms 216 (2004) 355–366.
[41] J.A.V. Pomoell, et al., Ion ranges and irradiation-induced defects in multiwalled carbon nanotubes, Journal of Applied Physics 96 (5) (2004) 2864–2871.
[42] A.V. Krasheninnikov, K. Nordlund, Ion and electron irradiation-induced effects in nanostructured materials, Journal of Applied Physics 107 (7) (2010), 071301.
[43] N.P. O'Brien, M.A. McCarthy, W.A. Curtin, A theoretical quantification of the possible improvement in the mechanical properties of carbon nanotube bundles by carbon ion irradiation, Carbon Vol. 53 (2013). Pg. 346–356.

[44] N.P. O'Brien, M.A. McCarthy, W.A. Curtin, Improved inter-tube coupling in CNT bundles through carbon ion irradiation, Carbon 51 (2013) 173–184.

[45] E. Salonen, A.V. Krasheninnikov, K. Nordlund, Ion-irradiation-induced defects in bundles of carbon nanotubes, Nuclear Instruments and Methods in Physics Research Section B-beam Interactions with Materials and Atoms 193 (2002) 603–608.

[46] P.M. Ajayan, F. Banhart, Strong bundles, Nature Materials 3 (3) (2004) 135–136.

[47] Effective stiffness and strength refer to measuring strength and stiffness from stress calculations that consider the entire cross section of all CNT shells and tubes within the MWNT or CNT bundle and considering a CNT shell thickness of ∼0.33 nm.

[48] T. Filleter, H.D. Espinosa, Multi-scale mechanical improvement produced in carbon nanotube fibers by irradiation cross-linking, Carbon 56 (2013) 1–11.

[49] Z. An, et al., Bio-inspired borate cross-linking in ultra-stiff graphene oxide thin films, Advanced Materials 23 (33) (2011) 3842–3846.

[50] S.H. Qin, et al., Covalent cross-linked polymer/single-wall carbon nanotube multilayer films, Chemistry of Materials 17 (8) (2005) 2131–2135.

[51] C. Miko, et al., Effect of ultraviolet light irradiation on macroscopic single-walled carbon nanotube bundles, Applied Physics Letters 88 (15) (2006) 151905.

[52] C.Y. Wang, et al., Strong carbon-nanotube-polymer bonding by microwave irradiation, Advanced Functional Materials 17 (12) (2007) 1979–1983.

[53] M.H. Miao, et al., Effect of gamma-irradiation on the mechanical properties of carbon nanotube yarns, Carbon 49 (14) (2011) 4940–4947.

[54] T. Kyotani, et al., Chemical modification of the inner walls of carbon nanotubes by HNO oxidation, Carbon 39 (2001) 771–785.

[55] M.V. Naseh, et al., Functionalization of carbon nanotubes using nitric acid oxidation and DBD plasma, World Academy of Science, Engineering and Technology 27 (2009) 567–569.

[56] A.S. Wu, et al., Carbon nanotube fibers as torsion sensors, Applied Physics Letters 100 (20) (2012), 201908.

[57] T. Xu, et al., Surface modification of multi-walled carbon nanotubes by O-2 plasma, Applied Surface Science 253 (22) (2007) 8945–8951.

[58] A.S. Wu, et al., Strain rate-dependent tensile properties and dynamic electromechanical response of carbon nanotube fibers, Carbon 50 (10) (2012) 3876–3881.

[59] A.S. Wu, T.W. Chou, Carbon nanotube fibers for advanced composites, Materials Today 15 (7–8) (2012) 302–310.

[60] N.G. Sahoo, et al., Improvement of mechanical and thermal properties of carbon nanotube composites through nanotube functionalization and processing methods, Materials Chemistry and Physics 117 (1) (2009) 313–320.

[61] X.W. Pei, W.M. Liu, J.C. Hao, Functionalization of multiwalled carbon nanotube via surface reversible addition fragmentation chain transfer polymerization and as lubricant additives, Journal of Polymer Science Part A–Polymer Chemistry 46 (9) (2008) 3014–3023.

[62] C. Cooper, et al., Detachment of nanotubes from a polymer matrix, Applied Physics Letters 81 (2002) 3873–3876.

[63] A. Barber, et al., Interfacial fracture energy measurements for multi-walled carbon nanotubes pulled from a polymer matrix, Composites Science and Technology 64 (2004) 2283–2289.

[64] A.H. Barber, S.R. Cohen, H.D. Wagner, Measurement of carbon nanotube-polymer interfacial strength, Applied Physics Letters 82 (23) (2003) 4140–4142.
[65] A.M. Beese, et al., Bio-inspired carbon nanotube-polymer composite yarns with hydrogen bond-mediated lateral interactions. (2013) ACS Nano, 7 (4) 3434-3446.
[66] F. Vollrath, D.P. Knight, Liquid crystalline spinning of spider silk, Nature 410 (6828) (2001) 541–548.
[67] F. Vollrath, B. Madsen, Z.Z. Shao, The effect of spinning conditions on the mechanics of a spider's dragline silk, Proceedings of the Royal Society B-Biological Sciences 268 (1483) (2001) 2339–2346.
[68] M.J. Buehler, Nature designs tough collagen: explaining the nanostructure of collagen fibrils, Proceedings of the National Academy of Sciences 103 (2006) 12285–12290.
[69] S. Keten, et al., Nanoconfinement controls stiffness, strength and mechanical toughness of beta-sheet crystals in silk, Nature Materials 9 (4) (2010) 359–367.
[70] M. Meyers, et al., Biological materials: structure and mechanical properties, Progress in Materials Science 53 (2008) 1–206.
[71] F.G. Omenetto, D.L. Kaplan, New opportunities for an ancient material, Science 329 (5991) (2010) 528–531.
[72] U. Wegst, M. Ashby, The mechanical efficiency of natural materials, Philosophical Magazine 84 (21) (2004) 2167–2181.
[73] A. Nova, et al., Molecular and nanostructural mechanisms of deformation, strength and toughness of spider silk fibrils, Nano Letters 10 (7) (2010) 2626–2634.
[74] T. Giesa, et al., Nanoconfinement of spider silk fibrils begets superior strength, extensibility, and toughness, Nano Letters 11 (11) (2011) 5038–5046.
[75] J.D. Currey, Mechanical properties of mother of pearl in tension, Proceedings of the Royal Society of London 196 (1125) (1977) 443–463.
[76] S. Keten, M.J. Buehler, Nanostructure and molecular mechanics of spider dragline silk protein assemblies, Journal of the Royal Society Interface 7 (53) (2010) 1709–1721.
[77] S.W. Cranford, et al., Nonlinear material behaviour of spider silk yields robust webs, Nature 482 (7383) (2012) 72–76.
[78] A. Tarakanova, M.J. Buehler, A materiomics approach to spider silk: protein molecules to webs, JOM 64 (2) (2012) 214–225.
[79] X. Zhang, et al., Strong carbon-nanotube fibers spun from long carbon-nanotube arrays, Small 3 (2) (2007) 244–248.
[80] T. Belytschko, et al., Atomistic simulations of nanotube fracture, Physical Review B 65 (23) (2002) 235430.
[81] M. Motta, et al., High performance fibres from 'dog bone' carbon nanotubes, Advanced Materials 19 (21) (2007) 3721–3726.
[82] J. Jia, et al., A comparison of the mechanical properties of fibers spun from different carbon nanotubes, Carbon 49 (4) (2011) 1333–1339.
[83] S. Fang, et al., Structure and process-dependent properties of solid-state spun carbon nanotube yarns, Journal of Physics: Condensed Matter 22 (33) (2010) 334221.
[84] M. Miao, et al., Poisson's ratio and porosity of carbon nanotube dry-spun yarns, Carbon 48 (10) (2010) 2802–2811.
[85] J.W.S. Hearle, P. Grosberg, S. Backer, Structural Mechanics of Fibers, Yarns, and Fabrics (1969).
[86] M. Zhang, K.R. Atkinson, R.H. Baughman, Multifunctional carbon nanotube yarns by downsizing an ancient technology, Science 306 (5700) (2004) 1358–1361.

[87] K. Koziol, et al., High-performance carbon nanotube fiber, Science 318 (5858) (2007) 1892–1895.
[88] W. Lu, et al., State of the art of carbon nanotube fibers: opportunities and challenges, Advanced Materials 24 (2012) 1805–1833.
[89] H. Cox, The elasticity and strength of paper and other fibrous materials, British Journal of Applied Physics 3 (3) (1952) 72.
[90] J. Zhao, et al., Double-peak mechanical properties of carbon-nanotube fibers, Small 6 (22) (2010) 2612–2617.
[91] P.J.F. Harris, Carbon Nanotube Science: Synthesis, Properties and Applications, Cambridge University Press, Cambridge, 2009.
[92] J.C. Charlier, Defects in carbon nanotubes, Accounts of Chemical Research 35 (12) (2002) 1063–1069.
[93] J. Kotakoski, A. Krasheninnikov, K. Nordlund, Energetics, structure, and long-range interaction of vacancy-type defects in carbon nanotubes: atomistic simulations, Physical Review B 74 (24) (2006) 245420.
[94] V. Datsyuk, et al., Chemical oxidation of multiwalled carbon nanotubes, Carbon 46 (6) (2008) 833–840.
[95] M.S. Dresselhaus, et al., Raman spectroscopy of carbon nanotubes, Physics Reports 409 (2) (2005) 47–99.
[96] M. Sammalkorpi, et al., Irradiation-induced stiffening of carbon nanotube bundles, Nuclear Instruments and Methods in Physics Research Section B: Beam Interactions with Materials and Atoms 228 (1) (2005) 142–145.
[97] H.Z. Jin, R.R. He, J. Zhu, Helicity and inter-tube bonding in bundles of single-walled carbon nanotubes, Journal of Electron Microscopy 48 (4) (1999) 339–343.
[98] K. Liu, et al., Carbon nanotube yarns with high tensile strength made by a twisting and shrinking method, Nanotechnology 21 (4) (2009) 045708.

CHAPTER 4

Synthesis and Properties of Ultralong Carbon Nanotubes

Rufan Zhang, Yingying Zhang, Fei Wei

Beijing Key Laboratory of Chemical Engineering and Technology, Department of Chemical Engineering, Tsinghua University, China

CHAPTER OUTLINE

- 4.1 Introduction 87
  - 4.1.1 Structure and growth mechanism of ultralong CNTs 88
    - 4.1.1.1 Structure of ultralong CNTs 88
    - 4.1.1.2 Growth mechanism of ultralong CNTs 90
- 4.2 Synthesis of Ultralong CNTs by CVD 101
  - 4.2.1 Catalysts and CVD process 101
  - 4.2.2 Carbon sources 102
  - 4.2.3 Substrates 102
- 4.3 Tuning the Structure of Ultralong CNTs 102
  - 4.3.1 Tuning the number of walls and diameters of ultralong CNTs 104
  - 4.3.2 Tuning the chiral consistency of ultralong CNTs 106
  - 4.3.3 Tuning the electrical properties of ultralong CNTs 108
  - 4.3.4 Tuning the morphology of ultralong CNTs 113
    - 4.3.4.1 Tuning of orientation and arrangement of ultralong CNT arrays ... 113
    - 4.3.4.2 Tuning the length of ultralong CNT arrays 116
    - 4.3.4.3 Tuning the areal density of ultralong CNT arrays 119
    - 4.3.4.4 Schulz–Flory distribution-controlled growth — of half-meter-long carbon nanotubes 119
  - 4.3.5 Mechanical properties of ultralong CNTs 122
  - 4.3.6 Potential applications of ultralong CNTs 127
- 4.4 Conclusions 129
- References 130

4.1 INTRODUCTION

Modern society has benefited from the advent of nanotechnology and nanomaterials. One of the key issues of obtaining further benefit from nanotechnology is the controlled mass production of low-dimensional nanomaterials. Carbon nanotubes (CNTs), one of the strongest materials ever found, are in great demand due to their superior mechanical and electronic properties. Perfect CNTs are tubular structures composed of many hexagons of carbon atoms in which there are no topological defects (TDs) such as pentagons or heptagons. Because of their high-aspect-ratio structure formed from strong sp^2 hybrid C–C bonds, they have been regarded as the strongest material ever discovered. Due to these ideal structures and properties, perfect CNTs are the most promising candidates in applications such as nanoelectronics, quantum lines, field emission transistors, superstrong materials, chemical sensors and detectors, etc. Especially, benefitting from their perfect structures and extraordinary mechanical properties, they show great potential as promising building blocks for superstrong fibers [1,2], and are even a candidate for building a cable for space elevators [3]. For instance, they have extremely high tensile strength, high Young's modulus and high breaking strain [1,2,4]. For the practical applications of CNTs, the key is to realize their bulk production with controlled structures and unlimited lengths. Ultralong CNTs usually refer to the horizontally aligned CNTs arrays with length up to centimeters or even decimeters grown on the surface of substrates by chemical vapor deposition (CVD) methods [1,5–12]. To date, the only effective method to prepare centimeter-long CNTs is gas flow-directed CVD on silicon substrates and the reported longest CNT was 20 cm [9]. However, the main obstacle for the bulk production of ultralong CNTs is their extremely low areal density. The areal density of ultralong CNTs is usually several CNTs per 100 μm [8–13], which is far lower than that of agglomerated CNTs [14] and vertically aligned CNT arrays [15]. In addition to the limited length and low number density, another barrier for the mass production of ultralong CNTs is the ultralow portion of long CNTs [9,12,16] among all the obtained CNTs.

To date, stirring progresses have been made on synthesizing ultralong CNTs and continuous research is ongoing. The geometry and structure of ultralong CNTs can be tuned by varying the processing conditions, such as temperature, catalyst, feeding gas, reaction time, etc. This provides a wide space for their application in micro-/nanoelectronics, sensors, field emission transistors, etc. Besides, half-meter-long CNTs with perfect structure have been synthesized by Wei's group of Tsinghua University. However, the question of how to improve the areal density of the ultralong CNTs has been baffling researchers for many years. The real difficulty lies in the lack of understanding of the selective growth mechanism for ultralong nanotubes and the corresponding techniques for their bulk production. Therefore, obtaining insights into the selective growth mechanism of ultralong CNTs is of both scientific and technical importance for designing future experiments to achieve better control of the growth as well as the bulk production of ultralong CNTs. This chapter will give a brief introduction of the structure, synthesis, structure tuning, properties, bulk production and application in superstrong fibers of ultralong CNTs.

4.1.1 Structure and growth mechanism of ultralong CNTs
4.1.1.1 Structure of ultralong CNTs
CNTs can be viewed as coaxial hollow cylinders curled up from single or multiple graphene sheets. In graphene, three out of the four outer electrons of each carbon

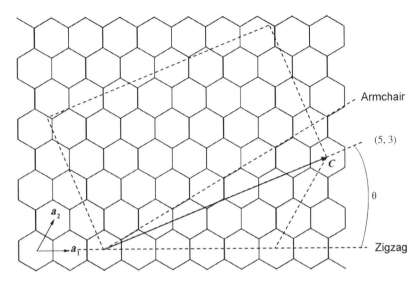

FIGURE 4.1

Illustration of curving a graphene sheet into a CNT shell with chiral index (n, m) along the vector $C = na_1 + ma_2$, where a_1 and a_2 are the unit vectors of graphene sheet and θ is the chiral angle of the CNT. We usually designate $n \geq m$.

atom take up three planar sp^2 hybrid orbits forming three σ bonds, while the remaining electron takes up the x-plane π orbit forming a vertical π bond. In each shell of CNTs, three σ bonds slightly deviate from the planar surface and the π bonds orient to the outside due to the quantum restriction effect and the σ–π rehybrid, which makes CNTs have higher mechanical strength and better electrical and thermal conductivity.

Since each shell of the CNTs is curved from a graphene sheet, its structure can be expressed by a vector C (Fig. 4.1), which has the following relationship with the unit vector of graphene, a_1 and a_2:

$$C = na_1 + ma_2 \qquad (4.1)$$

The chiral index (n, m) has a direct relationship with the chiral angle and electrical properties of CNTs. The chiral angle θ can be defined as the angle between vector C and a_1. According to the difference of chiral indexes, CNTs can be divided into three different types: armchair CNTs when $n = m$ ($\theta = 30°$), zigzag CNTs when $n > 0$ and $m = 0$ ($\theta = 0$), and chiral CNTs when $n > m \neq 0$ ($0 < \theta < 30°$). CNTs show different electrical properties for different chiral indexes. When $n - m = 3\ k$, CNTs show metallic characteristics; while for $n - m = 3k \pm 1$ (where $k = 0, 1, 2, 3...$), CNTs show semiconducting characteristics.

The atomic structures of CNTs have a direct influence on their electrical properties. The nano-sized dimensions of the CNTs, together with the unique electronic structure of a graphene sheet, make the electronic properties of them highly unusual. Both theoretical and experimental results show that the electronic properties of the CNTs are very sensitive to their geometric structures. As stated above, the chiral index (n, m) determines the conductive type of CNTs to be metallic or semiconducting. However, most of the CNTs do not have perfect geometric structures. Defects like pentagons, heptagons, vacancies or dopant species are found to modify their electronic properties drastically. The electronic properties of defective nanotube-based structures are then more complex than infinitely long, perfect CNTs. The introduction of defects into the CNTs is thus an effective way to tailor their intrinsic properties, in order to introduce new properties and to propose new potential applications of CNTs in nanoelectronics.

The carbon—carbon bond in a graphene layer is probably the strongest chemical bond in an extended system known in nature. Since CNTs are seamlessly rolled-up graphene layers, they have been speculated to possess exceptional mechanical properties, and it has become a topic of great interest to quantify these mechanical properties. One of the unusual features of CNTs is that they simultaneously involve widely varying scales: their length can be macroscopic, up to centimeters and even decimeters, while their width falls in the nanoscale. In addition, CNTs with the properties of 1—2 TPa modulus and 100—200 GPa strength, the strongest material ever known, have shown promising potential for the storage of mechanical energy, either by their deformation in the composite materials or by their elastic deformation produced by stretching or compressing the pristine tubes or tube arrays. Theoretical calculation suggested that the energy storage capacity, in the latter case, can be at least three orders higher than that of steel spring and several times that of the flywheels and lithium-ion batteries. The mechanical energy storage capacity of CNTs depends on their mechanical properties, which directly depend on their molecular structures. Just as stated above, the defects not only affect the electronic properties of as-grown CNTs but also give rise to a deleterious effect-deterioration of the axial mechanical properties of CNTs. For instance, theoretical calculation predicts that the tensile strength and critical strain of single-walled CNTs (SWCNTs) decrease by nearly a factor of 2 if an unreconstructed vacancy is present. Thus, it is of great significance to synthesize CNTs with perfect atomic structures.

Compared with agglomerated CNTs and vertically aligned CNT arrays, ultralong CNTs usually have fewer walls (usually less than five walls) and mostly are SWCNTs, with diameters typically smaller than 5 nm. For SWCNTs, they usually have diameters of 0.8—2.5 nm and most of the reported ultralong SWCNTs are semiconducting ones. Because of the free growth mode of individual ultralong CNTs, they can be free of the interaction with the substrate as well as the entanglement and twist themselves during the growing process, which makes them grow into centimeters and even decimeters long with perfect structures.

4.1.1.2 Growth mechanism of ultralong CNTs

To date, a vapor—liquid—solid (VLS) mechanism has been widely accepted to interpret the growth mechanism of CNTs [17—21]. According to this mechanism, the carbon source decomposes on the metal catalyst nanoparticles (NPs) which are in the state of liquid at high temperature. The carbon atoms decomposed from the carbon source will dissolve into the liquid metal NPs and then precipitate out of the metal NPs after reaching the saturated condition and form CNTs through self-assembly. This mechanism provides a simple explanation for CNT growth but cannot interpret the detailed behavior of how the CNTs are formed at the atomic scale, such as how to keep the open ends of CNTs during growth, how to repair their structural defects, and how to control their diameter and chirality, etc.

As for the growth mechanisms of ultralong CNT, various mechanisms were proposed. One of them is the "kite mechanism" [22] to interpret the tip-growth mode of ultralong CNTs. During a CNT growth process, one end of the CNT is floating over the substrate because of the thermal buoyancy caused by the temperature difference between the substrate and the flow gas. The catalyst NP is at the floating end of the CNT, which follows a tip-growth mechanism. The kite mechanism successfully explained many facts, such as that ultralong CNTs could continuously grow up to 20 cm long [9] and cross over obstacles and trenches on substrates [8,22]. However, for short CNTs, they were considered to follow a base-growth mode [23], in which the catalyst NPs anchored on substrate during CNT growth and the CNTs extend their length from the base ends. In this section, we will give a brief introduction to the growth mechanism and modes of ultralong CNTs.

4.1.1.2.1 Growth mechanism of CNTs

Maruyama [24–26] and Ding et al. [27–30] have made systematic studies on the detailed growth process and mechanism of CNTs based on VLS mechanism using molecular dynamic (MD) simulation methods. The dissolution–precipitation process of carbon atoms on melted metal NPs can be divided into three stages according to Ding's model (Fig. 4.2). The first stage is called the *unsaturation stage* during which carbon atoms continuously dissolve into the melted metal NPs and after a short time, the concentration of the carbon atoms reaches a highly supersaturated stage, which is the second stage. During this stage, the carbon atoms begin to precipitate on the metal NP surface and form carbon atom islands (small graphene sheet), which will then form CNTs at proper temperature. After that, the concentration of carbon atoms will keep supersaturated, which is the third stage.

Temperature plays a very important role in the formation of CNTs [30]. At temperature <600 °C, the carbon atom islands (small graphene sheets) precipitated from the metal NPs will not have enough energy to overcome the interaction between graphene sheets and the surface of metal NPs, thus they will then form large graphene sheets and eventually cover the whole surface of metal NPs and block the dissolution–precipitation process of other carbon atoms on the metal NPs (Fig. 4.3). However, if the temperature is too high (such as higher than 1500 K), the precipitation process of carbon atoms from the metal NPs will occur too fast and the graphene sheets will accumulate at some points on the metal NP surface, eventually forming soot-like structures. Only at proper temperature, the graphene sheets precipitated from the metal NPs can form stable tube-like structures, i.e. CNTs.

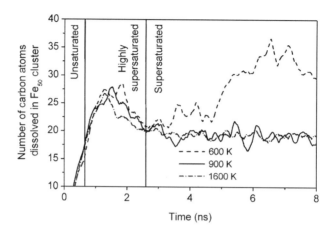

FIGURE 4.2

Dissolved carbon content in the $Fe_{50}C$ cluster during the growth of a graphene sheet at 600 K (dashed line), an SWCNT at 900 K (solid line), and the soot-like structure at 1600 K (dot dashed line).

Source: Reproduced with permission from Ref. [30]. Copyright (2004) American Chemical Society.

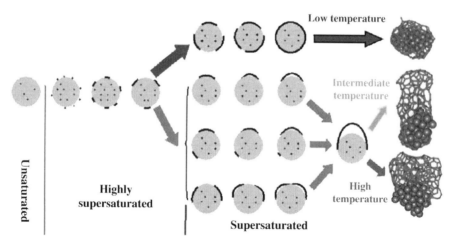

FIGURE 4.3

Detailed vapor–liquid–solid model of SWCNT growth at different temperatures. (For color version of this figure, the reader is referred to the online version of this book.)

Source: Reproduced with permission from Ref. [30]. Copyright 2004, American Chemical Society.

In the above process, both the diameters of metal NPs and the interaction between the carbon atoms and the metal NPs have significant influence on the diameters and structures of CNTs [28,29]. Generally speaking, when the atom number in the metal clusters is higher than 20, the catalyzed CNTs have relatively good

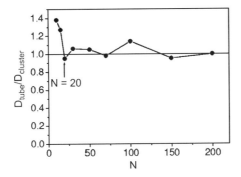

FIGURE 4.4

Dependence of the SCWNT diameter, D_{Tube}, on the Fe_N cluster diameter, $D_{Cluster}$, for $10 < N < 200$.

Source: Reproduced with permission from Ref. [29]. Copyright 2004, American Institute of Physics.

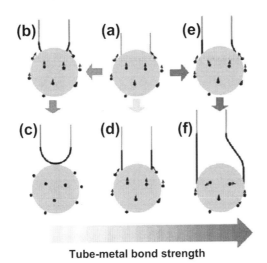

FIGURE 4.5

Three possible SWCNT growth scenarios from a catalyst particle: I (a–c), the nanotube forms a closed cap; II (a–d), the nanotube elongates without a change in diameter; and III (a–f), the nanotube elongates and the diameter approaches that of the catalyst particle. The arrow shows the SWCNT catalyst interaction strength giving rise to the three growth scenarios. (For color version of this figure, the reader is referred to the online version of this book.)

Source: Reproduced with permission from Ref. [28]. Copyright 2008, American Chemical Society.

structures and their diameter is consistent with that of metal clusters. But when the diameter of metal clusters decreases, the diameter of the corresponding CNTs cannot decrease accordingly because its diameter limit is about 0.6–0.7 nm. When the diameter of a CNT decreases, the curvature of their graphene sheet

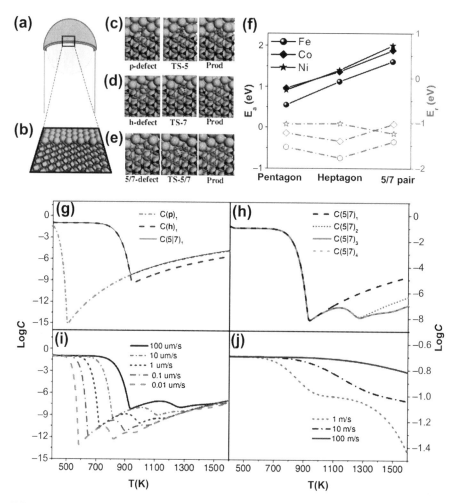

FIGURE 4.6

(a) Defect healing at the SWCNT–catalyst particle interface, where a circular tube's open end is attached to a step edge on the catalyst particle. (b) A fraction of the SWCNT–catalyst step [a step along the (211) direction on the (111) surface of the fcc crystal] interface is modeled as an interface of the graphene-stepped metal surface. (c)–(e). Healing of the pentagon (**p** defect), heptagon (**h** defect), and pentagon–heptagon pair (5|7) and the corresponding geometries (original defect formations, transition states and products after healing) involved. (f) The energy barrier (E^*) (black real line with symbols) and the reaction energy (E_r) (red dashed line with symbols) for the **p**, **h**, and 5|7 defects healing on stepped Fe(111)/Co(111)/Ni(111) surfaces. (g) The concentrations of pentagon $[C(\mathbf{p})_1]$, heptagon $[C(\mathbf{h})_1]$, and their summation $C(5|7)_1$ on the front most edge of a growing SWCNT as functions of temperature (T) at the growth rate of 100 μm = s; (h) The concentrations of $C(5|7)_i$ vs T at different depths from tube edge at the growth rate of 100 μm = s; (i) Concentrations of survived (5|7) in the increases accordingly and this will make the structure of the CNTs unstable because of the extremely high curvature energy. The relationship of CNT diameter and atom number of metal clusters can be seen in Fig. 4.4.

The influence of interaction between catalyst metal NPs and carbon atoms on the diameters and structures of CNTs is mainly focused on the control of open ends of CNTs during growing process. The holding of open ends (contacted with the metal clusters) of CNTs directly influences their structures and lengths during the growth process. There are dangling bonds at the bottom carbon atoms of the newly formed CNTs from the precipitated carbon atoms on metal NPs. The dangling bonds have a tendency to enclose each other to reach a stable state (Fig. 4.5(a)–(c) process). If the adhesion between the metal NPs and carbon atoms is not strong enough to prevent the dangling bonds to enclose each other, the open ends of CNTs will easily close up. But if the adhesion between the metal NPs and carbon atoms is too strong to prevent the dangling bonds to keep a stable state, the open ends of the CNTs will be enlarged by the adhesion and the diameter of the CNTs will increase as high as consistent with that of the metal NPs (Fig. 4.5(a)–(d)). Only with proper adhesion with metal NPs, can the open ends of the CNTs keep a stable state, thus making the structure of the CNTs stably extend (Fig. 4.5(a)–(f)).

Yuan et al. has theoretically studied the energetics of TDs in CNTs and their kinetic healing during the catalytic growth [31]. They indicated that with the assistance of a metal catalyst, TDs formed during the addition of C atoms can be efficiently healed at the CNT–catalyst interface. Theoretically, a TD-free CNT wall with 10^8–10^{11} carbon atoms can be achievable, and, the growth of perfect CNTs up to 0.1–100 cm long is possible since the linear density of a CNT is ~ 100 carbon atoms per nanometer. Their calculation also showed that, among catalysts most often used, Fe had the highest efficiency for defect healing (Fig. 4.6).

4.1.1.2.2 Tip-growth mode of ultralong CNTs

In 2003, Huang et al. proposed for the first time the gas flow-directed growth of CNTs. They found that SWCNTs could grow from the catalyst region and then kept floating in the gas flow in CVD process, during which the catalysts kept on the growing tips of the SWCNTs. The CNTs are parallel with each other during the gas flow-directed CVD growth process and it is easy for catalysts to keep a long life of activity, which benefits the growth of CNTs. In 2004, they proposed a kite mechanism to interpret the ultralong CNTs growth based on their tip-growth mode (Fig. 4.7(a)) [22]. During the CNT growth process, there is a temperature difference between the gas flow and the substrate, which accordingly produces

CNT wall, C(5|7)$_w$ vs T at different growth rates; (j) Concentrations of survived 5|7 in the CNT wall at very fast growth rates. Fe was considered as the catalyst in this study. (For interpretation of the references to color in this figure legend, the reader is referred to the online version of this book.)

Source: Reproduced with permission from Ref. [31]. Copyright 2008, American Physical Society.

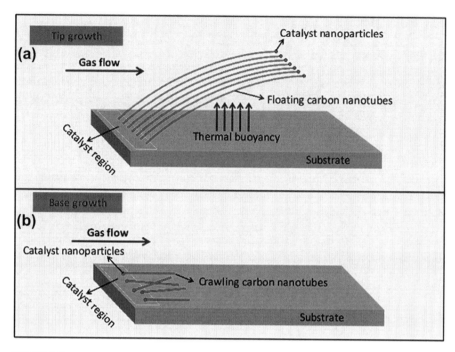

FIGURE 4.7

Tip-growth (a) and base-growth (b) mechanisms of ultralong CNTs. (For color version of this figure, the reader is referred to the online version of this book.) Flowchart describing the formation of possible morphologies of carbon nanotube graphoepitaxy by miscut of C-plane sapphire, annealing, and CVD. (a) Equilibrium shape of -Al2O3, with facets C{0001}, R{1102},S{1011},P{1123}, and A{1120}, in order of increasing surface energy. The same drawing is used to show the different miscut directions. (b) Miscut toward [1100] produces a vicinal-Al2O3 (0001) surface with atomic steps along [1120] (c) Annealing leads to R-faceted nanosteps. (d) SWNTs grow straight along[1120] (the ball represents the catalyst nanoparticle). (e) Miscut toward [1210] produces a vicinal Al2O3 (0001) with atomic steps along [1010]. (f) Annealing initially leads to metastable P-faceted nanosteps. (g) SWNTs grow straight along [1010]. (h) Further annealing from (f) leads to sawtooth-shaped-R-/R-faceted nanosteps. (i) SWNTs grown loosely conformal to the sawtooth-shaped nanosteps, with segment along [1120] and [2110].

Source: Reproduced with permission from Ref. [102]. Copyright 2013, Elsevier.

a thermal buoyancy vertical to the substrate. Some of the CNTs will be lifted up by the buoyancy and float in the laminar gas flow. The catalyst NPs stay at the tip ends of the floating CNTs and continuously catalyze the CNTs growth. They found that metal NPs were kept at the tip ends of as-grown CNTs, which gave direct evidence of the tip-growth mode. They pointed out that the growth speed could be as high as 3 μm/s when using CO/H_2 as the carbon source while it could reach 20 μm/s when using CH_4/H_2 as the carbon source. In the same year, Zheng et al. fabricated 4.8-cm-long SWCNTs at 900 °C using ethanol as the carbon source and using 0.1 mol/l $FeCl_3$ as catalyst in the CVD process. The growth speed was 3 μm/s and the length was the longest one at that time. In 2009, Wang et al. synthesized 18.5-cm-long SWCNTs using the gas flow-directed CVD process. In 2010, Wen et al. obtained 20-cm-long double-/triple-walled CNTs (TWCNTs) using the gas flow-directed CVD by adding a little water to increase the catalyst activity. The growth speed was as high as 80–90 μm/s. Many groups [6,8,22,32–36] reported that ultralong CNTs can cross over hundreds of micron-wide trenches and dozens of micron-high barriers on substrates. All the above facts confirmed that the ultralong CNTs are floating in the gas flow during the growth process.

Besides, the tip-growth mode has also been confirmed by the isotope labeling method. Wen et al. studied the growth of ultralong CNTs on Si/SiO_2 substrates using $^{12}CH_4$ and $^{13}CH_4$ as the carbon source, which can be easily distinguished by Raman spectroscopy. They first used $^{12}CH_4$ as the carbon source for growing ultralong CNTs and then switched to $^{13}CH_4$ after several minutes. They found that the ^{12}C were located at the starting ends of ultralong CNTs (near the catalyst region) while the ^{13}C were located at the other ends (far away from the catalyst region). If the ultralong CNTs followed the base-growth mode, the catalysts would stay fixed on the substrate. Thus, the newly formed CNTs containing ^{13}C should be near the catalyst region, not far away from the catalyst region, which was contrary to the experimental results. Therefore, the ultralong CNTs should follow the tip-growth mode.

4.1.1.2.3 Base-growth mode of ultralong CNTs

Although the tip-growth mode of ultralong CNTs successfully explained many facts, there were still many results which could not be interpreted by the tip-growth mode. Many facts have proven the base-growth mode (Fig. 4.7(b)). Han et al. [37] realized template-free directional growth of SWCNTs on a- and r-plane sapphire. Ismach et al. [38] reported the CNT highly oriented graphoepitaxy growth on faceted nanosteps (Fig. 4.8). Ago et al. [39] found that the lattice-directed ultralong CNT growth on quartz substrates mostly followed the base-growth mode and only a few of them followed the tip-growth mode. He et al. [23] confirmed the base-growth mode of ultralong CNTs by reactivation of iron catalyst NPs on Si/SiO_2 substrates. Huang et al. [32] found that although the ultralong CNTs could cross over trenches and barriers on substrates, they did not find the catalyst NPs at the tip ends of as-grown ultralong CNTs. They believed that the as-grown ultralong CNTs followed the base-growth mode and this kind of CNT could also float in the gas flow and cross over trenches and barriers.

4.1.1.2.4 Chirality selectivity of CNT growth

Many reports [40–43] have shown that for CNTs synthesized with different methods, those with larger chiral angles have a higher abundance. That is to say, the larger the chiral angles (the closer to armchair CNTs), the easier the growth

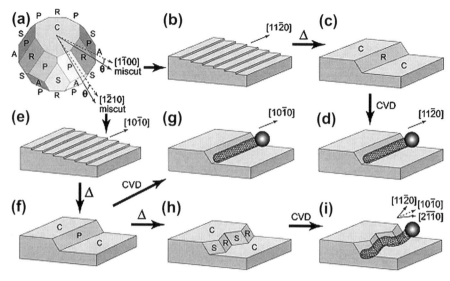

FIGURE 4.8

Flowchart describing the formation of possible morphologies of carbon nanotube graphoepitaxy by miscut of C-plane sapphire, annealing, and CVD. (a) Equilibrium shape of al2O3, with facets C{0001},R{1(¯1)02},S{10(¯1)1},P{11(¯2)3}, and A{11(¯2)0}, in order of increasing surface energy. The same drawing is used to show the different miscut directions. (b) Miscut toward [1(¯1)00] produces a vicinal -A12O3 (0001) surface with atomic steps along [11(¯2)0]. (c) Annealing leads to R-faceted nanosteps. (d) SWNTs grow straight along [11(¯2)0] (the ball represents the catalyst nanoparticle). (e) Miscut toward [1(¯2)10] produces a vicinal -Al2O3 (001) with atomic steps along [10(¯1)0]. (f) Annealing initially leads to metastable P-faceted nanosteps. (g) SWNTs grow straight along [10(¯1)0]. (h) Further annealing from (f) leads to sawtooth-shaped S-/R-faceted nanosteps. (i) SWNTs grow loosely conformal to the sawtooth-shaped nanosteps, with segments along [11(¯2)0] and [2(¯1)(¯1)0].

Source: Reproduced with permission from Ref. [38]. Copyright 2008, American Chemical Society.

of CNTs is. Ding et al. [27] proposed a dislocation theory of chirality-controlled nanotube growth based on many experimental results as well as MD studies.

If we define the zigzag CNTs (achiral CNTs, the chiral angle = 0) as a crystal with perfect structure, all the CNTs with chiral angle ≠ 0 (chiral CNTs) can be viewed as having a screw dislocation along their axial direction (Fig. 4.9). The screw dislocation is in proportion to the chiral angle. The growth of CNTs is just like the increase of their tubal end-edges. For achiral CNTs (zigzag CNTs), a high amount of energy is needed to add a new carbon atom to the completely closed tubal end-edges, which means starting a new tube end-edge. With the increase in carbon atom number, the energy for adding each atom gradually decreases. For adding the last carbon atom to form a complete new tubal end-edge, the corresponding energy is

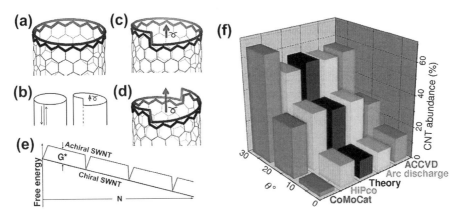

FIGURE 4.9

An axial screw dislocation in the CNT. An achiral zigzag (n, 0) tube (a) can be viewed as a perfect crystal, and transformed into a chiral one by cutting, shifting by a Burgers vector b (red arrows in (b)–(d)), and resealing a tube cylinder (b). The chiral (n, 1) in (c) and (n, 2) in (d) tubes contain the axial screw dislocations with a single and double value of b, accordingly; the corresponding kinks at the open tube-end are marked in red. (e) Free energy profile during the growth of a chiral or achiral nanotube. (f) The distribution of CNT product as a function of chiral angle. Experimental data of CoMoCat [41,42], HiPco [41], arc discharge [40], and ACCVD [40,41,43] are extracted from literature. Theoretical results come from Ding's MD study [27]. θ is the chiral angle of CNTs. (For interpretation of the references to color in this figure legend, the reader is referred to the online version of this book.)

Source: Reproduced with permission from Ref. [27]. Copyright 2008, The National Academy of Sciences of the United States of America.

the lowest. However, for the next new tubal end-edge, the energy for adding the first carbon atom will then be much higher than that for adding the last carbon atom. But for the chiral CNTs, the case is different. The energy for adding the first atom to a new tubal end-edge in chiral CNTs is much lower than that in achiral CNTs because of the screw dislocation of chiral CNTs, which makes the growth of chiral CNTs much easier than that of achiral CNTs. The larger the helix angles, the easier the growth for CNTs. Wen et al. found that most of the as-grown 10-cm-long TWCNTs show a preference of large chiral angles (Table 4.1) [12], which also confirmed the dislocation-controlled nanotube growth.

4.1.1.2.5 Summary

The above facts show that both the tip-growth and base-growth modes coexist during the ultralong CNTs growth process. The dominant growth mode depends on the experimental conditions such as substrates. But what to be mentioned is that no matter what growth mode works, to obtain ultralong CNTs, part or whole CNTs must be floating in the gas flow during the growth process. Otherwise, the ultralong

Table 4.1 Chirality Distribution of Ultralong TWCNTs

No.	Length (mm)	Diameter (nm)	Chirality (n, m)	Chirality Angle α (°)	$\Delta\alpha$ (°)	Semiconducting(s)/Metallic(m)
1	60	1.36	11, 9	26.7	4.1	s
		2.12	19, 12	22.6	0.4	s
		2.87	26, 16	22.2		s
2	40	2.71	21, 19	28.3	2.7	s
		3.60	30, 23	25.6	3.3	s
		4.34	33, 31	29.0		s
3	30	1.44	13, 8	22.2	2.4	s
		2.21	21, 11	19.8	4.9	s
		2.93	25, 18	24.6		s
4	10	3.43	32, 18	20.8	3.4	s
		4.31	43, 19	17.4	1.1	m
		5.12	50, 24	18.5		s
5	7	3.56	30, 20	23.4	6.3	s
		4.38	44, 19	17.1	2.6	s
		5.03	48, 25	19.7		s
6	6	1.64	18, 5	11.9	2.0	s
		2.26	24, 8	13.9	3.8	s
		3.10	35, 8	10.1		m
7	2	3.78	35, 20	21.1	4.3	m
		4.46	45, 19	16.8	1.5	s
		5.19	51, 24	18.3		m
8	2	1.72	19, 5	11.4	1.9	s
		2.46	28, 6	9.5	5.0	s
		3.23	34, 12	14.6		s

Source: Reproduced with permission from Ref. [12]. Copyright 2010, Wiley-VCH Verlag GmbH & Co. KGaA, Weinheim.

CNTs cannot cross over the trenches and barriers on substrates and cannot grow into ultralong ones.

In comparison, the tip-growth mode shows more advantages than the base-growth mode for the high-speed growth of defect-free ultralong CNTs. That is because a high growth speed needs a large amount of carbon to decompose on the surface of catalyst NPs. Taking the perfect TWCNTs growing at 1000 °C with methane as the carbon source as an example, the number of methane molecules colliding onto a 4-nm metal catalyst NP by thermal movement is about 5×10^9/s at the reaction temperature, pressure and carbon concentration. The number of carbon atoms needed is about 5×10^7/s if the TWCNTs grow with a speed of 24 μm/s. Considering the influence of the effective collision ratio and effective surface adsorption ratio of methane molecules, we can say that the growth of ultralong TWCNTs is a process that has reached the limit of the chemical equilibrium between the consumption and effective supply of carbon atoms. It is reported that the catalyst NPs can rotate 180 rounds per 11 min during the CNTs growing process [44]. For the fast CNT growth near the speed limit in this example, the catalysts must have a very large rotational inertia and tremendous mass and heat transfer. It is very hard for catalysts that follow the base-growth mode to meet such a rigorous demand. Only those catalysts that float in the gas flow and can freely move can meet such a rigorous requirement.

4.2 SYNTHESIS OF ULTRALONG CNTs BY CVD

Arc discharge, laser ablation, and CVD are the three main methods for synthesizing CNTs. The as-grown CNTs have different structures, morphologies and yields with different synthesis methods. The arc discharge and laser ablation methods are hard to precisely control for the self-assembly of the CNTs during the high energy input process. While for the CVD methods, they have shown advantages for the synthesis of ultralong CNTs, such as easy control of parameters, lower reaction temperature and easy realization of mass production, etc. The CVD methods have been widely used in ultralong CNTs' synthesis. In order to obtain better ultralong CNT arrays, external forces are usually used to guide the CNTs' growth during their growing process. These forces can be divided into two types: the first type can be called *field force*, such as gas flow force, magnetic field force and electrical field force, while the second one is *substrate surface force*, such as lattice, atom steps and edges.

4.2.1 Catalysts and CVD process

Transition metal catalyst is indispensable for synthesizing ultralong CNTs using CVD. The commonly used catalysts are Fe, Mo, Co, Cu, and Cr NPs. The carbon source can decompose into individual carbon atoms or atom pairs on the surface of these metal NPs at 700–1100 °C, and then form CNTs. The single-dispersive NPs are a key factor for the growth of ultralong CNTs. The diameter of CNTs are proportionally related to the size of the catalyst NPs [45,46]. Controlling the structure of catalytic NPs allows the variation of diameter, length and even the chirality of CNTs [46]. The metal NPs are usually obtained from the decomposition of catalyst precursor by thermal reduction. The following shows a typical method as an example to fabricate catalyst NPs and synthesize ultralong CNTs.

A quartz tube (with length of 0.5–1.5 m and inner diameter of 20–50 mm) is used as the reactor. Single-crystal silicon wafers (1–10 cm long, 0.5–1 cm wide and 0.5–1.0 mm thick) with a 300–800-nm-thick SiO_2 layer on the surface are employed as substrates. An ethanol solution of $FeCl_3$ (or $CuCl_2$, 0.001–0.1 mol/l) is coated onto the upstream end of the substrates by microcontact printing. After reduction in H_2 and Ar (H_2:Ar = 1:1–4:1 in volume with a total flow of 100–400 sccm) at 900 °C for 10–40 min, the precursor $FeCl_3$ becomes Fe NPs which work as the catalysts for CNT growth. Then, the sample is heated from 900 to 1000 °C in 2 min and the inlet gas is switched to CH_4 and H_2 (CH_4:H_2 = 1:2 in volume with a total flow of 60–100 sccm) for growing ultralong CNTs. The growth duration for the CNTs is usually 10–20 min, which depends on the length of CNTs desired.

4.2.2 Carbon sources

There are many kinds of carbon sources for growing ultralong CNTs, such as methane, ethane, carbon monoxide, and ethanol. Compared with ethene and propylene, all the above carbon sources decompose slowly at high temperature. Methane, ethanol and carbon monoxide are the most widely used carbon sources. To better restrict the decomposition of the carbon source, a high concentration of hydrogen is also used during the CNTs' growing process, which can keep the activity of catalysts for a long time to get macroscale long CNTs. This is different from the synthesis of agglomerated CNTs and vertically aligned CNT arrays which exploit a high concentration of ethene and propylene as the carbon source.

4.2.3 Substrates

The most widely used substrates for making ultralong CNTs are silicon wafers with hundreds of nanometer-thick SiO_2 layer. Silicon is very stable at high temperature and the surface of silicon wafer is very flat, which is beneficial for the characterization and pose application of as-grown CNTs. Besides, the SiO_2 layer can prevent metal NPs from acting with the silicon and losing activity. Meanwhile, it also works as an insulation layer so as to make in-situ characterization of CNTs' electrical properties.

Sapphire and single-crystal quartz can also be used as substrates to prepare ultralong CNTs and arrays with parallel arrangement. When cutting off along the Y direction, a series of atom steps will form in the direction vertical to the cutting direction ($2\bar{1}0$). There is a guiding effect when preparing CNTs on the quartz substrate cut along the Y direction (Y-cut). This guiding effect will arrange CNTs along the atom steps direction, but the as-grown CNTs are usually shorter than centimeters. Table 4.2 is a summary of synthesis methods for ultralong CNTs.

4.3 TUNING THE STRUCTURE OF ULTRALONG CNTs

Controlled synthesis of ultralong CNTs is very important for their applications. Due to the fact that the diameter and wall numbers of CNTs are greatly affected by the reacting parameters, it is feasible to tune CNTs' structure by varying the growing parameters.

Table 4.2 Methods for Synthesizing Ultralong CNTs

Methods[1]	Catalyst	Temperature (°C)	Substrate	Growth Mode[4]	Literature
EFCVD	Fe	900	Wafer	T	[33]
EFCVD	Fe	800	Wafer	T	[103]
GCVD	Fe/Mo	900	Wafer	T	[36,47,104]
GCVD	$FeCl_3$	900	Wafer	T	[6]
CVD	Fe	900	Quartz	B	[105]
GCVD	Fe/Mo	900	Wafer	T	[106,107]
GECVD	Ferritin	800	Sapphire	B	[71]
GECVD	Ferritin	900	Sapphire	B	[37,108]
GCVD	Co(Mo)	850	Si	T	[6,32]
MFCVD	Fe	750	SiO_2	T	[109]
GCVD[2]	$FeCl_3$	950	Wafer	T	[8]
GECVD	Ferritin	800	α-Al_2O_3	B	[38]
GCVD	Cu	925	Wafer	T	[48]
GECVD	Fe(Mo)	900	Sapphire	B	[72,110]
GECVD	Ferritin	900, 925	ST-cut quartz	B	[69]
GCVD	$FeCl_3$/$CoCl_2$	900–950	Wafer	T	[52]
GECVD	Fe, Co, Ni	900	Quartz	B	[111]
GCVD	Cu	900	ST-cut quartz	B	[66]
CVD	SWCNTs	975	Wafer/quartz	B	[54]
GCVD	FeMo	950	Wafer	T	[11]
GECVD	Cu	900	ST-cut quartz	B	[67]
CVD	SiO_2, Al_2O_3, TiO_2, Er_2O_3	900	Wafer	B	[112]
GECVD	FeMo	900	SiO_2/Si wafer	B	[113]
GCVD[3]	$FeCl_3$	1000	SiO_2/Si wafer	T	[12]
GECVD	Ferritin	750–900	SiO_2/Si wafer	B	[114]
GECVD	Fe line	925	ST-cut quartz	B	[115]
GCVD	$DyCl_3$	900	SiO_2/Si wafer	T	[116]
GCVD[4]	$FeCl_3$	1000	SiO_2/Si wafer	T	[12]

T: tip-growth mode; B: base-growth mode.
[1]EFCVD: electric field-assisted CVD; MFCVD: magnetic field-assisted CVD; GCVD: gas flow-assisted CVD; GECVD: graphoepitaxy CVD. If not specified, the CNT products were single-wall CNTs.
[2]Single-/double-/multi-wall CNTs.
[3]Triple-wall CNTs.
[4]Double-/triple-wall CNTs.

4.3.1 Tuning the number of walls and diameters of ultralong CNTs

As mentioned above, the number of walls and diameters of ultralong CNTs are greatly affected by temperature and catalysts. There are many methods to control the number of walls of ultralong CNTs which can be seen in Table 4.2. For the control of CNT diameter, it is important to keep the discrete dispersion of catalyst NPs and their narrow distribution of diameters. There are already several successful practices about how to prepare discrete dispersive catalyst NPs. Li et al. [46] obtained discrete catalytic NPs with diameters in the range of 1–2 nm and 3–5 nm, respectively, by placing controllable numbers of metal atoms into the cores of apoferritin, and used these NPs for growing SWCNTs on substrates by CVD. Huang et al. [47] designed another method to make discrete catalyst NPs. They first spin-coated the substrate with a layer of photoresist (1813 photoresist) followed by drying at 100 °C for 5 min, a high-contract black–white film with desired patterns was used as a photomask. After exposure to UV light for 3–5 min, the substrate was developed in 2–3% tetramethylammonium hydroxide aqueous solution for 1–2 min. Then hexane solution containing catalyst NPs was dropped on the patterned substrate. After drying at room temperature, the substrate was further developed in 1 M NaOH aqueous solution to remove all the photoresist on the substrate, leaving a catalyst pattern on the substrate defined by the pattern on the photomask. A thermally stable silicon material with ordered mesopores was also used as a substrate to disperse Co and Mo NPs [6]. The as-obtained catalyst NPs could effectively produce well-aligned SWCNTs with a narrow distribution of diameters. The most widely used method to prepare catalyst NPs was to directly coat $FeCl_3$ or $CuCl_2$ ethanol solution (0.001–0.01 M) on the substrates using the micro-printing method, and then anneal the catalyst precursor protected by H_2 and Ar at 700–900 °C to obtain well-dispersed catalyst NPs [9,12,48–51].

Temperature directly affects the thermodynamic and kinetic behavior of ultralong CNTs during their growth process. Thus, the diameter and number of walls of ultralong CNTs can be tuned by changing the temperature. Yao et al. [51,52] found that the diameter of SWCNTs became smaller with increasing temperature (Fig. 4.10). Generally, the diameter of CNTs is determined by the diameter of the catalyst NPs. However, Yao et al. predicted that the diameter of catalyst NPs did not change when the diameter of SWCNTs changed. It was believed that the solubility of carbon atoms in metal NPs could vary with temperature. The interaction between CNTs and metal NPs could also vary with temperature. Both the solubility of carbon atoms and the interaction between CNTs and metal NPs can cause a change in CNT diameter. CNTs can be viewed as a seamless graphite tube and the curved graphite sheets have strain energy, which increases accordingly with temperature. Thus, when temperature increases, the curvature of graphite sheets increases too, resulting in the diameter of CNTs becoming smaller. For the ultralong multiwalled CNTs (MWCNTs), there is a similar phenomenon. In addition, the temperature also affects the number of walls of MWCNTs. Wen et al. [12] found that both the diameter and the number of walls of as-grown ultralong CNTs catalyzed by Fe NPs decreased with increasing temperature (Fig. 4.11). The selectivity of TWCNTs

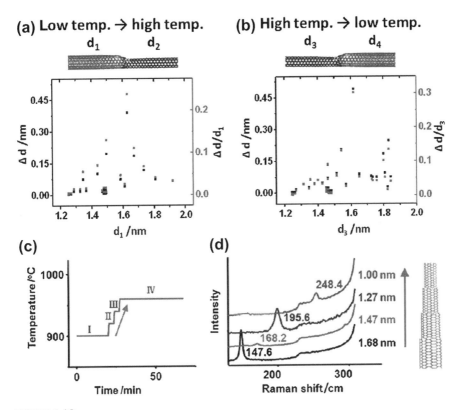

FIGURE 4.10

Effect of temperature on the diameter of SWCNTs. The relationship between the change of diameter and the initial diameter: (a) from low temperature to high temperature ($\Delta d = |d_2 - d_1|$); (b) from high temperature to low temperature ($\Delta d = |d_4 - d_3|$). (c) The ascending temperature time curve and (d) typical Raman result of an ultralong SWCNT grown under the temperature time process in panel c. The locations where the red, blue and green curves are detected are 300, 1800, and 7850 μm, respectively, away from the place where the black curve is detected. The inset shows a schematic tapered structure of this SWCNT. (For interpretation of the references to color in this figure legend, the reader is referred to the online version of this book.)

Source: Reproduced with permission from Ref. [51]. Copyright 2008, American Chemical Society.

(with diameter of 3–5 nm) could be as high as 90% at 1000 °C. Moreover, by adding a little water during the growth process, the proportion of DWCNTs (with diameter of 1.7–1.9 nm) can be increased to 54% and the proportion of TWCNTs (with diameter of 2.6–3.3 nm) decreased to 44%. This can be used as a good method to synthesize ultralong CNTs with a desired diameter and number of walls.

Hong et al. [8] synthesized ultralong CNTs using three different kinds of $FeCl_3$ ethanol solution and found that the diameter and number of walls of as-grown CNTs increased with increasing the solution concentration.

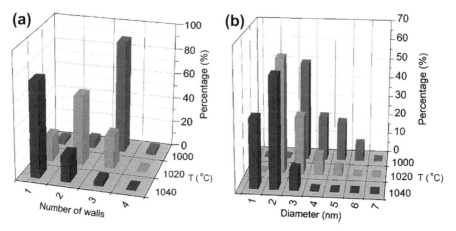

FIGURE 4.11

Structure distribution of CNTs grown at different temperatures [12]. (a) Distribution of number of walls of centimeter-long CNTs. (b) Distribution of outer diameters of centimeter-long CNTs. (For color version of this figure, the reader is referred to the online version of this book.)

Source: Reproduced with permission from Ref. [12]. Copyright 2010, Wiley-VCH Verlag GmbH & Co. KGaA, Weinheim.

The impact of catalyst on the growth of ultralong CNTs mainly works in the following ways. The activity of catalysts influences the structure, defects and length of ultralong CNTs. The diameter of catalyst NPs determines the diameter and number of walls of ultralong CNTs. The distribution of catalyst NPs determines the areal density of ultralong CNTs. Generally, the larger the diameter of the catalyst, the larger the diameter of the CNTs. An et al. [53] used identical metal-containing molecular nanoclusters as catalyst to grow SWCNTs and found that the diameter of SWCNTs could be tuned by varying the reacting conditions. To prevent the agglomeration of catalyst NPs at high temperature, many porous inorganic materials (such as Al_2O_3, SiO_2, mesoporous silicon and molecular sieves) are used to support the catalyst NPs. These methods often use metal NPs such as Co, Co–Mo and Fe. Besides, different kinds of metal NPs usually produce CNTs with different diameters.

4.3.2 Tuning the chiral consistency of ultralong CNTs

CNTs' electrical properties are largely determined by their chirality, so it is of great importance to keep the CNTs' chiral consistency for their application in integrated circuits. For centimeters-long CNT with the same chirality, it is possible to fabricate thousands of nanoelectronic devices on the same ultralong CNTs. However, it is a challenge to tune the CNTs' chirality at will. There are very few reports about chirality tuning of ultralong CNTs.

For ultralong CNT chirality tuning, Yao et al. [54] first proposed cloning growth of SWCNTs on Si/SiO$_2$ substrates. As shown in Fig. 4.12, they cut off an ultralong CNT on the same substrate into many small sections and used each of the open ends as seeds/catalysts to grow new SWCNTs which had the same chirality as the parent ones. This method shows a new way to fabrication of CNTs with desired structures but has a very low efficiency. Only 9% of the cutoff SWCNT sections can effectively grow new ones on Si/SiO$_2$ substrates and the efficiency can be up to 40% on quartz substrates.

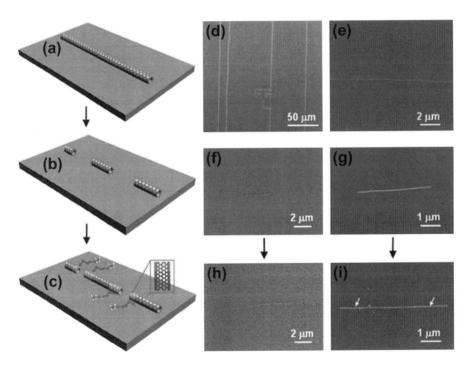

FIGURE 4.12

(a, c) Schematic diagrams of as-grown ultralong SWCNTs (a) on substrate by low feeding gas flow, which could be cut into short segments, open-end SWCNTs seeds (b), and the duplicate SWCNTs (red in color) which could be continued to grow from the parent SWCNTs segments via directly adding C_x (C_2 and/or C_3) radicals to the open-end seed (c). (d, e) SEM image and AFM image of ultralong SWCNTs used for preparing open-end SWCNTs seeds, respectively. (f, g) Representative SEM image and AFM image of short parent SWCNTs segments for the second growth. (h, i) SEM image and AFM image of duplicate SWCNTs continued grown from the SWCNTs in panels f and g, respectively. (For interpretation of the references to color in this figure legend, the reader is referred to the online version of this book.)

Source: Reproduced with permission from Ref. [54].Copyright 2009, American Chemical Society.

Wen et al. [12] performed a systematic study of the chirality tuning of ultralong TWCNTs. The ratio of TWCNTs in as-grown CNTs can be up to 90%. The chirality of these TWCNTs was characterized by transmission electron microscope (TEM) electron deflection patterns. It was found that the chirality of a 10-cm-long TWCNT can be maintained for at least 35 mm. As seen in Fig. 4.13, for the section between 25 mm and 60 mm from the starting end of the TWCNT, the chirality is unchanged and showed semiconducting properties (Fig. 4.14). At the 70 mm position, the chirality varied a little. This may result from the temperature change at the 70 mm position.

The chiral angle for long TWCNTs (centimeters long) usually ranged from $20°-30°$, while for short TWCNTs (shorter than centimeters), the chiral angles were usually smaller than $20°$ and randomly distributed (Fig. 4.15). The difference between the chiral angles for the two adjacent walls in TWCNTs is usually smaller than $6.3°$. According to Yakobson's theory, there is a difference between the growing speeds for CNTs with different chirality [27]. The larger the chiral angle, the higher the growing speed is. In this work, the growth speed for ultralong TWCNTs (24 μm/s) is higher than the typical growth speed. Thus, the distribution of the TWCNTs' chiral angle is close to the large value. This method provides a good choice for growing narrow-distributed chiral ultralong CNTs.

The carbon source is another important parameter for controlling the diameter and chirality of ultralong CNTs. At present, the most widely used carbon sources are those with high decomposition temperature, such as methane, ethane, ethanol and CO, which are not prone to produce amorphous carbon on catalyst NPs at the CNTs' synthesizing temperature. This is beneficial for the growth of perfect ultralong CNTs. Many groups have investigated the effect of carbon source on the structure of CNTs. Wang et al. believed that the carbon source was the key factor influencing the diameter and chirality of CNTs [55]. They used Co—Mo as the catalyst and CO, ethanol, methanol or ethene as the carbon source, respectively. The as-grown SWCNTs have different chirality distributions (Fig. 4.16). For instance, the narrow-distributed chirality of SWCNTs needs a high pressure of CO as the carbon source, while for ethanol or methanol, the pressure is very low. Lu et al. [56] found that the chiral distribution of SWCNTs' chirality could be tuned by controlling the flux of the carbon source. They also found that even for the widely distributed catalyst sizes, the diameter distribution of SWCNTs could be tuned just by controlling the flux of the carbon source. The reason is that for a given flux of the carbon source, the carbon source can selectively choose the catalyst NPs which have relevant sizes to grow CNTs.

4.3.3 Tuning the electrical properties of ultralong CNTs

It is of great significance to obtain well-aligned metallic or semiconducting ultralong CNT with high selectivity for their applications in the field of nanoelectronics. The typical application of CNTs is to fabricate FETs based on semiconducting CNTs [57]. There are several ways to obtain highly selective semiconducting CNTs,

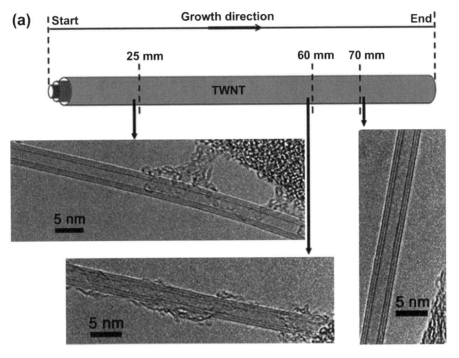
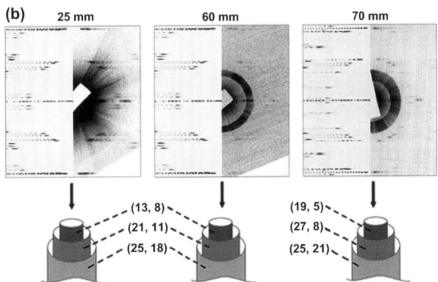

FIGURE 4.13

Long-range homogeneous atomic structure of TWCNTs. (a) Illustration of a 100-mm-long TWCNT and HRTEM images on three positions that are 25, 60 and 70 mm away from the growth-starting point. (b) Electron diffraction patterns recorded on the TWCNT at these three positions and the chiral indices were assigned to all three shells at each location.

Source: Reproduced with permission from Ref. [12]. Copyright 2010, Wiley-VCH Verlag GmbH & Co. KGaA, Weinheim.

CHAPTER 4 Synthesis and Properties of Ultralong Carbon Nanotubes

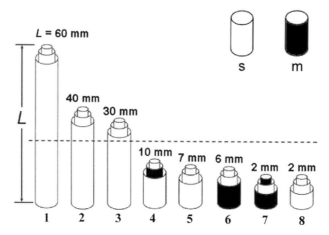

FIGURE 4.14

Illustration of eight TWCNTs showing the selective growth of long (>30 mm), semiconducting TWCNTs; "s" and "m" indicate semiconducting and metallic shells, respectively.

Source: Reproduced with permission from Ref. [12]. Copyright 2010, Wiley-VCH Verlag GmbH & Co. KGaA, Weinheim.

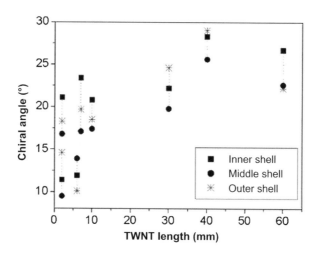

FIGURE 4.15

The chiral angle distribution of TWCNTs with different lengths.

Source: Reproduced with permission from Ref. [12]. Copyright 2010, Wiley-VCH Verlag GmbH & Co. KGaA, Weinheim.

such as selectively separate semiconducting and metallic CNTs from each other [58–60] or selectively removing metallic CNTs from semiconducting ones [61–64]. However, all the above methods could bring pollution or destruction to the CNTs. The best way is to selectively synthesize semiconducting ultralong

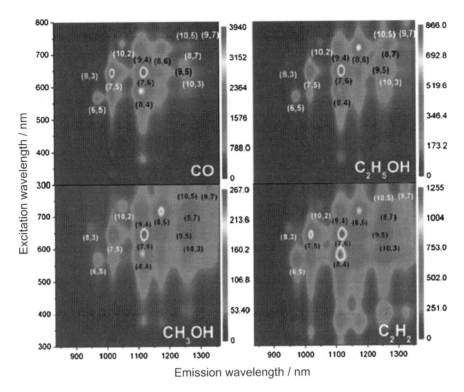

FIGURE 4.16

Photoluminescence excitation (PLE) intensity map as a function of excitation and emission wavelength for SDBS micellarized SWCNTs in D_2O produced from four different carbon precursors on Co—Mo catalysts. (For color version of this figure, the reader is referred to the online version of this book.)

Source: Reproduced with permission from Ref. [55].Copyright 2007, American Chemical Society.

CNTs (or metallic ones). The earliest report about highly selective semiconducting CNT growth was in 2002, when Kim et al. [65] obtained semiconducting CNTs with a selectivity of 70%. But these were randomly oriented and short CNTs (hundreds of micrometers long). In recent years, there has been significant progress in the fabrication of semiconducting CNTs with high selectivity. Ding et al. [66] selectively synthesized SWCNTs where 95% are semiconducting. Quartz substrates were used some methanol as added to ethanol as the carbon source. They believed that there were two important factors for this highly selective growth mechanism. The first one is the addition of methanol. Although methanol cannot be decomposed at 900 °C, the hydroxyl of methanol can selectively etch part of metallic CNTs. This is because the ionization energy of metallic CNTs is much lower than that of semiconducting ones. The above etching effect can also remove some of the amorphous carbon during the CNT growing process. The second factor is that the Y-cut quartz substrates have a selective effect. Wang et al. [11] recently made more than 100 FETs on an 18.5-cm-long semiconducting CNT and found that they had identical electrical properties, showing the electrical consistency of centimeters-long CNTs.

Hong et al. [67] used an ultraviolet-assisted CVD method to grow CNTs using methane as the carbon source on single-crystal quartz substrates and obtained 95% selective semiconducting CNTs. The on/off ratio of such semiconducting CNTs was lower than that of FETs made by individual semiconducting ones, due to the existence of a small portion of metallic CNTs. However, the well-aligned ultralong CNT arrays with a high areal density were very convenient to be directly used for fabrication of complicated circuits using microprocessing techniques, which was a breakthrough for the development of integrated circuits (Fig. 4.17(a) and (b)). In addition, Huang et al. [68] deposited Ag NPs on the as-grown CNTs and could tell the metallic and semiconducting CNTs from each other. But this method could only identify them but not separate them from each other. Wen et al. reported the on/off ratio as high as 10^6 on TWCNTs (Fig. 4.17(c)) [12].

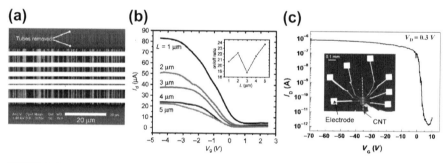

FIGURE 4.17

High on/off ratio FETs fabricated with as-grown aligned CNT arrays top-gated by solid electrolyte polymer films. (a) Large magnification view of the set of electrodes of a test device. The widths of the gaps between the 40-nm-thick gold contact lines correspond to the channel lengths of FETs (from top to bottom): 4, 3, 2, 1, and 5 μm. (b) Family of transfer characteristics (drain current, I_d, vs top gate voltage, V_g) of one of the test devices under a bias voltage V_{ds} 80 = mV. The variable parameter of the family is the transistor channel length L. The channel width is 0.5 mm. (Inset) The on/off ratios obtained for the device as a function of the channel length, L. (c) Current versus gate voltage (VG) characteristics at a constant bias of 1 V of a typical FET device with an on/off ratio of 10^6 fabricated on the semiconducting TWNT. (For color version of this figure, the reader is referred to the online version of this book.)

Sources: (a) Reproduced with permission from Ref. [67]. Copyright 2010, Wiley-VCH Verlag GmbH & Co. KGaA, Weinheim. (b) Reproduced with permission from Ref. [12]. Copyright 2009, American Chemical Society.

4.3.4 Tuning the morphology of ultralong CNTs

As mentioned above, ultralong CNTs have perfect structures and low defect densities thus showing great potential for nanoelectronics fabrication, superstrong fiber synthesis, etc. To realize these applications, it is necessary to tune the morphology of ultralong CNT arrays in addition to the structural tuning of individual CNTs.

4.3.4.1 Tuning of orientation and arrangement of ultralong CNT arrays

The tuning of orientation and arrangement of ultralong CNT arrays can be divided into three kinds of methods: the first one is using gas flow to orient the floating ultralong CNTs, the second one is using substrate surface force, and the last one is using an electrical field or magnetic field to orient the ultralong CNTs.

4.3.4.1.1 Gas flow-directed growth of ultralong CNTs

For ultralong CNTs growing on an Si/SiO$_2$ substrate, part of the growing CNTs are floating in the gas flow, which has significant influence on the orientation and arrangement of these ultralong CNTs. Many groups have investigated the effect of gas flow on the morphology of ultralong CNTs [8,10,16,22,36,47,50,69] (Fig. 4.18). Generally, the stable flow (laminar flow) is favorable for the arrangement of ultralong CNTs and the orientation of the CNTs is controlled by the flow direction. The disturbance of the gas flow will make the ultralong CNTs wave with the flow. To obtain horizontally aligned ultralong CNT arrays, it is necessary to cause the growing CNTs to lift off the substrate and be directed by the gas flow. In order to obtain well-aligned ultralong CNT arrays, it is very important to maintain a stable gas flow. The flow must be a laminar one; otherwise the arrangement of the ultralong CNTs will be disturbed. Hong et al. [8] put a small quartz tube into a larger one to decrease the Reynolds number of gas flow and eventually to make the flow more stable. Jin et al. [5] obtained more horizontally parallel ultralong CNT arrays with a high areal density using ultralow gas velocity. Peng et al. [50] systematically studied the effect of Reynolds number and Richardson number on the growth of ultralong CNTs. They found that a small Reynolds number and a large Richardson number are favorable for growing ultralong CNTs. In addition to using gas flow to direct the normal growth of ultralong CNTs, we can also use the gas flow-directing effect to fabricate various CNT patterns by changing the gas flow direction or the substrate orientation.

4.3.4.1.2 Substrate surface-directed growth of ultralong CNTs

Although gas flow tuning can effectively direct the ultralong CNTs' orientation, yet it is poorly controllable. A more precise nanosized directing method is in great demand.

Su et al. [70] studied the crystal lattice-directed growth of SWCNTs on quartz substrates. They found that the as-grown SWCNTs would follow the (100) and (111) direction during growth. They believed that this resulted from the interaction

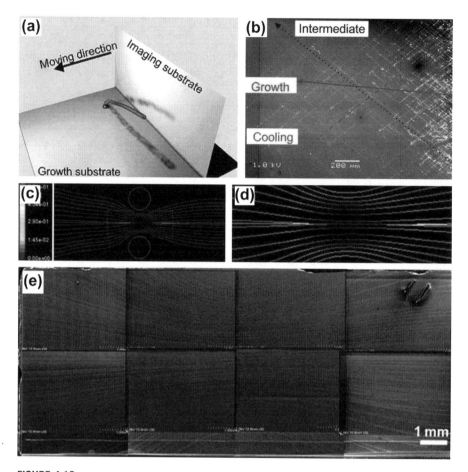

FIGURE 4.18

(a) Illustration of sample rotation for fabrication CNT to cross structures. (b) SEM picture of the resulting aligned nanotubes with an indication of the gas flow direction for each of the corresponding phases. (c) The hydromechanical simulation result when two cylindrical barriers were placed each on one side of the substrate. (d) The magnified image of (c) shows the streamline pattern at the place where the silicon wafer substrate was located. (e) Merged SEM images show the panorama of the obtained SWCNTs. (For color version of this figure, the reader is referred to the online version of this book.)

Sources: (a) Reproduced with permission from Ref. [10]. Copyright 2009, IOP Publishing. (b) Reproduced with permission from Ref. [16]. Copyright 2007, American Chemical Society

between the CNTs and crystal lattice of the substrate, which directed the orientation of ultralong CNTs.

Ismach et al. [38,71] found that CNTs would follow the miscut direction when growing on a miscut C-plane sapphire. These CNTs were not oriented by the gas flow or the surface lattices. A wake-growth mechanism was proposed to explain such a phenomenon [71]. There were many atomic steps on the miscut C-plane sapphires. The catalysts of CNTs would slide along the atomic steps during the CNT growth process, making the CNTs lie along the atomic steps, just like a long wake. This process involved three main factors: (1) higher nanotube-surface van der Waals (vdW) interactions near the step that result from increased contact area; (2) electrostatic interaction between the local electric fields created by uncompensated dipoles at the atomic steps and the induced dipoles across the SWCNTs; and (3) better wetting of the atomic steps by Fe NPs because of capillarity and higher coordination. Besides, the vdW interaction of CNTs and silicon substrates is about 2.2 eV/nm, while the electrostatic interaction of CNTs and silicon substrates can be up to 50 eV/nm, which has more effect on the arrangement of CNTs.

Ago et al. [72] obtained well-aligned and isolated SWCNTs grown on the R-plane ($1\bar{1}02$) and A-plane ($11\bar{2}0$) surfaces of sapphire (Al_2O_3) substrates by CVD, not on the C-plane surface as observed by Ismach et al. [38,71]. Although there were also atomic steps on the R-plane of sapphires, the SWCNTs did not align with these atomic steps, but with an inclination. It was deemed that the reason for CNTs aligning with atomic steps in Ismach's work was that the sapphires were miscut and the miscut inclination should be larger than 5°. Only in this case, there could be a local electric field which could then enhance the interaction between the CNTs and atomic steps. If the miscut inclination was smaller than 5°, CNTs would not follow the atomic steps. An atomic arrangement programmed growth was proposed for this plane-selective arrangement of CNTs on sapphire. It was believed that there were many one-dimensional arrays of Al atoms on the R- and A-plane surfaces of sapphire. Because of the strong interaction of Al atoms and SWCNTs, the SWCNTs could be aligned by the Al atom arrays. However, on the C-plane surface of sapphire, there were no such Al arrays because of the threefold symmetry in the atomic arrangement of the Al atoms.

Kocabas et al. [34,73,74] studied the orientation tuning of CNTs on Y-cut quartz substrates. They found that when cutting quartz substrates along the Y direction, atomic steps would form vertical to the cutting direction, which could be called the lattice direction. There was a lattice-inducing effect when growing CNTs on the Y-cut quartz substrates, which would make the CNTs align with the atomic steps. Some researchers combined the lattice-directed tuning with the gas flow-directed tuning and successfully fabricated Serpentine-like ultralong CNTs.

No matter what is the lattice orientation or the atomic steps orientation, even the crystal-edge orientation, all the above orientations belong to the graphoepitaxy methods. Both the lattices and atomic steps are nanosized and thus can precisely control the orientation of ultralong CNTs.

4.3.4.1.3 Electric field-directed growth of ultralong CNTs

When synthesized within an electric field, the CNTs will be affected by an electrical force. We can orient the arrangement of ultralong CNTs by this method. Many studies have been reported on the electric field-directed arrangement tuning of ultralong CNTs [33,75,76].

4.3.4.2 Tuning the length of ultralong CNT arrays

When the concept of "ultralong CNTs" was first proposed, the length of as-grown ultralong CNTs was only 0.6 mm [65]. To date, the longest ultralong CNTs can be up to 20 cm [9]. The key for growing ultralong CNTs is to keep the long lifetime of catalysts. Besides, it is necessary to keep the gas flow stable during the ultralong CNTs growth process as any disturbance of the gas flow would make the floating catalysts contact the substrate and then lose its activity. Once the metal catalyst NPs have contacted the substrate, they will combine with heteroatoms on the substrates at high temperature and then their size and structures will change greatly, resulting in losing their activity. There are many reasons for deactivity of catalysts. One of the many reasons is the deposition of amorphous carbon on the surface of catalysts. An effective route to keep catalysts' activity is by removing the amorphous carbon from the catalysts.

Wen et al. [9] found that adding some water could significantly improve the lifetime of catalysts and obtained ultralong CNTs with length up to 20 cm. Except for adding water, adding some CO has a similar effect on the improvement of CNTs' growth [77,78]. Adding CO can also effectively improve the CNT growth speed [9]. The highest growth speed can be 90 μm/s (Fig. 4.19).

Hata et al. [79] made some investigation on the effect of water on the improvement of CNTs' growth speed in 2004. They believed that water can remove the amorphous carbon deposited on catalysts and then improve their activities. From Fig. 4.20, we can see that the effect of adding water varied with water amount and time period, in which water is added. For instance, if the added water takes up 0.4% of the total feedstock, the ultralong CNTs can grow into 20 cm in 40 min, resulting in an average growth speed of 80 μm/s. While for ultralong CNTs synthesized without water, their growth speed is only 21 μm/s, which is much lower than the former one.

From Fig. 4.19, we can also find that there is a time-dependent characteristic for the effect of adding water on the CNTs' growth. Adding water at the start of the reaction has the best effect. The later the time water is added, the weaker the effect is. There are three reasons for this phenomenon. The first one is that the small amount of water can effectively remove the amorphous carbon on the catalysts, which makes their activity improve greatly. The second one is that at the initial stage of feeding carbon, there is a fast process of catalysts' deactivity due to the amorphous carbon, which mainly occurs in the initial 2–3 min. If there is no water, the deactivity can be as high as 70% which makes the activity of catalysts decrease greatly and thus makes the growth speed of ultralong CNTs decrease greatly. The third one is that there is a balance between the carbon source decomposition speed and the carbon consumption speed for growing CNTs, which makes the deactivity become weak after 2 min.

4.3 Tuning the Structure of Ultralong CNTs

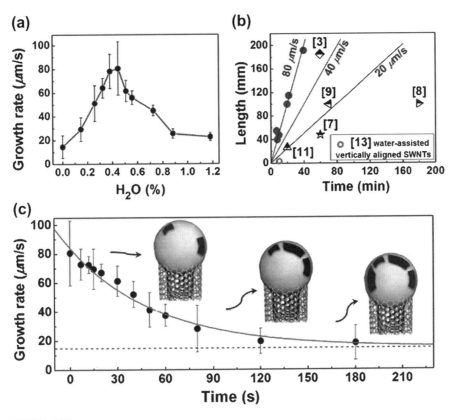

FIGURE 4.19

(a) Effect of water concentration on the growth rate of CNTs. (b) Comparison of the length and growth rate of CNTs in the references. (c) Effect of the time of water addition (after the feeding of the carbon source) on the growth rate of CNTs. (For color version of this figure, the reader is referred to the online version of this book.)

Source: Reproduced with permission from Ref. [9]. Copyright 2010, American Chemical Society.

Yuan et al. [31] theoretically explored the energetics of TDs in CNTs and their kinetic healing during the catalytic growth. Their study indicated that, with the assistance of a metal catalyst, TDs formed during the addition of C atoms can be efficiently healed at the CNT–catalyst interface. Theoretically, a TD-free CNT wall with 10^8–10^{11} carbon atoms is achievable, and, as a consequence, the growth of perfect CNTs up to 0.1–100 cm long was possible since the linear density of a CNT is ~100 carbon atoms per nanometer. At present, the main factor limiting the length of ultralong CNTs is the length of furnace heating zone which is usually 20–30 cm long. According to the kite mechanism [22], the catalyst NP keeps floating in the gas flow during CNTs' growth process. Therefore, the key point for growing ultralong

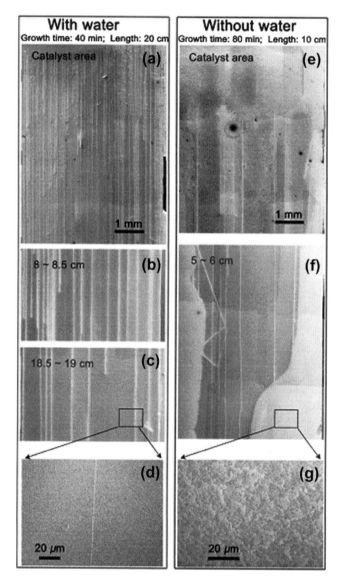

FIGURE 4.20

SEM pictures of long CNTs grown on Si substrates with water (left) and without water (right). The white lines in the left column are CNTs. The white areas in the right column are amorphous carbon, which is shown in panel g. (For color version of this figure, the reader is referred to the online version of this book.)

Source: Reproduced with permission from Ref. [9]. Copyright 2010, American Chemical Society.

CNTs is to make their tip end with catalyst NPs stay in the heating zone, in which the temperature should be maintained constant. One way to realize this idea is to make the furnace move along the reactor tube at the speed which is identical with the growth speed of ultralong CNTs, keeping the growth end of ultralong CNTs always in the heating zone of the furnace during the CNTs' growth process.

4.3.4.3 Tuning the areal density of ultralong CNT arrays

For the practical application of ultralong CNTs, the key is to realize the bulk production of ultralong CNTs with controlled structures. However, the main obstacle for the bulk production of ultralong CNTs is their extremely low areal density. The areal density of ultralong CNTs is usually several CNTs per 100 μm [8–13], which is far lower than that of agglomerated CNTs and vertically aligned CNT arrays.

Due to the agglomeration of catalyst NPs on the substrate at high temperature, small NPs tend to merge into large ones, which accordingly produce MWCNTs. These MWCNTs tend to follow the base-growth mechanism due to the strong interaction between the large catalyst NPs and the substrate, which hinders their extension into ultralong ones. The entanglement of MWCNTs is another obstacle blocking their growth. Only those few-walled CNTs (FWCNTs) with small diameters and with length longer than several hundred microns can be effectively directed by the gas flow. In general, for effective growth of ultralong CNTs, one of the three key factors is to prevent the merging of small catalyst NPs into large ones during the process, another one is to create an upward force to lift up more CNTs, and the last one is to provide a stable growth environment for ultralong CNTs, such as stable temperature, gas flow, and improved catalyst lifetime.

In order to improve the areal density of ultralong CNT arrays, it is important to maintain the discrete dispersive state of catalyst NPs during the CNT growth process. In addition to catalyst, substrates are another important factor influencing the CNTs' areal density. Generally speaking, the CNT arrays grown on silicon substrates have a much lower areal density than that of CNT arrays grown on quartz substrates. To date, the highest areal density of ultralong CNT arrays grown on silicon substrates is about 1–2 CNTs per 10 μm [48] (Fig. 4.21). While on a quartz substrate, the areal density of ultralong CNT arrays can be as high as 50 CNTs/μm [13].

Adding water can also improve the areal density of ultralong CNT arrays [9]. From Fig. 4.22, we can see that the average space between ultralong CNTs decreased from 500 to 100 μm compared with that without adding water. A little water can effectively remove the amorphous carbon, thus providing more growth sites for ultralong CNTs.

4.3.4.4 Schulz–Flory distribution-controlled growth — of half-meter-long carbon nanotubes

The Schulz–Flory (SF) distribution [81–85] is a mathematical equation that describes the relative ratios of linear condensation polymers of different length after a kinetics-controlled polymerization process, where the monomers are assumed to be equally reactive. The SF distribution can be expressed as

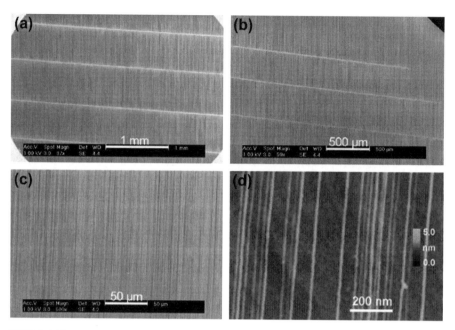

FIGURE 4.21

SEM and AFM images of high-density and perfectly aligned arrays of long SWNTs along the [72] direction on the ST-cut quartz substrate using patterned copper catalyst. The bright stripes in (a) and (b) correspond to copper catalyst. (c) High-magnification SEM image of the arrays of SWNTs. (d) AFM image of 1 μm × 0.75 μm area, in which 22 SWNTs have been found. (For color version of this figure, the reader is referred to the online version of this book.)

Source: Reproduced with permission from Ref. [13]. Copyright 2008, American Chemical Society.

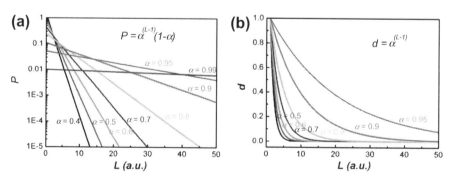

FIGURE 4.22

Theoretical number distribution of ultralong CNTs. (a) Theoretical percentage of ultralong CNTs. (b) Theoretical number density of ultralong CNTs [80]. (For color version of this figure, the reader is referred to the online version of this book.)

$P_x = p^{x-1}(1-p)$, where P_x is the mole fraction of polymers containing x segments (x-mers) and p is the probability that each monomer reacts to link the growing polymer chains while $(1-p)$ is the probability that no linkage exists. The SF distribution has been widely accepted to study the distribution of linear condensation polymers.

CNTs are one-dimensional carbon molecules and are highly desired due to their extraordinary properties resulting from their perfect atomic structures. It has been found that the number of long CNTs decreases rapidly with increased length [9–11,69,86,87]. Zhang et al. [80] found that the distribution of CNTs with different length, which can be regarded as spiral linear "polymers" of different length, can be well interpreted using the SF distribution. The distribution of long CNTs is controlled by the catalyst activity probability, which can be tuned by varying the process parameters. Based on the SF distribution, $P_x = p^{x-1}(1-p)$, p was redefined as the probability that a catalyst stays active enough to maintain the CNT adding a unit length (we could choose "1 mm" as a unit length in the following). α can be called **catalyst activity probability**, while $(1-\alpha)$ can be called *catalyst deactivation probability*. The percentage of CNTs with length L can be expressed as

$$P_L = \alpha^{(L-1)}(1-\alpha). \tag{4.2}$$

The CNT number density at the distance L from the starting position on substrates can be defined as the sum of the percentage of CNTs with length $\geq L$, which is

$$d_L = \sum_{L}^{\infty} P_L = \alpha^{(L-1)}. \tag{4.3}$$

The theoretical number distribution of ultralong CNTs is shown in Fig. 4.22. A low α indicates that the growth of long CNTs is easy to be terminated, resulting in a low percentage of long CNTs (Fig. 4.22(a)). With the increase of α, the percentage of long CNTs increases accordingly. The number density of ultralong CNTs is shown in Fig. 4.22(b). No matter what α, the number density of ultralong CNTs decreases with their increasing length. The higher the α, the slower the decrease in number density.

The catalyst activity is influenced by many factors, such as temperature, feedstock, substrates, gas velocities and so on (Fig. 4.23). Both the length and number density of ultralong CNTs can be improved by optimizing growth conditions. With an α as high as 0.995/mm and a "furnace-moving" method, 550-mm-long CNTs were synthesized. A "furnace-moving" method (Fig. 4.24(a)) was proposed to break the limitation induced by the physical length of the furnace. CNTs with length up to 550 mm were synthesized with 2 h growth time (Fig. 4.24(b)). α in this process was as high as 0.995/mm (Fig. 4.24(c)), enabling the growth of such long CNTs with a relatively high number density.

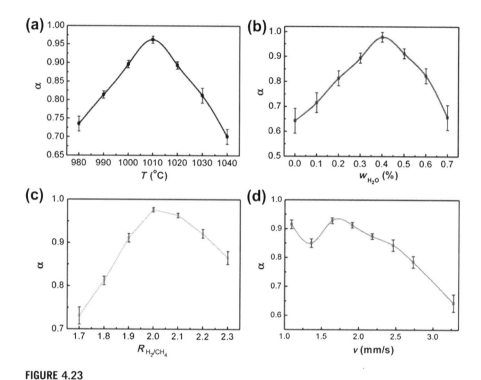

FIGURE 4.23

Relationship of catalyst activity probability (α) with different synthesis conditions. (a) Relationship of α with growth temperature. (b) Relationship of α with water content. (c) Relationship of α with H_2/CH_4 ratio. (d) Relationship of α with gas velocity [80]. (For color version of this figure, the reader is referred to the online version of this book.)

4.3.5 Mechanical properties of ultralong CNTs

One of the most exciting properties of CNTs is their extraordinary mechanical performance. CNTs with a high-aspect-ratio structure formed from strong sp^2 hybrid C—C bonds have been regarded as the strongest material ever discovered [4,88,89], and the only candidate for space elevators [3]. Theoretically, the tensile strength of individual CNTs can be higher than 100 GPa, Young's modulus higher than 1 TPa, and breaking strain up to 20%. Due to their high strength and strain, they have also shown promising potential for the storage of mechanical energy [90–93]. However, it is difficult to obtain the experimental values of these mechanical properties of CNTs because of their nanosized diameter. Ever since their discovery, the measurement of CNTs' mechanical properties has attracted extensive interest. Various methods were designed for this purpose and different data were obtained.

Treacy et al. [89] estimated the Young's modulus of isolated CNTs by measuring the amplitude of their intrinsic thermal vibrations in the TEM and obtained the

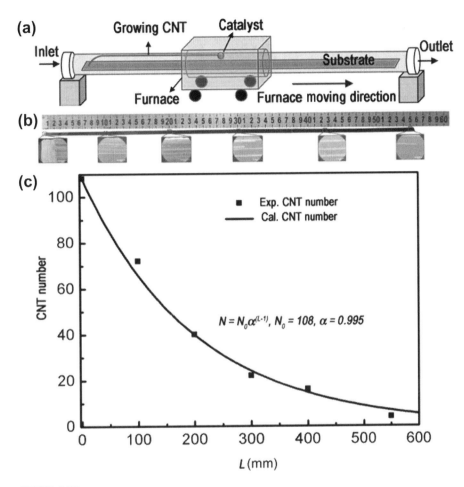

FIGURE 4.24

Synthesis of 550 mm long CNTs. (a) Schematic illustration of the "furnace-moving" method. (b) SEM image of 550 mm long CNTs. (c) Number of CNTs at different length on the substrate [80]. (For color version of this figure, the reader is referred to the online version of this book.)

values for Young's modulus ranging from 0.40 to 4.15 TPa. Although it was an interesting method to measure the Young's modulus at the nanoscale, there were many experimental uncertainties and errors in this process. Yu et al. [2,94] measured the tensile strengths of individual MWCNTs with a "nanostressing stage" located within a scanning electron microscope (SEM) (Fig. 4.25). The MWCNTs broke in the outermost layer ("sword-in-sheath" failure), and the tensile strength of this layer ranged from 11 to 63 GPa for the set of 19 MWCNTs that were loaded (Fig. 4.26). Analysis of the stress–strain curves for individual MWCNTs indicated that the Young's modulus of the outermost layer varied from 270 to 950 GPa. They

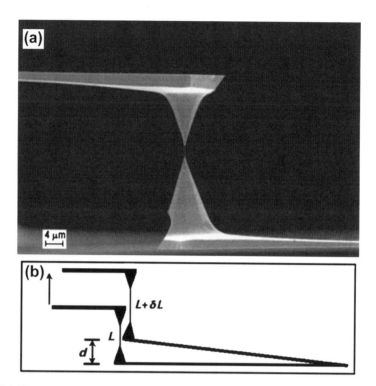

FIGURE 4.25

(a) Individual MWCNT is clamped in place and stretched by two opposing AFM tips.
(b) Schematic of the tensile loading experiment.

Source: Reproduced with permission from Ref. [94]. Copyright 2002, ASME Publications.

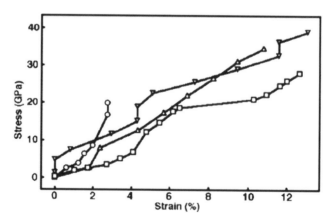

FIGURE 4.26

Plot of stress versus strain curves for five individual MWCNTs.

Source: Reproduced with permission from Ref. [94]. Copyright 2002, ASME Publications.

also measured the mechanical response of 15 SWCNT ropes under tensile load [95]. They found that eight of the as-used ropes broke at strain values of 5.3% or lower. The breaking strength of 15 SWCNTs range from 13 to 52 GPa (mean 30 GPa) based on a model. According to the same model, the eight average Young's modulus values determined range from 320 to 1470 GPa (mean 1002 GPa). Demczyk et al. [96] conducted pulling and bending tests on individual CNTs in situ in a TEM and computed a tensile strength of 150 GPa based on the observation of the force required to break the CNTs. From corresponding bending studies on such CNTs, the Young's modulus was estimated to be 0.9 TPa (0.8 TPa after sub-continuum corrections). Wu et al. [97] measured the stiffness of SWCNTs of defined crystal structure by combing optical characterization with a magnetic actuation technique and obtained the Young's modulus of $E = 0.91 \pm 0.16$ TPa. Peng et al. [4] measured the tensile strength of defect-free MWCNTs higher than 100 GPa.

From the above data, we can see that few of them could reach the theoretical values simultaneously. Except for the experimental errors and uncertainties, a main reason is that the as-used CNTs have structural defects. The extraordinary mechanical performance of CNTs depends on their perfect molecular structure. However, if there are defects in the atomic structures of as-grown CNTs, it is easy to envision that even a small number of defects will result in some degradation of their mechanical performance due to their quasi-one-dimensional atomic structures. The defects can appear both at the stage of CNT growth and at the later processing stage [98]. The defects, such as pentagon–heptagon pairs, vacancies, or dopant, are found to drastically influence the mechanical properties of CNTs, i.e. Young's modulus, tensile strength, critical strain, etc. (Fig. 4.27). The more defects the CNTs have, the worse the mechanical perforce is. Compared with the agglomerated CNTs and vertically aligned CNT arrays, the horizontally well-aligned ultralong CNTs have fewer defects and should have better mechanical properties. However, the length of individual ultralong CNTs has reached 20 cm [9], which is far beyond the characterization scale of the normally used instruments such as SEM, atomic force microscope, TEM, etc. Manipulation of individual ultralong CNTs is not an easy task due to their nanoscale diameters and macroscale lengths, which hampered the measurement of mechanical properties of ultralong CNTs.

Zhang et al. [1] developed a unique method to measure the mechanical properties of ultralong individual CNTs. They made individual ultralong CNTs visible under optical microscopes by depositing some TiO_2 particles onto the suspended CNTs, which also made the manipulation of individual CNTs feasible at the macroscale (Fig. 4.28). Through a special device, they experimentally characterized the as-grown CNTs with length over 10 cm, had a breaking strain up to 17.5%, tensile strength up to 200 GPa and Young's modulus up to 1.34 TPa (Fig. 4.29). The ultralong CNTs could endure a continuously repeated mechanical strain-release test for over 1.8×10^8 times. These CNTs also exhibited high potential for the storage of mechanical energy. They could store mechanical energy with a density as high as 1125 Wh/kg, which is much higher than that in flywheels and lithium-ion batteries,

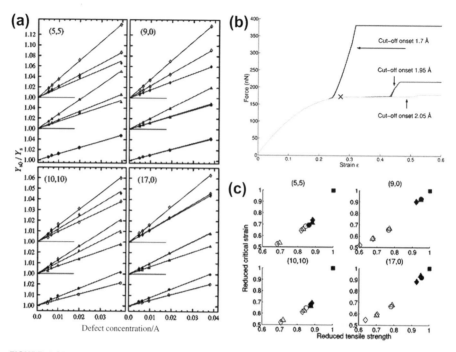

FIGURE 4.27

(a) Inverse of the scaled Young's modulus Y_{s0}/Y_s (where Y_{s0} is the surface-based Young's modulus of a perfect tube) plotted as a function of the defect concentration (or $1/L$) for four different nanotubes with single, double and triple vacancies. Circles refer to single vacancies, triangles to double vacancies and diamonds to triple vacancies. Filled symbols stand for reconstructed defects, whereas open symbols are for nonreconstructed defects. Nonreconstructed double and triple vacancies have two possible axial orientations which result in different Y_s behavior and two separate data sets. (b) Force–strain curves for an ideal (5,5)-nanotube at 10 K temperature with three different cutoff radii. The onset of the cutoff function can be perceived as an artificial peak. The flat regime depicts a sudden bond elongation to a length corresponding to the cutoff. Thus, the beginning of this plateau, "X", corresponds to the largest value of force outside the cutoff peak and will be interpreted as bond rupture. (c) Tensile strength and critical strain of nanotubes with defects. Filled symbols correspond to reconstructed vacancies and open symbols to nonreconstructed vacancies. Squares are the reference values for perfect tubes, circles stand for tubes with a monovacancy, triangles with a double vacancy and diamonds for tubes with a triple vacancy.

Source: Reproduced with permission from Ref. [31]. Copyright 2012, American Physical Society.

and they have a very high conversion efficiency, indicating the CNTs can be a promising medium for the storage of mechanical energy as well as a promising candidate for superstrong fibers. All these mechanical properties have reached the theoretical values. The basic reason is the as-grown ultralong CNTs have perfect structures.

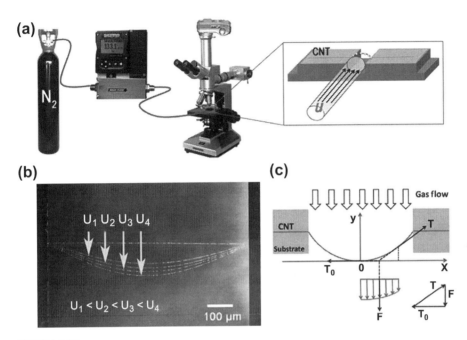

FIGURE 4.28

(a) Illustration of the gas-flow-blowing system used for testing the mechanical strength of CNT/TiO$_2$ string. The suspended CNT/TiO$_2$ chain was observed using an optical microscope. The elongation of the string was recorded by a digital camera when gas flow was introduced onto the suspended tube section. **U** in the inset was the gas velocity. (b) Elongation of the CNT/TiO$_2$ string with the increasing gas velocity. U_1, U_2, U_3 and U_4 represent different gas velocities. The two ends of the suspended CNTs remained fixed during the process. (c) Force analysis illustration for a stretched CNT. **T** is the pull exerted along the axis of the CNT, **T$_0$** is the pull exerted at the middle point of the suspended CNT, and **F** is the drag force exerted on the TiO$_2$ particles. (For color version of this figure, the reader is referred to the online version of this book.)

Source: Reproduced with permission from Ref. [1]. Copyright 2011, Wiley-VCH Verlag GmbH & Co. KGaA, Weinheim.

4.3.6 Potential applications of ultralong CNTs

The ultralong CNTs have shown perfect structures and extraordinary mechanical properties. The most promising application is to use them as superstrong fibers, even for space elevators [3]. The space elevator was first proposed in the 1960s, which was used as a method of getting into space. When the space elevator was first proposed, there was no material strong enough to build a cable to lift the elevator. Now, due to the emergence of ultralong CNTs, it is possible to realistically discuss the construction of a space elevator. Ultralong CNTs have the strength-to-mass ratio required for this endeavor. Much work has been done to design space elevators

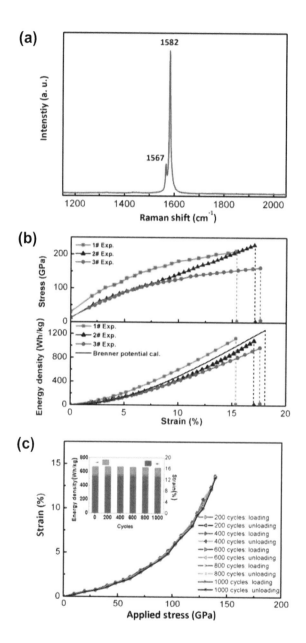

FIGURE 4.29

(a) A typical Raman spectrum from the as-grown CNTs. (b) Stress–strain curves and energy density–strain curves for three CNTs with high strain. The black line represents the energy density based on the second Brenner potential method. (c) Strain–stress behavior of a CNT after 200, 400, 600, 800 and 1000 strain–relaxation cycles. Inset: Energy density and strain property. (For color version of this figure, the reader is referred to the online version of this book.)

Source: Reproduced with permission from Ref. [1]. Copyright 2011, Wiley-VCH Verlag GmbH & Co. KGaA, Weinheim.

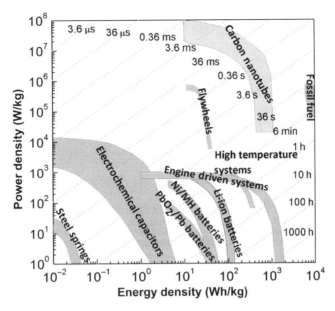

FIGURE 4.30

Ragone plot of several energy storage systems. (For color version of this figure, the reader is referred to the online version of this book.)

Source: Reproduced with permission from Ref. [1]. Copyright 2011, Wiley-VCH Verlag GmbH & Co. KGaA, Weinheim.

[3,99–101]. We have also demonstrated that super long CNTs with experimental strain up to 17% can store mechanical energy with a density as high as 1125 Wh/kg, which is much higher than that in lithium-ion batteries and flywheels, and has very high conversion efficiency (Fig. 4.30). This shows that nanomaterials can be an effective storage medium with mechanical energy as an alternative to electrochemical and electrostatic energy storage. In addition, the scale-up of the CNT systems can have other astonishing applications due to the strength. For instance, a rope with 4.5 cm in diameter and 3 km in length that used this perfect tube would have the capability for strain and recovery that can provide the large capacity/force to stop a 5000 ton train with a velocity of 350 km/h and reversibly absorb all the kinetic energy.

4.4 CONCLUSIONS

The growth of half-meter-long individual CNTs that are of high quality and pure has been demonstrated. If the long CNTs could be mass produced at a competitive cost, this superfiber material could displace many existing fibers in commercial use and change engineering design. Higher performance existing applications and new applications would come from the use of this ultralong CNT superfiber material.

References

[1] R. Zhang, et al., Superstrong ultralong carbon nanotubes for mechanical energy storage, Advanced Materials 23 (2011) 3387–3391.
[2] M.F. Yu, et al., Strength and breaking mechanism of multiwalled carbon nanotubes under tensile load, Science 287 (2000) 637–640.
[3] B.C. Edwards, Design and deployment of a space elevator, Acta Astronomica 47 (2000) 735–744.
[4] B. Peng, et al., Measurements of near-ultimate strength for multiwalled carbon nanotubes and irradiation-induced crosslinking improvements, Nature Nanotechnology 3 (2008) 626–631.
[5] Z. Jin, et al., Ultralow feeding gas flow guiding growth of large-scale horizontally aligned single-walled carbon nanotube arrays, Nano Letters 7 (2007) 2073–2079.
[6] L.X. Zheng, et al., Ultralong single-wall carbon nanotubes, Nature Materials 3 (2004) 673–676.
[7] X.S. Wang, et al., Selective fabrication of quasi-parallel single-walled carbon nanotubes on silicon substrates, Nanotechnology 21 (2010) 395602.
[8] B.H. Hong, et al., Quasi-continuous growth of ultralong carbon nanotube arrays, Journal of the American Chemical Society 127 (2005) 15336–15337.
[9] Q. Wen, et al., Growing 20 cm long DWNTs/TWNTs at a rapid growth rate of 80–90 μm/s, Chemistry of Materials 22 (2010) 1294–1296.
[10] Y. Liu, et al., Flexible orientation control of ultralong single-walled carbon nanotubes by gas flow, Nanotechnology 20 (2009) 185601.
[11] X.S. Wang, et al., Fabrication of ultralong and electrically uniform single-walled carbon nanotubes on clean substrates, Nano Letters 9 (2009) 3137–3141.
[12] Q. Wen, et al., 100 mm long, semiconducting triple-walled carbon nanotubes, Advanced Materials 22 (2010) 1867–1871.
[13] L. Ding, D. Yuan, J. Liu, Growth of high-density parallel arrays of long single-walled carbon nanotubes on quartz substrates, Journal of the American Chemical Society 130 (2008) 5428–5429.
[14] Y. Wang, F. Wei, G. Luo, H. Yu, G. Gu, The large-scale production of carbon nanotubes in a nano-agglomerate fluidized-bed reactor, Chemical Physics Letters 364 (2002) 568–572.
[15] Z. Ren, et al., Synthesis of large arrays of well-aligned carbon nanotubes on glass, Science 282 (1998) 1105–1107.
[16] M. Hofmann, D. Nezich, A. Reina, J. Kong, In-situ sample rotation as a tool to understand chemical vapor deposition growth of long aligned carbon nanotubes, Nano Letters 8 (2008) 4122–4127.
[17] S. Yahachi, Nanoparticles and filled nanocapsules, Carbon 33 (1995) 979–988.
[18] R.S. Wagner, W.C. Ellis, Vapor-liquid-solid mechanism of single crystal growth, Applied Physics Letters 4 (1964) 89–90.
[19] S. Robert, et al., Chemical vapor deposition growth of single-walled carbon nanotubes at 600°C and a simple growth model, Journal of Physical Chemistry B 108 (2004) 1888–1893.
[20] J. Gavillet, et al., Root-growth mechanism for single-walled carbon nanotubes, Physical Review B 87 (2001) 275504.

[21] G.Y. Zhang, X.C. Ma, D.Y. Zhong, E.G. Wang, Polymerized carbon nitride nanobells, Journal of Applied Physics 91 (2002) 9324–9332.

[22] S.M. Huang, M. Woodson, R. Smalley, J. Liu, Growth mechanism of oriented long single walled carbon nanotubes using "fast-heating" chemical vapor deposition process, Nano Letters 4 (2004) 1025–1028.

[23] M. He, et al., Iron catalysts reactivation for efficient CVD growth of SWNT with base-growth mode on surface, Journal of Physical Chemistry B 108 (2004) 12665–12668.

[24] Y. Shibuta, S. Maruyama, Molecular dynamics simulation of generation process of SWNTs, Physica B: Condensed Matter 323 (2002) 187–189.

[25] Y. Shibuta, S. Maruyama, Molecular dynamics simulation of formation process of single-walled carbon nanotubes by CCVD method, Chemical Physics Letters 382 (2003) 381–386.

[26] S. Maruyama, Y. Shibuta, Molecular dynamics in formation process of SWNTs, Molecular Crystals and Liquid Crystals 387 (2002) 87–92.

[27] B.I. Yakobson, A.R. Harutyunyan, D. Feng, Dislocation theory of chirality-controlled nanotube growth, Proceedings of the National Academy of Sciences U S A 106 (2009) 2506–2509.

[28] D. Feng, et al., The importance of strong carbon-metal adhesion for catalytic nucleation of single-walled carbon nanotubes, Nano Letters 8 (2008) 463–468.

[29] D. Feng, A. Rosen, K. Bolton, Molecular dynamics study of the catalyst particle size dependence on carbon nanotube growth, Journal of Chemical Physics 121 (2004) 2775–2779.

[30] D. Feng, B. Kim, A. Rosen, Nucleation and growth of single-walled carbon nanotubes: a molecular dynamics study, Journal of Physical Chemistry B 108 (2004) 17369–17377.

[31] Q. Yuan, Z. Xu, B.I. Yakobson, F. Ding, Efficient defect healing in catalytic carbon nanotube growth, Physical Review Letters 108 (2012) 245505.

[32] L.M. Huang, et al., Cobalt ultrathin film catalyzed ethanol chemical vapor deposition of single-walled carbon nanotubes, Journal of Physical Chemistry B 110 (2006) 11103–11109.

[33] Y. Zhang, et al., Electric-field-directed growth of aligned single-walled carbon nanotubes, Applied Physics Letters 79 (2001) 3155–3157.

[34] C. Kocabas, et al., Guided growth of large-scale, horizontally aligned arrays of single-walled carbon nanotubes and their use in thin-film transistors, Small 1 (2005) 1110–1116.

[35] Y.Y. Zhang, J. Zhang, H.B. Son, J. Kong, Z.F. Liu, Substrate-induced Raman frequency variation for single-walled carbon nanotubes, Journal of the American Chemical Society 127 (2005) 17156–17157.

[36] S. Huang, B. Maynor, X. Cai, J. Liu, Ultralong, well-aligned single-walled carbon nanotube architectures on surfaces, Advanced Materials 15 (2003) 1651–1655.

[37] S. Han, X. Liu, C. Zhou, Template-free directional growth of single-walled carbon nanotubes on a- and r-plane sapphire, Journal of the American Chemical Society 127 (2005) 5294–5295.

[38] A. Ismach, D. Kantorovich, E. Joselevich, Carbon nanotube graphoepitaxy: highly oriented growth by faceted nanosteps, Journal of the American Chemical Society 127 (2005) 11554–11555.

[39] H. Ago, et al., Visualization of horizontally-aligned single-walled carbon nanotube growth with $^{13}C/^{12}C$ isotopes, Journal of Physical Chemistry C 112 (2008) 1735–1738.

[40] S.M. Bachilo, et al., Structure-assigned optical spectra of single-walled carbon nanotubes, Science 298 (2002) 2361–2366.

[41] S.M. Bachilo, et al., Narrow (n, m)-distribution of single-walled carbon nanotubes grown using a solid supported catalyst, Journal of the American Chemical Society 125 (2003) 11186–11187.

[42] Y. Miyauchi, S. Chiashi, Y. Murakami, Y. Hayashida, S. Maruyama, Fluorescence spectroscopy of single-walled carbon nanotubes synthesized from alcohol, Chemical Physics Letters 387 (2004) 198–203.

[43] K. Hirahara, et al., Chirality correlation in double-wall carbon nanotubes as studied by electron diffraction, Physical Review B 73 (2006) 195420.

[44] M. Marchand, et al., Growing a carbon nanotube atom by atom: "and yet it does turn!" Nano Letters 9 (2009) 2961–2966.

[45] C.L. Cheung, A. Kurtz, H. Park, C.M. Lieber, Diameter-controlled synthesis of carbon nanotubes, Journal of Physical Chemistry B 106 (2002) 2429–2433.

[46] Y. Li, et al., Growth of single-walled carbon nanotubes from discrete catalytic nanoparticles of various sizes, Journal of Physical Chemistry B 105 (2001) 11424–11431.

[47] S. Huang, X. Cai, J. Liu, Growth of millimeter-long and horizontally aligned single-walled carbon nanotubes on flat substrates, Journal of the American Chemical Society 125 (2003) 5636–5637.

[48] W.W. Zhou, et al., Copper catalyzing growth of single-walled carbon nanotubes on substrates, Nano Letters 6 (2006) 2987–2990.

[49] Y.G. Yao, et al., Crinkling ultralong carbon nanotubes into serpentines by a controlled landing process, Advanced Materials 21 (2009) 4158–4162.

[50] B.H. Peng, Y.G. Yao, J. Zhang, Effect of the Reynolds and Richardson numbers on the growth of well-aligned ultra long single-walled carbon nanotubes, Journal of Physical Chemistry C 114 (2010) 12960–12965.

[51] Y.G. Yao, X.C. Dai, R. Liu, J. Zhang, Z.F. Liu, Tuning the diameter of single-walled carbon nanotubes by temperature-mediated chemical vapor deposition, Journal of Physical Chemistry C 113 (2009) 13051–13059.

[52] Y.G. Yao, et al., Temperature-mediated growth of single-walled carbon-nanotube intramolecular junctions, Nature Materials 6 (2007) 283–286.

[53] L. An, J. Owens, L. McNeil, J. Liu, Synthesis of nearly uniform single-walled carbon nanotubes using identical metal-containing molecular nanoclusters as catalysts, Journal of the American Chemical Society 124 (2002) 13688–13689.

[54] Y.G. Yao, C.Q. Feng, J. Zhang, Z.F. Liu, "Cloning" of single-walled carbon nanotubes via open-end growth mechanism, Nano Letters 9 (2009) 1673–1677.

[55] B. Wang, et al., (n, m) selectivity of single-walled carbon nanotubes by different carbon precursors on Co–Mo catalysts, Journal of the American Chemical Society 129 (2007) 9014–9019.

[56] C. Lu, J. Liu, Controlling the diameter of carbon nanotubes in chemical vapor deposition method by carbon feeding, Journal of Physical Chemistry C 110 (2006) 20254–20257.

[57] R. Martel, T. Schmidt, H.R. Shea, T. Hertel, P. Avouris, Single- and multi-wall carbon nanotube field-effect transistors, Applied Physics Letters 73 (1998) 2447–2449.

[58] M.S. Arnold, A.A. Green, J.F. Hulvat, S.I. Stupp, M.C. Hersam, Sorting carbon nanotubes by electronic structure using density differentiation, Nature Nanotechnology 1 (2006) 60–65.

[59] R. Krupke, F. Hennrich, H. von Lohneysen, M.M. Kappes, Separation of metallic from semiconducting single-walled carbon nanotubes, Science 301 (2003) 344–347.

[60] S. Banerjee, T. Hemraj-Benny, S.S. Wong, Routes towards separating metallic and semiconducting nanotubes, Journal of Nanoscience and Nanotechnology 5 (2005) 841–855.

[61] Y.Y. Zhang, Y. Zhang, X.J. Xian, J. Zhang, Z.F. Liu, Sorting out semiconducting single-walled carbon nanotube arrays by preferential destruction of metallic tubes using xenon-lamp irradiation, Journal of Physical Chemistry C 112 (2008) 3849–3856.

[62] G.Y. Zhang, et al., Selective etching of metallic carbon nanotubes by gas-phase reaction, Science 314 (2006) 974–977.

[63] L.M. Gomez, et al., Scalable light-induced metal to semiconductor conversion of carbon nanotubes, Nano Letters 9 (2009) 3592–3598.

[64] H.J. Huang, R. Maruyama, K. Noda, H. Kajiura, K. Kadono, Preferential destruction of metallic single-walled carbon nanotubes by laser irradiation, Journal of Physical Chemistry B 110 (2006) 7316–7320.

[65] W. Kim, et al., Synthesis of ultralong and high percentage of semiconducting single-walled carbon nanotubes, Nano Letters 2 (2002) 703–708.

[66] L. Ding, et al., Selective growth of well-aligned semiconducting single-walled carbon nanotubes, Nano Letters 9 (2009) 800–805.

[67] G. Hong, et al., Direct growth of semiconducting single-walled carbon nanotube array, Journal of the American Chemical Society 131 (2009) 14642–14643.

[68] S.M. Huang, et al., Identification of the structures of superlong oriented single-walled carbon nanotube arrays by electrodeposition of metal and Raman spectroscopy, Journal of the American Chemical Society 130 (2008) 11860–11861.

[69] A. Reina, M. Hofmann, D. Zhu, J. Kong, Growth mechanism of long and horizontally aligned carbon nanotubes by chemical vapor deposition, Journal of Physical Chemistry C 111 (2007) 7292–7297.

[70] M. Su, et al., Lattice-oriented growth of single-walled carbon nanotubes, Journal of Physical Chemistry B 104 (2000) 6505–6508.

[71] A. Ismach, L. Segev, E. Wachtel, E. Joselevich, Atomic-step-templated formation of single wall carbon nanotube patterns, Angewandte Chemie International Edition 116 (2004) 6266–6269.

[72] H. Ago, et al., Aligned growth of isolated single-walled carbon nanotubes programmed by atomic arrangement of substrate surface, Chemical Physics Letters 408 (2005) 433–438.

[73] C. Kocabas, et al., Experimental and theoretical studies of transport through large scale, partially aligned arrays of single-walled carbon nanotubes in thin film type transistors, Nano Letters 7 (2007) 1195–1202.

[74] C. Kocabas, M. Shim, J. Rogers, Spatially selective guided growth of high-coverage arrays and random networks of single-walled carbon nanotubes and their integration into electronic devices, Journal of the American Chemical Society 128 (2006) 4540–4541.

[75] C.C. Chiu, et al., Tip-to-tip growth of aligned single-walled carbon nanotubes under an electric field, Journal of Crystal Growth 290 (2006) 171–175.

[76] H. Hongo, F. Nihey, Y. Ochiai, Horizontally directional single-wall carbon nanotubes grown by chemical vapor deposition with a local electric field, Journal of Applied Physics 101 (2007) 024325.

[77] Q. Wen, et al., CO_2-assisted SWNT growth on porous catalysts, Chemistry of Materials 19 (2007) 1226–1230.

[78] A.G. Nasibulin, et al., An essential role of CO_2 and H_2O during single-walled CNT synthesis from carbon monoxide, Chemical Physics Letters 417 (2006) 179–184.

[79] K. Hata, et al., Water-assisted highly efficient synthesis of impurity-free single-walled carbon nanotubes, Science 306 (2004) 1362–1364.

[80] R. Zhang, Y. Zhang, H. Xie, F. Wei. Growth of half-meter long carbon nanotubes based on Schultz-Floury distribution. ACS Nano, under review, 2013.

[81] P.J. Flory, Molecular size distribution in linear condensation polymers, Journal of the American Chemical Society 58 (1936) 1877–1885.

[82] P.J. Flory, Principles of Polymer Chemistry, vol. 1, Cornell University Press, 1953.

[83] C. Bianchini, et al., Simultaneous polymerization and Schulz-Flory oligomerization of ethylene made possible by activation with MAO of a C1-symmetric [2, 6-bis (arylimino) pyridyl] iron dichloride precursor, Organometallics 23 (2004) 6087–6089.

[84] D.S. McGuinness, P. Wasserscheid, D.H. Morgan, J.T. Dixon, Ethylene trimerization with mixed-donor ligand (N, P, S) chromium complexes: effect of ligand structure on activity and selectivity, Organometallics 24 (2005) 552–556.

[85] X. Zhang, J. Wang, An alternative deductive method for molecular weight distribution function of linear polymers, Acta Polymerica Sinica 5 (2005) 725–730.

[86] L.X. Zheng, B.C. Satishkumar, P.Q. Gao, Q. Zhang, Kinetics studies of ultralong single-walled carbon nanotubes, Journal of Physical Chemistry C 113 (2009) 10896–10900.

[87] H.P. Liu, D. Takagi, S. Chiashi, Y. Homma, The controlled growth of horizontally aligned single-walled carbon nanotube arrays by a gas flow process, Nanotechnology 20 (2009) 345604.

[88] D. Walters, et al., Elastic strain of freely suspended single-wall carbon nanotube ropes, Applied Physics Letters 74 (1999) 3803.

[89] M. Treacy, T. Ebbesen, J. Gibson, Exceptionally high Young's modulus observed for individual carbon nanotubes, Nature 381 (1996) 678–680.

[90] A.B. Dalton, et al., Super-tough carbon-nanotube fibres, Nature 423 (2003) 703.

[91] K. Koziol, et al., High-performance carbon nanotube fiber, Science 318 (2007) 1892–1895.

[92] A. Cao, P.L. Dickrell, W.G. Sawyer, M.N. Ghasemi-Nejhad, P.M. Ajayan, Super-compressible foamlike carbon nanotube films, Science 310 (2005) 1307–1310.

[93] F. Hill, T. Havel, C. Livermore, Modeling mechanical energy storage in springs based on carbon nanotubes, Nanotechnology 20 (2009) 255704.

[94] D. Qian, G.J. Wagner, W.K. Liu, M.F. Yu, R.S. Ruoff, Mechanics of carbon nanotubes, Applied Mechanics Reviews 55 (2002) 495–533.

[95] M.F. Yu, B.S. Files, S. Arepalli, R.S. Ruoff, Tensile loading of ropes of single wall carbon nanotubes and their mechanical properties, Physical Review Letters 84 (2000) 5552–5555.

[96] B. Demczyk, et al., Direct mechanical measurement of the tensile strength and elastic modulus of multiwalled carbon nanotubes, Materials Science and Engineering A 334 (2002) 173–178.

[97] Y. Wu, et al., Determination of the Young's modulus of structurally defined carbon nanotubes, Nano Letters 8 (2008) 4158–4161.

[98] M. Sammalkorpi, A. Krasheninnikov, A. Kuronen, K. Nordlund, K. Kaski, Mechanical properties of carbon nanotubes with vacancies and related defects, Physical Review B 70 (2004) 245416.

[99] N.M. Pugno, On the strength of the carbon nanotube-based space elevator cable: from nanomechanics to megamechanics, Journal of Physics: Condensed Matter 18 (2006) S1971.

[100] N.M. Pugno, Space elevator: out of order? Nano Today 2 (2007) 44–47.

[101] J. Pearson, E. Levin, J. Oldson, H. Wykes. The Lunar Space Elevator, pp. 1–11, 2004.

[102] R. Zhang, et al., The reason for the low density of horizontally aligned ultralong carbon nanotube arrays, Carbon 52 (2012) 232–238.

[103] E. Joselevich, C.M. Lieber, Vectorial growth of metallic and semiconducting single-wall carbon nanotubes, Nano Letters 2 (2002) 1137–1141.

[104] S. Huang, X. Cai, C. Du, J. Liu, Oriented long single walled carbon nanotubes on substrates from floating catalysts, Journal of Physical Chemistry B 107 (2003) 13251–13254.

[105] C. Kocabas, M.A. Meitl, A. Gaur, M. Shim, J. Rogers, Aligned arrays of single-walled carbon nanotubes generated from random networks by orientationally selective laser ablation, Nano Letters 4 (2004) 2421–2426.

[106] S. Li, Z. Yu, C. Rutherglen, P.J. Burke, Electrical properties of 0.4 cm long single-walled carbon nanotubes, Nano Letters 4 (2004) 2003–2007.

[107] Z. Yu, S. Li, P.J. Burke, Synthesis of aligned arrays of millimeter long, straight single-walled carbon nanotubes, Chemistry of Materials 16 (2004) 3414–3416.

[108] X. Liu, K. Ryu, A. Badmaev, S. Han, C. Zhou, Diameter dependence of aligned growth of carbon nanotubes on a-plane sapphire substrates, Journal of Physical Chemistry C 112 (2008) 15929–15933.

[109] N. Kumar, W. Curtis, J. Hahm, Laterally aligned, multiwalled carbon nanotube growth using *Magnetospirillum magnetotacticum*, Applied Physics Letters 86 (2005) 173101.

[110] H. Ago, et al., Synthesis of horizontally-aligned single-walled carbon nanotubes with controllable density on sapphire surface and polarized Raman spectroscopy, Chemical Physics Letters 421 (2006) 399–403.

[111] K. Ryu, et al., Synthesis of aligned single-walled nanotubes using catalysts defined by nanosphere lithography, Journal of the American Chemical Society 129 (2007) 10104–10105.

[112] S. Huang, Q. Cai, J. Chen, Y. Qian, L. Zhang, Metal-catalyst-free growth of single-walled carbon nanotubes on substrates, Journal of the American Chemical Society 131 (2009) 2094–2095.

[113] N. Ishigami, et al., Crystal plane dependent growth of aligned single-walled carbon nanotubes on sapphire, Journal of the American Chemical Society 130 (2008) 9918–9924.

[114] A. Rutkowska, D. Walker, S. Gorfman, P.A. Thomas, J.V. Macpherson, Horizontal alignment of chemical vapor-deposited SWNTs on single-crystal quartz surfaces: further evidence for epitaxial alignment, Journal of Physical Chemistry C 113 (2009) 17087−17096.

[115] S.W. Hong, T. Banks, J.A. Rogers, Improved density in aligned arrays of single-walled carbon nanotubes by sequential chemical vapor deposition on quartz, Advanced Materials 22 (2010) 1826−1830.

[116] Q. Yong, W. Chunyan, H. Bin, Dysprosium-catalyzed growth of single-walled carbon nanotube arrays on substrates, Nanoscale Research Letters 5 (2009) 442−447.

[117] Q. Zhang, J.Q. Huang, M.Q. Zhao, W.Z. Qian, F. Wei, Carbon nanotube mass production: principles and processes, ChemSusChem 4 (2011) 864−889.

CHAPTER 5

Alloy Hybrid Carbon Nanotube Yarn for Multifunctionality

Lakshman K. Randeniya

CSIRO Materials Science and Engineering, Lindfield, NSW, Australia

CHAPTER OUTLINE

5.1 Introduction 137
    5.1.1 Interface between metal nanoparticles and CNTs 138
5.2 Electrical Conductivity of CNT Yarns.. 140
5.3 Metal Deposition on CNT Macrostructures 142
    5.3.1 Metal nanoparticle deposition on CNT yarns 143
5.4 Gas Sensing Applications 150
    5.4.1 Sensing mechanisms 152
    5.4.2 Gas sensing using CNT macrostructures 154
5.5 Summary 161
References 161

5.1 INTRODUCTION

Heterostructures of carbon nanotubes (CNTs) and metal nanoparticles (MNP)/polymers possess unique properties and have potential applicability in diverse industries [1,2]. Hybrid or composite materials take advantage of the important properties of the constituent materials. In some cases, totally new and useful properties can also emerge. CNTs have very useful intrinsic properties such as small size, large surface area, mechanical strength, environmental stability and high electrical conductivity. Their inherent size and hollow geometry can make them attractive as supports for heterogeneous catalysts. While CNTs in powder form are useful for a great variety of applications, macrostructures of CNT materials such as ropes, webs and yarns have their own unique properties that have been explored [3–6]. A number of laboratories have demonstrated the production of CNT yarns from vertically aligned "drawable" multiwalled CNT (MWCNT) forests using chemical vapor deposition methods (Fig. 5.1(a)). Precise control over the catalyst, gas purity, temperature and time are required for the production of continuously drawable products [3,7]. Using concepts from textile production, a yarn is spun from vertically aligned MWCNT forests (Fig. 5.1 (b)). These mechanically robust (tensile strengths of 1–2 GPa) and light-weight structures (~ seven times lighter than Cu) may now

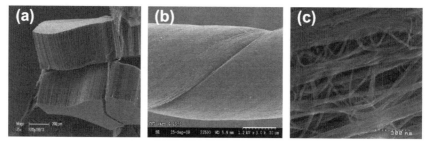

FIGURE 5.1

CNT yarn fabrication. (a) Drawable, vertically aligned MWCNT forest made by chemical vapor deposition using Fe catalysts [3,8]; (b) CNT yarn spun from a drawable MWCNT forest; (c) high-resolution image showing individual strands/fibers of nanotubes.

be produced on a large scale due to recent developments in spinning technologies [8]. Continual improvements to the spinning processes and to the preparation of well-aligned MWCNT webs are being made in order to obtain yarns with better mechanical and physical properties.

This chapter particularly focuses on the fabrication and applications of heterostructures of CNTs and MNP. In this context, we classify nanoparticles of metals as typically having diameters less than 50 nm. Such particles contain a large fraction of their atoms on the surface. Due to spatial confinement of the excitations, these particles show unique properties that may be tuned by size selection. The formation of heterojunctions from CNT macrostructures and metal nanocrystal is therefore an interesting proposition. The unique properties of these hybrids may be explored including the electrical resistance of nanotube—nanoparticle networks. Electrical properties depend on the nanoscale junctions that exist between the constituent materials and on the microscale and macroscale connectivity of the tubes and MNP.

Metals are important materials in CNT devices. Nanolayers of metal generally catalyze the synthesis of CNT and metal—CNT connections are central for devices. In addition, the interface properties can be useful in catalysis, sensing and other applications. Therefore, a good understanding of the interaction between metal and CNT is important for further progress in these fields.

5.1.1 Interface between metal nanoparticles and CNTs

The nature of bonding between metals and CNTs will determine the charge transfer kinetics, which is central to practical applications. In this chapter, we focus not on the growth of CNT on metal surfaces but on the aspects of metal particle attachment to the sidewalls of CNTs mainly through van der Waals bonding. Zhang et al. studied the sticking of a number of electron beam-evaporated metals on suspended single-walled carbon nanotubes (SWCNTs) using transmission electron microscopy [9]. Metals such as Ni and Ti formed continuous films and strong covalent bonding through curvature-induced hybridization of carbon sp^2 orbitals and metal d orbitals (see Fig. 5.2). These metals have a strong tendency to form carbides. Au, Al and Fe formed large islands indicating their relatively low condensation/sticking coefficients on nanotubes. Pd formed quasi-continuous films on SWCNTs. The strength of van der Waals bonding of metal atoms and CNT depends on a number of factors including surface wetting. Surface wetting is greater when the metal–metal bonding is weaker compared to metal–CNT bonding. When the metal–CNT bonding is weaker, the higher mobility of metal atoms on the surface leads to coalescence of metal particles [10]. The strength of the bonding, however, does not necessarily determine the electrical properties of the contact. For example, Ti wets the CNT surface better than Pd, but the electrical contact is poorer in case of Ti. In the case of Ti, the interlayer with CNT is depopulated from electrical charge [11].

The introduction of metal particles has given rise to large increases in the sensing response of CNTs to a large number of industrially and environmentally important gases. These gases include NH_3, NO_2, CO, H_2S, NO and H_2 [12–20]. It has been

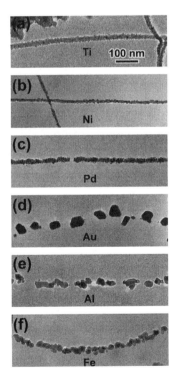

FIGURE 5.2

Transmission electron microscopic images of (a) Ti, (b) Ni, (c) Pd, (d) Au, (e) Al, and (f) Fe.
Source: From Ref. [9].

argued that the electrical response to gas adsorption is greatly affected by the interfacial potential barriers [21]. For example, when Au particles are first deposited on SWCNTs, there is a charge transfer from the tubes to MNP, which enhances the electrical conductance of the p-type CNTs [22]. When CO molecules are adsorbed on the metal surface, the charge density reverts onto the CNTs leading to an increase in resistance. Electronic structure calculations indicate dramatic variations in the charge density at the MNP–SWCNT interface. The increase in the potential barrier as a function of exposure time eventually leads to a saturation of response as the process of charge transfer is no longer possible [22].

5.2 ELECTRICAL CONDUCTIVITY OF CNT YARNS

We now examine the electrical conductance in CNT yarns and their hybrids with metals. Due to their light weight and tolerance to high current density, advanced electrically conducting CNT composite yarns are well suited for use as, for example, low-impedance, high-capacity and more robust microelectrodes in medical implants; low-resistant, high-strength wires in microelectronic; or light-weight conducting alternatives for complex systems in aerospace industry [6]. Large cost savings and lower carbon footprints can result in the use of such yarns in the aviation and aerospace industry where the reduction of weight is a significant advantage. They also have the potential to be the next-generation power transmission lines. The sagging of lines, common to metal wire conductors, is unlikely to be a problem with CNT-based conductors. However, the increase of conductivity in the yarns is a significant challenge and some approaches have been proposed in the literature.

The electrical conductivity of the MWCNT-based macrostructures such as CNT yarns is much lower than that of defect-free individual CNTs due to the presence of amorphous carbon and other impurities, which cause scattering and contact resistances. Table 5.1 lists the measured conductivities of single CNTs and bundles of tubes (e.g. yarns and fibers). Under ballistic conditions, the SWCNTs display very high electrical conductivity of about 10^6 S/cm and the multiwalled tubes have conductivities of approximately 3×10^4 S/cm [23–26]. Our measurements and previous measurements by Li et al. [27,28] indicate that pristine CNT yarns prepared from long CNT fibers behave typically as semiconductors and have electrical conductivities in the range of $5-6 \times 10^2$ S/cm at room temperature. Similarly, lower conductivity due to contact resistances are also present in the fibers made from SWCNTs. Ericson et al. prepared fibers solely containing SWCNT fibers by using fuming H_2SO_4 to support alignment [29]. The measured electrical conductivity of ∼5000 S/cm is about 10 times larger than that of the MWCNT counterparts, but significantly smaller than that of corresponding isolated tubes.

Li et al. studied two possible approaches for improving electrical conductivity in the yarns: (1) minimize the contact resistances between nanotubes by improving the alignment and the lengths of the nanotubes and (2) use postsynthesis treatments [27]. These authors synthesized well-aligned millimeter-long nanotubes using chemical

Table 5.1 Conductivity (σ) of Individual CNT and CNT-Based Macrostructures. The Last Letters in the Notation Refer to Different Structures, T, Tubes (Individual); F, Film; Y, Yarn

Material	σ (Scm-1)	Remarks	References
MWCNT	$\sim 3 \times 10^4$	Mean free path = 30 μm implying ballistic conduction	[24,26]
SWCNT	$\sim 1 \times 10^6$	Diameter 1 nm; length 90 nm, ballistic conduction	[30]
MWCNT	$\sim 10^3 - 10^2$	Defects decrease conduction; diameter 8–18 nm	[25]
SWCNF	$\sim 1 - 3 \times 10^4$	Diameter > 10 nm and lengths > 10 μm; metallic above 50 K	[31]
MWCNY	$\sim 3 \times 10^2$	Diameter 3–10 μm	[32]
MWCNY	$\sim 6 \times 10^2$	Diameter 3 μm; semiconducting	[27]
MWCNY	$\sim 5 - 6 \times 10^2$	Diameter 13–27 μm; semiconducting	[28]

vapor deposition. They used ethylene as the source gas and Fe as a catalyst. They then investigated the effects of a variety of mild postsynthesis techniques on the conductivity of the yarns. The authors investigated annealing, acid treatment and incorporation of Au nanoparticles (via spontaneous deposition from 0.01 M solution of $HAuCl_4$) on the conductivity of the yarns. The proposed methods made some small improvements to electrical conductivity of the yarns. The spontaneous deposition of Au only sparsely decorated the yarns.

Zhao studied the electrical conductivity in double-walled carbon nanotube (DWNT) cables and found that iodine doping increased the conductivity remarkably [33]. The raw DWNT cables were placed in the iodine vapor at 200 °C for 12 h. Following the treatment, the authors found that the morphology and the physical size (diameter) of the cables were unchanged but that the electrical conductivity had increased. They found evidence to suggest that iodine is uniformly present within the cable. Theoretical investigations suggested that iodine atoms are less likely to penetrate into the interlayer spacing or inside of the tubes, especially when they are well capped. The iodine atoms most probably penetrated into interstitial spaces between tubes and formed intercalated structures. They found that about 2.3% of the iodine is strongly chemically bonded to the nanotubes. The cables exhibited high current-carrying capacity of 10^5 A/cm^2 and could be joined together into arbitrary length and diameter, without degradation of their electrical properties. Interestingly, both raw and iodine-doped cables showed metallic behavior in terms of variation of resistance with temperature. The temperature coefficient of resistivity for iodine-doped cables was found to be much smaller compared to that of Cu, even smaller than that of the undoped (raw) cables.

SWCNTs prepared from different methods lead to different proportions of metallic and semiconducting tubes. From the commercial preparation techniques, the HIPco® method generally leads to about 60% semiconducting tubes, whereas the CoMoCAT® method yields more than 90% semiconducting tubes [34]. Increasing the fraction of metallic tubes in the original samples therefore will result in macrostructures with higher electrical conductivity. Sundaram et al. has shown that introducing sulfur soon after the cracking of ferrocene catalyst leads to nucleation of SWCNTs, which can be drawn out as continuous fibers [35]. The fibers had a large percentage of metallic tubes as confirmed by Raman spectroscopy. These authors, however, did not report the electrical conductivity of the resulting fibers.

The utility of the CNT yarns can be facilitated by providing electrical conducting pathways in them. Postsynthesis treatment of CNT yarn is one way of achieving this goal [27]. Choi et al. showed that Au and Pt particles spontaneously deposited onto SWCNTs enhanced the electrical conductivity [36]. Kong et al. [37] obtained hybrids of Au and SWCNTs and the conductivity increased up to 2×10^3 S/cm. Li et al. [27] attempted the incorporation of gold particles into MWCNT yarns using galvanic deposition in an ethanol solution of $HAuCl_4$ in order to increase the electrical conductivity. However, the gold nanoparticles were found to be sparsely distributed on the fibers and the increase in the electrical conductivity, which could be attributed to the presence of gold particles on the nanofibers, was less than 20%.

5.3 METAL DEPOSITION ON CNT MACROSTRUCTURES

A large number of techniques were developed for depositing MNP on CNTs. Extensive reviews of these techniques are available in the literature [1,2,38]. Two approaches seem common wherein the metal particles are incorporated directly on to the sidewalls of the nanotubes or attached via organic or biolinker molecules. The use of linker molecules allows covalent and noncovalent bonding of MNP and CNTs. For direct deposition, salts of noble metals are commonly used as precursors of nanoparticles and the metals are released by a chemical reduction process. If the reduction occurs in the presence of CNTs, the van der Waals forces allow adhesion of metal particles on the tubes. Various methods, for example, the application of heat and light and the use of chemical agents, have been used to reduce the metal cations. In some cases, spontaneous reduction of metal salts has been observed [27,36].

Electrochemical and electroless deposition methods have also been used extensively to achieve deposition of MNP on to CNTs. Precious and noble metals (Pt, Au, Pd, Ag, Rh, Ru) as well as Ni, Co, and Cu have been deposited and the resulting hybrids were proposed for various applications. These metals are commonly used in heterogeneous catalytic reactions, molecule detection (gas phase and solution phase) [1,2,38,39] and other situations where their properties can be enhanced when CNTs are employed as support materials. Some of the proposed methods are not easily used for metal deposition on CNT macrostructures. Electroless deposition is one method with substantial merit for use with CNT macrostructures.

Unlike electrodeposition, the electroless deposition process does not involve the use of an external power source. Therefore, the method is limited by the electrochemical redox potentials of metal systems and CNT. In cases of substrate-enhanced electro deposition [40] and self-fueled electro deposition (SFED) [28] techniques, metals such as Au, Pt, Pd, Cu, Ni and Ag can be deposited using a sacrificial metal with a lower reduction potential. It is noteworthy that metals like Cu and Ag have redox potentials smaller than those of CNTs. In both methods, CNTs act merely as electrodes where the charge transfer is facilitated. For example, while Cu and Ag can be used to deposit Au or Pd on CNT macrostructures, a metal such as Al, with a very low reduction potential, is required to deposit Cu.

5.3.1 Metal nanoparticle deposition on CNT yarns

MNP can be incorporated into CNT yarns using a simple electrochemical method [28]. The method is called SFED and is essentially electrochemical in nature and has no external power source. The method provides a contamination-free route for incorporating metals into CNT yarns and into other CNT-based macrostructures leading to composites with enhanced electrical conductivity. The method does not require the use of oxidative pretreatments, which has the potential to damage the surface of the fibers. By a simple attachment of a reducing agent (a metal) to the end of the yarns and immersion into a solution of selected electrolytes, MNP can be deposited all along the lengths of the yarns or other macrostructures. The authors deposited Cu, Au, Ni, Pd, Pt and Ag onto CNT yarns using this method.

Figure 5.3 shows simple arrangements that can be used to deposit MNP onto a yarn. The arrangement in Fig. 5.3(a) is most suitable for longer lengths and is suitable for coating Au, Cu and Pd. The two ends of the yarn are connected to small wires of the reducing metal (M_R) through a galvanic contact. Silver epoxy was used to secure the contacts. A suitable glass structure with end hooks can be used to hold the yarn loosely in place. The yarn is immersed in a suitable salt solution of the metal to be deposited (M_D). The galvanic contact stays just above the level of the electrolyte. Typical concentrations of the metal salt solution that were used in the experiments ranged from 5 mM to 20 mM. Once the reducing metal is in contact with the electrolyte solution, the deposition process starts and continues until the reducing metal is taken out of the solution. For longer depositions, the cleaning of the reducing metal wire at regular intervals may be required. Two porous barriers can be employed to prevent the migration of particles generated at the reducing metal toward the yarn. The rate of deposition depends on the nature of M_D and M_R and the concentration of the solution. It was also found that the exposure of the yarn to ultraviolet (UV) light for a few minutes accelerates deposition and leads to stronger attachment of Pd and Pt metal particles onto the yarn. The extent of coating and the size of the particles can be controlled by a variety of means, including the solution concentration, time of deposition, temperature of the bath and pretreatment of yarns with chemical and plasma techniques. The details of

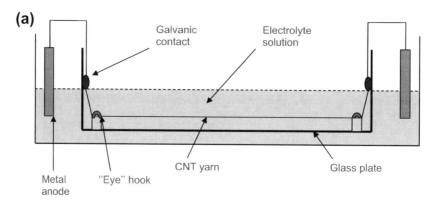

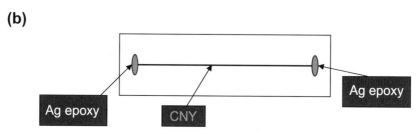

FIGURE 5.3

(a,b) The arrangement used for SFED deposition of metals onto CNT yarns. CNY, carbon nanoyarn. (For color version of this figure, the reader is referred to the online version of this book.)

the electrolytes and the conditions used for preparing conducting yarns are shown in Table 5.2.

For depositing metals on shorter lengths of yarns, the simple arrangement shown in Fig. 5.3(b) may be used. Here the CNT yarn is fixed on to a microscope glass slide using Ag epoxy. When the epoxy has dried, immersing the whole of the glass slide in

Table 5.2 Parameters Used for SFED Deposition of Metal Nanoparticles on CNT Yarns

Metal to be Deposited (M_D)	Metal Salt Used	Examples of Suitable Reducing Metals (M_R)
Au	$HAuCl_4 \cdot 3H_2O$	Ag, Cu
Pd	K_2PdCl_4	Ag, Cu
Pt	K_2PtCl_4	Cu, Zn
Cu	$CuSO_4/H_2SO_4$	Zn
Ag	$AgNO_3/HNO_3$	Cu, Zn
Ni	$NiCl_2 \cdot 6H_2O$	Zn

Source: From Randeniya et al. [28].

the electrolyte is sufficient for the deposition of Au and Pd. Ag in the contact sites acts as the reducing agent. For thicker coatings, Ag contacts may need to be redone to ensure that Ag is available for the continuous reduction and that the yarn is held in place. This method can be used to deposit metal particles on yarns up to about 7 cm long. The Au- and Pd-coated CNT yarn chemiresistors used for gas sensing experiments were prepared by this method.

As the metal deposits on the yarn, the electrical conductivity goes up. The change in electrical conductivity is a measure of the metal content on the yarns. For practical applications like gas sensing where a deposition of small amount of particles is required, the deposition process can be completed within a few minutes. The particle size and distribution can be affected by the pretreatment procedures, which will be described later. Dense distributions of Au particles as small as 2–3 nm on the CNT yarns were achieved using this method [17]. For metal-incorporated yarns, Randeniya et al. found metal-like electrical conductivity for Cu- and Au-coated yarns (see Figure 5.9 for temperature dependence of raw and hybrid yarn resistivity).

Optical images of CNT yarns coated with different metals are shown in Fig. 5.4. Cu and Au yield the most robust coatings where the metal particles are strongly

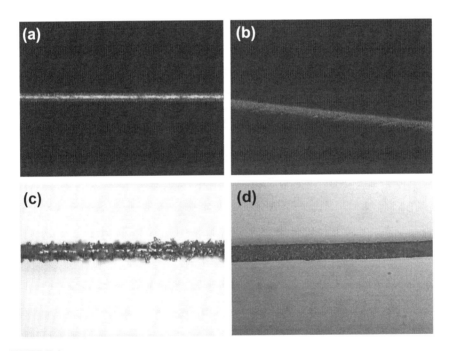

FIGURE 5.4

CNT yarn coated with (a) Cu, (b) Au, (c) Ag and (d) Pd. Very smooth metal layers can be seen for Au-, Cu- and Pd-deposited yarns. Ag particles agglomerate and do not form uniform layers. (For color version of this figure, the reader is referred to the online version of this book.)

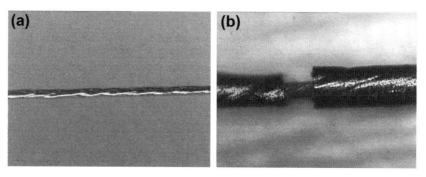

FIGURE 5.5

Optical images of CNT yarns. (a) Pure yarn. (b) A Pt-coated yarn with the metal layer developed a crack. The lightly coated yarn can be seen in the middle. (For color version of this figure, the reader is referred to the online version of this book.)

attached to the yarn. Stringent tests with tape were not able to remove the metal layers from the hybrid yarn. For Pd- and Pt-deposited yarns, continued tape tests led to the removal of some metal particles resulting in the reduction of electrical conductivity. Ag does not deposit as a smooth film and easily peels off from the CNT yarn. No increase in electrical conductivity could be measured due to the fragility of the hybrid yarn confirming that the metal particles are very loosely attached to the yarns.

Figure 5.5(b) shows the difficulty with Pt–CNT hybrid yarns with thick metal layers. When the thickness of the film increases, it cracks and the electrical connectivity is lost. Under these circumstances, the net increase in the electrical conductivity resulting from Pt deposition is small and results through a very thin and continuous film remaining on the yarn.

The average size of the MNP depositing on the CNT yarn was estimated from X-ray diffraction (XRD) spectroscopy using the Scherrer equation. This equation relates the full width at half maximum of the XRD peaks to the average crystal size. The results for different metals are shown in Table 5.3. The depositions for the yarns used for the XRD study were made at room temperature using 10 mM solutions of electrolyte. In general, Au, Cu, Pd and Ni formed similar-sized particles,

Table 5.3 Average Size of Metal Particles Depositing on the Yarn Determined from X-Ray Diffraction Spectroscopy

Metal	Average Size (nm)
Au	20–40
Cu	20–35
Pd	15–35
Pt	4–6
Ni	15–30

whereas Pt deposition led to much smaller sized particles (4–6 nm). When the Pt–CNT hybrid yarn was exposed to high temperatures, the average crystal size and the lattice parameter increased as shown in Fig. 5.6.

Figure 5.7 shows the scanning electron microscopic (SEM) images of composite yarns of CNT with Au (Au-CNT), Cu (Cu-CNT) and Pt (Pt-CNT) prepared by the SFED technique. Figure 5.7(a) shows smooth distribution of gold in the Au-CNT yarns. This morphology is not affected by heating the sample up to 400 °C. The surface of Cu-CNT yarn, which is shown in Fig. 5.7(b), shows a distribution of particles that is less smooth compared to that of Au-CNT. Although the surface of Pt-incorporated CNT yarn is much smoother (consisting of smaller nanocrystallites as discussed below), some discontinuities (micron-sized cracks) can be observed on the surface (Fig. 5.7(c)). These cracks become wider following heating at 400 °C. For Pd-CNT yarns, such cracks start to appear only following heating in air at 400 °C (Fig. 5.7(d)).

Au and Cu deposition were most successful due to the high electrical conductivity and nucleation properties of the metals. As the metal layer becomes thicker, the conductivity of the yarn increases as expected. Figure 5.8 shows the change in conductance as a function of thickness of the metal-deposited yarn. The conductivity rises rapidly at first and then saturates. Table 5.4 lists the measured terminal

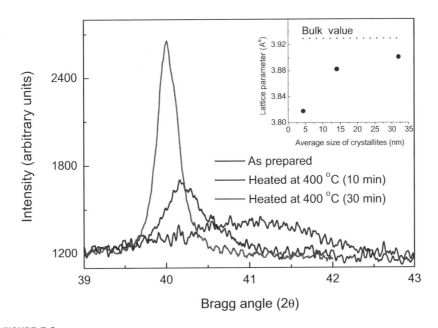

FIGURE 5.6

XRD spectra for Pt-CNT hybrid yarns as deposited and treated at high temperatures. (For color version of this figure, the reader is referred to the online version of this book.)

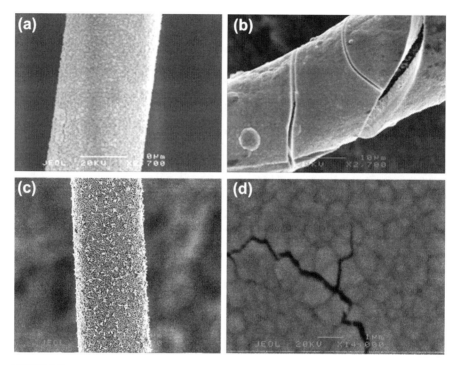

FIGURE 5.7

Low-resolution SEM diagrams of metal-carbon-nanotube hybrid yarns. (a) Au-coated yarn, (b) Pt-coated yarns showing delamination and cracking, (c) Pd-coated yarn and (d) some small cracking in very thick Pd-coated yarn.

electrical conductivities for individual doped/coated yarns. For Au- and Cu-coated yarns, the measured conductivity is about 50% of that of the bulk metal.

For metal-incorporated yarns, Randeniya et al. found metal-like electrical conductivity for Cu- and Au-coated yarns. The estimated temperature coefficient of resistivity is 3.2×10^{-3}/K for Au-CNT and 3.9×10^{-3}/K for Cu-CNT. The corresponding values for pure gold and pure copper are 3.2×10^{-3}/K and 3.9×10^{-3}/K, respectively. This further confirmed the prominent role of nanocrystalline gold and copper in conducting electricity through the corresponding composite yarns. The results showed that Au-CNT and Cu-CNT yarns have unique properties compared to other metal-CNT yarns (e.g. Pd and Pt) studied. The mechanical robustness and the nature of the electrical conductivity were found to be particularly remarkable. The authors concluded that Au and Cu particles are strongly attached to the CNT fibers compared to Pd and Pt. The high ductility of Au and Cu minimizes the development of structural defects and provides the continuity of particle interconnectedness through these yarns. Pt-CNT yarns showed a general trend in the temperature dependence of the resistivity similar to that of pure semiconducting CNT yarn;

5.3 Metal Deposition on CNT Macrostructures

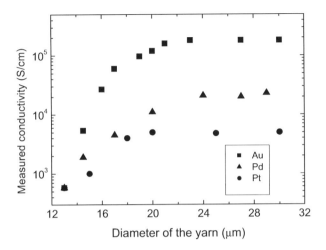

FIGURE 5.8

Electrical conductivity of the Au-, Pd- and Pt-coated CNT yarns as function of thickness. Original thickness of the yarns was 13 μm.

Table 5.4 Conductivity of Doped and Coated Macrostructures of CNT

Coated and Doped Yarns/Cables	Conductivity at Room Temperature (10^5 Scm^{-1})	Reference
I_2-DWCNY	~100	Zhao et al.[33]
Cu-MWCNY	3.0	Randeniya et al.[28]
Au-MWCNY	2.0	Randeniya et al.[28]
Pd-MWCNY	0.40	Randeniya et al.[28]
Pt-MWCNY	0.03	Randeniya et al.[28]

DWCNY, double-walled carbon nanoyarn; MWCNY, multiwalled carbon nanoyarn.

however, the influence of temperature was found to be smaller for the case of Pt-CNT yarn. The Pd-CNT yarns showed behavior of a semiconductor with distinct temperature regimes where the resistivity is relatively constant. Although there is a reduction by a factor of 40 in the room-temperature resistivity for a Pd-CNT yarn compared to a pure CNT yarn, the temperature dependence of resistivity still resembled that of a semiconductor-like material. For both Pt-CNT and Pd-CNT yarns, there are temperature regimes of metal-like behavior, where the resistivity increased with temperature [28]. Further investigations are required to understand the interplay between the metallic and semiconducting behaviors of the composite yarns. Table 5.4 lists the room-temperature conductivities of metal- and iodine-incorporated yarns.

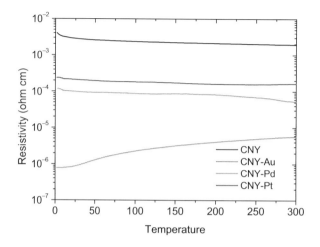

FIGURE 5.9

Temperature dependence of the resistivity of metal-coated yarns. (For color version of this figure, the reader is referred to the online version of this book.)

5.4 GAS SENSING APPLICATIONS

In addition to the applications that require the production of continuous long CNT yarns, reports on other possible applications have emerged. Biological and gas sensing applications require much smaller lengths of these materials and make use of the semiconducting properties of CNT, flexibility, high tensile strengths and the relative ease of handling and functionalization for device manufacturing process. For example, simple and straightforward methods for the deposition of a variety of noble-MNP on CNT yarns have been proposed. The yarns can also be easily treated with acids or in plasma to activate the sites for metal deposition. These developments should pave the way for future use of CNT yarns and ropes for the manufacture of flexible room-temperature gas sensing devices.

The detection of biological and chemical species in the atmosphere is important in relation to the control of industrial process, monitoring of environmental pollution, medical diagnosis, public security and agriculture. The development of inexpensive miniature sensors that can detect these gases in real time with good sensitivity and selectivity can have a large positive impact on human lives. Trace-level detection of NO_2 and NH_3 at room temperature remains a great need and a challenge. NO_2 is a hazardous gas in the environment and detection at trace levels is important. It plays a major role in the formation of ozone and acid rain. Continued or frequent exposure to concentrations higher than the air quality standard (53 parts per billion (ppb)) may cause increased incidences of acute respiratory illness in children and even death [41]. Similarly, ammonia is a toxic gas that is widely used in industries, produced naturally through nitrogen fixation and present in high concentrations in farming areas. The concentration of ammonia in exhaled breath is a useful indicator in the diagnosis of kidney and gastrointestinal diseases. Measurements of concentrations less than 2 ppb for environmental monitoring and 50 ppb for diagnostic breath analysis are required [42]. No device yet exists that can repeatedly detect such levels of gas concentrations at room temperature to fulfill the needs of safety, health and medical diagnostic requirements of communities. The unpleasant odors from landfill sites, some agricultural areas and industry need to be monitored and assessed to prevent health effects and the degradation of quality of life in surrounding communities. The biodegraded products of landfills include methane, carbon dioxide, ammonia, hydrogen, hydrogen sulfide and nitrous oxide. Therefore, repeated trace-level detection of these gases using an inexpensive, room-temperature device remains a challenge and a great need [43]. Room-temperature polymer-based sensors are not sensitive enough to be useful for environmental and medical diagnostic applications [44]. Conventional metal-oxide sensors operate at high temperatures (200–600 °C) and have high power consumption, low selectivity, high sensitivity to humidity and long-term drift [45].

The use of CNTs in sensing has been extensively studied since 2000 when Kong et al. showed the high sensitivity of a CNT wire transistor to nitrogen dioxide and ammonia at room temperature [46]. CNTs, and in particular SWCNTs, have exceptional sensitivity to gases at room temperature and have good environmental stability and excellent mechanical and electronic properties; however, practical gas sensing applications are hindered by excessively slow recovery and nonselectivity [46–48]. When devices made from CNTs (chemiresistors and field-effect transistors (FETs)) are exposed to reactive gases, the conductivity of the nanotubes change from the baseline value. In an ideal gas sensor, the conductance is expected to return to the baseline value (i.e. recover) when the reactive gas concentration is returned to zero. Rapid recovery of conductance is an essential element of a gas sensor as it allows repeated measurements of concentration changes. Kong et al. reported spontaneous recovery times of 12 h for NO_2 and NH_3 detections [46], which are extremely long. Heating of the devices was required to return to baseline conductance within a reasonable time. Very high sensitivity (e.g. detection of 100 parts per trillion of NO_2) and selectivity were achieved with CNTs combined with polymers. The recovery of the devices was achieved by the use of UV radiation [47,48]. The use of UV radiation (259 nm) in the presence of air produces ozone, which rapidly degrades the quality of sensors [49], limiting their use in inert buffer gases such as argon. In addition, polymer composite CNTs often used for achieving selectivity and higher sensitivity [48] become unstable when repeatedly exposed to UV radiation even in inert atmospheres. Therefore, improving the spontaneous recovery rates is a challenge and as shown later may be achieved through the use of MNP.

In addition to the gases mentioned above, CNT has been used to detect many other gases (e.g. H_2S, CO, H_2O, O_2, NO) and organic vapors (e.g. nitrotoluene, ethanol, and cyclohexane), which are important in industrial and environmental monitoring [16,50–53]. The disadvantage here is that substantial modifications to

CNTs are needed to achieve selectivity toward a given gas. Incorporation of MNP has been the choice of many scientists to achieve selectivity.

As mentioned earlier, the strong interaction of MNP with CNTs leads to significant charge reorganization and variations in charge density at the interface. The metal-CNT heterostructures can be used for addressing a number of important requirements in the sensors. First, they can be used to improve the sensitivity toward a given gas. It has already been shown that the use of CNT-MNP heterostructures increases the sensitivity toward certain gases. The use of Pd and Pt has been shown to be successful in enhancing the response toward hydrogen and methane [18,54–56]. Pt, Pd, Au, Rh, and Ag have been used with the CNTs for the detection of NH_3, NO_2, NO and H_2S [13,15,17,21,54]. The other advantage of the approach is that the choice of metal can give selectivity. This is quite important as CNT indiscriminately responds to many industrially and environmentally important gases [15]. The judicious use of different MNP on CNTs allows selectivity to be achieved.

In recent years, several reviews have discussed the progress in CNT-based sensor development for gas detection [16,51,52,57,58]. CNTs have attracted interest in gas sensing applications because of the lower operating temperatures compared to conventional oxide thin-film devices. In general, CNT gas sensors are based on individual strands, assemblies or dispersed solutions of single or multiwalled tubes that may be functionalized by the addition of suitable metal or metal-oxide nanoparticles. They can also be composited with suitable polymers to obtain responses to a wide variety of gases. In the simplest chemiresistor configuration, a single CNT or mesh of CNTs bridges electrodes and the binding of the analyte on the CNT surface results in charge transfer between the gaseous adsorbate and the CNT resulting in a change in electrical resistance. In general, CNT has p-type semiconducting properties. Therefore, the resistance decreases in response to adsorption of a strongly electron-deficient gas like NO_2 as charge transfer occurs from the nanotubes to the adsorbed molecules. Similarly, the adsorption of electron-rich gases, such as ammonia, increases the resistance of CNTs. Ab initio calculations can be used to predict the adsorption energy and the extent of charge transfer (normally presented as the fraction of charge transferred per adsorbed molecule). The presence of defect structures complicates the problem [59,60]. The production method and postproduction methods used for device fabrication (e.g. plasma and acid treatments, ultrasonic treatment) can induce various structural and dangling bond defects. The adsorption energy and the charge transfer normally increase due to the presence of the defects. In addition to adsorption at structural defects on each tube, the adsorption on interstitial sites and grooves must also be included in theoretical treatments. These considerations are particularly relevant for devices that use ropes and yarn structures where CNTs are all in perpetual contact with many surrounding tubes.

5.4.1 Sensing mechanisms

An understanding of mechanisms leading to the molecule-induced change in resistance observed in metal-decorated CNT assemblies is still developing. The selectivity and response enhancement obtained for certain gases need to be explained considering the changes occurring at the interface between metal particles and CNT. For CNT-based gas sensors (without metal decoration) used in FET configuration (carbon nanotube field-effect transistor (CNTFET)), there has been a long debate on the mechanisms responsible for the observed response [58]. Does the response occur due to charge transfer between the gas and the CNT occurring on the channel of the CNT or is it a result of the modulation of Schottky barrier between CNT and the bulk metal contact due to adsorption at the interface? For gases like NO_2 which are strong electron acceptors, there is theoretical evidence that the bonding between the gas molecules and the CNT is strong enough for charge transfer to be effective. Because CNTs are p-type semiconductors, the charge transfer to NO_2 enhances their conductance, which is consistent with the experimental observations. For weakly bound molecules such as NH_3, the charge transfer is unlikely to be efficient and additional mechanisms have been proposed [46]. There is growing evidence now to suggest that the Schottky barrier modulation at the interface between CNT and the metal contacts contributes predominantly to the sensing mechanism [21,22,61−63].

The early experiments carried out to differentiate between the two mechanisms gave inconsistent results. Liu et al. used polymethyl methacrylate (PMMA) for alternate passivation of the CNT channel and the CNT−metal contact regions [64]. Their results suggested that both NH_3 and NO_2 responded well to the CNTFET devices with passivated CNT−metal contact regions. Zhang et al. [63] also used PMMA to passivate the metal−CNT contact region of a CNTFET device and found that the response to NO_2 gas was delayed. The delay occurred due to the time taken by the NO_2 gas to diffuse through the PMMA layer to the metal−CNT contact region. These results highlighted the inadequacies of using PMMA as the passivation agent. Later, Peng et al. also examined the response of ammonia gas to CNTFET from room temperature up to 200 °C [62]. In their experiments, passivation was achieved by the deposition of 500-nm-thick layer of Si_3N_4. Peng et al. argued that NH_3 gas is unlikely to penetrate the Si_3N_4 layer and therefore the experiments should provide conclusive evidence for the mechanisms responsible. Three configurations of the device were examined: (1) an as-prepared CNTFET device in which both the CNT channel and the Au−CNT contact regions were exposed (device A), (2) in which Au−CNT contact regions were passivated (device B) and (3) in which only the CNT channel region was passivated (device C). At room temperature, both devices A and C showed good sensitivity to NH_3. However, device B did not respond to concentrations up to 500 ppm. Therefore, the authors concluded that their experiment demonstrated unambiguously the importance of adsorption at the contact region for the overall sensitivity of the device. Therefore, the Schottky barrier modulation is the key mechanism for the sensitivity of the CNTFET for ammonia gas. Further experiments at 150 °C and 200 °C suggested that weak charge transfer between ammonia and CNT is possible at higher temperatures, and small sensitivity toward the target gas was obtained for device B.

The previous discussion concerned the interface between the CNT and the bulk metal contact and further efforts are required to obtain a sound understanding of gas sensing mechanisms in metal nanoparticle-decorated CNTs (MNPCNTs). The interface phenomenon is inherently more complicated for nanosize particles and some controversies remain. Kaufmann and colleagues have examined the mechanisms responsible for gas sensing in MNPCNT devices. In one study, they examined the relationship between the metal work function and the charge transfer to SWCNT valence band by the nitric oxide gas [21]. They compared the responses of FET devices consisting of bare SWCNT and metal-decorated SWCNT. When exposed to 10 ppm NO (a weak electron donor), a larger shift in gate voltage was observed for the metal-decorated SWCNT devices compared to the bare SWCNT devices, which indicated increased electron transfer into p-type SWCNT when metal particles were present. In order to explain the results, the authors pointed to the apparent proportionality between metal work function and the measured absolute value of gate voltage shift in the metal-decorated SWCNT devices. The alignment of Fermi levels occur in contact regions between MNP and the semiconducting CNT. Charge redistribution occurs during this alignment and depletion regions are formed. The conductance decreased as these depletion regions acted as scattering sites for the charge carriers, reducing their mobility. This was seen as a "tilt" in the conductance versus gate voltage curves obtained for metal-decorated devices. Following adsorption of NO, electron transfer occurs from the metal to the SWCNT across a small potential barrier. The recombination leads to depletion of holes, which shows as a negative shift in device gate voltage. The authors argued that the potential barrier and therefore the negative shift in gate voltage does not depend on the size of the nanoparticles but on the work function of the metal.

Kauffman et al. also investigated the sensor response of carbon monoxide gas to Au-decorated SWCNTs [22]. Using optical spectroscopic technique and electrical transport mechanisms, they obtained evidence for the charge transfer events between gas molecules, Au nanoparticles and the CNTs. Further evidence for charge density variations at the metal-CNT resulting from gas molecule adsorption on nanoparticles was obtained.

5.4.2 Gas sensing using CNT macrostructures

More recently, it has been shown that MWCNTs in the form of these yarn- or rope-like structures can also be used to detect gases such as NH_3 and NO_2 [12,65]. The ropes were prepared by a floating catalyst Chemical vapour deposition (CVD) method (Fig. 5.10). They had good mechanical strength and were manually mounted onto SiO_2 substrates for gas sensing measurements. The change in electrical conductance as a function of target gas concentrations was monitored using a four-point probe method. Figure 5.11 shows the response and recovery cycles for ammonia measurements. While the ropes exhibited a good response to parts per million concentrations of NH_3 (50–1000 ppm), desorption and hence recovery of the sensor required heating the ropes to temperatures of the order of 180 °C. Here the response

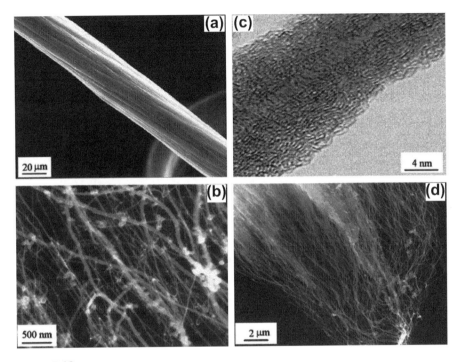

FIGURE 5.10

(a) SEM diagram of a CNT rope. (b) SEM image of the rope showing most of the CNTs aligning along the axial direction. (c) High-resolution image of a CNT. (d) Typical appearance of a broken rope after tensile test.

Source: Adapted from Li et al. [12].

is defined as the percentage change in resistance due to gas exposure relative to the baseline resistance. The response of the ropes to NO_2 also was examined [65]. When the ropes were treated with HNO_3, the sensitivity increased by 100% (Fig. 5.12). The recovery in the case of NO_2 was found to be much slower and incomplete and was attributed to the bundled nature of the rope structure creating adsorption states in the pores between the tubes inside the ropes.

Randeniya et al. examined the gas sensing characteristics of MWCNT yarns decorated with Au, Pd and Pt [17,56]. The deposition of metal particles was achieved using the SFED technique described earlier. For these experiments, the authors employed low-twist yarns (2000 twists per meter). This allowed CNT yarns with a more open structure and a larger surface area to be obtained compared to full-strength yarns (6000–10,000 twists per meter). The mechanical strength of the low-twist yarns remained at a level where they were easily handled and mounted onto substrates and filament posts. Due to the lower number of twists used to prepare the yarn, the diameter across the lengths of the yarns varied between 20 μm and

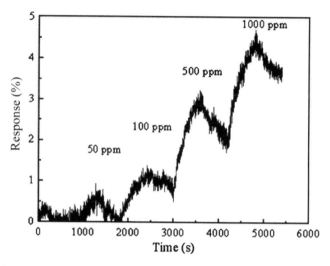

FIGURE 5.11

Dynamic adsorption–desorption response of an MWCNT rope sensor at room temperature with different concentrations.

Source: Adapted from Li et al. [12].

40 μm. The average resistance of the samples was found to be 500–600 Ω/cm. The yarns were mounted on glass slides (for acid and plasma treatments) by stretching and gluing the ends of the yarn to the substrate using Ag epoxy. The yarns were then cut into approximately 0.4–0.8 cm lengths and attached to filament posts (distances between filament posts were variable) to form simple chemiresistors and were inserted into the gas sensing apparatus.

Au-coated yarns showed much enhanced response to ammonia gas. The increase in the yarn electrical resistance was consistent with the previous experiments using individual CNTs, mats and ropes. Figure 5.13 shows that a sensitivity enhancement by a factor of six can be obtained for 10-min exposures using Au-decorated yarns compared to bare yarns. These authors also examined the recovery characteristics of the MWCNT yarn chemiresistors. By measuring the response to 550 ppm NH_3, the authors also found that the change in resistance in Au-coated CNT yarn sensors depended on the embedded amount of gold particles. The sensitivity to NH_3 increased at first with the Au loading and reached a maximum. The sensitivity decreased when Au amounts on the CNT yarns were increased further. A possible explanation for this observation is that at higher Au loadings, the conductivity of the yarn occurs mainly through Au network. The composite then behaves as a metal-like material. The charge scattering at the metal–CNT interface becomes less important for the overall charge migration through the yarn.

Randeniya et al. also examined the recovery rate of the Au-decorated yarns as a function of particle size. They used different surface treatments to obtain distinctly

5.4 Gas Sensing Applications

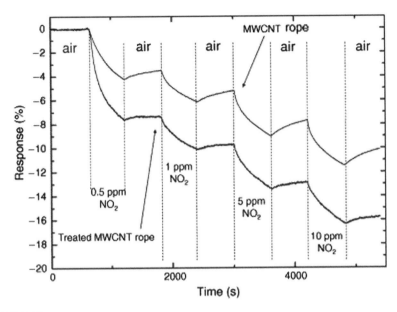

FIGURE 5.12

Dynamic adsorption–desorption responses at room temperature for a pristine MWCNT rope and an acid-treated MWCNT rope for different concentrations of NO_2. The dashed lines separate the cycles of adsorption (injection of NO_2) and desorption (injection of dry air).

Source: Adapted from Mendoza et al. [65].

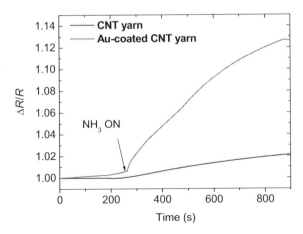

FIGURE 5.13

Increase in electrical resistance in MWCNT yarns in response to 500 ppm ammonia gas. The yarn decorated with Au nanoparticles shows enhanced sensitivity. Here R is the baseline resistance and ΔR is the change in resistance. (For color version of this figure, the reader is referred to the online version of this book.)

different Au particle distributions on the yarns. Figure 5.14 shows the impact of surface treatments on the size distribution of the Au particles on the CNT surfaces. Here the plasma treatment involved the use of 1:1 mixture of either Ar/O_2 or Ar/H_2. The acid treatment used a 1:1 mixture of concentrated HNO_3/H_2SO_4. Examination of SEM figures showed that the plasma-treated samples had two size distributions. First, there is a prominent dense distribution of particles smaller than 3 nm. Second, there is a sparse distribution of 10–20 nm particles. The acid-treated samples formed Au particles similar to that of the larger particle distribution on the plasma-treated sample. The smaller particle population is completely absent on the acid-treated (Fig. 5.14(b)) and untreated samples.

Figure 5.15 shows the response and recovery cycles from the measurements of ammonia gas using an Ar/O_2 plasma-pretreated and Au-coated CNT yarn chemiresistor. The device could be used to detect NH_3 levels down to about 500 ppb. Except in the case for 500 ppb sample, the exposure time was 300 s. For the case of 500 ppb, a longer exposure time (600 s) was necessary to detect the presence of NH_3. The lowest concentrations reported to be detected using CNT ropes at room temperature (not Au-coated) previously is 50 ppm [8]. A fixed exposure time was required because the signal saturation was not reached within a reasonable period. The resistance kept increasing slowly even after exposure to ammonia for 1 h.

In general, due to the high mobility of the Au atoms on the surface of CNTs and the larger binding energy for Au–Au than that of Au-CNT, the clustering of Au atoms is energetically preferred over forming Au-CNT bonds [16]. This can explain the formation of a larger particle distribution in nontreated and acid-treated yarn samples. However, for plasma-treated samples, the formation of defects appears to have substantially increased the binding energy between Au and CNTs. This led to a much denser population of smaller Au nanoparticles as seen in the SEM.

The impact of the size and the shape of nanoparticles on the physical and chemical properties are well known [18,19]. The smaller particles offer more surface area and therefore more active sites for adsorption of NH_3. For small particle sizes (<3 nm), great majority of the atoms will be found on the surface of the

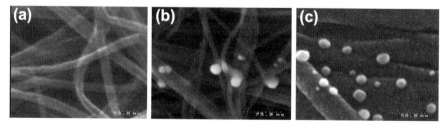

FIGURE 5.14

SEM diagrams of CNT yarns (a) with no metal deposition, (b) with Au deposited after acid treatment and (c) with Au deposited after Ar/O_2 plasma treatment.

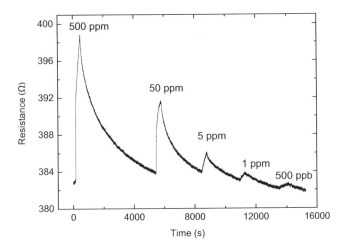

FIGURE 5.15

Response (change of resistance) cycles for Ar/O$_2$ plasma-treated and Au-coated CNT yarn samples for different concentrations of NH$_3$. The exposure time in each case is 300 s except in the case for 500 ppb where an exposure time of 600 s was used.

Source: Adapted from Randeniya et al. [17].

nanocrystallites. Therefore, desorption of NH$_3$ gas may be easier from smaller Au particles as the distance for diffusion is shorter. In particular, as shown here, the recovery characteristics improved when smaller (<3 nm) Au particles were present on the CNTs. The addition of Au particles also gave greater selectivity for ammonia sensing using CNT yarns. As mentioned earlier, the sensitivity for ammonia in Au-coated CNT yarn sensors was higher by factors of up to 10 in comparison to that of pure CNT yarn sensors. The sensitivity toward other gases such as methane, hydrogen and carbon dioxide remained nearly the same for pure CNT sensors and for Au-coated CNT sensors.

Randeniya et al. found that when the MWCNT yarns are decorated with Pd, the electrical conductivity of the yarn goes up [56]. This confirms that the Pd-decorated yarns consist of sufficient connectivity between individual CNTs and the Pd clusters (30–50 nm) leading to continuous electrical paths. When a chemiresistor made from the Pd-decorated yarn was first exposed to hydrogen, a large drop of resistance was observed [56]. This drop was interpreted as the reorganization of the Pd particles on the surface leading to better electrical connectivity between metal particles. The dissolution of hydrogen leads to a change in crystal parameters of Pd (expansion) and causes a rapid and reversible decrease in the resistance in meso-wire arrays on graphite [66]. On CNT yarns, the affect is not reversible except for small concentrations below 0.1% of hydrogen. It is likely that a breakup and reorganization of Pd particles occurs and the connectivity remains even after hydrogen has desorbed.

Following this initial drop of resistance, which was irreversible, the sensing of hydrogen was possible through a mechanism where the resistance of the hybrid yarn increased as a function of hydrogen concentration [56]. Figure 5.16 shows the measurement of hydrogen in nitrogen buffer gas using Pt-Pd- and Pd-deposited yarns. In all cases, good sensitivity together with fast response and recovery were obtained. The authors detected hydrogen concentrations down to 20 ppm using the Pd-decorated CNT yarns. When a layer of Pt was introduced on top, the lower limit of detection could be extended to 5 ppm.

Here the mechanism for hydrogen detection may be explained as follows. The dissolution of hydrogen atoms in Pd leads to the formation of hydrides of Pd and lowers the work function of the metal. The electrons are now more easily transferred to CNT, which has p-type semiconducting properties. The carrier density of the yarn is lowered and the resistance of the hybrid increases [18]. The picture may not be so simple as this model does not take into consideration the properties of metal−CNT

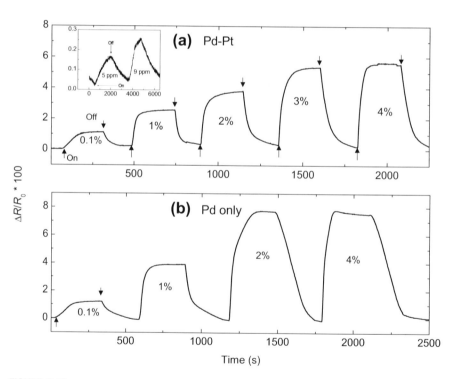

FIGURE 5.16

Detection of hydrogen using the Pd-Pt-decorated (a) and Pd-decorated (b) CNT yarns. The response is expressed as a percentage change in baseline resistance. Upward pointing arrows denote the time when hydrogen gas was turned on and downward pointing arrows denote the time when hydrogen gas was turned off.

interface [21]. The formation of potential barriers at the interface has the ability to interfere in the charge transfer process. These considerations could be more important in the detection of lower levels of hydrogen and when the detection is carried out in air.

5.5 SUMMARY

Macroscopic structures of CNTs combined with metal alloy nanoparticles provide exciting prospects for several areas of applications. For longer lengths of the yarns, the possibility of manufacturing light-weight electrical conductors will provide real commercial benefits. The direct incorporation of metals and nonmetals for increasing electrical conductivity has been demonstrated. The significant increase in weight of the hybrid material may be a disadvantage for certain applications.

Shorter lengths of CNT yarns show potential as microelectrodes and robust and flexible chemiresistors for molecule detection. The decoration of metal nanoclusters provides real possibilities for achieving high sensitivity and selectivity toward many gases that are present in industrial and landfill emissions. Currently, the use of macrostructures for gas sensing has been limited to the use of MWCNTs. The use of single-walled nanotubes can provide enhanced sensitivity and future opportunities for metal-CNT gas sensors.

References

[1] X. Peng, J. Chen, J.A. Misewich, S.S. Wong, Carbon nanotube–nanocrystal heterostructures, Chemical Society Reviews 38 (4) (2009) 1076–1098.

[2] G. Wildgoose, C. Banks, R. Compton, Metal nanoparticles and related materials supported on carbon nanotubes: methods and applications, Small 2 (2) (2006) 182–193.

[3] K.R. Atkinson, S.C. Hawkins, C. Huynh, C. Skourtis, J. Dai, M. Zhang, et al., Multifunctional carbon nanotube yarns and transparent sheets: fabrication, properties, and applications, Physica B: Physics of Condensed Matter 394 (2) (2007) 339–343.

[4] E. Muñoz, A.B. Dalton, S. Collins, M. Kozlov, J. Razal, J.N. Coleman, et al., Multifunctional carbon nanotube composite fibers, Advanced Engineering Materials 6 (10) (2004) 801–804.

[5] A.B. Dalton, S. Collins, J. Razal, E. Munoz, V.H. Ebron, B.G. Kim, et al., Continuous carbon nanotube composite fibers: properties, potential applications, and problems, Journal of Materials Chemistry 14 (1) (2004) 1–3.

[6] W. Lu, M. Zu, J.-H. Byun, B.-S. Kim, T.-W. Chou, State of the art of carbon nanotube fibers: opportunities and challenges, Advanced Materials 24 (14) (2012) 1805–1833.

[7] C.P. Huynh, S.C. Hawkins, Understanding the synthesis of directly spinnable carbon nanotube forests, Carbon 48 (4) (2010) 1105–1115.

[8] T. Canh-Dung, S.M. Smith, G. Higgerson, B. Anh, L.K. Randeniya, T-C. Thanh. Spinning CNT based composite yarns using a dry spinning process, Nanoscience and Nanotechnology (ICONN), (2010) International Conference on, 17–20.

[9] Y. Zhang, N.W. Franklin, R.J. Chen, H. Dai, Metal coating on suspended carbon nanotubes and its implication to metal–tube interaction, Chemical Physics Letters 331 (1) (2000) 35–41.

[10] F. Banhart, Interactions between metals and carbon nanotubes: at the interface between old and new materials, Nanoscale 1 (2) (2009) 201–213.

[11] N. Nemec, D. Tománek, G. Cuniberti, Contact dependence of carrier injection in carbon nanotubes: an ab initio study, Physical Review Letters 96 (7) (2006) 076802.

[12] Y.H. Li, Y.M. Zhao, Y.Q. Zhu, J. Rodriguez, J.R. Morante, E. Mendoza, et al., Mechanical and NH3 sensing properties of long multi-walled carbon nanotube ropes, Carbon 44 (9) (2006) 1821–1825.

[13] M. Penza, G. Cassano, R. Rossi, M. Alvisi, A. Rizzo, M.A. Signore, et al., Enhancement of sensitivity in gas chemiresistors based on carbon nanotube surface functionalized with noble metal (Au, Pt) nanoclusters, Applied Physics Letters 90 (17) (2007), 173123–1, -3.

[14] M. Penza, R. Rossi, M. Alvisi, G. Cassano, E. Serra, Functional characterization of carbon nanotube networked films functionalized with tuned loading of Au nanoclusters for gas sensing applications, Sensors and Actuators B: Chemical 140 (1) (2009) 176–184.

[15] M. Penza, R. Rossi, M. Alvisi, E. Serra, Metal-modified and vertically aligned carbon nanotube sensors array for landfill gas monitoring applications, Nanotechnology 21 (10) (2010) 105501.

[16] D. Kauffman, A. Star, Carbon nanotube gas and vapor sensors, Angewandte Chemie International Edition 47 (35) (2008) 6550–6570.

[17] L.K. Randeniya, P.J. Martin, A. Bendavid, J. McDonnell, Ammonia sensing characteristics of carbon-nanotube yarns decorated with nanocrystalline gold, Carbon 49 (15) (2011), 173123-1-3.

[18] J. Kong, M.G. Chapline, H. Dai, Functionalized carbon nanotubes for molecular hydrogen sensors, Advanced Materials 13 (18) (2001) 1384–1386.

[19] J.C. Charlier, L. Arnaud, I.V. Avilov, M. Delgado, F. Demoisson, E.H. Espinosa, et al., Carbon nanotubes randomly decorated with gold clusters: from nano2hybrid atomic structures to gas sensing prototypes, Nanotechnology 20 (37) (2009) 375501.

[20] S. Mubeen, T. Zhang, B. Yoo, M.A. Deshusses, N.V. Myung, Palladium nanoparticles decorated single-walled carbon nanotube hydrogen sensor, The Journal of Physical Chemistry C 111 (17) (2007) 6321–6327.

[21] D.R. Kauffman, A. Star, Chemically induced potential barriers at the carbon nanotube–metal nanoparticle interface, Nano Letters 7 (7) (2007) 1863–1868.

[22] D.R. Kauffman, D.C. Sorescu, D.P. Schofield, B.L. Allen, K.D. Jordan, A. Star, Understanding the sensor response of metal-decorated carbon nanotubes, Nano Letters 10 (3) (2010) 958–963.

[23] S.N. Song, X.K. Wang, R.P. Chang, J.B. Ketterson, Electronic properties of graphite nanotubules from galvanomagnetic effects, Physical Review Letters 72 (5) (1994) 697–700.

[24] A. Bachtold, M. Henny, C. Terrier, C. Strunk, C. Schönenberger, J.P. Salvetat, et al., Contacting carbon nanotubes selectively with low-ohmic contacts for four-probe electric measurements, Applied Physics Letters 73 (2) (1998) 274–276.

[25] H. Dai, E.W. Wong, C.M. Lieber, Probing electrical transport in nanomaterials: conductivity of individual carbon nanotubes, Science 272 (1996) 523–526.

[26] C. Berger, Y. Yi, Z.L. Wang, W.A. de Heer, Multiwalled carbon nanotubes are ballistic conductors at room temperature, Applied Physics A 74 (3) (2002) 363–365.
[27] Q.W. Li, Y. Li, X.F. Zhang, S.B. Chikkannanavar, Y.H. Zhao, A.M. Dangelewicz, et al., Structure-dependent electrical properties of carbon nanotube fibers, Advanced Materials 19 (20) (2007) 3358–3363.
[28] L.K. Randeniya, A. Bendavid, P.J. Martin, C. Tran, Composite yarns of multiwalled carbon nanotubes with metallic electrical conductivity, Small 6 (16) (2010) 1806–1811.
[29] L.M. Ericson, H. Fan, H. Peng, V.A. Davis, W. Zhou, J. Sulpizio, et al., Macroscopic, neat, single-walled carbon nanotube fibers, Science 305 (5689) (2004) 1447–1450.
[30] B. Gao, Y.F. Chen, M.S. Fuhrer, D.C. Glattli, A. Bachtold, Four-point resistance of individual single-wall carbon nanotubes, Physical Review Letters 95 (19) (2005) 196802–196805.
[31] R.S. Lee, H.J. Kim, J.E. Fischer, A. T, E. SR, Conductivity enhancement in single-walled carbon nanotube bundles doped with K and Br, Nature 388 (1997) 255–257.
[32] M. Zhang, K.R. Atkinson, R.H. Baughman, Multifunctional carbon nanotube yarns by downsizing an ancient technology, Science 306 (2004) 1358–1361.
[33] Y. Zhao, J. Wei, R. Vajtai, P.M. Ajayan, E.V. Barrera, Iodine doped carbon nanotube cables exceeding specific electrical conductivity of metals, Science Reports 1 (2011).
[34] A.V. Naumov, O.A. Kuznetsov, A.R. Harutyunyan, A.A. Green, M.C. Hersam, D.E. Resasco, et al., Quantifying the semiconducting fraction in single-walled carbon nanotube samples through comparative atomic force and photoluminescence microscopies, Nano Letters 9 (9) (2009) 3203–3208.
[35] R.M. Sundaram, K.K.K. Koziol, A.H. Windle, Continuous direct spinning of fibers of single-walled carbon nanotubes with metallic chirality, Advanced Materials 23 (43) (2011) 5064–5068.
[36] H.C. Choi, M. Shim, S. Bangsaruntip, H. Dai, Spontaneous reduction of metal ions on the sidewalls of carbon nanotubes, Journal of the American Chemical Society 124 (31) (2002) 9058–9059.
[37] B.S. Kong, D.H. Jung, S.K. Oh, C.S. Han, H.T. Jung, Single-walled carbon nanotube gold nanohybrids: application in highly effective transparent and conductive films, Journal of Physical Chemistry C 111 (23) (2007) 8377–8382.
[38] V. Georgakilas, D. Gournis, V. Tzitzios, L. Pasquato, D.M. Guldi, M. Prato, Decorating carbon nanotubes with metal or semiconductor nanoparticles, Journal of Materials Chemistry 17 (26) (2007) 2679–2694.
[39] C. Gao, Z. Guo, J.-H. Liu, X.-J. Huang, The new age of carbon nanotubes: an updated review of functionalized carbon nanotubes in electrochemical sensors, Nanoscale 4 (6) (2012) 1948–1963.
[40] L. Qu, L. Dai, Substrate-enhanced electroless deposition of metal nanoparticles on carbon nanotubes, Journal of the American Chemical Society 127 (31) (2005) 10806–10807.
[41] A. Afzal, N. Cioffi, L. Sabbatini, L. Torsi, NOx sensors based on semiconducting metal oxide nanostructures: progress and perspectives, Sensors and Actuators B: Chemical 171–172 (0) (2012) 25–42.
[42] B. Timmer, W. Olthuis, A.V.D. Berg, Ammonia sensors and their applications—a review, Sensors and Actuators B: Chemical 107 (2) (2005) 666–677.

[43] F. Yavari, Z. Chen, A.V. Thomas, W. Ren, H.-M. Cheng, N. Koratkar, High sensitivity gas detection using a macroscopic three-dimensional graphene foam network, Science Reports 1 (2011) (10.1038/srep00166).
[44] H. Bai, G. Shi, Gas sensors based on conducting polymers, Sensors 7 (3) (2007) 267–307.
[45] G. Korotcenkov, Gas response control through structural and chemical modification of metal oxide films: state of the art and approaches, Sensors and Actuators B: Chemical 107 (1) (2005) 209–232.
[46] J. Kong, N.R. Franklin, C. Zhou, M.G. Chapline, S. Peng, K. Cho, et al., Nanotube molecular wires as chemical sensors, Science 287 (5453) (2000) 622–625.
[47] J. Li, Y. Lu, Q. Ye, M. Cinke, J. Han, M. Meyyappan, Carbon nanotube sensors for gas and organic vapor detection, Nano Letters 3 (7) (2003) 929–933.
[48] P. Qi, O. Vermesh, M. Grecu, A. Javey, Q. Wang, H. Dai, et al., Toward large arrays of multiplex functionalized carbon nanotube sensors for highly sensitive and selective molecular detection, Nano Letters 3 (3) (2003) 347–351.
[49] G. Chen, T.M. Paronyan, E.M. Pigos, A.R. Harutyunyan, Enhanced gas sensing in pristine carbon nanotubes under continuous ultraviolet light illumination, Science Reports 2 (343) (2012).
[50] E.S. Snow, F.K. Perkins, J.A. Robinson, Chemical vapor detection using single-walled carbon nanotubes, Chemical Society Reviews 35 (9) (2006) 790–798.
[51] T. Zhang, S. Mubeen, N.V. Myung, M.A. Deshusses, Topical review: recent progress in carbon nanotube-based gas sensors, Nanotechnology 19 (33) (2008) 332001.
[52] B. Mahar, C. Laslau, R. Yip, Y. Sun, Development of carbon nanotube-based sensors: a review, Sensors Journal, IEEE 7 (2) (2007) 266–284.
[53] A. Goldoni, et al., Sensing gases with carbon nanotubes: a review of the actual situation, Journal of Physics: Condensed Matter 22 (1) (2010) 013001.
[54] A. Star, V. Joshi, S. Skarupo, D. Thomas, J.-C.P. Gabriel, Gas sensor array based on metal-decorated carbon nanotubes, The Journal of Physical Chemistry B 110 (42) (2006) 21014–21020.
[55] Y. Lu, J. Li, J. Han, H.T. Ng, C. Binder, C. Partridge, et al., Room temperature methane detection using palladium loaded single-walled carbon nanotube sensors, Chemical Physics Letters 391 (4–6) (2004) 344–348.
[56] L.K. Randeniya, P.J. Martin, A. Bendavid, Detection of hydrogen using multi-walled carbon-nanotube yarns coated with nanocrystalline Pd and Pd/Pt layered structures, Carbon 50 (5) (2012) 1786–1792.
[57] D.R. Kauffman, A. Star, Graphene versus carbon nanotubes for chemical sensor and fuel cell applications, Analyst 135 (11) (2010) 2790–2797.
[58] P. Bondavalli, P. Legagneux, D. Pribat, Carbon nanotubes based transistors as gas sensors: state of the art and critical review, Sensors and Actuators B: Chemical 140 (1) (2009) 304–318.
[59] J.A. Robinson, E.S. Snow, SC. Bădescu, T.L. Reinecke, F.K. Perkins, Role of defects in single-walled carbon nanotube chemical sensors, Nano Letters 6 (8) (2006) 1747–1751.
[60] J. Andzelm, N. Govind, A. Maiti, Nanotube-based gas sensors – role of structural defects, Chemical Physics Letters 421 (1–3) (2006) 58–62.

[61] T. Yamada, Equivalent circuit model for carbon nanotube Schottky barrier: Influence of neutral polarized gas molecules, Applied Physics Letters 88 (8) (2006) 083106−83113.

[62] N. Peng, Q. Zhang, C.L. Chow, O.K. Tan, N. Marzari, Sensing mechanisms for carbon nanotube based NH3 gas detection, Nano Letters 9 (4) (2009) 1626−1630.

[63] J. Zhang, A. Boyd, A. Tselev, M. Paranjape, P. Barbara, Mechanism of NO_2 detection in carbon nanotube field effect transistor chemical sensors, Applied Physics Letters 88 (12) (2006), 123112−1-3.

[64] X. Liu, Z. Luo, S. Han, T. Tang, D. Zhang, C. Zhou, Band engineering of carbon nanotube field-effect transistors via selected area chemical gating, Applied Physics Letters 86 (24) (2005) 243501−243503.

[65] E. Mendoza, J. Rodriguez, Y. Li, Y.Q. Zhu, C.H.P. Poa, S.J. Henley, et al., Effect of the nanostructure and surface chemistry on the gas adsorption properties of macroscopic multiwalled carbon nanotube ropes, Carbon 45 (1) (2007) 83−88.

[66] F. Favier, E.C. Walter, M.P. Zach, T. Benter, R.M. Penner, Hydrogen sensors and switches from electrodeposited palladium mesowire arrays, Science 293 (5538) (2001) 2227−2231.

CHAPTER

Wet Spinning of CNT-based Fibers

6

Simon Jestin, Philippe Poulin

Centre de Recherche Paul Pascal - CNRS, University of Bordeaux, 115 Avenue Schweitzer, 33600 Pessac, France

CHAPTER OUTLINE

6.1 Introduction to Wet Spinning	167
6.2 Fibers Obtained from the Coagulation of Carbon Nanotubes	170
6.2.1 Direct coagulation in bad solvents	170
6.2.1.1 Fibers spun from CNT solutions in superacids	170
6.2.1.2 Fibers spun from CNT dispersions	171
6.2.2 Polymer-induced coagulation	173
6.2.2.1 Fiber formation and physicochemical mechanisms	173
6.2.2.2 Composition, structure and properties	177
6.3 Fibers Obtained from the Coagulation of CNT–Polymer Mixtures	185
6.3.1 High-performance synthetic polymers	186
6.3.1.1 PVA–CNT fibers	186
6.3.1.2 PAN–CNT fibers	190
6.3.1.3 PBO–CNT fibers	193
6.3.1.4 UHMW PE–CNT fibers	194
6.3.2 CNT–conductive polymer composite fibers	195
6.3.3 CNT–natural polymer composite fibers	198
6.4 Conclusions	201
References	202

6.1 INTRODUCTION TO WET SPINNING

Wet spinning is an alternative to other fiber manufacturing techniques and enables a wide variety of polymers to be spun. It has been used over the past 12 years to produce various types of composite fibers composed of polymers loaded with carbon nanotubes (CNTs) and even fibers solely composed of CNTs. As for other fiber materials, fibers obtained by wet-spinning approaches offer a significant chance to achieve materials with highly aligned CNTs that can be easily manipulated and used in a variety of textile, cable and composite applications. Recent developments of CNT-based wet-spun fibers are the topic of the present chapter. From a general point of view, wet-spinning processes are particularly appropriate for polymers that cannot be melted or for avoiding any degradation of the material due to heating. As sketched in Fig. 6.1, the generic process consists of injecting a concentrated solution of a dissolved macromolecular material through a spinneret immersed in a liquid bath, hence the name of wet spinning, in contrast to dry spinning or melt spinning. A derivative and associated method is called dry-jet wet spinning. In this case, the dope material is injected in air slightly above the coagulating medium. The air gap distance between the spinneret and the coagulation bath depends on the chemicals used. A solid filament is obtained by precipitation, coagulation or gelation of the dissolved macromolecular material. The solvent is then removed by evaporation or circulation in a liquid medium into which the solvent diffuses out from the fiber. These general features are met in most of wet-spinning technologies. Nevertheless, the details of each process depend of the formulations and the desired properties of a given fiber material.

Wet spinning was first developed for cellulose by Chardonnet at the end of the nineteenth century. His patent granted in 1884 led to the development of artificial silk (also known as viscose or rayon from 1924). In 1902, Charles, Edward and Clayton [1] improved this method to produce fibers via a chemically safer process. Extraction of cellulose is achieved from cotton or wood by means of caustic soda and carbon disulfide giving a highly viscous solution that can be extruded into fibers. This material was named viscose due to this initial solution state prior to spinning.

Other natural materials such as milk casein are used to produce wet-spun fibers [2]. Spider silk is another example of natural fibers obtained by wet spinning. A high-protein-concentration solution is extruded through the spider natural "spinnerets" and the combination of several complex biochemical reactions leads to the coagulation of the involved proteins into solid silk fiber.

Many other fibers have been produced by wet spinning and derived processes over the past decades, mainly from synthetic polymeric materials such as polyacrylonitrile (PAN), polyaramid, acrylic, ultrahigh molecular weight polyethylene (UHMW PE), polyvinyl alcohol (PVA) or more recently conductive polymers such

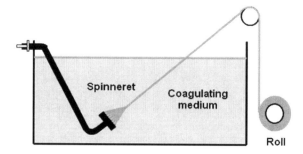

FIGURE 6.1

Schematic of a generic wet-spinning process for the production of polymer fibers. (For color version of this figure, the reader is referred to the online version of this book.)

as poly(3,4-ethylenedioxythiophene) (PEDOT)/poly(styrenesulfonic acid) (PSS) [3], polyaniline (PANi) [4,5] or polypyrrole (PPy) [6,7].

The general features of wet-spinning technologies and their extensions, such as dry-jet wet spinning, to modified processes are well documented in the literature [8,9] and beyond the scope of the present chapter, which is devoted to the case of CNT materials. Nevertheless, some of these features remain essential for the specific case of CNT fibers or composite wet-spun CNT—polymer fibers. In particular, for neat polymer solutions, it is necessary to achieve homogeneous and relatively concentrated dispersions in liquid media to obtain spinnable dopes. This is one of the first challenges met in the wet spinning of CNT-based fibers. The rate of fiber production is lower than that of melt spinning but remains viable for industrial applications. In addition, because wet-spinning technologies have been industrially developed for decades, it is likely a promising approach for future CNT-based fibers that would be produced on large scale at a reasonable cost. Indeed, knowledge already acquired for the wet spinning of neat polymers provides a valuable basis for the development of new fibers composed of CNTs.

Wet-spinning technologies are of great interest to produce composite fibers made of polymers and CNTs and also fibers solely composed of CNTs. In contrast to other processes, wet-spinning processes can be carried out with almost any type of CNT since the material is introduced in a liquid dispersion before being processed. The process is not limited by the production of a carpet or foams from which fibers are directly spun. In addition, the use of a liquid medium allows various treatments of the CNTs before fiber spinning. In particular, the CNTs can be purified or covalently functionalized before the production of fibers. The wet-spinning approach is versatile but has some downsides. Indeed, dispersing CNTs in aqueous or organic solvent often requires mechanical energy to disentangle and unbundle the nanotubes after their synthesis. Mechanical energy can be supplied, for example, by high-shear mixing, sonication or ball milling. These methods allow homogeneous and spinnable dispersions to be achieved. Nevertheless, the CNTs are shortened during the dispersion process [10,11]. The scission of nanotubes can be detrimental for the achievement of high mechanical or electrical properties.

The wet spinning of polymer—CNT fibers shares the advantages and drawbacks of wet-spinning technologies used for neat polymers. Wet spinning allows the use of polymers that cannot be melted or heated at high temperature, as required for melt spinning. This is particularly interesting for the development of composite fibers with fragile polymers such as conducting polymers or biopolymers. Fiber wet spinning displays another significant advantage over melt spinning methods with the possibility to achieve large fractions of CNTs in composite fibers. In melt spinning processes, the fraction of CNTs cannot be increased up to large levels because increases in viscosity that hinder the extrusion of polymer—CNT mixtures. However, the relative fraction of CNTs can be much greater when the CNTs and polymers are dispersed or dissolved in a common solvent. In this case, the use of a solvent allows the viscosity difficulties to be circumvented. Another specific feature of wet spinning is the possibility to spin fibers not only from isotropic liquids but also from liquid crystalline phases. This is a key feature for example for the production of high-strength polyaramid fibers, which are spun from liquid crystalline phases of rigid polymers. Similar concepts have been explored for CNTs, which can also form liquid crystalline phases at sufficiently high concentration. Fiber wet spinning allows a high degree of alignment and thereby superior mechanical properties to be achieved. Finally and as already indicated, even if the production rate of fiber wet spinning is slower than that of melt spinning, the technology remains viable for industrial production. It is therefore a promising field of research that can lead to new fibers with unique properties made by potentially scalable methods.

The specific and exciting opportunities of CNT-based wet-spun fibers are described below. The present chapter is subdivided into two main parts. In the first part, we consider fibers that are obtained by the coagulation or precipitation of CNTs initially dispersed in a given liquid and then injected into a bad solvent. This approach allows the formation of fibers that contain high fractions of CNTs and even fibers solely composed of CNTs. In the second part, we consider fibers spun from liquid mixtures of CNT and polymers. These methods are often directly derived from methods already used to spin fibers made of neat polymers. In this case, the fiber solidification results from the solidification of the polymer matrix. The fractions of CNTs can still be relatively high. Particularly, original fibers including conductive polymers or biopolymers can be used. The properties of the fibers strongly depend on the details of the process used to spin them and on the polymers used in composite fibers. Consequently, general comparisons of properties are not always relevant. This is why the properties of the achieved fibers are discussed in the chapter for each different type of produced fibers.

6.2 FIBERS OBTAINED FROM THE COAGULATION OF CARBON NANOTUBES

6.2.1 Direct coagulation in bad solvents

6.2.1.1 Fibers spun from CNT solutions in superacids

As indicated in the introduction, wet spinning requires starting materials that are homogeneous fluids. Unfortunately, CNTs are insoluble in most solvents. They generally have to be dispersed by supplying mechanical energy to disentangle and unbundle raw CNTs. In addition, the particles have to be stabilized against aggregation. This can be achieved by the addition of amphiphilic agents such as surfactants or polymers adsorbed at the interfaces of the CNTs. The adsorbed molecules provide steric or electrostatic repulsive interactions that prevent the CNTs from rebundling. It is also possible to covalently functionalize the CNTs and make them soluble in a given medium. Nevertheless, covalent functionalization can be quite tedious and can downgrade several properties of the CNTs. A particularly exciting alternative has been developed over the past years by M. Pasquali's group at the Rice University. This alternative is based on the discovery that CNTs are spontaneously soluble in superacids [12–15]. These solutions are achieved without supplying mechanical energy and are thermodynamically stable in contrast to surfactant-stabilized dispersions. The CNTs are undamaged during the dissolution process. At high concentration, CNTs dissolved in super acids exhibit liquid crystalline phases. CNT solutions in superacids can be injected in a solvent in which the CNTs are not soluble such as diethyl ether or water. The CNTs directly coagulate in such a bad solvent and form fibers that can be drawn and dried. The achieved fibers are solely composed of CNTs and exhibit very high electrical conductivity ($>10^5$ S/m). Nevertheless, because the CNTs used in the first experiments were slightly defective and relatively short, the mechanical properties of such fibers were still inferior to those of high-performance synthetic polymeric fibers available in the market. Nevertheless, wet spinning from superacids is probably one of the most promising approaches to achieve superfiber materials in the future by a scalable method. Indeed, the production of CNT fibers solely composed of CNTs dissolved without supplying mechanical energy is a significant advantage over other technologies. It is anticipated that the use of this method with high-quality and ultralong CNTs could lead to new fibers with extraordinary mechanical and electrical properties. A full and comprehensive description of fibers spun from CNT solutions in superacids is given in [13,15] and is not detailed in the present chapter. A recent paper validating the above expectations has appeared during the preparation of the present chapter [16]. The reader is refered to this paper for the latest advances concerning the wet-spinning of nanotube fibers from solutions in super acids.

6.2.1.2 Fibers spun from CNT dispersions

Also with the aim of developing fibers solely composed of CNTs, Zhang et al. [17] in A. Windle's group at the Cambridge University have achieved fibers from multi-walled carbon nanotubes (MWNTs) and nitrogen-doped multiwalled carbon nanotubes (N-MWNTs). The CNTs are not dissolved in thermodynamically stable forms but are instead dispersed by sonication in ethylene glycol. The obtained suspensions exhibit a sufficient kinetic metastability to be spun into long continuous fibers made of highly aligned CNTs. High alignment results not only from the spinning process but also from the fact that the suspensions are in liquid crystalline states. The fibers are obtained by injecting the dispersions in a coagulating bath of diethyl ether. Ethylene glycol rapidly diffuses out the extruded CNT fiber into the ether, and ether is back-diffusing inside the fiber. Fibers can be heated to fully remove the ether by evaporation. The MWNT and N-MWNT fibers exhibit high Young's modulus of 69 and 142 GPa, respectively, on an average. These values are still below the Young's modulus of individual CNTs but compare already well with the Young's modulus of high-performance synthetic polymer fibers such as polyaramid, PVA or UHMW PE fibers. However, the tensile strength of the achieved fibers is still relatively low, on an average 150 and 170 MPa, respectively, for MWNT and N-MWNT fibers. This can be understood by the fact that the nanotubes are not bound to each other. As a result, creep can occur and the fibers can break even for small strain deformation. It is interesting to note that N-MWNT fibers exhibit better properties than MWNT fibers. This difference is ascribed to the less-defective structure of N-MWNTs compared to MWNTs. The electrical conductivity of neat CNT fibers is particularly high, on the order of 10^4 S/m. This value does not yet compare with metallic fibers but is already well sufficient for a variety of functional textiles with antistatic, heating and sensing capabilities.

The achievement of homogeneous dispersions suitable for fiber spinning is quite challenging. Homogeneity can be assisted by the use of surfactants. Steinmetz et al. [18], for example, used dispersions of single-walled nanotubes (SWNTs) stabilized in water by the addition of sodium dodecyl sulfate (SDS), a surfactant commonly used to stabilize CNTs in water. The dispersions are homogenized by sonication. The dispersions are injected into the cylindrical coflowing stream of a coagulating mixture composed either of ethanol/glycerol or ethanol/glycol. Fibers are washed in water and ethanol to remove SDS and coagulating agents and obtain thereby fiber solely composed of CNTs. The achieved fibers are mechanically weak, presumably because of the presence of impurities in the used CNT materials. But the fibers exhibited good electrical properties with conductivity on the order of 10^2 S/m. Kozlov et al. [19] also achieved polymer-free fibers using a wet-spinning approach. The authors used an acid-induced flocculation of surfactant-stabilized SWNT dispersions. Flocculation occurs near the solution spinning point in the acidic bath and induces the formation of a gel fiber. The fibers are washed in methanol to remove all the hydrochloric acid. Both hollow and solid fibers are obtained by varying the injection rates of the CNT dispersion in the acidic bath. The pure SWNT hollow fibers exhibit the following density-normalized mechanical properties: a Young's modulus of 12 GPa/g/cm^3, a specific strength of 65 MPa/g/cm^3 and a strain to failure of 1%. Electrical conductivity can exceed 10^4 S/m for fibers annealed at 1000 °C under argon, and the electrochemical capacitances were found to lie between 50 and 100 F/g, which is greater than the capacitance of common "buckypapers", which are mats of randomly orientated CNTs.

Several important conclusions can be drawn from the above contributions. The production of fibers by the direct coagulation of the CNTs in a nonsolvent medium leads to fibers solely composed of nanotubes. As a consequence, these fibers exhibit high electrical conductivity. Unfortunately, because the CNTs have to be dispersed, and often shortened during the dispersion process, they tend to be too short for the achievement of excellent mechanical properties. Indeed, the entanglement of long nanotubes should promote the transfer of mechanical stress between the CNTs and lead thereby to stronger fibers. The challenge in the field for the next years will consist in optimizing dispersions and solutions with straight and long nanotubes. Typical lengths above 1 or 2 μm should lead to substantial improvements of properties. Dispersion in superacids is a spontaneous process and therefore one of the most promising approaches to achieve solutions of long nanotubes. Nevertheless, dispersions in other solvents remain interesting, in particular when water is used as solvent. Water-based dispersions stabilized by surfactants or polymers may still be improved in the future and have the advantage of being more easily processable and more environment-friendly than dispersions in other solvents.

6.2.2 Polymer-induced coagulation

Fibers formed by the precipitation in a nonsolvent medium have been presented in the previous section. In the present one, we discuss the formation of fibers that result from the adsorption of polymer chains at the surface of the CNTs. The production of such fibers is achieved by injecting a nanotube dispersion into a polymer solution. Polymer bridging leads to a strong adhesion of the CNTs, which become bound by the polymer chains. The CNT assembly, which does not only involve intertube van der Waals interactions as in the previous case, exhibits improved mechanical properties because of the polymer-induced adhesion of the CNTs. More details on the manufacturing of such fibers and on their properties are given below.

6.2.2.1 Fiber formation and physicochemical mechanisms

In 2000, Vigolo et al. [20] proposed the first semicontinuous wet-spinning process to obtain CNT fibers. This method was called particle coagulation spinning process. The method consists in dispersing nanotubes in water using SDS as surfactant and tip sonication for homogenization. The composition of these dispersions was optimized, and a maximum content of 0.35 wt% SWNT with 0.9 wt% SDS has been used in the first experiments. Homogeneity of the dispersion is the key factor to achieve good fiber spinning. As shown in Fig. 6.2, the dispersion is injected in the coflowing stream of a rotating bath that contains 5 wt% of PVA. PVA, due to its amphiphilic character, adsorbs at the CNT interface and displaces the surfactant molecules. Polymer chains adsorbed onto two or more CNTs create molecular bridges. The resultant coagulation process is called bridging flocculation in the field

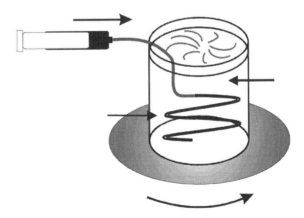

FIGURE 6.2

Schematic of a laboratory-scale experimental setup used to make nanotube fibers by wet spinning in a coflow configuration. The spinning dope is injected via a dosing pump or syringe pump into a rotating bath that contains the coagulating medium. The formed prefiber can then be extracted from the bath to be washed dried and thermally or mechanically treated [20]. (For color version of this figure, the reader is referred to the online version of this book.)

of colloids [21]. This process does not involve any chemical reaction; no covalent bonds are formed or disrupted during the process. It results in the formation of a prefiber, which can be viewed as an interconnected network of nanotubes bound by PVA chains.

The process leads to flow-induced alignment of the nanotubes along the fiber axis. The prefibers can be removed from the coagulating bath and washed with water to remove excess of PVA and SDS. During drying, capillary forces, water evaporation and drainage cause the prefiber to collapse into a dry and thin cylindrical or ribbonlike fiber. By varying spinning parameters, it is possible to obtain fibers with thicknesses that vary from a few microns to 100 µm with a density of about 1.3 g/cm^3 and a length of several tens of centimeters. M.H. Jee et al. have recently investigated [23] the effect of the wet-spinning conditions on the structural, mechanical and electrical properties of wet-spun PVA–CNT fibers spun in coaxial flows. They have clearly demonstrated that nanotube alignment and physical properties are improved by the shear stress of the wet-spinning process. As shown further in the present chapter, postsynthesis treatment such as hot-drawing can also have a strong influence on the structure and properties of PVA–CNT fibers.

Nanotube dispersions that have been used for this process have been stabilized not only by SDS but also by lithium dodecyl sulfate [24] or even by amphiphilic polymers such as denatured DNA [25]. The dispersant used has to be displaced by the PVA during fiber spinning, so that the bridging coagulation mechanism can take place. This means that the adsorption energy of the dispersant at the CNT interface must be lower than that of the PVA chains. It is worth noting that this wet-spinning process can also be achieved from homogenous solutions of CNTs without any dispersant. For example carboxylated CNTs are water soluble and can be directly spun via the PCS process. Néri et al. realized dispersions of oxidized MWNTs in basic aqueous solutions [26]. The oxidized CNTs are soluble because of the presence of ionized carboxylic groups at their surface in a basic medium. The dispersion is injected in the coflowing stream of an acidified PVA (pH < 3) solution to neutralize the carboxylic groups and achieve coagulation. The achieved fibers exhibit properties that compare well with the properties of fibers spun from surfactant-stabilized dispersions. Pénicaud et al. [27] proposed a fiber spinning process of composite CNT/PVA fibers without the use of any surfactant. They used dimethyl sulfoxide (DMSO) solutions of CNTs reduced with alkali metals [28]. Such nanotubes are spontaneously soluble in polar organic solvents. A great advantage of this approach is the possibility to spin fibers with CNTs that have not been sonicated and therefore not damaged or shortened during the dispersion process. The same advantage is found for fibers obtained from solutions in superacids as described in the above section.

Another example of wet-spun CNT fiber using polymer-induced coagulation was proposed by Muñoz et al. [29]. The authors used polyethyleneimine (PEI) instead of PVA as the coagulating agent. The process is similar to the process developed for composite CNT–PVA fibers. CNTs are dispersed in an aqueous surfactant solution, which is injected into the coflowing stream of a mixture of 40 wt% PEI–60 wt%

methanol rotating bath. The authors obtained thin fibers with a skin-core structure and a diameter ranging from about 15 to 50 μm. The resulting fibers contained only 25 wt% PEI. Their density-normalized modulus, tensile strength and strain to failure are, respectively, 6 GPa cm^3/g, 115 MPa cm^3/g and 3%. More interestingly, the PEI–CNT fibers exhibited high electrical conductivity above 2×10^4 S/m. This high conductivity compared to raw PVA–CNT fibers is presumably due to the low content of PEI in the fibers. This lower content reflects weaker interactions between CNTs and PEI than between CNTs and PVA. The strong interactions of PVA and CNTs make polymer-induced coagulation particularly efficient. This is an advantage for the robustness of the process. Nevertheless, it also generally leads to large fractions of PVA within the fibers. Posttreatments such as extensive washing or thermal degradation are necessary to remove fully or partially the PVA.

Manufacturing of fibers in the coflowing stream of a rotating bath can be easily achieved on the laboratory scale. It allows tests of various chemical compositions and types of fibers to be screened with small amounts of materials in a short time. But the length of as-produced fibers does not exceed 1 m typically. For most future applications of CNT-based fibers, it is clear that a continuous production of indefinitely long fibers would be more suitable. Fortunately, coflowing streams can be achieved via many ways. The simplest method consists in using coaxial pipes [23,30–33]. As sketched in Fig. 6.3, the dope is injected via a spinneret into a pipe into which a coagulating medium is circulating [32,33]. A third fluid can even be used to delay the coagulation along the line and avoid clogging of the spinneret [22,33]. Dalton et al. and Razal et al., for example [24,30,31], used a cylindrical glass pipe in which a PVA solution is flowing. The schematic of their setup is shown in Fig. 6.4. The fibers obtained are several meters long and are collected on a rotating mandrel immersed in a washing bath. This method cannot yet be considered as a true continuous process because the collection of prefibers on the rotating mandrel adds an intermediate stage before the achievement of the dried final fibers. Indeed, after

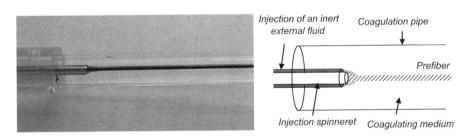

FIGURE 6.3

Wet spinning in coflow conditions can also be achieved in coaxial pipes. The dope is injected in the stream of a coagulating medium that flows in a larger pipe. As schematically shown, a third inert fluid can be injected near the spinneret to delay the coagulation slightly further along the spinning line and thereby avoiding clogging at the tip of the spinneret [22]. (For color version of this figure, the reader is referred to the online version of this book.)

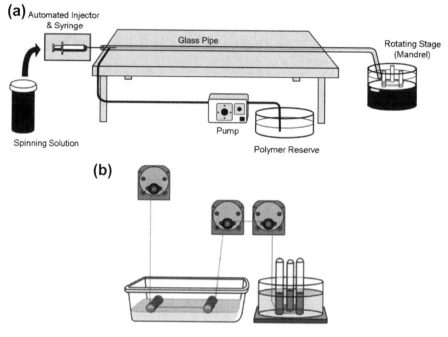

FIGURE 6.4

Details of the continuous spinning system, showing schematic representations of (a) the spinning apparatus and (b) the apparatus for postspinning solvent treatment and draw [31].

having been collected onto the mandrel, the fibers have still to be washed and dried by unwinding them from the mandrel.

From a general point of view, the robustness, efficiency and production rate of the process depends on the coagulation kinetics of the fibers. This is also true for any wet-spinning process. Wet-spun fibers are produced from a liquid dope that is injected in a particular media into which the fiber material solidifies. The mechanisms of solidification can strongly differ as a function of the considered chemicals and types of fibers. Nevertheless, regardless the exact mechanisms, the solidification has to be fast enough so that fibers can be drawn and treated at a reasonable rate. Studying the kinetics of solidification is not straightforward but Mercader et al. [32] recently proposed an experimental methodology to evaluate the kinetics of fiber solidification. The method consists in flowing forming fibers into a cylindrical pipe. The fibers progressively solidify as they circulate along the pipe. At some point, a diameter constriction of the pipe induces an elongational flow field in the flowing direction. Drag forces experienced by the fiber can lead to their scission if they are not yet sufficiently solidified. By contrast, the fiber can pass through the constriction without breaking if it is sufficiently strong. Varying the location of the diameter constriction therefore allows the strength of the fiber to be tested for different resident times of the fiber in the coagulating medium. This offers a simple way to evaluate the kinetics of solidification of the fiber. The method was used to study the coagulation of composite PVA–CNT fibers [32]. It was shown that the fiber solidifies more quickly with polymers of high molecular weight. The better efficiency of large macromolecules at quickly solidifying CNT fibers is ascribed to the greater number of physical bonds formed between a polymer chain and a CNT. Polymers of smaller molecular weight diffuse more quickly but do not create strong bonding between the CNTs. As a consequence, the fiber solidifies more slowly in the presence of short polymers. The methodology proposed by Mercader et al. could be used in the future to optimize several other factors such as addition of cross-linkers, salts, temperature, etc. It could also be used for the development and optimization of other types of wet-spun fibers.

6.2.2.2 Composition, structure and properties
6.2.2.2.1 From composite to neat CNT fibers

The composition, structure and properties of PVA–CNT fibers achieved by polymer-induced coagulation have been investigated in depth over the past years. Typically, raw CNT–PVA fibers contain a weight fraction of CNT of about 20%. Of course the exact amount of CNT depends on the nature of the CNT and on the details of the process. Washing the fibers allows the fraction of CNT to be raised above 50%. But the polymer-induced coagulation process does not directly lead to fibers solely composed of CNTs as achieved via other processes described in previous sections and other chapters of the present book. Nevertheless, PVA–CNT fibers can be thermally treated at high temperature in inert atmosphere to achieve neat CNT fibers. Indeed, above 400 °C, the PVA is almost entirely degraded leaving voids within fibers, which remain solely composed of CNTs. Without further treatment, baked fibers become brittle because the CNTs are only held together by van der Waals forces and entanglements. They are not bonded anymore by PVA chains. Nevertheless, the resultant baked fibers are porous and, more importantly, highly conductive as other fibers solely composed of CNTs. They exhibit thereby a great potential for applications in which porosity and conductivity are sought after. These applications include microelectrodes, electrochemical biosensors, microelectromechanical actuators and supercapacitors. Such applications are briefly described at the end of this section, after the description of the structure and properties of PVA–CNT composite fibers.

6.2.2.2.2 Structure and alignment of CNTs

PVA–CNT fibers produced by polymer-induced coagulation exhibit a hierarchical structure as evidenced in a study by Neimark et al. [34]. This hierarchical structure is reminiscent of the structure of several biomaterials and can perhaps explain some of their surprising properties presented below. The outer surface of the fiber is composed (1) of aligned elementary filaments, each of about 0.2–2 μm diameter built of packed CNT bundles and (2) of a core filled by a nanofelt of SWNT bundles. After thermal degradation of the PVA and based on nitrogen vapor sorption/desorption measurements, these fibers became lightweight and highly porous materials with a density of 0.2 cm^3/g with an average pore diameter of 8 nm.

Alignment of the CNTs within the fibers has been characterized by X-ray diffraction [35–37] and polarized Raman spectroscopy [38]. Both methods confirmed the alignment of the CNT along the main fiber axis. Nevertheless, the degree of alignment is relatively poor for raw fibers. For example, assuming that the distribution of orientations follows a Gaussian distribution, the full width at half maximum of the distribution for raw fibers often exceeds 70°. This is why postsynthesis treatments are needed to increase the alignment of the CNTs. These treatments can in particular include drawing of fibers [39] at room temperature in wet states [36,40] or hot-drawing of dried fibers [40,41]. The latter can be more easily implemented in a continuous production line and actually exists in several industrial processes of fiber manufacturing. Hot-drawing treatments have been shown to be particularly efficient at improving the degree of alignment of CNTs along the axis of wet-spun fibers. The same holds of course for composite fibers made by melt spinning approaches. But these other types of fibers usually contain a small fraction of nanotubes and are out of the scope of the present chapter.

6.2.2.2.3 Mechanical properties

The mechanical properties of PVA−CNT composites fibers depend largely on the treatments and composition of the fibers. The first fibers produced in 2000 by Vigolo et al. exhibit a Young's modulus between 9 and 15 GPa, a strain to failure up to 3% and a tensile strength of about 150 MPa. The properties are relatively poor in comparison to the properties of individual nanotubes and even in comparison with the properties of high-performance commercial polymer fibers. It is believed that the weakness of the first PVA−CNT fibers is partially due to the type of CNTs used at that time. Indeed, Vigolo et al. used CNTs produced by the electric arc method. This method is known to produce CNTs of high quality with few structural defects; but the CNTs are produced along with a significant fraction of impurities, which include particles of amorphous carbon. These particles present in large amounts weaken the fibers. In 2003, A. Dalton et al. in R. Baughman's group achieved "supertough" fibers by using CNTs produced by the so-called HiPco process [24]. The main impurities formed in this synthesis process are iron particles, which can be easily removed in acid solutions. As a result, purified samples contain a very small fraction of impurities. This improvement is reflected in the significant increase of the mechanical properties. The fibers are produced in a coaxial coflowing CNT dispersion/PVA solution system, from SWNTs stabilized using lithium dodeycl sulfate as surfactant, and are washed in an acetone bath before being dried. These fibers contain a high weight fraction of CNTs of 60%, for only 40% of PVA. As shown in the stress vs strain curves of Fig. 6.5, the fibers exhibit high mechanical performances: a tensile strength of about 1.8 GPa and a strain to failure of about 100%. The most surprising property is the energy needed to break the fiber, sometimes referred to as "toughness". Indeed, the energy to failure of PVA−CNT fibers is

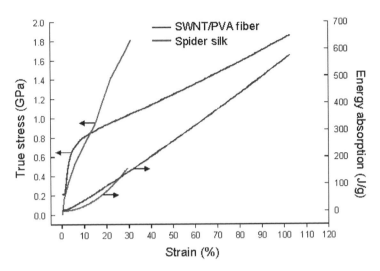

FIGURE 6.5

Stress and energy absorption vs strain for high-toughness spider silk and a nanotube composite fiber. The energy to break during the 104% elongation of nanotube composite fiber was 570 J/g, vs a total energy to break of 165 J/g for the spider silk fiber [24]. (For color version of this figure, the reader is referred to the online version of this book.)

near 600 J/g. The authors call such fibers "supertough fibers" because their energy to failure exceeds by far the energy to failure of polyaramid fibers used in bulletproof vests and even the energy to failure of spider silk, which was considered before this report as the toughest material on earth.

In 2005, Miaudet et al. [41] obtained even higher toughness values, up to 870 J/g with a strain to failure up to 430%. In this case, the spinning solution is composed of 0.35% SWNT and 1% SDS in water. The fibers are spun in a rotating bath. Similar fibers have been made with spinning solutions containing 0.9% MWNT and 1.2% SDS in water. These first CNT–PVA fibers made of MWNTs exhibit a strain to failure and a high toughness that compares well with the properties of fibers made of SWNTs. Although these energies to failure are very high, the interest of such fibers for energy adsorption application is not obvious. For example, bulletproof vests applications require energy absorption at a low strain. In order to improve the stiffness of the fibers and therefore their capability to absorb energy at low strain, Miaudet et al. [41] carried out hot-drawing treatments of PVA/CNT composite fibers. The SWNT- and MWNT-based fibers are hot drawn by 850% at 180 °C (a temperature between the glass transition temperature ~80 °C and melting temperature ~220 °C of PVA). These thermomechanical treatments strongly improve the PVA crystallinity and the alignment of the CNTs and polymer chains. Hot-stretched fibers exhibit a Young's modulus up to 45 GPa and 35 GPa, and a tensile strength up to 1.8 and 1.4 GPa for SWNT and MWNT, respectively, with a strain to failure reduced to 10%. The energy to failure becomes reduced as a consequence of the decrease of the strain to failure. But such fibers still exhibit a large toughness, on the order of 60 J/g, that exceeds that of polyaramid fibers. It was also shown that the properties of the fibers could be modulated by varying the temperature and draw ratio.

6.2.2.2.4 Thermomechanical properties and shape memory effects

Among actively moving materials, CNTs can be used as fillers of shape memory polymers (SMPs). SMPs have applications in packaging, biomedical devices, heat shrink tubing, deployable structures, microdevices, etc. SMPs are usually deformed at high temperature (T_d) and then cooled down under fixed strain to trap the deformed polymer chains, thus storing mechanical energy. Upon reheating, typically in the vicinity of the glass transition temperature (T_g), the polymer chains become mobile and the material can relax by reverting toward its original and more stable shape. The efficiency of SMPs is controlled by the composition of the polymer, as defined by its chemical structure, molecular weight, degree of cross-linking and fraction of amorphous and crystalline domains [42–49]. The energy that is restored on shape recovery is a growing function of the energy supplied during the deformation at high temperature. SMPs can exhibit large strain when they revert to their initial shape. Unfortunately, this large strain is usually associated with a low stress recovery from a few tenths of megapascals to a few tens of megaPascals. Consequently, the energy density, which results from a combination of stress and strain, is rather low. Combining large stress and large strain recovery as well as finding more controlled programming procedures remain critical challenges for the development of smarter and stronger shape memory materials. CNT fibers exhibit several properties potentially useful for such objectives. Indeed, as seen in the above section, CNT fibers can absorb a large amount of mechanical energy, which is a necessary, but not yet sufficient, condition to achieve large energy densities in shape memory phenomena. CNT improves the stiffness of SMPs, which is critical for large stress recovery. Their electrical conductivity is of particular interest to engineering materials, which can be heated via Joule heating and directly stimulated by an electrical current. Nevertheless, the manifestation of these properties is here again expected to strongly depend on the fraction of nanotubes and on their ordering. This is why wet-spun CNT fibers are of particular interest in the field. They indeed allow large fractions of aligned CNTs to be embedded in a polymer. These advantages have been validated in the recent years with PVA–CNT composite fibers [40,50]. A qualitative demonstration of the shape memory effect of PVA–CNT fibers is shown in Fig. 6.6. This observation confirms that, in contrast to metallic alloys, CNT–PVA fibers can exhibit a large recovery strain, as most SMPs.

Figure 6.7 shows more quantitatively the stress needed to stretch CNT–PVA composite fibers up to 800% at different temperatures. A greater stress is needed to deform the fibers at low T_d. At higher T_d, the fibers become softer and can be more easily deformed. This softness is associated with a lower supply of mechanical energy. This can be estimated in Fig. 6.7, where the area under each curve

6.2 Fibers Obtained from the Coagulation of Carbon Nanotubes

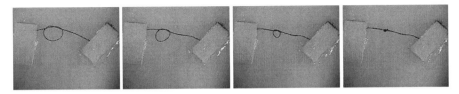

FIGURE 6.6

Qualitative evidence of the shape memory behavior of a CNT–PVA composite fiber. The fiber shown has been stretched at 150 °C. It has then been cooled down to room temperature under tensile load. A knot was made with the stretched fibers. The fiber shrinks and the knot tightens when the fiber is reheated. Reheating is here simply achieved by blowing hot air toward the fibers. The series of pictures shows the fiber shrinking as a function of time. The time interval between each picture is 3 s [40]. (For color version of this figure, the reader is referred to the online version of this book.)

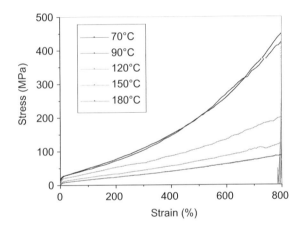

FIGURE 6.7

Stress vs strain curves of nanotube composite fibers. The fibers are stretched up to 800% at different temperatures (T_d) [50]. (For color version of this figure, the reader is referred to the online version of this book.)

corresponds to the energy supplied to the fibers at different T_d, from 70 °C to 180 °C, and on mechanical stretching.

As shown in Fig. 6.8, when reheated at fixed strain, the fibers generate a strong stress with a maximum at a well-defined temperature (T_s). The occurrence of a peak recovery stress in conditions of fixed strain has already been observed for other SMPs and nanocomposites [46,51], but with no direct link between T_s and T_d. In conventional materials, the peak recovery stress occurs in the vicinity of the glass transition of the neat polymer. When the materials are initially deformed above

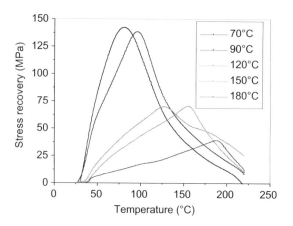

FIGURE 6.8

Stress generated by a nanocomposite fiber when it is reheated. The strain is kept fixed and the temperature is increased from room temperature to 230 °C at a rate of 5 °C per minute. The different colors correspond to the temperatures T_d at which the fibers have been initially deformed. A peak is observed in each case for a temperature T_s roughly equal to T_d [50]. (For color version of this figure, the reader is referred to the online version of this book.)

the glass transition temperature, the peak often disappears and the stress generated on shape recovery substantially decreases. This is due to the fact that polymer chains can relax when deformed at temperatures well above T_g, thus decreasing the potential for stored mechanical energy.

Here the peak is preserved well above the T_g of the neat PVA. Strikingly, T_s and T_d are actually roughly equal. This near equality means that the fibers memorize the temperature at which they have been deformed. The peak of stress generated can be observed up to 180 °C, which is about 100 °C above the T_g of the neat PVA. This distinctive feature provides an opportunity to rationally control T_s without varying the chemical structure of the material. In addition, it is observed that the maximum stress generated by the fiber is close to 150 MPa. This value is from one to two orders of magnitude greater than the stress generated by conventional SMPs. It is obtained for fibers that have been deformed at 70 °C and 90 °C; temperatures that are in the vicinity of the T_g of the neat PVA. They correspond to the conditions for which the greatest energy is supplied during initial deformation. Additionally, because CNT fibers are electrically conductive, thermal shape memory effects can be triggered by Joule heating when an electrical current is passed through the fiber. This can be useful for the direct use in microdevices where heating via an external source can be difficult. CNTs substantially alter the properties of the composite fibers in several ways. First, they act as reinforcements with an increase of one order of magnitude of the storage modulus [50]. Second and as already reported [41,52], they favor the stabilization of crystalline domains. This can contribute to the locking of mechanical constraints. Therefore, CNTs by increasing the stiffness of the polymer can allow more energy to be absorbed and restored. This can explain the very large stress measured in the present experiments. However, the origin of the temperature memory is not yet very clear. It could likely be arising from a broad glass transition with the contribution of confined polymers at the interface of nanotubes or crystalline domains. It has been shown that significant gradients of T_g can develop at the interface of nanoparticles [51,53]. Amorphous polymer shells around the CNTs or around crystalline domains largely overlap and percolate, such as the CNTs themselves, meaning that there is a distribution of polymer−CNT or amorphous polymer−crystallites distances that ranges from molecular contact to several nanometers. This distribution of confinement results in a wide broadening of the relaxation time spectrum and specifically the glass transition through a distribution of polymer fractions that exhibit different T_g. This property could be responsible for peaks of stress recovery well above the T_g of the neat polymer. Indeed, when the material is stretched at T_d, the polymer fractions that have lower glass transition temperatures (far from the interface) can quickly relax and do not efficiently participate in the storage of mechanical energy. In contrast, polymer fractions with glass transition temperatures close to T_d dominate the behavior by storing and restoring mechanical energy. We also note that composites treated in the vicinity of T_g still exhibit higher toughness and generate greater stress recovery. This indicates that the fractions of amorphous polymer with unshifted or slightly shifted glass transition temperatures remain the major components of the composite. This scenario is still speculative and further research is needed to clarify the microscopic origin of the temperature memory. In particular, temperature memory should take place in other nanocomposites and even in neat polymers since they exhibit a broad glass transition. This topic is currently a particularly active field of research, and some of the above expectations have been beautifully demonstrated in A. Lendlein [54−56] and X. Tie's groups [57,58]. The temperature memory has indeed been observed in other polymers including biocompatible materials and extended to multishape memory effects.

6.2.2.2.5 Electrochemical and electromechanical properties

As indicated above, CNT-PVA fibers or other wet-spun fibers that contain a large fraction of CNTs can be used to make fibers solely composed of CNTs. This can be achieved by thermal treatments that induce the degradation of the polymer. The resultant fibers are porous and exhibit a large electrical conductivity. These are ideal features for electrochemical and electromechanical applications. CNT fibers offer therefore a unique opportunity to develop microelectrodes with enhanced activity potentially useful in a variety of applications such as biosensors, biofuel cells and electromechanical actuators. Actually, some of these applications have been validated over the past years. In particular, it was shown that CNT fibers made from baked PVA−CNT fibers can be used as efficient microelectrodes for biosensing applications with high spatiotemporal resolution and improved selectivity [59,60]. Such microelectrodes have already been tested for the sensing of glucose

[60] and dopamine, for example [59,61]. Because of their nanoporosity, CNT fibers exhibit an excellent potential to absorb redox mediators and various biocatalysts. This particular feature, compared to more conventional electrodes, make CNT-based fibers particularly appealing for the development of novel microelectrodes for biofuel cell applications. In such applications, redox reactions are catalyzed by enzymes adsorbed at the interfaces of the electrodes [62]. Gao et al. [63] have recently compared the efficiency of microelectrodes made from carbon fibers or from porous CNT fibers. The current densities, power generation and stability are found to be significantly improved by using CNT fibers. Power densities up to 740 µW/cm^2 are achieved, making such biofuel cells among the most efficient ones ever achieved. It is believed that these improvements are due to the efficient adsorption of enzymes at the surface of the CNTs. Actually, it is even possible that because of their nanodiameter CNTs can almost literally wire the active core of the enzyme and thereby promote better electron transfers. While a lot still needs to be done to clarify the origin of such improvements, it is clear that CNT-based fibers offer a great potential in the field of microelectrodes.

Last, CNTs can also be used for electromechanical actuators. Their charge density in a given electrolyte can be varied by applying a low voltage with respect to a counter electrode. A double layer forms at the nanotube interface and the material expands or contracts in response to quantum mechanical and electrostatic effects [64−75]. CNTs operate at low voltage and can generate a large stress because of their stiffness. The first macroscopic manifestation of this phenomenon was reported by Baughman et al. in 1999 [64]. These first actuators were made of the so-called "buckypapers", which are mats of randomly orientated CNTs. These first actuators could generate a stress of about 0.75 MPa when stimulated in an aqueous electrolyte. This stress is far from the potential theoretically expected for an individual nanotube but exceeds the stress generated by biological muscles [64]. Several groups have theoretically and experimentally investigated the involved mechanisms and attempted to improve the performances of such nanotube-based actuators [65−75]. Nevertheless, optimization of nanotube structures for such applications remains challenging. Indeed, an optimal actuator has to combine high mechanical strength, good electrical conductivity and a large surface specific area to maximize the interface exposed toward the electrolyte. In addition, alignment of the nanotubes is expected to be critical since it could promote macroscopic dimensional changes along a given direction. This is why CNT fibers again could exhibit several advantages over random assemblies of CNTs for applications as electromechanical actuators. This was actually confirmed with the measurements of generated stress that exceeds 10 MPa [40], which is an order of magnitude greater than the stress achieved with mats of randomly assembled CNTs. Nevertheless, in spite of these promising improvements, further research is still necessary to make this new technology viable. Indeed, CNT-based actuators made from buckypapers and also from CNT fibers often exhibit creep because the nanotubes are not bound together. The structures are just held by van der Waals forces and entanglements. The lifetime of such structures after repeated deformations is rather low. It is therefore critical to develop materials that still combine porosity and electrical conductivity but with improved interactions between the CNTs so that the actuators can sustain repeated cycles of deformation without degradation of properties.

In conclusion, it can be stressed that the polymer-induced coagulation of CNTs has allowed over the past years the achievement of new fibers with particularly exciting properties. These exciting properties include a giant toughness, a great potential for novel microelectrodes or shape memory effects with unprecedented energy densities. Nevertheless, if these exciting properties are of great fundamental interest and raised a number of scientific questions, it is far from obvious that such fibers will be available in the market in the near future. Indeed, those fibers are wet spun in coflowing conditions. The scale-up of coflow experiments achieved in the laboratory is not obvious in particular if multifilament spinning is considered. Monofilament production for high-value applications such as biomedical applications (biofuel cells and biosensors) would likely be useful and commercially viable in niche markets. However, it is difficult to anticipate applications of textiles or composites with giant toughness or unusual shape memory phenomena considering the large amount of fibers needed in such fields. In fact, other processes using static coagulating baths, as commonly employed in fiber wet-spinning industrial technologies, would be preferable for these large-scale applications.

6.3 FIBERS OBTAINED FROM THE COAGULATION OF CNT–POLYMER MIXTURES

A general and more direct approach to produce polymer–CNT fibers consists simply in adding CNTs to polymer solutions that are spinnable in static coagulation baths, as commonly achieved for several industrial polymer fibers. This simple concept can be easily achieved as long as the fraction of CNTs in the fibers remains low, typically below 10 wt%. In this case, the CNTs play essentially the role of an additive that does not strongly affect the wet-spinning process. Unfortunately, a low fraction of CNT is a limitation to significant changes of the fiber properties. Nevertheless, even at low weight fraction of CNTs, the approach can still be useful to provide electrical conductivity or improvements of mechanical properties in polymers that are difficult to be processed via melt spinning. These polymers include in particular some conducting polymers, high-molecular-weight polymers and biopolymers.

Developments of fibers that contain greater weight fractions of CNTs, typically above 10 wt%, require more innovative processes. Major challenges are faced in this case because it is necessary to disperse a large fraction of CNTs in a polymer solution. The dispersion has to be sufficiently homogeneous so that it remains spinnable. Even though more challenging, research in the field has been recently carried out and has successfully led to scalable processes reminiscent of industrial wet-spinning processes that still allow large fractions of nanotubes to be embedded in polymer fibers. Examples of recent achievements, with high or low weight fractions of CNTs, are given in the present section. Wet-spinning processes are often characteristic of a given type of polymer. The chemical and flow conditions differ depending on the considered material. The fiber properties and targeted applications also differ strongly depending on the type of polymer. The present section is therefore subdivided into sections that specifically deal with different classes of polymers.

6.3.1 High-performance synthetic polymers

Several synthetic polymers are used to develop high-performance fibers. These polymers include, but are not limited to, UHMW PE, polyaramid, PVA, poly(p-phenylene-2,6-benzobisoxazole) (PBO) and PAN, which can be further converted into carbon fibers. These polymers can be wet spun to form fibers that exhibit excellent mechanical properties. Several investigations have been carried out over the past years to improve further the properties of these fibers by including CNTs.

6.3.1.1 PVA—CNT fibers
6.3.1.1.1 CNT—PVA fibers at low weight fractions of CNTs

PVA is among the main polymers that are industrially spun via wet- or gel-spinning approaches [76]. This polymer combines good mechanical properties and chemical stability and is therefore used in a large variety of applications [76]. In addition, PVA can be spun from aqueous solutions; this is a significant advantage in terms of production cost and environmental impact. It is known that PVA induces the aggregation of nanotubes in aqueous solutions [20,32]. The aggregation of the nanotubes in the presence of PVA results from the adsorption of polymer chains onto two or more CNTs, this mechanism being known as bridging coagulation. PVA indeed interacts strongly with the nanotubes to displace weakly adsorbed surfactant molecules. This phenomenon was previously used on purpose to spin PVA—CNT fibers with large fractions of CNTs in coflowing configurations as described in the previous section [20,32]. The injection of solutions in coflowing configurations can be easily achieved on a laboratory-scale using devices with rotating baths or coaxial pipes. Unfortunately, controlling coflowing streams on large scale and in multifilaments configurations, as needed for most applications, remains particularly challenging. Fiber spinning in static media would be more suitable. Indeed, coagulation of a polymer in a static bath would be reminiscent of conventional industrial wet-spinning processes and therefore easily scalable. Successful efforts to directly spin fibers from a mixture of PVA and CNTs in static coagulating baths have been recently reported in the literature.

Zhang and collaborators [77] studied the gel-spinning of PVA—SWNT composites fibers. The nanotubes and the polymer are mixed in a solution of water and DMSO. The mixture is homogenized by sonication and mechanical stirring. It is believed that PVA contributes to the stabilization of nanotube dispersions in DMSO-based solvents. This is a key feature for the gel-spinning process of the present composite fibers. The SWNT weight fraction is nevertheless kept relatively low in order to achieve homogeneous and spinnable dispersions. The relative weight

6.3 Fibers Obtained from the Coagulation of CNT–Polymer Mixtures

fraction of SWNT to PVA is 3 wt%. The dispersions are injected into a dry ice-cooled methanol bath at -25 ± 5 °C. As soon as the dispersion contacts the cold methanol, a gel fiber is formed and is continuously collected on a fiber take-up unit. This gel fiber is then kept immersed in methanol for 2 days and is dried in a vacuum oven at 70 °C for 3 days before being drawn on a hot plate at 200 °C. Even though the fraction of nanotubes in the achieved fibers is of only 3 wt%, clear improvements of mechanical properties are achieved compared to neat PVA fibers. In particular, the Young's modulus is raised from 26 GPa to 36 GPa via the addition of nanotubes. Minus et al. have further studied this class of gel-spun fibers and investigated the effects of thermal and mechanical treatments [78]. They observed that the presence of 1 wt% of SWNT could lead to enhancements of 49% and 63% of the Young's modulus and tensile strength, respectively, compared to control neat PVA fibers spun and treated in similar conditions. Uddin et al. [79] achieved even greater levels of relative improvements of fibers gel-spun from dispersions in DMSO. The authors could indeed produce fibers with a tensile strength of about 2.2 GPa and a Young's modulus of 36 GPa for fibers containing only 0.3 wt% of CNTs. Neat PVA fibers spun in similar conditions exhibit a tensile strength of 1.7 GPa and a Young's modulus of about 28 GPa. The achieved fibers also exhibit an excellent energy to failure associated to a large elongation at break of about 10%.

PVA-based composite fibers have also been produced from aqueous dispersions of CNTs in PVA solutions. In contrast to the gel-spinning process from DMSO-based solutions, coagulation is achieved here in a solution of salts, often sodium sulfate (Na_2SO_4), in water. Sodium sulfate has a dehydrating effect. The precipitation of PVA can be promoted by the addition of tetraborate and tetra borate ions. Su et al. [80] for example developed PVA/gelatin (GE)/MWNT composite fibers. MWNTs with diameters of 10–30 nm and lengths ranging from 5 to 15 μm are functionalized with carboxylic groups for better dispersion in water. Acid-treated MWNTs are dispersed in a basic aqueous solution using high-power tip sonication for 10 min. This dispersion is then mixed with a solution of PVA solution and GE. The PVA/GE ratio is kept at 100/10, and the MWNT weight fraction in the dried fibers is varied from 0.25% to 1%. Fiber spinning is achieved using a saturated Na_2SO_4 solution as the coagulating medium. The fibers obtained are drawn at high temperature and then washed in water. The addition of nanotubes allowed improvements of more than 20% of the fiber's Young's modulus for a weight fraction of only 1 wt%.

The above contributions, either in gel-spinning from DMSO-based dispersions or wet spinning from aqueous dispersions, demonstrated excellent reinforcement of the PVA fibers at low weight fraction of nanotubes. CNTs exhibit particularly strong interactions with PVA. They play therefore the role of efficient direct reinforcement with a good mechanical stress transfer from the polymer to the nanotubes. Nevertheless, the achieved levels of reinforcement are above the generally observed reinforcements of polymers by CNTs. The particular case of PVA fibers thus raises the question of the origin of the efficiency of CNTs at reinforcing such fibers. In fact, two specific effects, which are not always met in other CNT–polymer nanocomposites, can come into play. First, CNTs can be well aligned in fiber materials, in particular if the fibers are drawn at high temperature. According to micromechanical models [81–84], the alignment of the CNTs is essential to achieve a significant reinforcement. Second, in addition to orientation effects, it has been shown that the presence of the nanotubes increases the crystallinity of the polymer [41,50,78–80]. This effect is most often characterized by X-ray diffraction experiments and shown to be enhanced by drawing the fibers at high temperature. Improvements of crystallinity are reflected by an increase of the stiffness of the fibers. Improvements of the mechanical fiber properties can be quite pronounced even at low fraction of CNTs.

6.3.1.1.2 CNT–PVA fibers at high weight fractions of CNTs

New properties are expected if the fraction of nanotubes could be increased above a few wt%. Indeed, fibers that contain only a few wt%, or less, of CNTs are not likely to be electrically conductive. Some groups recently investigated new approaches to produce PVA fibers that contain CNTs at weight fractions typically above 10 wt%. For example, Xue et al. [85] proposed a method to spin electrically conductive yarns made of PVA and CNTs. In this approach, CNTs are directly mixed with PVA in aqueous media with a weight ratio of CNT to PVA up to 40%. But in such conditions, the nanotubes form aggregates and the dispersions are inhomogeneous. Nevertheless, these dispersions could still be injected in a static coagulation bath made of an aqueous solution of Na_2SO_4. The fibers produced had an average diameter of 500 μm and exhibited nonuniform properties and structure due to the presence of aggregates in the spinning dispersions. They were conductive and showed an average resistivity on the order of several tens of kΩ/cm. But such fibers are fragile and their conductivity varies along the filament [85]. The main challenge to achieve more uniform fibers consists in stabilizing CNTs in aqueous solutions of PVA to prevent the formation of aggregates before fiber spinning. The spinning solutions have to contain a large ratio of CNT to PVA and need to be sufficiently concentrated. In addition, the stabilization should be achieved by modifying the nanotubes without completely screening or downgrading their properties. Mercader et al. [86] found that these requirements can be met by using polyoxyethylene glycol octadecyl ether, a nonionic surfactant additive that coats the nanotube and prevents the PVA chains from adsorbing at their interface. The selected surfactant avoids bridging coagulation and destabilization of the dispersion. The choice of the appropriate surfactant is particularly delicate because it has to be strongly adsorbed onto the nanotubes via a large hydrophobic group and still be water soluble. The fibers are spun by the injection of the PVA–CNT dispersion in a static bath of an aqueous solution of Na_2SO_4. The coagulated fibers are continuously circulated in a bath of water to be washed and are then dried using infrared heating before being wound up. All these stages match the basic phenomena used in the continuous spinning of industrial PVA fibers. The proposed approach allows the spinning of fibers that contain a ratio of CNT to PVA above 10 wt% and that are electrically conductive. The use of

MWNTs, that all have a metallic behavior, enables the production of fibers that exhibit a resistivity of about 35 kΩ/cm. This resistivity can be decreased to 300 Ω/cm by annealing the fibers at 180 °C for 1 h [86]. The observed improvement, by two orders of magnitude, is ascribed to the relaxation of mechanical stress of the polymer chains that allows better intertube contacts. Examples of mechanical properties of the achieved fibers are shown in Figs. 6.9 and 6.10 and Table 6.1.

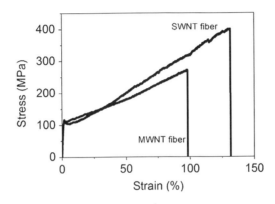

FIGURE 6.9

Stress vs strain of PVA–carbon nanotube composite fibers made using the process described in [86]. Fibers made of SWNT exhibit a greater strain to failure than fibers made of MWNT [86].

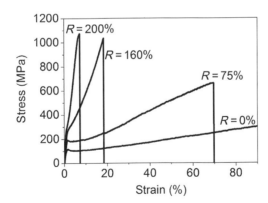

FIGURE 6.10

Stress vs strain of PVA–SWNT composite fibers made using the process described in [86] and prestretched at different draw ratios specified by R [86]. $R = (l/l_0 - 1) * 100$, where l_0 is the initial length of the fiber, and l the length of the fiber after drawing.

Table 6.1 Mechanical properties of carbon nanotube–PVA composite fibers made using the process described in Ref. [86]. The last line, fiber Type F (fiber made of PVA and SWNT), has been hot-stretched by 200% at 140 °C before characterization.

Type	Young's Modulus	Tensile Strength	Strain to Failure	Energy to Failure
Unit	GPa	MPa	%	J/g
PVA/MWNT	10	270	95	130
PVA/SWNT = F	12	400	130	230
F + 200% @140 °C	38	1100	8	–

SWNT-based fibers exhibit better properties than MWNT-based fibers. In addition, drawing at high temperature leads to significant improvements of tensile strength and Young's modulus.

6.3.1.2 PAN–CNT fibers

PAN fibers, also called acrylic fibers, can be obtained by wet spinning or gel spinning. Such fibers can be used as spun for clothing or can also be carbonized under inert atmosphere to produce carbon fibers. PAN precursors fibers must be stabilized before carbonization. The aim of stabilization is to cross-link the polymer and therefore to prevent relaxation and chain scission during carbonization. This reaction is carried out under air at temperatures of 200–400 °C. Stabilization causes both cyclization and dehydrogenation and the fiber must be kept under tension to prevent relaxation. Carbonization converts organic polymeric material into pure carbon. It is performed by pyrolysis under inert atmosphere. Solubilization of pure PAN powder is usually performed in dimethylformamide (DMF) for dry-spinning and in dimethylacetamide (DMAc) or DMF and DMSO for wet spinning. Solid matter concentration and solubilization temperatures are about 23–30 wt% and 110 °C–140 °C for dry spinning and 18–25 wt% and 20 °C–60 °C for wet spinning. Dry spinning is better suited for fine titer fiber, whereas wet spinning is more suitable for coarser filaments.

Weisenberg et al. [87] produced PAN–CNT composite fibers using the dry-jet wet-spinning technique. The authors used MWNTs produced by chemical vapor deposition. The CNTs are dispersed in DMAc by using sonication at concentration of 1, 3 and 5 wt%. The dope is heated to 110 °C and is injected via a 305-μm-diameter needle into a 10 wt% DMAc/H_2O solution at 2.5 °C. A 10-mm air gap is maintained between the end of the die and the coagulation bath. Observations by electron microscopy of broken fibers showed MWNT pulling out from the PAN matrix. Such a pullout phenomenon is known to enhance energy absorption of composite materials. Weisenberg et al. [87] observed that for a volume fraction of 1.8 vol% in the final fibers, highly graphitized MWNTs increased the tensile strength and the tensile modulus of PAN fibers by 31% and 36%, respectively, and the energy to failure by 80%.

S. Kumar and his group extensively investigated PAN−CNT composite fibers [88−97] over the past years. As for the case of PVA, several experimental observations suggest that CNTs strongly interact with PAN. Indeed excellent levels or reinforcements have been achieved. In a first study [88,89] by S. Kumar's group, PAN−SWNT fibers are produced with weight fractions of SWNTs up to 10 wt%. The fibers are dry-jet wet spun from dispersions in DMAc or DMF. As-produced fibers, before any oxidative or carbonization treatments, already demonstrate a strong influence of the CNTs. Indeed, the Young's modulus of PAN fibers is more than doubled and the tensile strength increased by 43% in the presence of CNTs [88]. The glass transition temperature is shown to increase from 103 °C for control PAN fiber to 143 °C for PAN−SWNT (90:10) composite fiber. Shrinkage measurements made on both PAN and PAN−SWNT fibers also demonstrate significant modifications of the fiber properties. In the presence of CNTs, the shrinkage measured at 200 °C is half the shrinkage of the PAN control fiber. All these observations suggest strong interactions between PAN and CNTs. The PAN and composite PAN−SWNT fibers are further thermally oxidized in a furnace at 250 °C at constant length and for 10 h in air atmosphere [89]. As shown in the electron micrographs of Fig. 6.11 and in contrast to neat PAN fibers, a fibrillar-type structure is observed in oxidized composite fiber. Large 50-nm-diameter fibrils observed in oxidized composite fibers surface suggest that the oxidized PAN strongly adheres to the SWNTs [89].

Further studies were performed using the gel-spinning technique in methanol [91]. The fibers are processed with 1 wt% of SWNTs. The fibers are drawn and then stabilized under air atmosphere between 285 °C and 330 °C at various stress levels. Last, the fibers are carbonized under argon atmosphere at 1100 °C. The addition of nanotubes leads to improvements of the tensile strength from 0.72 N/tex to 0.89 N/tex and of the tensile modulus from 17.8 N/tex−22.5 N/tex. X-ray diffraction experiments show enhancements of the orientation of PAN molecules and greater crystal sizes in the presence of CNTs. More surprisingly, the carbonization reaction temperature is only 1100 °C for PAN−SWNT fibers, which is far below the temperatures usually employed for the carbonization of neat PAN fibers [91]. Even at this low temperature, SWNTs lead to a more graphitic structure compared to neat PAN fibers. PAN−SWNT fibers exhibit a nanofibrillar structure after carbonization. The nanofibrils consist of SWNTs surrounded by well-developed graphitic structures. PAN matrix carbonized at the same temperature is disordered and forms mostly amorphous carbon. The addition of SWNT in such PAN-based carbon fibers can result in an increase of 64% of the tensile strength and of 49% of the tensile modulus. PAN-CNT- composite(99/1)-based carbon fibers of 1 μm or less have also been processed using island-in-a-sea bicomponent cross-sectional geometry and gel spinning [93]. Here again improvements of mechanical properties are observed on addition of CNTs. As in the case of PVA fibers, it appears that the role of nanotubes in the improvement of

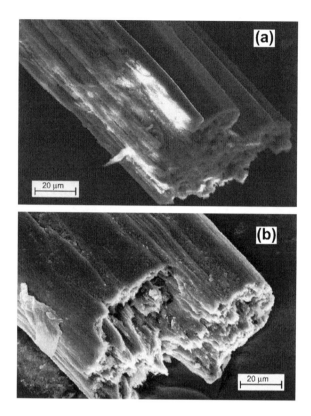

FIGURE 6.11

Scanning electron micrographs of the section of oxidized fibers: (a) PAN and (b) PAN—SWNT composite fiber (90:10). Fibrillar structure evidence of strong interaction between CNTs and PAN is observed in (b) [89].

mechanical properties is twofold. It was indeed suggested that the increase in tensile strength of the PAN-CNT-based carbon fiber over PAN-based carbon fiber comes from two effects: the classical filler reinforcement of CNTs and the presence of graphitic layers surrounding the CNTs. The formation of these surrounding layers results from the ability of CNTs to act as templates for the orientation and crystallization of polymer chains.

Optimization of fiber spinning conditions developed more recently allowed fibers with greater fractions of nanotubes to be produced [94]. Nevertheless, it was observed that above 5 wt% of MWNTs, mechanical properties of the composite fibers tend to decrease. This is due to the presence of large MWNT aggregates that create voids and stress concentrations. However, increasing the fraction of nanotubes enables significant improvements of the electrical conductivity of the PNA—CNT composite fibers.

Last, deeper studies of the stabilization mechanisms of PAN—CNT fibers have been reported and demonstrated subtle and new effects provided by CNTs. In particular, it was shown that CNTs enhance the stabilization reaction in PAN fibers [95]. A less-porous and less-defective structure is observed when CNTs are present. CNTs reduce thermal shrinkage [96] and also enable the maximum tension that can be applied to the fiber to be increased during the stabilization reaction [95,96]. Optimization of all the involved factors leads in the end to better mechanical properties [97].

6.3.1.3 PBO—CNT fibers

PBO is a rigid-rod polymer characterized by a very high tensile strength, high stiffness and high thermal stability. PBO fibers clearly rank among the best high-performance fibers. Various approaches have been proposed to attempt further improving the properties of PBO fibers by including CNTs [98—100]. A great challenge is the achievement of spinnable solutions that contain both PBO and CNTs. Kumar et al. [98,99] proposed a technique in which PBO synthesis is carried out in the presence of SWNT. The resulting mixture is a lyotropic liquid crystalline solution containing PBO or PBO/SWNT. Zhou et al. [100] used a grafting-from approach to covalently functionalize MWNT with PBO and also used in situ polymerization to achieve PBO—CNT solutions. More recently, Hu et al. [101] also used covalently bonded CNTs to PBO to form composite fibers via a dry-jet wet-spinning process. The main objective is to improve the interfacial adhesion of the polymer to the nanotubes.

Another example of PBO/CNT fiber spinning is given in Ref. [98]. PBO and PBO—SWNT are dry-jet wet spun using a 250-μm-diameter spinneret. A 10-cm air gap distance between the spinneret and the distilled water coagulation bath is maintained for all the solutions. Draw ratios up to 10 can be achieved, and fibers are washed in running water for a week, prior to drying in a vacuum oven set at 80 °C. The fibers obtained are heat-treated at 400 °C under a tension of 40 MPa for 2 min. Fibers with large fractions of nanotubes were reported [98,99]. Quite spectacular improvements of mechanical properties of PBO fibers have been achieved for weight fractions of CNTs of about 10 wt%. Typical stress vs strain curves are shown in Fig. 6.12.

Neat PBO fibers already exhibit excellent mechanical properties with a tensile strength and a Young's modulus in the ranges of, respectively, 1.8—2.6 GPa and 230—160 GPa. Fibers with a weight fraction of 10 wt% exhibit a tensile strength and Young's modulus in the ranges of, respectively, 2.9—4.2 GPa and 150—180 GPa. These values are remarkably high in the world of synthetic fibers and confirm that CNTs can indeed be used to achieve superfiber materials. Composite fibers made with covalently bonded CNTs also exhibit significant improvements of mechanical properties [101].

The investigated PBO—CNT fibers exhibit remarkable mechanical properties but are not electrically conductive in spite of the large fraction of embedded nanotubes. This effect can be ascribed to the alignment of the nanotubes and to the losses of

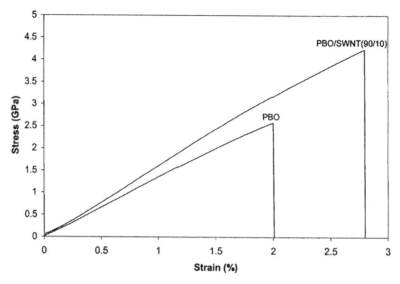

FIGURE 6.12

Stress vs strain curves for PBO and PBO—SWNT (90/10) fibers [98].

intertube contacts during fiber drawing. It has been indeed shown in melt-spun fibers that drawing reduces the electrical conductivity of the fibers and increases the percolation threshold [102]. The loss of conductivity is essentially dominated by the relative translation of the nanotubes with respect to each other during fiber drawing [102].

6.3.1.4 UHMW PE—CNT fibers

UHMW PE is widely used in fiber technology and enables the spinning of high-tenacity fibers for various applications including bulletproof vests. In spite of its relatively low melting temperature of about 150 °C, its chemical nature leads to a highly hydrophobic material with superior chemical resistance. The first UHWM PE—CNT composite fibers were made in 2005 [103] using MWNTs. After purification in potassium permanganate and sulfuric acid at reflux at 120 °C and subsequent washing with deionized water, MWNTs are functionalized with a titanate coupling agent. The alkyl part of the titanate agent is able to react with the carboxyl groups of MWNTs, introducing a steric hindrance, therefore preventing CNT from agglomeration in UHMW PE. The authors could achieve fibers with weight fractions of nanotubes ranging from 0.25 wt% to 3 wt%. Slight improvements of mechanical properties could thereby be obtained.

Ruan et al. performed gel spinning of UHMW PE—MWNT fibers containing 5 wt% of MWNTs [104]. Nanotubes are oxidized and dispersed in ethanol using tip sonication. This dispersion is added to a mixture of 3 wt% UHMW PE in decalin. The solution is injected in a water bath at room temperature to form gel fibers, which are hot pressed at 120 °C. The obtained precursor fibers are then hot-drawn twice at 120 and 130 °C at draw ratios, respectively, of 5 and 30. The achieved fibers exhibit excellent mechanical properties with improvements of 34% in ductility and 64% in energy to failure. Enhancements of 11.6, 18.8 and 15.4% of tensile modulus, tensile strength and strain at break, respectively, are observed for UHMW PE–MWNT in comparison to neat UHMW PE. Even though these levels of improvements can appear lower than those achieved with other polymers, one has to keep in mind that neat UHMW PE, such as PBO, has already excellent intrinsic mechanical properties. Improving them further is therefore particularly challenging.

In 2008, Yeh et al. proposed an improved process [105] still using decalin as solvent for the polymer–CNT mixture. The extruded material enters a quenching water bath at 25 °C placed at 520 mm or 810 mm from the spinneret. This technique is similar to dry-jet wet spinning. The gel fibers are then extracted from the water bath and rinsed in an *n*-hexane bath to remove residual decalin. Finally, the fibers are dried in air to remove any remaining solvents. Various concentrations of polymers and CNTs have been tested. Increasing the CNT concentration was found to strongly increase the viscosity of the spinning dope. Nevertheless, solutions containing 2.0 wt% of UHMW PE with 0.005 wt% of CNT are found to be still spinnable. Large draw ratios up to 110 could be achieved with the 2.0 wt% UHMW PE–0.002 wt% CNT optimum formulation. Such fibers exhibit strong birefringence under cross-polarized microscopy. Because of their strong alignment, mechanical properties of these ultradrawn composite fibers are quite high. Their Young's modulus is indeed 90 GPa and their tensile strength is about 3 GPa [105]. More recently, Mahfuz et al. reported a related method to spin UHMW PE–CNTs composite fibers [106] but investigated different experimental conditions and treatments. The authors observed that the dope could not be spun, because of viscosity enhancement, if the solution contained more than 2 wt% of MWNT. The fiber's Young's modulus increased by 114% and the tensile strength increased by 62% with the addition of 2 wt% of MWNT. The strain to failure is reduced by 30%. Mechanical properties are found to be enhanced via the strain hardening process. Due to the presence of the MWNTs, a 15% increase in crystallinity is reported. This effect shows again that the mechanical properties of the polymer are not only modified by the presence of stiff inclusions and transfer of mechanical load but also modified because the presence of the CNTs affects the structure of the polymer.

6.3.2 CNT–conductive polymer composite fibers

Conducting polymers are promising materials for future organic electronics. They are also promising materials under the form of fibers for novel functional textiles, microelectrodes, actuators or biomedical applications. Nevertheless, fiber spinning of conducting polymers is rather delicate for several reasons. The chemical stability of the polymer is the first issue. This is why fibers from these polymers are most often spun by wet-spinning approaches. Melt spinning at high temperature could degrade the polymer. Second, fibers solely composed of conducting polymers are sometimes mechanically weak and cannot be easily processed. Developments of new composite fibers with CNTs and conducting polymers have been performed over the past years. The presence of the nanotubes is expected to facilitate the spinning of the fibers and to improve their mechanical and electrical properties. The main research efforts have been devoted to PANi, PPy and PEDOT and are discussed below.

G. Wallace and collaborators extensively worked on conducting polymer—CNT fibers. Already in 2005, V. Mottaghitalab et al. [107] proposed a method to prepare the first PANi—CNT composite fibers. SWNTs are dispersed in dimethyl propylene urea. Emeraldine base (EB) powder, the neutral form of PANi, is mixed with the CNT dispersion and the mixture is homogenized by sonication to achieve a spinnable dope. This spinning dope is deaerated under vacuum and injected through a single-hole spinneret of 250 μm diameter into a coagulation bath. This bath is composed of 10 wt% of N-methyl-pyrrolidone in water. After coagulation, the fiber is kept in water to reduce the solvent content and then dried at room temperature for 12 h. Hot-drawing is then carried out to improve the structure and properties of the fibers. The last step of the process is the PANi doping. Doping is mandatory to convert the EB into its conductive protonated form called the emeraldine salt. Doping is achieved in an aqueous solution of methane sulfonic acid. The addition of CNT at a weight fraction of 2% was found to increase the Young's modulus by a factor of 3. More strikingly, the conductivity at the same weight fraction of CNT is increased by more than one order of magnitude. This large improvement of conductivity was explained by a homogeneous dispersion of the CNTs in the PANi matrix that forms an interconnected network with good bonding to the polymer. This process was further improved using 2-acrylamido-2-methyl-1propane sulfonic acid (AMPSA) as dispersant for SWNTs in dichlororacetic acid (DCAA) and PANi [108]. The combination of DCAA and AMPSA resulted in an efficient solvent system for dissolution of PANi. Moreover, dynamic light scattering measurement showed that DCAA—SWNT samples containing AMPSA have smaller hydrodynamic radius compared to neat DCAA—SWNT, indicating that AMPSA assists in debundling of SWNTs, as well as in stabilizing SWNTs. The solution is injected through a single-hole spinneret into an acetone coagulation bath at room temperature. The mechanical and electrical properties of the fibers were found to be substantially improved by the addition of CNTs. Conductivity values as high as 7.5×10^4 S/m are measured with only 0.76 wt% of SWNTs.

From a general point of view, combined improvements of mechanical and electrical properties are not always straightforward. For example, fiber drawing is known to increase the mechanical properties of fiber materials. However, fiber drawing also results in the disruption of internanotube contacts. This is reflected by a decrease of the electrical conductivity [102]. Here the addition of CNTs leads to a combined improvement of mechanical and electrical properties. This combined improvement is of particular interest for functional materials and in particular for electromechanical actuators. Indeed, such materials should ideally exhibit a large electrical

6.3 Fibers Obtained from the Coagulation of CNT–Polymer Mixtures

conductivity and high mechanical strength to have a large energy density. Neat conducting polymers have been extensively investigated for applications in this field. But their performance was often limited by a low Young's modulus. Composite CNT–PANi fibers exhibit enhanced mechanical and electrical properties [107–112], which led to fiber materials becoming particularly promising for novel actuators with enhanced performances [110–112]. This new class of fibers could also be potentially useful for other electrochemical applications such as battery materials [113].

More recently, G. Wallace's group investigated wet-spun PPy/alginate/SWNT composite fibers [114]. PPy is a conductive polymer that cannot be easily dissolved because of strong interchain interactions. Composite fibers investigated in [114] are made of alginate host fibers containing pyrrole monomer and SWNT. The pyrrole monomer is polymerized inside the fiber. The final weight ratio of pyrrole monomer to alginate to SWNT is 2.63/1/0.04 in the fiber-spinning dope. The dope is injected through a single-hole spinneret of 100 μm into a coagulation bath composed of 5 wt% of $CaCl_2$ in a mixture of methanol/water (70/30). After coagulation, the fibers are transferred to an oxidant/doping bath. Several formulations were tested as described in Ref. [114]. After spinning, the fibers are kept for 1 hour and 30 minutes in a methanol/water (70/30) bath containing 15 wt% of $CaCl_2$.

The presence of the nanotubes was found to strongly affect the properties of the fibers. Swelling tests were carried out by immersion in water. The swelling ratio of the PPy-alginate fibers decreases from 200% to 50% in the presence of nanotubes. For the formulation and doping conditions used, the Young's modulus and tensile strength are increased in the presence of CNTs. The electrical conductivity is multiplied by a factor greater than 5. As a consequence, the investigated fibers exhibited interesting properties as electromechanical actuators. Indeed, it was found that PPy-alginate-CNT fibers possess a reversible actuation strain of 0.7% at a high scan rate of 100 mV/s.

Among the latest efforts in the field, R. Jalili et al. [115] developed composite fibers made of SWNT and PEDOT:PSS. The authors used highly exfoliated nanotubes and high-quality dispersions to obtain wet-spinnable composite formulations at various nanotube volume fractions. As observed with other conductive polymers, the addition of well-dispersed SWNTs resulted in large enhancement of modulus, tensile strength, and electrical conductivity. In addition, a significant increase of the electrochemical capacitance was observed. The work reported in [115] is of particular interest because, as shown in Figs. 6.13 and 6.14, it presents comprehensive measurements of both mechanical and electrical properties of wet-spun fibers as a function of the volume fraction of nanotubes. At low volume fraction, up to 2%, the Young's modulus, tensile strength and electrical conductivity increase almost linearly with the fraction of CNTs.

From a general point of view, the achieved exciting properties of CNT–conducting polymer composite fibers provide novel opportunities to develop electrodes, fabrics and other textile structures potentially useful in the fields of energy storage, energy conversion and biomedical applications.

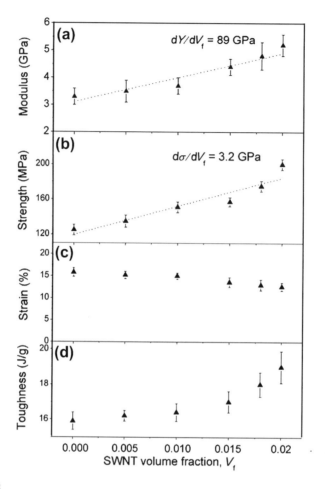

FIGURE 6.13

(a–d) Mechanical properties of PEDOT:PSS–SWNT composite fibers as a function of SWNT loading [115].

6.3.3 CNT–natural polymer composite fibers

Wet-spinning technologies have been used for a long time to produce fibers from natural polymers. The field is currently finding a regrowth of interest for two main reasons. First of all, natural polymers are preferred over synthetic polymers made from oil for environmental issues. Second, natural polymers exhibit often a better biocompatibility compared to synthetic polymers. They are therefore more appropriate for applications in the field of biomedicine and even simple wearable clothing. The inclusion of CNTs in fibers made of natural polymers is an exciting challenge to develop stronger fibers or fibers that exhibit novel

6.3 Fibers Obtained from the Coagulation of CNT—Polymer Mixtures

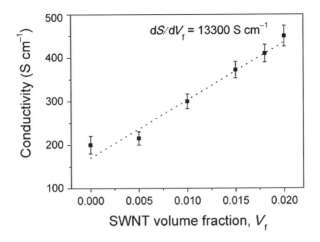

FIGURE 6.14

Electrical properties of PEDOT:PSS-SWNT composite fibers as a function of SWNT loading [115].

functionalities provided by the electronic properties of the nanotubes. In this context, several groups have investigated over the past years CNT—natural polymer composite fibers. The used polymers include in particular cellulose, alginate and chitosan.

Several formulations have been proposed to develop nanotube/cellulose composite fibers [116—119]. Ionic liquids, considered as green solvents, are often used as solvents for the nanotubes and the cellulose. The high viscosity of the solutions makes it difficult for the spinning of fibers with large fractions of nanotubes. Nevertheless, CNT—cellulose fibers were found to be electrically conductive even for small fractions of nanotubes on the order of 0.1 wt%. Unfortunately, at such low volume fraction, the mechanical properties are not substantially affected by the presence of the nanotubes. Using an optimized dry-jet wet-spinning process, Zhang et al. [116] could nevertheless achieve cellulose fibers with a weight fraction as high as 9 wt%. Electron microscopy observation of the fiber cross-sections showed that the nanotubes are well dispersed in the cellulose matrix for composite containing less than 4 wt% of CNTs. Aggregation of the nanotubes is observed at greater weight fractions. Therefore, CNT reinforcement capability is reduced, and top mechanical properties are achieved for composite fibers containing 4 wt% of nanotubes. In such fibers at room temperature, the Young's modulus is about 18 GPa, whereas the Young's modulus of neat cellulose fibers is about 5 GPa. Also, the cellulose fiber containing 4 wt% CNTs showed a high electrical conductivity of about 10^{-2} S cm. Large improvements of mechanical properties and electrical conductivity were also observed in CNT—cellulose composite fibers made with cellulose produced by bacteria [117].

Polymers other than cellulose have been used to make CNT—natural composite fibers. For example, chitosan/CNT composite fibers can also be spun via wet spinning [120—122]. Chito (1 → 4)-2-amino-2-deoxy-β-D-glucose is a biocompatible polymer, derived by the alkaline deacetylation of chitin, a polysaccharide found in the exoskeletons of shrimps and crabs. Cross-linked chitosan forms hydrogels that swell in acidic media, due to the protonation of amino groups. The composite fibers could be useful for bioactuator applications like artificial muscles, with the capacity of retaining a liquid. The addition of CNTs led to a composite reinforcement, doubling the tensile modulus up to 10 GPa approximately. The swelling behavior of chitosan is reduced by the presence of nanotubes. The equilibrium water content of investigated SWNT—chitosan composite fibers is about 10%. This amount of water is lower than that retained in neat chitosan in similar conditions of pH. The reduced swelling in the presence of nanotubes was ascribed to the formation of a rigid composite microgel, which provides elastic resistance against swelling.

Alginate fibers containing CNTs are another example of biopolymer—CNT composite materials. Sodium alginate is a biocompatible anionic polysaccharide found in brown algae. K. Kornev et al. [123,124] used this natural polymer in association with SWNTs and surfactants to realize aqueous spinning dope. SWNTs are dispersed in water using SDS. Different amounts of this dispersion are added to a 1.25 wt% sodium alginate (120—190 kDa) solution. SWNT concentrations range from 0 wt% to 0.3 wt%. The spinning dope is injected into a coagulation bath consisting of an aqueous solution of $CaCl_2$ at 15 wt%. This coagulation bath is placed in a rotating stage to produce coflow conditions. Fibers with low CNT weight fractions (from 0.6 to 2.4 wt%) showed an electrical resistivity greater than 0.5 Ω m. However, the resistivity was found to significantly decrease with the CNT loading. Fibers that contain 12 and 23 wt% of CNTs exhibited resistivities of only 3.7×10^{-2} and 3.10×10^{-3} Ω m, respectively.

Several other biopolymers have been used to make nanocomposite fibers using the PCS in coflowing configurations (see Section 6.2.2.1). The biopolymers cannot only be used as coagulating medium or as the matrix of composite fibers, but, as shown in Ref. [122], they can also be used as efficient dispersants for the CNTs. In particular, dispersions achieved with deoxyribonucleic acid (DNA), hyaluronic acid (HA), heparin and chondroitin sulfate have been tested to make CNT-chitosan-based fibers. The dispersions are homogenized with sonication and spinning is achieved using a rotating coagulating bath. Sodium salts of DNA and HA are used as dispersants for SWNT and create a suspension of negatively charged particles. In contrast, chitosan is positively charged and forms an effective coagulant for these dispersions. The mechanical properties and electrical properties of the achieved fibers depend strongly on the polymer used as dispersant [122]. For example, fibers made of nanotubes dispersed with DNA exhibit a Young's modulus of 65 GPa, whereas fibers made in similar conditions of nanotubes dispersed with HA exhibit a Young's modulus of only 4 GPa. However, the latter fibers exhibit a greater electrical conductivity. The exact mechanisms of reinforcement and electrical conductivity are not yet clear as they involve complex interactions between CNTs and natural macromolecules. Nevertheless, it can be concluded that fibers solely composed of biomaterials and CNTs can be processed via water-based and environment-friendly processes. The achieved fibers exhibit electrical, electrochemical and mechanical properties potentially useful for future biomedical applications [122].

6.4 CONCLUSIONS

CNTs are exceptional particles for the development of novel fibers by wet-spinning approaches. First of all, they can be used as neat materials solubilized or dispersed in various types of solvents and directly spun to form fibers solely composed of CNTs. The achieved fibers have been shown to exhibit an excellent electrical conductivity. Nevertheless, improvements of their mechanical properties are still needed to enlarge their spectrum of potential applications. Challenges at different levels are faced toward these improvements. It will first be necessary to produce long and defect-free CNTs. In addition, it will be necessary to develop specific approaches to process such long nanotubes in liquid media and spinnable dope to form fibers. CNTs can also be used as fillers in polymer fiber made by wet spinning. They exhibit ideal dimensions and shapes to fit within fiber structures. Wet spinning is particularly well suited for polymers that cannot be easily molten such as conductive polymers, natural polymers or ultrahigh molecular weight polymers. In addition, in contrast to melt spinning, wet-spinning approaches allow large fractions of CNTs to be included in polymer fibers. Nevertheless, wet spinning of composite fibers requires, as for neat CNT fibers, the achievement of perfectly homogeneous dispersions. This point is critical in particular when large fractions of nanotubes are targeted. Research of new dispersants and conditions of coagulation are in this case of great importance. When successfully formed, it has been shown for a variety of polymers that the fiber properties can be substantially altered by the presence of the nanotubes. For almost all the polymers investigated, the presence of the nanotubes induces an increase of the Young's modulus and provides electrical conductivity provided that the CNT concentration is large enough. Conductive fibers can be used for antistatic textiles, conductive clothing and more advanced applications such as sensing textiles and microelectrodes for biological or energy applications. The increase of the Young's modulus can be simply understood by considering the direct mechanical transfer of tensile load from the soft matrix to the stiff inclusions. Nevertheless, the influence of the nanotubes goes well beyond this simple reinforcement mechanism. Indeed, it has been shown that nanotubes can also improve the strain to failure of polymer fibers and lead thereby to the toughest materials ever reported. CNTs affect in some cases the structure of the polymer and modify its thermal properties. It has been indeed observed for several polymers that CNTs can increase the crystallinity of the polymer and shift the glass or melting transition temperatures. Increasing the crystallinity of the polymer has a strong effect on the mechanical properties, well beyond the direct reinforcement by transfer of mechanical load. Other positive side effects of nanotubes have been reported. In particular, the presence of nanotubes allows the tension of PAN fibers to be increased before they are converted in carbon fibers. This indirect effect leads again to substantial improvements of the produced fibers. More advanced functions are achieved in fibers composed of nanotubes and bio- or conductive polymers. Such fibers exhibit electrical and electromechanical properties potentially useful for various energy storage and conversion applications including actuators with enhanced properties.

Exciting and promising results have been achieved at the laboratory scale. The fibers produced are most often prepared using monofilament fiber spinning instruments. It is today critical to consider the scale-up of production of such fibers so that applications can be fully validated. Scale-up will not just necessitate producing fibers via multifilament fiber spinning instruments. It will also be necessary to optimize the rate of production and develop sufficiently robust processes. Fundamental investigations of the kinetics of coagulation can be particularly useful for this goal as it can help in identifying the conditions for a quick and efficient solidification of the fibers. Another area for future research and technological development is the investigation of fibers composed of other novel forms of carbon nanoparticles. In particular, graphene [125,126] is currently the topic of strong interest in various fields including nanocomposites. It has been recently shown that the combination of graphene and CNTs can be used to produce by wet-spinning approaches fibers that are tougher than fibers solely composed of nanotubes [125,126]. There is no doubt that investigations of new fibers made by wet-spinning processes and including graphene materials could also lead to novel properties.

References

[1] F.C. Charles, J.B. Edward, B. Clayton, Treatment of Viscose for the Preparation of Useful Products Therefrom, U.S. Patent GB 190103592 (A), 1902.
[2] E.O. Whittier, S.P. Gould, Making casein fiber, Industrial & Engineering Chemistry 32 (1940) 906–907.
[3] H. Okuzaki, Y. Harashina, H. Yan, Highly conductive PEDOT/PSS microfibers fabricated by wet-spinning and dip-treatment in ethylene glycol, European Polymer Journal 45 (2009) 256–261.
[4] F. Zhang, P.A. Halverson, B. Lunt, M.R. Linford, Wet spinning of pre-doped polyaniline into an aqueous solution of a polyelectrolyte, Synthetic Metals 156 (2006) 932–937.
[5] S. Pomfret, P. Adams, N. Comfort, A. Monkman, Electrical and mechanical properties of polyaniline fibres produced by a one-step wet spinning process, Polymer 41 (2000) 2265–2269.
[6] J. Foroughi, G.M. Spinks, G.G. Wallace, P.G. Whitten, Production of polypyrrole fibres by wet spinning, Synthetic Metals 158 (2008) 104–107.

[7] J. Foroughi, G.M. Spinks, G.G. Wallace, Effect of synthesis conditions on the properties of wet spun polypyrrole fibres, Synthetic Metals 159 (2009) 1837–1843.

[8] Fourné, Synthetic fibers, s. d. 1999.

[9] Z.K. Walczak, Processes of Fiber Formation, first ed., Elsevier Science, 2002.

[10] A. Lucas, C. Zakri, M. Maugey, M. Pasquali, P. van der Schoot, P. Poulin, Kinetics of nanotube and microfiber scission under sonication, Journal of Physical Chemistry C 113 (2009) 20599–20605.

[11] G. Pagani, M.J. Green, P. Poulin, M. Pasquali, Competing mechanisms and scaling laws for carbon nanotube scission by ultrasonication, Proceedings of the National Academy of Sciences of the United States of America (2012).

[12] L.M. Ericson, H. Fan, H. Peng, V.A. Davis, W. Zhou, J. Sulpizio, et al., Macroscopic, neat, single-walled carbon nanotube fibers, Science 305 (2004) 1447–1450.

[13] W. Zhou, Single wall carbon nanotube fibers extruded from super-acid suspensions: preferred orientation, electrical, and thermal transport, Journal of Applied Physics 95 (2004) 649.

[14] Y. Wang, L.M. Ericson, C. Kittrell, M.J. Kim, H. Shan, H. Fan, et al., Revealing the substructure of single-walled carbon nanotube fibers, Chemistry of Materials 17 (2005) 6361–6368.

[15] R.D. Booker, M.J. Green, H. Fan, A.N.G. Parra-Vasquez, N. Behabtu, C.C. Young, et al., High-shear treatment of single-walled carbon nanotube—superacid solutions as a pre-processing technique for the assembly of fibres and films, Proceedings of the Institution of Mechanical Engineers, Part N: Journal of Nanoengineering and Nanosystems 222 (2008) 101–109.

[16] N. Behabtu, C.C. Young, D.E. Tsentalovich, O. Kleinerman, X. Wang, A.W.K. Ma, E.A. Bengio, R.F. ter Waarbeek, J.J. de Jong, R.E. Hoogerwerf, S.B. Fairchild, J.B. Ferguson, B. Maruyama, J. Kono, Y. Talmon, Y. Cohen, M.J. Otto, M. Pasquali, Strong, Light, Multifunctional Fibers of Carbon Nanotubes with Ultrahigh Conductivity, Science 339 (6116) (2013) 182–186.

[17] S. Zhang, K.K.K. Koziol, I.A. Kinloch, A.H. Windle, Macroscopic fibers of well-aligned carbon nanotubes by wet spinning, Small 4 (2008) 1217–1222.

[18] J. Steinmetz, M. Glerup, M. Paillet, P. Bernier, M. Holzinger, Production of pure nanotube fibers using a modified wet-spinning method, Carbon 43 (2005) 2397–2400.

[19] M.E. Kozlov, R.C. Capps, W.M. Sampson, V.H. Ebron, J.P. Ferraris, R.H. Baughman, Spinning solid and hollow polymer-free carbon nanotube fibers, Advanced Materials 17 (2005) 614–617.

[20] B. Vigolo, A. Pénicaud, C. Coulon, C. Sauder, R. Pailler, C. Journet, et al., Macroscopic fibers and ribbons of oriented carbon nanotubes, Science 290 (2000) 1331–1334.

[21] J. Swenson, M.V. Smalley, H.L.M. Hatharasinghe, Mechanism and strength of polymer bridging flocculation, Physical Review Letters 81 (1998) 5840–5843.

[22] C. Mercader, Filage continu de fibres de nanotubes de carbone: de la solidification aux propriétés finales, Thesis 2010.

[23] M.H. Jee, S.H. Park, J.U. Choi, Y.G. Jeong, D.H. Baik, Effects of wet-spinning conditions on structures, mechanical and electrical properties of multi-walled carbon nanotube composite fibers, Fibers and Polymers 13 (2012) 443–449.

[24] A.B. Dalton, S. Collins, E. Muñoz, J.M. Razal, V.H. Ebron, J.P. Ferraris, et al., Super-tough carbon-nanotube fibres, Nature 423 (2003) 703.

[25] J.N. Barisci, M. Tahhan, G.G. Wallace, S. Badaire, T. Vaugien, M. Maugey, et al., Properties of carbon nanotube fibers spun from DNA-stabilized dispersions, Advanced Functional Materials 14 (2004) 133–138.

[26] W. Néri, M. Maugey, P. Miaudet, A. Derré, C. Zakri, P. Poulin, Surfactant-free spinning of composite carbon nanotube fibers, Macromolecular Rapid Communications 27 (2006) 1035–1038.

[27] A. Pénicaud, L. Valat, A. Derré, P. Poulin, C. Zakri, O. Roubeau, et al., Mild dissolution of carbon nanotubes: composite carbon nanotube fibres from polyelectrolyte solutions, Composites Science and Technology 67 (2007) 795–797.

[28] A. Pénicaud, P. Poulin, A. Derré, E. Anglaret, P. Petit, Spontaneous dissolution of a single-wall carbon nanotube salt, Journal of the American Chemical Society 127 (2005) 8–9.

[29] E. Muñoz, D.S. Suh, S. Collins, M. Selvidge, A.B. Dalton, B.G. Kim, et al., Highly conducting carbon nanotube/polyethyleneimine composite fibers, Advanced Materials 17 (2005) 1064–1067.

[30] A.B. Dalton, S. Collins, J. Razal, E. Munoz, V.H. Ebron, B.G. Kim, et al., Continuous carbon nanotube composite fibers: properties, potential applications, and problems, Journal of Materials Chemistry 14 (2003) 1–3.

[31] J.M. Razal, J.N. Coleman, E. Muñoz, B. Lund, Y. Gogotsi, H. Ye, et al., Arbitrarily shaped fiber assemblies from spun carbon nanotube gel fibers, Advanced Functional Materials 17 (2007) 2918–2924.

[32] C. Mercader, A. Lucas, A. Derré, C. Zakri, S. Moisan, M. Maugey, et al., Kinetics of fiber solidification, Proceedings of the National Academy of Sciences of the United States of America (2010).

[33] O. Bonhomme, J. Leng, A. Colin, Microfluidic wet-spinning of alginate microfibers: a theoretical analysis of fiber formation, Soft Matter 8 (2012) 10641–10649.

[34] A.V. Neimark, S. Ruetsch, K.G. Kornev, P.I. Ravikovitch, P. Poulin, S. Badaire, et al., Hierarchical pore structure and wetting properties of single-wall carbon nanotube fibers, Nano Letters 3 (2003) 419–423.

[35] P. Launois, A. Marucci, B. Vigolo, P. Bernier, A. Derré, P. Poulin, Structural characterization of nanotube fibers by X-ray scattering, Journal of Nanoscience and Nanotechnology 1 (2001) 125–128.

[36] B. Vigolo, P. Poulin, M. Lucas, P. Launois, P. Bernier, Improved structure and properties of single-wall carbon nanotube spun fibers, Applied Physics Letters 81 (2002) 1210.

[37] V. Pichot, S. Badaire, P.A. Albouy, C. Zakri, P. Poulin, P. Launois, Structural and mechanical properties of single-wall carbon nanotube fibers, Physical Review B – Condensed Matter and Materials Physics 74 (2006).

[38] E. Anglaret, A. Righi, J.L. Sauvajol, P. Bernier, B. Vigolo, P. Poulin, Raman resonance and orientational order in fibers of single-wall carbon nanotubes, Physical Review B 65 (2002) 165426.

[39] M.H. Jee, J.U. Choi, S.H. Park, Y.G. Jeong, D.H. Baik, Influences of tensile drawing on structures, mechanical, and electrical properties of wet-spun multi-walled carbon nanotube composite fiber, Macromolecular Research 20 (2012) 650–657.

[40] L. Viry, C. Mercader, P. Miaudet, C. Zakri, A. Derré, A. Kuhn, et al., Nanotube fibers for electromechanical and shape memory actuators, Journal of Materials Chemistry 20 (2010) 3487−3495.

[41] P. Miaudet, S. Badaire, M. Maugey, A. Derré, V. Pichot, P. Launois, et al., Hot-drawing of single and multiwall carbon nanotube fibers for high toughness and alignment, Nano Letters 5 (2005) 2212−2215.

[42] A. Lendlein, S. Kelch, Shape-memory polymers, Angewandte Chemie International Edition 41 (2002) 2034−2057.

[43] C. Liu, H. Qin, P.T. Mather, Review of progress in shape-memory polymers, Journal of Materials Chemistry 17 (2007) 1543−1558.

[44] B.K. Kim, S.Y. Lee, M. Xu, Polyurethanes having shape memory effects, Polymer 37 (1996) 5781−5793.

[45] T. Ohki, Q.-Q. Ni, N. Ohsako, M. Iwamoto, Mechanical and shape memory behavior of composites with shape memory polymer, Composites Part A: Applied Science and Manufacturing 35 (2004) 1065−1073.

[46] J. Hu, Z. Yang, L. Yeung, F. Ji, Y. Liu, Crosslinked polyurethanes with shape memory properties, Polymer International 54 (2005) 854−859.

[47] J. Morshedian, H.A. Khonakdar, M. Mehrabzadeh, H. Eslami, Preparation and properties of heat-shrinkable cross-linked low-density polyethylene, Advances in Polymer Technology 22 (2003) 112−119.

[48] H. Qin, P.T. Mather, Combined one-way and two-way shape memory in a glass-forming nematic network, Macromolecules 42 (2009) 273−280.

[49] T. Chung, A. Romo-Uribe, P.T. Mather, Two-way reversible shape memory in a semicrystalline network, Macromolecules 41 (2008) 184−192.

[50] P. Miaudet, A. Derré, M. Maugey, C. Zakri, P.M. Piccione, R. Inoubli, et al., Shape and temperature memory of nanocomposites with broadened glass transition, Science 318 (2007) 1294−1296.

[51] Y. Miyamoto, K. Fukao, H. Yamao, K. Sekimoto, Memory effect on the glass transition in vulcanized rubber, Physical Review Letters 88 (2002) 225504.

[52] M. Cadek, J.N. Coleman, K.P. Ryan, V. Nicolosi, G. Bister, A. Fonseca, et al., Reinforcement of polymers with carbon nanotubes: the role of nanotube surface area, Nano Letters 4 (2011) 353−356.

[53] J. Berriot, H. Montes, F. Lequeux, D. Long, P. Sotta, Gradient of glass transition temperature in filled elastomers, Europhysics Letters 64 (2003) 50−56.

[54] K. Kratz, S.A. Madbouly, W. Wagermaier, A. Lendlein, Temperature-memory polymer networks with crystallizable controlling units, Advanced Materials 23 (2011) 4058−4062.

[55] K. Kratz, U. Voigt, A. Lendlein, Temperature-memory effect of copolyesterurethanes and their application potential in minimally invasive medical technologies, Advanced Functional Materials 22 (2012) 3057−3065.

[56] U.N. Kumar, K. Kratz, M. Heuchel, M. Behl, A. Lendlein, Shape-memory nanocomposites with magnetically adjustable apparent switching temperatures, Advanced Materials 23 (2011) 4157−4162.

[57] T. Xie, Tunable polymer multi-shape memory effect, Nature 464 (2010) 267−270.

[58] T. Xie, Recent advances in polymer shape memory, Polymer 52 (2011) 4985−5000.

[59] J. Wang, R.P. Deo, P. Poulin, M. Mangey, Carbon nanotube fiber microelectrodes, Journal of the American Chemical Society 125 (2003) 14706−14707.

[60] L. Viry, A. Derré, P. Garrigue, N. Sojic, P. Poulin, A. Kuhn, Optimized carbon nanotube fiber microelectrodes as potential analytical tools, Analytical and Bioanalytical Chemistry 389 (2007) 499–505.
[61] L. Viry, A. Derré, P. Poulin, A. Kuhn, Discrimination of dopamine and ascorbic acid using carbon nanotube fiber microelectrodes, Physical Chemistry Chemical Physics 12 (2010) 9993–9995.
[62] N. Mano, F. Mao, A. Heller, Characteristics of a miniature compartment-less glucose-O_2 biofuel cell and its operation in a living plant, Journal of the American Chemical Society 125 (2003) 6588–6594.
[63] F. Gao, L. Viry, M. Maugey, P. Poulin, N. Mano, Engineering hybrid nanotube wires for high-power biofuel cells, Nature Communications 1 (2010) 2.
[64] R.H. Baughman, C. Cui, A.A. Zakhidov, Z. Iqbal, J.N. Barisci, G.M. Spinks, et al., Carbon nanotube actuators, Science 284 (1999) 1340–1344.
[65] J. Fraysse, A. Minett, O. Jaschinski, G. Duesberg, S. Roth, Carbon nanotubes acting like actuators, Carbon 40 (2002) 1735–1739.
[66] G. Sun, J. Kürti, M. Kertesz, R.H. Baughman, Dimensional changes as a function of charge injection in single-walled carbon nanotubes, Journal of the American Chemical Society 124 (2002) 15076–15080.
[67] S. Ghosh, V. Gadagkar, A.K. Sood, Strains induced in carbon nanotubes due to the presence of ions: ab initio restricted Hartree–Fock calculations, Chemical Physics Letters 406 (2005) 10–14.
[68] S. Gupta, M. Hughes, A.H. Windle, J. Robertson, Charge transfer in carbon nanotube actuators investigated using in situ Raman spectroscopy, Journal of Applied Physics 95 (2004) 2038–2048.
[69] M. Hughes, G.M. Spinks, Multiwalled carbon nanotube actuators, Advanced Materials 17 (2005) 443–446.
[70] J. Riemenschneider, S. Opitz, M. Sinapius, H.P. Monner, Modeling of carbon nanotube actuators: part I — modeling and electrical properties, Journal of Intelligent Material Systems and Structures 20 (2009) 245–250.
[71] J. Riemenschneider, S. Opitz, M. Sinapius, H.P. Monner, Modeling of carbon nanotube actuators: part II — mechanical properties, electro mechanical coupling and validation of the model, Journal of Intelligent Material Systems and Structures 20 (2009) 253–263.
[72] C. Bartholome, A. Derré, O. Roubeau, C. Zakri, P. Poulin, Electromechanical properties of nanotube-PVA composite actuator bimorphs, Nanotechnology 19 (2008).
[73] J. Barisci, G. Spinks, G. Wallace, J. Madden, R. Baughman, Increased actuation rate of electromechanical carbon nanotube actuators using potential pulses with resistance compensation, Smart materials and Structure, Volume 12, (2003) 549–555.
[74] J.D.W. Madden, J.N. Barisci, P.A. Anquetil, G.M. Spinks, G.G. Wallace, R.H. Baughman, et al., Fast carbon nanotube charging and actuation, Advanced Materials 18 (2006) 870–873.
[75] Y. Yun, V. Shanov, Y. Tu, M.J. Schulz, S. Yarmolenko, S. Neralla, et al., A multiwall carbon nanotube tower electrochemical actuator, Nano Letters 6 (2006) 689–693.
[76] I. Sakurada, Polyvinyl Alcohol Fibers, Marcel Dekker, 1985.
[77] X. Zhang, T. Liu, T.V. Sreekumar, S. Kumar, X. Hu, K. Smith, Gel spinning of PVA/SWNT composite fiber, Polymer 45 (2004) 8801–8807.

[78] M.L. Minus, H.G. Chae, S. Kumar, Interfacial crystallization in gel-spun poly(vinyl alcohol)/single-wall carbon nanotube composite fibers, Macromolecular Chemistry and Physics 210 (2009) 1799–1808.

[79] X. Xu, A.J. Uddin, K. Aoki, Y. Gotoh, T. Saito, M. Yumura, Fabrication of high strength PVA/SWCNT composite fibers by gel spinning, Carbon 48 (2010) 1977–1984.

[80] J. Su, Q. Wang, R. Su, K. Wang, Q. Zhang, Q. Fu, Enhanced compatibilization and orientation of polyvinyl alcohol/gelatin composite fibers using carbon nanotubes, Journal of Applied Polymer Science 107 (2008) 4070–4075.

[81] J.C.H. Affdl, J.L. Kardos, The Halpin-Tsai equations: a review, Polymer Engineering & Science 16 (1976) 344–352.

[82] H.L. Cox, The elasticity and strength of paper and other fibrous materials, British Journal of Applied Physics 3 (1952) 72–79.

[83] T. Liu, S. Kumar, Effect of orientation on the modulus of SWNT films and fibers, Nano Letters 3 (2003) 647–650.

[84] F.M. Blighe, K. Young, J.J. Vilatela, A.H. Windle, I.A. Kinloch, L. Deng, et al., The effect of nanotube content and orientation on the mechanical properties of polymer–nanotube composite fibers: separating intrinsic reinforcement from orientational effects, Advanced Functional Materials 21 (2011) 364–371.

[85] P. Xue, K.H. Park, X.M. Tao, W. Chen, X.Y. Cheng, Electrically conductive yarns based on PVA/carbon nanotubes, Composite Structures 78 (2007) 271–277.

[86] C. Mercader, V. Denis-Lutard, S. Jestin, M. Maugey, A. Derré, C. Zakri, et al., Scalable process for the spinning of PVA-carbon nanotube composite fibers, Journal of Applied Polymer Science 125 (2012) E191–E196.

[87] M.C. Weisenberger, E.A. Grulke, D. Jacques, A.T. Rantell, R. Andrewsa, Enhanced mechanical properties of polyacrylonitrile/multiwall carbon nanotube composite fibers, Journal of Nanoscience and Nanotechnology 3 (2003) 535–539.

[88] T.V. Sreekumar, T. Liu, B.G. Min, H. Guo, S. Kumar, R.H. Hauge, et al., Polyacrylonitrile single-walled carbon nanotube composite fibers, Advanced Materials 16 (2004) 58–61.

[89] B.G. Min, T.V. Sreekumar, T. Uchida, S. Kumar, Oxidative stabilization of PAN/SWNT composite fiber, Carbon 43 (2005) 599–604.

[90] H.G. Chae, M.L. Minus, S. Kumar, Oriented and exfoliated single wall carbon nanotubes in polyacrylonitrile, Polymer 47 (2006) 3494–3504.

[91] H.G. Chae, M.L. Minus, A. Rasheed, S. Kumar, Stabilization and carbonization of gel spun polyacrylonitrile/single wall carbon nanotube composite fibers, Polymer 48 (2007) 3781–3789.

[92] W. Wang, N.S. Murthy, H.G. Chae, S. Kumar, Structural changes during deformation in carbon nanotube-reinforced polyacrylonitrile fibers, Polymer 49 (2008) 2133–2145.

[93] H.G. Chae, Y.H. Choi, M.L. Minus, S. Kumar, Carbon nanotube reinforced small diameter polyacrylonitrile based carbon fiber, Composites Science and Technology 69 (2009) 406–413.

[94] R. Jain, M.L. Minus, H.G. Chae, S. Kumar, Processing, structure, and properties of PAN/MWNT composite fibers, Macromolecular Materials and Engineering 295 (2010) 742–749.

[95] Y. Liu, H.G. Chae, S. Kumar, Gel-spun carbon nanotubes/polyacrylonitrile composite fibers. Part I: effect of carbon nanotubes on stabilization, Carbon 49 (2011) 4466–4476.

[96] Y. Liu, H.G. Chae, S. Kumar, Gel-spun carbon nanotubes/polyacrylonitrile composite fibers. Part II: stabilization reaction kinetics and effect of gas environment, Carbon 49 (2011) 4477–4486.

[97] Y. Liu, H.G. Chae, S. Kumar, Gel-spun carbon nanotubes/polyacrylonitrile composite fibers. Part III: effect of stabilization conditions on carbon fiber properties, Carbon 49 (2011) 4487–4496.

[98] S. Kumar, T.D. Dang, F.E. Arnold, A.R. Bhattacharyya, B.G. Min, X. Zhang, et al., Synthesis, structure, and properties of PBO/SWNT composites &, Macromolecules 35 (2002) 9039–9043.

[99] T. Uchida, S. Kumar, Single wall carbon nanotube dispersion and exfoliation in polymers, Journal of Applied Polymer Science 98 (2005) 985–989.

[100] C. Zhou, S. Wang, Y. Zhang, Q. Zhuang, Z. Han, In situ preparation and continuous fiber spinning of poly(p-phenylene benzobisoxazole) composites with oligo-hydroxyamide-functionalized multi-walled carbon nanotubes, Polymer 49 (2008) 2520–2530.

[101] Z. Hu, J. Li, P. Tang, D. Li, Y. Song, Y. Li, et al., One-pot preparation and continuous spinning of carbon nanotube/poly(p-phenylene benzobisoxazole) copolymer fibers, Journal of Materials Chemistry 22 (2012) 19863–19871.

[102] F. Grillard, C. Jaillet, C. Zakri, P. Miaudet, A. Derré, A. Korzhenko, et al., Conductivity and percolation of nanotube based polymer composites in extensional deformations, Polymer 53 (2012) 183–187.

[103] Y. Wang, R. Cheng, L. Liang, Y. Wang, Study on the preparation and characterization of ultra-high molecular weight polyethylene / carbon nanotubes composite fiber, Composites Science and Technology 65 (2005) 793–797.

[104] S. Ruan, P. Gao, T.X. Yu, Ultra-strong gel-spun UHMWPE fibers reinforced using multiwalled carbon nanotubes, Polymer 47 (2006) 1604–1611.

[105] J. Yeh, S. Lin, K. Chen, K. Huang, Investigation of the ultradrawing properties of gel spun fibers of ultra-high molecular weight polyethylene/carbon nanotube blends, Journal of Applied Polymer Science 110 (2008) 2538–2548.

[106] H. Mahfuz, M.R. Khan, T. Leventouri, E. Liarokapis, Investigation of MWCNT reinforcement on the strain hardening behavior of ultrahigh molecular weight polyethylene, Journal of Nanotechnology 2011 (2011) 1–9.

[107] V. Mottaghitalab, G.M. Spinks, G.G. Wallace, The influence of carbon nanotubes on mechanical and electrical properties of polyaniline fibers, Synthetic Metals 152 (2005) 77–80.

[108] V. Mottaghitalab, G.M. Spinks, G.G. Wallace, The development and characterisation of polyaniline–single walled carbon nanotube composite fibres using 2-acrylamido-2 methyl-1-propane sulfonic acid (AMPSA) through one step wet spinning process, Polymer 47 (2006) 4996–5002.

[109] Y.J. Kim, M.K. Shin, S.J. Kim, S.-K. Kim, H. Lee, J.-S. Park, et al., Electrical properties of polyaniline and multi-walled carbon nanotube hybrid fibers, Journal of Nanoscience and Nanotechnology 7 (2007) 4185–4189.

[110] V. Mottaghitalab, B. Xi, G.M. Spinks, G.G. Wallace, Polyaniline fibres containing single walled carbon nanotubes: enhanced performance artificial muscles, Synthetic Metals 156 (2006) 796–803.

[111] G.M. Spinks, V. Mottaghitalab, M. Bahrami-Samani, P.G. Whitten, G.G. Wallace, Carbon-nanotube-reinforced polyaniline fibers for high-strength artificial muscles, Advanced Materials 18 (2006) 637–640.

[112] G.M. Spinks, S.R. Shin, G.G. Wallace, P.G. Whitten, I.Y. Kim, S.I. Kim, et al., A novel "dual mode" actuation in chitosan/polyaniline/carbon nanotube fibers, Sensors and Actuators B: Chemical 121 (2007) 616–621.

[113] C.Y. Wang, V. Mottaghitalab, C.O. Too, G.M. Spinks, G.G. Wallace, Polyaniline and polyaniline–carbon nanotube composite fibres as battery materials in ionic liquid electrolyte, Journal of Power Sources 163 (2007) 1105–1109.

[114] J. Foroughi, G.M. Spinks, G.G. Wallace, A reactive wet spinning approach to polypyrrole fibres, Journal of Materials Chemistry 21 (2011) 6421.

[115] R. Jalili, J.M. Razal, G.G. Wallace, Exploiting high quality PEDOT: PSS–SWNT composite formulations for wet-spinning multifunctional fibers, Journal of Materials Chemistry 22 (2012) 25174–25182.

[116] H. Zhang, Z.G. Wang, Z.N. Zhang, J. Wu, J. Zhang, J.S. He, Regenerated-cellulose/multiwalled-carbon-nanotube composite fibers with enhanced mechanical properties prepared with the ionic liquid 1-Allyl-3-methylimidazolium chloride, Advanced Materials 19 (2007) 698–704.

[117] P. Chen, H.-S. Kim, S.-M. Kwon, Y.S. Yun, H.-J. Jin, Regenerated bacterial cellulose/multi-walled carbon nanotubes composite fibers prepared by wet-spinning, Current Applied Physics 9 (2009) e96–e99.

[118] S.S. Rahatekar, A. Rasheed, R. Jain, M. Zammarano, K.K. Koziol, A.H. Windle, et al., Solution spinning of cellulose carbon nanotube composites using room temperature ionic liquids, Polymer 50 (2009) 4577–4583.

[119] S. Rahatekar, A. Rasheed, R. Jain, M. Zammaranoa, K.K. Koziol, A.H. Windle, et al., in: Processing of Natural Fibers Nanocomposites Using Ionic Liquids, ECS, 2009, pp. 119–127.

[120] G.M. Spinks, S.R. Shin, G.G. Wallace, P.G. Whitten, S.I. Kim, S.J. Kim, Mechanical properties of chitosan/CNT microfibers obtained with improved dispersion, Sensors and Actuators B: Chemical 115 (2006) 678–684.

[121] A.J. Granero, J.M. Razal, G.G. Wallace, M. in het Panhuis, Conducting gel-fibres based on carrageenan, chitosan and carbon nanotubes, Journal of Materials Chemistry 20 (2010) 7953–7956.

[122] C. Lynam, S.E. Moulton, G.G. Wallace, Carbon-nanotube biofibers, Advanced Materials 19 (2007) 1244–1248.

[123] V. Sa, K.G. Kornev, A route toward wet spinning of single walled carbon nanotube fibers: sodium alginate – SWCNT fibers, in: Materials Research Society Symposium Proceedings (2009), pp. 127–132.

[124] V. Sa, K.G. Kornev, A method for wet spinning of alginate fibers with a high concentration of single-walled carbon nanotubes, Carbon 49 (2011) 1859–1868.

[125] M.K. Shin, B. Lee, S.H. Kim, J.A. Lee, G.M. Spinks, S. Gambhir, et al., Synergistic toughening of composite fibres by self-alignment of reduced graphene oxide and carbon nanotubes, Nature Communications 3 (2012) 650.

[126] R.R. Wang, J. Sun, L. Gao, C.H. Xu, J. Zhang, Fibrous nanocomposites of carbon nanotubes and graphene-oxide with synergetic mechanical and actuative performance, Chemical Communications 47 (30) (2011) 8650–8652.

CHAPTER 7

Dry Spinning Carbon Nanotubes into Continuous Yarn: Progress, Processing and Applications

Canh-Dung Tran

Computational Engineering & Scientific Research Centre (CESRC), Faculty of Engineering and Surveying, University of Southern Queensland, Toowoomba, Qld, Australia

CHAPTER OUTLINE

7.1 Introduction	212
7.1.1 Wet spinning	212
7.1.2 Direct spinning of CNT aerogel	213
7.1.3 Dry spinning	213
7.2 Basis of CNT Assembly in Macroscopic Structures	213
7.2.1 Interrelations of CNTs in macroscopic structures	214
7.2.2 Potential energy function for the interaction between CNTs	215
7.3 From Textile Spinning Technology to Dry CNT Spinning	216
7.3.1 CNT forest/matrix	217
7.3.2 CNT web: formation and model	217
7.3.2.1 Formation of CNT web	217
7.3.2.2 Dual slip-link-based CNT web model	218
7.3.3 Aligning and tensioning system	219
7.3.4 Packing fibers in CNT yarn	219
7.4 Multistep Spinning Process Using a Drafting System	222
7.4.1 Separating a CNT web using an electrostatic approach	222
7.4.1.1 Dilatation of CNT fibers	222
7.4.1.2 Dilatation of CNT silver/web using electrostatic field	223
7.4.2 Multistep CNT dry spinning process and results	224
7.5 Several Treatments for CNT Yarn Improvement	225
7.5.1 Compaction of a CNT yarn	225
7.5.2 Heat treatment of a web	226
7.5.3 Improving the strength of CNT yarn using functionalization	226
7.5.4 Multiply twisted yarns	227
7.6 CNT-Based Composite Yarns	227
7.6.1 CNT–polymer composite yarns	227
7.6.2 Biscrolling CNT web and particles into yarn	230
7.6.3 CNT–metal composite yarns	230

7.7 Applications of CNT Yarns..	231
7.7.1 Textile electrodes and supercapacitors	232
7.7.2 CNT yarns for actuators..	232
7.7.3 CNT yarn sensors..	232
7.7.4 CNT yarns in bioengineering...	233
7.7.5 CNT-based high-performance yarns...	234
7.8 Conclusion...	234
Acknowledgments	234
References	235

7.1 INTRODUCTION

The carbon nanotube (CNT) atomic structure with unique properties, such as high tensile strength and Young's modulus, high electrical and thermal conductivities and very high aspect ratio [1−3], has underpinned potentially novel applications, for example, nanomechanics [4], advanced electronics [5], biotechnology [6] and materials science. A number of investigations have shown that CNTs can enhance the properties of polymer composites at relatively low levels of CNTs owing to the excellent mechanical properties [7−10]. Recently, in spite of the emergence of graphene, CNT is still a potential material for many important applications requiring slender fiber as a priority, such as the advanced fibrous materials. The prerequisite for these applications of CNT is based on its very high aspect ratio (1000−10,000) [11].

Between 1991 and 2010, significant effort has been spent on producing CNT yarns to make use of the CNT's unique properties [12−17]. Methods for manufacturing CNT yarns include the direct synthesis of CNT yarns [18]; producing yarn from CNT aerogel [19]; spinning yarn from vertically aligned CNT arrays [14,17]; and wet spinning method of CNT yarns [20,21]. These methods were based on traditional textile spinning principles and can be classified into three principal approaches: (1) the wet spinning or solution-based spinning method; (2) the direct spinning from CNT aerogel; and (3) the dry spinning or solid state-based spinning method.

7.1.1 Wet spinning

Owing to the chemical inertness and agglomeration of CNTs, yarns are formed by the extrusion of acid/alkaline media or polymer solutions containing dispersed CNTs instead of directly spinning pristine CNTs [20−23]. In this approach, CNTs are dispersed in superacids which provide charge to the CNTs and promote their ordering into an aligned phase of individual CNTs surrounded by acid anions [20]. The solution is then extruded into a coagulation solution to produce continuous CNT yarn. Since concentrated CNT solutions would favor the formation of liquid-crystal domains, CNTs can be processed into yarns with controlled morphologies. Different from the just mentioned technique, several works [24−26] reported the use of coagulation spinning approach to assemble CNTs into yarn. In this way, CNTs are homogeneously dispersed in aqueous solutions of sodium dodecyl sulfate [26] or lithium dodecyl sulfate [27] which can be adsorbed onto the surface of the CNT fibers and neutralize the van der Waals force. The CNT solution is then extruded into the second flowing stream of polymer solution containing very low concentration of polyvinyl alcohol. The elongational flow via the extrusion yields the flow-induced alignment and a preferential orientation of the CNTs in a sliver. After being washed and dried, the surfactants and polymers are removed and CNT yarns are then formed. Recently, another wet spinning method was developed to produce CNT yarn without using superacid. In this method, CNTs are dispersed in ethylene glycol. While this crystalline dispersion liquid is injected into a diethyl ether, the ethylene glycol diffuses out of the CNT yarn [28]. The report claimed that CNT fibers are highly aligned due to the combination of shear forces and the liquid-crystalline phase. The yarn produced using wet spinning method has good electrical conductivity (5000 S/cm) and thermal conductivity (21 W/km) [24,26].

7.1.2 Direct spinning of CNT aerogel

The assembly of CNTs into a sliver is carried out in a furnace chamber, where CNT as an aerogel is synthesized from a precursor material mixed with hydrogen (as a carrier gas) during a chemical vapor decomposition [18,19,29,30]. The CNT aerogel is then pulled out of the furnace to form a yarn which is wound on a bobbin. Hence, the method is known as the direct spinning or one-step spinning method.

7.1.3 Dry spinning

Since Jiang et al. [14] reported that yarns can also be assembled from a vertically aligned CNT forest on a substrate, significant efforts have been devoted to improve the performance of CNT yarns using the dry spinning method [17,31–34]. In spite of several recent works on dry CNT yarn spinning, the knowledge on yarn assembly is still open. In this chapter, we present an overview of this CNT spinning approach together with recent progress and applications.

7.2 BASIS OF CNT ASSEMBLY IN MACROSCOPIC STRUCTURES

In a macroscopic structure (forest, yarn, web and three-dimensional structures such as bioengineering filters), there is a hierarchical structure composed of two levels: (1) individual CNTs at the molecular level and (2) bundles of aggregated CNTs (called secondary fibers, Fig. 7.1(a),(d)) held by van der Waals forces [34,35]. For the sake of presentation in this review, the individual CNT is described as a CNT or nanotube, a bundle of CNTs as CNT fiber, and sliver or web describes fibers of low density and without twist. When gathered from a CNT forest, the CNT fibers form a continuous network called the web (Fig. 7.1(a)).

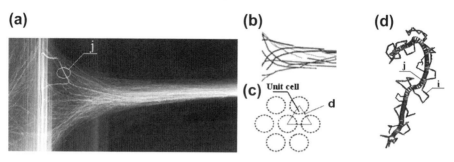

FIGURE 7.1

The structure of several stages of the CNT spinning process: (a) CNT forest structure; (b) schema of pulling out a CNT web; (c) schema of the cross-section of a hexagonal array of CNTs and (d) the model of a CNT fiber. (For color version of this figure, the reader is referred to the online version of this book.)

7.2.1 Interrelations of CNTs in macroscopic structures

Experience shows that during synthesis, CNTs usually form fibers consisting of hundreds of CNTs [36,37]. Indeed, the micrograph (Fig. 7.1(a)) shows that the CNTs tend to agglomerate in the form of close-packed fibers in macroscopic structures (Fig. 7.1(c),(d)). The fibers form a continuous network, called the web, with a preferred orientation along the yarn axis. Tran et al. [34] postulate that the dry spinning of CNTs brings together the characteristics of both traditional staple spun yarn and synthetic filament spinning, in which the web and yarn structures are the key factors affecting the yarn's physical and mechanical properties. In addition, CNT fibers in web/yarns have similarities to the staple fibers of a spun yarn. As a fibrous material, the study of CNT macroscopic structures in general and CNT yarn in particular is necessary because there is an interrelation between the structure and properties of CNT fibers and yarns. This interrelation is given in Fig. 7.2.

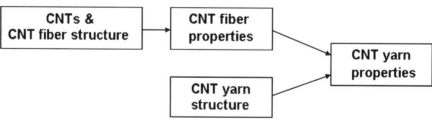

FIGURE 7.2

Interrelations of CNTs, CNT fiber and yarn structure and properties.

It can be appreciated that fibers in a web vary in diameter and length and this causes difficulties in spinning staple spun yarn from the point of view of conventional textiles [38].

7.2.2 Potential energy function for the interaction between CNTs

The CNTs interaction force is based on van der Waals force, in which the interaction energy between pairs of carbon atoms ϕ can be approximated using the Lennard Jones potential (LJP) as follows [39,40]:

$$\phi(d) = 4\varepsilon\left[\left(\frac{\sigma}{d_a}\right)^{12} - \left(\frac{\sigma}{d_a}\right)^{6}\right] \quad (7.1)$$

where d_a is the distance between atomic centers; ε, the depth of the potential energy well; σ, the spacing where the potential energy is zero.

Recently, Girifalco et al. [41] have developed a continuum approach to calculate the interaction potential $\overline{\phi}$ between two nanostructures (for example, CNT–CNT, CNT ropes, graphene–graphene) by integrating the LJP over the surface of a structure as follows:

$$\phi(\overline{d}) = -\frac{|\phi(d_0)|}{0.6}\left(\left(\frac{3.41}{3.13\overline{d} + 0.28}\right)^{4} - 0.4\left(\frac{3.41}{3.13\overline{d} + 0.28}\right)^{10}\right) \quad (7.2)$$

where d is the center to center spacing of two CNTs, d_0, the equilibrium distance (the distance between CNT centers in the unload state), ρ, the distance characteristic of the specific geometries in the interaction and $|\phi(d_0)|$, the energy well depth and $\overline{d}$, the normalization of d is given by

$$\overline{d} = \frac{d - \rho}{d_0 - \rho} \quad (7.3)$$

The parameters d_0, ρ and $|\phi(d_0)|$ are dependent on the chiral pair of CNTs and are given in Table 7.1.

Several theoretical and experimental studies [42,43] reported that the breaking force for fibers/yarns was related to the sliding of CNTs rather than breakage of individual CNTs and can be approximated as

$$F = F_1 + F_2 = F_1 + lb \quad (7.4)$$

Table 7.1 Universal Potential Function Parameters of Several Chiral Pairs of CNTs [41]

System	Energy Well Depth $|\phi(d_0)|$ (nj/m)	Equilibrium Spacing d_0 (nm)	Distance Parameter ρ (nm)
(6,6)–(6,6) CNTs	0.1176	1.1281	0.8142
(10,10)–(10,10) CNTs	0.1525	1.6732	1.3570
(12,12)–(12,12) CNTs	0.1674	1.9441	1.6284

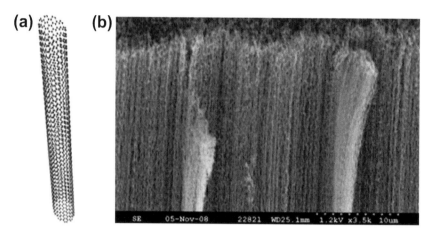

FIGURE 7.3

CNT structure and CNTs forest (matrix): (a) a schema of CNT and (b) a forest of wavy CNTs.

where F_1, the sliding force, depends on the interactive energy of CNT fibers (Eqn (7.1)); F_2, the force to overcome the effect of CNTs' corrugation (Fig. 7.3(b)), scales linearly with the length of contact between neighboring CNTs (l) and b, a constant value [43–45].

Thus, based on Eqn (7.4), there are two interaction energies in yarn: (1) van der Waals-based microcohesion between CNTs and (2) the macrofriction cohesion between CNT fibers.

7.3 FROM TEXTILE SPINNING TECHNOLOGY TO DRY CNT SPINNING

This section discusses the similarities between traditional textile spinning and dry spinning of CNT into yarn.

For dry spinning method, yarns are spun from a substrate-based forest of CNTs which is a forest of vertically aligned CNTs (Fig. 7.4). In spite of efforts to improve the properties of CNT yarns, these are still much inferior to the properties of an individual CNT. Several studies reported that while textile yarns can achieve up to 70% of the specific strength of their constituent fibers [46], CNT yarns only reach few percent of the tensile strength of an individual CNT [17,34].

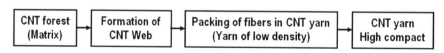

FIGURE 7.4

CNT dry spinning method.

7.3 From Textile Spinning Technology to Dry CNT Spinning

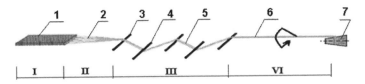

FIGURE 7.5

A schema of the modified dry spinning process. I: CNT forest/matrix; II: web formation; III: aligning and tensioning system and IV: highly compact yarn; CNT forest (1); CNT web (2); capstan rods (3&4); CNT sliver (5); CNT yarn (6); Bobbin (7) [34]. (For color version of this figure, the reader is referred to the online version of this book.)

The ability to dry spin was attributed in part to the extent of packing of fibers applied during spinning, i.e. packing reduced the inter-CNT fiber distance and increased the cohesion energy between CNT fibers within yarn. Recently, a modified dry CNT spinning process has been developed [34]. The process is based on the principle of a staple spun yarn spinning and includes four separate stages with four corresponding structures, namely CNT forest, CNT web, CNT sliver, and CNT yarn. For this process, the system was partitioned into four distinct zones (Fig. 7.5): CNT forest/matrix (1), web formation zone (2), tensioning/drafting zone (3), and compacting zone (4). More details can be found in Ref. [34].

7.3.1 CNT forest/matrix

A spinnable CNT forest is the first condition for the dry spinning method. Several publications claimed that not all CNT arrays can be spun into yarn [47] and the spinnability of CNTs has a strong relationship with the morphology of CNT arrays including length and shape and density of CNTs. Indeed, one approach for improving the properties of CNT yarns is to develop longer and better aligned CNT forests on wafers [48,49].

7.3.2 CNT web: formation and model

7.3.2.1 Formation of CNT web

The dry spinning process starts by forming CNT web, whereby a forest of vertically oriented CNT forest converts to a horizontally continuous CNT web in which CNT fibers are oriented along the web axis in a variety of ways, some of them are parallel and others are poorly aligned or coiled [34] (Fig. 7.6).

The web formation influences the properties of CNT yarns. Hence, its mechanism has been simulated in detail as reported in recent publications [49–51]. For example, an elegant model by Kunznetsov et al. [50] proposed that the connecting CNTs between CNT fibers were the basis for the continuous spinning of a CNT yarn from a CNT forest. Furthermore, they reported that during the process, the densification effect at the top and bottom of the forest strengthens the fibers interconnections.

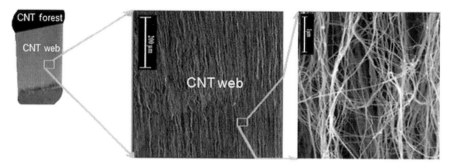

FIGURE 7.6

A structure of CNT web. (For color version of this figure, the reader is referred to the online version of this book.)

7.3.2.2 Dual slip-link-based CNT web model

From the observation of the CNT dry-drawing process, the scanning electron microscopy (SEM) analysis of webs showed that CNT fibers entangle with each other. This entanglement is a key factor because it allows the array of parallel fibers to unfold continuously into a CNTs mat. The entanglement of CNTs can be regarded as network of CNTs made of fibers "entanglement junctions" (Fig. 7.7(b)) and have a strong relation with the properties of CNT yarns. We postulate that CNT fibers are movable because the constraint imposed by the surrounding CNTs can change due to their own motion and a constraint will disappear if surrounding CNTs shift.

Thus, the entanglement of CNT fibers (or web) can be modeled using the dual slip-link theory which was introduced by Doi and Takimoto [52] in modeling polymeric materials. For the model, the entanglement junction is represented by a slip link through which fiber chains can pass freely. The CNT fibers are linked together by slip links which are destroyed if fibers slide off the slip links. This model corresponds to the formation of a network of random CNT fibers with nonuniform tension

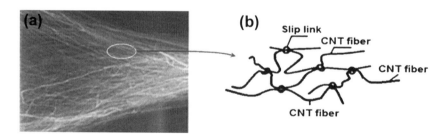

FIGURE 7.7

Model of CNT web formation: (a) web formation from a CNT forest and (b) schema of CNT web using the dual slip-link model. (For color version of this figure, the reader is referred to the online version of this book.)

pulled out of a CNT forest. Geometrical parameters of CNTs such as length, diameter, waviness, and the distance between CNTs can be controlled independently, so that the effects of these parameters on the production and the properties of CNT yarns can be investigated.

7.3.3 Aligning and tensioning system

A key issue in fabricating high-performance CNT yarns is to align the CNT fibers along the yarn axis. Indeed, a web with a poorly oriented structure results in yarns with less than optimal packing of the fibers. A system consisting of several rods forms an aligning and tensioning system (ATS) system to improve the orientation and control the nonuniform tension of fibers in sliver/yarn. Indeed as sliver passes through the ATS, a capstan effect rod system (CERS) [34], the increased tension extends and aligns the fibers due to the slip links as presented in the dual slip-link model (Fig. 7.7(b)) and then some strain can occur. The ATS yields two effects: "elongation and alignment" of fibers in the sliver due to tension and the condensing of the fibers together due to the radial pressure. Both effects lead to an improvement of the properties of a CNT yarn including mechanical properties and electrical conductivity [34,53]. Experimental works showed an obvious alignment of CNT fibers in sliver (Fig. 7.8).

7.3.4 Packing fibers in CNT yarn

As mentioned earlier, the dry spinning depends on the packing density of fibers. The role of packing was to increase the van der Waals-based microinteraction between CNTs and the macrofriction between CNT fibers and yields the required interbundle lateral cohesion of the yarn (Section 7.2.2). Since the tenacity of a CNT yarn depends on stress transfer in shear mode between neighboring CNT fibers, a better contact results in a stronger stress transfer between neighboring CNT fibers under tensile loading conditions. In traditional textile, there are several methods to condense fibers into yarn and classified into two groups: twisting (Fig. 7.9(a)) and rubbing/false-twisting (Fig. 7.9(b)).

FIGURE 7.8

SEM image showing the microstructure of the fibers in CNT web/sliver; (a, b) CNT web SEM images and (c) sliver SEM image under high tension between two initial rods (3) and (4) on the CERS (Fig. 7.5).

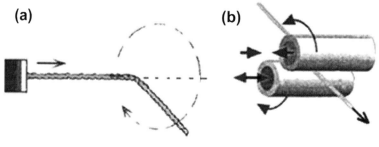

FIGURE 7.9

Packing CNT fibers in yarns: (a) a schema of twist method and (b) a schema of rubbing method. (For color version of this figure, the reader is referred to the online version of this book.)

To achieve a higher tensile strength, the CNTs fibers need not only to have a minimum length but also to share maximum contact area. Besides the increase of van der Waals forces between CNTs, CNT fibers are held together by the fiber-to-fiber interception formed by twist. Thus, twisting leads to an interlocked CNT yarn and can enhance stress transfer between CNT bundles under tensile forces [42]. The tensile strength of CNT yarns can be effectively improved by a high frictional force between the fibers with a higher twist level [17,31]. Traditional textile mechanics would also predict loss of strength as twist levels increase above an optimum value [46].

The loss of strength of CNT yarns at high twist levels has been noted in Ref. [31]; and it is necessary to determine an optimum twist for CNT yarns to prevent fiber slippage [34]. Several interesting studies on the relation between the twist levels and tensile strength of CNT yarns using SEM and focused ion beam (FIB) have been recently reported in Ref. [54]. Furthermore, the highly twisted CNT yarn tends to snarl or coil when unconstrained and causes many difficulties in using and processing CNT yarns [55,56]. Figure 7.10 shows a microsnarling on a CNT yarn [34].

Experimental results showed that the tensile strength of the CNT yarn prepared by the dry spinning process with ATS increased significantly (two times higher) compared with yarn spun without ATS [34], for given CNT parameters (i.e. length, diameter and number of walls) and engineering parameters such as take-up speed and twisting level. The tensile strength and strain data cover a range from 970 MPa to 1.4 GPa and 5–8%, respectively (Fig. 7.11).

For the dry spinning process, the improvement in the yarn structure was obtained by stretching CNT fibers and enabling more effective yarn compaction. While the densification treatment in the process was not only from the twist but also by the radial pressure through the ATS which can be considered as a method to produce a highly densified CNT yarn without twist. Several other traditional textile twistless methods can be used to produce densified compact CNT [57] such as rubbing method. However, since the aspect ratio of CNT fibers is very high ($\sim 10{,}000$),

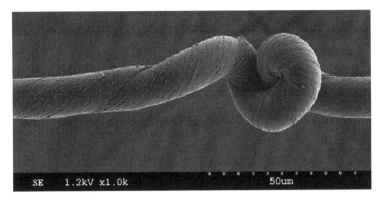

FIGURE 7.10

Yarn of high twist with densification: a snarl of yarn.

Source: Courtesy of Ref. [34].

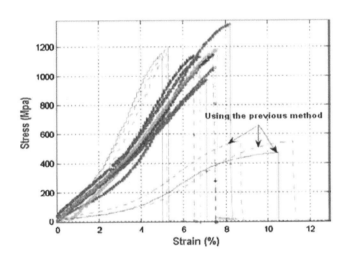

FIGURE 7.11

The improvement of the tensile strength using the modified spinning process for CNT yarns: A comparison of the tensile strength of the CNT yarns. (For color version of this figure, the reader is referred to the online version of this book.)

Source: Courtesy of Ref. [34].

the rubbing could make CNT fibers into a coil (called nep in textile mechanics) or produce a random structure within CNT yarn ([58], Fig. 7.2). Thus, the issue requires further investigation.

Although the modified spinning process leads to a significant improvement of the properties of CNT yarns, it was not possible to conduct an efficient elongation process in one step due to some limits in optimizing the CNT alignment in a web. Furthermore, since CNT sliver is too weak to stretch in one go without causing fiber damage, a new dry spinning process using a multistep drafting system is required to overcome disadvantages [59].

7.4 MULTISTEP SPINNING PROCESS USING A DRAFTING SYSTEM

The separation of CNT fibers using an electric field [60] forms a multistep CNT spinning process. The process is a further development of the modified process and is described by a schema in Fig. 7.12.

7.4.1 Separating a CNT web using an electrostatic approach
7.4.1.1 Dilatation of CNT fibers

The alignment and tensioning will be improved if fiber interaction is firstly reduced in the sliver/web structure. To separate CNTs, a transverse extensional force will be used for the dilatation of CNT fibers. Under the transverse load, there will be a displacement δ ($\delta = d - d_0$). From Eqn (7.2), the force and energy per unit length required to reach a given deformation are determined, respectively, as follows [41]:

$$F(\bar{\delta}) = 6.119 \frac{|\phi(d_0)|}{d_0 - \rho} \left[\left(\left(\frac{1}{1 + 0.9179\bar{\delta}} \right)^5 - \left(\frac{1}{1 + 0.9179\bar{\delta}} \right)^{11} \right) \right] \quad (7.5)$$

$$U(\bar{\delta}) = |\phi(d_0)| \left[1 - \frac{1}{0.6} \left(\left(\frac{1}{1 + 0.9179\bar{\delta}} \right)^4 - 0.4 \left(\frac{1}{1 + 0.9179\bar{\delta}} \right)^{10} \right) \right] \quad (7.6)$$

where $\bar{\delta} = \delta/(d_0 - \rho)$.

The dilatation of CNT fibers (bundle of CNTs) is carried out by a transverse extensional force. Considering a hexagonal bundle of a finite number of CNTs

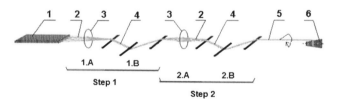

FIGURE 7.12

Model of two-step spinning process: (1.A, 2.A): separation; (1.B, 2B): aligning and tensioning; CNT forest (1); CNT webs (2); Rings of electric field (3); CNT slivers (4); CNT yarn (5) and Bobbin (6). (For color version of this figure, the reader is referred to the online version of this book.)

(N) (Fig. 7.1(c)), the energy density (per unit volume) needed to split a fiber into individual CNTs is given by [39]:

$$E = \frac{4\phi(d_0)}{\sqrt{3}d_0^2} \frac{\sum_{i=1}^{N}(3i-1)}{\sum_{i=1}^{N}(2i-1)} \qquad (7.7)$$

where i is the characteristic coefficient of the hexagonal array of CNT fibers and N, the number of CNT center-to-center lengths that make up the length of a side of the hexagon encompassing all unit cells in the array.

7.4.1.2 Dilatation of CNT silver/web using electrostatic field

For our work, the dilatation of CNT web/sliver/yarn is carried out using a dielectrophoresis [61], in which the energy for separating the fibers is set up using a direct current (DC) electric field. The electrical potential causes a repulsive force between CNTs which helps separate CNT fibers (Fig. 7.13(a),(b)) [60]. The process is set up as shown in Fig. 7.13(c),(d), where a web/sliver of CNT (2) at first goes through a ring (3). A high voltage is applied to the ring such that at a critical voltage

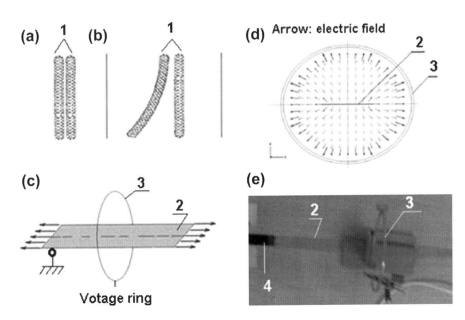

FIGURE 7.13

Dilatational separation of CNT fibers and control of CNT web density using the electrostatic field approach: (a) & (b) a schema of two CTNs without and with a dilatational separation; (c) a schema of the dilatational separation of CNT bundles; (d) electric field depicted by "arrows" and (e) an actual operational process of the dilatational separation of CNT fibers. CNTs (1), CNT web (2), ring of high voltage (3) and CNT forest (4). (For color version of this figure, the reader is referred to the online version of this book.)

(Fig. 7.13(d)), the repulsive force within the charged CNTs' web acts against the van der Waals force between CNTs and the web/sliver is then spread out. The voltage imposed in rings is determined based on the desired dilatation force which is approximated by Eqns (7.5)–(7.7). Stretching of CNT sliver/web depends on the electrostatic forces caused by the voltage imposed in the ring.

These effects provide the basis to develop multistep spinning process for producing CNT yarns whose structure is controllable.

7.4.2 Multistep CNT dry spinning process and results

For the process, the ATS includes a repeating multistep drafting system, in which each step of drafting consists of two operations: (1) separation of CNT fibers ($k.A$ in Fig. 7.12); and (2) compact CNTs web into a sliver and then align and strengthen the sliver by a smaller tension force using a CERS ($k.B$) ($k = 1,2,...,n$). The number of steps n depends on the quality of the forest (i.e. the geometrical parameters of CNTs as well as the alignment of CNT forest) and the desired performance of yarns.

At a step k, the separation ($k.A$) supports the efficient and effective alignment and stretch of CNT fibers with a much smaller tension force and limits the damage to CNT fibers. Indeed, based on the dual slip-link-based CNT web model, under the dilation, the slack CNT fibers are elongated through the slip links without breaking (Fig. 7.14).

Furthermore, the role of the electrostatic field imposed at the CNT web when pulled out of the CNT forest is not only to separate CNT fibers but also to align CNTs as reported in several works [62,63]. Especially, the separation force allows reducing the entanglement while transferring vertically aligned CNTs into the horizontal CNT web.

It has been experimentally shown that the alignment of CNTs in a web is gradually improved via the ATS steps and the damage to CNT bundles is significantly reduced [59]. Figure 7.15 depicts the morphology and tension of CNT fibers at the two steps using two rings of 400 and 500 V DC at step 1 and step 2, respectively.

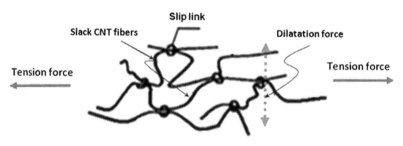

FIGURE 7.14

A schema of the CNT fibers' elongation under the dilatation and tensile forces based on the dual slip-link-based CNT web model. (For color version of this figure, the reader is referred to the online version of this book.)

FIGURE 7.15

Multistep spinning process; (a) initial CNT web; (b) CNT sliver at the first step and (c) CNT sliver at the second step in the ATS with an electrostatic field.

Owing to the use of much smaller tension forces, the multistep spinning process allows for spinning ultrafine yarns (e.g. we achieved 4-μm-diameter yarns). The elongation ratio of the two-step spinning process can reach four times higher value in comparison with the ones without voltage rings. The elongation ratio depends on the voltage imposed in rings, the tensile force of the ATS at the steps and the width of CNT strip.

Several experimental studies have been carried out and showing some promising results on both the alignment of CNTs in the yarn structure and the mechanical properties of yarns using the multistep dry process [59].

7.5 SEVERAL TREATMENTS FOR CNT YARN IMPROVEMENT

Treatments for CNT yarns can be carried out during the yarn formation (spinning) process or as a further postspinning processing. The separation of distinctive mechanical and structural features at the different stages of the web and yarn formation in both the modified spinning process and the multistep spinning process allows for different relevant treatments to be performed at individual stages of the spinning process. Several typical treatments to (1) improve the bulk properties of yarns and/or (2) create CNT yarns of desired features have been recently reported and are described in the following sections.

7.5.1 Compaction of a CNT yarn

A yarn of CNT fibers consists of thousands of individual CNTs. The van der Waals forces between these CNTs hold them together and are responsible for the strength of the fiber as a monolithic entity. The van der Waals force decreases with the distance between the CNTs, and the strength of a CNT fiber increases with the total contact area among the CNTs. Therefore, compaction of a CNT fiber is an efficient way to increase the strength of the CNT fiber. In addition to using the twisting and nontwisting methods to condense CNT yarn, the compaction of CNT yarns can be done by various methods including mechanical, thermal, physical and chemical approaches.

For mechanical processing, based on the principles of Sirospun and multisliver single yarn [64,65], a CNT yarn has been spun from two or many webs split on one single wafer [34]. The method forms many intercept points between CNT fibers

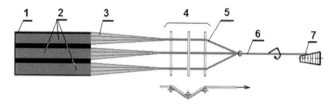

FIGURE 7.16

Schema of a multisliver yarn based on the Sirospun principle: wafer (1); strips of CNT forest (2); CNT webs (3); ATS (4); CNT slivers (5); CNT yarn (6) and bobbin (7).

inside yarns and leads to a dramatic improvement of the mechanical properties of yarns (Fig. 7.16). A similar installation as described in Fig. 7.16 can be done using different sources (wafers) and in such cases, a CNT-based composite yarn can be produced in order to satisfy certain requirements.

The compaction of a yarn is increased by introducing a volatile solvent to the CNT fibers and then evaporating it because the capillary force attracts adjacent CNTs to each other during evaporation. Indeed, several recent publications [66,67] reported that a sliver/yarn pulled out from an aligned CNT forest and twisted was then passed through acetone or ethanol liquid for shrinking to increase the linear density and then improve the tensile strength of yarns. Finally, the compaction of CNT yarns can also be improved via an annealing processing [34].

7.5.2 Heat treatment of a web

It has been shown that the van der Waals force causes the intertubular CNT interactions related to the bundle and web formation and yarn compaction and that the magnitude of the force depends on thermal load and temperature [68–70]. Hence, temperature can be used as a method to control the CNT interaction during the CNT spinning process. Heat treatment of the web has been considered as a factor in decreasing the van der Waals forces as the web is drafted through the ATS, thus increasing the drafting efficiency of the ATS ([34], Fig. 14).

After extraction from the forest, the CNT web goes through a furnace whose temperature is adjustable from 200 to 600 °C. The results showed that the temperature affects the mechanical properties of yarn via its effect on the van der Waals interaction between CNTs. It is important to appreciate that the process of web heat treatment is carried out to improve the drafting properties of the CNT web ([34], Fig. 15).

7.5.3 Improving the strength of CNT yarn using functionalization

The mechanical properties of CNT yarn can be improved by functionalizing CNTs (introducing covalent bonds between the CNTs) in yarns using chemical or physical methods. Recently, Cai et al. [71] reported functionalization for spun CNT yarns using aryldiazonium salts that involves the pH-controlled application of the

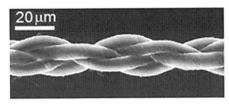
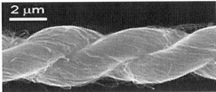

FIGURE 7.17

SEM images of multiply twisted CNT yarns.

Source: Courtesy of AFM, CSIRO.

diazonium salts to CNTs. The process was carried out using the modified CNT spinning process (i.e. between the two rods of the ATS, see Fig. 7.5) during the yarn formation process. The work led to the formation of oligomerizes polyene structures on the CNT surfaces and resulted in a significant improvement of mechanical properties for CNT yarns. The tensile strength of functionalized CNT yarns which was reported by Cai et al. [71] is 60% stronger than that without functionalization.

7.5.4 Multiply twisted yarns

Using conventional textile principles, multiply twisted yarns have been spun to prohibit the slippage of nanotubes and then enhance mechanical properties of CNT yarns (Atkinson et al., 2007) [72] and to design CNT yarns with desired features [73]. Furthermore, the plying is a useful method to balance highly twisted CNT yarns [56]. Figure 7.17 shows several CNT multiply yarns.

7.6 CNT-BASED COMPOSITE YARNS

Recently, CNT-based composite yarns have attracted significant attention across many research fields [11,74–76]. The introduction of additives allows for (1) improving properties of yarns and (2) introducing new functionalities into CNT yarns, for example, CNT composite yarns with high electrical and thermal conductivities and higher tensile strength. Several different techniques for manufacturing CNT-based composite yarns, which have been recently reported, include biscrolling nanotube sheets and functional guests into yarns [75]; CNT–polymer composite yarns [11,77] and CNT–metal composite yarns [76]. These techniques are classified into two groups, namely CNT–polymer composite and CNT–metal composite, and can be directly implemented using the dry spinning process.

7.6.1 CNT–polymer composite yarns

While CNTs served as a filler and were dispersed in a polymer solution in certain composite manufacturing processes [78–81], in the present method of forming, a CNT–polymer composite yarn polymer filled the pores between the stretched

FIGURE 7.18

Schema of the polymer/CNT composite manufacturing process: CNT forest (1), CNT web (2), CNT sliver (3), squeeze rollers (4), furnace (5), guide rods (6), yarn (7), bobbin (8) and polymer applicator (9) [11]. (For color version of this figure, the reader is referred to the online version of this book.)

and aligned CNT fibers in a web/sliver/yarn. The direct process is based on the modified dry CNT spinning process and consists of the following steps: (1) web formation from CNT forests together with the arrangement and alignment of CNT fibers, (2) application of polymer onto CNT web or sliver, (3) compressing and squeezing polymer in the CNT reinforcing layer and (4) curing the composite (Fig. 7.18).

In several studies [77,82], resins were applied by immersing CNT yarns in a polymer solution. In a different approach, Tran et al. [11] have developed the resin impregnation methods for achieving high alignment of CNT fibers in the composite yarn. Applying resin on the CNT web before it passes through the ATS ensures an even distribution of polymer on the CNT web and subsequently in the yarn cross-section. Indeed, a morphological observation of web, sliver and yarn, using an FIB combined with SEM, confirmed the even distribution of polymer on the CNT web and yarn cross-section (Fig. 7.19) as well as the fact that the PU has filled the spaces between the CNTs fibers in a yarn (Fig. 7.20).

Experimental results showed that the tensile strength of CNT−polymer composite yarns (CNT-PVA/DMSO [77] and CNT-PU [11]) was much higher

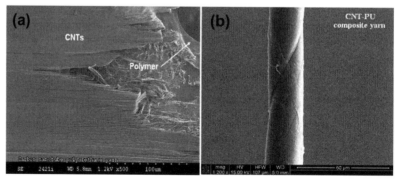

FIGURE 7.19

SEM images of PU/CNT composite sliver: (a) CNT sliver impregnated with PU is obtained by rupturing the PU/CNT composite sliver and (b) the surface layer of PU−CNT composite. PU: polyurethane; DMSO; dimethyl sulfoxide.

Source: Courtesy of Ref. [11].

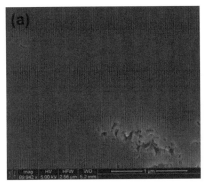 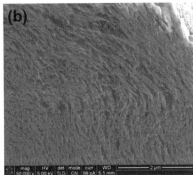

FIGURE 7.20

Morphology of a part of the cross-sections of pure CNT and PU/CNT composite yarns using SEM–FIB: (a) cross-section of a PU/CNT composite yarn with some defects and (b) cross-sections of a pure CNT yarn.

Source: Courtesy of Ref. [11].

than that of pure CNT yarns. In order to further improve the properties of CNT polymer composite yarns, the functionalization can be an efficient method to improve the interfacial interaction and to enhance the covalent bonding between a polymer and CNTs [83–85].

Furthermore, a multiweb spinning from multiwafers (Fig. 7.21) allows the development of multilayer composite yarns and multilayer or sandwich composite constructions which can satisfy various requirements [11]. It should be noted that

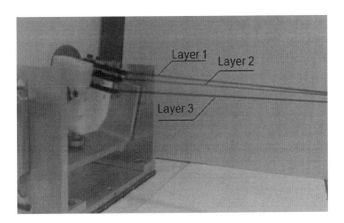

FIGURE 7.21

CNT multiweb spinning from multiwafers [11]. (For color version of this figure, the reader is referred to the online version of this book.)

these processes can be carried out directly on the dry spinning process for CNT—polymer composite yarns.

7.6.2 Biscrolling CNT web and particles into yarn

Recently, Baughman's research group at the Alan G. MacDiarmid NanoTech Institute, University of Texas at Dallas, USA has successfully developed the technique of biscrolling CNT webs and functional guest into yarns [75]. A schema of the process is given in Fig. 7.22. Instead of impregnating polymer liquid/foam in CNT web/sliver [11], powder of particles or nanofibers of desired properties were sprayed onto a CNT web. The web was then scrolled into yarns with various complex structures (Archimedean spirals, Fermat spirals, or Spiral pairs). The technique opened potential for manufacturing of advanced multifunctional yarns such as superconducting yarns, high-performance Li-ion battery electrode yarns, and catalytic fuel cell yarns. Based on this innovation, Baughman's research group has developed many significant electrochemical devices based on CNT yarns (see Section 7.7 for details).

7.6.3 CNT—metal composite yarns

The electrical and thermal conductivities of pure CNT yarns are much lower than that of defect-free individual CNTs due to the presence of amorphous carbon and other impurities, which cause electron scattering and contact resistances [86,87]. Experiment showed that multiwalled CNT yarns have electrical conductivity in the range of $5-6 \times 10^2$ S/m at room temperature (Randeniva et al., 2010). In order to enhance electrical and thermal conductivities, manufacturing of CNT—metal composites with physical vapor decomposition-based methods has recently attracted considerable attention. For example, the incorporation of gold particles into CNT yarns using galvanic deposition in an ethanol solution of $HAuCl_4$ has increased the electrical conductivity of yarns [88]. Choi et al. [89] reported that a spontaneous deposition of Au and Pt particles on single-walled CNTs enhanced the electrical conductivity. Kong et al. [90] obtained hybrids of Au and single-walled CNTs with conductivities increased up to 2×10^3 S/cm.

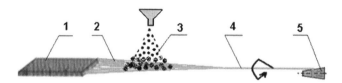

FIGURE 7.22

A schema of CNT biscrolling process: CNT forest (1); CNT web (2); particles spray (3); CNT yarn (4) and bobbin (5). (For color version of this figure, the reader is referred to the online version of this book.)

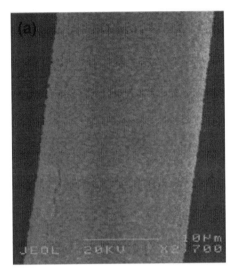 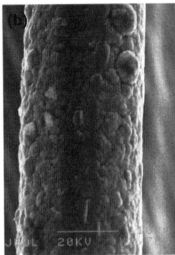

FIGURE 7.23

SEM images for Au–CNT yarn (a) and Cu–CNT yarn (b).

Source: Courtesy of Ref. [76].

Recently, Randeniva et al. (2010) have employed a new technique, the self-fueled electrodeposition (SFED), for incorporating metals (gold, silver and platinum) into CNT yarns. The technique leads to the manufacturing of CNT-based composite yarns with metallic electrical conductivity. The method does not require the use of oxidative pretreatments, which cause degradation of the surface of the CNT fibers and allows CNT–metal (Cu and Au) to reach metallic electrical conductivity owing to the penetration of metal particles into the CNT yarns (Randeniva et al., 2010). Furthermore, the process can be operated both on CNT yarns and during the spinning process. Experimental results showed that Au–CNT and Cu–CNT yarns have unique properties in comparison to other metal–CNT yarns. Several SEM images of CNT–metal (Au and Cu) composite yarns manufactured with the SFED technique are shown in Fig. 7.23. However, a drawback of the technique is a resultant slight loss of tensile strength of CNT yarn. More recently, Zhao et al. [91] have recorded the electrical resistivity of iodine-doped, double-walled nanotube fibers to be around 10^{-7} Ω m.

7.7 APPLICATIONS OF CNT YARNS

A broad range of applications of CNTs have been proposed between 1991 and 2010. These applications encompass various areas from composite materials, electronics, aerospace to medical and bioengineering. Owing to the limit of this chapter, only applications of CNT-based yarns spun by any spinning method are presented.

Potential applications of CNT yarns and CNT-based composite yarns have been identified in the areas of electrochemical devices, high-performance materials and bioengineering. Examples include specialty composites; high-power-density flexible batteries and supercapacitors [21,24,92–94]; flexible CNT-woven textile electrodes [95,96]; CNT textile antennas [97,98]; heat storage CNT textiles for clothing [99]; magnetic CNT yarns for actuators and transformers [100,101]; thermal electrochemical and mechanical–electrical energy harvesting [102]; flexible stretchable loudspeakers [103–105]; and self-repairing structures and artificial muscles [73,106,107]. Several typical applications are highlighted as follows.

7.7.1 Textile electrodes and supercapacitors

Based on the capability of inducing high capacitance of CNT yarns [12], significant efforts have been recently spent to develop CNT yarn electrodes and CNT yarn supercapacitors [21,24,92,94,108,109]. For example, Dalton et al. [24] have used CNT yarns to manufacture yarn supercapacitors which consist of two CNT/PVA composite yarns. The yarns are coated with electrolyte by dipping in aqueous PVA/phosphoric acid, then twisted together and recoated with electrolyte. Kozlov et al. [21] determined the specific capacitance of CNT yarns using a three-electrode cell, of which silver is the reference electrode, carbon felt the counter electrode, and the ionic liquid ethyl-methylimidazolium trifluoromethylsulfonyl imide, the electrolyte.

7.7.2 CNT yarns for actuators

Recently, CNT yarns have been used to develop electromechanical actuators [100,101,110,111]. The CNT yarn electrode of an electrochemical cell containing an ion solution can behave as an actuator with expanding and shrinking actions. When a voltage is applied between an actuating CNT yarn electrode and a counter electrode, charges are injected into the yarn. The predominant cause of actuation is the electrostatic forces that are repulsive interactions between like charges in the nanotubes leading to expansion and elongation of the nanotubes. When the voltage is off, the yarn shrinks back to the original size. On the other hand, when a force is applied and varied under constant potential, the yarn electrode generates a change in current. Since CNT fibers are known to withstand stresses above 800 MPa and temperatures up to 450 °C in air, they can potentially be used as sensors for high-stress and high-temperature applications [112,113].

7.7.3 CNT yarn sensors

Owing to their repeatable, stable resistance–strain behavior and low density, CNT yarns can be used as piezo-resistive sensors to measure strain of composite materials via the electrical resistance of yarns [114,115]. Especially, by permanent integration within a composite material during the fabrication, the embedded yarn sensors can monitor the composite deformation in real time or crack propagation in composite structures [115,116]. Wang et al. [96] have reported the use of CNT/PVA composite yarn as a microelectrode for detecting biomolecules, such as NADH, hydrogen peroxide and dopamine. Experiments showed that CNT yarns displayed higher electrocatalytic activities than carbon fibers. Recently, Zhu et al. [117] have designed a brush-like electrode based on a CNT yarn for electrochemical biosensor applications for sensing glucose. Glucose oxidase enzyme is immobilized at the brush-like end of the CNT fiber, and the enzyme layer is encapsulated by the epoxy-polyurethane semipermeable membrane. It is worthy of notice that this application can extend to environmental monitoring. Also for this application, Randeniya et al. [118] have developed a chemiresistor using metal composites with CNT yarns (Pt−Pd−CNT), which allows for detecting hydrogen of a very low concentration (5 ppm).

7.7.4 CNT yarns in bioengineering

Recently reported were several interesting applications of CNT-based yarns in the medical and bioengineering including artificial muscles and self-repairing structures. For example, Edwards et al. [73] have produced a CNT-based three-dimensional knitted scaffold from multiply CNT yarns and composite scaffolds (Fig. 7.24) which were created through the deposition of poly(lactic-co-glycolic acid) nanofibers onto the knitted tubes. The scaffold supports cell growth and promotes a uniform cell distribution and can be used for nerve regeneration in tissue engineering (bioengineering), where the ability to electrically stimulate nerve cells could prove beneficial. Another typical application of CNT yarns is artificial muscles [106]. According to their work, CNT actuators can provide up to a few percentage of corresponding natural muscle stroke and generate 100 times higher stress level than natural muscles. Foroughi et al. [107] have also used an electrolyte-filled twist-spun CNT yarn as a torsional artificial muscle in a simple three-electrode electrochemical system.

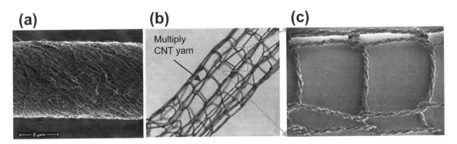

FIGURE 7.24

Three-dimensional knitted scaffold from multiply CNT yarns: (a) CNT yarn; (b) and (c) tube of knitted scaffold. (For color version of this figure, the reader is referred to the online version of this book.)

7.7.5 CNT-based high-performance yarns

Based on the excellent and unique mechanical and physical properties of CNTs (high tensile strength and Young's modulus, high electrical and thermal conductivities, high interfacial load transfer efficiency, and very high aspect ratio), CNTs have been a candidate for applications requiring high strength and lightweight materials and can be a superb choice of reinforcements for high-performance composites [11,27,77,119]. For example, the toughness and lightweight of CNT yarns and CNT–polymer composite yarns could enable their applications in antiballistic vests, safety belts, and explosion-proof blankets for aircraft cargo bays.

Many potential applications of CNT-based yarns were recently presented under different aspects in several valuable technical reviews, including Refs [120–122].

7.8 CONCLUSION

This chapter provides an overview of the dry spinning method that processes a vertically aligned CNT forest into continuous yarns. Aspects discussed include the interaction between CNTs, web formation, interrelation between structure and yarn properties, fabrication methods as well as some treatments of CNT yarns. The idea of the dual slip-link model helps explain the CNT web formation and formulate the new multistep process for dry spinning of CNTs into a yarn. The spinning process is based on the introduction of dielectrophoresis in the ATS and allows effectively and efficiently handling the tensioning and alignment of the CNT fibers in a web/sliver. The new process allows developing different types of CNT yarns, such as CNT-metal/composite yarns in a one-step process. Indeed, potential applications of pure CNT yarns as well as CNT-based composite yarns have been presented in several multifunctional products such as composite materials, biomedical engineering, strain-damage sensing, electrochemical devices and microelectrodes. Experiments have shown that with a suitable process (the voltages imposed in rings, tension forces in the ATS, twisting level, and other treatments at individual stages of the process), CNT yarns can potentially exploit the excellent properties of individual CNTs to enhance micro- and macrostructural performance. Thus, the dry spinning of yarns is producing a new material called *superfiber* that could have wide applications in many fields.

Acknowledgments

The author would like to thank MSE, FMF (CSIRO, Australia) and CESRC (USQ, Australia); and Drs R. Hore (FMF, CSIRO) and W. Humphries (AFM, CSIRO) for direct and indirect supports. The author also notes the assistance of Drs S. Smith, C. Huynh, S. Lucas, L.K. Randeniya (MSE, CSIRO) and T. Truong (DSTO, Au). The author highly appreciates the help by Dr D.G. Phillips (MSE and TFT, CSIRO) and Professor T. Tran-Cong (CESRC, USQ) in critically reading the manuscript and correcting errors. All support is gratefully acknowledged.

References

[1] M.S. Dresselhaus, G. Dresselhaus, P. Avouris, Carbon Nanotubes, Synthesis, Structure, Properties and Applications, Springer, Berlin, 2001.

[2] T.W. Ebbesen, Carbon Nanotubes: Preparation and Properties, CRC, New York, 1997.

[3] M.M.J. Treacy, T.W. Ebbesen, J.M. Gibson, Exceptionally high Young's modulus observed for individual carbon nanotubes, Nature 381 (1996) 680–687.

[4] C.Y. Li, T.W. Chou, Strain and pressure sensing using SWCNTs, Nanotechnology 15 (2004a) 1493–1496.

[5] S.J. Tan, A.R.M. Verschueren, C. Dekker, Room-temperature transistor based on a single carbon nanotube, Nature 393 (1998) 49–52.

[6] E.B. Malarkey, V. Parpura, Applications of carbon nanotubes in neurobiology, Neurodegenerative Disease 4 (2007) 292–299.

[7] S.J.V. Frankland, V.M. Harik, Analysis of carbon nanotube pull-out from a polymer matrix, Surface Science 525 (2003) L103–L108.

[8] P. Potschke, S.M. Dudkin, I. Alig, Dielectric spectroscopy on melt processed polycarbonate multi-walled carbon nanotube composites, Polymer 44 (2003) 5023–5030.

[9] E.T. Thostenson, C.Y. Li, T.W. Chou, Nano-composites in context, Composites Science and Technology 65 (3–4) (2005) 491–516.

[10] E.T. Thostenson, Z.F. Ren, T.W. Chou, Advances in the science and technology of carbon nanotubes and their composites: a review, Composites Science and Technology 61 (13) (2001) 1899–1912.

[11] C.D. Tran, S. Lucas, D.G. Phillips, L.K. Randeniya, R.H. Baughman, T. Tran-Cong, Manufacturing polymer-CNT composite using a novel direct process, Nanotechnology 22 (2011) 145302.

[12] R.H. Baughman, A.A. Zakhidov, W.A. de Heer, Carbon nanotubes—the route towards applications, Science 297 (2002) 787.

[13] A. Goho, Nice threads: the golden secret behind spinning carbon nanotube fibers, Science News 165 (23) (2004) 363–365.

[14] K. Jiang, Q. Li, S. Fan, Nanotechnology: spinning continuous carbon nanotube yarns, Nature 419 (2002) 801.

[15] K. Koziol, J. Vilatela, A. Moisala, M. Motta, P. Cunniff, M. Sennett, A. Windle, High-performance carbon nanotube fiber, Science 318 (2007) 1892–1895.

[16] M. Motta, Y.L. Li, I.A. Kinloch, A.H. Windle, Mechanical properties of continuously spun fibers of carbon nanotubes, Nano Letters 5 (2005) 1529–1533.

[17] M. Zhang, K.R. Atkinson, R.H. Baughman, Multifunctional carbon nanotube yarns by downsizing an ancient technology, Science 306 (2004) 1358–1361.

[18] H.W. Zhu, C.L. Xu, D.H. Wu, B.Q. Wei, R. Vajtai, P.M. Ajayan, Direct synthesis of long single-walled carbon nanotube strands, Science 296 (2002) 884–886.

[19] Y.L. Li, I.A. Kinloch, A.H. Windle, Direct spinning of carbon nanotube fibers from chemical vapor deposition synthesis, Science 304 (2004b) 276–278.

[20] L.M. Ericson, H. Fan, H.Q. Peng, V.A. Davis, W. Zhou, J. Sulpizio, Y.H. Wang, R. Booker, J. Vavro, C. Guthy, A.N.G. Parra-Vasquez, M.J. Kim, S. Ramesh, R.K. Saini, C. Kittrell, G. Lavin, H. Schmidt, W.W. Adams, W.E. Billups, M. Pasquali, W.F. Hwang, R.H. Hauge, J.E. Fischer, R.H. Smalley, Macroscopic, neat, single-walled carbon nanotube fibers, Science 305 (2004) 1447–1450.

[21] M.E. Kozlov, R.C. Capps, W.M. Sampson, V.H. Ebron, J.P. Ferraris, R.H. Baughman, Spinning solid and hollow polymer-free carbon nanotube fibers, Advanced Materials 17 (2005) 614–617.

[22] V.A. Davis, A.N.G. Parra-Vasquez, M.J. Green, P.K. Rai, N. Behabtu, V. Prieto, R.D. Booker, J. Schmidt, E. Kesselman, W. Zhou, H. Fan, W.W. Adams, R.H. Hauge, J.E. Fischer, Y. Cohen, Y. Talmon, R.E. Smalley, M. Pasquali, True solutions of single-walled carbon nanotubes for assembly into macroscopic materials, Nature Nanotechnology 4 (12) (2009) 830–834.

[23] J. Steinmetz, M. Glerup, M. Paillet, P. Bernier, M. Holzinger, Production of pure nanotube fibers using a modified wet-spinning method, Carbon 43 (11) (2005) 2397–2400.

[24] A.B. Dalton, S. Collins, E. Munoz, J.M. Razal, V.H. Ebron, J.P. Ferraris, J.N. Coleman, B.G. Kim, R.H. Baughman, Super-tough carbon nanotube fibres, Nature 423 (2003) 703.

[25] E. Munoz, D.S. Suh, S. Collins, M. Selvidge, A.B. Dalton, B.G. Kim, J.M. Razal, G. Ussery, A.G. Rinzler, M.T. Martinez, R.H. Baughman, Highly conducting carbon nanotube/polyethylenimine, Composite Fibers Advanced Materials 17 (2005) 1064–1067.

[26] B. Vigolo, A. Penicaud, C. Coulon, C. Saunder, R. Pailler, C. Journet, P. Bernier, P. Poulin, Macroscopic fibers and ribbons of oriented CNT, Science 290 (2000) 1331–1334.

[27] A.B. Dalton, S. Collins, E. Munoz, J.M. Razal, V.H. Ebron, J.P. Ferraris, J.N. Coleman, B.G. Kim, R.H. Baughman, Continuous carbon nanotube composite fibers: properties, potential applications, and problems, Journal of Materials Chemistry 14 (2004) 1–3.

[28] S.J. Zhang, K.K. Koziol, I.A. Kinloch, A.H. Windle, Macroscopic fibers of well-aligned carbon nanotubes by wet spinning, Small 4 (8) (2008a) 1217–1222.

[29] K.L. Stano, K. Koziol, M. Pick, M.S. Motta, A. Moisala, J.J. Vilatela, S. Frasier, A.H. Windle, Direct spinning of carbon nanotube fibres from liquid feedstock, International Journal of Material Forming 1 (2) (2008) 59–62.

[30] J.J. Vilatela, A.H. Windle, You have full text access to this content, Yarn-like carbon nanotube fibers, Advanced Materials 22 (44) (2010) 4959–4963.

[31] K.R. Atkinson, S.C. Hawkins, C. Huynh, C. Skourtis, Zhang M Dai, S. Fang, A.A. Zakhidov, S.B. Le, A.E. Aliev, C.D. Williams, R.H. Baughman, Multifunctional carbon nanotube yarns and transparent sheets: fabrication, properties, and applications, Physica B: Condensed Matter 15 (2007) 339–343.

[32] S. Fang, M. Zhang, A.A. Zakhidov, R.H. Baughman, Structure and process dependent properties of solid-state spun carbon nanotube yarns, Journal of Physics. Condensed Matter 22 (2010) 334221.

[33] Q. Li, X. Zhang, R.F. DePaula, L. Zheng, Y. Zhao, L. Stan, T.G. Holesinger, P.N. Arendt, D.E. Peterson, Y.T. Zhu, Sustained growth of ultra-long carbon nanotube arrays for fiber spinning, Advanced Materials 18 (23) (2006) 3160–3163.

[34] C.D. Tran, W. Humphries, S.M. Smith, C. Huynh, S. Lucas, Improving the tensile strength of carbon nanotube spun yarns using a modified spinning process, Carbon 47 (2009) 2662.

[35] J. Tersoff, R.S. Ruoff, Structural properties of a carbon nanotube crystal, Physical Review Letters 73 (1994) 676–679.

[36] D. Bethune, C.H. Kiang, M. de Vries, G. Gorman, R. Savoy, J. Vazquez, et al., Cobalt catalysed growth of carbon nanotubes with single-atomic layer walls, Nature 363 (1993) 605−607.
[37] S. Iijima, T. Ichihashi, Single-shell carbon nanotubes of 1-nm diameter, Nature 363 (1993) 603−605.
[38] P.R. Lord, Handbook of Yarn Production: Technology, Science and Economics, CRC, New York, 2003.
[39] D.W. Coffin, L.A. Carlsson, R.B. Pipes, On the separation of carbon nanotubes, Composites Science and Technology 66 (2006) 1132−1140.
[40] L.A. Girifalco, Molecular-properties of C-60 in the gas and solid phases, Journal of Physical Chemistry 96 (2) (1992) 858−861.
[41] L.A. Girifalco, M. Hodak, R.S. Lee, Carbon nanotubes, buckyballs, ropes and a universal graphitic potential, Physical Review B 62 (19) (2000) 13104−13110.
[42] P.M. Ajayan, L.S. Schadler, C. Giannaris, A. Rubio, Single-walled carbon nanotube polymer composites: strength and weakness, Advanced Materials 12 (10) (2000) 750−753.
[43] D. Qian, G.J. Wagner, W.K. Liu, M.F. Yu, R.S. Ruo, Mechanics of carbon nanotubes, Applied Mechanics Reviews 55 (6) (2002) 495−533.
[44] F. Bobaru, Influence of van der Waals forces on increasing the strength and toughness in dynamic fracture of nano-fibre networks: a peri-dynamic approach, Modelling and Simulation in Materials Science and Engineering 15 (2007) 397−417.
[45] C. Thomsen, S. Reich, A.R. Goni, H. Jantoljak, P.M. Rafailov, I. Loa, et al., Intermolecular interaction in carbon nanotube ropes, Physica Status Solidi B 215 (1999) 435−441.
[46] J.W.S. Hearle, P. Grosberg, S. Backer, Structural Mechanics of Fibers, Yarns and Fabrics, Wiley-Interscience, New York, 1969.
[47] C.P. Huynh, S.C. Hawkins, Understanding the synthesis of directly spinnable carbon nanotube forests, Carbon 48 (2010) 1105.
[48] K. Jiang, J. Wang, Q. Li, L. Liu, C. Liu, S. Fan, Super-aligned carbon nanotube arrays, films, and yarns: a road to application, Advanced Materials 23 (2011) 1154−1161.
[49] X.B. Zhang, K.L. Jiang, C. Teng, P. Liu, L. Zhang, J. Kong, T.H. Zhang, Q.Q. Li, S.S. Fan, Spinning and processing continuous yarns from 4-inch wafer scale super-aligned carbon nanotube arrays, Advanced Materials 18 (12) (2006) 1505.
[50] A.A. Kuznetsov, A.F. Fonseca, R.H. Baughman, A.A. Zakhidov, Structural model for dry-drawing of sheets and yarns from carbon nanotube forests, ACS Nano 5 (2011) 985−993.
[51] J.J. Vilatela, J.A. Elliott, A.H. Windle, A model for the strength of yarn-like carbon nanotube fibers, ACS Nano 5 (3) (2011) 1921−1927.
[52] M. Doi, J. Takimoto, Molecular modelling of entanglement, Philosophical Transactions of the Royal Society of London. Series A 361 (2003) 641−652.
[53] W. Guo, C. Liu, X. Sun, Z. Yang, H.G. Kia, H. Peng, Aligned carbon nanotube/polymer composite fibers with improved mechanical strength and electrical conductivity, Journal of Materials Chemistry 22 (3) (2012) 903−908.
[54] K. Sears, C. Skourtis, K. Atkinson, N. Finn, W. Humphries, Focused ion beam milling of carbon nanotube yarns to study the relationship between structure and strength, Carbon 48 (2010) 4450−4456.

[55] D.G. Phillips, C.D. Tran, W.B. Fraser, G.H.D. van der Heijden, Torsional properties of staple fibre plied yarns, Journal of the Textile Institute 101 (7) (2010) 595–612.

[56] C.D. Tran, G.H.D. van der Heijden, D.G. Phillips, Application of topological conservation to model key features of zeros torque multiply yarns, Journal of Textile Institute 99 (4) (2008b) 325–337.

[57] C.A. Lawrence, Fundamentals of Spun Yarn Technology, CRC, London, 2003.

[58] M. Miao, Production, structure and properties of twist-less carbon nanotube yarns with a high density sheath, Carbon 50 (2012) 4973–4983.

[59] C.D. Tran, T. Tran-Cong, S. Hawking, C. Huynh, Dry spinning CNT yarn using a multi-step drafting system, submitted.

[60] C.D. Tran, T. Tran-Cong, K. Le-Cao, D. Ho-Minh. Processing the CNTs' interaction in web using an electrostatic field based process, 2012 International Conference On Nano-science and Nano-technology (ICONN 2012), CD proceedings, 6–9 Feb 2012b, Perth, Australia.

[61] H.A. Pohl, Dielectrophoresis, Cambridge University Press, Cambridge, 1978.

[62] T. Kimura, K. Ago, M. Tobita, S. Ohshima, S. Kyotani, M. Yumura, Polymer composites of carbon nanotubes aligned by a magnetic field, Advanced Materials 14 (19) (2002) 1380–1383.

[63] M. Monti, M. Natali, L. Torre, J.M. Kenny, The alignment of single walled carbon nanotubes in an epoxy resin by applying a DC electric field, Carbon 50 (2012) 2453–2464.

[64] D.E.A. Plate, Sirospun new spinning technique for worsted weaving yarn, Journal of Australasian Textiles 2 (1) (1982) 10–12.

[65] C.D. Tran, D.G. Phillips, Modelling and performance of multi-strand spun yarns including two-strand siro-spun. The Fiber Society Conference, May 2008a, CD proceedings, ENSISA Uni., Mulhouse, France.

[66] K. Liu, Y.H. Sun, R.F. Zhou, H. Zhu, J. Wang, L. Liu, S.S. Fan, K. Jiang, Carbon nanotube yarns with high tensile strength made by a twisting and shrinking method, Nanotechnology 21 (4) (2010b) 045708.

[67] M. Motta, A. Moisala, I.A. Kinloch, A.H. Windle, High performance fibres from dog bone carbon nanotubes, Advanced Materials 19 (2007) 3721–3726.

[68] V.A. Parsegian, B.W. Ninham, Temperature-dependent van der Waals forces, Biophysical Journal 10 (7) (1970) 664–674.

[69] N.R. Raravikar, P. Keblinski, A.M. Rao, S. Rao Mildred, M.S. Dresselhaus, L.S. Schadler, et al., Temperature dependence of radial breathing mode Raman frequency of single-walled carbon nanotubes, Physical Review 66 (2002) 235424.1–9.

[70] Y.C. Zhang, X. Chen, X. Wang, Effects of temperature on mechanical properties of multi-walled carbon nanotubes, Composites Science and Technology 68 (2008b) 572–581.

[71] J.Y. Cai, J. Min, J. McDonnell, S. Jeffrey, J.S. Church, D. Christopher, C.D. Easton, W. William Humphries, S. Lucas, L. Andrea, A.L. Woodhead, An improved method for functionalisation of carbon nanotube spun yarns with aryldiazonium compounds, Carbon 50 (12) (2012) 4655–4662.

[72] A. Ghemes, Y. Minami, J. Muramatsu, M. Okada, H. Mimura, Y. Inoue, Fabrication and mechanical properties of carbon nanotube yarns spun from ultra-long multi-walled carbon nanotube arrays, Carbon 50 (2012) 4579–4587.

[73] S.L. Edwards, J.S. Church, J.A. Werkmeister, J.A.M. Ramshaw, Tubular micro-scale multi-walled carbon nanotube-based scaffolds for tissue engineering, Biomaterials 30 (2009) 1725–1731.

[74] F. Deng, W. Lu, H. Zhao, Y. Zhu, B.S. Kim, T.W. Chou, The properties of dry-spun carbon nanotube fibers and their interfacial shear strength in an epoxy composite, Carbon 49 (5) (2011) 1752–1757.

[75] M.D. Lima, S. Fang, X. Lepro, C. Lewis, R.O. Robles, J.C. Gonzalez, E.C. Martinez, M.E. Kozlov, J. Oh, N. Rawat, C.S. Haines, M.H. Haque, V. Aare, S. Stoughton, A.A. Zakhidov, R.H. Baughman, Biscrolling nanotube sheets and functional guests into yarns, Science 331 (2011) 51–55.

[76] L.K. Randeniya, A. Bendavid, P.J. Martin, C.D. Tran, Composite yarns of multi-walled carbon nanotubes with metallic electrical conductivity, Small 6 (16) (2010) 1806–1811.

[77] K. Liu, Y.H. Sun, X. Lin, R.F. Zhou, J.P. Wang, S.S. Fan, K.L. Jiang, Scratch-resistant, highly conductive, and high-strength carbon nanotube-based composite yarns, ACS Nano 4 (2010a) 5827–5834.

[78] M.J. Biercuk, M.C. Llaguno, M. Radosavljevic, J.K. Hyun, A.T. Johnson, J.E. Fischer, Carbon nanotube composites for thermal management, Applied Physics Letters 80 (15) (2002) 2767–2769.

[79] Y. Bin, M. Kitanaka, D. Zhu, M. Matsuo, Development of highly oriented polyethylene filled with aligned carbon nanotubes by gelation/crystallization from solutions, Macromolecules 36 (16) (2003) 6213–6219.

[80] F. Du, J.E. Fischer, K.I. Winey, Coagulation method for preparing single-walled carbon nanotube/poly(methyl-methacrylate) composites and their modulus, electrical conductivity, and thermal stability, Journal of Polymer Science. Part B, Polymer Physics 41 (24) (2003) 3333–3338.

[81] P. Poulin, B. Vigolo, P. Launois, Films and fibers of oriented single wall nanotubes, Carbon 40 (10) (2002) 1741–1749.

[82] S. Ryu, Y. Lee, J.W. Hwang, S. Hong, C. Kim, T.G. Park, H. Lee, S.H. Hong, High-strength carbon nanotube fibers fabricated by infiltration and curing of mussel-inspired catecholamine polymer, Advanced Materials 23 (17) (2011) 1971–1975.

[83] J.L. Bahr, J.P. Yang, D.V. Kosynkin, M.J. Bronikowski, R.E. Smalley, J.M. Tour, Functionalization of carbon nanotubes by electrochemical reduction of aryl diazonium salts: a bucky paper electrode, Journal of the American Chemical Society 123 (27) (2001) 6536–6542.

[84] S. Banerjee, T. Hemraj-Benny, S.S. Wong, Covalent surface chemistry of single-walled carbon nanotubes, Advanced Materials 17 (1) (2005) 17–29.

[85] C.J. Yong, G.S. Nanda, W.C. Jae, Polymeric nano composites of polyurethane block copolymers and functionalized multi-walled carbon nanotubes as crosslinkers, Macromolecular Rapid Communications 27 (2) (2006) 126–131.

[86] A. Bachtold, M. Henney, C. Terrier, C. Strunk, C. Schonenberger, J.P. Salvetat, J.M. Bonard, L. Forro, Contacting carbon nanotubes selectively with low-ohmic contacts for four-probe electric measurements, Applied Physics Letters 73 (1998) 274–276.

[87] C. Berger, Y. Yi, Z.L. Wang, W.A. de Heer, Multi-walled carbon nanotubes are ballistic conductors at room temperature, Applied Physics A 74 (2002) 363–365.

[88] Q. Li, Y. Li, X. Zhang, S.B. Cikkannanavar, Y. Zhao, A.M. Dangelewicz, L. Zheng, S.K. Doorn, Q. Jia, D.E. Peterson, P.N. Arendt, Y. Zhu, Structure-dependent electrical properties of carbon nanotube fibers, Advanced Materials 19 (20) (2007) 3358−3363.

[89] H.C. Choi, M. Shim, S. Bangsaruntip, H. Dai, Spontaneous reduction of metal ions on the sidewalls of carbon nanotubes, Journal of the American Chemical Society 124 (31) (2002) 9058−9059.

[90] B.S. Kong, D.H. Jung, S.K. Oh, C.S. Han, H.T. Jung, Single-walled carbon nanotube gold nanohybrids: application in highly effective transparent and conductive films, Journal of Physical Chemistry C 111 (23) (2007) 8377−8382.

[91] Y. Zhao, J. Wei, R. Vajtai, P.M. Ajayan, E.V. Barrera, Iodine doped carbon nanotube cables exceeding specific electrical conductivity of metals, Scientific Reports 1 (2011) 83.

[92] E. Munoz, A.B. Dalton, S. Collins, M. Kozlov, J. Razal, J.N. Coleman, B.G. Kim, V.H. Ebron, M. Selvidge, J.P. Ferraris, R.H. Baughman, Multifunctional carbon nanotube composite fibers, Advanced Engineering Materials 10 (2004) 801−804.

[93] H.X. Zhang, C. Feng, Y.C. Zhai, K.L. Jiang, Q.Q. Li, S.S. Fan, Cross-stacked carbon nanotube sheets uniformly loaded with SnO_2 nano-particles: a novel binder-free and high-capacity anode material for lithium-ion batteries, Advanced Materials 21 (22) (2009) 2299−2304.

[94] X.H. Zhong, Y.L. Li, Y.K. Liu, X.H. Qiao, Y. Feng, J. Liang, J. Jin, L. Zhu, F. Hou, J.Y. Li, Continuous multilayered carbon nanotube yarns, Advanced Materials 22 (6) (2010) 692−696.

[95] L. Viry, A. Derre, P. Garrigue, N. Sojic, P. Poulin, A. Kuhn, Optimized carbon nanotube fiber microelectrodes as potential analytical tools, Analytical and Bioanalytical Chemistry 389 (2) (2007) 499−505.

[96] J. Wang, R.P. Deo, P. Poulin, M. Mangey, Carbon nanotube fiber microelectrodes, Journal of the American Chemical Society 125 (2003) 14706.

[97] P.J. Burke, S. Li, Z. Yu, Quantitative theory of nanowire and nanotube antenna performance, IEEE Transactions on Nanotechnology 5 (4) (2006) 314−334.

[98] G.W. Hanson, Fundamental transmitting properties of carbon nanotube antennas, IEEE Transactions on Antennas and Propagation 53 (11) (2005) 3426−3435.

[99] L. Hu, M. Pasta, F.L. Mantia, L.F. Cui, S. Jeong, H.D. Deshazer, J.W. Choi, S.M. Han, Y. Cui, Stretchable, porous, and conductive energy textiles, Nano Letters 10 (2010a) 708−714.

[100] R.H. Baughman, C.X. Cui, A.A. Zakhidov, Z. Lqbal, J.N. Barisci, G.M. Spinks, G.G. Wallace, A. Mazzoldi, D.D. Rossi, A.G. Rinzler, O. Jaschinski, S. Roth, M. Kertesz, Carbon nanotube actuators, Science 284 (1999) 1340−1344.

[101] L. Viry, C. Mercader, P. Miaudet, C. Zakri, A. Derre, A. Kuhn, M. Maugey, P. Poulin, Nanotube fibers for electromechanical and shape memory actuators, Journal of Materials Chemistry 20 (2010) 3487−3495.

[102] R. Hu, B.A. Cola, N. Haram, N. Joseph, J.N. Barisci, S. Lee, S. Stoughton, G. Wallace, C. Too, M. Thomas, A. Gestos, M.E. dela Cruz, J.P. Ferraris, A.A. Zakhidov, R.H. Baughman, Harvesting waste thermal energy using a carbon nanotube-based thermo-electrochemical cell, Nano Letters 10 (2010b) 838−846.

[103] A.E. Aliev, M.D. Lima, S. Fang, R.H. Baughman, Underwater sound generation using carbon nanotube projectors, Nano Letters 10 (2010) 2374−2380.

[104] K. Suzuki, S. Sakakibara, M. Okada, Y. Neo, H. Mimura, Y. Inoue, T. Murata, Study of carbon nanotube web thermo-acoustic loud speakers, Japanese Journal of Applied Physics 50 (2011) 01BJ10.

[105] L. Xiao, Z. Chen, C. Feng, L. Liu, Z.Q. Bai, Y. Wang, L. Qian, Y. Zhang, Q. Li, K. Jiang, S. Fan, Flexible, stretchable, transparent carbon nanotube thin film loudspeakers, Nano Letters 8 (12) (2008a) 4539−4545.

[106] A.E. Aliev, J. Oh, M.E. Kozlov, A.A. Kuznetsov, S. Fang, A.F. Fonseca, R. Ovalle, M.D. Lima, M.H. Haque, Y.N. Gartstein, M. Zhang, A.A. Zakhidov, R.H. Baughman, Giant stroke, super-elastic carbon nanotube aerogel muscles, Science 323 (2009) 1575−1578.

[107] J. Foroughi, G.M. Spinks, G.G. Wallace, J. Oh, M.E. Kozlov, S.L. Fang, T. Mirfakhrai, J.D.W. Madden, M.K. Shin, S.J. Kim, R.H. Baughman, Torsional carbon nanotube artificial muscles, Science 334 (6055) (2011) 494−497.

[108] J.A. Lee, M.K. Shin, S.H. Kim, S.J. Kim, G.M. Spinks, G.G. Wallace, R. Ovalle-Robles, M.D. Lima, M.E. Kozlov, R.H. Baughman, Hybrid nano-membranes for high power and high energy density super-capacitors and their yarn application, ACS Nano 6 (2012) 327−334.

[109] L. Xiao, P. Liu, L. Liu, K. Jiang, X. Feng, Y. Wei, L. Qian, S. Fan, T. Zhang, Barium-functionalized multi-walled carbon nanotube yarns as low-work-function thermionic cathodes, Applied Physics Letters 92 (15) (2008b) 153108.

[110] T. Mirfakhrai, J.Y. Oh, M. Kozlov, S.L. Fang, M. Zhang, R.H. Baughman, J.D. Madden, Carbon nanotube yarn actuators: an electrochemical impedance model, Journal of the Electrochemical Society 156 (2009) K97−K103.

[111] T. Mirfakhrai, J.Y. Oh, M.E. Kozlov, E.C. Fok, M. Zhang, S.L. Fang, R.H. Baughman, J.D. Madden, Electrochemical actuation of carbon nanotube yarns, Smart Materials and Structures 16 (2) (2007) 243−249.

[112] T. Mirfakhrai, M. Kozlov, S. Fang, M. Zhang, R.H. Baughman, J.D. Madden, Carbon nanotube yarns: sensors, actuators, and current carriers, Proceedings of SPIE 6927 (2008a) 692708.

[113] T. Mirfakhrai, J. Oh, M. Kozlov, S. Fang, M. Zhang, R.H. Baughman, J.D. Madden, Carbon nanotube yarns as high load actuators and sensors, Advances in Science and Technology 61 (2008b) 65−74.

[114] D.Y. Khang, J.L. Xiao, C. Kocabas, S. MacLaren, T. Banks, H.Q. Jiang, Y.Y.G. Huang, J.A. Rogers, Molecular scale buckling mechanics in individual aligned single-wall carbon nanotubes on elastomeric substrates, Nano Letters 8 (1) (2008) 124−130.

[115] H. Zhao, Y. Zhang, P.D. Bradford, Q. Zhou, Q. Jia, F.-G. Yuan, et al., Carbon nanotube yarn strain sensors, Nanotechnology 21 (2010) 305502.

[116] J.L. Abot, Y. Song, M.S. Vatsavaya, S. Medikonda, Z. Kier, C. Jayasinghe, N. Rooy, V.N. Shanov, M.J. Schulz, Delamination detection with carbon nanotube thread in self-sensing composite materials, Composites Science and Technology 70 (7) (2010) 1113−1119.

[117] Z.G. Zhu, W. Song, K. Burugapalli, F. Moussy, Y.L. Li, X.H. Zhong, Nano-yarn carbon nanotube fiber based enzymatic glucose biosensor, Nanotechnology 21 (16) (2010) 165501.

[118] L.K. Randeniya, P.J. Martin, A. Bendavid, Detection of hydrogen using multi-walled carbon nanotube yarns coated with nano-crystalline Pd and Pd/Pt layered structures, Carbon 50 (2012) 1786−1792.

[119] R.J. Mora, J.J. Vilatela, A.H. Windle, Properties of composites of carbon nanotube fibres, Composites Science and Technology 69 (10) (2009) 1558–1563.
[120] L. Liu, W. Ma, Z. Zhang, Macroscopic carbon nanotube assemblies: preparation, properties, and potential applications, Small 7 (11) (2011) 1504–1520.
[121] W. Lu, M. Zu, J.H. Byun, B.S. Kim, T.W. Chou, State of the art of carbon nanotube fibers: opportunities and challenges, Advanced Materials 24 (14) (2012) 1805–1833.
[122] J. Park, K.H. Lee, Carbon nanotube yarns, Korean Journal of Chemical Engineering 29 (3) (2012) 277–287.

CHAPTER 8

Synthesis and Properties of Boron Nitride Nanotubes

N. Govindaraju, R.N. Singh

School of Materials Science and Engineering, Helmerich Advanced Technology Research Center, Oklahoma State University, Tulsa, OK, USA

CHAPTER OUTLINE

8.1 Introduction	243
8.2 Nanotubes: Basic Structure	246
8.3 Synthesis of BNNTs	249
8.3.1 Arc discharge	249
8.3.2 Ball milling	249
8.3.3 Carbothermal synthesis	250
8.3.4 Chemical vapor deposition	251
8.3.4.1 Plasma-based CVD synthesis	*252*
8.3.4.2 In situ synthesis of BN precursors	*257*
8.3.4.3 Other CVD methods of synthesis of BNNTs	*258*
8.3.5 Laser heating and ablation	258
8.4 Properties of Boron Nitride Nanotubes	259
8.4.1 Electrical properties	259
8.4.2 Mechanical properties	259
8.4.3 Optical properties	260
8.4.4 Thermal properties, thermal stability, and wetting behavior	260
8.5 Comparison of BNNTs and CNTs	261
8.6 Summary	263
Acknowledgments	263
References	263

8.1 INTRODUCTION

Structure–property relationships form the core of materials science research. The development of atomistic and quantum theories in the early part of the twentieth century revolutionized our understanding of materials and the fundamental role that atomic structure plays in a wide spectrum of material properties we observe in our everyday lives. This fundamental understanding is a cornerstone for the exponential development in technology since 1910. Magnetic resonance imaging, the microprocessor, and the laser are among the devices and technologies whose successful development hinged on our enhanced knowledge of fundamental material properties.

Materials in the universe encompass a staggering range of scales, from the nanometer regime (10^{-9} m) to the regime of a light year ($\sim 10^{16}$ m). The physical laws, which describe the behavior of matter at both ends of the length spectrum, are very different. For instance, while Newtonian mechanics describes the motion of planetary bodies well, at the scale of the atom, it is inadequate, and quantum mechanics comes into play. Consequently, physical properties of a given material can drastically change from the macro- to the micro- and nanoscale.

In this context, nanoscale materials (<100 nm in length scale) bridge the atomic world and the micro- and macroworld and exhibit a host of interesting material properties, which are useful for technological applications. A useful way of looking at the position occupied by nanoscale materials and their utility in developing new technologies is to look at the evolution of the energy band structure in a solid. It is known that an isolated individual atom, for instance, carbon, exhibits discrete atomic levels. However, as the carbon atoms are brought together to form a diamond crystal structure, the outermost electrons of the individual carbon atoms begin to interact and result in the formation of a continuous band of electron energy levels. This band structure is the result of discrete electron energy levels being so closely spaced apart that they essentially form a continuous band of energies. If N numbers of atoms are brought together in a lattice and ΔE_α is the average spacing between atomic levels, the electron energy spacing in the band can be approximated by Eqn (8.1) [1].

$$\delta = \frac{\Delta E_\alpha}{N} \tag{8.1}$$

At room temperature, the spacing between electron energy levels in a band can be considered to be negligible (and hence the band is deemed continuous) when the following condition is satisfied.

$$\delta \ll kT \approx \frac{1}{40} \tag{8.2}$$

Therefore, when a few hundred atoms are brought together (assuming $\Delta E_\alpha \approx 1$ eV), the material may be considered to be a "regular" solid, which may be described by "macroscopic" laws. In the case of carbon atoms in a diamond lattice, 100 atoms would lead to a crystal size of approximately 36 nm. Therefore, for diamond crystal sizes significantly >36 nm, we may safely assume that the material behaves as a solid continuum, and its behavior can be described well by classical physics. However, for crystals sizes close to or below 36 nm, quantum mechanical effects begin to play an important role. Under such circumstances, it is possible to predict the material properties from first principles calculations. Also, because physical laws at this length scale are very different from classical continuum descriptions, it is possible to realize extreme values for physical properties such as electron mobilities and saturation velocities. It is for this reason that nanostructured materials such as nanotubes and nanowires have attracted significant attention recently.

As mentioned above, the effects of nanoscale dimensions on the observed material properties can be significant. Several critical fundamental material property parameters such as the melting point, Curie temperature, Debye temperature, and the superconductive temperature decrease as the size of a nanoparticle reduces [2]. Figure 8.1 shows the change in melting point for Si tetrahedral nanoparticles as a function of the nanoparticle size. The solid line is a theoretical fit using Eqn (8.3).

$$\frac{T_x}{T_{x,\infty}} = \left[1 - \frac{\alpha_{shape}}{D}\right]^{S^{-1/2}} \quad (8.3)$$

where,

$$\alpha_{shape} = \left[\frac{D(\gamma_s - \gamma_l)}{\Delta H_{m,\infty}}\right]\left(\frac{A}{V}\right) \quad (8.4)$$

Here, T_x is the size-dependent temperature parameter (x: melting point, Curie temperature, Debye temperature, and superconducting temperature), $T_{x,\infty}$ is the corresponding bulk temperature, α_{shape} is the shape parameter, D is the diameter of the nanostructure, A is the surface area of the nanostructure, V is its volume, $\Delta H_{m,\infty}$ is the enthalpy of melting for the bulk material, γ_l is the liquid surface energy, and γ_s

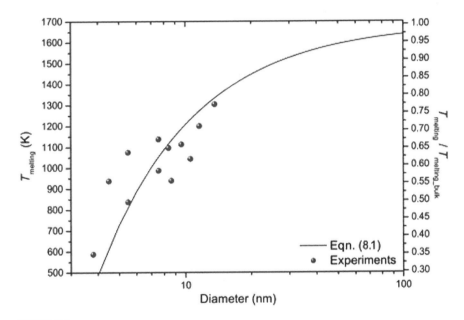

FIGURE 8.1

Plot of Si tetrahedral nanoparticle melting point as a function of particle size. The solid blue dots are experimental data while the solid line is a fit using Eqn (8.3). Significant depression in the melting point is evident with decreasing particle size. (For interpretation of the references to color in this figure legend, the reader is referred to the online version of this book.)

Source: Reprinted from Ref. [2]. Copyright (2009), with permission from Elsevier.

is the solid surface energy. Equations (8.3) and (8.4) and Fig. 8.1 clearly show the stark effect that size can have on fundamental material properties. In Fig. 8.1, the melting point drops by more than 50% (as compared to the bulk value) when the size of the Si nanoparticle is $\sim$2 nm.

Similarly, other physical properties such as electrical conductivity, thermal conductivity, and mechanical strength are found to differ, and in some cases are enhanced significantly as compared to their bulk values. These drastic changes in physical properties have led to an aggressive research effort to exploit nanostructured materials for advanced technological applications. One such nanostructure, which has received significant attention in recent years, is the nanotube.

8.2 NANOTUBES: BASIC STRUCTURE

Nanotubes, as the name implies, are hollow tubular nanostructures with diameters in the nanometer regime and lengths ranging, in general, from the nanometer to the micrometer regime. A nanotube either can consist of a single sheet of atoms wrapped to form a tube (a single-wall nanotube) or may consist of multiple layers wrapped to form a hollow core (a multiwall nanotube).

Nanotubes are formed by a variety of materials, but are easier to synthesize in two-dimensional (2-D) layered compounds such as graphite, hexagonal boron nitride (h-BN), and WS_2 [3]. While it is possible to form nanotubes from three-dimensionally (3-D) structured materials such as GaN, it was found that such nanostructures exhibited a faceted morphology and were more reactive [3]. The preferential formation and stability of nanotubes in the case of 2-D structured materials may be attributed to weak (van der Waals) bonding between 2-D sheets and the ability to "stitch" together incomplete bonds by rolling up a sheet of 2-D material. In the case of 3-D structured materials such as GaN, the presence of strong bonding in all three spatial directions makes it difficult to achieve a stable structure. Among nanotube materials, carbon nanotubes (CNTs) have received significant attention in recent years. Boron nitride nanotubes (BNNTs) are structural counterparts of CNTs. Figure 8.2 illustrates the similarities in structure between BNNTs and CNTs.

A CNT is formed by rolling up a single sheet of carbon atoms (graphene). Similarly, a BNNT is formed by rolling up a single sheet of h-BN. Both these materials consist of a 2-D hexagonal arrangement of atoms—in the case of CNTs, the hexagonal lattice is entirely made up of C atoms, while in the case of BNNTs, the lattice is made up of alternating B and N atoms. Before discussing the synthesis and properties of BNNTs, it is useful to briefly discuss the basic parameters used to describe a CNT or a BNNT. Given the structural simplicity of the CNT, we will use it as the basis for the following discussion.

Figure 8.3 shows the lattice unit vectors ($\vec{a}_1$, $\vec{a}_2$), which are used to describe the structure of a CNT.

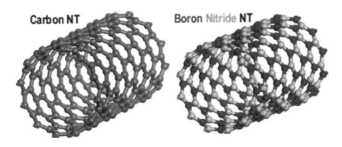

FIGURE 8.2

The atomic structures of CNTs and BNNTs are analogous. B and N atoms are illustrated in blue and pink, respectively [4]. (For interpretation of the references to color in this figure legend, the reader is referred to the online version of this book.)

Source: Reprinted with permission from John Wiley and Sons, Copyright (2007), Wiley-VCH Verlag GmbH & Co. KGaA, Weinheim.

The *chiral vector* (Eqn (8.5); shorthand notation for the chiral vector: (n, m)) describes the structure of the nanotube and defines its circumference. Specifically, it determines the two crystallographically equivalent lattice sites, which when joined together result in a nanotube. The diameter of a nanotube is given by Eqn (8.6), and angle between the chiral vector and the lattice vector $\vec{a}_1$ is defined as the *chiral angle* or the *helicity* of the nanotube (Eqn (8.7)). The chiral angle defines the tilt angle of the hexagonal lattice with respect to the nanotube axis.

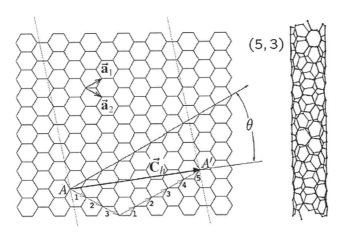

FIGURE 8.3

Schematic diagram showing the basic parameters, which are used to define the structure of a CNT. The figure shows the chiral vector ($\vec{C}_h$) and the chiral angle (or helicity, θ). For illustration, a (5, 3) CNT is also shown. The same structural parameters can be used to define a BNNT.

Source: Reprinted with permission from Ref. [30]. Copyright (2007) by the American Physical Society.

$$\vec{C}_h = n\vec{a}_1 + m\vec{a}_2 \tag{8.5}$$

$$d_t = \frac{|\vec{C}_h|}{\pi} = \frac{a}{\pi}\sqrt{n^2 + nm + m^2} \tag{8.6}$$

$$\cos\theta = \frac{\vec{C}_h \cdot \vec{a}_1}{|\vec{C}_h||\vec{a}_1|} = \frac{2n + m}{2\sqrt{n^2 + nm + m^2}} \tag{8.7}$$

While there are a variety of combinations of (n, m) vectors, which can result in a nanotube, the following specialized chiral vectors are given specific names.

1. Nanotubes with a chiral vector $(n, 0)$ and chiral angle of $0°$ are defined as *zigzag nanotubes*. These nanotubes exhibit a zigzag pattern of atomic configuration along the circumference.
2. *Armchair nanotubes* have a chiral vector (n, n) and a corresponding chiral angle of $30°$. In this case, the atoms along the tube circumference form an armchair pattern.

All other nanotubes with a general (n, m) chiral vector ($n \neq m$ and $n \neq 0$) are defined as *chiral nanotubes*. Figure 8.4 shows schematic diagrams for all three kinds of nanotube structures for the case of BN. Nanotubes can consist of a single sheet of atoms, in which case, they are called *single-walled* (SW) nanotubes as opposed to *multiwall* (MW) nanotubes, which can consist of several concentric nanotubes. In contrast to CNTs, BNNTs usually tend to form multiwall structures easily and the yield of single-walled BNNTs (SWBNNTs) is usually low [4]. Also, SWBNNTs are usually found to be mixed with multiwall BNNTS (MWBNNTs). Therefore, in the discussion in the following sections, the term BNNT refers to MWBNNTs unless otherwise noted.

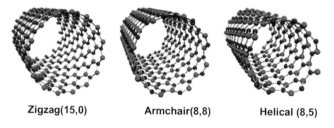

Zigzag(15,0)   Armchair(8,8)   Helical (8,5)

FIGURE 8.4

Illustrations of the three main types of BNNTs: zigzag, armchair, and helical (or chiral) nanotubes. The boron and nitrogen atoms are shown in blue and red, respectively. (For interpretation of the references to color in this figure legend, the reader is referred to the online version of this book.)

Source: Reprinted with permission from Ref. [6]. Copyright (2010) American Chemical Society.

8.3 SYNTHESIS OF BNNTs

A variety of synthesis methods have been used to produce BNNTs. In contrast to graphite, h-BN is insulating which makes the synthesis of BNNTs challenging. Therefore, large-scale (several kilogram range) synthesis of BNNTs is yet to be achieved. Some of the main methods of BNNT synthesis are briefly described below. The purpose of the following discussion is to provide an overview of the major BNNT synthesis methods and is by no means a comprehensive review of the topic. Detailed information on the numerous BNNT synthesis techniques is available in the articles published by Arenal, Goldberg, and Zhi [4–7].

8.3.1 Arc discharge

The arc discharge method was one of the earliest approaches used for the synthesis of BNNTs. Electrical discharge between a cooled copper cathode and an h-BN rod inserted into a hollow tungsten anode resulted in the deposition of "sooty" material on the cathode, which was found to contain BNNTs [8]. Other electrode combinations such as graphite and HfB_2, a tantalum tube filled with BN powder and Cu, and ZrB_2 electrodes were used. In general, it was found that the arc discharge method results in low yield of BNNTs and the purity of the nanotubes was also found to be poor. Modified versions of this process may provide higher yields and better quality [4,5]. Figure 8.5 shows a BNNT synthesized by this method.

8.3.2 Ball milling

High-energy ball milling, a predominantly mechanical process, nevertheless results in significant structural and chemical changes in the material. Nonequilibrium synthesis of materials at low temperatures via ball milling is possible through a combination of multiple processes, which occur during milling. These processes include thermal shock, high-speed plastic deformation, mechanical grinding and fracturing, cold welding, and intimate mixing [9].

BNNTs were typically synthesized by the prolonged (approximately 150 h) high-energy milling of pure boron or h-BN powder using stainless-steel milling vessels and hardened steel balls in a pressurized ($\sim 2.3 \times 10^3$ Torr) NH_3 atmosphere. The milled material was then annealed at high temperature (>1000 °C) in an N_2 atmosphere for $\sim$10 h. It was found that large quantities of BNNTs can be synthesized using this method. The yield of the BNNTs depended on the duration of the milling treatment [11]. It was proposed that nanotube formation by this method was caused by two different mechanisms. The first mechanism being the nitridation of B nanoparticles in the NH_3 atmosphere, which in turn served as nucleation sites for the formation of BNNTs. The second mechanism proposed was that the Fe (and other metals such as Cr and Ni) from the milling process was incorporated into the B powder during high-energy milling and that the metal particles then served as catalysts for BNNT growth [11]. In order for these two mechanisms to operate

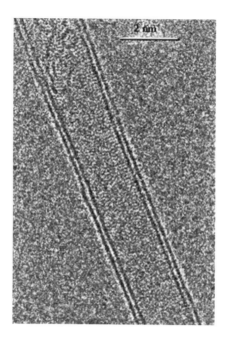

FIGURE 8.5

Transmission electron micrograph of a BNNT synthesized by the arc discharge method [31].
Source: Reprinted with permission from Ref. [31]. Copyright (2001), American Institute of Physics.

effectively, it is necessary that both the ball milling and annealing steps be carried out for long times. Other variations of this technique have been reported including the use of tungsten carbide (WC) balls, and a mixture of NiB and alumina [5,12]. Even though the yield of BNNTs can be very high using this method, the resultant nanotubes can suffer from contamination and structural defects. Figure 8.6 shows a micrograph of BNNTs synthesized using this technique.

8.3.3 Carbothermal synthesis

This method of synthesis of BNNTs is based on the carbothermal reaction used for producing bulk h-BN. The carbothermal reaction (Eqn (8.8)) involves the reduction of boron oxide by a "carbon-like" reducing agent in N_2 with the reaction temperature being maintained between 1000 and 1400 °C [5].

$$B_2O_3 + 3C + N_2 \rightarrow 2\,BN + 3\,CO \tag{8.8}$$

CNTs have been used as templates for synthesizing BNNTs using the carbothermal reaction method. In this approach, the C atoms in the CNTs are substituted by BN atoms resulting in BNNTs. The typical process for this method of synthesis involves the heating of B_2O_3 powder along with CNTs in a graphite crucible [13,14] at high temperatures (>1200 °C) in an N_2 atmosphere. Temperature was

FIGURE 8.6

Transmission electron microscope image of BNNTs synthesized by the ball milling process. The growth of the nanotubes from the milled material is clearly evident. The largest nanotube imaged has a "bamboo"-like morphology.

Source: Reprinted with permission from Ref. [32]. Copyright (1999), American Institute of Physics.

found to be a critical parameter for this method. Temperatures close to 1200 °C resulted in the absence of BNNTs and temperatures close to 1500 °C resulted in the formation of other BN nanostructures such as nanocones and some MWBNNTs [13]. The use of MoO_3 along with the CNT and B_2O_3 was found to significantly enhance the yield of pure MWBNNTs, however, Mo clusters were found to be encapsulated into BNNT tubular shells [15].

Variations of this method include the use of B_4C as a carbon source and boric acid as a boron source [5]. Since CNTs can be synthesized in large quantities, this method offers a good way to synthesize large quantities of SWBNNTs [13] and MWBNNTs [14] from single-walled CNTs and multiwall CNTs. However, the purity of the resultant material can be poor since there can be a significant fraction of B—C—N nanotubes present in the final product [4,13,14]. On the other hand, carbon is a dopant atom in BNNTs, and this method may prove to be a useful way to produce doped BNNTs, albeit with limited control on the dopant concentration. Figure 8.7 shows micrographs of BNNTs synthesized by this method.

8.3.4 Chemical vapor deposition

Chemical vapor deposition (CVD) relies on gas-phase precursors to undergo chemical reactions to synthesize different kinds of materials. Based on the technique and precursors used, it is possible to produce large-area, high-quality materials with low contamination using CVD. A variety of CVD methods have been used for BNNT synthesis. Some of the notable results are presented here, while further references can be found elsewhere [4–7].

In order to reduce contamination in BNNTs, it is desirable to use low-contamination gas-phase precursors. One way to realize a pure gas-phase boron precursor is to synthesize it *in situ*. Another approach involves the use of

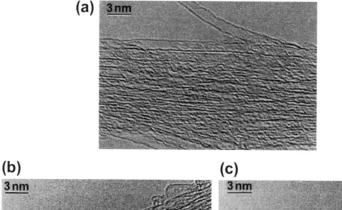
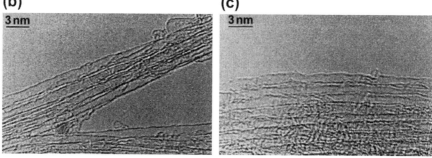

FIGURE 8.7

Transmission electron micrographs of (a) initial SWCNT bundles (before carbothermal substitution reaction), (b) B_xC_{1-x} bundles formed by substitution reaction done at 1250 °C, and (c) BNNT bundles formed by a substitution reaction done at 1350 °C.

Source: Reprinted from Ref. [13]. Copyright (1999), with permission from Elsevier.

plasma-based synthesis techniques. Both approaches result in the formation of high-purity BNNTs. In the discussion below, plasma-based synthesis techniques are discussed first followed by *in situ* precursor synthesis approaches.

8.3.4.1 Plasma-based CVD synthesis

Plasma-based synthesis methods provide unique opportunities to synthesize novel materials under nonequilibrium conditions. The plasma consists of a highly reactive mixture of ions, electrons, and unstable neutral gas molecules and atoms. This highly dynamic and reactive environment facilitates gas-phase reactions, which result in precursor species that normally do not form under equilibrium conditions. Therefore, plasma-based methods enable materials synthesis under unique conditions such as low temperatures. Plasma-assisted CVD techniques have been used to synthesize a variety of materials including ceramics and semiconductors. Two approaches for plasma-based synthesis of BNNTs are discussed below.

8.3.4.1.1 Microwave plasma-enhanced chemical vapor deposition

Microwave plasma-enhanced CVD (MPECVD) process enables the generation of highly reactive BN gas-phase precursors in a controlled, high-purity environment resulting in high-quality BNNT synthesis with very low contamination. Further, due to the highly active nature of the plasma, reactions can be facilitated at temperatures significantly lower than traditional BN synthesis processes (for example, temperatures >1200 °C for carbothermal synthesis vs 800 °C for MPECVD). Low-temperature synthesis not only drives down the cost of production but also broadens the potential applications for BNNTs by facilitating BNNT deposition on temperature-sensitive substrate materials.

An ASTEX MPECVD reactor (1.5 kW; 2.45 GHz) was used to successfully synthesize SW- and MWBNNTs [16,17] in our group. The MPECVD reactor was equipped with an RF substrate heater. Temperature was monitored by means of an optical pyrometer and a thermocouple embedded in the graphite susceptor directly below the substrate. *In situ* diagnostic instrumentation including optical emission spectroscopy (OES) and quadrupole mass spectroscopy (QMS) was used to monitor the plasma chemistry. The gas-phase precursors consisted of 5% B_2H_6 (diborane) in H_2, NH_3, and H_2. Electron beam evaporation was used to deposit Ni and Co catalyst layers of different thicknesses on oxidized Si(111) substrates.

The following steps were involved in the synthesis of BNNTs on catalyst-coated oxidized Si(111) substrates by the MPECVD technique. The reaction chamber was first evacuated to a base pressure of 10^{-6} Torr and heated to 350 °C to ensure a clean environment for deposition. The temperature was then ramped up to 800 °C and held in an NH_3 atmosphere for 10 min. This step transformed the metal catalyst thin films into nanosize islands. Plasma initiation was done using H_2 at 15 Torr and 600 W. Once the plasma was stabilized, H_2 was partially replaced by NH_3, and the deposition process was initiated by the introduction of B_2H_6 into the gas mixture. The effect of temperature on BNNT synthesis was studied by systematically varying the deposition temperature between 600 and 950 °C in different experiments. The effect of catalyst layer thickness on BNNT synthesis was also studied. Scanning electron microscopy (SEM), transmission electron micron microscopy (TEM), selected area diffraction, Raman spectroscopy, and energy-dispersive X-ray spectroscopy were used to characterize the BNNTs.

The catalyst layer was found to be critical for the synthesis of BNNTs—there was no BNNT formation in areas without a catalyst film. Further, the thickness of the catalyst had a significant effect on BNNT synthesis. BNNTs were obtained for catalyst thicknesses <10 nm, while higher thicknesses inhibited BNNT formation. High BNNT yields were obtained when the catalyst thickness was below 2 nm. Figure 8.8 shows an SEM image of a dense mass of BNNTs grown on a 1-nm-thick Ni catalyst layer.

TEM micrographs revealed the presence of BNNT bundles (bundle width: 50–70 nm) consisting of individual BNNTs. A TEM image of a single BNNT is shown in Fig. 8.9. The BNNT diameter was found to vary between 5 and 20 nm with a wall thickness of 3–4 nm.

Electron diffraction analysis (Fig. 8.10) with the electron beam perpendicular to the BNNT axis of two aligned BNNTs revealed that they are zigzag nanotubes of crystalline nature with the <100> direction parallel to the tube axis. Figure 8.10(b)

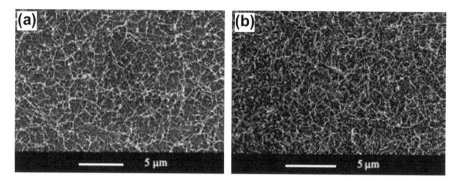

FIGURE 8.8

SEM images of BNNTs grown on (a) Ni catalyst (1 nm) and (b) Co catalyst (1 nm). Dense structures of BNNTs are clearly evident. The BNNTs were synthesized at 800 °C for 1 h. Gas composition: B_2H_6 (5 sccm); NH_3 (27.5 sccm); H_2 (10 sccm).

Source: Reprinted from Ref. [17]. Copyright (2009), with permission from Elsevier.

shows a pictorial representation of the electron diffraction pattern, wherein the solid circles represent diffraction spots from (0 0 2) planes, which constitute the side walls of the nanotubes while the open circles represent electron diffraction from the top and bottom walls. Raman spectra measurements (λ: 532 nm; Nd:YVO$_4$ laser) on the BNNTs showed the presence of the 1368 cm^{-1} peak, which is characteristic of h-BN, thereby confirming that BNNTs were indeed being synthesized by this process.

Systematic experimentation demonstrated that in addition to optimum catalyst thickness, the correct combination of gas-phase precursor ratios and temperature was critical for BNNT synthesis. For instance, low temperatures (500 °C) resulted in the formation of flakes with very few BNNTs. High temperatures, on the other hand, resulted in the formation of large metal islands ($\sim$100 nm) covered with BN.

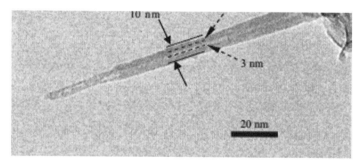

FIGURE 8.9

TEM micrograph of an individual BNNT. The wall thickness of this particular BNNT is 3 nm and the diameter of the BNNT is 10 nm.

Source: Reprinted from Ref. [17]. Copyright (2009), with permission from Elsevier.

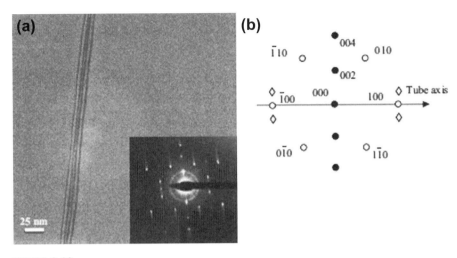

FIGURE 8.10

(a) TEM image of two aligned BNNTs. The inset shows their corresponding electron diffraction patterns. (b) Pictorial representation of the diffraction pattern. The BNNTs are crystalline in nature with a zigzag structure the <100> directions are parallel to the tube axis.

Source: Reprinted from Ref. [17]. Copyright (2009), with permission from Elsevier.

Based on the evaluation of the OES and QMS data, the following processes and mechanisms were proposed for the formation of BNNTs by the MPECVD process using B_2H_6, NH_3, and H_2. Under a plasma environment, the following four reactions resulted in the formation of a BN precursor, which in turn formed a BNNT in the presence of a catalyst.

$$\text{Decomposition of } B_2H_6: B_2H_6 \rightarrow B_xH_y + H_2 (x = 1 - 2; y = 1 - 2) \quad (8.9)$$

$$\text{Dissociation of } NH_3: NH_3 \rightarrow NH_z + H_2 \ (z = 0 - 2) \quad (8.10)$$

$$\text{Formation of } BNH_\gamma \text{ from (9) and (10): } B_xH_y + NH_z \rightarrow BNH_\gamma \ (\gamma = 2) \quad (8.11)$$

$$\text{Dissociation of } BNH_\gamma \text{ to form BN precursor: } BNH_\gamma \rightarrow BN + H_2 \quad (8.12)$$

It was proposed that the BN precursor from reaction 8.12 resulted in the formation of BNNTs in the catalyst-coated areas via a modified vapor–liquid–solid (VLS) mechanism. According to the proposed mechanism, the catalyst melted under the influence of plasma and substrate heating. The plasma generated gas-phase BN precursor then dissolved in the melted catalyst droplet, and upon supersaturation, precipitated to form BNNTs.

The fact that BNNT formation can be mediated by a catalyst opens up the interesting possibility of selective growth of BNNTs for applications such as nanoelectromechanical systems. It was demonstrated that BNNTs can indeed be synthesized selectively over large areas using the MPECVD technique as follows. Electron beam lithography was used to define different diameter Ni (2 nm thick) catalyst islands on a substrate. The substrate was exposed to BNNT deposition conditions described earlier. BNNTs were deposited selectively on the catalyst islands (Fig. 8.11) and not elsewhere, thereby demonstrating the feasibility of the process.

Therefore, the MPECVD technique offers a powerful way to synthesize high-quality, low-contamination BNNTs over large areas. Other boron nanostructures were also synthesized using a similar process [18].

8.3.4.1.2 Plasma-enhanced pulsed laser deposition

The second plasma-based approach for BNNT synthesis used a combination of an RF plasma and pulsed laser deposition (PLD); in other words, a plasma-enhanced pulsed laser deposition (PE-PLD) technique [19]. Pure iron was deposited on an oxidized Si substrate by PLD. The Fe-coated substrates were heated to 600 °C

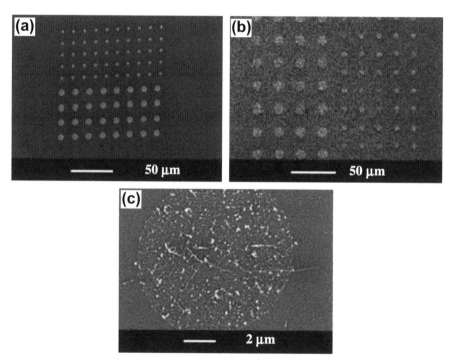

FIGURE 8.11

SEM micrographs demonstrating that selected area deposition of BNNTs is possible using a patterned catalyst as a template. (a): Ni catalyst islands before deposition. (b) and (c): Ni catalyst islands selectively covered with BNNTs.

Source: Reprinted from Ref. [17]. Copyright (2009), with permission from Elsevier.

and exposed to a pure N_2 RF plasma. Under these conditions, h-BN was ablated from a target using PLD, which led to the formation of BNNTs. It was proposed that the RF plasma induces a substrate bias voltage, which accelerates the ions and the BN vapor to the substrate resulting in the formation of BNNTs preferentially on Fe. It was found that a combination of RF plasma and PLD was required to successfully synthesize BNNTs. The thickness of the Fe layer was also found to be important.

8.3.4.2 In situ synthesis of BN precursors

Several different carbon-free boron precursors have been synthesized *in situ*. Two such precursors are discussed here. The first precursor, borazine ($B_3N_3H_6$), was synthesized in situ as a result of the reaction of $(NH_4)_2SO_4$, $NaBH_4$, and Co_3O_4 [20]. The resulting borazine vapors were transported by N_2 carrier gas into a tube-furnace reactor maintained at 1000–1100 °C. It was found that good-quality BNNTs were synthesized when NiB was used as the catalyst material (deposited on oxidized Si wafers). The second *in situ* precursor synthesis approach involved producing $B_4N_3O_2H$, from a combination of H_3BO_3 (boric acid) and $C_3N_6H_6$ (melamine) [21,22]. $B_4N_3O_2H$ powder was placed in a graphite crucible, which in turn was placed into a graphite susceptor, and heated to 1700 °C using an RF heater. $B_4N_3O_2H$ decomposed to form B_2O_3 (in vapor form), which was transported by N_2, and reduced into BN, which deposited on the graphite walls in the form of nanotubes. It was proposed that B–N–O nanoclusters initially nucleated on the walls of the graphite crucible. Subsequently, BN was believed to have precipitated and diffused on the B–N–O nanoclusters to result in the formation of BNNTs. Other BN nanostructures were also synthesized using this method [22]. Figure 8.12 shows BNNTs synthesized by using $B_4N_3O_2H$ as a precursor.

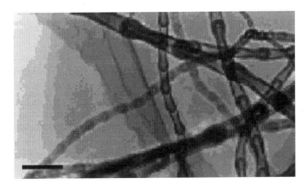

FIGURE 8.12

Transmission electron microscopy micrographs for BNNTs synthesized by the CVD technique using in situ synthesis of $B_4N_3O_2H$ (scale bar on lower left corner; 100 nm).

Source: Reprinted with permission from Ref. [33]. Copyright (2002) American Chemical Society.

8.3.4.3 Other CVD methods of synthesis of BNNTs

Other CVD-based methods have been used to synthesize BNNTs. These include RF induction heating of a mixture of boron oxide and magnesium oxide in a BN reaction tube [23]. The reaction between these components resulted in the formation of Mg and B_2O_2 vapors, which were transported by an argon carrier gas into a BN-walled reaction chamber that was maintained at 1100 °C under a continuous flow of NH_3. The B_2O_2 reacted with NH_3 to result in BNNT deposition on the BN walls of the reaction chamber. This method was found to yield large quantities of pure BNNTs without the use of a catalyst.

Another technique which may not be strictly classified as CVD but is nevertheless discussed here since it may lead to large-scale synthesis of BNNTs. In this approach, powders of β-rhombohedral boron and h-BN were mixed together and exposed to Li vapors at high temperatures (1200 °C) in an inert He environment for 10–20 h [24]. BNNTs and BN nanocones were found in the resulting powder. Numerous other CVD-based techniques for BNNT synthesis have been reported, further details can be found elsewhere [4,5]. It should be emphasized that CVD-based techniques provide a variety of routes for synthesizing high-quality BNNTs over large areas.

8.3.5 Laser heating and ablation

This method of synthesis uses pulsed or continuous wave (CW) lasers to synthesize BNNTs. The earliest version of this technique utilized a diamond anvil cell (DAC) to synthesize MWBNNTs under high pressures [25]. Single-crystal cubic BN was compressed to 5–15 GPa under an N_2 atmosphere and subjected to high-power CO_2 laser (240 W; 80 μm spot size) heating for 1 min. It was estimated that the temperatures in the DAC reached ~5000 °C resulting in BNNT formation. Another approach used CW laser heating to synthesize BNNTs but did not employ high pressures [26]. Instead, hot-pressed and thermally shocked h-BN microplatelets were heated by a high-power CO_2 laser ($\lambda = 10.6$ μm; 70 W; and 160 μm spot size) for 3 min in an N_2 atmosphere (~75 Torr). BNNTs were found in the heated area.

Instead of CW laser heating, laser ablation using pulsed lasers has also been utilized to synthesize BNNTs. In one approach, BN was mixed with Ni and Co powder and hot pressed to form a target [27]. The target was mounted in a quartz tube maintained at a temperature of 1200 °C with a carrier gas flow of Ar, N_2, or He. Subsequently the target was ablated by an excimer laser (248 nm; spot size: 1×3 mm²; pulse frequency: 10 Hz; pulse duration: 34 ns; energy: 400 mJ). The ablated material resulted in BNNT formation on a water-cooled copper block held downstream from the target. It was proposed that Co and Ni metals may have served as catalysts for the BNNT growth. A catalyst-free laser ablation approach was also developed for the synthesis of BNNTs [28]. A BN target was ablated with a CO_2 laser ($\lambda = 10.6$ μm; 1000 W; 7.5 mm spot size) in an N_2 atmosphere (~750 Torr). The ablation products were transported by a N_2 carrier gas and collected on a trap and filter. BNNTs were found on the ablated products collected on the trap and filter.

Laser ablation is one of the few techniques, which results in a high yield of SWBNNTs. However, laser ablation and laser heating are prohibitive in terms of cost, and the equipment cannot be operated continuously, thereby limiting large-scale synthesis of BNNTs by this method [5].

8.4 PROPERTIES OF BORON NITRIDE NANOTUBES

In contrast to CNTs, BNNTs are not good electrical conductors, thereby making the measurement of physical properties of these materials very challenging. Nevertheless, significant research has been performed on elucidating the fundamental physical properties of BNNTs. A few important physical properties of BNNTs are described below; further details can be found in the articles by Arenal, Goldberg, and Zhi [4–7].

8.4.1 Electrical properties

BNNTs are estimated to have a bandgap of ~ 5 eV [4,6] and are wide-bandgap semiconductors. Owing to the significant ionic nature of the bonds in BNNTs, their bandgap is independent of the tube chirality. Given the high-energy bandgap, it is also difficult to find elements which will act as dopants in a BNNT. Some encouraging results were obtained by using carbon as dopant in BNNTs. It was found that the addition of C to BNNT field emitters (FEs) enables the smooth tuning of the bandgap of the BNNTs down to 1 eV [4,29].

Current–voltage (I–V) measurements were performed using a low-energy electron point source microscope on a pure CNT, a pure BNNT, and a B–C–N nanotube synthesized by the carbothermal substitution method (Fig. 8.13) [29]. The pure CNT exhibited metallic I–V characteristics while the pure BNNT exhibited insulating behavior. Current–voltage characteristics of the B–C–N nanotube were found to be intermediate between the BNNT and CNT I–V curves thereby indicating that carbon addition to the BNNT had modified the electronic transport in the BNNT [4,29]. B–C–N NTs and ropes were also found to be good FEs with current densities close to that of CNTs. Since BNNTs have better thermal stability (Section 8.4.4) as compared to the CNTs, these results may indicate that B–C–N NTs may serve as stable long-life FE sources. Piezoelectric effects and BNNT-based field effect transistors have also been demonstrated [4].

8.4.2 Mechanical properties

Experimental measurements of the Young's modulus of BNNTs yielded values between 0.5 and 1.5 TPa for MWBNNTs [4,6,7]. While there is a paucity of experimental data on the mechanical properties of BNNTs, there is a wealth of theoretical predictions. The following predictions have been made with regard to the mechanical properties of BNNTs.

Owing to the differences in the activation and formation energy of defects in BNNTs vs CNTs, a crossover phenomenon is predicted for the yield strength of these two systems with respect to temperature. At moderate temperatures, CNTs

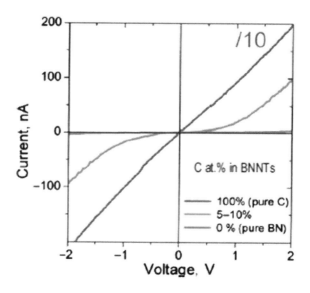

FIGURE 8.13

Current–voltage characteristics for CNTs, B—C—N nanotubes, and BNNTs. The plots are linear for CNTs (metallic behavior), while negligible currents flow through the BNNTs (insulating behavior). B—C—N nanotubes show intermediate I—V characteristics showing the effect of C doping on BNNT electronic transport. (For color version of this figure, the reader is referred to the online version of this book.)

Source: Reprinted with permission from Ref. [29]. Copyright (2004) Cambridge University Press.

are expected to be stronger while at high temperatures or for long deformation times, BNNTs are expected to outperform CNTs. It is also predicted that under plastic deformation, zigzag BNNTs are expected to be significantly weaker than the armchair nanotubes. Armchair BNNTs are expected to perform comparably to CNTs under plastic deformation [7].

8.4.3 Optical properties

Majority of experimental work on BNNTs has focused on the study of photoluminescence (PL), cathodoluminescence, and absorption spectra [7]. PL spectra indicate that BNNTs are indirect bandgap materials [4]. The emission spectra of BNNTs exhibit peaks at ~ 230 nm, ~ 279 nm, ~ 338 nm, and ~ 460 nm [7]. The PL yield for BNNTs is expected to exceed that of CNTs [4]. This fact coupled with the observed emission peaks in the UV regime indicates that BNNTs may be useful UV light sources.

8.4.4 Thermal properties, thermal stability, and wetting behavior

One-dimensional nanostructures can exhibit significantly enhanced thermal conductivities as compared to the bulk materials. Thermal transport in BNNTs is due to

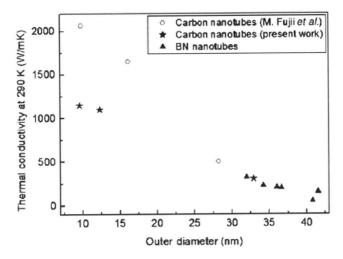

FIGURE 8.14

Experimentally measured thermal conductivity values for BNNTs and CNTs. For nanotube diameters close to 40 nm, BNNTs and CNTs exhibit similar behavior, data at lower BNNT diameters were not measured [4,7]. CNT data in the figure were obtained from the article published by Fujii et al. [10].

Source: Reprinted figure with permission from Ref. [34]. Copyright (2006) American Physical Society.

phonons and experimental measurements performed on BNNTs and CNTs show that for BNNT diameters close to 40 nm, the thermal conductivity values for BNNTs and CNTs are comparable [4,7]. However, data for smaller BNNT diameters (which should exhibit significantly higher thermal conductivity) were lacking (Fig. 8.14).

BNNTs have better thermal stability as compared to CNTs. Figure 8.15 shows thermogravimetric plots for BNNTs and CNTs. It can be seen from the figure that BNNTs, in this particular case, are stable up to ~900 °C as compared to CNTs which began to oxidize at ~500 °C [7].

As compared to bulk BN, BNNTs exhibit significant hydrophobicity with water contact angles in the range of 145°–160° [7]. The contact angles of other liquids on BNNTs were found to be similar to that of CNTs [4]. Figure 8.16 illustrates the significant hydrophobicity of BNNTs.

8.5 COMPARISON OF BNNTs AND CNTs

Given the structural similarities between BNNTs and CNTs, it is useful to briefly compare and contrast these two interesting nanostructures. The discussion below is based on the articles published by Arenal, Goldberg, and Zhi [4–7]. Hexagonal BN and graphene have similar lattice structures and comparable values for the elastic folding energies indicating that they both may be able to form nanotubes.

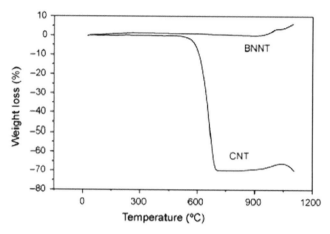

FIGURE 8.15

Thermogravimetric analysis data for BNNTs and CNTs showing that BNNTs have superior thermal stability up to temperatures as high as 900 °C.

Source: Reprinted from Ref. [7]. Copyright (2010), with permission from Elsevier.

FIGURE 8.16

Water has a high contact angle on BNNTs. This photograph illustrates the hydrophobicity of BNNT layers. (For color version of this figure, the reader is referred to the online version of this book.)

Source: Reprinted with permission from Ref. [35]. Copyright (2009) American Chemical Society.

However, once BNNTs and CNTs are formed, they can exhibit very different properties. These differences stem from the fact that the carbon atoms in CNTs are bonded covalently while in the case of BNNTs, the B—N bonds have a significant ionic component.

Therefore, BNNTs are wide-bandgap semiconductors while CNTs can exhibit metallic or semiconducting behavior. In addition, due to the differences in bonding for the two materials, unlike CNTs, the electronic and optical properties of BNNTs are largely independent of its chirality. The high bandgap of BNNTs results in high resistivity as compared to CNTs making BNNT synthesis and characterization very challenging. In terms of optical properties, BNNTs emit in the violet and ultraviolet regime (220–460 nm) while CNTs emit in the infrared wavelengths (800–1700 nm).

As discussed in earlier sections, the mechanical properties and thermal conductivity of BNNTs and CNTs are comparable, but BNNTs offer significant advantages in terms of thermal stability. Also, BNNTs exhibit significant hydrophobicity, which can be useful for certain water-repellant applications. Another difference between BNNTs and CNTs, which may have significant technological implications, is the difference in their gas adsorption behavior. Owing to the variation in electronic properties in CNTs (based on chirality), the adsorption properties of CNTs can vary by wide margins. Given the difficulty in controlling the structure of CNTs, it makes the synthesis of CNT-based gas adsorbers difficult. Since BNNT electronic properties are largely independent of their structures, it is possible to realize stable gas adsorbers using these materials. Experimental measurements indicate that BNNTs can rival or exceed CNTs in adsorbing hydrogen.

8.6 SUMMARY

Although challenging to synthesize, BNNTs are wide-bandgap semiconductor nanostructures exhibiting a unique combination of properties. They exhibit high resistivity, are amenable to doping which leads to continuous tuning of the bandgap, emit UV radiation, possess high strength and thermal stability, and are expected to show high values of thermal conductivity. Moreover, in contrast to CNTs, the electronic and optical properties of BNNTs are largely independent of their chirality. Given the difficulties in controlling atomic structure at the nanoscale (especially for materials synthesized in large quantities), BNNTs offer a stable platform for various applications. Therefore, with further research and development on large-scale synthesis and characterization of BNNTs, it is possible to envisage a range of technological applications for these materials including BNNT-based composites, FEs, gas storage materials, UV emission sources, and nanoelectromechanical systems.

Acknowledgments

The authors are grateful for the support of the National Science Foundation (NSF; Grant No. ECCS1237959 and CBET-1227788) for the present work. Any opinions, findings, conclusions or recommendations expressed in this material are those of the authors and do not necessarily reflect the views of the National Science Foundation.

References

[1] J.C. Phillips, G. Lucovsky, Bonds and Bands in Semiconductors, Momentum Press, 2009.
[2] G. Guisbiers, L. Buchaillot, Universal size/shape-dependent law for characteristic temperatures, Physics Letters A 374 (2009) 305–308.
[3] R. Tenne, Inorganic nanotubes and fullerene-like nanoparticles, Nature Nanotechnology 1 (2006) 103–111.

[4] D. Golberg, Y. Bando, C. Tang, C. Zhi, Boron nitride nanotubes, Advanced Materials 19 (2007) 2413–2432.
[5] R. Arenal, X. Blasé, A. Loiseau, Boron-nitride and boron-carbonitride nanotubes: synthesis, characterization and theory, Advances in Physics 59 (2) (2010) 101–179.
[6] D. Golberg, Y. Bando, Y. Huang, T. Terao, M. Mitome, C. Tang, C. Zhi, Boron nitride nanotubes and nanosheets, ACS Nano 4 (6) (2010) 2979–2993.
[7] C. Zhi, Y. Bando, C. Tang, D. Golberg, Boron nitride nanotubes, Materials Science and Engineering R 70 (2010) 92–111.
[8] N.G. Chopra, R.J. Luyken, K. Cherrey, V.H. Crespi, M.L. Cohen, S.G. Louie, A. Zettl, Boron nitride nanotubes, Science 269 (5226) (1995) 966–967.
[9] Y. Chen, J. Fitz Gerald, J.S. Williams, S. Bulcock, Synthesis of boron nitride nanotubes at low temperatures using reactive ball milling, Chemical Physics Letters 299 (1999) 260–264.
[10] M. Fujii, X. Zhang, H.Q. Xie, H. Ago, K. Takahashi, T. Ikuta, H. Abe, T. Shimizu, Physical Review Letters 95 (2005) 065502.
[11] Y. Chen, M. Conway, J.S. Williams, J. Zou, Large-quantity production of high-yield boron nitride nanotubes, Journal of Materials Research 17 (8) (2002) 1896–1899.
[12] J.D. Fitz Gerald, Y. Chen, M.J. Conway, Nanotube growth during annealing of mechanically milled boron, Applied Physics A 76 (2003) 107–110.
[13] D. Golberg, Y. Bando, W. Han, K. Kurashima, T. Sato, Single-walled B-doped carbon, B/N-doped carbon and BN nanotubes synthesized from single-walled carbon nanotubes through a substitution reaction, Chemical Physics Letters 308 (1999) 337–342.
[14] W. Han, Y. Bando, K. Kurashima, T. Sato, Synthesis of boron nitride nanotubes from carbon nanotubes by a substitution reaction, Applied Physics Letters 73 (1998) 3085–3087.
[15] D. Golberg, Y. Bando, K. Kurashima, T. Sato, MoO_3-promoted synthesis of multi-walled BN nanotubes from C nanotube templates, Chemical Physics Letters 323 (2000) 185–191.
[16] L. Guo, R.N. Singh, Selective growth of boron nitride nanotubes by plasma-enhanced chemical vapor deposition at low substrate temperature, Nanotechnology 19 (2008) 065601.
[17] L. Guo, R.N. Singh, Catalytic growth of boron nitride nanotubes using gas precursors, Physica E 41 (2009) 448–453.
[18] L. Guo, R.N. Singh, H.J. Kleebe, Growth of boron-rich nanowires by chemical vapor deposition (CVD), Journal of Nanomaterials (2006) 1–6, 58237.
[19] J. Wang, V.K. Kayastha, Y.K. Yap, Z. Fan, J.G. Lu, Z. Pan, I.N. Ivanov, A.A. Puretzky, D.B. Geohegan, Low temperature growth of boron nitride nanotubes on substrates, Nano Letters 5 (12) (2005) 2528–2532.
[20] O.R. Lourie, C.R. Jones, B.M. Bartlett, P.C. Gibbons, R.S. Ruoff, W.E. Buhro, CVD growth of boron nitride nanotubes, Chemistry of Materials 12 (7) (2000) 1808–1810.
[21] R. Ma, Y. Bando, T. Sato, K. Kurashima, Growth, morphology, and structure of boron nitride nanotubes, Chemistry of Materials 13 (2001) 2965–2971.
[22] R. Ma, Y. Bando, T. Sato, Controlled synthesis of BN nanotubes, nanobamboos, and nanocables, Advanced Materials 14 (5) (2002) 366–368.
[23] C. Tang, Y. Bando, T. Sato, K. Kurashima, A novel precursor for synthesis of pure boron nitride nanotubes, Chemical Communications (2002) 1290–1291.

[24] M. Terauchi, M. Tanaka, K. Suzuki, A. Ogino, K. Kimura, Production of zigzag-type BN nanotubes and BN cones by thermal annealing, Chemical Physics Letters 324 (2000) 359–364.

[25] D. Golberg, Y. Bando, M. Eremets, K. Takemura, K. Kurashima, H. Yusa, Nanotubes in boron nitride laser heated at high pressure, Applied Physics Letters 69 (14) (1996) 2045–2047, 30.

[26] T. Laude, Y. Matsui, A. Marraud, B. Jouffrey, Long ropes of boron nitride nanotubes grown by a continuous laser heating, Applied Physics Letters 76 (2000) 3239–3241.

[27] D.P. Yu, X.S. Sun, C.S. Lee, I. Bello, S.T. Lee, H.D. Gu, K.M. Leung, Synthesis of boron nitride nanotubes by means of excimer laser ablation at high temperature, Applied Physics Letters 72 (16) (1998) 1966–1968.

[28] R.S. Lee, J. Gavillet, M. Lamy de la Chapelle, A. Loiseau, J.-L. Cochon, D. Pigache, J. Thibault, F. Willaime, Catalyst-free synthesis of boron nitride single-wall nanotubes with a preferred zig-zag configuration, Physical Review B 64 (2001) 121405.

[29] D. Golberg, Y. Bando, P. Dorozhkin, Z.-C. Dong, Synthesis, analysis, and electrical property measurements of compound nanotubes in the B-C-N ceramic system, MRS Bulletin 29 (2004) 38–42.

[30] J.-C. Charlier, X. Blase, S. Roche, Electronic and transport properties of nanotubes, Reviews of Modern Physics 79 (2007) 677–732.

[31] B.G. Demczyk, J. Cumings, A. Zettl, R.O. Ritchie, Structure of boron nitride nanotubules, Applied Physics Letters 78 (18) (2001) 2772–2774.

[32] Y. Chen, L.T. Chadderton, J. Fitz Gerald, J.S. Williams, A solid-state process for formation of boron nitride nanotubes, Applied Physics Letters 74 (20) (1999) 2960–2962.

[33] R. Ma, Y. Bando, H. Zhu, T. Sato, C. Xu, D. Wu, Hydrogen uptake in boron nitride nanotubes at room temperature, Journal of the American Chemical Society 124 (2002) 7672–7673.

[34] C.W. Chang, A.M. Fennimore, A. Afanasiev, D. Okawa, T. Ikuno, H. Garcia, D. Li, A. Majumdar, A. Zettl, Isotope effect on the thermal conductivity of boron nitride nanotubes, Physical Review Letters 97 (2006) 085901.

[35] C.H. Lee, J. Drelich, Y.K. Yap, Superhydrophobicity of boron nitride nanotubes grown on silicon substrates, Langmuir 25 (9) (2009) 4853–4860.

CHAPTER

Boron Nitride Nanotubes, Silicon Carbide Nanotubes, and Carbon Nanotubes—A Comparison of Properties and Applications

9

Janet Hurst

National Aeronautics and Space Administration (NASA), Glenn Research Center, Cleveland, OH

CHAPTER OUTLINE

- 9.1 Introduction 267
- 9.2 BNNT and SiCNT Structure and Synthesis 268
  - 9.2.1 Structure 268
  - 9.2.2 Synthesis 270
    - 9.2.2.1 BNNTs 270
    - 9.2.2.2 SiCNT 271
  - 9.2.3 Property comparison of SiCNT and BNNT 274
- 9.3 Composites Reinforced with High-Temperature Nanotubes 276
  - 9.3.1 BNNT composites 276
  - 9.3.2 SiCNT composites 281
- 9.4 Applications of High-Temperature Nanotubes 281
- 9.5 Concluding Remarks 285
- References 285

9.1 INTRODUCTION

Development of noncarbon nanotubes (CNTs) for use in aggressive environments has been pursued by the National Aeronautics and Space Administration (NASA) Glenn Research Center (GRC) for nearly 10 years. While carbon nanomaterials offer tremendous potential, noncarbon materials also have an important place in the vision embraced by NASA. This vision ranges from exploration conducted in the cold of the interstellar space and the hot toxic environment of Venus to designing of future generations of commercial aircraft. The wide range of mission profiles requires a broad investment in new materials and structures to meet varied requirements ranging from commercial aeronautics to deep space missions. Although each mission has specific materials requirements, broad themes still emerge. The materials development driver is for increasing specific strength, specific energy and power. The NASA Nanotechnology Roadmap [1] attempts to address these requirements. This plan has targeted goals such as a 25% reduction in weight for aerospace structures and 50% lower density materials to help meet needs such as reduced fuel consumption and emissions. Significant weight savings are anticipated via CNT reinforcement of structures such as airframes; however, there are also important applications where the properties of CNTs may not be sufficient to meet the challenges. New material systems based on high-temperature nanotubes are needed to solve issues including weight reduction, higher engine operating temperatures, radiation protection, higher thermal conductivity materials to improve thermal management and higher temperature and/or higher power electrical systems and electronics.

In the past decade, the promising properties and applications of CNT have been pursued throughout the world. Only to a much lesser degree have higher temperature nanotubes been explored. Boron nitride nanotubes (BNNTs) and silicon carbide nanotubes (SiCNTs) are of interest to NASA due to their superior thermo-oxidative stability relative to CNTs, which suggests potential use in hostile environments. Interest ranges from utilizing BNNT insulation properties for thermal management to radiation protection for future manned space missions. Interest in SiCNT covers a wide range of applications from structural to electronic. In these and perhaps applications not yet considered, non-CNT compositions, SiCNT and BNNT, may bring nanoengineered property improvements to demanding power and propulsion applications as well as enable more aggressive space missions.

9.2 BNNT AND SiCNT STRUCTURE AND SYNTHESIS
9.2.1 Structure

As remarkably inert and oxidatively stable materials, nanotubes composed of BN and/or SiC are interesting and yet have presented processing challenges. Bulk boron nitride is a man-made material and is synthesized in four polytypes. Commercially, both cubic and hexagonal (h-BN) polytypes are important. Rhombohedral BN and wurtzite BN (w-BN) polytypes can also be synthesized. The wide range of uses for BN includes cosmetics to thermal management to high-temperature lubricants and release agents to military armor [2,3]. Similarly, bulk SiC has been of interest for many thermal and structural applications ranging from gas turbine components and heating elements to nuclear fuel cladding and even jewelry [4,5]. High-temperature and/or high-power electronic and electrical applications are particularly important due to the high electric field breakdown strength and high maximum current density of SiC [6–8]. SiC has two common polytypes, cubic SiC (β-SiC) and hexagonal SiC (α-SiC). Both are composed of a tetrahedron as the building block for over 200 polymorphs. These polymorphs have the same structure in two dimensions but have variations in stacking sequence in the third dimension [9]. The structures of hexagonal C, h-BN and α-SiC are shown in Fig. 9.1.

The bulk properties of h-BN have generated considerable interest in BNNT as a high-temperature analog to CNTs. Both BN and C can be synthesized in similar

9.2 BNNT and SiCNT Structure and Synthesis

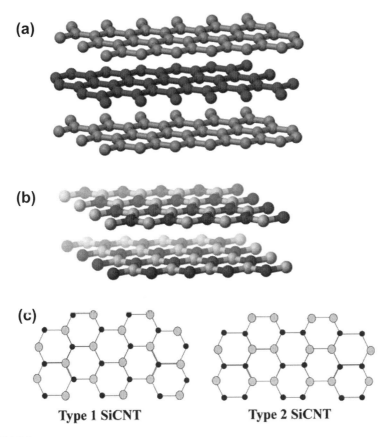

FIGURE 9.1

Hexagonal sheets of (a) carbon, (b) boron nitride, (c) silicon carbide, two arrangements with silicon in gray, carbon in black. (For interpretation of the references to colour in this figure legend, the reader is referred to the online version of this book.)

Source: (c) After Ref. [46]. M. Menon, et al., Physical Review B, 69, 115322.

graphene structures. Hexagonal layers of primarily in-plane sp^2 bonding [10–12] are held together by van Der Waals forces. In h-BN sheets, the B atoms sit directly above the N atoms in an adjacent layer, so hexagons are stacked directly on top of one another, as seen in Fig. 9.1. However in carbon, C atoms fit between C atoms in adjacent sheets, and so are shifted by a half hexagon. The B—N bonds reduce the number of chiral variants available in BNNTs to zigzag and armchair relative to CNTs with their numerous chiral variants. While h-BN layers were first synthesized in nanotube form by Chopra in 1995 [10], the existence of BNNTs was anticipated earlier due to these structural similarities [11,12] with carbon.

SiCNT is composed of hexagonal sheets as well, and like BNNT, both zigzag and armchair types are predicted to be energetically stable [13]. A two-dimensional graphene structure similar to CNT is more difficult to achieve in SiC as the larger radii of Si relative to B, N and C leads to higher coordination numbers and sp^3 bonding. However, this can be accommodated by localized distortions of the lattice. This "rippled" structure predicted for SiCNT [13] has more electropositive C atoms rolling outward and more electronegative Si rolling inward relative to the tube diameter. A similar rippled effect also occurs in BNNTs. Two potential atomic arrangements are predicted. Type 1 has only alternating Si—C bonds whereas type 2 contains some C—C bonds as well. Both are shown in Fig. 9.1 and each can be expected to exhibit zigzag and armchair forms. These structural variations cause property differences as well.

9.2.2 Synthesis

9.2.2.1 BNNTs

The high-temperature oxidative stability advantage of BNNTs over CNTs comes at the price of somewhat more difficult synthesis. However, there have been many successful synthesis demonstrations of this material by a variety of methods, mostly similar to CNT production methods. Some of these include arc discharge [10,14,15], chemical vapor deposition (CVD) [16—19], reactive milling [20—22], laser methods [23,24] and chemical templating from CNT reaction [25—27]. Still commercial development of good-quality BNNTs has not occurred. Limited availability exists through research groups. Synthesis methods have an effect on BNNT product including nanotube diameter and length as well as defects and impurities; this in turn affects BNNT properties such as thermal conductivity and strength. As NASA-GRC is primarily an engineering organization, synthesis of BNNT has been pursued with an eye toward suitability of product for power, propulsion and space applications including ease of future commercialization. Current synthesis methods were tried and downselected based on reproducibility, safety, flexibility and relevance to NASA-GRC applications. Thermal CVD met these criteria. Using this process, laboratory-scale quantities of BNNT on the order of 2 g/day/rig are produced by batch process, although continuous processing should be possible. BNNT is synthesized either in powder form or as surface coatings on substrates including super alloys, alumina, carbon, Si and SiC wafers and other SiC surfaces. Some of these are shown in Fig. 9.2. This material is being used for property investigations by NASA-GRC and its partners. Control of nanotube diameter and form can be accomplished through selection of appropriate catalyst details and adjusting processing parameters such as time, temperature, gas composition, gas pressure and flow rate. Due to both safety concerns regarding nanoparticles and ease of subsequent dispersion of larger diameter BNNT in matrices, powder is generally synthesized in the 80—100 nm range, although smaller diameters are occasionally produced as well. BNNT coatings synthesized on substrates typically have diameters of about 40 nm. As there are several excellent reviews of BNNT properties and processing published elsewhere [28—31], these are not further examined here.

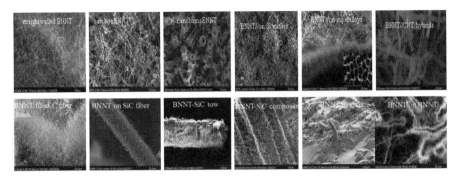

FIGURE 9.2

Several BNNT morphologies and infiltrations performed at the NASA-GRC.

9.2.2.2 SiCNT

Synthesis of both SiCNT and SiC nanorods has been demonstrated in small quantities [32–46]. Processing approaches include CVD [42–44] and carbothermal reduction into porous sacrificial templates such as silica and carbon [45,46]. SiC whiskers, whose sizes are on the order of microns, have also been utilized for many years for composite reinforcement [47,48]. The most successful approach for synthesis of SiCNT has been chemical templating, which uses CNT as a starting material [34–41]. As noted above, this technique has been used for successful BNNT synthesis as well.

SiCNT production using the chemical templating approach has been reported by several groups [34–41]. This method scales up easily and is used at NASA-GRC for SiCNT production in 20-g batches. A simplified illustration of the technique is shown in Fig. 9.3. CNTs are exposed to Si, which is either vaporized from fine Si metal or from gaseous sources of Si such as $SiCl_4$. Conversion of individual CNT to SiCNT occurs at temperatures of 1200–1400 °C. Typical output from this production method is shown in Fig. 9.4. Nanotubes with an outer layer of SiC and an

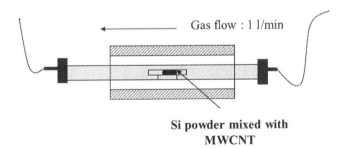

FIGURE 9.3

Schematic for furnace arrangement for CNT to SiCNT conversion. (For color version of this figure, the reader is referred to the online version of this book.)

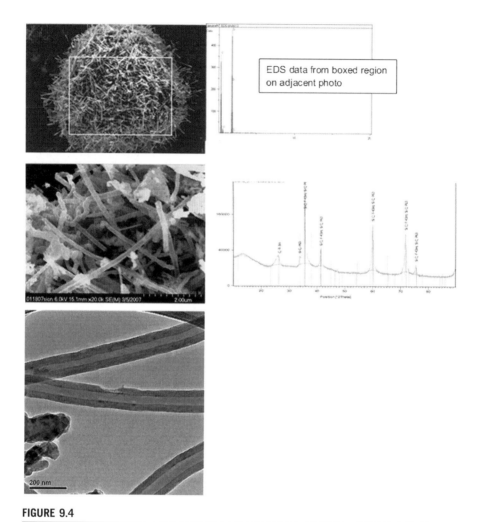

FIGURE 9.4

Typical SEM, TEM images, as well as EDS and X-ray data for SiCNT. (For color version of this figure, the reader is referred to the online version of this book.)

interior layer of CNT are also synthesized, in a coaxial cable affect. The somewhat disordered appearance of nanotube walls for the outer portion of a coaxial cable type of SiCNT is illustrated in Fig. 9.5. This is the result of stacking faults, a common occurrence in SiC structures [9]. Controlling the extent of the conversion reaction is somewhat difficult and grains of β-SiC are often nucleated on the nanotube surfaces as seen in Fig. 9.6. SiC nanorods are found as well, although these are less than 10% of the product. About 50–90 weight% conversion to SiC occurs, depending on the processing parameters and starting materials. The final form of the SiCNT product is very similar to the CNT starting materials. The CNT inner

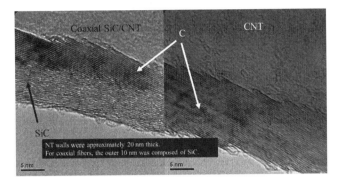

FIGURE 9.5

SiCNT and CNT wall structures. Stacking faults are common in SiCNT walls. This is typical in bulk SiC as well. (For color version of this figure, the reader is referred to the online version of this book.)

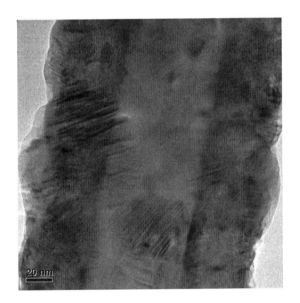

FIGURE 9.6

Isolated SiC grains on SiCNT surfaces causing a roughened surface.

core can be removed via oxidation at temperatures of 600–700 °C, leaving behind the SiCNT sheath. This technique extends the temperature stability range of CNT to higher temperatures. The improved thermal stability of SiCNT relative to CNT allows reinforcement into matrices, which might otherwise be difficult for CNT due to temperature or chemistry. This method may compare favorably with multistep purification techniques required to extend the stability range of CNT to higher

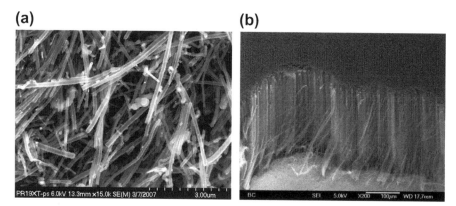

FIGURE 9.7

SiCNT can be synthesized via a high-temperature conversion process from a variety of CNT products including inexpensive MWCNT (a) or nanoforests (b).

Source: From NanoLab.

temperatures [49]. Another advantage of this method is that a wide array of commercially available CNT products such as nanotubes of various diameters, number of walls, chiral variants, lengths, as well as aligned nanoforests may be successfully used as starting materials; some of these are shown in Fig. 9.7. The improvement in thermal stability as measured by thermogravimetric analysis (TGA) that is achieved for an inexpensive multiwalled carbon nanotube (MWCNT) starting material is shown relative to the SiCNT product following conversion in Fig. 9.8.

9.2.3 Property comparison of SiCNT and BNNT

Many properties of BNNT and SiCNT are similar to those of CNTs, as shown in Table 9.1. The elastic modulus of BNNT has been found to be the same as that of CNT, with 1.22 TPa [50] and more recently 1.3 TPa [51] being reported for BNNT. Tensile strength of individual nanotubes of up to 33 GPa has been reported [52]. Robust buckling characteristics within individual single-walled BNNT was observed by Zheng [52], which further increases interest in BNNT for structural reinforcement.

BNNT shares high thermal conductivity with CNT although unlike CNT, BNNT is an electrical insulator [53−55], an unusual combination of properties making it potentially useful in thermal management. BNNT and CNT thermal conductivities have been measured as about 350 W/m K 300 W/m K, respectively (both for outer tube diameters of 30−40 nm). The thermal conductivities of both BNNT and CNT are predicted to be in excess of 3000 W/m K for 14 nm tube diameters [56].

Unlike CNT, the band gap of BNNT has been found to be very stable at 5−6 eV, [12,57−59], independent of chirality and diameter [12]. However, there has been success in increasing the metallic character of BNNT as well [60]. Also unlike

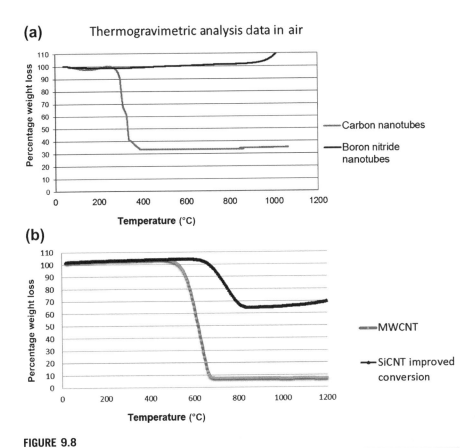

FIGURE 9.8

Thermogravimetric analysis measured in air for BNNT (a) and SiCNT (b).

CNT, BNNT is very thermo-oxidatively stable. BNNT stability far exceeds CNT stability with demonstrated thermal stability from 900 °C [61] to over 1000 °C [62,63] for BNNT as compared to only 400–600 °C for CNT (Fig. 9.8).

Fewer property measurements have been conducted on SiCNT, so most available properties are mainly theoretical predictions. Table 9.1 shows several significant properties of SiCNT and BNNT. The elastic modulus of SiCNT is expected to be only half that of either BNNT or CNT, at about 0.61 TPa. [64]. This value is similar to those of SiC nanorods and whiskers and is nearly twice that of the β-SiC phase.

Modeling suggests that thermal conductivity may be very poor for both SiCNT and SiC nanorods with values being two orders of magnitude lower than that of bulk SiC [65]. Both SiCNT and BNNT are potentially wide band gap semiconductors. Notably, very different band gaps are predicted for the two atomic arrangements of SiCNT previously discussed, type 1 and type 2. Type 1, with alternating Si–C bonds, is predicted to have a semiconductor band gap of 3.2 eV and 3.6 eV for zigzag

Table 9.1 Properties of CNT, BNNT and SiCNT

Material	CNT	BNNT	SiCNT
Elastic modulus	1.3 TPa	~1.3 Pa [51]	~0.61 TPa (equivalent to those of SiC nanorods and whiskers. Nearly double β-SiC[64]
Strength	11–150 GPa	33 GPa [51]	?
Band gap	Varies with chirality and diameter [12]	5 eV [12,60–62]	3.4–3.6 eV[46] 0.7–0.5 eV [46]
Thermal conductivity (W/(m K))	3000 [56]	350 [53] experimental >600 [54] theoretical	<10 [65]
Thermal stability	400–600 °C	>1000 °C [61–63]	>800–1000 °C

and armchair structures, respectively. On the other hand, type 2 is an electrical conductor with a band gap of only 0.7 eV and 0.5 eV for zigzag and armchair structures, respectively [13].

The thermal stability as found by TGA in air is compared for BNNT synthesized by CVD, SiCNT synthesized by chemical templating (NASA-GRC), and a commercially available MWCNT in Fig. 9.8. While CNT rapidly oxidizes and disappears above 500 °C, both BNNT and SiCNT slowly oxidize at 1100 °C and 800 °C, respectively. It is clear that both BNNT and SiCNT offer significant advantages over CNT for the high-temperature processing required for glasses, metals and ceramics.

9.3 COMPOSITES REINFORCED WITH HIGH-TEMPERATURE NANOTUBES

9.3.1 BNNT composites

In the past decade, BNNT as a structural reinforcement has been demonstrated in a handful of efforts. As BNNT becomes more generally available, its utilization in composites for strength enhancement as well as for multifunctional uses (i.e. strength and thermal conductivity or strength and radiation protection) will increase. Recent research has demonstrated BNNT-reinforced polymer films of polyaniline [66] and polystyrene [67] with a 20% increase in elastic modulus while maintaining transparency of the films. BNNT has also been demonstrated as reinforcement in aluminum ribbons [68,69] and aluminum composites [70]. In aluminum ribbons, tensile strength was more than doubled as a result of BNNT addition. BNNT has also been added in small amounts to Al_2O_3 and Si_3N_4.

9.3 Composites Reinforced with High-Temperature Nanotubes

Both materials became more deformable at high temperatures with additions of 2.5–5 wt% BNNT [71].

The effort at NASA-GRC has been focused on higher temperature matrices such as glass, aluminum and ceramics where typically CNT thermo-oxidative stability is problematic. The first successful demonstration of BNNT as a structural reinforcement was in glass–ceramic composites in 2006 by Bansal, Hurst and Choi [72]. In their procedure, 4 wt% BNNT (synthesized at NASA-GRC by mechanical milling similar to Chen [20,21]) was dispersed in a barium calcium aluminosilicate glass powder and subsequently hot pressed to 630 °C. Flexural strength was nearly doubled and composite fracture toughness was increased by 35%, see Fig. 9.9. The BNNT phase demonstrated "fiber" pullout on the fracture surface, also in Fig. 9.9, typical of fiber-reinforced composite behavior. This suggested that standard composite mechanical modeling theory can be used to understand nanotube-reinforced material behavior.

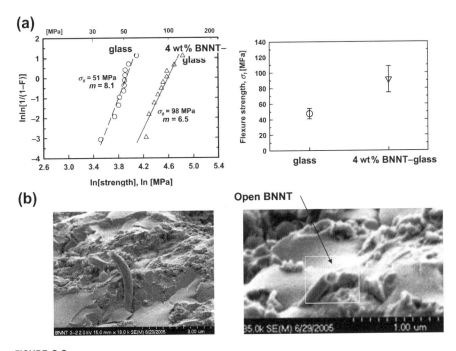

FIGURE 9.9

4 wt% BNNT reinforcement in a barium calcium aluminosilicate glass composite. (a) Weibull strength distribution (left) for an unreinforced glass and BNNT-reinforced glass showing nearly twice the strength for BNNT reinforcement. A moderate increase of 35% in fracture toughness was also demonstrated (right). (b) Fracture surfaces of BNNT-reinforced glass composites showed well-distributed BNNT and BNNT pull-out from fracture surfaces. (For color version of this figure, the reader is referred to the online version of this book.)

FIGURE 9.10

Fuzzy fiber concept; nanotubes grown on individual fibers to provide enhanced interlaminar properties.

Randomly oriented dispersions of nanotubes are a useful first step in composite reinforcement with BNNT. However, directional growth of nanotubes as in the case of nanoforests of fuzzy fibers, offers promising nanoengineering opportunities. The "fuzzy" fiber concept, Fig. 9.10, has previously been demonstrated with CNT grown on carbon fibers reinforcing a thermoset resin system. A 69% improvement in interlaminar shear was demonstrated [73–75]. This approach was recently investigated by the author with BNNT grown on tows and inside SiC-woven preforms to synthesize "fuzzy" fibers of silicon carbide fiber for reinforcement within silicon carbide matrix composites (SiC_f/SiC_m ceramic matrix composites (CMCs)) [76,77].

In the current generations of SiC_f/SiC_m composites, industry has favored laminate or two-dimensional fiber layups for ease of manufacturing. However, this approach produces composites with insufficient shear properties for the high-stress conditions likely to be encountered in many hot section components for gas turbine engines such as blades and vanes. Three-dimensional weaving approaches could generate improvements in interlaminar properties but are expensive and may be impossible to carry out with high-modulus SiC tows. Additionally, component fatigue properties may suffer with reduced fiber in loading directions. In SiC_f/SiC_m composites, SiC fibers must be coated with an interfacial debonding layer of planar BN, which is required SiC composites to provide graceful failure and to provide good high-temperature mechanical properties [78]. A typical SiC_f/SiC_m composite micrograph is shown in Fig. 9.11. Carbon layers are also used in this application rather than BN, but use of these composites is limited in temperature and oxidation exposure. Summarizing, it is anticipated that improvements in interlaminar strength and localized toughening can be achieved with "fuzzy" fibers of BNNT on SiC fibers.

BNNT growth on SiC tows and woven fabric has been demonstrated with good infiltration of tows and both 8-harness satin and pinweave preforms. A thermal CVD method, as previously discussed, was used to synthesize BNNT on fiber surfaces. This is an ideal application for the CVD BNNT synthesis method. CVD BNNT synthesis uses similar equipment and processing gases as those used to deposit planar BN coatings for SiC tows and plies are shown in Fig. 9.12 with extensive BNNT growth. Also shown in Fig. 9.12 is a fiber fracture surface taken from an

9.3 Composites Reinforced with High-Temperature Nanotubes

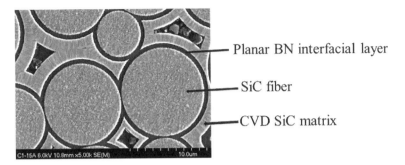

FIGURE 9.11

Typical microstructure of a SiC_f/SiC_m composite demonstrating the planar BN interface. (For color version of this figure, the reader is referred to the online version of this book.)

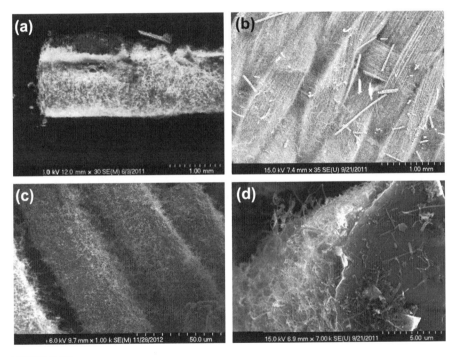

FIGURE 9.12

(a) BNNT-coated SiC tow, (b) BNNT-coated ply surface, and (c, d) interior fiber surface coated with BNNT within woven 8-harness satin fabric.

interior tow of a woven fabric with abundant BNNT growth coating the fiber surface. BNNT coats nearly all SiC surfaces similar to the way planar BN is currently infiltrated into woven SIC fabric as an interfacial layer in typical SiC composites. Four BNNT-infiltrated SiC plies were assembled into simple composites, 1.25 cm wide by

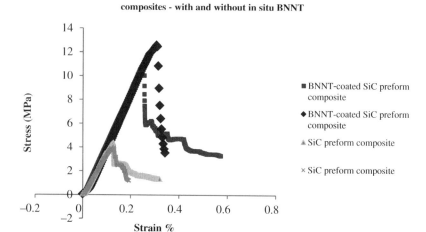

FIGURE 9.13

Addition of BNNT to SiC composites resulted in a nearly three times significant increase in tensile strength. (For color version of this figure, the reader is referred to the online version of this book.)

0.055 cm by 10 cm length, and hot pressed at 2000 °C and 41 MPa. Coupon testing of BNNT-reinforced SiC composites demonstrated a three times improvement in room temperature tensile strength relative to unreinforced SiC, Fig. 9.13. Examination of the fracture surfaces found BNNT undamaged by hot pressing and intact on the SiC surfaces following fracture, Fig. 9.14. Ongoing work is currently focused on

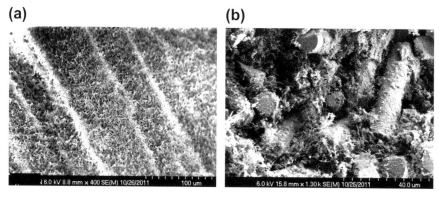

FIGURE 9.14

BNNT survived hot pressing at 2000 °C and are present on the surface of the composite (a). BNNT was also found on fractured surfaces within composites following tensile testing (b).

measurement of interlaminar properties and development of dense SiC/BNNT/SiC composites. High-temperature property investigation is also necessary with thorough investigation of mechanical properties including crack growth, high-temperature tensile strength, and fatigue.

9.3.2 SiCNT composites

Little has been published on SiCNT-reinforced composites. SiCNT has significant thermo-oxidative stability and a more reactive surface than BNNT. At NASA-GRC, SiCNT has been used to reinforce environment barrier coating (ebc) compositions with promising results. As spallation of an ebc would be catastrophic to many gas turbine components, high toughness and strength are important. A two to three times improvement in strength was found for SiCNT-reinforced ebc bond coatings of either SiCNT-HfO_2-Si or SiCNT-$Yb_2Si_2O_7$-Si [41]. Another significant finding is that SiCNT was stable in air to 1450 °C after 100 h exposure when protected by a matrix. Thermal conductivities of these SiCNT-reinforced coating compositions were measured by the laser flux method. Thermal conductivity was found to be low, in the range of a few W/mK. This was similar to predicted values. Oxidation kinetics for SiCNT was determined by both TGA studies and scale thickening measurements. Activation energies were in good agreement for the two methods at 155 kJ/mol and 142 kJ/mol.

SiCNT was also added to a $BaO \cdot SrO \cdot Si_2Al_2O_8$ glass and tested in thermal cycling up to 1500 °C. Improved thermal resistance and a 20% increase in thermal conductivity for SiCNT-reinforced specimens was found, as shown in Fig. 9.15.

9.4 APPLICATIONS OF HIGH-TEMPERATURE NANOTUBES

While CNTs are suitable for many inert applications at high temperature or for short exposures in other environments, BNNTs, SiCNTs and SiC nanorods are promising for more aggressive environments. A summary of some of the many applications for high temperature nanotube/rods that are being considered at NASA-GRC is included in Table 9.2. As SiCNTs and BNNTs become more widely available, additional applications will be suggested by those requiring materials solutions for specific problems. As BNNTs and SiCNTs are unlikely to compete in price with CNTs for some time, the property improvements achieved with these must be compelling.

A good example of an extreme environment application for BNNT and SiCNT is found within gas turbine engines, which demand material properties such as high strength and extreme chemical inertness in the presence of high-temperature combustion gases with material temperatures of well over 1300 °C for hot section turbomachinery. Future engine designs may require hot section turbomachinery, which can operate for thousands of hours at temperatures up to 1700 °C. CNT reinforcement will not suffice due to poor oxidative stability. A few of the potential high-temperature applications within gas turbine engines for BNNT investigated by the

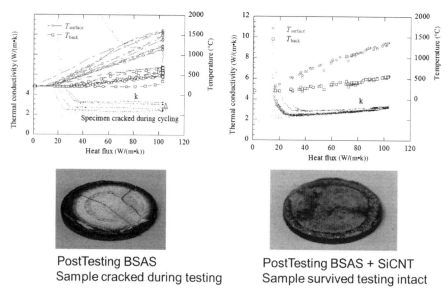

FIGURE 9.15

SiCNT added to $BaO \cdot SrO \cdot Si_2Al_2O_8$ (BSAS) glass demonstrated a 20% higher thermal conductivity relative to the baseline glass as well as superior thermal gradient cyclic stress resistance at temperatures up to 1500 °C. No cracking was observed for the SiCNT–BSAS specimens. The BSAS specimens did not survive testing. (For color version of this figure, the reader is referred to the online version of this book.)

author are shown in Fig. 9.16. Many of these applications are also appropriate for SiCNT reinforcement.

Also, there is a growing trend toward hybrid electric propulsion from cars to commercial aircraft to optimize energy efficiency. In the case of commercial aircraft, future concepts often include electric motors, which are driven by either power generated from turbines or energy storage systems, which will drive the fan. Thermal management becomes a major challenge for electric (or more electric) aircraft that require many high power density components, such as electric motors, power electronics, and high-voltage transmission. To achieve high power density in electric motors, removal of heat is necessary to avoid increased temperature within electric coils and permanent magnets. Similarly, elevated temperatures in power electronics are limiting efforts to improve their power density. Battery life is also limited by elevated temperatures during rapid discharge cycles. Clearly, effective thermal management would increase battery life in vehicles from electric cars to aircraft. One approach for thermal management is to transfer the heat to a heat sink through a thermal interface material (TIM). Aligned CNT forests are under development as TIM for thermal management in numerous applications such as batteries and power electronics because of their very high thermal conductivity.

9.4 Applications of High-Temperature Nanotubes

Table 9.2 Potential NASA Application of High-Temperature Nanotubes and Nanorods

Nanoreinforcement	Matrix	NASA Applications
BNNT	Ceramics such as SiC, glass, metal, polymers	Metals: aluminum reinforcement for low-weight high strength structure Ceramics: SiC/SiC composites for hot-section turbomachinery Glass fibers: high strength and/or high temperature glass fibers, flywheels Polymers: thermal management structures Radiation shielding
SiCNT, SiC nanorods	SiC, glass, metal, polymers	Environmental barrier coatings, SiC, composites SiC fibers for >3000 F turbomachinery
CNT	Polymer, glass	Polymer: low-weight, high-strength structures such as airframes, etc. Glass fibers: flywheels, etc.
B_4C	Polymer	Radiation shielding

But for applications where the heat sink is the primary structure itself or the heat sink is attached to the primary structure, it would be highly beneficial to have a TIM that is thermally conductive, but electrically insulating. Additionally, as there are 171 miles of wiring in a Boeing 747 [79], lighter weight insulation could translate into saving of over 4000 pounds through reduced fuel consumption and increased payload. Similar weight savings are important for space-based technology as well.

Another unique application is radiation shielding for long duration space missions. Compositions containing boron can provide both structural reinforcement as well as protection from cosmic galactic radiation (CGR). This radiation becomes a significant concern with longer mission profiles beyond the protective confines of the Earth's atmosphere. Manned missions to Mars or asteroids will require new radiation protection solutions that are effective and very light weight due to the detrimental effects of CGR as well as numerous additional ionizing sources, on both astronaut health and equipment reliability. Effective solutions such as lead or water are simply too heavy to consider for launched systems. Therefore, both BNNT and B_4C may be a part of a light-weight multiphase structural solution for this complex issue due to the very high neutron capture cross-section of ^{10}B. A demonstration of BNNT with a high percentage of ^{10}B has been published [80]. Also, recent work has shown h-BN membranes as being more resistant to electron beam irradiation than graphene [81].

If nanotube-reinforced aluminum could be made at reasonable costs, aluminum could more easily compete with composites in many aerospace structural applications where weight reduction is a prime motivation for new materials development [1]. As

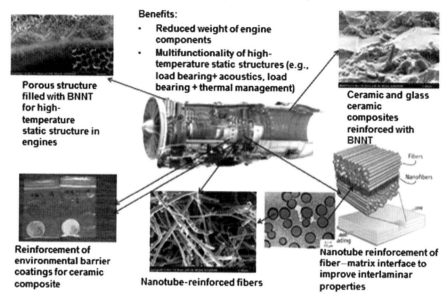

FIGURE 9.16

Potential applications of BNNT within a gas turbine engine. (For color version of this figure, the reader is referred to the online version of this book.)

strength improvements of three times appears to be achievable in several systems, including aluminum, one could envision using one-third as much material in an airframe or an Al, which was as strong as steel but with a much lower weight. BNNT may also be useful in PMCs where better dielectric properties, higher thermal conductivity or optical properties achievable with BNNT are needed. Both BNNT and SiCNT may be useful for high-temperature wide band gap semiconductor applications for nanoelectronics. Additional applications include medicine [81] and ion emitters [82] and there are likely numerous others including multifunctional applications.

The remaining issue for application of BNNT and SiCNT is the availability of successful commercial sources of material. Potential applications of high-temperature nanotubes are not sufficient cause for commercial development. Near-term and significant applications are necessary to spur commercialization of BNNT and SiCNT; otherwise they remain of academic interest alone. Early applications must exist which must meet the following three criterions:

1. There are no adequate, less-costly alternatives available
2. Additions of nanotubes provide high performance
3. A near-term market for the specific application exists.

High-temperature nanotubes are actually a family of products of various tube lengths, diameters, chiral variants and qualities available. Application requirements, costs and handling issues will dictate the product needed. For example, while fine-diameter BNNT of 8 nm may be useful for high thermal conductivity applications, nanomaterial with this diameter is both difficult to handle and disperse in composite matrices. Larger diameter, multiwalled products are often better suited for composite applications and have the additional advantage of simplified handling. It has been found that this material also provides better reinforcement properties within composites [67,68,72,83], even though individual nanotube strength may be lower [84]. Early commercialization efforts are likely to focus on these materials, similar to CNT development paths.

An important practical matter is that nanotubes with diameters over 100 microns may be treated as a fine powder as indicated by current National Institute for Occupational Safety and Health (NIOSH) regulations [85] whereas those below 100 microns require additional safety procedures, which are still largely under development for nanomaterials. Commercial development for larger diameter nanotubes may initially be more practical.

9.5 CONCLUDING REMARKS

Tremendous strides continue to be made throughout the field of nanotechnology. With increasing maturity, nanotechnology is no longer merely of academic interest; it is rapidly becoming an engineering tool to tailor materials, devices and/or surfaces to create improved specific properties or to provide several properties. With this shift, nanotechnology directed toward high-temperature applications and other specific needs is of increasing interest. The significant progress that has been made in synthesis of BNNT and SiCNT has resulted in sufficient material to begin investigation into composite development and applications. As more thorough property investigations begin, new applications will continue to be discovered. Improvements in strength of three times are being made with very little effort at optimization. Strength improvements of 5–10 times may be possible, opening entirely new avenues for higher specific strength materials. With other property improvements, higher specific energy and power applications will follow.

References

[1] M. Meador, B. Files, J. Li, H. Manohara, D. Powell, E. Siochi, NASA Nanotechnology Roadmap, http://www.nasa.gov/pdf/501325main_TA10-Nanotech-DRAFT-Nov2010-A.pdf.

[2] A. Lipp, K. Schwetz, K. Hunold, Journal of the American Ceramic Society 5 (1) (1989) 3.

[3] R. Paine, C. Narula, Chemical Reviews 90 (1) (1990) 73.

[4] V.A. Izhevskyi, et al., Cerâmica 46 (2000) 297.

[5] A.R. Powell, L.B. Rowland, Proceedings of the IEEE 90 (6) (2002) 942.
[6] G. Hunter, et al., Silicon carbide, Proceedings of the Materials Research Society Symposium (2004) 815. (Eds. M. Dudly, P. Gouma, T. Kimolo, P. Neudeck, Saddows).
[7] P. Neudeck, R.S. Okojie, L.-Y. Chen, Proceedings of the IEEE 90 (6) (2002) 1065.
[8] A. Elasser, et al., Proceedings of the IEEE 90 (6) (2002) 969.
[9] A.P. Mirgorodsky, et al., Physical Review B52 (1995) 3993.
[10] N.G. Chopra, et al., Science 269 (1995) 966.
[11] A. Rubio, et al., Physical Review B 49 (1994) 5081.
[12] X. Blase, A. Rubio, S. Louie, M. Cohen, Europhysics Letters 28 (1994) 335.
[13] M. Menon, et al., Physical Review B 69 (2004) 115322.
[14] J. Cumings, Chemical Physics Letters 316 (3–4) (2000) 211.
[15] C.M. Lee, et al., Current Applied Physics 6 (2006) 166.
[16] O.R. Lourie, et al., Chemistry of Materials 12 (200) (1808).
[17] C.H. Lee, et al., Nanotechnology 19 (2008) 455605.
[18] Y. Huang, et al., Nanotechnology 22 (2011) 145602.
[19] A. Pakdel, et al., Nanotechnology 23 (2012) 215601.
[20] Y. Chen, et al., Chemical Physics Letters 299 (1999) 260.
[21] Y. Chen, et al., Applied Physics Letters 74 (1999) 2960.
[22] J. Yu, et al., Chemistry of Materials 17 (2005) 5172.
[23] D. Golberg, et al., Applied Physics Letters 69 (1996) 2045.
[24] M. Smith, et al., Nanotechnology 20 (2009) 505604.
[25] W. Han, et al., Applied Physics Letters 73 (1998) 3085.
[26] E. Borowiak-Palen, et al., AIP Conf. Proc. 723, p. 141.
[27] D. Golberg, et al., Chemical Physics Letters 323 (2000) 185.
[28] D. Golberg, Y. Bando, C. Tang, C. Zhi, Advanced Materials 19 (18) (2007) 2413.
[29] A. Pakdel, et al., Materials Today 15 (6) (2012) 256.
[30] J. Wang, et al., Nanoscale 2 (2010) 2028.
[31] D. Golberg, et al., Israel Journal of Chemistry 50 (2010) 405.
[32] H.J. Dai, et al., Nature 375 (1995) 769.
[33] Z. Pan, et al., Advanced Materials 12 (2000) 1186.
[34] C. Pham-Huu, et al., Journal of Catalysis 2 (10) (2001) 400.
[35] X. Sun, et al., Journal of the American Chemical Society 124 (2002) 14464.
[36] T. Taguchiw, et al., Journal of the American Ceramic Society 88 (2) (2005) 459.
[37] J. Keller, et al., Carbon 41 (11) (2003) 2131.
[38] G. Mpourmpakis, et al., Nano Letters 6 (8) (2006) 1581.
[39] N. Keller, et al., Journal of the Brazilian Chemical Society 16 (3B) (2005) 514.
[40] J. Hurst, Conf. Nano Materials for Defense Appl., San Diego, CA, 2007.
[41] D. Zhu, J. Hurst, Presentation MS&T, Columbus, OH, 2011.
[42] B. Wei, et al., Chemical Physics Letters 354 (3) (2002) 264.
[43] Z. Xie, et al., Journal of Nanoscience and Nanotechnology 7 (2) (2007) 647.
[44] H. Peng, Journal of Materials Research 15 (2000) 9.
[45] Z. Yang, et al., Chemistry of Materials 16 (2004) 3877.
[46] G. Meng, et al., Chinese Physics Letters 15 (1998) 689.
[47] I. Ibrahim, et al., Journal of Materials Science 26 (1991) 1137.
[48] C. Calow, A. Moore, Composites 2 (4) (1971) 231.
[49] P. Hou, et al., Carbon 46 (2008) 2003.
[50] N.G. Chopra, et al., Solid State Communications 105 (1998) 297.

[51] X.L. Wei, et al., Advanced Materials 22 (2010) 4895.
[52] M. Zheng, et al., Small 8 (2012) 116.
[53] Y. Xia, et al., Applied Physics Letters 84 (2004) 4626.
[54] C.W. Chang, et al., Physical Review Letters 97 (2006) 85901.
[55] L. Lindsay, D.A. Broido, Physical Review B 84 (2011) 155421.
[56] S. Berber, et al., Physical Review Letters 84 (2000) 4613.
[57] R. Arenal, et al., Advances in Physics 59 (2010) 101.
[58] A. Pakdel, et al., Journal of Materials Chemistry 22 (2012) 4818.
[59] C.H. Lee, et al., Chemistry of Materials 2 (2010) 1782.
[60] M. Terrones, et al., Nano Letters 8 (2008) 1026.
[61] Y. Chen, et al., Applied Physics Letters 84 (2004) 2430.
[62] J. Hurst, et al., Developments in Advanced Ceramics and Composites 26 (8) (2005) 355.
[63] D. Golberg, et al., Scripta Materialia 44 (2001) 1561.
[64] W.H. Moon, et al., Nanotechnology 3 (2003) 158.
[65] J.W. Lyver, et al., Journal of Computational and Theoretical Nanoscience 8 (2011) 529.
[66] C. Zhi, et al., Angewandte Chemie International Edition 44 (2005) 7929.
[67] C. Zhi, et al., Journal of Materials Research 21 (2006) 2794.
[68] M. Yamaguchi, et al., Acta Materialia 60 (2012) 6213.
[69] M. Yamaguchi, et al., Nanoscale Research Letters 8 (2013) 3.
[70] J. Hurst, et al., Unpublished Work.
[71] Q. Huang, et al., Proc. 16th Int. Conf. on Comp. Mater., Kyoto, Japan, 2007, 810.
[72] N.P. Bansal, J.B. Hurst, R. Choi, Journal of the American Ceramic Society 89 (1) (2006) 388.
[73] K. Liew, Nanotechnology 22 (2011) 085701.
[74] S.S. Wicks, et al., Composites Science and Technology 70 (2010) 20.
[75] E. Garcia, et al., Composites Science and Technology 68 (9) (2008) 2034.
[76] Hurst, et al., NT13.
[77] Hurst, et al., ICACC13.
[78] R. Naslain, Composites Science and Technology 64 (2) (2004) 155.
[79] www.boeing.com/.
[80] J. Yu, Y. Chen, et al., Advanced Materials 18 (2006) 2157.
[81] G. Ciofani, et al., Nanoscale Research Letters 4 (12) (2009) 113.
[82] C. Su, et al., Diamond and Related Material 16 (4–7) (2007) 1393.
[83] J. Hurst, N. Bansal, D. Zhu, Unpublished Work.
[84] B. Peng, et al., Nature Nano 3 (2008) 626.
[85] http://www.cdc.gov/niosh/docs/2012-147/pdfs/2012-147.pdf.

CHAPTER 10

Carbon Nanotube Fiber Doping

Noe T. Alvarez[1,2], Vesselin N. Shanov[1,2], Tim Ochmann[1,2], Brad Ruff[1,3]

[1] *Nanoworld Laboratories,* [2] *Chemical and Materials Engineering,* [3] *Mechanical Engineering and School of Dynamics, University of Cincinnati, Cincinnati, OH 45221-0072*

CHAPTER OUTLINE

10.1 Introduction	289
10.2 Doping	290
10.2.1 Types of doping	291
10.2.1.1 Endohedral doping or encapsulation	*291*
10.2.1.2 Exohedral doping or chemisorption	*292*
10.2.1.3 Substitutional doping	*293*
10.2.2 Functionalization	296
10.3 Single-Walled Carbon Nanotube Doping	296
10.3.1 Redox potential	297
10.3.2 Work functions	298
10.4 Multiwalled Carbon Nanotube Doping	299
10.5 Characterization of Doped CNTs	300
10.5.1 Electrical conductivity of doped MWCNT fibers	300
10.5.2 Raman, XPS, ^{13}C-nuclear magnetic resonance	301
10.6 Experimental Challenges in Characterization	303
10.6.1 Electrical contact	303
10.6.2 Diameter measurements	304
10.7 Summary	305
Acknowledgments	306
References	306

10.1 INTRODUCTION

It has been more than two decades since carbon nanotubes (CNTs) were observed by Ijima [1,2], and the most promising physical properties of CNTs are still not present in bulk materials that are used for commercial products. Despite the blossoming research and significant improvements related to synthesis, type, and chirality separations, the search for strong CNT ropes, ballistic conducting fibers, and optically and thermally highly conducting CNT-based materials continues [3–6]. Control over chirality, type, diameter and length are highly desired if we intend to take advantage of individual CNT physical properties. Fortunately, some researchers have already made significant improvements in synthesis; improvements in selective synthesis of metallic or semiconducting nanotubes have been reported [7–9]. Additionally, researchers around the world have reported different benchmarks in CNT length. Up to 22-mm-long vertical arrays of CNTs have been synthesized at the University of Cincinnati [10]. Postprocessing methods have also been applied to separate metallic from semiconducting CNTs, or vice versa [11,12]. Further studies and application of these techniques has allowed chiral separations from solution mixtures. Density gradient separations have allowed separating CNTs not only by type but also by diameter, chirality and mirror images [11–14]. However, all these efforts have not been enough to bring CNTs into commercial products. Some fibers and thin films manufactured from these posttreated materials still fall below expectations compared to materials in commercial products.

Besides synthesis of pure metallic or semiconducting CNTs, another approach to improve electrical conductivity of the CNTs is chemically doping them [15–17]. By using electron withdrawing or donating atoms, ions or molecules, it is possible to substantially increase the density of free carriers. This increases the electrical and thermal properties of the CNT bundles and fibers. Furthermore, it has been suggested that dopant molecules on the CNT surface may assist the tube–tube contact within the CNT bundle or fiber, which would subsequently improve their electrical conductivity [18,19].

Although doping is a well-established technique generally used to improve and engineer electrical properties of materials in the semiconductor industry, this idea still creates some misunderstanding within material scientists and chemists when it is applied to CNTs. Doping, as conceived in the semiconductor industry, implies an atomic substitution of matrix elements by other similar atoms, with the intention of increasing electron charge carriers or donors (n doping) or increasing the number of free charge carriers or holes (p doping). Due to the quasi-one-dimensional nature of CNTs, doping in CNTs has broad varieties in a similar manner to doping in fullerenes. Unlike the semiconductor industry, confusing information is often found in the literature regarding doping, intercalation, physisorption, substitution and functionalization. In this chapter, we will discuss doping types that are normally not employed for bulk Si doping and will provide a more comprehensive definition and classification of CNT materials doping. Also, within the chapter, we use the terms fiber, thread and yarn indistinctly to refer to all CNT fibrous assemblies.

10.2 DOPING

In order to come up with a more precise CNT doping definition, we consider doping to be any physisorption or chemisorption of foreign atoms or molecules on the CNT outer or inner surface or atoms within the lattice of the CNT that would not modify the hexagonal lattice of the carbon atoms on the nanotube. Excess dopant accumulation such as particle formation should probably be considered as a form of a composite material rather than doping.

10.2.1 Types of doping

A more comprehensive classification of CNT doping types can be borrowed from the similarity of CNTs to fullerenes. Fullerenes can be doped with atoms, molecules, or ions by substituting a C atom on the fullerene surface, by external chemisorption or by internally loading them. This doping classification, based on the position of the dopant molecule with respect to the fullerene, brings a more comprehensive scenario for CNT doping. While dopants can be chemisorbed onto the CNT surface, be encapsulated within the CNT shell, and substitute carbon atoms from the hexagonal lattice, as shown in Fig. 10.1(a), (b) and (c), respectively [20–22], in fullerene science, each of these doping types have received a more generic name such as exohedral, endohedral and substitutional doping [16,23,24]. Due to the similarity of CNT to fullerenes, application of the same definition provides a general nomenclature for the variety of CNT doping.

10.2.1.1 Endohedral doping or encapsulation

This type of doping implies that the dopant molecule, ion, or atom is located within the inner side of the CNT shell. Several reports claim the insertion of organic and inorganic molecules within the CNT; this type of doping is also known as amphoteric doping in the literature [25–28]. Fullerene and metallofullerene encapsulation within the CNT has been theoretically predicted [20,29,30] and experimentally demonstrated [31–35]. Figure 10.2 illustrates endohedral doping by fullerenes (a) and reveals transmission electron microscopic images that show the result of experimental efforts (b) [25,31]. Local electronic band gap modification by endohedral doping of single-walled CNTs (SWCNTs) has been reported. Encapsulating fullerenes within the ~1.3-nm-diameter CNTs has modified the E_{11} and E_{22} electronic transitions up to 0.050 and 0.116 eV, respectively, as reported by Okasaki et al. [31].

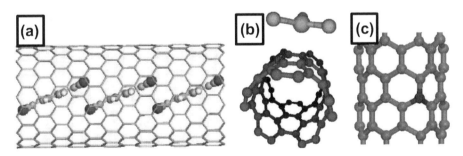

FIGURE 10.1

Types of CNT doping (a) endohedral, (b) exohedral [21], and (c) substitutional doping [22]. (For color version of this figure, the reader is referred to the online version of this book.)

Source: (a) Reproduced with permission from Ref. [20] Copyright 2004

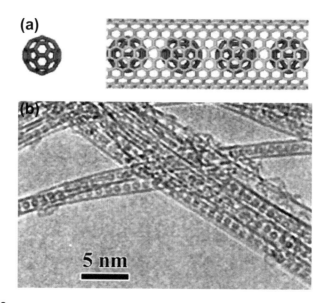

FIGURE 10.2

Scheme that illustrates fullerene encapsulation within the CNT (a) [25] and transmission electron microscopic images of experimental peapods, fullerenes within the CNTs (b) [31].

The advantage of endohedral doping compared to other types of doping is the stability of the dopants to air and environmental exposure. Molecules trapped within the CNT walls are almost isolated from the environment; therefore, any sudden leakage or reactions with O_2 from the atmosphere are unexpected. This approach should be considered as a robust doping approach for CNT applications where the nanotube temperature rises or long-term exposures to the environment occur. However, loading dopants within the CNTs could be challenging.

10.2.1.2 Exohedral doping or chemisorption

Another type of doping, perhaps the most common and simple way to modify the electronic structure of the CNTs, is the exohedral doping, also called π-stacking, noncovalent functionalization, chemisorption, in-plane doping or intercalation [16,36–39]. Basically, any chemisorption of atoms or molecules on the outer surface will disturb CNT electronic density [40]. Furthermore, the as-grown CNTs are considered unintentionally p-type doped. Most likely, this is due to the chemisorbed O_2 molecules that are collected by their exposure to the environment, air. Figure 10.3(a), illustrates the intercalation and chemisorbed molecules on the CNT surface [41]. A simple CNT baking under vacuum conditions at moderately high temperatures has been revealed to be sufficient to remove the chemisorbed O_2 molecules from the CNT surface [42].

Extensive intentional exohedral p-type doping experiments have been performed during the past decade. Typical p-type dopants are any electron withdrawing atoms

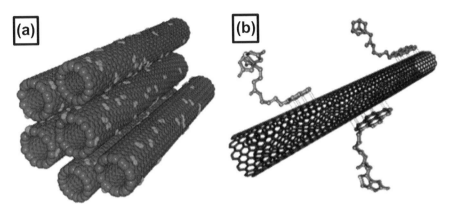

FIGURE 10.3

(a) Illustration of exohedral doping [41] also called (b) noncovalent functionalization or π-stacking [51]. (For color version of this figure, the reader is referred to the online version of this book.)

or molecules that interact with the CNT surface; acids, transition metals, and other organic molecules have been studied [15,43–45]. Furthermore, p-type-doped CNTs have been shown to be, by far, more stable than n-type-doped CNTs at atmospheric conditions in open air [19]. Arguably, HNO_3 has been considered as the most stable p-type dopant at ambient temperatures [18,46]. Although n-type dopants have shown to be more sensitive to air and the environment, the results reported in the literature are clearly interesting. Early on, K doping of CNTs had shown significant improvement in their electrical properties [15]. Lately, hydrazine doping has been demonstrated and higher stability of doped CNTs compared to alkali metals has been reported [47].

Additionally, exohedral doping is strictly related to the reactivity of individual CNTs and the dopant molecules. Regarding SWCNTs, it has been shown that among a variety of band gap semiconductor CNTs, they react differently to dopant molecules or ions, suggesting that electron donations to dopants are different between semiconducting SWCNTs [48,49]. Theoretical studies on exohedral doping have shown that the affinity of the dopant molecule depends on other factors such as diameter of the CNT, redox potential, and defects on the CNT [50]. As shown in Fig. 10.3(b), molecules that promote π-stacking are another type of exohedral doping [39,51].

10.2.1.3 Substitutional doping

CNT substitutional doping is the closest doping type to traditional semiconductor industry doping. It is an atom-by-atom substitution of C by a dopant atom within the CNT lattice. In Si technology, doping with electron donors and acceptors shifts the valence and conduction bands of Si, which leads to a well-controlled Fermi level energy. Depending on the dopant concentrations, typically in parts per million, it is possible to tune the electrical conductivity of the Si. In a process similar to Si

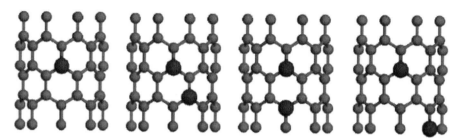

FIGURE 10.4

Nitrogen substitutional doping on a (8,0) CNT that illustrates the atom-by-atom substitution and some possible option between a single and double atomic substitution [53]. (For color version of this figure, the reader is referred to the online version of this book.)

doping, increasing the substitutional doping concentration increases the conductivity of CNTs. Also, due to the large activation energy required for a carbon atom substitution, substitutional doping in CNTs is done during synthesis [52].

Due to their size and valence band electronic structure, B and N have been considered as the most suitable elements for CNT substitutional doping. Small concentrations of B- or N-containing precursors are introduced into the reactor at the same time as the carbon precursor, so the dopant atom (B or N) is inserted into the CNT lattice instead of carbon. Figure 10.4 illustrates four types of substitutional doping with a single- and double-atom arrangement on the CNT surface [53]. Although CNT synthesis by chemical vapor deposition (CVD) is considered as the most suitable preparation technique for B- and N-doped CNTs, other synthesis methods such as arc discharge [54,55], laser ablation [56] and plasma-assisted CVD [57] have been explored for this type of doping.

Usually, B is considered a p-type dopant, while N is an n-type dopant; however, it has been theoretically postulated and experimentally demonstrated that depending on the arrangement of N atoms within the CNT lattice and their concentration, N can give either p- or n-type doping [24,58,59]. N atom connected to three C atoms in a similar geometry to C in hexagonal lattice has shown to act as an n-type dopant, while N substitution in a pyridine-like configuration can act as an n- or a p-type dopant [16,59–61]. In general, several experimental reports claim that p-type (B) and n-type (N) substitutional doping create quasi-bond states below or above the Fermi energy in close analogy to acceptor and donors in semiconductors. Surprisingly, researchers have found that substitutional doping with B on an armchair CNT (10,10) does not scatter electrons at the Fermi energy, suggesting that the conductance remains constant as reported by theoretical studies [62]. An individual nitrogen atom within the CNT lattice increases the Fermi energy by c. 1.21 eV as reported by Terrones et al., [59] besides generating additional states in the conduction band above the Fermi level. Typically, substitutional doping changes the semiconducting nature of the CNTs to a quasi-metallic state such as the confirmed case of nitrogen doping.

Due to the similar size of N and B to C atoms, N or B doping allows retaining the hexagonal atom configuration on a cylindrical geometry of pristine CNTs. Usually, N and B dopant atoms are located in the plane of the CNT wall, while in case of CNT doping with P, Ni, Pd and other transition metals, small dislocations toward the inner or outer side from the CNT plane have been observed [63–65]. For obvious reasons, they have been called endo- and exo-substitutional doping, Fig. 10.5 [63]. This behavior results in largely localized or delocalized electron density in the endo- and exo-substitutional doping, respectively [63].

Furthermore, B-N codoping affecting the electronic transport of CNTs has been reported [36]. Theoretical studies reveal that B-N codoping is transparent to the electron transport and does not modify the conductivity of the codoped CNTs; however, larger concentrations can generate segregations where domains of B-N are localized and reduce the electron transport almost linearly with the concentration. Interestingly, charge localizations were observed that behave like a typical dopant when B and N are far apart from each other [36]. Another intriguing doping approach is the combination of substitutional and exohedral doping [22]. Theoretical studies reveal that substitutionally doped CNTs with B and N can assimilate higher concentrations of exohedrally doped Pd atoms [22]. Another report of CNT codoping describes the introduction of N and P atoms simultaneously into the CNT lattice [66].

Carbon replacement by B, N and P is a technique that has been practiced extensively in graphite structures, fullerenes, and carbon fibers [60,67]. Studies have demonstrated that N-substitutional doping allows up to 4.5 at. % atomic concentration. However, significant distortion of the lattice has been observed for higher concentrations [52,60,68].

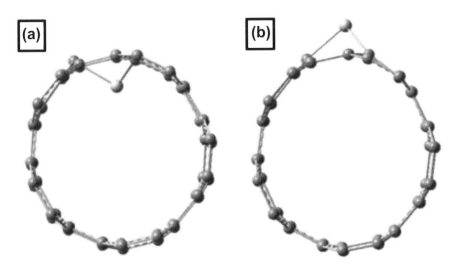

FIGURE 10.5

(a) Endo- and (b) exo-substitutional doping of a (8,0) CNT [63].

10.2.2 Functionalization

There is a subtle difference between doping and functionalization, with their purpose constituting the main difference between them. While doping intends to modify the electronic nature of the CNT, functionalization intends to add chemically reactive sites on the CNT surface. Besides adding a functional group, functionalization also modifies the electronic structure of CNTs, since charges in a functionalized CNT are localized around the functional group, as demonstrated by Veloso et al. [69]. In general, while doping is the addition of a foreign molecule/atom without changing the hexagonal lattice of the CNT, functionalization is the addition of a foreign species typically by covalent bond formation on the CNT lattice defects. Both functionalization and doping may look quite similar but they differ substantially in their purpose. Although substitutional doping is related to covalent bond formation, in order to be considered true doping, it still needs to preserve the hexagonal lattice of the CNTs. In our opinion, a carbon atom substitution on defect sites of CNTs should still be considered as functionalization rather than doping.

Another approach to distinguish functionalization from doping is to set some minimum bonding energy as the threshold for functionalization. However, the variety of functional groups and doping agents makes this approach inapplicable. In the case of doping with transition metals, the bonding energies are shared within more than a single carbon atom, which reduces the bonding energy of an individual carbon atom to the dopant. For clarification purposes, we will consider functionalization as the localized bond formation between a foreign molecule/atom and a C atom on the CNT surface. Functionalization produces a covalent bond with a reactive functional group. Generally, covalent bond formation requires high enough energy to break the sp^2 C—C bonds; therefore typical functionalization is performed under harsher conditions to promote the chemical reaction between the CNT atoms and the foreign element or molecule.

10.3 SINGLE-WALLED CARBON NANOTUBE DOPING

Relevant work on SWCNT doping has been reported, mostly involving SWCNTs assembled on thin films and bucky papers, although there is work done on individually dispersed SWCNTs that has provided insightful understanding of doping. The availability of sensitive analytical tools to study doping, such as Raman spectroscopy, has allowed tracking very small concentrations of dopants and monitoring peak shifts as a function of concentrations. Since most SWCNTs are small in diameter, frequently less than 2.2 nm, commercial Raman tools can easily track their radial breathing mode (RBM) shifts. SWCNTs with diameters above ~2.2 nm are not easily accessible by most of the commercial Raman setups. Physical properties of SWCNTs such as electrical conductivity are directly affected by changes in their redox potential and Fermi level. Other tools such as Vis—near-IR absorption and photoluminescence spectroscopy can also detect SWCNT electronic structure changes as a function of dopant concentrations [70,71]. The lowest resistivity of doped SWCNTs reported is 1.2×10^{-6} ($\Omega*m$), most likely as a result of exohedral doping [72].

10.3.1 Redox potential

The redox potential of SWCNTs is defined as its ability to donate and withdraw electrons, which basically identifies the chemical reactivity or stability of each individual CNT. Several attempts to determine SWCNT redox potentials have been reported, most of them supported by Raman spectroscopy, photoluminescence and UV–Vis absorption data. Researchers have quantified the redox potential for specific tubes and have extrapolated this data to provide a more general understanding of their redox potential. The difficulty about accurate redox potential determination lies in the large mixture of diameters and chiralities within the synthesized SWCNT batch. Additionally, some limitations of Raman spectroscopy affect the ability to understand the SWCNT redox potentials. Each excitation source wavelength can only excite a few nanotubes from the whole sample. Therefore, a continuous Raman excitation source is required to study broader diameter range and large chiral diversity of CNTs, instead of the discrete wavelength sources available on most of the commercial Raman setups.

Several literature reports have claimed selective reactions that support different redox potentials in SWCNTs with different diameters or chiralities [49,73]. Perhaps, one of the most complete diagrams of the redox potential of CNTs originally reported by Kim et al. [74] is shown in Fig. 10.6. Equation (10.1) has been proposed to calculate the redox potential (V) of the corresponding van Hove singularities (vHs) where a and b are the fitting parameters and d_t is the diameter of an SWCNT.

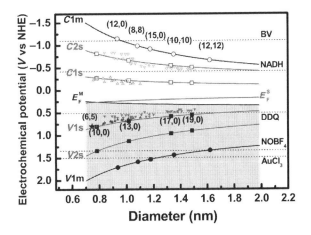

FIGURE 10.6

Redox potential of CNTs as a function of their diameter [74]. (For color version of this figure, the reader is referred to the online version of this book.)

$$E_{(SHE)}^{SWCNT}[V] = \frac{a}{d_t[nm]} + b \qquad (10.1)$$

Based on data collected from photoluminescence spectroscopy, the plot in Fig. 10.6 summarizes the redox potential of different diameter SWCNTs, where **Cis** and **Cim** are the *i* unoccupied vHs for semiconducting and metallic SWCNTs, respectively, while **Vis** and **Vim** are the *i* occupied vHs of semiconducting and metallic SWCNTs, respectively. The values within parentheses are the chiral index of each individual SWCNT. Benzyl viologen and β-nicotinamide adenine dinucleotide are the electron donating molecules reported by Kim et al. [75] and Kang et al. [76], while 2,3-dichloro-5,6-dicyano-*p*-benzoquinone and nitrosyl tetrafluoroborate ($NOBF_4$) are electron withdrawing molecules used to study SWCNT redox potentials [74].

10.3.2 Work functions

Besides diameter, chirality and band gap, work functions of SWCNTs are important physical parameters in electron transport between nanotubes and their contacts. Work functions (φ) are normally defined in metals as the difference in the potential energy of an electron between the vacuum level (ϕ) and the Fermi level (E_F) ($\varphi = \phi - E_F$). The vacuum level is the energy of an electron at rest at a point sufficiently far outside the surface such that the electrostatic image force on the electron may be neglected [77]. Reported experimental data on CNT work functions demonstrate an increase of the work function as the dopant concentration increases as displayed in Fig. 10.7 [78]. For Au^{3+} (p-type) doping of SWCNT sheets, an

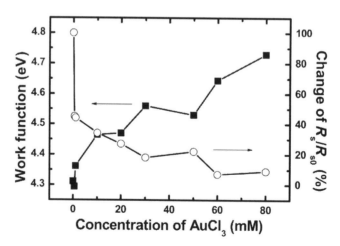

FIGURE 10.7

Work functions of CNT sheets and electrical resistance as a function of $AuCl_3$ concentration doping. The work function of the CNTs increases, while electrical resistance decreases [78].

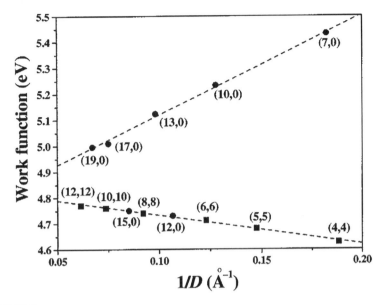

FIGURE 10.8

Work functions (electron volts) of individual metallic (squares) and semiconducting (dots) SWCNTs as a function of the inverse diameter (1/D) [80].

increment of 0.42 eV was observed, starting at 4.31 eV for a pristine SWCNT up to 4.73 eV for the highest dopant concentration (80 mM) [78]. Similar reduction of SWCNT work functions by p-doping has been reported in the past [79].

The relationship between the work function (φ) by the unit charge (e) and the reduction potential (V) versus a normal hydrogen electrode (NHE) is shown in Eqn (10.2) [78].

$$\frac{\varphi}{e} = V(\text{vs NHE}) + 4.44 \qquad (10.2)$$

Although doping of the CNT sheet shows a simple trend, the work function is directly proportional to the doping concentrations. In reality, particularly for small-diameter SWCNTs, the work function of zigzag nanotubes could be up to 1.0 eV different from that of their armchair counterparts, as seen in Fig. 10.8 [80]. It has also been suggested that the work function of the metallic tubes is smaller than that of semiconducting tubes [78].

10.4 MULTIWALLED CARBON NANOTUBE DOPING

Unlike SWCNTs where characterization tools are more suitable to their diameter, multiwalled CNT (MWCNT) doping studies are more complex. For instance, active modes such as the RBMs of CNTs with diameters larger than ~2.2 nm are no longer active since they are out of the range of what most commercial Raman systems can detect. Additionally, the large diversity of diameters, numbers of walls, and wall chiralities make MWCNT doping studies difficult, and systematic studies with such a huge variety becomes almost impossible. Most MWCNT doping studies are limited to characterization of a mixture of MWCNTs assembled into macroscopic bundles. X-ray photoelectron spectroscopy (XPS), electron energy loss spectroscopy, and X-ray diffraction have been used as characterization tools [58,81,82]. Researchers have reported an increase in electrical conductivity by more than an order of magnitude for doped CNT fibers. It is worthwhile to mention that the lowest electrical resistivity of 1.5×10^{-7} (Ω*m) reported is the result of exohedral doping of MWCNT fibers with iodine [83].

10.5 CHARACTERIZATION OF DOPED CNTs
10.5.1 Electrical conductivity of doped MWCNT fibers

Electrical conductivity or resistivity changes of CNTs is probably the most relevant testing tool for CNT doping [42,84]. Small traces of chemisorbed molecules affect the resistivity of the CNTs; therefore, CNTs have long been studied as gas sensing components of sensor devices [84]. Basically, any molecule upon adsorption on the CNT surface will modify the electron density of the CNT surface, which will be traced by the electrical conductivity of the CNT assemblies [40]. In case of doping, the chemisorbed molecules seem to improve the contact between CNTs as well, since the electrical conductivity in a CNT network is junction—junction dependent [19,85]. Electrical conductivity is probably the best option to determine effective doping of MWCNTs.

Doping of CNT fibers has progressed with the ability to manufacture them. K and Br doping of 8-mm-long CNT bundles has been demonstrated over a decade ago [15]. Other results for CNT ropes and fibers have followed; lately, significant improvements for CNT fiber doping have been reported [83,86,87]. Experiments performed at the Nanoworld Laboratories at the University of Cincinnati have shown an increase of conductivity by up to one order of magnitude over the undoped CNT fibers, as shown in Fig. 10.9. These dopings were the result of 30 min soaking of as-spun CNT fibers in 0.05 M aqueous solutions of Au (KAuCl$_4$) according to the protocol published by Jarosz et al. [88]. The results of multiple samples have consistently shown a decrease in resistivity of about one order of magnitude. Figure 10.9 compares these improvements with the electrical resistivity of Au and Cu wires, which are about two orders of magnitude lower than currently doped CNT fibers. Also, it contains some relevant resistivity measurements reported in the literature for SWCNTs and MWCNTs [72,83,89,90]. The resistivity measurements were performed using a four-point probe setup with four electrodes positioned in parallel. Current is supplied by the external connectors and voltage is measured between the internal connectors. The internal connectors are 4 mm apart, while the distance between the internal and external electrodes is 2 mm; all connectors are 500 μm pins.

10.5 Characterization of Doped CNTs

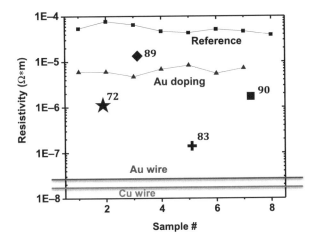

FIGURE 10.9

Electrical resistivity of Au-doped CNT fibers compared to Au and Cu wire as well as low-resistivity CNT fibers reported in the literature [72,83,89,90]. Reference samples correspond to the resistivity of as-spun undoped CNT fibers. The Au doping represents the resistivity of multiple CNT fibers after Au doping. (For color version of this figure, the reader is referred to the online version of this book.)

10.5.2 Raman, XPS, ^{13}C-nuclear magnetic resonance

Raman spectroscopy has been used extensively to characterize SWCNT doping. Due to its high sensitivity to the geometry, electronic structure and vibrational modes [17,91], it is considered an ideal tool for studying exohedral, endohedral and substitutional doping [47,71,92]. In the case of n-type substitutional doping, it has been reported that Raman spectroscopy is able to detect significantly small dopant concentrations, as low as 10^{-5} at. % [93]. It is commonly accepted that CNT doping with electron withdrawing molecules will promote a charge transfer from the CNTs to the dopant (p-type), which will stiffen the C—C bonds [16,94]. As a result, one should expect peak blue shifts to higher energy (Raman upshifts) [45,47,78]. On the contrary, CNT doping with electron donating molecules will soften the C—C bonds on the CNT surface and will promote red peak shifts (Raman downshifts) [47]. With just a few exceptions in the case of exohedral doping, a massive number of literature reports support this trend and have been commonly accepted. Figure 10.10 illustrates G band peak shifts upon doping: benzene doping is located almost at the center, while electron withdrawing groups tetracyanoquinodimethane and tetracyanoethylene drive Raman upshifts [95]. Electron donating groups tetrathiafulvalene and aniline promote Raman downshifts. Identification of p-type and n-type doping by Raman is simple and true for SWCNTs and MWCNTs when they are based on G peaks. Regardless of the nanotube type, the blue shift of G is

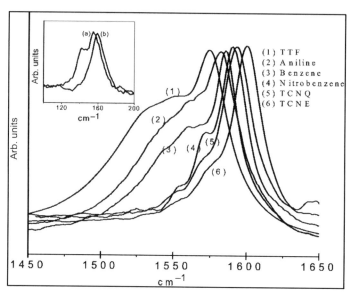

FIGURE 10.10

Typical Raman spectroscopy shifts observed on the G band of CNTs after doping. Raman upshifts promoted by electron withdrawing groups nitrobenzene, TQCN (tetracyanoquinodimethane) and TCNE (tetracyanoethylene), and Raman downshifts produced by donating groups TTF (tetrathiafulvalene) and aniline [95].Imet shows RBM bands of SW-CNTs after interaction with (a) TTF and (b) TCNE.

typical for p-type and the red shift of G is typical of n-type. This phenomenon has been reported by several researchers [78,79].

More recently, the G′ peak and peaks nearby have been established as valuable tools for p- or n-type doping identification by Raman spectroscopy [71]. Maciel and colleagues have identified another peak adjacent to the G′ peak. Depending on the position of these peaks, it is possible to determine the doping type. The G′ peak is an important feature of sp^2 carbon materials and it has been referred to as highly sensitive to π-electronic structures. It has also been used to identify the metallic nature of SWCNTs [96], and is considered the main feature in graphene studies for identifying the number of layers [97].

XPS is another valuable tool used to determine the doping in CNTs. Due to its sensitivity; it has been used extensively to study most exohedral and substitutional doping. The binding energy determined by XPS is related to the core electrons and the Fermi level, so changes in the Fermi level can be observed by binding energy peak shifts [46]. Additionally, by deconvoluting the binding energy curve, it is possible to determine the nature of the dopant and estimate the dopant concentration on the CNT surface [47].

Solid-state ^{13}C-nuclear magnetic resonance (^{13}C-NMR) has been used extensively for studying functionalized SWCNTs [98,99]; however, ^{13}C-NMR can also determine chemical doping of CNTs where no covalent bond was present [47]. Chemical shits of ~8 ppm and peak broadenings were observed for n-type and p-type doped CNTs. Furthermore, NMR has also been used to study charge transfer-induced changes at the Fermi level, where a continuous increase of the density of states at the Fermi level was observed as a function of Li intercalation increase [100].

10.6 EXPERIMENTAL CHALLENGES IN CHARACTERIZATION
10.6.1 Electrical contact

Due to the miniaturization of electronic devices, electrical contacts are becoming more highly critical components than before and they are more relevant when related to nanotechnology. A large amount of work has been dedicated to find the ideal material as a contact element between CNTs and metals. As in many bulk semiconducting and metal contacts, the Fermi level determines the contact quality that results in a Schottky barrier between semiconducting materials and metals. Contact of CNTs and CNT fibers with metal electrodes, as shown in Fig. 10.11(a), has long been reported as Schottky barriers between CNTs and the metal electrodes, which may affect the electron transport. Furthermore, the physical properties of CNTs are much different than those of bulk semiconductor materials, and as such, many of the concepts developed for traditional contacts may need to be revisited [101].

Besides Schottky barriers between the CNT fiber/yarn and the metal contacts, the unintentional doping of CNTs as a result of exposure to air and other sample handling issues create systematic errors during electrical resistivity measurements of CNT fibers. Perhaps one of the most common difficulties is related to the CNT fiber diameter and length measurements. Common connectors from the electronic industry have electrode dimensions of hundreds of microns, which are significantly larger than a CNT and their typical CNT fiber diameters (10–30 μm). As a result of the size differences, there is an inherent difficulty in determining the point of contact between the electrode and the CNT thread. As shown in Fig. 10.11, the state-of-the-art contacts created between the CNT thread and the electrode do not precisely point out the length of the analyzed thread. Therefore, the errors associated with measuring the length of the samples are large compared to the CNT fiber diameter. Furthermore, the contacts are normally made of Ag paste, which is a colloidal solution consisting of micron-sized Ag particles in an organic solvent. During the contact preparation, the solvent usually travels along the CNT thread for over 100 μm in some cases, wetting the CNT fiber and carrying Ag particles with it. Unfortunately, even by scanning electron microscopy it is not easy to determine where the real metal contact ends or where the pure CNT fiber begins as shown in Fig. 10.11(b) and (c). Some of these Ag particles are located on the CNT thread surface as nonuniform bubbles and aggregates and there are cases where a Ag aggregate has traveled far along the CNT thread such that it may actually stay

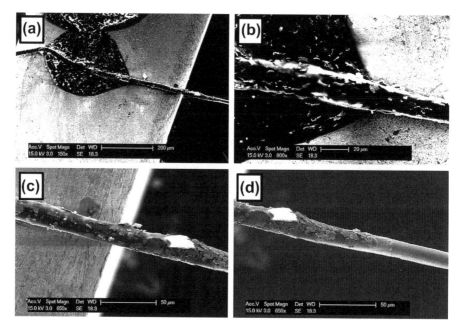

FIGURE 10.11

Scanning electron microscopic images of a typical CNT thread and the Ag paste at the contact point with the metal electrode. Ag paste particles do not wet CNT thread uniformly and are not continuous along the thread. (a) Typical metal contact to Ag paste and CNT fiber; (b) higher magnification of contact area; (c) Ag particles along the fibers, nonuniform wetting and (d) wetting interface between the CNT fiber and Ag particles.

isolated from the bulk Ag particles that make contact with the metal electrode (Fig.10.11(d)). To overcome this issue, analysis of a large number of samples is necessary. Also, a longer separation distance between the contact electrodes and fast drying conducting paste is recommended.

Most contacts in the electronic industry are typically made by soldering components together. Unfortunately, the traditional metals used in soldering do not wet the CNT thread well enough to guarantee contact with the electrode. To overcome this limitation, CNT surface functionalization is suggested. By adding functional groups on the CNTs, it is possible to improve the CNT fiber wetting by the soldering metal.

10.6.2 Diameter measurements

As-synthesized and dry-spun CNT fibers or threads are normally seen as a cylindrical fiber (Fig. 10.12(a)). These cylindrical fibers have structure similar to any thread made from animal wool or synthetic fiber, although in the CNT threads it is more

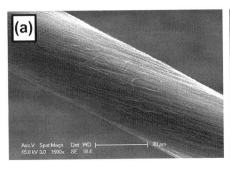
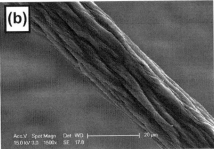

FIGURE 10.12

Scanning electron microscopic images of CNT threads: (a) as-spun and (b) after densification and doping. The images reveal a change in morphology of the cylindrical CNT fibers.

difficult to observe void spaces between the individual CNTs on the thread surface. The similarity to natural fiber threads is more evident once these CNT threads are submerged into densifying organic solvents that increase the density of CNTs within the fiber. As shown in Fig. 10.12(b), the CNT thread diameter decreases and its surface texture changes after immersing in acetone solvent. Unfortunately, the smooth-looking fiber surface changes and looks cracked, with voids along its surface. Therefore, the original circular-looking CNT fiber may not have the same cross-sectional area as a solid wire. Since the resistivity of circular wires depend on the square of diameter, small errors in estimating the diameter of CNT fibers can significantly affect the resistivity calculations. A clever solution to keep the CNT fiber circular has been practiced by Alvarenga et al. [102], in which CNT thread has been drawn through a tungsten carbide die to maintain its circular shape through compacting.

10.7 SUMMARY

Doping has been shown to improve the electrical conductivity of nanotube fibers. The improved conductivity is needed to open up more applications of nanotube materials. Among the variety of doping methods, exohedral doping has produced CNT fiber with the highest electrical conductivity, according to the literature. Endohedral doping is considered the simplest. Substitutional doping in CNTs is done during synthesis typically using B and N elements. A significant amount of work is still needed to understand all these doping processes and control them at the atomic level. The redox potential and work functions define the chemical reactivity of CNTs and are useful in predicting the properties of doped CNT materials. Doping SWCNTs has the advantage of more readily available characterization tools as compared to doping MWCNTs, although the broad diameter range of CNTs is still a limiting factor in accurately characterizing the level of doping.

Acknowledgments

We are grateful to the funding agencies: NSF through grant CMMI-07272500 from NCA&T, a DURIP-ONR grant, NSF SNM grant 1120382, and ERC supplemental STTR funded by NSF (2010–2012) "Translational Research to Commercialize CNT and Mg Based Materials from Biomedical and Industrial Nanotechnology Applications". We would also like to acknowledge Chaminda Jayasinghe, Ge Li, Feng Wang and Christopher Katuscak for their assistance in the lab; Doug Hurd for his help in the machine shop; and General Nano LLC for granting access to their e-Beam evaporator.

References

[1] I. Iijima, Helical microtubules of graphitic carbon, Nature 354 (1991) 56–58.
[2] S. Iijima, T. Ichihashi, Single-shell carbon nanotubes of 1-nm diameter, Nature 363 (1993) 603–605.
[3] K. Jiang, et al., Superaligned carbon nanotube arrays, films, and yarns: a road to applications, Advanced Materials 23 (2011) 1154–1161.
[4] W. Lu, et al., State of the art of carbon nanotube fibers: opportunities and challenges, Advanced Materials 24 (2012) 1805–1833.
[5] N. Grobert, Carbon nanotubes becoming clean, Materials Today 10 (2007) 28–35.
[6] P. Avouris, J. Chen, Nanotube electronics and optoelectronics, Materials Today 9 (2006) 46–54.
[7] R.M. Sundaram, K.K. Koziol, A.H. Windle, Continuous direct spinning of fibers of single-walled carbon nanotubes with metallic chirality, Advanced Materials 23 (2011) 5064–5068.
[8] A.R. Harutyunyan, et al., Preferential growth of single-walled carbon nanotubes with metallic conductivity, Science 326 (2009) 116–120.
[9] Y. Li, et al., Preferential growth of semiconducting single-walled carbon nanotubes by a plasma enhanced CVD method, Nano Letters 4 (2004) 317–321.
[10] V. Shanov, M. Schulz. (cited 2012, 24.12.12), Available from: http://www.min.uc.edu/nanoworldsmart/.
[11] X. Tu, et al., DNA sequence motifs for structure-specific recognition and separation of carbon nanotubes, Nature 460 (2009) 250–253.
[12] M.S. Arnold, et al., Sorting carbon nanotubes by electronic structure using density differentiation, Nature Nanotechnology 1 (2006) 60–65.
[13] S. Ghosh, S.M. Bachilo, R.B. Weisman, Advanced sorting of single-walled carbon nanotubes by nonlinear density-gradient ultracentrifugation, Nature Nanotechnology 5 (2010) 443–450.
[14] E.H. Haroz, et al., Enrichment of armchair carbon nanotubes via density gradient ultracentrifugation: Raman spectroscopy evidence, ACS Nano 4 (2010) 1955–1962.
[15] R.S. Lee, et al., Conductivity enhancement in single-walled carbon nanotube bundles doped with K and Br, Nature 388 (1997) 255–257.
[16] M. Terrones, A.G. Souza Filho, A.M. Rao, Doped carbon nanotubes: synthesis, characterization and applications, in: S.T.i.A. Phys (Ed.), Carbon Nanotubes: Advanced Topics in The Synthesis, Structure, Properties and Applications, Springer-Verlag, Berlin, Heidelberg, 2008, pp. 531–566.

[17] A.G. Souza Filho, M. Terrones, Properties and applications of doped carbon nanotubes, in: Y.K. Yap (Ed.), B-C-N Nanotubes and Related Nanostructures, Springer, Houghton, MI, USA, 2009.

[18] T.M. Barnes, et al., Reversibility, dopant desorption, and tunneling in the temperature-dependent conductivity of type-separated, conductive carbon nanotube networks, ACS Nano 2 (9) (2008) 1968–1976.

[19] A.G. Rinzler, E.P. Donoghue, All the dope on nanotube films, ACS Nano 5 (5) (2011) 3425–3427.

[20] J. Lu, et al., Amphoteric and controllable doping of carbon nanotubes by encapsulation of organic and organometallic molecules, Physical Review Letters 93 (2004) 116804.

[21] S. Guerini, et al., Electronic properties of FeCl3-adsorbed single-wall carbon nanotubes, Physical Review B 72 (2005) 233401.

[22] G.X. Chen, et al., First-principles study of palladium atom adsorption on the boron- or nitrogen-doped carbon nanotubes, Physica B 404 (2009) 4173–4177.

[23] M.S. Dresselhaus, G. Dresselhaus, P.C. Eklund, in: A. Press (Ed.), Science of Fullerenes and Carbon Nanotubes, Elsevier Science, USA, 1996.

[24] P. Ayala, et al., The doping of carbon nanotubes with nitrogen and their potential applications, Carbon 48 (2010) 575–586.

[25] T. Takenobu, et al., Stable and controlled amphoteric doping by encapsulation of organic molecules inside carbon nanotubes, Nature Materials 2 (2003) 683–688.

[26] L.J. Li, et al., Diameter-selective encapsulation of metallocenes in single-walled carbon nanotubes, Nature Materials 4 (2005) 481–485.

[27] A. Ilie, et al., Effects of KI encapsulation in single-walled carbon nanotubes by Raman and optical absorption spectroscopy, Journal of Physical Chemistry B 110 (2006) 13848–13857.

[28] D. Tasis, et al., Chemistry of carbon nanotubes, Chemical Reviews 106 (2006) 1105.

[29] J. Lu, et al., Interplay of single-walled carbon nanotubes and encapsulated La@C82, La2@C80 and Sc3N@C80, Physical Review B: Condensed Matter and Materials Physics 71 (2005) 235417.

[30] Y. Cho, et al., Orbital hybridization and charge transfer in carbon nanopeapods, Physical Review Letters 90 (2003) 106402.

[31] T. Okazaki, et al., Optical band gap modification of single-walled carbon nanotubes by encapsulated fullerenes, Journal of the American Chemical Society 130 (2008) 4122–4128.

[32] H. Kataura, et al., High-yield fullerene encapsulation in single-wall carbon nanotubes, Synthetic Metals 121 (2001) 1195–1196.

[33] D.J. Hornbaker, et al., Mapping the one-dimensional electronic states of nanotube peapod structures, Science 295 (2002) 828–831.

[34] J. Lee, et al., Bandgap modulation of nanotubes by encapsulated metallofullerenes, Nature 415 (2002) 1005–1008.

[35] B.W. Smith, M. Monthioux, D.E. Luzzi, Encapsulated C60 in carbon nanotubes, Nature 396 (1998) 323–324.

[36] H. Khalfoun, P. Hermet, L. Henrard, B and N codoping effect on electronic transport in carbon nanotubes, Physical Review B 81 (2010) 193411.

[37] S. Latil, S. Roche, J.C. Charlier, Electronic transport in carbon nanotubes with random coverage of physisorbed molecules, Nano Letters 5 (2005) 2216–2219.

[38] C. Bower, et al., Intercalation and partial exfoliation of single-walled carbon nanotubes by nitric acid, Chemical Physics Letters 288 (1998) 481–486.
[39] M. Holzinger, et al., Multiple functionalization of single-walled carbon nanotubes by dip coating, Chemical Communications 47 (2011) 2450–2452.
[40] B.L. Allen, P.D. Kichambare, A. Star, Carbon nanotube field-effect-transistor based biosensors, Advanced Materials 19 (2007) 1439–1451.
[41] A. Zettl, (27.12.2012). Available from: http://www.lbl.gov/Science-Articles/Archive/zettl-nanotubes.html.
[42] P.G. Collins, et al., Extreme oxygen sensitivity of electronic properties of carbon nanotubes, Science 287 (2000) 1801–1804.
[43] W. Zhou, et al., Charge transfer and Fermi level shift in p-doped single-walled carbon nanotubes, Physical Review B 71 (2005) 205423.
[44] S.B. Fagan, et al., 1,2-Dichlorobenzene interacting with carbon nanotubes, Nano Letters 4 (2004) 1285–1288.
[45] A.M. Rao, et al., Evidence for charge transfer in doped carbon nanotube bundles from Raman scattering, Nature 388 (1997) 257–259.
[46] R. Graupner, et al., Doping of single-walled carbon nanotube bundles by Brønsted acids, Physical Chemistry Chemical Physics 5 (2003) 5472–5476.
[47] K.S. Mistry, et al., n-Type transparent conducting films of small molecule and polymer amine doped single-walled carbon nanotubes, ACS Nano 5 (5) (2011) 3714–3723.
[48] M.S. Strano, et al., Reversible, band-gap-selective protonation of single-walled carbon nanotubes in solution, Journal of Physical Chemistry B 107 (2003) 6979–6985.
[49] M.J. O'Connell, E.E. Eibergen, S.K. Doorn, Chiral selectivity in the charge-transfer bleaching of single-walled carbon-nanotube spectra, Nature Materials 4 (2005) 412–418.
[50] S.B. Fagan, et al., Ab initio study of 2,3,7,8-tetrachlorinated dibenzo-p-dioxin adsorption on single wall carbon nanotubes, Chemical Physics Letters 437 (2007) 79–82.
[51] M. Holzinger, Supramolecular Assemblies. (cited 2012, 11.10.12), Available from: http://dcm.ujf-grenoble.fr/site/site.php?style=spip&id_rubrique=176&id_article=292&scrolling=no&dossier=.
[52] F. Alibart, et al., Comparison and semiconductor properties of nitrogen doped carbon thin films grown by different techniques, Applied Surface Science 254 (2008) 5564–5568.
[53] J. Wei, et al., Effects of nitrogen substitutional doping on the electronic transport of carbon nanotube, Physica E 40 (2008) 462–466.
[54] M. Glerup, et al., Synthesis of N-doped SWNT using the arc-discharge procedure, Chemical Physics Letters 387 (2004) 193–197.
[55] O. Stephan, et al., Doping graphitic and carbon nanotube structures with boron and nitrogen, Science 266 (1994) 1683–1685.
[56] Y. Zhang, et al., Heterogeneous growth of B-C-N nanotubes by laser ablation, Chemical Physics Letters 279 (1997) 264–269.
[57] E.G. Wang, et al., Optical emission spectroscopy study of the influence of nitrogen on carbon nanotube growth, Carbon 41 (2003) 1827–1831.

[58] P. Ayala, et al., Tailoring N-doped single and double wall carbon nanotubes from a nondiluted carbon/nitrogen feedstock, Journal of Physical Chemistry C 111 (2007) 2879–2884.
[59] M. Terrones, et al., N-doping and coalescence of carbon nanotubes: synthesis and electronic properties, Applied Physics A 74 (2002) 355–361.
[60] T. Belz, et al., Structural and chemical characterization of n-doped nanocarbons, Carbon 36 (1998) 731–741.
[61] M. Terrones, et al., New direction in nanotube science, Materials Today 7 (2004) 30–45.
[62] H.J. Choi, et al., Defects, quasibound states, and quantum conductance in metallic carbon nanotubes, Physical Review Letters 84 (13) (2000) 2917–2920.
[63] C.S. Yeung, Y.K. Chen, Y.A. Wang, Theoretical studies of substitutionally doped single-walled nanotubes, Journal of Nanotechnology 2010 (2010) 801789.
[64] W.Q. Tian, et al., Electronic properties and reactivities of perfect, defected, and doped single-walled carbon nanotubes, in: T. Dumitrica (Ed.), Trends in Computational Nanomechanics, Springer, 2010, pp. 421–471.
[65] I.O. Maciel, et al., Synthesis, electronic structure, and Raman scattering of phosphorus-doped single-wall carbon nanotubes, Nano Letters 9 (2009) 2267–2272.
[66] E. Cruz-Silva, et al., Heterodoped nanotubes: theory, synthesis, and characterization of phosphorus-nitrogen doped multiwalled carbon nanotubes, ACS Nano 2 (2008) 441–448.
[67] M. Endo, et al., Anode performance of a Li ion battery based on graphitized and B-doped milled mesophase pitch-based carbon fibers, Carbon 37 (1999) 561–568.
[68] J. Mandumpal, S. Gemming, G. Seifert, Curvature effects of nitrogen on graphitic sheets: structures and energetics, Chemical Physics Letters 447 (2007) 115–120.
[69] M.V. Veloso, et al., Ab Initio study of covalent functionalized carbon nanotubes, Chemical Physics Letters 430 (2006) 71–74.
[70] L.J. Li, R.J. Nicholas, Bandgap-selective chemical doping of semiconducting single-walled carbon nanotubes, Nanotechnology 15 (2004) 1844–1847.
[71] I.O. Maciel, et al., Electron and phonon renormalization near charged defects in carbon nanotubes, Nature Materials 7 (2008) 878–883.
[72] V.A. Davis, et al., True solutions of single-walled carbon nanotubes for assembly into macroscopic materials, Nature Nanotechnology 4 (2009) 830–834.
[73] K. Okazaki, Y. Nakato, K. Murakoshi, Absolute potential of the Fermi level of isolated single-walled carbon nanotubes, Physical Review B 68 (2003) 354341–354345.
[74] K.K. Kim, et al., Doping strategy of carbon nanotubes with redox chemistry, New Journal of Chemistry 34 (2010) 2183–2188.
[75] S.M. Kim, et al., Reduction-controlled viologen in bisolvent as an environmentally stable n-type dopant for carbon nanotubes, Journal of the American Chemical Society 131 (2009) 327–331.
[76] B.R. Kang, et al., Restorable type conversion of carbon nanotube transistor using pyrolitically controlled antioxidizing photosynthesis coenzyme, Advanced Functional Materials 19 (2009) 2553–2559.
[77] C. Kittel, Introduction to Solid State Physics, eighth ed., John Wiley & Sons, 2005.
[78] K.K. Kim, et al., Fermi level engineering of single-walled carbon nanotubes by $AuCl_3$ doping, Journal of the American Chemical Society 130 (2008) 12757–12761.

[79] B.S. Kong, et al., Single-walled carbon nanotube gold nanohybrids: application in highly effective transparent and conductive films, Journal of Physical Chemistry C 111 (2007) 8377–8382.

[80] J. Zhao, J. Han, J.P. Lu, Work functions of pristine and alkali-metal intercalated carbon nanotubes and bundles, Physical Review B 65 (2002), 193401(1–4).

[81] L. Duclaux, Review of the doping of carbon nanotubes (multiwalled and single-walled), Carbon 40 (2002) 1751–1764.

[82] R. Che, et al., Electron energy-loss spectroscopy characterization and microwave absorption of iron filled carbon-nitrogen nanotubes, Nanotechnology 18 (2007) 355705.

[83] Y. Zhao, et al., Iodine doped carbon nanotube cables exceeding specific electrical conductivity of metals, Scientific Reports 1 (2011) 83.

[84] J.A. Robinson, E.S. Snow, F.K. Perkins, Improved chemical detection using single-walled carbon nanotube network capacitors, Sensors and Actuators B 135 (2007) 309–314.

[85] J.L. Blackburn, et al., Transparent conductive single-walled carbon nanotube networks with precisely tunable ratios of semiconducting and metallic nanotubes, ACS Nano 2 (2008) 1266–1274.

[86] L. Grigorian, et al., Reversible intercalation of charged iodine chains into carbon nanotube ropes, Physical Review Letters 80 (1998) 5560–5563.

[87] N. Bendiab, et al., Structural determination of iodine localization in single-walled carbon nanotube bundles by diffraction methods, Physical Review B 69 (2004) 195415.

[88] P. Jarosz, et al., Carbon nanotube wires and cables: near-term applications and future perspectives, Nanoscale 3 (2011) 4542–4553.

[89] K. Liu, et al., Scratch-resistant, highly conductive and high strength carbon nanotube based composite yarns, ACS Nano 4 (2010) 5827–5834.

[90] X.H. Zhong, et al., Continuous multilayered carbon nanotube yarns, Advanced Materials 22 (2010) 692–696.

[91] R. Saito, G. Dresselhaus, M.S. Dresselhaus, Physical Properties of Carbon Nanotubes, Imperial College Press, London, UK, 1998.

[92] P. Corio, et al., Characterization of single wall carbon nanotubes filled with silver and with chromium compounds, Chemical Physics Letters 383 (2004) 475–480.

[93] C.L. Pint, et al., Supergrowth of nitrogen-doped wingle-walled carbon nanotube arrays: active species, dopant characterization, and doped/undoped heterojunctions, ACS Nano 5 (9) (2011) 6925–6934.

[94] M.S. Dresselhaus, G. Dresselhaus, Intercalation compounds of graphite, Advances in Physics 51 (2002) 1–186.

[95] R. Voggu, et al., Extraordinary sensitivity of the electronic structure and properties of single-walled carbon nanotubes to molecular charge-transfer, Physical Chemistry C 112 (2008) 13053–13056.

[96] K.K. Kim, et al., Dependence of Raman spectra G band intensity on metallicity of single-wall carbon nanotubes, Physical Review B 76 (2007) 205426.

[97] J.S. Park, et al., G0 band Raman spectra of single, double and triple layer graphene, Carbon 47 (2009) 1303–1310.

[98] L.B. Alemany, et al., Solid-state NMR analysis of fluorinated single-walled carbon nanotubes: assessing the extent of fluorination, Chemistry of Materials 19 (2007) 735–744.

[99] L. Zeng, et al., Demonstration of covalent sidewall functionalization of single wall carbon nanotubes by NMR spectroscopy: side chain length dependence on the observation of the sidewall sp3 carbons, Nano Research 1 (2008) 72–88.

[100] M. Schmid, et al., Metallic properties of Li-intercalated carbon nanotubes investigated by NMR, Physical Chemistry B 74 (2006) 073416.

[101] F. Leonard, The Physics of Carbon Nanotube Devices, William Andrew Inc., 2009.

[102] J. Alvarenga, et al., High conductivity carbon nanotube wires from radial densification and ionic doping, Applied Physics Letters 97 (2010) 182106.

CHAPTER 11

Carbon Nanofiber Multifunctional Mat

Carla L. Lake, Patrick D. Lake
Applied Sciences Inc, Cedarville, OH, USA

CHAPTER OUTLINE

11.1 Introduction	313
11.2 Development of Carbon Nanofiber Mat	315
11.2.1 Carbon nanofiber	315
11.2.2 Carbon nanofiber mat	317
11.2.3 Fabrication of CNF mat-reinforced composites	321
11.2.4 Short beam shear testing	322
11.3 Conclusion	328
Acknowledgments	328
References	328

11.1 INTRODUCTION

Multifunctional structural systems that utilize nanoscale reinforcing materials have been of interest to the scientific community for the past 20 years [1–8]. Structural components built using polymer matrix composites (PMCs) are attractive for a wide range of applications due to their combination of light weight and high stiffness. PMCs allow for the design of complex unitized structures, which can significantly reduce the number of parts and fasteners in an aerospace structure. PMCs when combined with carbon nanomaterials (CNMs) are regarded as ideal candidates to replace conventional metallic components to reduce weight and potentially impart multifunctionality to the component.

Various CNMs have been incorporated into polymers to form nanocomposites [4–13]. The potential benefits of having carbon nanotubes (CNT) and carbon nanofibers (CNFs) in nanocomposites include weight reduction, improved mechanical properties (modulus, strength, fracture toughness, fatigue resistance [14–18], delamination resistance [19,20], impact strength [21,22], and structural damping [23–27]), improved electrical [28] and thermal conductivity, increased thermal stability [29], improved flame retardancy [30–32], enhanced barrier properties [33], and reduced environmental effects such as moisture absorption [34] and degradation by irradiation [35].

Recent results show significant improvements of PMC fracture toughness with the addition of CNFs, as a result of crack deflection by the dispersed nanostructures [36–38]. Small quantities of tailored CNFs can efficiently reinforce the matrix without any adverse impact on the overall composite. Improvements in tensile strength and tensile modulus by 11% and 22.3%, respectively, have been reported in carbon fabric–epoxy composites with incorporation of 2 wt.% of CNF in the matrix, serving to significantly increase the interphase volume and result in microcrack mitigation [39–41].

Figure 11.1 shows an example of CNF bridging a crack that has formed in epoxy. This ability to hinder crack propagation has been identified as the primary mechanism by which CNF is able to improve interlaminar shear strength (ILSS) and resin mechanical properties in general.

CNMs are a new class of multifunctional additives that are becoming commercially available at practical production volumes and prices. This class of materials includes single-walled carbon nanotubes (SWNT), multiwalled carbon nanotubes (MWNT), nanographene platelets (NGP), and stacked-cup carbon nanotubes, also known as CNFs. The advantage of using CNMs in applications that require multifunctionality comes from their extremely large surface area, which enables them to reinforce and interact with the polymer matrix at the molecular level. By being able to control the loading and the dispersion of CNMs, several functionalities can be translated into lighter composite parts.

As in all cases where nanosized additives are involved, the development of high-performance composites requires homogeneous dispersion of their nano

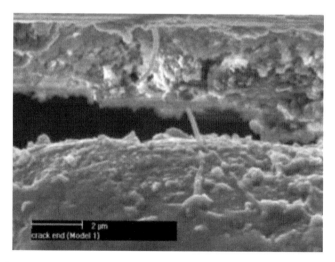

FIGURE 11.1

ASI's carbon nanofibers can bridge cracks in epoxy resin composites providing structural support.

counterparts. Well-documented success stories, in both theoretical and experimental levels, show clear benefits of nanotechnology in PMCs to improve their structural performance and impart multifunctionality. Typically these findings have been limited to small laboratory-scale environments, which have not proceeded to commercially reality. This has much to do with the wide spectrum of nanoparticle species, qualities and purity grades available in the market today; the absence of quality control; and the nonapplication of standard manufacturing approaches. There are a large number of raw material manufacturers using different composite synthesis methods, and their standards for product quality and consistency are not uniform. Thus design engineers for spacecraft and aircraft are faced with comparing a multitude of nanoparticles that respond differently to processing schemes and are often difficult to handle and qualify.

Filtration of nanoparticles by larger carbon fibers/glass fibers used as the primary reinforcing phase is typically a problem when incorporating nanofibers or nanotubes into laminated composites. A larger loading of nanoparticles occurs at the resin injection site and a lower loading occurs at the outlet side. Miller et al. showed evidence of nanoparticle filtration on a resin transfer molding-processed epoxy/carbon fiber composite [35]. Factors that contributed to nanoparticle filtration were poor dispersion prior to infiltration and the braid architecture of the structural reinforcement. Ultimately, achieving multifunctional properties in composites depends on a proper choice of fiber type and/or geometry, mastering the composite processing techniques, effective placement of the nanofillers in the polymeric continuous phase, and developing innovative product designs that overcome processing and interface challenges.

Nanomaterials hold extraordinary promise for improving composite structures. Considering the technical and processing hurdles associated with these materials, there is need for development of a continuous sheet good based on carbon nanoparticles. The sheet material should be easy to handle and integrate into traditional composite processing techniques. A useful form of a continuous sheet good would be a mat or veil. Continuous sheets of nonwoven veils of various materials have been widely used in industry for surface engineering of composite properties ranging from surface quality to adding functionality. CNF mat, or nanomat, produced in high volumes could serve as a new veil material with a wide range of applications including lightning strike protection [42], electromagnetic interference shielding [43], electrostatic dissipation, thermal management [44] and vibration damping [25,26]. This chapter describes fabrication of CNF mat, new processing techniques developed, and initial composite property results.

11.2 DEVELOPMENT OF CARBON NANOFIBER MAT

11.2.1 Carbon nanofiber

Applied Sciences, Inc.'s (ASI's) "Pyrograf® III" CNFs, shown in Fig. 11.2, have demonstrated their value as additives to improve the performance of polymer

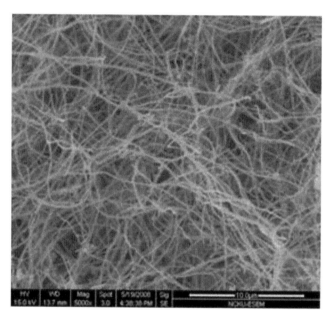

FIGURE 11.2

Scanning electron microscope image of Pyrograf III material.

composites in various studies. Pyrograf III CNFs are graphitic filaments having a diameter of approximately 100 nm and an aspect ratio greater than 1000. In terms of physical size, and production cost, CNF completes a continuum bounded by carbon black, fullerenes, and SWNT to MWNT on one end and continuous carbon fiber on the other end. Relative to other CNMs, CNFs are more easily dispersible than MWNT. The van der Waals forces are lower due to the larger diameter of CNFs, which means that less energy is required to disperse them into individual fibers. Although MWCNTs are becoming commercially available at low prices, the raw materials require extensive processing steps, such as purification and functionalization, to make the materials effective for use. These postprocessing steps have yet to be demonstrated at a level higher than the laboratory scale and pose significant challenges for scale-up efforts associated with commercialization.

CNF can be manufactured with a distinctly different structure than CNT. As shown in Fig. 11.3, CNF can show a "stacked-cup" morphology where graphene appears to be a stack of conic sections or a scroll structure with edges that terminate on the surface of the fiber. The stacked-cup structure is markedly different than nanotubes, which are composed of sheets of graphene arranged in a structure resembling concentric cylinders. The CNF also has a larger average outer diameter (~ 100 nm), a resultant lower specific surface area (~ 20 m^2/g), and a longer average length (~ 100 μm) than CNT. Despite these differences in scale and morphology, composites produced from CNF exhibit properties similar to

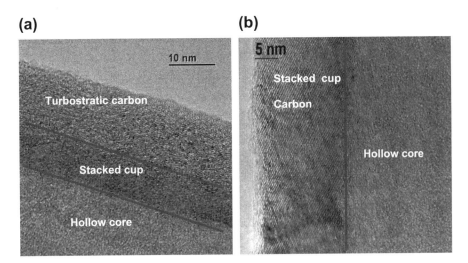

FIGURE 11.3

(a) The transmission electron microscope (TEM) image represents a typical PR-19 carbon nanofiber with a large fraction of turbostatic carbon deposited on the catalytic layer (stacked-cup carbon). The catalytic carbon layer is carbon precipitated from the catalyst particle, while the turbostatic layer is added through chemical vapor deposition techniques. (b) The image is typical of PR-25 fiber that only has the catalytic carbon layer. (For color version of this figure, the reader is referred to the online version of this book.)

Source: Figures courtesy of Jane Howe, Oak Ridge National Laboratory, TN, USA.

composites made from MWNT. The difference is that CNF composites often cost less and there are fewer processing issues.

11.2.2 Carbon nanofiber mat

CNF mat was produced by integrating CNF with polyacrylonitrile (PAN)-derived carbon fiber veil to form a continuous nanomaterial sheet good that can be incorporated into composite systems using traditional composite processing methods and equipment. CNF and carbon fiber veil are commercially available at low cost and high volume. This is a proprietary process developed by ASI and there can be a wide variety of applications for this material. CNF used in the process is a highly graphitic multifunctional nanomaterial and is available in tons at less than $0.75/g. Several product forms of CNF are available with an index of graphitization, surface state, and other properties that can be tailored for specific applications. The CNF mat can be produced using carbon veils from suppliers such as Hollingsworth & Vose Company, Technical Fibre Products, Inc., or Southeast Nonwovens, Inc. However, other noncarbon veils, such as glass veils or polymer veils, can also be used in the fabrication of CNF mat. Figures 11.4 and 11.5 compare optical and scanning electron micrographs of CNF mat produced according to the traditional method of paper making (Fig. 11.4) and ASI. CNF mat produced following its proprietary method (Fig. 11.5).

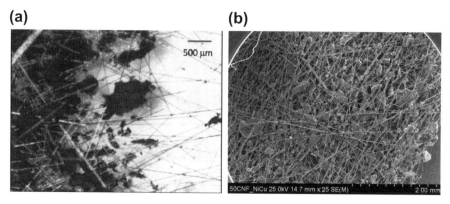

FIGURE 11.4

(a) Optical and (b) SEM images of nanofiber mat produced via filtration, a process typically used in papermaking. The filtration process generates a distribution of agglomerated nanofiber within the body of the veil and greatly limits the strength and durability of the sheet good. (For color version of this figure, the reader is referred to the online version of this book.)

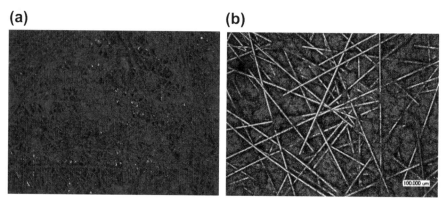

FIGURE 11.5

(a) Optical and (b) SEM images of nanofiber mat produced via ASI's proprietary process. The process developed by ASI generates a uniform sheet good with a well-developed network of nanofiber supported by the carbon fiber veil that is well dispersed and is durable enough to process on reel-to-reel processing operations. (For color version of this figure, the reader is referred to the online version of this book.)

The veil shown in Fig. 11.4 was produced using a traditional wet-laid paper-making process. One can readily see that the papermaking process yields a poor dispersion of the nanomaterials and contains agglomerates of CNF nested between PAN carbon fibers. These sandwiched agglomerates are virtually impossible to penetrate with resin during processing to incorporate the sheets into a composite.

Furthermore, the sandwiched agglomerates are inefficient in forming a conductive network. In contrast, the veil produced using ASI's proprietary process results in an excellent dispersion of CNFs.

For this study, long lengths of CNF mat were produced, with widths from 12 in. to 30 in. Blends of different CNF types, lengths, and surface treatments were used to fabricate the mats. The areal weight of the mat material can be tailored based on application requirements. The CNF mat was produced without a backing paper since the material has sufficient strength to be handled without damage. In addition, the process used can incorporate other nanomaterials including CNT and NGP. Figure 11.6 shows a 14 in. wide by 80 ft long section of a CNF mat ready for prepreg processing.

The CNF mat is easy to handle and overcomes all the primary difficulties derived from dispersion, processing, and handling typically associated with CNMs.

Small-scale impregnation trials were performed on CNF mat samples to determine the material's wettability and durability. Cytec 5250-4 bismaleimide (BMI) resin film (prepreg grade) with an areal weight of 34 gsm was used to impregnate the CNF mat samples. The small-scale tests showed the CNF mat samples impregnated well. The successfully impregnated veil samples were fully cured. A sample piece of cured CNF mat laminate is shown in Fig. 11.7.

In addition, over 100 linear ft of CNF nanomat in two rolls were produced for prepregging trials in Cytec's Anaheim facility (Fig. 11.8). Prepregging trials were conducted at Cytec by sandwiching the nanomat between two 150 gsm films of BMI resin 5250-4 provided by Cytec. Cytec's choice of the 150 gsm weight for the films was based on the estimated quantity of resin required to wet out the assumed weight of the CNF nanomat. Cytec reported that the nanomat exhibited good strength and was easy to pull from the roll.

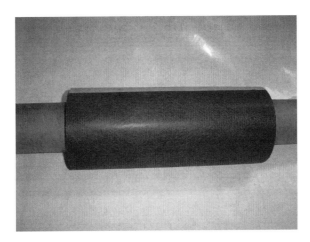

FIGURE 11.6

Image of 80-ft-long roll of nanofiber mat. (For color version of this figure, the reader is referred to the online version of this book.)

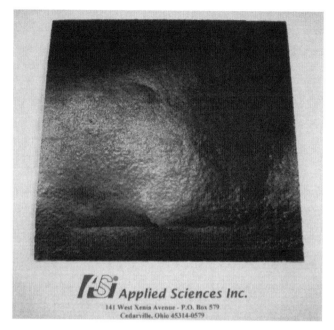

FIGURE 11.7

Image of fully cured laminate. The laminate was produced by impregnating a CNF mat using BMI resin film (Cytec Engineered Materials, Anaheim, CA). The CNF mat impregnated well and is fully consolidated. (For color version of this figure, the reader is referred to the online version of this book.)

FIGURE 11.8

CNF mat being processed on a 12″ pilot line at Cytec.

The nanomat material was prepregged at 210 °F with 100 psi for all nips and with a sled speed of 5 fpm. The resulting material was flexible and had a small amount of tack. The color of the material was consistently black and showed no sign of significant imperfections. Cytec reported that, overall, this material was relatively easy to prepreg. Impregnation trials were performed using handsheets and laminates that were cocured to a glass fabric using Cytec 5240-4 BMI resin film. The material was processed with no difficulty.

Approximately 100 ft^2, in a continuous length, of CNF nanomat was processed on Renegade Materials industrial-scale prepreg line. A roll of 35 gsm BMI film was used in the trial. The nanomat was impregnated with resin using films on either side of the nanomat. The material processed well with no special handling or equipment required. Figure 11.9 shows images of the CNF nanomat processed at Renegade Materials facility in Springboro, OH, USA.

11.2.3 Fabrication of CNF mat-reinforced composites

Laminates were produced via the vacuum-assisted resin transfer molding (VARTM) technique. A filtration effect has occasionally been reported when manufacturing nanostructured composite laminates by VARTM—especially at filler concentrations above 1 wt%. Low-viscosity resin and proper use of resin distribution media has been shown to greatly reduce the negative effects of this phenomenon. ASI avoided the problems associated with filtration by utilizing novel CNF product forms, namely, CNF mat, a continuous sheet good composed mainly of CNF, as interleaves for the woven carbon fiber fabric in the composite. In other composites, ASI applied CNF directly to the woven carbon fiber reinforcement and proceeded with the composite layup.

Test coupons for each mode of testing were subsequently cut by water jet from the panels. The water-lubrication provided smooth specimen surfaces and prevented an undesirable temperature increase.

The following four composite configurations were fabricated:

FIGURE 11.9

CNF nanomat processing on industrial-scale prepreg line at Renegade Materials. (For color version of this figure, the reader is referred to the online version of this book.)

1. A baseline two-dimensional woven laminated composite consisting of six layers of the Qiso triaxial braided fiber architecture of $0° \pm 60°$ (from A&P Technology) with an SC-79 epoxy matrix (from Applied Poleramics). The layers were stacked in a (0°/90°) sequence. This composite configuration will be referred as BL in the subsequent discussions.
2. A laminated composite panel consisting of five layers of the Qiso triaxial braided fiber interleaved with four layers of CNF mat; this configuration represents a 16.6% weight reduction in comparison to the BL panel. This reduced weight composite configuration, designated as RW, was fabricated to determine if the panel weight could be reduced while maintaining the same mechanical performance as the BL, by interleaving the CNF mat plies.
3. A laminated composite panel with the same layup as the BL, to which a 5 wt% CNF-reinforced resin was applied to each of the woven fabric plies. The presence of the CNF in the fiber—matrix interface of the composite laminates is expected to enhance material shear strength properties. This composite panel is designated as 5CNF.
4. A laminated composite with the same layup as the BL, to which CNF mat was added to outer surfaces and interleaved between each ply. This composite configuration, designated as CNF mat, is expected to add mechanical reinforcement to the composite and surface electrical conductivity.

Figure 11.10 shows schematic drawings representing each of the laminated composite configurations.

11.2.4 Short beam shear testing

In order to determine the effect of adding the CNF, CNF mat, and nanocapsules, short beam shear (SBS) tests were performed as per American Society for Testing and Materials (ASTM) D2344, at two different temperatures, 23 °C and 200 °C. The effect of different material configurations on the interlaminar shear behavior of the composite laminates was also studied.

As shown in Fig. 11.11, the composite specimen is placed on two cylindrical supports and a cylindrical head is moved down to apply a force at the center and generate an increasing transverse load until the first failure is recorded. The load at failure is then used to determine the apparent interlaminar shear strength of the composite. The results are average values of five individually tested specimens for each sample. The experiment measures the effectiveness of the CNFs to reinforce the interface of the laminated composites.

In the SBS test, the determination of ILSS is based on classical (Bernoulli—Euler) beam theory. For a beam of rectangular cross-section loaded in three-point bending, the maximum interlaminar shear stress occurs at the midthickness of the beam between the center and end supports and is calculated according to Eqn (11.1).

$$ILSS = \frac{0.75 \times P_m}{b \times h} \qquad (11.1)$$

11.2 Development of Carbon Nanofiber Mat

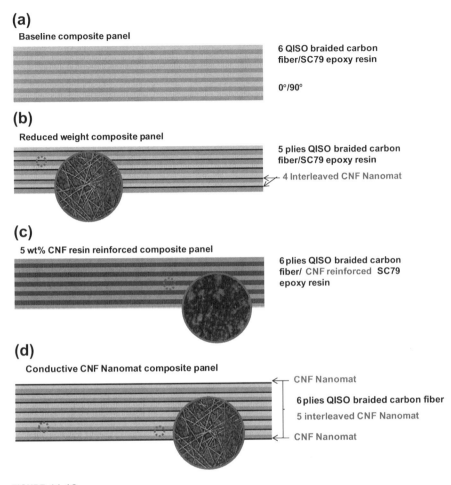

FIGURE 11.10

(a) Baseline two-dimensional woven laminated composite (BL) consisting of six layers of triaxial braided fiber architecture and SP-79 epoxy matrix; (b) laminated composite panel consisting of five layers of triaxial braided fiber interleaved with four layers of CNF mat (RW); (c) laminated composite panel with the same layup as (a), to which a 5 wt% CNF-reinforced resin was applied to each of the woven fabric plies (5CNF); (d) laminated composite with the same layup as (a), to which the CNF mat was added to outer surfaces and interleaved between each ply.

where

P_m is the maximum load during test (lbf)
b is the measured specimen width (in.)
h is the measured specimen thickness (in.)

The ILSS is an important material property associated with composite laminates and defines when individual plies fail in shear. Conventional laminated composites

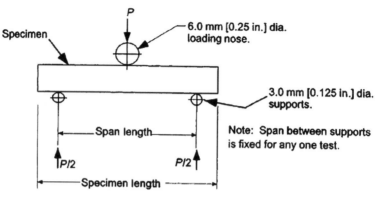

FIGURE 11.11

Short beam shear test loading configuration (ASTM D2344).

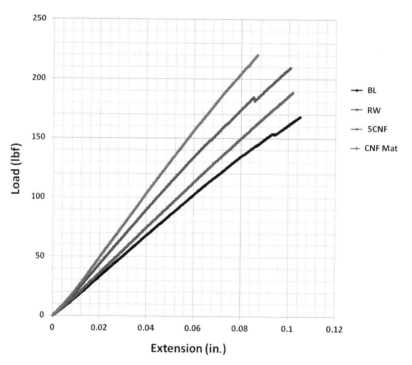

FIGURE 11.12

Representative load–displacement curves for all the composites tested at 23 °C. (For color version of this figure, the reader is referred to the online version of this book.)

have an inherent weakness that manifests itself in poor resistance to interlaminar shear. Because CNFs have exceptional stiffness and tensile strength, it is anticipated that adding them to the fiber—matrix interface of composite laminates will enhance material shear strength properties.

For each composite configuration tested, a load—displacement curve was selected that best represents the average of five separate test specimens. The typical load—displacement curves for each of the five composite panels tested at room and high temperature are shown in Figs 11.12 and 11.13.

The material response at room temperature shows a typical brittle failure mode. As anticipated, all the samples containing CNF (5CNF) or CNF mat (RW and CNF mat), show failure at higher loads when compared to the BL specimen. The RW configuration shows a higher load to failure when compared to the BL, even though this specific configuration has less one ply of carbon fiber, totaling a reduction of over 16% in the weight of the composite panel. This result alone shows the great promise of CNF-reinforced composites for lightweight structural composites.

The material response at elevated temperature shows a nearly linear elastic trend during the early stage of loading. This regime continues until an apparent elastic

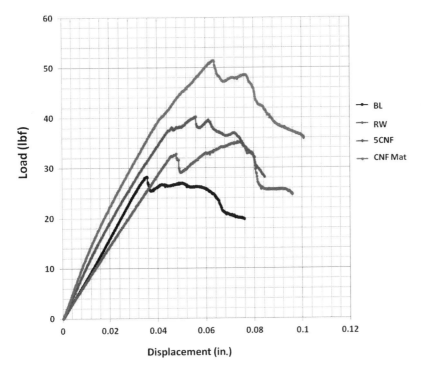

FIGURE 11.13

Representative load—displacement curves for all the composites tested at 200 °C. (For color version of this figure, the reader is referred to the online version of this book.)

CHAPTER 11 Carbon Nanofiber Multifunctional Mat

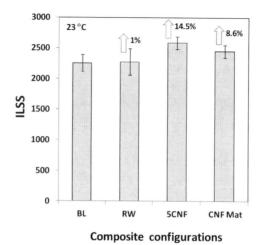

FIGURE 11.14

ILSS average values for the five different composite configurations tested by SBS at room temperature (23 °C). (For color version of this figure, the reader is referred to the online version of this book.)

limit is reached. At this point, the behavior of the material follows one of the two trends: the load decreases or increases. By increasing the temperature, the failure behavior changes from "almost brittle" to "almost plastic".

The BL composite reaches the elastic limit at the lowest load among the five composites. The CNF mat composite achieved the highest peak load among all the composites and maintained a higher load at 0.1″ displacement compared to all others.

All the composites, with exception of RW, showed a typical behavior of load decrease after the peak load was reached. The RW reaches a peak load higher than that of the BL and lower than that of the rest of the composites, but after this point, the load continues to increase, with the highest peak load observed at a displacement of 0.074. This result indicates that while the damage initiation occurred at a lower load, the damage tolerance of the material increases with the presence of the CNF mat interleaved between each ply.

From the load histories, the SBS strength was calculated for each composite using Eqn (11.1). The average ILSS values were determined from five tests for each of the composite configurations tested at room and elevated temperatures and are shown below in Figs 11.14 and 11.15, respectively.

Replacing one of the braided Qiso plies by four interleaved lightweight CNF mats (RW) resulted in a composite panel with 16.6% lower weight and a similar ILSS as the unreinforced laminate (BL). This result shows the potential of adding nanomaterials to decrease the overall weight of Carbon Fiber Reinforced Plastic (CFRP) composites. At room temperature, the highest increase in ILSS, 14.5%, was recorded for the composite laminated containing 5 wt% of CNFs dispersed in the resin. The increased ILSS suggests that the improved shear performance may

11.2 Development of Carbon Nanofiber Mat

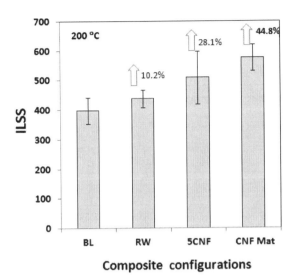

FIGURE 11.15

ILSS average values for the five different composite configurations tested by SBS at elevated temperature (200 °C). (For color version of this figure, the reader is referred to the online version of this book.)

Table 11.1 Summary of ILSS of the Different Composite Configurations Tested at Room Temperature (23 °C)

Specimen	Average (psi)	Standard Deviation (psi)	Coefficient of Variation (%)	Improvement (%)
BL	2259.2	137.8	6.1	–
RW	2281.3	217.1	9.5	1.0
5CNF	2587.8	102.4	3.96	14.5
CNF mat	2452.6	102.6	4.18	8.56

Table 11.2 Summary of Short Beam Shear Strength of the Different Composite Configurations Tested at High Temperature (200 °C)

Specimen	Average (psi)	Standard Deviation (psi)	Coefficient of Variation (%)	Improvement (%)
BL	397.7	44.8	11.3	–
RW	438.4	29.1	6.6	10.2
5CNF	509.3	88.8	17.4	28.1
CNF mat	575.8	44.9	7.8	44.8

be related to the ability of the high aspect ratio CNFs to effectively "anchor" the microcracks together, delaying their growth, coalescence and thus laminate failure. Also, CNFs are able to deflect and stop the growth of the microcracks.

At high temperatures, the ILSS values decrease substantially, which is associated with the softening of the matrix. This was evidenced as well by the "almost plastic" failure behavior. Also, the presence of the CNFs, either dispersed in the resin (5CNF) or in the nanomat (RW and CNF mat), leads to higher improvement when compared to the room-temperature test results. Another interesting aspect of these results is the fact that the presence of the interleaved CNF mat, in the CNF mat sample, increases the ILSS by almost 45%, compared to BL. This was expected, since in a plastic regime, the presence of the nanomaterials at the fiber/matrix interface prevents delamination, which allows better performance of the composite. The presence of well-dispersed CNFs in resin prevents microcrack formation and/or propagation, thus showing the highest ILSS improvement. A summary of the ILSS results for both room- and elevated temperature testing are reported in Tables 11.1 and 11.2.

11.3 CONCLUSION

A method of producing a nanofiber mat composed of highly graphitic CNFs in an isotropic array embedded in a carbon fiber veil has been developed. Areal weight can be tailored using combinations of CNF having different aspect ratios and degrees of graphitization. An optimum formulation was used to generate over 250 ft of nanomat for use in prepregging trials. Sufficient strength to provide handling ease and straightforward production of prepreg rolls using conventional methods has been observed from the nanofiber mat. CNF mat is useful for imparting electrical conductivity to structural composites and can be produced with conventional commercial materials.

The most significant finding from this study is that through use of low loadings of CNFs, traditional composites can be made stronger and more resistant to damage and fatigue. This study demonstrated the feasibility of use of CNFs and carbon nanomats in composite laminates to either increase the envelope of mechanical properties at a given weight or reduce the number of plies of conventional carbon fiber reinforcement while retaining the mechanical properties of the composite. Interlaminar shear stress (ILSS) results at both room temperature and elevated temperature (200 °C) showed that the addition of well-dispersed CNFs improves the shear properties of the composite.

Acknowledgments

This work was supported by USAF SBIR Contract FA8650-09-M-5021.

References

[1] M. Rahmat, P. Hubert, Carbon nanotube-polymer interactions in nanocomposites — a review, Composites Science and Technology 72 (2011) 72—84.

[2] G. Pandey, E. Thostenson, Carbon nanotube-based multifunctional polymer nanocomposites, Polymer Reviews 52 (2012) 355–416.

[3] M. Kessler, Polymer matrix composites: a perspective for a special issue of polymer reviews, 52 (2012) 229–233.

[4] G.G. Tibbetts, et al., Composites Science and Technology 67 (7–8) (June 2007) 1709–1718.

[5] P.M. Ajayan, L.S. Schadler, P.V. Braun, Nanocomposite Science and Technology, John Wiley & Sons, New York, 2003.

[6] O. Breuer, U. Sundaraj, Big return from small fibers: a review of polymer/carbon nanotube composites, Polymer Composites 25 (2005) 630–645.

[7] E.T. Thostenson, Z.F. Ren, T.W. Chou, Advances in the science and technology of carbon nanotubes and their composites: a review 61 (2001) 1899–1912.

[8] K.T. Lau, D. Hui, The revolutionary creation of new advanced materials — carbon nanotube composites, Composites Part B: Engineering 33 (2002) 263–277.

[9] S.C. Tjong, Structural and mechanical properties of polymer nanocomposites, Materials Science and Engineering: R: Reports 53 (3–4) (2006) 73–197.

[10] K.T. Lau, C. Gu, D. Hui, A critical review on nanotube and nanotube/nanoclay related polymer composite materials, Composites Part B: Engineering 37 (6) (2006) 425–436.

[11] X.L. Xie, Y.W. Mai, X.P. Zhou, Dispersion and alignment of carbon nanotubes in polymer matrix: a review, Materials Science and Engineering: R: Reports 49 (4) (2005) 89–112.

[12] J. Jordan, K.I. Jacob, R. Tannenbaum, M.A. Sharaf, I. Jasiuk, Experimental trends in polymer nanocomposites — a review, Materials Science and Engineering A 393 (1–2) (2005) 1–11.

[13] G.G. Tibbetts, J.J. McHugh, Mechanical properties of vapor-grown carbon fiber composites with thermoplastic matrices, Journal of Materials Research 14 (7) (1999) 1–10.

[14] I.C. Finegan, G.G. Tibbetts, D.G. Glasgow, J.-M. Ting, M.L. Lake, Surface treatments for improving the mechanical properties of carbon nanofiber/thermoplastic composites, Journal of Materials Science 38 (2003) 3485–3490.

[15] R.J. Kruiger, K. Alam, D.P. Anderson, Strength prediction of partially aligned discontinuous fiber-reinforced composites, Journal of Materials Research 16 (1) (2001) 226–232.

[16] K. Lafdi, M. Matzek, Carbon nanofibers as a nano-reinforcement for polymeric nanocomposites. In: 48th International SAMPE symposium proceedings, Long Beach, USA, 2003.

[17] B. Wetzel, P. Rosso, F. Haupert, K. Friedrich, Epoxy nanocomposites — fracture and toughening mechanisms, Engineering Fracture Mechanics 73 (16) (2006) 2375–2398.

[18] Y.X. Zhou, V. Rangari, H. Mahfuz, S. Jeelani, P.K. Mallick, Experimental study on thermal and mechanical behavior of polypropylene, talc/polypropylene and polypropylene/clay nanocomposites, Materials Science and Engineering: A 402 (1–2) (2005) 109–117.

[19] R. Sadeghian, B. Minaie, S. Gangireddy, K-T. Hsiao, Mode-1 delamination characterization for carbon nanofiber toughened polyester/glassfiber composites. In: 50th International SAMPE Symposium Proceedings, Long Beach, May 2005.

[20] N.A. Siddiqui, R.S.C. Woo, J.K. Kim, C.C.K. Leung, A. Munir, Mode I interlaminar fracture behavior and mechanical properties of CFRPs with nanoclay-filled epoxy

matrix, Composites Part A: Applied Science and Manufacturing 38 (2) (2007) 449–460.

[21] J.C. Lin, L.C. Chang, M.H. Nien, H.L. Ho, Mechanical behavior of various nanoparticle filled composites at low-velocity impact, Composite Structures 74 (1) (2006) 30–36.

[22] Q. Yuan, R.D.K. Misra, Impact fracture behavior of clay–reinforced polypropylene nanocomposites, Polymer 47 (12) (2006) 4421–4433.

[23] I.C. Finegan, G.G. Tibbetts, R.F. Gibson, Modeling and characterization of damping in carbon nanofiber/polypropylene composites, Composites Science and Technology 63 (2003) 1629–1635.

[24] J. Zeng, B. Saltysiak, W.S. Johnson, D.A. Schiraldi, S. Kumar, Processing and properties of poly(methyl methacrylate)/carbon nano fiber composites, Composites: Part B 35 (2004) 173–178.

[25] H. Rajoria, N. Jalili, Passive vibration damping enhancement using carbon nanotube-epoxy reinforced composites, Composites Science and Technology 65 (14) (2005) 2079–2093.

[26] J. Gou, S. O'Braint, H. Gu, G. Song, Damping augmentation of nanocomposites using carbon nanofiber paper, Journal of Nanomaterials (2006) 1–7. Article ID 32803.

[27] N.A. Koratkar, B.Q. Wei, P.M. Ajayan, Multifunctional structural reinforcement featuring carbon nanotube films, Composites Science and Technology 63 (11) (2003) 1525–1531.

[28] I.C. Finegan, G.G. Tibbetts, Electrical conductivity of vapor-grown carbon fiber/thermoplastic composites, Journal of Materials Research 16 (6) (2001) 1668–1674.

[29] J.W. Gilman, Flammability and thermal stability studies of polymer-layered silicate (clay) nanocomposites, Applied Clay Science 15 (1999) 31–49.

[30] T. Kashiwagi, E. Grulke, J. Hilding, K. Groth, R. Harris, K. Butler, J. Shields, S. Kharchenko, J. Douglas, Thermal and flammability properties of polypropylene/carbon nanotube nanocomposites, Polymer 45 (2004) 4227–4239.

[31] T. Kashiwagi, E. Grulke, J. Hilding, R. Harris, W. Awad, J. Douglas, Thermal and flammability properties of polypropylene/carbon nanotube nanocomposites, Macromolecular Rapid Communications 23 (2002) 761–765.

[32] T. Kashiwagi, F. Du, K.I. Winey, K.M. Groth, J.R. Shields, S.P. Bellayer, H. Kim, J.F. Douglas, Flammability properties of polymer nanocomposites with single-walled carbon nanotubes: effects of nanotube dispersion and concentration, Polymer 46 (2005) 471–481.

[33] B. Xu, Q. Zheng, Y.H. Song, et al., Calculating barrier properties of polymer/clay nanocomposites: effects of clay layers, Polymer 47 (8) (2006) 2904–2910.

[34] J.K. Kim, C.G. Hu, R.S.C. Woo, M.L. Sham, Moisture barrier characteristics of organoclay–epoxy nanocomposites, Composites Science and Technology 65 (5) (2005) 805–813.

[35] S.G. Miller, L. Micham, C.C. Copa, J.M. Criss Jr., E.A. Mintz, Nanoparticle Filtration in a RTM Process Epoxy/Carbon Fiber Composite. SAMPE 2011, Long Beach, CA, 23–6 May 2011.

[36] J.L. Abot, Y. Song, M.J. Schulz, V.N. Shanov, Composites Science and Technology 68 (2008) 2755–2760.

[37] T. Yokozeki, Y. Iwahori, S. Ishiwata, Composites A 38 (2007) 917–924.

[38] T. Yokozeki, Y. Iwahori, S. Ishiwata, K. Enomoto, Composites A 38 (2007) 2121–2130.

[39] Y. Zhou, I. Pervin, S. Jeelani, P.K. Mallick, Journal of Materials Processing Technology 198 (2008) 445.
[40] Y. Iwahori, S. Ishiwata, T. Sumizawa, T. Ishikawa, Composites Part A 36 (10) (2005) 1430.
[41] M.J. Palmeri, K.W. Putz, T. Ramanathan, L.C. Brinson, Multi-scale reinforcement of CFRPs using carbon nanofibers, Composites Science and Technology 71 (2011) 79–96.
[42] J. Gou, Y. Tang, F. Liang, Z. Zhao, D. Firsich, J. Fielding, Carbon nanofiber paper for lightning strike protection of composite materials, Composites: Part B 41 (2010) 192–198.
[43] P. Lake, et al., Carbon Nanofiber Mat, US Patent Application Number 61/238337, 31 August 2010.
[44] N.K. Mahanta, A.R. Abramson, M.L. Lake, D.J. Burton, J.C. Chang, H.K. Mayer, J.L. Ravine, Thermal conductivity of carbon nanofiber mats, Carbon 48 (2010) 4457–4465.

CHAPTER

Direct Synthesis of Long Nanotube Yarns for Commercial Fiber Products

12

Miao Zhu, Hongwei Zhu

Center for Nano and Micro Mechanics (CNMM), Key Laboratory for Advanced Manufacturing by Material Processing Technology, Ministry of Education and School of Materials Science and Engineering, Tsinghua University, Beijing, P. R. China

CHAPTER OUTLINE

12.1 Introduction	333
12.2 Direct Synthesis of Long CNT Yarns	335
12.3 Growth of High-Quality CNTs	340
12.4 Applications of CNT Yarns/Fibers	341
12.5 Conclusions	345
Acknowledgments	345
References	346

12.1 INTRODUCTION

Several structures produced by nanotechnology are scientifically and technologically interesting but they are submicroscopic in size. These structures or materials can be even more useful if they can be constructed and retain their properties at the macroscale. Carbon nanotubes (CNTs) are the strongest materials ever synthesized [1]. CNTs have extremely high tensile strength, high modulus, large aspect ratio (up to 10^8), low density, good chemical and environmental stability, high thermal conductivity, and electrical conductivity. These unique properties make CNTs very attractive not only for the fabrication of nanoscale devices but also for broad potential applications at the macroscale, such as strong fibers for body armor, lightweight aerospace structures, and electronic fibers for smart textiles. CNT yarn, the one-dimensional (1D) assembly of CNTs, is of great academic and commercial value due to its inherent properties and the capability to assemble into fibers (Fig. 12.1). The outstanding properties of CNTs can be exploited in a fiber, with the alignment of CNTs parallel to each other and to the fiber axis. Moreover, CNTs have the thermal and chemical stability of graphene and therefore do not degrade under ambient conditions, and they preserve their mechanical properties up to even very high temperatures. Scaling up fiber manufacturing for commercialization is a goal of current

CHAPTER 12 Direct Synthesis of Long Nanotube Yarns

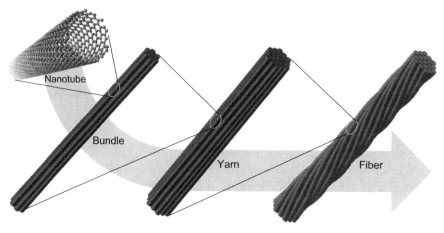

FIGURE 12.1

From CNTs to CNT yarns/fibers. (For color version of this figure, the reader is referred to the online version of this book.)

research in order to fully develop carbon-based fibers with properties that exceed those of existing fiber materials [2].

CNT fibers match the highest reported strengths for a couple of the strongest commercially available fibers. Tensile tests have been performed on CNT fibers directly spun from an aerogel, showing a maximum tensile strength of 1.46 GPa [3] and the highest strength reported to date is 8.8 GPa [3]. Figure 12.2 shows the excellent strength performance of CNT fibers compared with other materials [4]. It should be noted that the specific strength shown in Fig. 12.2(a) is only the high-strength peak of the tested fibers. Further experiments showed the presence of low-strength peaks, which would be predominant as the length increases due to the presence of defects (Fig. 12.2(b)) [4]. It is a significant difficulty that needs to be overcome when it comes to large-scale commercial application.

Meanwhile, the excellent conductivity combined with high strength/stiffness, low density and low cost of production make CNT yarn an attractive substitute to metallic conductor, although it is not always the best one in conductivity performance [5]. For example, field emission properties of CNT yarns were measured and a larger emission constant than traditional thermionic cathodes was achieved [6]. The electrical transporting performance strongly depends on the fiber structure, which provides a practical way to modify or improve their electrical performance by doping or annealing it in oxidizing and reducing environments [7].

This chapter gives an overview of the production, properties and prospects of high-performance CNTs yarns and fibers that generally may be called nanotube superfibers. In general, CNT yarns/fibers can be produced by three ways: (1) wet spinning from solution of CNTs [8,9], (2) dry spinning from arrays of aligned CNTs [10] and (3) direct synthesis by chemical vapor deposition (CVD) [11] or

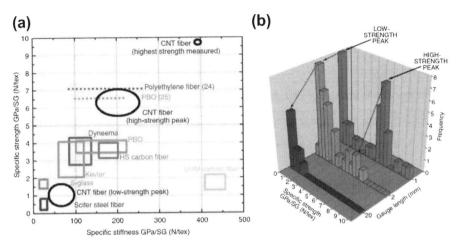

FIGURE 12.2

(a) Comparison of CNT fibers and other materials in strength. (b) Specific strength distribution of CNT fibers [4]. PBO, poly(p-phenylene-2,6-benzobisoxazole) (For color version of this figure, the reader is referred to the online version of this book.)

spinning from the aerogel of CNTs by CVD [12]. We will focus on the third one and discuss the direct synthesis techniques of long CNT yarns by CVD, which is continuous and can currently spin fibers at rates of hundreds of meters per minute, allowing manufacturing scale-up and commercialization.

12.2 DIRECT SYNTHESIS OF LONG CNT YARNS

In all the competing processes that are used to produce single-walled nanotubes (SWNTs), the typical lengths of tangled nanotube bundles reach several tens of micrometers. CNT yarns by postspinning cannot achieve the strength and conductivity of individual nanotubes. The direct synthesis technique is highly preferred to develop the ultimate CNT superfiber materials. Long CNT yarns fabricated directly from CVD were first reported by H.W. Zhu et al. in 2002 [11,13]. CNT yarns ~20 cm long were obtained by a floating catalytic technique using an enhanced vertical furnace as shown in Fig. 12.3(a). The long yarns of nanotubes assemble continuously from arrays of nanotubes, which are intrinsically long (Fig. 12.3(b)). Electrical resistivity of the as-grown yarn was measured via the four-probe method at temperature ranging from 5 K to room temperature. When the temperature was higher than 90 K, CNT yarn showed metallic behavior with its resistivity $\rho = 5-7 \times 10^{-6} \, \Omega \cdot m$. Additionally, Young's modulus of the CNT yarn was measured to be ranging from 49 to 77 GPa according to the tensile tests (Fig. 12.3(c)). X-ray diffraction studies on the yarns focusing on the low-frequency regions showed the well-defined peak at $Q = 0.51/\text{Å}$ corresponding to

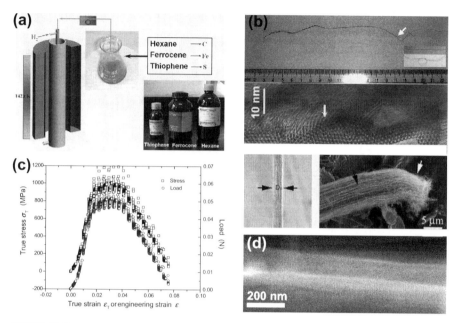

FIGURE 12.3

(a–d) Direct synthesis of CNT yarns [11]. (For color version of this figure, the reader is referred to the online version of this book.)

$d(1,0)$ of the nanotube triangular lattice [14]. The lattice parameter calculated from this peak position is 1.42 nm, which comes from a lattice assembled from 1.1-nm-diameter nanotubes.

The salient feature of this synthesis method was in the use of *n*-hexane in combination with thiophene and hydrogen. The use of thiophene (sulfur additive) has been shown to increase the yield of CNTs. The temperature and the hydrogen flow rates (which also had a positive influence on CNT yield) were optimized in this technique to create a continuous process in which large portions of long CNTs were formed and assembled into macroscopic yarns. Without hydrogen flow, the yield of nanotube growth rapidly decreased and no long yarns were produced.

Since the CNT yarns were of macroscopic lengths and could be manipulated quite easily, direct tensile tests were performed on individual yarns to provide lower bound estimates for the mechanical properties (e.g. modulus), suggesting the robustness of these macroscopically long nanotube assemblies. These very long nanotube yarns were handled and manipulated easily. The Young's modulus estimates for these structures from the direct tensile tests fell short of values expected for individual nanotubes or small nanotube bundles. But the numbers obtained for modulus were lower bound estimates due to the uncertainty in knowing the exact cross-sectional area of the yarns supporting the load. The yarns were not monolithic structures and consisted of parallel nanotube ropes separated by interstitial space. If one considered only the cross-sectional area supporting the load (less than 48%) during the tensile test, the modulus values for the yarns jumped from 49–77 GPa to ~100–150 GPa, consistent with the modulus of large CNT bundles. Although an individual CNT has elastic modulus of about 1 TPa, the value could fall to ~100 GPa for nanotube bundles, due to the internanotube defects (for example, imperfect lattice of nanotube bundles due to different nanotube diameters) present along the bundles.

The long nanotube yarn created by the direct synthesis technique was certainly a good alternative to the fibers and filaments spun from nanotube slurries. The mechanical and electrical properties of these yarns were indeed superior to those of these fibers. The yarns could be produced in high yield and continuously. The thickness of the yarns and their length could be further optimized, by tuning the processing conditions, to produce practically useful nanotube-based macroscale cables and fibers. For example, when increasing the addition of thiophene and the introduction rate of the solution, thicker yarns (1 mm) consisting of densely bundled structures (super bundles) were formed in the as-grown materials with an average length of 20 cm [15]. The superbundles exhibited good alignment and polygonization (Fig. 12.3(d)).

Based on this method, in late 2004, Y.L. Li et al. reported a modified floating technique to directly spin the CNT fibers formed in CVD [12], which was more efficient and economical for commercial application. Two wind-up devices with different geometry were placed at different positions of the furnace to produce CNT yarns in large scale in one step. Figure 12.4(a) shows the schematic diagram of the CVD apparatus and the formation of CNT fibers. The best conductivity of as-produced yarns was 8.3×10^5 S/m, while the strength was within the range of typical carbon fibers. The strain to failure could exceed 100% on initial loading.

In the continuous spinning process, sulfur was also identified as segregating to the surface of the iron catalyst particles and the interface between the particle and the nanotube. The nanotubes produced by the process were exceptionally long, and their growth to this length over a few seconds was associated with the promoting effect of added sulfur. If the sulfur was made available soon after the ferrocene was first cracked to nucleate the floating iron catalyst particles, then it retarded the growth of the particles so that they nucleated SWNTs, which were then extracted as continuous ~10-μm-diameter fiber [16]. Additionally, the fact that armchair (and thus metallic) nanotubes were dominant in the fiber was surprising and added further piquancy to their potential applicability. It retarded the growth of iron catalyst size by collision, so that the particles were still of appropriate dimensions for SWNTs at the point where the carbon became available for nanotube growth. It was also possible that it played a key role in enabling the motion of Stone-Thrower-Wales dislocations under the driving force of curvature energy minimization at the temperature of the process.

It was proved that a shrinking process would occur when as-grown CNTs were wetted in an organic solution like acetone, which was extremely advantageous to enhance the strength of the obtained CNT yarn [4]. Based on this, CNT yarns were continuously spun to a length of several kilometers (Fig. 12.4(b)) [16]. The specific strength of as-spun fibers was 0.4–1.25 GPa and the electrical conductivity

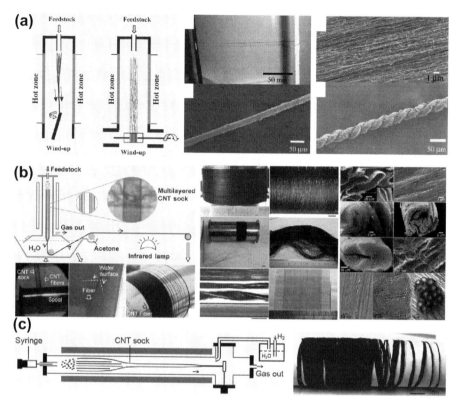

FIGURE 12.4

(a–c) Direct spinning of CNT yarns [12,16,17]. (For color version of this figure, the reader is referred to the online version of this book.)

reached 5.0×10^5 S/m. The specific surface area was measured by the Brunauer–Emmett–Teller method and the result (194 m^2/g) was in the range of CNT films or bucky papers.

Furthermore, the CVD spinning apparatus was modified to be made of a horizontal furnace and an on-site water vapor densification system, using which "cotton"-like yarns were synthesized directly (Fig. 12.4(c)) [17]. One of the advantages of using the horizontal furnace was that the growth of CNTs will be more stable and reliable because of the relative low convection in the reactor.

The typical parameters of preparation and the characteristics of as-spun CNT fibers are summarized in Table 12.1.

Future efforts will focus on the adaptation of CNT yarn spinning techniques, currently under development in laboratory facilities, to mass production at the industrial level. Increasing emphasis will be placed on analysis and test that provide insights on how various parameters, such as twist angle and fiber size, can influence the mechanical properties of CNT fibers. CNT production methods, based on CVD,

Table 12.1 Typical Parameters to Synthesize CNT Fibers by CVD

No	Ref.	Precursor	Reaction Temperature	Catalyst	CNT Formation	Yarn Strength	Resistivity (10^{-6} $\Omega \cdot m$)
1	[11]	n-Hexane	1423 K	Ferrocene/thiophene	SW: $d = 1.1$–1.7 nm	1.2 GPa av 49–77 GPa (E)	5–7
2	[12]	Ethanol	1323–1473 K	Ferrocene/thiophene	SW: $d = 1.1$–1.7 nm, MW: $d = 30$ nm	0.1–1 GPa	1.2
3	[3]	Ethanol, ethylene glycol, hexane	1453 K	Ferrocene/thiophene	SW/MW	0.14–1.46 GPa	–
4	[4]	Ethanol	1323–1473 K	Ferrocene/thiophene	DW: $d = 4$–10 nm	1.3–8.8 GPa	–
5	[18]	Ethanol	1573 K	Ferrocene/thiophene	DW: $d = 5$–10 nm	1.2–2.2 GPa	–
6	[16]	Ethanol/acetone	1443 K	Ferrocene/thiophene	DW: $d = 8$–10 nm	0.4–1.25 GPa	2.0
7	[19]	Ethanol	1423 K	Ferrocene/thiophene	DW: $d = 6$ nm	74 MPa	5555.6

SW, single walled; DW, double walled; MW, multi walled.

are to be evaluated and CVD reactors optimized, so that production output may reach industrial levels. CNT yarn synthesis and spinning techniques are to be modified so that vertical integration into the process chains of carbon fiber production facilities is straightforward and with a minimum cost and risk. Design and optimization of CNT yarn production equipment will be based on computer modeling (e.g. computational fluid dynamics for the CVD reactor) and demonstration prototypes. Computer modeling, based on molecular theory and molecular mechanics/dynamics, validated by experimental measurements, is to be used for CNT characterization regarding physical, mechanical and tribological properties and for investigating interactions between CNTs in a bundle. Based on these, equivalent macroscale models of CNTs will be developed for simulating yarn behavior with an acceptable computational cost.

12.3 GROWTH OF HIGH-QUALITY CNTS

The use of well-defined nanostructured materials, such as CNTs, for the precursor of the spun yarn/fiber, will require reliable synthesis methods [20]. It should also be noted that a CNT yarn is not only a structure but also a material, and as such each new variation of yarn design will produce a different set of properties. The ability to control the structure and arrangement of CNTs is a prerequisite to realizing many practical applications. Newly developed approaches for high-quality CNT synthesis and fast and precise structural characterizations enable us to fulfill their potential.

Atomic configurations of individual CNTs have been obtained by high-resolution transmission electron microscopy (HRTEM) with atomic sensitivity [21]. A structural reconstruction was carried out by Fourier-filtered analysis of Moiré patterns, and it was possible to acquire the carbon honeycomb lattice images through the entire periphery of individual nanotubes [22] (Fig. 12.5(a)). This visualization technique provided supplementary access in nanoscale characterizations by combining with other techniques.

The atomic structures of nucleation points for CNTs could also be imaged with atomic resolution to reveal the effects of SWNT–catalyst interactions on SWNT growth [23]. A novel and rational approach to visualize the nucleation points of CNTs on the atomic scale was proposed. The atomic structure of both the catalysts and the tubes growing out were observed by HRTEM (Fig. 12.5(b)). The growth pathway extracted from the transmission electron microscopic image sequence demonstrated a root mechanism in SWNT growth. This result represented a significant first step toward understanding, and ultimately controlling, the chirality of a nanotube.

Based on this observation, a strategy to control the chirality of SWNTs was proposed [24]. Atomic-resolution imaging of nucleation points made it possible to explore the structural correlation between the catalysts and the corresponding nanotubes. It was found that the angle of the step edge on the (111) plane of the catalytic particle with respect to the nanotube growth direction was the key factor in determining the tube chirality (Fig. 12.5(c)). The result showed the potential in promoting

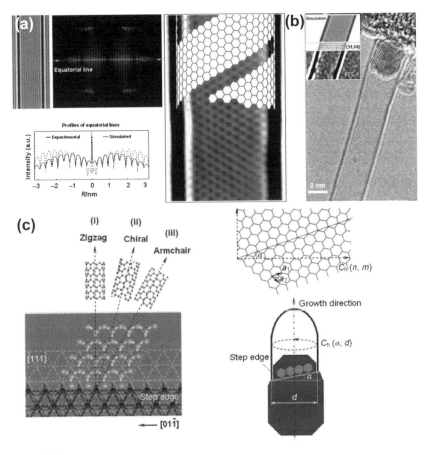

FIGURE 12.5

(a–c) Growth mechanism and chirality control of CNTs [22–24]. (For color version of this figure, the reader is referred to the online version of this book.)

some experimental advances in synthesis of nanotubes with the desired structure. Similar studies could be extended and used to obtain a more detailed understanding of the role of crystal structure in other catalyst systems.

12.4 APPLICATIONS OF CNT YARNS/FIBERS

CNT-based yarns and fibers have elicited interest for multifunctional applications such as strain sensing combined with structural reinforcement. Due to CNTs' unique properties, industrialization of CNT yarn production will open new markets for carbon fiber yarn and textile industries, with positive economic impact and new job openings.

One of the most important applications of long nanotube strands based on their properties will be as reinforcements in composite materials. Composite yarns comprised of CNT fibers and nanocrystalline metals were synthesized to improve the conductivity of CNT yarns [25]. The conductivity of composite yarns reached $2-3 \times 10^5$ S/cm, compared to the 500 S/cm of CNT yarns. However, the tensile strength decreased to less than half of the strength of CNT yarns. In fact, in most cases, composite was considered as an important method to improve the strength of CNT yarn. W.H. Guo et al. proposed a CNT/polymer composite fiber with both increase of the tensile strength and electrical conductivity [26]. Poly(vinyl alcohol) and poly(amic acid) were infiltrated to CNT yarn to increase their strength, and after a curing process, an additional increase about 300 MPa could be obtained [27–30].

The ultralong CNT yarns are quite stiff and exceptionally strong and could sustain large strains in tension without showing signs of fracture [31]. In other directions, nanotubes were highly flexible (Fig. 12.6(a)). They have a high Young's modulus of ~150 GPa (specific modulus: 188 GPa/SG) and tensile strength of 2.4 GPa (specific strength: 3.0 GPa/SG) based on the nanocantilever beam and direct tensile tests. The average Young's modulus of 1D nanotube–polymer composite was 17% higher than the theoretical value, with an elastic elongation of ~7%. The main advantage of using long CNT yarns for structural polymer composites (Fig. 12.6(b)) was that nanotube reinforcements would increase the toughness of the composites by absorbing energy during their highly flexible elastic behavior. Other advantages were low density of the nanotubes, increased electrical conduction and better performance during compressive load. Another possibility, which was an example of a nonstructural application, is filling of photoactive polymers with nanotubes. These composites show a large increase in conductivity with only a little loss in photoluminescence and electroluminescence yields. Nanotube–polymer composites could also be used in the biochemical field as membranes for molecular separations. There are also a multitude of foreseen consumer products: light-weight, multi-functional, highly customized and personalized, durable and efficient clothing and apparel with antiballistic, antiradiation, antiodor,

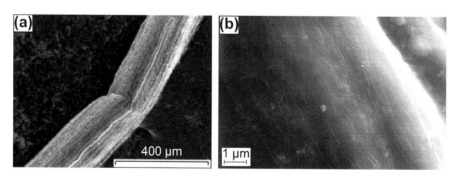

FIGURE 12.6

(a–b) Applications of CNTs in composite fibers.

antimicrobial, heat, fire- and wrinkle-resistant properties; textiles acting as electronic components capable of storing energy, emitting radiation, transmitting signals, sensing biometrics and even moving as artificial muscles [32]; sensors in construction materials capable of sensing cracks or temperature changes; and lightweight wiring in aerospace applications.

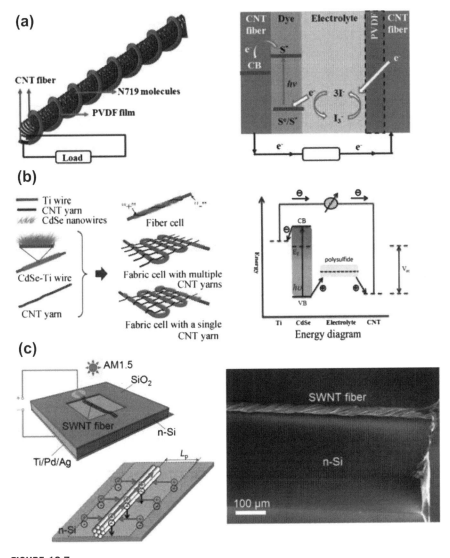

FIGURE 12.7

(a–c) Applications of CNTs in photovoltaics [33,36,37]. CB, conduction band; VB, valence band. (For color version of this figure, the reader is referred to the online version of this book.)

CNT yarns/fibers have been widely used in large-scale fields such as photovoltaic devices, mechanical and biochemical sensors, field emission devices, composite materials and so on. For example, CNT-fiber-electrode-based dye-sensitized photovoltaic wire was recently reported, as shown in Fig. 12.7(a) [33]. Two CNT fibers were twisted together to form a photovoltaic wire. The inner one absorbed with dye (N719) was used as the working electrode and another coated with poly(Vinylidene Fluoride) film was used as the counter electrode. With the introduction of aligned titanium dioxide nanoparticles in the photovoltaic wire, the power conversion efficiency could reach up to 3.9% [34,35]. Some nanowires, e.g. CdSe/Ti wires, when combined with CNT fibers, can be used as electrodes to fabricate solar cells (Fig. 12.7(b)) [36]. Y. Jia et al. also fabricated solar cells using CNT fibers as hole collectors, as shown in Fig. 12.7(c) [37]. The highest efficiency (nominal) of the tested samples is 10.6%, indicating the promising prospect of applying CNT yarns in the energy conversion field.

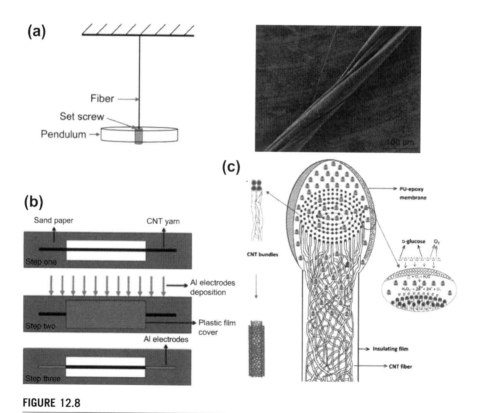

FIGURE 12.8

(a–c) Applications of CNTs in sensors [38–40]. (For color version of this figure, the reader is referred to the online version of this book.)

Due to the excellent strength of CNT yarns, a mechanical sensor was considered as yet another important application. A torsion sensor was designed based on the increase in electrical contact between CNTs (see Fig. 12.8(a)) [38]. The result showed that shear strain of over 24% could be obtained for neat fibers without permanent changes in electrical conductivity. An earlier resistance—strain sensor was reported in 2010, as shown in Fig. 12.8(b) [39]. Al electrode was deposited on the CNT fiber and a tensile strain was applied on it; then an electrical resistance response was given by the sensor. CNT yarn was also investigated and actively used in other sensors. An electrochemical biosensor was designed based on CNT fibers with better performance in sensitivity, linear detection range and linearity than the traditional Pt-Ir coil electrode (Fig. 12.8(c)) [40]. Furthermore, the geometry of the CNT fiber electrode is much smaller and the anodic peak current is much larger than the one obtained by Pt-Ir electrode. In a field electron emitter application [41], continuous CNT yarn was cut into several segments to obtain open-ended cross-sections. The field emitter was of great stability with continuous working over 7 h without emission current degradation. A thermionic cathode was further designed via functionalizing the yarns with barium [42]. Barium doping reduced the work function of CNT yarns greatly, and the emission current density of 185 mA/cm^2 was obtained under the electric field of 850 V/5 mm at 1317 K. The field emission properties of hafnium carbide-decorated CNT yarns were also studied [43]. The emissive material should have a low threshold emission field and large stability at high current density. Furthermore, an ideal emitter was required to have a nanometer-size diameter, a structural integrity, a high electrical conductivity, a small energy spread and a large chemical stability. Long CNT yarns possessed all these properties.

12.5 CONCLUSIONS

In summary, the main methods reported thus far to directly synthesize long CNT yarns by CVD and their potential applications were reviewed. All the direct approaches to grow CNT yarns are based on CVD and are applicable to commercial applications. High strength is the remarkable advantage of CNT yarns compared with traditional CNTs. Devices based on CNT yarns will not only have the benefits that CNTs bring to us, such as low cost and low density, but also will be strong enough to work in a harsh environment robustly and stably. Therefore, CNT yarns will be promising materials in the future and more and more applications will be developed. The superior properties and different material forms will open up new parameter spaces and allow designers to develop revolutionary engineering designs across many industry segments.

Acknowledgments

This work is supported by the National Science Foundation of China (50972067) and the Beijing Natural Science Foundation (2122027).

References

[1] R.H. Baughman, A.A. Zakhidov, W.A. de Heer, Carbon nanotubes — the route toward applications, Science 297 (5582) (2002) 787–792.

[2] J.J. Vilatela, A.H. Windle, A multifunctional yarn made of carbon nanotubes, Journal of Engineered Fibers and Fabrics 7 (2012) 23–28. (Special issue).

[3] M. Motta, et al., Mechanical properties of continuously spun fibers of carbon nanotubes, Nano Letters 5 (8) (2005) 1529–1533.

[4] K. Koziol, et al., High-performance carbon nanotube fiber, Science 318 (5858) (2007) 1892–1895.

[5] K.L. Stano, et al., Direct spinning of carbon nanotube fibres from liquid feedstock, International Journal of Material Forming 1 (2008) 59.

[6] P. Liu, et al., Thermionic emission and work function of multiwalled carbon nanotube yarns, Physical Review B 73 (2006) 235412.

[7] Q.W. Li, et al., Structure-dependent electrical properties of carbon nanotube fibers, Advanced Materials 19 (20) (2007) 3358–3363.

[8] L.M. Ericson, et al., Macroscopic, neat, single-walled carbon nanotube fibers, Science 305 (2004) 1447–1450.

[9] B. Vigolo, et al., Macroscopic fibers and ribbons of oriented carbon nanotubes, Science 290 (2000) 1331–1334.

[10] M. Zhang, K.R. Atkinson, R.H. Baughman, Multifunctional carbon nanotube yarns by downsizing an ancient technology, Science 306 (2004) 1356–1361.

[11] H.W. Zhu, et al., Direct synthesis of long single-walled carbon nanotube strands, Science 296 (5569) (2002) 884–886.

[12] Y.L. Li, I.A. Kinloch, A.H. Windle, Direct spinning of carbon nanotube fibers from chemical vapor deposition synthesis, Science 304 (5668) (2004) 276–278.

[13] Direct Synthesis of Long Single-Walled Carbon Nanotube Strands. US 7615204, 10.11.2009.

[14] B.Q. Wei, R. Vajtai, Y.Y. Choi, P.M. Ajayan, H.W. Zhu, C.L. Xu, D.H. Wu, Structural characterizations of long single-walled carbon nanotube strands, Nano Letters 2 (10) (2002) 1105–1107.

[15] H.W. Zhu, B. Jiang, C.L. Xu, D.H. Wu, Long super-bundles of single-walled carbon nanotubes, Chemical Communications 17 (2002) 1858–1859.

[16] Rajyashree M. Sundaram, Krzysztof K.K. Koziol, Alan H Windle, Continuous direct spinning of fibers of single-walled carbon nanotubes with metallic chirality, Advanced Materials 23 (43) (2011) 5064–5068.

[17] X.H. Zhong, et al., Continuous multilayered carbon nanotube yarns, Advanced Materials 22 (6) (2010) 692–696.

[18] M. Motta, et al., High performance fibres from 'dog bone' carbon nanotubes, Advanced Materials 19 (21) (2007), 3721–3726.

[19] Xiao-Hua Zhong, et al., Fabrication of a multifunctional carbon nanotube "cotton" yarn by the direct chemical vapor deposition spinning process, Nanoscale 4 (2012) 5614–5618.

[20] H.W. Zhu, B.Q. Wei, Macrostructures of carbon nanotubes. Encyclopedia of nanosci, Nanotechnology 16 (2011) 33–53. (American Scientific Publishers, Edited by Hari Singh Nalwa).

[21] H.W. Zhu, K. Suenaga, A. Hashimoto, K. Urita, S. Iijima, Structural identification of single & double-walled carbon nanotubes by high-resolution transmission electron microscopy, Chemical Physics Letters 412 (1–3) (2005) 116–120.

[22] H.W. Zhu, K. Suenaga, J.Q. Wei, K.L. Wang, D.H. Wu, Atom-resolved imaging of carbon hexagons of carbon nanotubes, Journal of Physical Chemistry C 112 (30) (2008) 11098–11101.

[23] H.W. Zhu, K. Suenaga, K. Mizuno, A. Hashimoto, K. Urita, K. Hata, S. Iijima, Atomic-resolution imaging of the nucleation points of single-walled carbon nanotubes, Small 1 (12) (2005) 1180–1183.

[24] H.W. Zhu, K. Suenaga, J.Q. Wei, K.L. Wang, D.H. Wu, A strategy to control the chirality of single-walled carbon nanotubes, Journal of Crystal Growth 310 (24) (2008) 5473–5476.

[25] L.K. Randeniya, et al., Composite yarns of multiwalled carbon nanotubes with metallic electrical conductivity, Small 6 (16) (2010) 1806–1811.

[26] W.H. Guo, et al., Aligned carbon nanotube/polymer composite fibers with improved mechanical strength and electrical conductivity, Journal of Materials Chemistry 22 (3) (2012) 903–908.

[27] C. Fang, et al., Enhanced carbon nanotube fibers by polyimide, Applied Physics Letters 97 (2010) (18190618).

[28] M. Kulkarni, et al., Elastic response of a carbon nanotube fiber reinforced polymeric composite: a numerical and experimental study, Composites Part B-Engineering 41 (5) (2010) 414–421.

[29] S. Wu Amanda, et al., Carbon nanotube fibers for advanced composites, Materials Today 15 (2012) 302–310.

[30] T. Chen, et al., Nitrogen-doped carbon nanotube composite fiber with a core-sheath structure for novel electrodes, Advanced Materials 23 (40) (2011) 4620–4625.

[31] H W. Zhu, Synthesis and Characterizations of Single-Walled Carbon Nanotube Macrostructures. Ph. D thesis, Tsinghua University, 2003.

[32] A.E. Aliev, et al., Giant-stroke, super elastic carbon nanotube aerogel muscles, Science 323 (5921) (2009) 1575–1578.

[33] F.J. Cai, T. Chen, H.S. Peng, All carbon nanotube fiber electrode-based dye-sensitized photovoltaic wire, Journal of Materials Chemistry 22 (30) (2012) 14856–14860.

[34] T. Chen, et al., Intertwined aligned carbon nanotube fiber based dye-sensitized solar cells, Nano Letters 12 (5) (2012) 2568–2572.

[35] Sen Zhang, et al., Porous, platinum nanoparticle-adsorbed carbon nanotube yarns for efficient fiber solar cells, 6 (8) (2012) 7191–7198.

[36] Luhui Zhang, et al., Fiber and fabric solar cells by directly weaving carbon nanotube yarns with CdSe nanowire-based electrodes, Nanoscale 4 (2012) 4954–4959.

[37] Yi Jia, et al., Strong, conductive carbon nanotube fibers as efficient hole collectors, Nanoscale Research Letters 7 (2012) 137.

[38] A.S. Wu, et al., Carbon nanotube fibers as torsion sensors, Applied Physics Letters 100 (2012) (20190820).

[39] H.B. Zhao, et al., Carbon nanotube yarn strain sensors, Nanotechnology 21 (2010) (30550230).

[40] Z.G. Zhu, et al., Nano-yarn carbon nanotube fiber based enzymatic glucose biosensor, Nanotechnology 21 (2010) (16550116).

[41] Y. Wei, et al., Efficient fabrication of field electron emitters from the multiwalled carbon nanotube yarns, Applied Physics Letters 89 (2006) (0631016).
[42] L. Xiao, et al., Barium-functionalized multiwalled carbon nanotube yarns as low-work-function thermionic cathodes, Applied Physics Letters 92 (2008) (15310815).
[43] Y.C. Yang, et al., In situ fabrication of HfC-decorated carbon nanotube yarns and their field-emission properties, Carbon 48 (2) (2010) 531–537.

CHAPTER 13

Carbon Nanotube Sheet: Processing, Characterization and Applications

Rachit Malik[1], Noe Alvarez[1], Mark Haase[1], Brad Ruff[2], Yi Song[2], Bolaji Suberu[2], Duke Shereen[3], David Mast[3], Andrew Gilpin[1], Mark Schulz[2], Vesselin Shanov[1]

[1] *Department of Chemical & Materials Engineering, University of Cincinnati, Cincinnati, OH USA 45221,* [2] *Department of Mechanical Engineering, University of Cincinnati, Cincinnati, OH USA 45221,* [3] *Department of Physics, University of Cincinnati, Cincinnati, OH USA 45221*

CHAPTER OUTLINE

13.1 Introduction	350
13.2 Two-Dimensional Films, "Buckypapers" and Sheets of Carbon Nanotubes	351
13.2.1 "Buckypaper" and thin films with limited/random nanotube alignment	351
13.2.2 Dry CNT sheets pulled from aligned arrays of MWNTs grown via CVD	352
13.2.3 Alternative techniques for CNT sheet production	353
13.3 Functionalization and Characterization of CNT Sheets	354
13.3.1 Techniques for functionalization of CNTs	355
13.3.1.1 Wet/solution functionalization	355
13.3.1.2 Plasma (dry) functionalization	356
13.4 CNT Sheet Products Manufacturing	361
13.4.1 CNT sheet manufacturing	361
13.4.2 Atmospheric pressure plasma functionalization	362
13.4.3 Characterization	363
13.4.3.1 Raman spectroscopy	363
13.4.3.2 X-ray photoelectron spectroscopy	367
13.4.3.3 FTIR Analysis	369
13.4.3.4 Contact angle testing	371
13.4.3.5 Optical characterization	371
13.4.4 Applications of CNT sheet	373
13.4.4.1 Electromagnetic interference shielding	373
13.4.4.2 Composites	377
13.5 Conclusions and Future Work	382
Acknowledgments	382
References	382

13.1 INTRODUCTION

Carbon nanotubes (CNTs) have come a long way since their discovery in 1991 [1]. The arc discharge evaporation method initially used to synthesize nanotubes has given way to large-scale production by chemical vapor deposition (CVD) [2]. The individual nanoscale tubules of carbon have exceptional mechanical strength, high Young's modulus [3], and superior specific conductivities [4]. The exact magnitude of these properties depends on the diameter and chirality of the nanotubes and whether they are single-walled, double-walled or multiwalled. During growth, depending on the conditions in which they are formed, nanotubes assemble either as double-walled/multiwalled coaxial tubules (DWNTs/MWNTs) or as bundles (ropes) consisting of individual tubes (single-walled nanotubes, SWNTs) packed in two-dimensional (2D) triangular lattices [5]. Table 13.1 shows the typical properties observed for individual CNTs by researchers over the years [1–6].

CNTs have been extensively employed in many applications ranging from electrodes for fuel cells [7], batteries [8], supercapacitors [9] and solar cells [10] to substrates for biological cell growth [11,12] and materials for electromagnetic interference (EMI) shielding [13]. Thin-CNT films and sheets have found many innovative applications that require 2D structures. Transparent and conducting thin films for touch screens [14,15], polarized light emitters [16], loudspeakers [17], and high-performance energy storage devices [18] have been demonstrated. Thin-film polymer composites with CNT sheets sandwiched between two layers of polymer have shown unique properties, such as optical transparency [19]. Strong and conductive CNT sheets have also been produced and studied [20]; however, there are still many interesting applications to explore and challenges to overcome.

Individual CNTs have demonstrated exceptional mechanical strength and superior electrical conductivities, but extrapolating these properties to macroscopic structures of CNTs has been a challenge for researchers all around the world. The efforts have been focused on bridging the gap between properties of these structures and those of individual nanotubes. The CNT dispersion approach is widely used in the industrial production of composite films and fibers/yarns, including in solution processing, melt processing, electrospinning, and coagulation spinning [21,22].

Table 13.1 Properties of Carbon Nanotubes [6]

Property	SWNT	DWNT	MWNT
Tensile Strength (GPa)	13–53	23–63	10–60
Elastic Modulus (TPa)	~1	0.2–1.	0.3–1
Density (g/cm^3)	1.3–1.5	1.5	1.8–2.0
Electrical Conductivity (S/m)	>10^6		
Thermal Stability (°C)	>600 (in air before oxidation)		
Typical Diameter (nm)	1	~5	~20
Specific Surface Area (m^2/g)	400		

However, the dispersion of long CNTs is hindered by their entanglement and aggregation, and the CNTs are often presented with a low fraction (<10 wt.%) and random orientation. Consequently, the reported strengths of the composites processed by this approach are usually low [23,24]. Alignment plays a crucial role in determining the properties of the final structure. CNTs have a very high aspect ratio and thus alignment of tubes in one direction results in anisotropy, which leads to significant improvement of properties in that direction [25,26]. This discovery regarding the anisotropy and a high aspect ratio (typically 1,000,000:1 [27]) has fostered the interest in manufacturing of aligned 2D films and sheets made of CNTs. This chapter highlights the developments in manufacturing CNT films over the years along with an insight into current techniques and applications of CNT films, "buckypapers" and sheets alike.

13.2 TWO-DIMENSIONAL FILMS, "BUCKYPAPERS" AND SHEETS OF CARBON NANOTUBES

13.2.1 "Buckypaper" and thin films with limited/random nanotube alignment

The first technique to prepare free-standing films of nanotubes involved production of powdered SWNTs and MWNTs via the arc discharge approach. They were then dispersed in a polymer resin which was then allowed to set. Later, the films of the polymer were cut and nanotubes randomly aligned in the polymer matrix were observed [28]. Deheer et al. [29] built upon the idea of dispersing nanotube powders and formed a film of nanotubes by filtering them on a ceramic filter, thus opened a pathway for researchers around the world to follow and explore properties of CNTs on a macroscopic level. SWNT films were made by Bonard et al. [30] to study their field emission properties and MWNTs were employed by Lourie et al. [31] in an epoxy-based matrix to study the buckling and collapse of the embedded nanotubes. Jin et al. [32] reported an attempt to align randomly dispersed nanotubes in a thermoplastic polymer matrix by mechanical stretching along with heating the composite. Tang and group [33] also performed a similar experiment to induce alignment among the nanotubes via mechanical shearing. CNT actuators made from SWNTs were reported by Baughman et al. [34] and were shown to generate higher stresses than natural muscle and higher strains than high-modulus ferroelectrics. Since then, the dispersion of CNTs in surfactants and polymer matrices has become a routine activity and scientists have not limited themselves to the preparation of 2D structures, but also three-dimensional structures such as yarns [35], coaxial nanofibers [36] (with core and sheath-type structure) and mats [37] through techniques such as electrospinning and others used by the textile industry. Recently, SWNTs dispersed in water with the help of surfactant have been coated onto cotton paper electrodes and a flexible solid-state supercapacitor [38] has been produced. There is a vast amount of literature available on the above-discussed topic; hence, it is impossible to cite all the efforts made in the field. The authors have tried to encompass the boundaries and would like to direct interested readers toward comprehensive reviews compiled on the use of CNTs in polymer composites and a plethora of other potential applications [6,21,39–43].

13.2.2 Dry CNT sheets pulled from aligned arrays of MWNTs grown via CVD

Since its first implementation for CNT growth in 1999 [44], CVD has become the most favored method for tube synthesis in large quantities. Significant advances in improving CVD [45] have been made to produce tailor-made nanotubes. In 2002, Jiang et al. [46] made the remarkable discovery of spinning of continuous yarns of CNTs from superaligned arrays produced by CVD. This innovation has formed the cornerstone for current and future CNT research and, in the authors' view, could be the most significant finding in the field since the discovery of nanotubes. The production of superaligned arrays via CVD has been the key toward fostering research into spinning CNTs. Liu et al. [2] demonstrated controlled growth of spinnable arrays and its co-relation with the thickness of catalyst film. Yun et al. [47] demonstrated a modified water-assisted CVD process to produce long aligned MWNT arrays. The process described by Jiang et al. has been employed in principle to produce yarns and sheets of nanotubes. Since the focus of this chapter is CNT sheets, the results presented shall be pertaining to dry-spun CNT sheets. Zhang et al. [16] improvized on Jiang's method and instead of twisting, the ribbons pulled from arrays. They laid them flat to produce highly aligned free-standing sheets of multiwalled CNTs.

Building on previous work by Zhang et al. [16], researchers at the University of Cincinnati (UC) [48] envisioned production of multilayered sheets of MWNTs by laying multiple layers of pulled ribbon on top of itself. The details of this approach are discussed in later sections. Sheets produced by this technique have demonstrated superior mechanical strength as reported by Di et al. [20] who claim the strongest pure CNT sheets with specific tensile strength higher than that of high-strength steel and comparable to that of rigid carbon fiber laminates (~ 2 GPa). These sheets are much stronger than CNT films and buckypapers (0.1 GPa) produced via dispersion and filtration of nanotubes. This solid-state process of making CNT sheets is suited for future industrial scale production of nanotube materials. Along with high strength, these sheets also possess high electrical conductivity as shown by Zhang et al. [49] in their publication on the manufacturing of stretchable and transparent conductors with sheets of nanotubes embedded in a polydimethylsiloxane matrix. The CNT sheets reveal superior thermal properties as reported by Aliev et al. [50] and Poehls et al. [51]. The development of the CNT sheet has led to a deluge of literature studies based around it, along with a plethora of applications such as in incandescent displays [52], anodes for Li-ion battery [53], banner-type sound wave generator [54] and transparent electrodes for liquid-crystal displays [55] demonstrating commercial applicability.

13.2 Two-Dimensional Films, "Buckypapers" and Sheets of Carbon Nanotubes

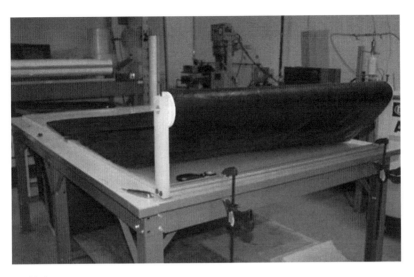

FIGURE 13.1

A long roll of carbon nanotube sheet, 1.2 m wide with a density of 15 g/cm^2. (For color version of this figure, the reader is referred to the online version of this book.)

Source: Reprinted from Ref. [89]. Copyright (2010) with permission from Materials Research Society.

13.2.3 Alternative techniques for CNT sheet production

This section discusses some proprietary and new techniques recently developed to mass produce CNT sheets. Lashmore et al. [56] at Nanocomp Technologies have developed a novel method to mass produce CNT sheets and threads. Iron particles with diameters equal to or slightly larger than nanotube diameter are delivered as a floating catalyst and utilized to carry out gas-phase catalytic decomposition of ethylene at high temperature. The process results in a cloud-type formation of CNTs which thickens over time and can be continuously drawn out in the form of large sheets and twisted to produce long threads. It is claimed that the process can be tuned to produce nanotubes of desired specifications such as diameter and number of walls. Though this process results in mass production of CNT sheets, the degree of alignment of the nanotubes within the sheet structure is still an area of research. Figure 13.1 shows an image of the large CNT sheets produced at Nanocomp Technologies employing the above described technique.

Wang et al. [57] describe a simple technique to produce CNT buckypapers with aligned nanotubes. This technique involves the growth of nanotubes on a silicon substrate like the previous techniques via CVD, but instead of drawing ribbons, the arrays are pushed flat with the help of a cylinder to produce aligned thick buckypapers. Figure 13.2 shows the illustration of the technique involving a microporous membrane to allow easy peel-off of the buckypapers.

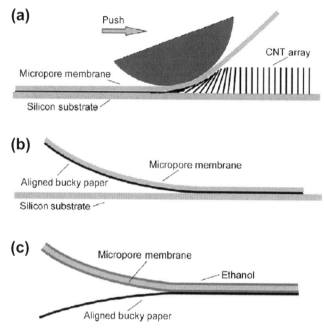

FIGURE 13.2

Schematics of the domino pushing method: (a) forming aligned buckypaper; (b) peel the buckypaper off from the silicon substrate; (c) peel the buckypaper off from the microporous membrane.

Source: Reprinted from Ref. [57]. Copyright (2008), with permission from IOP Publishing (free article).

13.3 FUNCTIONALIZATION AND CHARACTERIZATION OF CNT SHEETS

CNTs demonstrate high strength owing to a very stable structure consisting of a network of sp^2 bonded carbon atoms. This network of carbon–carbon double bonds lends stability to the nanotube, and also acts as a constraint toward the realization of many potential applications of CNTs, where the nanotubes need to associate themselves with other chemicals and materials. This brings the need for functionalization of CNTs.

An example of an application citing the requirement for functionalization of nanotubes is that in polymer composites. The challenge in developing polymer/CNT nanocomposites is to find ways to create macroscopic composites that benefit from the unique physical and mechanical properties of nanotubes within them. In order to achieve the desired results, it is crucial that there is a transfer of either mechanical load or electrical charge to individual nanotubes in the polymer composite. Efficient dispersion of individual nanotubes and/or the establishment of a strong chemical affinity (covalent or noncovalent) with the surrounding polymer matrix are essential to realize the above objective. Since pristine CNTs either agglomerate due to high Van der Waals forces between the tubes or do not interface chemically with the polymer matrix, they are not ideal for reinforcement of a polymer matrix. CNTs, because of their small size, make an excellent material of construction for microscale and nanoscale devices as was shown by Lin et al. [58] in the development of a glucose biosensor, and more recently by Chung et al. [59] in their engineering of a DNA biosensor with ionic liquid-multiwalled carbon nanotubes (IL-MWNTs) to detect *Salmonella typhi*. These findings and many more are perfect examples to demonstrate the need for functionalization/chemical modification of the surface of the nanotubes to tailor them as per the desired application.

13.3.1 Techniques for functionalization of CNTs

The two major techniques for functionalization of CNTs are covalent functionalization and noncovalent functionalization. The sp^2 hybridized π orbitals of CNTs provide the possibility for interactions with the long surfactant chains and also with conjugated polymer chains containing heteroatoms with free-electron pair [6]. Noncovalent CNT modification basically involves physical adsorption and/or wrapping of the long surfactant/polymer chains to the surface of the CNTs. The advantage of noncovalent functionalization is that it does not destroy the conjugated system of the CNT sidewalls, and therefore, it does not affect the final structural properties of the material.

Covalent functionalization involves chemical reaction where covalent bonds are formed between the CNTs and functional groups such as $-OH$ and $-COOH$. Carbon atoms in a CNT are sp^2 hybridized and hence can be reacted with concentrated acids, acid mixtures and peroxide solutions. In the case of covalent functionalization, the translational symmetry of CNTs is disrupted by changing sp^2 carbon atoms to sp^3 carbon atoms, and the properties of CNT, such as electronic and transport are influenced.

13.3.1.1 Wet/solution functionalization

Since the discovery of nanotubes, there has been immense interest in modifying nanotubes, and the majority of the study has pertained to functionalization of nanotubes in solution. There are several methods such as electrochemical [60] and chemical functionalization [61], fluorination [62,63], polymer wrapping [64], mechanochemical treatment [65] to modify the chemical composition of the CNTs dispersed in a solution by grafting functional groups to it. Recently, functionalized MWNTs were demonstrated to be used as contrast agents in the ultrasound technique [66]. The nanotubes were first oxidized and subsequently functionalized by 1,3-dipolar cycloaddition of azomethine ylides (ox-MWCNT-NH_3^+) to render them biocompatible and then were tested with ultrasound imagery by injection into a pig liver and heart, thereby successfully demonstrating another novel application of the special material we know as CNTs. Reports on new applications of specially functionalized nanotubes are published almost every month, thereby suggesting the wide applicability of CNTs as robust substrates and starting material for polymer growth as shown by Datsyuk et al. [67]. They carried out nitroxide-mediated polymerization of methyl methacrylate on surface of double-walled carbon nanotubes (DWNTs) with attached short polymer chains of hydrophobic polystyrene (PS) or hydrophilic poly(acrylic acid) via an in situ polymerization process. This form of functionalization has been extensively explored for powdered nanotubes but to a lesser extent for macrostructures made out of CNTs due to feasibility issues. Plus, the treatment of nanotube powders with strong acids, which is the first step for further functionalization, is shown to damage the CNT walls [68]. Furthermore, this method produces chemical waste and creates time and cost-efficiency issues which would become a major hindrance during scale-up.

13.3.1.2 Plasma (dry) functionalization

A technique which has recently gained popularity to modify nanotubes is treatment with plasma. The capability of functionalizing a large number of nanotubes in relatively shorter period of time can be cited as the main reasons for the growing interest in plasma; and the whole process being much more eco-friendly over acid treatment is a major advantage. Plasma has been utilized to provide a wide range of functional groups depending on the nature of the plasma gas and process parameters [69].

Low-pressure radio frequency (RF) plasma among other techniques has been shown to successfully bind oxygen [69], hydrogen [70], fluorine groups [71] and also deposit polymeric films [72] on CNT surface. Plasma treatment allows the user to exercise control over the amount of functional groups grafted at the nanotubes surface. This process can be tailored and saturation of functional groups on the surface can be prevented, thereby preserving the electronic structure and hence conductivity of nanotube material. There are a variety of techniques available to produce plasmas for CNT functionalization; hence, the scope of this section will be limited to low-pressure/atmospheric pressure glow discharge techniques due to the ease of use to treat CNT sheets and buckypapers.

Vohrer et al. [73] utilized glow discharge plasma treatments using N_2, O_2 and Ar gases to produce hydroxyl and carboxyl functional groups on SWNT and MWNT buckypapers. Figure 13.3 shows the X-ray photoelectron spectroscopy (XPS) analysis of N_2 plasma-treated MWNT buckypapers.

This group [73] carried out plasma treatments on MWNT and SWNT materials treated with nitrogen, N_2/O_2 (50:50) and oxygen plasma, respectively. The results of the XPS analysis to obtain the elemental compositions are given in Table 13.2. The same study reported a constant decrease in oxygen content with increasing storage time, which produced CNTs functionalized with oxygen and nitrogen plasmas and exhibiting improved hydrophilicity and biological cell adhesion proliferation.

Recently, an atmospheric pressure plasma jet was employed by Kolacyak et al. [74] using oxygen and nitrogen to carry out fast functionalization of MWNTs. The effectiveness of the plasma treatment was determined by characterizing the nanotubes using XPS and Fourier transform infrared (FTIR) spectroscopy.

13.3 Functionalization and Characterization of CNT Sheets 357

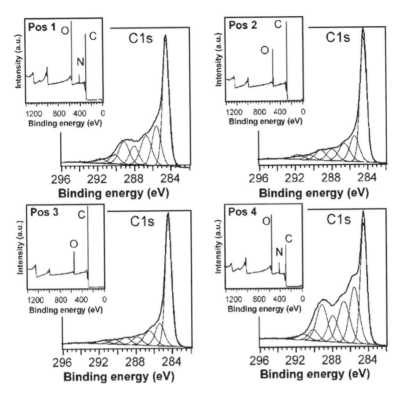

FIGURE 13.3

XPS survey scan and C1s spectra acquired at four different position of the same plasma-treated MWNT buckypaper. The plasma conditions were N_2, 10 W, and 10 min.

Source: Reprinted with permission from Ref. [73]. Copyright (2007), with permission from John Wiley and Sons.

Figure 13.4 shows the FTIR spectra of the nanotubes functionalized with oxygen and nitrogen.

The spectra show clear differences in the range of 1800–1000/cm compared to the untreated sample spectrum. This difference is attributed to carboxylic moieties with the two bands at ~1700/cm (−C═O) and 1350/cm (−C−O−) corresponding to the stretching of the carboxylic acid (−COOH) group. Plasma treatment induces structural changes in the structure, but FTIR itself is not conclusive and has to be supported with results from other techniques like Raman spectroscopy and XPS. Figure 13.5 presents the XPS survey spectra of the untreated and treated CNTs along with the elemental composition of the samples.

It was reported that oxygen plasma-functionalized nanotubes form stable dispersions in water. Figure 13.6 shows the stability of dispersions created with pristine nanotubes compared to those with treated nanotubes over time.

Table 13.2 Comparison of the Elemental Composition of SWNT and MWNT Material after Different Plasma Treatments with Increased Oxidation Potential and Two Different Treatment Times

	SWNT			MWNT		
	C	O	N	C	O	N
	Elemental Composition (atom %)					
Untreated	94.4	5.1	0.5	97.5	2.5	–
N_2 10 s	92.1	7.4	0.5	96.3	3.4	0.3
N_2 10 min	83.8	15.7	1.5	84.5	13.2	2.5
N_2/O_2 10 s	84.7	15.0	0.3	92.2	7.8	–
N_2/O_2 10 min	76.1	22.7	0.4	87.6	12.4	–
O_2 10 s	82.5	16.7	0.8	93.2	6.8	–
O_2 10 min	74.2	25.4	0.4	86.8	13.2	–

Source: Reprinted with permission from Ref. [73]. Copyright (2007), with permission from John Wiley and Sons.

Atmospheric pressure plasma presents an attractive alternative method to acidic treatment for surface functionalization of CNTs as a clean and easy method to use the process. The improved dispersibility is related to the intermolecular interaction of the functional groups of the CNTs with the water molecules. The plasma process allows fast functionalization of CNTs and can be operated continuously.

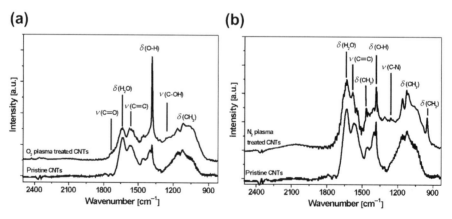

FIGURE 13.4

FTIR spectra (absorbance) of CNTs functionalized with (a) oxygen and (b) nitrogen plasma.
Source: Reprinted from Ref. [74]. Copyright (2011), with permission from Elsevier.

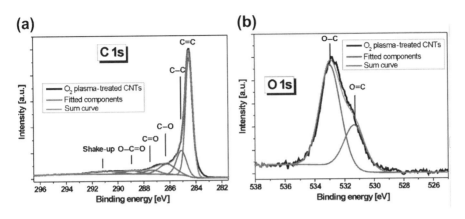

FIGURE 13.5

X-ray photoelectron spectra of CNTs after oxygen plasma treatment: (a) C(1s) detail scan, (b) O(1s) detail scan. (For color version of this figure, the reader is referred to the online version of this book.)

Source: Reprinted from Ref. [74]. Copyright (2011), with permission from Elsevier.

Naseh et al. did a comparative study on the effectiveness of plasma functionalization over the conventional chemical treatments to introduce carboxyl (−COOH) functionality to the nanotubes [75]. Using temperature programmed desorption (TPD) technique, they reported that acid-treated nanotubes have more functional groups produced on the surface, however, plasma makes the process shorter, cleaner

FIGURE 13.6

Sedimentation behavior of CNTs in water at pH 5. Images were taken 2 days after initial dispersion for (a) pristine MWCNTs and (b) oxygen plasma-treated CNTs. (For color version of this figure, the reader is referred to the online version of this book.)

Source: Reprinted from Ref. [74]. Copyright (2011), with permission from Elsevier.

and results in less damage to the CNT structure. Figure 13.7 shows the extent of damage to CNTs by plasma and acid treatments via SEM micrographs.

It is known that acid treatment under harsh conditions causes shortening of CNTs and nanotubes are shorter with the longer acid-treatment time [76]. Rosca and Hoa [77] reported that length of the nanotubes has an important influence on the conductivity of composites and in order to prepare a continuous and stable buckypaper, nanotubes should be adequately long. Therefore, plasma offers an attractive alternative to conventional acid functionalization techniques and establishes itself as the preferred technique for future industrial scale-up of CNT processing. Through the TPD technique, Naseh et al. [75] also studied how the plasma parameters affect the extent of oxygenated functional group formation by an air dielectric barrier discharge (DBD) plasma. The samples were heated in vacuum and the evolved gases from the decomposition/reaction of the functional groups present on the CNT surface were evaluated with an

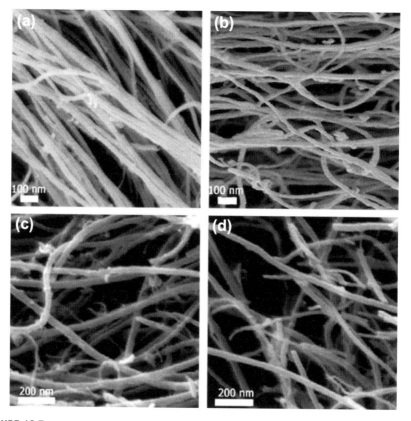

FIGURE 13.7

SEM micrograph of (a) as-prepared, (b) annealed, (c) plasma-functionalized and (d) acid-treated MWCNTs. Plasma conditions: power = 34.1 W and treatment time = 3 min.

Source: Reprinted from Ref. [75]. Copyright (2010) with permission from Elsevier.

FTIR attachment. They concluded that the extent of functionalization increases with increasing discharge power, provided the exposure time of the MWCNTs in the plasma atmosphere does not exceed a certain period of time. Further, plasma treatment using dry air produced functional groups with higher thermal stability (e.g. carbonyl and ether) compared to carboxyl groups being formed with acid treatment. Hence, we can establish that plasma is a useful tool to treat CNTs, especially in sheet and buckypaper form. The next section describes the progress made by UC in the production of CNT sheet, its functionalization with plasma and some applications.

13.4 CNT SHEET PRODUCTS MANUFACTURING
13.4.1 CNT sheet manufacturing

CNT sheets made at the UC are produced in a manner inspired from Zhang's [16] method building upon Jiang's innovation [46]. A recent report by Poehls et al. [51] describes the designing of a sheet machine manufactured at UC and employed to manufacture stronger, more handleable sheets of MWNTs. The setup used at UC to produce highly aligned CNT sheet was built in-house, drawing inspiration from the initial design of Jayasinghe and Shanov [48]. A thin teflon belt between two rollers was used as a substrate, on which the CNT ribbon was collected. When the teflon belt is rotated, a CNT ribbon is drawn from an array, and layers build up on top of each other. As the belt is rotated, the CNT array can be moved perpendicular to the drawing direction, in order to create a 2D sheet of highly aligned CNTs. The sheets are densified using volatile organic solvents like acetone and ethanol. Figure 13.8(A) shows a schematic of the sheet production along with a picture of the machine.

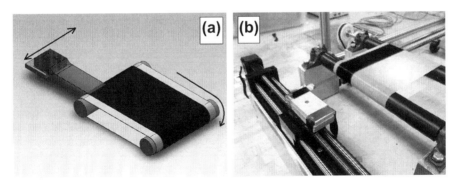

FIGURE 13.8

CNT sheet manufacturing process. (a) Schematic illustration of the setup. (b) CNT sheet preparation from vertically aligned CNT array [51]. (For color version of this figure, the reader is referred to the online version of this book.)

Table 13.3 The Effect of Solvent Densification on the Sheet Resistance of Layered CNT Sheet

No. of Layers	Sheet Resistance (Ω^{-2})	
	Undensified Sheet	Densified Sheet
1	7427	6869
2	4125	4603
4	2314	1370
10	865	471

The sheets consist of nanotubes aligned parallel to the drawing direction and thereby exhibit anisotropy with properties such as electrical resistance dependent on alignment of the nanotubes. Table 13.3 displays the effect of densification of the sheet on the electrical resistance value. Figure 13.9 shows a close-up optical image of the CNT sheet revealing its uniformity along with an SEM micrograph of the sheet illustrating a good degree of alignment of the CNTs within the sheet.

13.4.2 Atmospheric pressure plasma functionalization

Pristine CNT sheets were functionalized with atmospheric pressure plasma produced from a Surfx™ Atomflo 400-D reactor comprising oxygen and helium as the active and carrier gases, respectively. This type of plasma system has been previously employed to deposit glass coatings on aluminum [78] and other

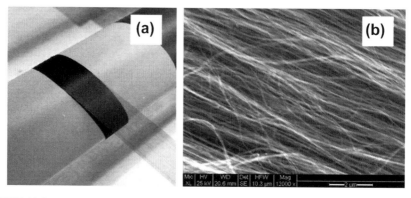

FIGURE 13.9

Optical image of CNT sheet accumulated on a teflon substrate—(a); SEM image of the sheet showing the degree of alignment of the nanotubes—(b). (For color version of this figure, the reader is referred to the online version of this book.)

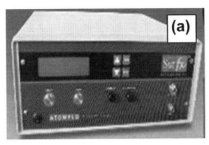 

FIGURE 13.10

Surfx Atomflo 400-D plasma controller—(a) and plasma head—(b) [79]. (For color version of this figure, the reader is referred to the online version of this book.)

materials. The base unit is supplied with an applicator with 25 mm in diameter active area and an air-cooled 300 W RF generator performing at 27.12 MHz. The RF atmospheric pressure plasma source Atomflo 400D is shown in Fig. 13.10. It produces a plasma environment at low temperature, which preserves the CNT materials from damaging.

The plasma is formed by feeding He at a constant flow rate of 30 L/min and the flow rate of O_2 (0.2–0.65 L/min) is adjusted as per the plasma power desired. CNT sheets were treated for different time intervals and with different plasma power to estimate how the plasma parameters affect the extent of functionalization of the sheet surface. In this section, both structural and chemical modifications induced by plasma treatments on the MWCNTs are discussed using Raman spectroscopy, XPS, FTIR spectroscopy and changes in hydrophobic character of the CNT material through contact angle testing is presented.

13.4.3 Characterization
13.4.3.1 Raman spectroscopy

Analysis of the Raman spectra of plasma-treated CNT sheets reveals that there are a significant number of defects that are created on the nanotube sidewall with the plasma. This has been inferred from the rise in intensity of the D-band or the disorder band peak in the spectra relative to the G-band or the graphitic band of the nanotubes. Figure 13.11 shows the Raman spectra obtained for sheets functionalized under plasma different conditions.

Figure 13.12 illustrates a possible mechanism for the modification of the MWNT walls by oxygen plasma [80].

The values of I_D/I_G ratios from the Raman spectra of samples prepared at UC are presented in Table 13.4 and are in accordance with data presented in other research articles [74]. It is also revealed that the nanotubes produced at UC are of high quality, which has been ascertained from the low I_D/I_G ratio for the untreated nanotube sheet sample (Table 13.4).

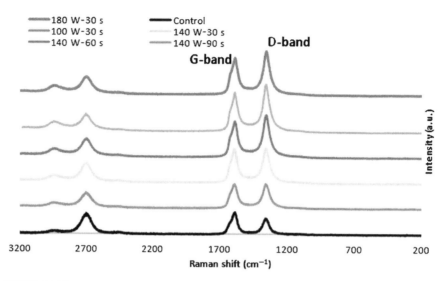

FIGURE 13.11

Raman spectra for untreated (control) and treated CNT sheet samples. Raman spectra were obtained using Renishaw inVia Raman Microscope with a 514 nm laser at 10% power level. (For color version of this figure, the reader is referred to the online version of this book.)

Figures 13.13 and 13.14 show plots of the observed I_D/I_G ratios as a function of plasma power and time of treatment. It is observed that the plasma effectively creates defects on the sidewalls of the nanotubes and the ratio increases with increase in plasma power. However, the change between 180 and 140 W treatment is significantly higher than that between 100 and 140 W, which could be attributed to higher oxidation power and temperature of the plasma at 180 W power.

A similar conclusion of increasing the number of defects due to a longer time of treatment can be drawn from Fig. 13.14. In this case, there is a stark increase in the I_D/I_G ratio as the treatment time goes from 30 to 60 s and an insignificant change between I_D/I_G ratios for 60 and 90 s.

Hence, two important conclusions can be drawn from the above results. First, plasma power in the high power range leads to the creation of a large number of defects, whereas in the medium and low power range, the difference is not significant. Therefore, high plasma power damages the nanotubes more readily albeit creating more functional groups but disturbing the CNT structure. Second, longer time of treatment tends to the creation of a larger number of defects but the rate of defects does not increase linearly with time and there is an upper limit to the extent of functionalization of the exposed nanotubes. Therefore, to achieve the desired functionalization with simultaneously preserving the CNT structure, it is necessary to have a short treatment time and reduced ($\sim$100–140 W) plasma power. This supposition is further supported via electrical characterization of the CNT sheet before and after plasma treatment as presented in Table 13.5.

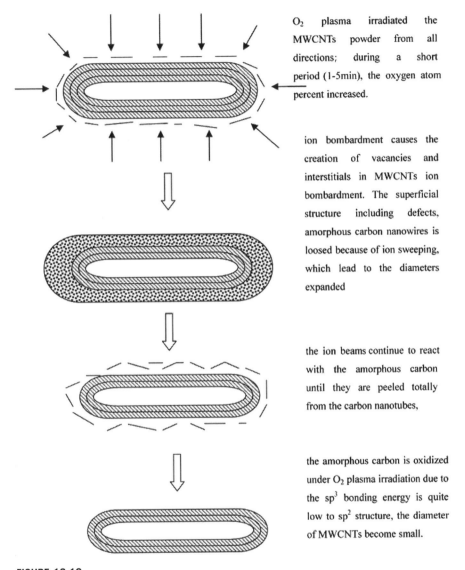

FIGURE 13.12

Schematic illustration of the possible mechanism of modification of CNTs by oxygen plasma, grafting of oxygenated moieties and ablation of carbonaceous products ($\sim CO_2$).

Source: Reprinted from Ref. [80]. Copyright (2007) with permission from Elsevier.

The increase in electrical resistance demonstrates that the electronic structure of the nanotubes is disturbed by the plasma treatment possibly by introduction of defects where the carboxyl groups (−COOH) are attached to the sidewalls. These results along with those from Raman spectroscopy confirm the change in the

Table 13.4 I_D/I_G Ratio from Raman Spectroscopy of CNT Sheet Samples Treated with He/O$_2$ Plasma for Different Times and Plasma Power

Plasma Power (W)	Treatment Time (s)	I_D/I_G Ratio
0	0	0.72
100	30	1.00
140	30	1.02
	60	1.17
	90	1.20
180	30	1.18

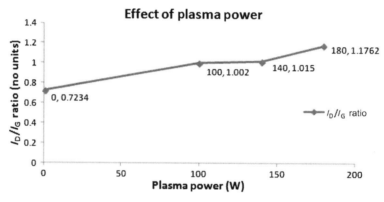

FIGURE 13.13

Effect of plasma power on the I_D/I_G ratio for treated and untreated CNT sheet. (For color version of this figure, the reader is referred to the online version of this book.)

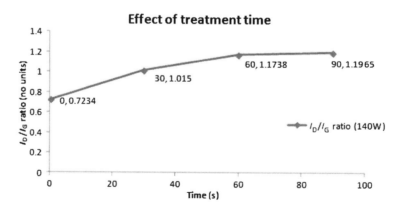

FIGURE 13.14

Effect of treatment time on the I_D/I_G ratio for treated and untreated CNT sheet. (For color version of this figure, the reader is referred to the online version of this book.)

Table 13.5 Effect of Plasma Treatment on the Electrical Resistance of the CNT Sheet

Sample (Plasma Power (W)–Treatment Time (s))	Resistance before Plasma Treatment (Ω)	Resistance after Plasma Treatment (Ω)
100–30	59.4	65.8
140–30	56.3	75.8
140–60	54.2	98.5
140–90	58.7	150.6
180–30	60.8	109.7

structure of the CNTs and are corroborated by results of XPS and FTIR analyses of the samples.

13.4.3.2 X-ray photoelectron spectroscopy

XPS spectra were acquired using a Phi 5300 X-ray photoelectron spectrometer with Mg K-alpha X-rays at an accelerating voltage of 15.0 kV, and charge compensation was provided by an electron flood gun. XPS analysis shows increasing oxygen concentration with increasing plasma power and time of exposure, as summarized in Table 13.6, and observed by the increasing relative intensity of the oxygen 1 s peak at 531 eV as in Fig. 13.15.

Oxygen incorporation into the nanotube backbone will disrupt the π conjugation of the backbone, which will cause an increase in the I_D/I_G ratio. Therefore, the increase in oxygen concentration determined by XPS analysis is in agreement with the increase in I_D/I_G ratio when raising plasma power as seen in the Raman analysis.

In addition to the determination of atomic composition, XPS analysis can be used to determine the degree of oxidation of the carbon atoms which comprise the sample. As a carbon atom becomes more heavily oxidized, its C(1s) peak position shifts to a higher binding energy; therefore, it is possible to identify and quantify the attachment of carbon to oxygen in high-resolution XPS spectra by a process

Table 13.6 Elemental Composition Results from XPS Analysis of Plasma-Treated CNT Sheet

Sample	C (%)	O (%)	N (%)
Control/untreated	95.9	3.8	0.3
100 W–30 s	87.8	11.3	0.9
140 W–30 s	85.0	13.8	0.9
140 W–60 s	82.2	16.8	1.0
180 W–30 s	85.0	14.1	1.3

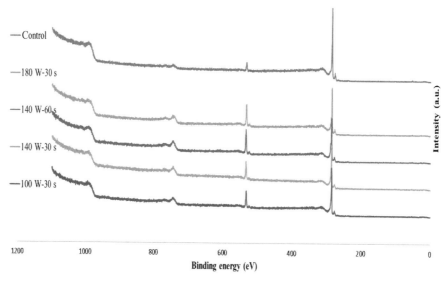

FIGURE 13.15

XPS survey results for CNT sheets treated by He–O_2 atmospheric pressure plasma with varying plasma power and time of treatment. (For color version of this figure, the reader is referred to the online version of this book.)

of curve fitting. Curve fits of plasma-treated (140 W/60 s) and pristine CNT sheets are shown in Fig. 13.16.

The spectra in Fig. 13.16 have been shifted so that their C–C peak occurs at 284.6 eV, an integrated (Shirley) background was subtracted, and every curve-fit component peak utilized the same asymmetric GL(30) (Gaussian/Lorentzian product) peak shape and full width at half maximum. The plasma-treated sample (Fig. 13.16(A)) shows the formation of peaks within the fitted curve which are attributed to –C=O, –C–O–X and O=C–O–X functionalities, where X is either a carbon or hydrogen atom. By comparing the area percentage of each component and the total amount of oxygen present in the sample (Table 13.6), X is postulated to be hydrogen, and therefore, carboxyl and hydroxyl functionalities predominate over ether and ester functionalities on the nanotube surface. The pristine nanotube sheet (Fig. 13.16(b)) shows a much less pronounced shoulder at higher binding energies; however, the curve fit does indicate that the pristine nanotubes have a small amount of defects or functional group sites. The π to π^* shake-up satellite (290.67 eV) is not a functional group, but it is instead evidence of conjugation existing within the pristine nanotubes. The conclusion from XPS analysis is that plasma treatment has created –C–OH, –C=O, and O=C–OH functionalities on the CNTs. This conclusion is supported by data published by other researchers who have functionalized CNTs with plasma [74].

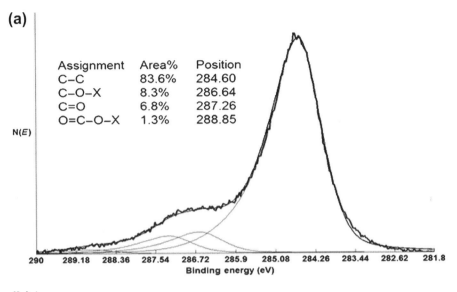
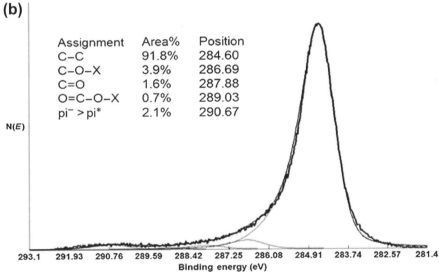

FIGURE 13.16

High-resolution XP spectra for plasma-functionalized—(a) and pristine—(b) CNT sheets.

13.4.3.3 FTIR Analysis

FTIR analysis of the sheet samples validates the results from the XPS analysis and suggests that indeed functionalization of the MWNTs has taken place with primarily carboxyl (—COOH), hydroxyl (—OH) and carbonyl (—C=O) groups being formed on the surface of the nanotubes. The samples were tested using Digilab Excalibur

FTS (Fourier Transform Spectrometer) 3000 FTIR (Fourier Transform Infrared Spectroscopy) spectrophotometer. Figure 13.17 shows the results from the FTIR analysis of the (1.) 100 W plasma-treated sample for 30 s compared with that of (2.) the control/pristine sheet sample. It can be seen that there are two distinct features or peaks that are observed for the plasma-treated samples in the 1714/cm and 1610–1650/cm regions, and these peaks can be attributed to the C=O stretching and water adsorbed on the surface, respectively. The stretching around the 1546/cm is common for both functionalized and unfunctionalized samples and can be attributed to C=C stretching of the CNT structure. A very broad peak in the 3000–3500/cm region is observed for all samples and more significantly for the treated samples. This region can be attributed to the —OH stretching from hydroxyl groups on the surface of the sheet and also to adsorbed water molecules. The adsorption of water or moisture from the environment on the sheet surface is critical as it provides hydrogen atoms and the treatment of the sheet with He—O_2 plasma makes the sheet more hydrophilic. A similar spectrum is also reported by Wang et al. [81] in their study of oxidative treatment of MWNTs with DBD plasma.

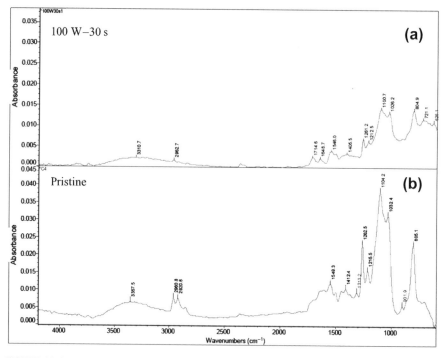

FIGURE 13.17

FTIR analysis of (a) 100 W plasma-treated sample for 30 s and (b) the control/pristine sheet sample. (For color version of this figure, the reader is referred to the online version of this book.)

13.4.3.4 Contact angle testing

The increase in the hydrophilicity as indicated by the water adsorption in the FTIR analysis is confirmed via contact angle testing. The contact angle visualized by a drop of deionized water was measured using Model 590 Rame-hart instrument contact angle goniometer. Figure 13.18 shows images of the droplet of water on pristine and plasma-treated sheet.

The measured contact angle for the CNT sheet drops from average 79.6° for the pristine sheet to 14.5° for the plasma-treated samples, showing the change from hydrophobic to hydrophilic nature of the sheet. This change can be explained on the premise that water droplet is now associated with the CNT sheet through H-bonding with —OH and —COOH groups on the nanotube surface which are generated via the plasma treatment.

13.4.3.5 Optical characterization

The CNT film drawn from the array grown on Si wafer is electrically conductive and also optically transparent. Hence, keeping the above two properties in mind, we made sheet samples of different number of layers and characterized them for transmittance with ultraviolet (UV), visible and infrared (IR) radiation. IR transparency is of particular importance for applications where heat dissipation is crucial, such as solar cells. CNT films of a varied number of layers were tested for percent transmittance (%T) as a function of wavelength.

The dependence of transmittance on the orientation of the nanotubes was also observed as multiple layered sheet samples were tested for three different orientations (parallel, perpendicular to the horizontal and cross-pattern). Characterization of the CNT sheet in the UV—visible region was carried out using Varian CARY 50 Bio spectrophotometer and the results obtained are shown in Fig. 13.19(B).

The effect of the alignment of nanotubes with respect to the direction of the incident radiation and the horizontal is observed with the change in %T as a function of the number of layers and the different orientation of the nanotubes in Fig. 13.20.

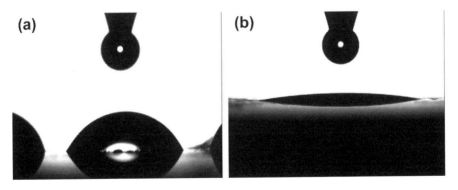

FIGURE 13.18

Droplet of D.I. water on pristine—(a) and plasma-treated CNT sheet—(b).

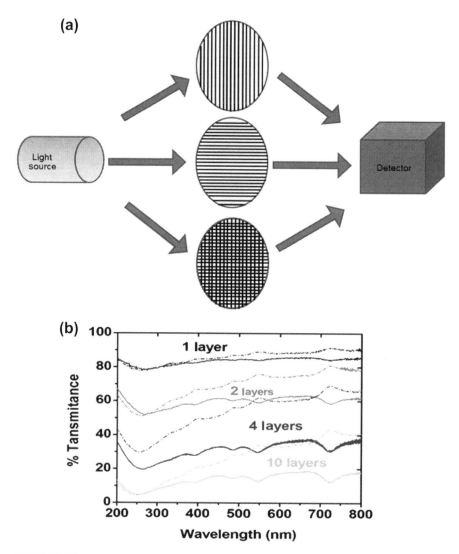

FIGURE 13.19

Illustration showing the different orientations of nanotubes used for UV–vis spectroscopy (solid line representing horizontal alignment and dashed line representing vertical alignment of nanotubes)—(a). Percent transmittance vs wavelength for CNT sheet toward UV–vis radiation as a function of number of layers of sheet—(b). (For interpretation of the references to color in this figure legend, the reader is referred to the online version of this book.)

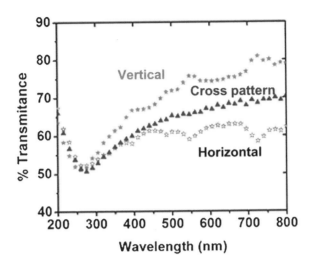

FIGURE 13.20

Effect of the three (vertical, horizontal and cross-pattern) orientations of the nanotubes in the CNT sheet sample on the percent transmittance as a function of wavelength for UV–vis radiation. (For color version of this figure, the reader is referred to the online version of this book.)

The observed trend of decreased transmittance (%T) at smaller UV wavelengths is logical and in accordance with a known fact that carbon materials are inherent absorbers of UV radiation. The CNT sheet produced and utilized in the current research was tested for IR transparency using Thermo Scientific Nicolet 6700 FTIR spectrometer, and the results are shown in Fig. 13.21.

The %T of the CNT sheet sample varies with wavelength and generally increases with a decrease in the wave number with that variation being more pronounced as the number of layers increases. A trend of increasing %T can be observed starting from UV (lowest) to far-IR (highest) wavelengths. Thus, the above results indicate the capability of the sheet to be used as a transparent conducting electrode.

13.4.4 Applications of CNT sheet

At UC, we are manufacturing CNT sheets comprising multiple layers and are exploring its different applications. Two of the studied applications of the CNT sheet are described below.

13.4.4.1 Electromagnetic interference shielding

An electromagnetic (EM) wave consists of mutually perpendicular components of electric and magnetic fields which are perpendicular to the direction of propagation of the wave, as shown in the illustration in Fig. 13.22.

The applied electric field on the surface of an ideal conductor induces a current that causes displacement of charge inside the conductor which cancels the applied

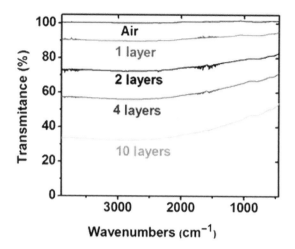

FIGURE 13.21

Percent transmittance vs wavelength for CNT sheet toward IR radiation as a function of number of layers of sheet. (For color version of this figure, the reader is referred to the online version of this book.)

field inside, at which point the current stops. Similarly, oscillating magnetic fields generate Eddy currents that act to cancel the applied magnetic field itself. Dispersed CNTs in a variety of polymer solutions and cast into composites [13,82–84] have demonstrated reasonable EMI shielding effectiveness. Fan et al. [85] have also developed a cladding of CNT as a shielding layer in a coaxial cable.

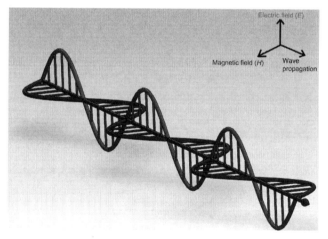

FIGURE 13.22

Illustration of the incident electromagnetic (EM) wave channeled through the waveguide with electric and magnetic field components. (For color version of this figure, the reader is referred to the online version of this book.)

The CNT sheet produced at UC is a freestanding, lightweight, robust and electrically conductive material thereby making it a perfect candidate for EMI shielding. The nanotube alignment in the sheet creates anisotropy in the conductivity and when the CNT main axis aligns parallel to the electric field component of the incident EM radiation, a significantly higher current is induced in the nanotubes. Therefore, a higher shielding effectiveness is observed. This is illustrated in Fig. 13.23 and Table 13.7 shows the results of shielding effectiveness from a 10 GHz incident wave.

The shielding measurements were made using two separate waveguide systems; one was X-band, 8.2–12.4 GHz and the other was KU-band, 12.4–18 GHz. The microwave signals were generated by a Hewlett–Packard 8671B generator which operates from 2 to 18 GHz. The signals were fed via low loss coaxial cables to a directional coupler (Krytar 4–20 GHz, −6 dB) connected to an SMA (SubMiniature version A) connector to waveguide coupler, which in turn was connected to a short section of straight waveguide. The output part of the waveguide system was identical to the input section but in reversed order: straight waveguide section, waveguide to SMA coupler connected to a Hewlett–Packard 437B microwave power meter with 8485 A power sensor (50 MHz–26.5 GHz, 1 µW–100 mW). The sample was placed in an air gap (~3 mm) between the input and output straight waveguide sections. The input directional coupler was connected, so that we could measure the amplitude of the signal that was reflected from the waveguide-sample system (using

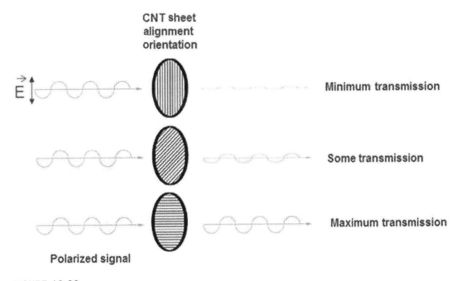

FIGURE 13.23

Illustration of the effect of nanotube orientation on the transmission of electric field through the sheet. (For color version of this figure, the reader is referred to the online version of this book.)

Table 13.7 Effect of Orientation of Nanotubes with Respect to Incident Electrical Field (E) on The Shielding Effectiveness of an 80-Layer-Thick CNT Sheet

Sample	Power Transmitted (dB)	Shielding Effectiveness (dB)
Air	−22.1	0
CNT sheet—perpendicular to E	−30.66	8.56
CNT sheet—parallel to E	−49.12	27.02

Hewlett–Packard 8485B power sensor, 100 pW–10 µW, 50 MHz–26.5 GHz), while the power meter allowed measurement of the amplitude of the transmitted signal. Measurements were first made without the sample (in air) and then made with the sample so that the relative change in the transmitted and reflected power was determined. The reflected and transmitted signals were also measured using a Tektronix 494P 10 kHz–21 GHz spectrum analyzer.

The effect of the frequency of the incident EM wave on the shielding effectiveness of the sheet was studied for the K_u (12.5–16 GHz) band of the EM spectrum. K_u band is primarily used for satellite communications, most notably for fixed and broadcast services. The results of the experiment are shown in Fig. 13.24.

It is observed that the shielding effectiveness of the nanotube sheet sample varies with frequency with peak effectiveness being observed at 16 GHz and the range

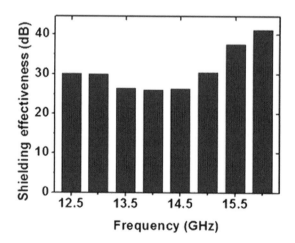

FIGURE 13.24

Effect of frequency of incident radiation on the shielding effectiveness of the CNT sheet (CNTs aligned parallel to electric field component of incident wave). (For color version of this figure, the reader is referred to the online version of this book.)

lying between 25 and 40 dB for CNT sheet samples with the nanotubes oriented in the direction of the oscillation of the electric field. These results demonstrate the ability of the CNT sheet to be effectively used as an EMI shielding material, especially for applications where weight of the material is a concern such as in aerospace applications.

13.4.4.2 Composites

Since their discovery, CNTs have been thought of as potential reinforcements for high-strength polymer composites. There is extensive literature available on CNTs dispersed in various polymer matrices and now the dry spun CNT sheet is also being explored as potential reinforcement. Recent progress made by Zhu et al. [86–88] indicates that dry drawn CNT sheets and sheets produced via floating catalytic CVD method [56] produce significantly stronger composites than those made from dispersed CNTs and could serve as a suitable reinforcement for strong composites in the future.

At UC, the effect of plasma functionalization of CNT sheets on the mechanical properties of prepregs made from CNT sheet was explored. CNT sheets are manufactured according to the process described earlier and are then densified with ethanol. Densified sheets are used to fabricate prepregs by a simple brush and paint-on technique employing solution of polyvinyl alcohol (PVA) in water.

These experiments were conducted to study the effect of various parameters on the final strength of the composite material. In the first experiment, a 100-layer CNT sheet sample with dimensions 0.25 inch by 6 inches was manufactured and cut in half. One half of it was plasma functionalized at 140 W plasma power for 30 s and the other half was left pristine. These sheet samples were then utilized to manufacture prepregs with 10 g/L PVA solution in water as matrix material. Then the CNT sheet/PVA prepregs were cured at 120 °C for 2 h in a hot press. Tensile testing of these and other composite materials has been carried out using Instron 5900 series testing instrument. The results of the test are shown in Fig. 13.25.

It is observed that prepregs with plasma-treated sheet samples demonstrate even lower strength than those containing pristine sheets. The data from the above experiment are tabulated in Table 13.8.

The decrease in Young's modulus and tensile strength can be attributed to the destructive effect of the plasma treatment as is observed in the increase of the I_D/I_G ratio obtained by Raman analysis and the increase of the electrical resistance presented earlier. Therefore, for the following experiment, the plasma power was reduced to 100 W and the treatment time was reduced to 10 s only. CNT/PVA prepregs were fabricated as described earlier with a 100-layer sheet sample cut into two equal halves, one of which was kept pristine and the other functionalized with plasma. The results are presented in Fig. 13.26.

An approximate 117% net increase in tensile strength is observed for samples treated with lower plasma power and less exposure time. This increase in strength

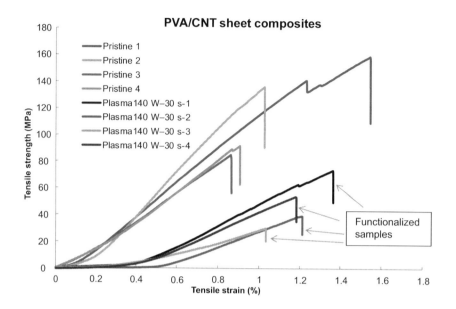

FIGURE 13.25

Stress—strain curves for PVA/CNT prepregs with pristine and plasma (140W-30 seconds) functionalized CNT sheets. (For color version of this figure, the reader is referred to the online version of this book.)

could be attributed to distinct phenomena working in synergy to improve the strength. First, a better impregnation of the sheet with the polymer solution is achieved due to increased hydrophilicity of the nanotubes. Second, hydrogen bonding interactions between the —OH groups of PVA and —COOH and —OH groups generated on the nanotube surface is created by the plasma treatment. Table 13.9 shows the data from the above described experiment.

Thus, plasma has presented itself as an attractive tool for functionalization of CNTs and could be instrumental in the development of strong composite materials of the future. But, there is potential to greatly increase the properties of CNT sheet composites by using several new processing techniques. First, the CNTs used are wavy. The waviness causes the CNT to be spaced apart which reduces the van der Waals forces and does not allow the CNT to all be loaded at the same time, thus reducing the strength of the sheet. Also the CNTs have defects which significantly reduce their strength. Thermally annealing the CNT at high temperature will make the CNT in the arrays straighter and stronger. Prestretching the sheet during drawing (as done by other groups) will also straighten the CNT and make the load transfer more uniform in the sheet. An aerospace resin to replace the PVA may increase the shear strength between nanotubes and increase the strength

Table 13.8 Mechanical Properties of Pristine and Plasma (140W-30 seconds)-Functionalized CNT Prepregs with PVA Matrix (1st Experiment)

		PVA CNT Sheet Composites				
	Young's Modulus (GPa)	Mean of Young's Modulus (GPa)	Change in Young's Modulus (%)	Tensile Strength (MPa)	Mean of Tensile Strength (MPa)	Change of Tensile Strength (%)
Pristine 1	12.7	11.6	0	159	116.7	0
Pristine 2	15.6			136.		
Pristine 3	11			84.9		
Pristine 4	6.9			86.7		
Functionalized 1	8	6.3	−45.9	73.6	49.1	−57.9
Functionalized 2	5.6			39.2		
Functionalized 3	4.3			30		
Functionalized 4	7.1			53.7		

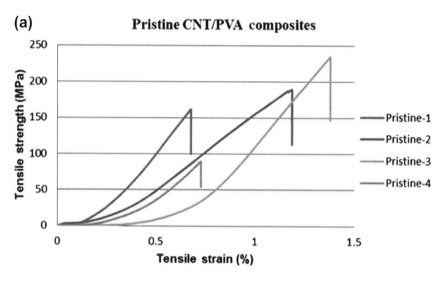
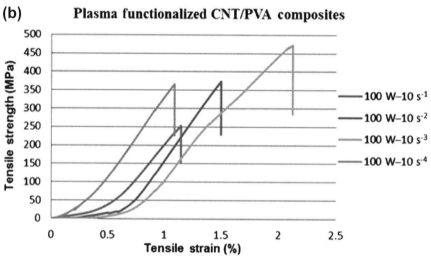

FIGURE 13.26

Stress–strain curves for CNT/PVA prepregs with (A) pristine and (B) plasma (100W-10 seconds)-treated sheets. (For color version of this figure, the reader is referred to the online version of this book.)

of the sheet. Finally, the use of longer nanotubes (e.g. 5–10 mm instead of 0.5 mm long) may increase the load transfer from nanotube to nanotube in the sheet. The combination of all these changes may greatly increase the strength of CNT sheet composites.

Table 13.9 Mechanical Properties of Pristine and Plasma (100W-10 seconds)-Functionalized CNT Prepregs with PVA Matrix (2nd Experiment)

	Young's Modulus (GPa)	Tensile Strength (MPa)	Mean of Young's Modulus (GPa)	Change in Young's Modulus (%)	Mean of Tensile Strength (MPa)	Change in Tensile Strength (%)
Pristine 1	35.6	161.7	23.3	0	168.8	0
Pristine 2	21.5	189.6				
Pristine 3	10.2	234.6				
Pristine 4	25.8	89.4				
Functionalized 1	36.4	253.2	39.4	69.2	366.5	117.1
Functionalized 2	43.6	373.8				
Functionalized 3	32.9	472.5				
Functionalized 4	44.6	366.6				

13.5 CONCLUSIONS AND FUTURE WORK

The manufacturing of dry spun sheets of MWNTs has been discussed. A review of the synthesis, processing, functionalization and characterization was presented. The application of atmospheric pressure plasma as a tool to carry out fast, efficient and clean functionalization of CNTs was demonstrated. Plasma-functionalized sheets were characterized by Raman spectroscopy, FTIR spectroscopy, XPS and contact angle measurement to determine the functional groups on the nanotube surface. CNT sheet has found many successful applications. Two applications, a lightweight, effective EMI shielding material and the reinforcement of strong polymer composites, were discussed. Plasma-functionalized sheet composites were shown to perform better than pristine sheet composites. Thus, plasma has been identified as a fast, clean and efficient tool to functionalize CNTs. For future work, it is crucial to optimize the plasma power and treatment to produce the best results. It will also be beneficial to understand the mechanism of plasma functionalization of the nanotubes by studying the active species in the plasma using optical emission spectroscopy. The use of other polymer matrix materials such as epoxy and bismaleimide resin will also be explored. Quantitative analysis of the functional groups on the sheet surface with the TPD technique should also help in optimizing the process. Lastly, consistent production of spin-capable arrays of CNTs by CVD is crucial for the advancement in this field of research. The latter will allow the manufacture and implementation of thicker, multi-layered CNT sheets which are expected to be easier to handle and more useful in manufacturing polymer composites. Such experiments with CNT sheets consisting of several thousand layers are in progress at UC and shall be reported in the future. Several new processing techniques will be combined and are expected to greatly increase the strength of CNT sheet composites.

Acknowledgments

We are grateful to the funding agencies: NSF through grant CMMI-07272500 from NCA&T, a DURIP-ONR grant, NSF SNM grant 1120382, and ERC supplemental STTR funded by NSF (2010–2012) "Translational Research to Commercialize CNT and Mg based Materials for Biomedical and Industrial Nanotechnology Applications". We would also like to acknowledge Feng Wang, Christopher Katuscak and Sandip Argekar for their assistance in the lab, Doug Hurd for his assistance in the machine shop, and General Nano LLC for granting access to their e-Beam evaporator.

References

[1] S. Iijima, Helical microtubules of graphitic carbon, Nature 354 (1991) (6348).
[2] K. Liu, et al., Controlled growth of super-aligned carbon nanotube arrays for spinning continuous unidirectional sheets with tunable physical properties, Nano Letters 8 (2) (2008).

[3] R.S. Ruoff, D. Qian, W.K. Liu, Mechanical properties of carbon nanotubes: theoretical predictions and experimental measurements, Comptes Rendus Physique 4 (9) (2003).
[4] T.W. Ebbesen, et al., Electrical conductivity of individual carbon nanotubes, Nature 382 (1996) (6586).
[5] P.M. Ajayan, et al., Single-walled carbon nanotube-polymer composites: strength and weakness, Advanced Materials 12 (10) (2000).
[6] N.G. Sahoo, et al., Polymer nanocomposites based on functionalized carbon nanotubes, Progress in Polymer Science 35 (7) (2010).
[7] C. Wang, et al., Proton exchange membrane fuel cells with carbon nanotube based electrodes, Nano Letters 4 (2) (2004).
[8] A.L.M. Reddy, et al., Coaxial MnO_2/carbon nanotube array electrodes for high-performance lithium batteries, Nano Letters 9 (3) (2009).
[9] D.N. Futaba, et al., Shape-engineerable and highly densely packed single-walled carbon nanotubes and their application as super-capacitor electrodes, Nature Materials 5 (12) (2006).
[10] M.W. Rowell, et al., Organic solar cells with carbon nanotube network electrodes, Applied Physics Letters 88 (23) (2006).
[11] M.P. Mattson, R.C. Haddon, A.M. Rao, Molecular functionalization of carbon nanotubes and use as substrates for neuronal growth, Journal of Molecular Neuroscience 14 (3) (2000).
[12] K.A. Crutcher, et al., Progress in the use of aligned carbon nanotubes to support neuronal attachment and directional neurite growth, in: M.J. Schulz, V.N. Shanov, Y. Yun (Eds.), Nanomedicine Design of Particles, Sensors, Motors, Implants, Robots, and Devices (2009).
[13] H.M. Kim, et al., Electrical conductivity and electromagnetic interference shielding of multiwalled carbon nanotube composites containing Fe catalyst, Applied Physics Letters 84 (4) (2004).
[14] Z.C. Wu, et al., Transparent, conductive carbon nanotube films, Science 305 (5688) (2004).
[15] C. Feng, et al., Flexible, stretchable, transparent conducting films made from super-aligned carbon nanotubes, Advanced Functional Materials 20 (6) (2010).
[16] M. Zhang, et al., Strong, transparent, multifunctional, carbon nanotube sheets, Science 309 (5738) (2005).
[17] L. Xiao, et al., Flexible, stretchable, transparent carbon nanotube thin film loudspeakers, Nano Letters 8 (12) (2008).
[18] J.-H. Kim, et al., Synthesis and electrochemical properties of spin-capable carbon nanotube sheet/MnO_x composites for high-performance energy storage devices, Nano Letters 11 (7) (2011).
[19] F. Meng, et al., Carbon nanotube composite films with switchable transparency, ACS Applied Materials & Interfaces 3 (3) (2011).
[20] J. Di, et al., Ultrastrong, foldable, and highly conductive carbon nanotube film, ACS Nano 6 (6) (2012).
[21] M.T. Byrne, Y.K. Gun'ko, Recent advances in research on carbon nanotube-polymer composites, Advanced Materials 22 (15) (2010).
[22] A.B. Dalton, et al., Super-tough carbon-nanotube fibres—these extraordinary composite fibres can be woven into electronic textiles, Nature 423 (6941) (2003).

[23] J.N. Coleman, et al., High-performance nanotube-reinforced plastics: understanding the mechanism of strength increase, Advanced Functional Materials 14 (8) (2004).
[24] H. Deng, et al., A novel concept for highly oriented carbon nanotube composite tapes or fibres with high strength and electrical conductivity, Macromolecular Materials and Engineering 294 (11) (2009).
[25] J. Hone, et al., Electrical and thermal transport properties of magnetically aligned single walled carbon nanotube films, Applied Physics Letters 77 (5) (2000).
[26] S. Badaire, et al., Correlation of properties with preferred orientation in coagulated and stretch-aligned single-wall carbon nanotubes, Journal of Applied Physics 96 (12) (2004).
[27] X. Wang, et al., Fabrication of ultralong and electrically uniform single-walled carbon nanotubes on clean substrates, Nano Letters 9 (9) (2009).
[28] P.M. Ajayan, et al., Aligned carbon nanotube arrays formed by cutting a polymer resin-nanotube composite, Science 265 (5176) (1994).
[29] W.A. Deheer, et al., Aligned carbon nanotube films—production and optical and electronic-properties, Science 268 (5212) (1995).
[30] J.M. Bonard, et al., Field emission from single-wall carbon nanotube films, Applied Physics Letters 73 (7) (1998).
[31] O. Lourie, D.M. Cox, H.D. Wagner, Buckling and collapse of embedded carbon nanotubes, Physical Review Letters 81 (8) (1998).
[32] L. Jin, C. Bower, O. Zhou, Alignment of carbon nanotubes in a polymer matrix by mechanical stretching, Applied Physics Letters 73 (9) (1998).
[33] B.Z. Tang, H.Y. Xu, Preparation, alignment, and optical properties of soluble poly(phenylacetylene)-wrapped carbon nanotubes, Macromolecules 32 (8) (1999).
[34] R.H. Baughman, et al., Carbon nanotube actuators, Science 284 (5418) (1999).
[35] F. Ko, et al., Electrospinning of continuous carbon nanotube-filled nanofiber yarns, Advanced Materials 15 (14) (2003).
[36] S.S. Ojha, et al., Characterization of electrical and mechanical properties for coaxial nanofibers with poly(ethylene oxide) (PEO) core and multiwalled carbon nanotube/PEO sheath, Macromolecules 41 (7) (2008).
[37] S.D. McCullen, et al., Morphological, electrical, and mechanical characterization of electrospun nanofiber mats containing multiwalled carbon nanotubes, Macromolecules 40 (4) (2007).
[38] S. Hu, R. Rajamani, X. Yu, Flexible solid-state paper based carbon nanotube supercapacitor, Applied Physics Letters 100 (10) (2012).
[39] X.L. Xie, Y.W. Mai, X.P. Zhou, Dispersion and alignment of carbon nanotubes in polymer matrix: a review, Materials Science & Engineering R: Reports 49 (4) (2005).
[40] J.N. Coleman, U. Khan, Y.K. Gun'ko, Mechanical reinforcement of polymers using carbon nanotubes, Advanced Materials 18 (6) (2006).
[41] J.N. Coleman, et al., Small but strong: a review of the mechanical properties of carbon nanotube-polymer composites, Carbon 44 (9) (2006).
[42] M. Endo, M.S. Strano, P.M. Ajayan, Potential applications of carbon nanotubes, Carbon Nanotubes 111 (2008).
[43] K. Jiang, et al., Superaligned carbon nanotube arrays, films, and yarns: a road to applications, Advanced Materials 23 (9) (2011).
[44] A.M. Cassell, et al., Large scale CVD synthesis of single-walled carbon nanotubes, Journal of Physical Chemistry B 103 (31) (1999).

[45] S. Vesselin, et al., Advances in chemical vapor deposition of carbon nanotubes, in: Nanoengineering of Structural, Functional and Smart Materials, CRC Press, 2005.

[46] K.L. Jiang, Q.Q. Li, S.S. Fan, Nanotechnology: spinning continuous carbon nanotube yarns—carbon nanotubes weave their way into a range of imaginative macroscopic applications, Nature 419 (6909) (2002).

[47] Y. Yun, et al., Growth mechanism of long aligned multiwall carbon nanotube arrays by water-assisted chemical vapor deposition, Journal of Physical Chemistry B 110 (47) (2006).

[48] Shanov, V., M. Schulz, C. Jayasinghe, Method for Making Multi-Layered Carbon Nanotube Sheet, USA, 2011.

[49] Y. Zhang, et al., Polymer-embedded carbon nanotube ribbons for stretchable conductors, Advanced Materials 22 (28) (2010).

[50] A.E. Aliev, et al., Thermal conductivity of multi-walled carbon nanotube sheets: radiation losses and quenching of phonon modes, Nanotechnology 21 (3) (2010).

[51] J.-H. Poehls, et al., Physical properties of carbon nanotube sheets drawn from nanotube arrays, Carbon 50 (11) (2012).

[52] P. Liu, et al., Fast high-temperature response of carbon nanotube film and its application as an incandescent display, Advanced Materials 21 (35) (2009).

[53] H.-X. Zhang, et al., Cross-stacked carbon nanotube sheets uniformly loaded with SnO_2 nanoparticles: a novel binder-free and high-capacity anode material for lithium-ion batteries, Advanced Materials 21 (22) (2009).

[54] Chen, Z., et al., Flexible Thermoacoustic Device for Use in Banner of Thermoacoustic Flag, Has Sound Wave Generator Comprising Carbon Nanotube Structure Having Carbon Nanotubes Combined by van der Waals Attractive Force, Univ Qinghua; Hon Hai Precision Ind Co Ltd.

[55] W. Fu, et al., Super-aligned carbon nanotube films as aligning layers and transparent electrodes for liquid crystal displays, Carbon 48 (7) (2010).

[56] J. Chaffee, et al., Direct synthesis of CNT yarns and sheets, NSTI Nanotech 2008, Vol 3, Nanotechnology Conference and Trade Show, Technical Proceedings 3 (2008).

[57] D. Wang, et al., Highly oriented carbon nanotube papers made of aligned carbon nanotubes, Nanotechnology 19 (7) (2008).

[58] Y.H. Lin, et al., Glucose biosensors based on carbon nanotube nanoelectrode ensembles, Nano Letters 4 (2) (2004).

[59] D.-J. Chung, A.K. Whittaker, S.-H. Choi, Electrochemical DNA biosensor based on IL-modified MWNTs electrode prepared by radiation-induced graft polymerization, Journal of Applied Polymer Science 126 (2012).

[60] J.L. Bahr, et al., Functionalization of carbon nanotubes by electrochemical reduction of aryl diazonium salts: a bucky paper electrode, Journal of the American Chemical Society 123 (27) (2001).

[61] K. Balasubramanian, M. Burghard, Chemically functionalized carbon nanotubes, Small 1 (2) (2005).

[62] K.H. An, et al., X-ray photoemission spectroscopy study of fluorinated single-walled carbon nanotubes, Applied Physics Letters 80 (22) (2002).

[63] E.T. Mickelson, et al., Fluorination of single-wall carbon nanotubes, Chemical Physics Letters 296 (1–2) (1998).

[64] M.J. O'Connell, et al., Reversible water-solubilization of single-walled carbon nanotubes by polymer wrapping, Chemical Physics Letters 342 (3–4) (2001).

[65] Z. Konya, et al., Large scale production of short functionalized carbon nanotubes, Chemical Physics Letters 360 (5–6) (2002).
[66] L.G. Delogu, et al., Functionalized multiwalled carbon nanotubes as ultrasound contrast agents, Proceedings of the National Academy of Sciences (2012).
[67] V. Datsyuk, et al., Double walled carbon nanotube/polymer composites via in-situ nitroxide mediated polymerisation of amphiphilic block copolymers, Carbon 43 (4) (2005).
[68] L. Dumitrescu, N.R. Wilson, J.V. Macpherson, Functionalizing single-walled carbon nanotube networks: effect on electrical and electrochemical properties, Journal of Physical Chemistry C 111 (35) (2007).
[69] A. Felten, et al., Radio-frequency plasma functionalization of carbon nanotubes surface O-2, NH_3, and CF_4 treatments, Journal of Applied Physics 98 (7) (2005).
[70] B.N. Khare, et al., Functionalization of carbon nanotubes using atomic hydrogen from a glow discharge, Nano Letters 2 (1) (2002).
[71] N.O.V. Plank, R. Cheung, Functionalisation of carbon nanotubes for molecular electronics, Microelectronic Engineering (2004) 73–74.
[72] P. He, et al., Plasma deposition of thin carbon fluorine films on aligned carbon nanotube, Applied Physics Letters 86 (4) (2005).
[73] U. Vohrer, et al., Plasma modification of carbon nanotubes and bucky papers, Plasma Processes and Polymers 4 (2007).
[74] D. Kolacyak, et al., Fast functionalization of multi-walled carbon nanotubes by an atmospheric pressure plasma jet, Journal of Colloid and Interface Science 359 (1) (2011).
[75] M.V. Naseh, et al., Fast and clean functionalization of carbon nanotubes by dielectric barrier discharge plasma in air compared to acid treatment, Carbon 48 (5) (2010).
[76] M.W. Marshall, S. Popa-Nita, J.G. Shapter, Measurement of functionalised carbon nanotube carboxylic acid groups using a simple chemical process, Carbon 44 (7) (2006).
[77] I.D. Rosca, S.V. Hoa, Highly conductive multiwall carbon nanotube and epoxy composites produced by three-roll milling, Carbon 47 (8) (2009).
[78] A. Ladwig, et al., Atmospheric plasma deposition of glass coatings on aluminum, Surface & Coatings Technology 201 (14) (2007).
[79] http://www.surfxtechnologies.com.
[80] T. Xu, et al., Surface modification of multi-walled carbon nanotubes by O_2 plasma, Applied Surface Science 253 (22) (2007) 8945–8951.
[81] W.-H. Wang, et al., Oxidative treatment of multi-wall carbon nanotubes with oxygen dielectric barrier discharge plasma, Surface & Coatings Technology 205 (21–22) (2011).
[82] Y. Yang, et al., Electromagnetic interference shielding characteristics of carbons nanofiber-polymer composites, Journal of Nanoscience and Nanotechnology 7 (2) (2007).
[83] N. Li, et al., Electromagnetic interference (EMI) shielding of single-walled carbon nanotube epoxy composites, Nano Letters 6 (6) (2006).
[84] Z. Liu, et al., Reflection and absorption contributions to the electromagnetic interference shielding of single-walled carbon nanotube/polyurethane composites, Carbon 45 (4) (2007).
[85] Jiang, K., et al., Coaxial Cable for Use as Carrier Used for Transferring Electrical Power and Signals, Comprises Core Comprising Carbon Nanotubes Having

Conductive Coating(s), Insulating Layer, Shielding Layer, and Sheathing Layer, Univ Tsinhua; Hon Hai Precision Ind Co Ltd; Univ Qinghua; Beijing Funa Techuang new Techn Co Ltd; Beijing Funate innovation Technology Co; Beijing Funate innovation Technology Co Ltd; Hongfujin Precision Ind Shenzhen Co Ltd.

[86] W. Liu, et al., Producing superior composites by winding carbon nanotubes onto a mandrel under a poly(vinyl alcohol) spray, Carbon 49 (14) (2011).

[87] X. Wang, et al., Mechanical and electrical property improvement in CNT/nylon composites through drawing and stretching, Composites Science and Technology 71 (14) (2011).

[88] Q. Cheng, et al., Functionalized carbon-nanotube sheet/bismaleimide nanocomposites: mechanical and electrical performance beyond carbon-fiber composites, Small 6 (6) (2010).

[89] Technology advances, MRS Bulletin 35 (03) (2010) 179–181.

CHAPTER 14

Direct Dry Spinning of Millimeter-long Carbon Nanotube Arrays for Aligned Sheet and Yarn

Yoku Inoue

Department of Electronics and Material Sciences, Shizuoka University, Naka, Hamamatsu, Shizuoka, Japan

CHAPTER OUTLINE

14.1 Introduction	390
14.2 Highly Spinnable MWCNT Arrays	391
14.2.1 Growth of millimeter-long spinnable MWCNT arrays	391
14.2.2 Characterization of MWCNT arrays	392
14.2.3 Dry spinning of webs from millimeter-long CNT arrays	393
14.3 Unidirectionally Aligned CNT Sheet	394
14.3.1 Fabrication of unidirectionally aligned CNT sheet	394
14.3.2 Electrical properties	395
14.3.3 Mechanical properties	396
14.3.4 Thermal properties	397
14.3.5 Discussions	398
14.4 Mechanical Properties of CNT Yarn	399
14.4.1 Fabrication of CNT yarn by dry spinning	399
14.4.2 Yarn evaluation methods	400
14.4.3 Mechanical property of a CNT web	401
14.4.4 Postspin twisting of as-spun yarns	402
14.4.4.1 Postspin twisting effect	402
14.4.4.2 CNT length dependence	405
14.4.4.3 Yarns' diameter dependence	405
14.4.5 Multiply twisted CNT yarns	407
14.4.5.1 Multiply twisting of as-spun CNT yarns	407
14.4.5.2 Twisting tension dependence	408
14.4.6 Pressed CNT yarns	409
14.4.7 Discussions	410
14.5 Conclusions	411
Acknowledgments	412
References	412

14.1 INTRODUCTION

Carbon nanotubes (CNTs) have outstanding material properties and have been studied extensively with the aim of being used in practical applications in industry. One of the most widely researched uses of CNTs is as filler material that is dispersed in a variety of polymers or solutions in order to promote electrical and/or thermal conductivity [1], or to reinforce the mechanical properties of the material being filled [2]. Since commercially available CNT products typically have lengths less than several tens of micrometers, the lengths of CNTs, when filled in the matrices of the materials they are being dispersed into, are further shortened to several micrometers because of fracturing that takes place during the dispersion processes. Therefore, in the case of CNT composites, even if the volume fraction of the CNTs in the composites is increased and their alignment improved, it is not given that these changes will result in significant improvements in the mechanical properties concomitantly. Ultralong CNTs have an advantage when used in composites as they can enhance load transfer. Longer CNTs are also useful when it comes to increasing the electrical conductivity of insulating materials. By bridging the insides of the matrices of the insulating materials, longer CNTs can help in increasing the electrical conductivity even as the quantity of the CNTs required is reduced.

Many researchers have attempted to grow longer vertically aligned CNT arrays [3–7]. In 2004, Hata et al. reported that adding a low water flow to a chemical vapor deposition (CVD) process increased the lifespan of the catalyst particles, resulting in an array of taller single-walled CNTs [8]. Ajayan et al. were able to achieve a 15 mm-tall multiwalled carbon nanotube (MWCNT) array by introducing air during the CVD process [7]. These ultralong growths resulted from the oxidation of the amorphous carbon that covers the surfaces of the catalyst particles and deactivates them. It has been reported that ultralong arrays with lengths of around 10 mm needed a growth period of longer than 10 h [7]. However, this increase in the growth period causes the morphologies of the CNTs to change. This, in turn, allows amorphous carbon to be deposited on the grown CNTs, thus resulting in changes in their diameters as well.

Recently, dry spinning of MWCNTs from vertically aligned arrays has been widely investigated in order to achieve lightweight, strong, functional large-scale CNT-based structures, including spun yarns [9–13] and sheets [14]. Connected by van der Waals forces, MWCNTs are drawn from the arrays while forming a CNT-based web, which is a two-dimensional CNT network formed from the three-dimensional array. The fact that MWCNTs are connected continuously over their lengths makes it easier to build large-scale CNT-based structures. The most noteworthy feature of MWCNT arrays is that all the MWCNTs are self-aligned in the drawing direction. By using these webs to assemble larger structures, this high degree of alignment can also be maintained in the resulting structures. Because of this alignment, these large-scale MWCNT-based structures exhibit mechanical properties that are highly desirable. Therefore, this spinning technology enables us to design new CNT-based materials that can be used in novel applications.

We have recently developed a rapid method for producing MWCNT arrays [14,15]. Millimeter-scale MWCNT arrays were grown by chloride-mediated chemical vapor deposition. The arrays had higher spinnability, as demonstrated by the fact that the MWCNT web could be drawn from 2.0-mm-high MWCNT arrays. We studied the spinnable CNT array structures and found that the highly straight and dense CNT forest structure is responsible for the high spinnability. Using the millimeter-long MWCNTs, material properties of unidirectionally aligned MWCNT sheet and twisted CNT yarns were investigated. It was found that outer walls of MWCNTs dominate as transfer channels for electron, phonon and load transfer, and so close packing of thinner and longer CNTs leads to high material properties.

14.2 HIGHLY SPINNABLE MWCNT ARRAYS
14.2.1 Growth of millimeter-long spinnable MWCNT arrays

Vertically aligned MWCNT arrays were synthesized using a conventional thermal chemical CVD system. A smooth quartz substrate was placed at the center of horizontal quartz tube furnace with iron chloride ($FeCl_2$) using a quartz boat. Typically, a thin metallic film deposited on a substrate is widely used as a catalyst; however, in our method, such a film need not be predeposited. During heating, the sample was maintained at a vacuum of 1×10^{-3} Torr, and once the optimal growth temperature was reached, it was purged with acetylene gas using a mass flow controller. CVD growth was carried out at furnace temperatures of 820–830 °C at low pressures <10 Torr. Figure 14.1 shows horizontal drawing of a web from a 2-mm-long

FIGURE 14.1

Two-millimeter-long spinnable MWCNT array. Transformation from a vertical array to a horizontal web is shown.

MWCNT array. MWCNTs with diameters of 30–50 nm are highly aligned perpendicular to the substrate and very highly bundled over the entire array. Since all the MWCNTs are strongly bundled to each other, the array can be easily peeled from the substrate and it is self-supporting; it resembles a thick black mat. It is essential that bundled MWCNTs spread across the entire substrate for high and steady spinnability.

14.2.2 Characterization of MWCNT arrays

Figure 14.2 shows a transmission electron microscopy (TEM) image of a typical CNT. The CNTs grown by the method have a multiwalled structure. The TEM image proved that MWCNTs have high crystal quality and little amorphous carbon deposition. The areal density is estimated by measuring the average distance between MWCNTs from side-view scanning electron microscopy (SEM) images of the arrays. Since the average distance of MWCNTs ranged from 200 to 300 nm, the areal density was estimated to be $1-3 \times 10^9/cm^2$. The high areal density is one of important issues for highly spinnable arrays. The high MWCNT quality was confirmed by Raman measurements.

Polarized Raman spectroscopy measurements were performed in backscattering geometry with incident light (laser light with a wavelength of 532 nm) normal to the side surface of the array (Fig. 14.3). The incident light was polarized parallel and perpendicular to the MWCNT alignment (see the inset of Fig. 14.3). For both polarizations, the ratio of the intensity of the graphite-like G band at 1580/cm to that of the disorder-induced D band at 1350/cm (I_G/I_D) is as high as 3.0. This high value indicates the high crystal quality of our MWCNTs and the low amount of amorphous

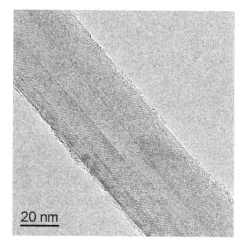

FIGURE 14.2

TEM image of a typical MWCNT.

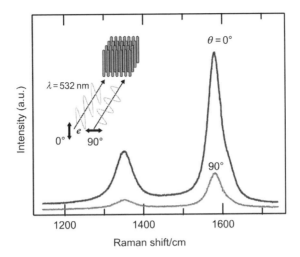

FIGURE 14.3

Polarized Raman spectra of an MWCNT array. Inset shows the measurement configuration. For the G band at 1580/cm, the parallel polarization intensity (red) is 4.4 times higher than the perpendicular polarization intensity. (For color version of this figure, the reader is referred to the online version of this book.)

carbon. The high quality of the MWCNTs is attributed to the high growth temperature (830 °C), which improves the crystalline quality of the graphene layers, and to the short growth time (5–16 min), which suppresses deposition of amorphous carbon. On the other hand, the G-band intensity ratio for the two polarizations, R ($= I_{parallel}/I_{perpendicular}$), provides information regarding the degree of MWCNT alignment in the array because Raman scattering is more intense when the polarization of the incident light is parallel to the axis of an MWCNT [16]. The obtained R is 4.4, which strongly indicates a high degree of alignment [17]. This alignment is due to the close proximity of growth sites.

14.2.3 Dry spinning of webs from millimeter-long CNT arrays

Due to high alignment with close distance (Fig. 14.4(a)), high crystal quality and small amount of amorphous carbon deposition, strong van der Waals attraction appears between individual CNTs and these vertically aligned CNT arrays can be continuously drawn into an aligned CNT web at a very high speed of more than 10 m/s [14]. A CNT web is a two-dimensional CNT network, in which all CNTs are well self-aligned in the direction of drawing. This self-alignment is a great advantage of dry spinning method. Excellent material properties of CNTs, having a cylindrical structure, are fully utilized when short CNTs are aligned in one direction in assembled materials and devices.

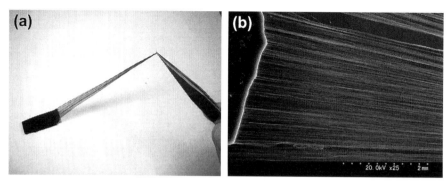

FIGURE 14.4

(a) MWCNT web initiation using tweezers. (b) MWCNT web. (For color version of this figure, the reader is referred to the online version of this book.)

Because the MWCNT arrays have high spinnability, no specialized equipment is required to form webs (Fig. 14.4(b)). They can be easily drawn by just pinching and pulling out an edge of the array using tweezers. During the spinning, drawn MWCNT bundles pull another one out from the array. This process repeats until all the MWCNTs have been removed from the substrate. The MWCNTs are aligned in the drawing direction in the web as shown in Fig. 14.4(c). The MWCNTs are automatically aligned in the drawing direction. This transformation, from a three-dimensional array to a two-dimensional web, appears to be a characteristic of CNTs. Previously reported methods for aligning a large number of CNTs require a lot of effort [16]. Forming webs of MWCNTs is a good method to obtain an aligned CNT structure. We believe this technique for forming MWCNT webs, which we term *CNT web technology*, will become a key CNT technology.

14.3 UNIDIRECTIONALLY ALIGNED CNT SHEET
14.3.1 Fabrication of unidirectionally aligned CNT sheet

To form aligned buckypapers, long-lasting MWCNT webs were wound on a drum (Fig. 14.5(a)). Because of the high spinnability and the high stickiness of the webs, large MWCNT paper sheets were easily formed in a short time [14]. To densify the as-stacked MWCNT paper, ethanol was sprayed on the sheets and then evaporated. The MWCNTs in the sheet were bonded to each other only by van der Waals forces; no binder material was needed. The thickness of the paper sheets could be controlled by varying the amount of webs rolled on the drum. Figure 14.5(b) shows the photograph of an A4 sheet of MWCNT paper, which was produced by cutting open a roll of paper. Owing to the high spinnability of the MWCNT array, an A4 sheet of paper can be fabricated in several minutes. The MWCNT paper produced in this study was 1.8 ± 0.1 μm thick and had a density

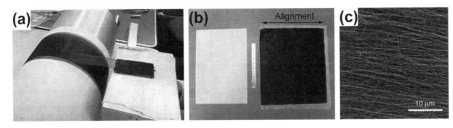

FIGURE 14.5

(a) Fabricating an aligned MWCNT paper sheet by winding MWCNT webs on a drum. (b) A4 sheet of aligned MWCNT buckypaper and (c) surface image of the MWCNT paper. (For color version of this figure, the reader is referred to the online version of this book.)

of 0.84 ± 0.08 g/cm^3. As a result of the high alignment of the MWCNTs in the webs, the MWCNTs are highly aligned in the paper (Fig. 14.5(c)). Therefore, the electrical, mechanical and thermal properties were significantly anisotropic.

In this study, all measurements of the MWCNT paper sheets were performed using conventional methods for macroscopic samples. The MWCNT paper sheets are not nanomaterials, but macrosized objects that can be handled just as easily as normal materials. Therefore, we consider that MWCNT paper sheets should be regarded as conventional sheets and should not be treated by techniques used for determining the physical parameters of CNTs (e.g. normalizing quantities by the density or subtracting the empty spaces in the material). Consequently, the parameters of the whole sample (i.e. the sample length, cross-sectional area and weight) were used when determining the material properties, which is conventional for macroscopic samples.

14.3.2 Electrical properties

The current–voltage characteristics were measured on a 1 cm × 1 cm area of a paper sheet parallel and perpendicular to the MWCNT alignment. Palladium plates were used as Ohmic electrodes. Linear current–voltage characteristics were observed (Fig. 14.6; see Table 14.1 for a summary). The sheet resistance in the parallel direction (13.8 Ω/sq) is lower than that in the perpendicular direction (100.1 Ω/sq) and the anisotropy ratio is 7.1. The resistivity of a 1.8-μm-thick paper sheet in the parallel direction is thus 2.5×10^{-3} Ω cm. The MWCNT paper is lightweight and highly electrically conductive with a high anisotropy. This macroscopic resistance of the MWCNT paper is mainly determined by the contact resistance between adjacent MWCNTs. Consequently, the parallel direction, which has fewer connecting points along current paths, has a lower resistance than the perpendicular direction. The high anisotropy is attributed to the MWCNTs having lengths of the order of millimeters and being highly aligned in the paper sheet.

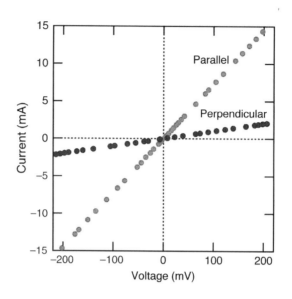

FIGURE 14.6

Current–voltage characteristics measured parallel and perpendicular to MWCNT alignment. (For color version of this figure, the reader is referred to the online version of this book.)

Table 14.1 Sheet Resistance and Resistivity Measured Parallel and Perpendicular to MWCNT Alignment

Direction	Sheet Resistance (Ω/sq)	Resistivity (Ω cm)
Parallel	13.8	2.5×10^{-3}
Perpendicular	100.1	1.8×10^{-2}

14.3.3 Mechanical properties

Figure 14.7 shows the stress–strain characteristics, which were measured using a tensile tester (Shimadzu, EZ-L) for MWCNT paper samples that were 1 cm × 1 cm in size and were mounted in a paper mount. The tensile strength, which was obtained by dividing the maximum load by the sample cross-sectional area, was 75.6 MPa parallel to the MWCNT alignment. Although the MWCNTs were bonded by van der Waals forces alone, the sheet strength is higher than that of disordered buckypaper [18], and is comparable to that of pure aluminum or nickel [19]. Unlike typical stress–strain curves for spun MWCNT fibers, which show that fracture occurs suddenly [9–13], the MWCNT paper fractures gradually due to sliding of MWCNTs, as indicated by the smooth peak structure in Fig. 14.7. In contrast, the tensile strength in the perpendicular direction is very low (see the inset in Fig. 14.7). The high strength in the parallel direction is attributed to the high

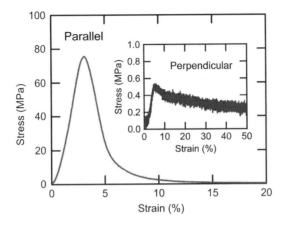

FIGURE 14.7

Stress–strain characteristics of the MWCNT paper sheet. (For color version of this figure, the reader is referred to the online version of this book.)

alignment of MWCNTs that have lengths of the order of millimeters and that are connected by very large surface areas, which increases the amount of van der Waals bonds that form between adjacent MWCNTs.

14.3.4 Thermal properties

Table 14.2 lists the highly anisotropic thermal properties. So far, for the aligned CNT sheet, electrical property was measured [20]. However, thermal conductivity has not been well investigated. The thermal diffusivities of the MWCNT paper sheets, α, were measured using a scanning laser heating thermal diffusivity meter (Ulvac-Riko, Inc., Laser PIT). The scanning laser heating provides a highly uniform and high-intensity energy source. An analysis method using both the amplitude decay and the phase shift was employed to eliminate the effects of heat loss. MWCNT paper sheets (5 mm × 25 mm in size) were measured at room temperature in a vacuum of <0.01 Pa. The experimental details of the measurement system are given in Ref. [21].

The thermal conductivities of the paper sheets, K, were calculated using $K = \alpha \rho C$, where ρ is the density and C is the specific heat of the MWCNT paper.

Table 14.2 Thermal Diffusivity and Thermal Conductivity Measured Parallel and Perpendicular to MWCNT Alignment

Direction	Thermal Diffusivity (m²/s)	Thermal Conductivity (W/m K)
Parallel	1.22×10^{-4}	69.6
Perpendicular	1.50×10^{-5}	8.6

The specific heat of an MWCNT was assumed to be 0.713 J/g K, which is almost identical to that of graphite at room temperature [22]. The thermal conductivity of the paper in the parallel direction was estimated to be 69.6 W/m K, which is much higher than that of reported buckypaper [23]. However, it is still lower than the thermal conductivity of single CNTs [24]. Since MWCNT paper consists of a huge number of MWCNTs, it is reasonable to consider that the large-scale thermal conductivity is governed by the heat resistance at MWCNT–MWCNT interfaces [25]. The anisotropy ratio for the thermal conductivity is as high as 8.1, which is due to the high alignment of ultralong MWCNTs.

14.3.5 Discussions

When MWCNTs are used as large-scale industrial materials, it is important to consider that irrespective of whether they are used for transferring electric current, mechanical stress or heat, the outermost walls of the MWCNTs form the main channel. Since electrons are injected generally from the outermost walls and the electrical conductivity in the graphene plane is much higher than that between the graphene planes, electron transport predominately occurs through the outermost wall [26]. In addition, when an electron transfers between two MWCNTs, it passes through the interface between the two surfaces. Consequently, macroscopic electrical conduction is governed by the outermost walls. For heat transport, in which phonons are the dominant carriers [27], similar reasoning can be used to predict that phonons are mainly transported through the outermost wall [25]. On the other hand, for load transfer in MWCNT structures such as buckypapers and spun yarns [9–13], the external load is balanced by shear forces at MWCNT–MWCNT interfaces. Since the total tensile strength of an MWCNT structure is much less than that of an individual MWCNT, fracture occurs due to sliding of MWCNTs. Therefore, only the outermost walls are responsible for the shear force.

Consequently, in addition to the hollow areas and interspaces between MWCNTs, the cross-sectional area of the inner walls does not participate in transfer of charge, load and heat. Nevertheless, when MWCNT structures are used as industrial materials, their total cross-sectional area should be taken into account, as is done with conventional macroscopic materials. Figure 14.8(b) shows the calculated relationship between the effective cross-sectional area and the MWCNT diameter. Assuming that the aligned MWCNTs are hexagonally close packed, the effective cross section is simply given by the cross-sectional area of the outermost wall that is contained in a hexagon (Fig. 14.8(a)). The ratio of the effective cross-sectional area to the total cross-sectional area, $X_{effective}$, is expressed by

$$X_{effective} = \frac{2\pi dD}{\sqrt{3}(d+D)^2}, \qquad (14.1)$$

where D is the diameter of the MWCNT. In this calculation, the spacing between the walls of an MWCNT is assumed to be identical to the distance between adjacent MWCNTs; this distance is denoted by d and set to be 0.34 nm, which is the spacing

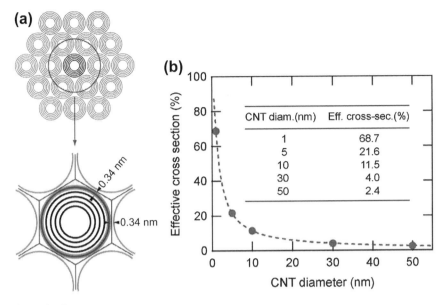

FIGURE 14.8

(a) Schematic of the effective cross-sectional area of an aligned MWCNT structure. Calculation of the effective cross-sectional area ratio in the total cross-sectional area can be reduced to that of a hexagon, which contains an MWCNT. The green area is the effective cross section in the red hexagon. The spacing between the walls in an MWCNT and the distance between adjacent MWCNTs are assumed to be equal to the spacing of graphite (002) planes (0.34 nm). (b) Calculated effective cross-sectional area as a function of MWCNT diameter. The inset tabulates the results. (For interpretation of the references to color in this figure legend, the reader is referred to the online version of this book.)

between graphite (002) planes. The effective cross-sectional area decreases drastically with increasing diameter. The calculation reveals that even for a 1-nm-diameter single-walled nanotube, 30% of the total cross-sectional area does not participate in transfer of charge, load and heat. In this study, the MWCNTs were 30 nm in diameter; thus, at most only 4% of the cross-sectional area is effective. Based on the density of the MWCNT paper sheets being lower than expected, it is obvious that they are not ideally packed. The actual effective cross-sectional area is speculated to be considerably <4%. To enhance the properties of CNT structures, it is critical to densely pack thin CNTs so as to reduce ineffective spaces.

14.4 MECHANICAL PROPERTIES OF CNT YARN

14.4.1 Fabrication of CNT yarn by dry spinning

CNT yarns were fabricated by dry spinning method, in which CNTs were drawn out from CNT arrays and the drawn web was twisted with no chemical binder

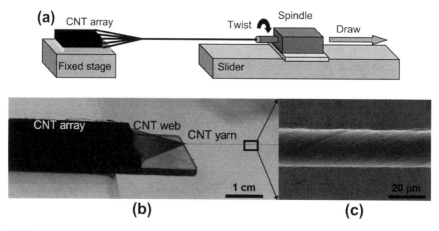

FIGURE 14.9

(a) Schematic representation of CNT yarns' spinning procedure. (b) Spinning from 1-cm-wide CNT array. (c) SEM image of as-spun CNT yarn. (For color version of this figure, the reader is referred to the online version of this book.)

materials [28]. For yarns spinning, we used a simple spinning system schematically illustrated in Fig. 14.9(a). First, a uniform CNT array was placed on a sample stage, and then a CNT web was drawn with tweezers and attached to a spindle. To spin a CNT yarn, a rotating spindle was drawn back along a motorized slider. The spinning system provides simultaneously constant twisting and drawing of a CNT web. We call the spun yarn as-spun CNT yarn. Typical spinning parameters were 32,000/min for spindle rotation and 120 mm/s for slider speed, which give a twisting angle of 25° for a 5-mm-wide web.

14.4.2 Yarn evaluation methods

For mechanical properties measurements, a tensile tester was used at a constant displacement rate of 1 mm/min and a gauge length of 10 mm. Testing specimens were prepared as follows: after spinning, a small piece of a CNT yarn was fixed on a paper mount using adhesive tape and cyanoacrylate adhesive as shown in Fig. 14.10. After fixing the paper mount into the testing machine, the paper was cut at the cutting line and stretching force has been applied to the yarns. The longitudinal strain was measured using a noncontacting video extensometer (Shimadzu TRViewX). For each yarn, five specimens have been tested and the results were averaged.

In order to determine the tensile strength of CNT yarns, accurate measurements of yarns cross-sectional areas are required. Cross-sectional area was assumed as a circular shape and calculated from yarn diameter. Yarn diameters were measured using a scanning electron microscope, in which scale was corrected by a length standard. To minimize errors, we measured the diameter in three different points of the yarn and the average value has been used for calculations. The weight density of the

14.4 Mechanical Properties of CNT Yarn

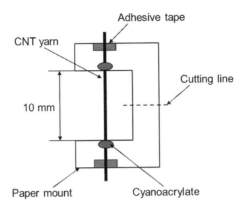

FIGURE 14.10

Specimen for tensile strength measurement. (For color version of this figure, the reader is referred to the online version of this book.)

yarns was calculated by determining the mass of a 5 cm piece of CNT yarn using a microbalance (Sartorius ME 5) with a resolution of 1 μg.

14.4.3 Mechanical property of a CNT web

First, we measured tensile mechanical property of a CNT web. An as-drawn web with a width of 5 mm was attached on a paper mount. A measured load–strain characteristic is shown in Fig. 14.11. We found that a very small tension of only 0.035 N would

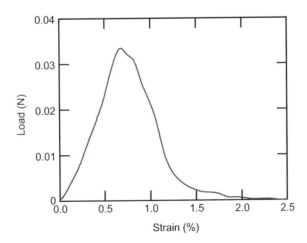

FIGURE 14.11

Load–strain curve of a CNT web. (For color version of this figure, the reader is referred to the online version of this book.)

easily break the CNT web. A peak-shaped structure of the load—strain curve indicates that CNTs are connected to each other with certain overlapping lengths and, under tensile stress, CNTs gradually slide in the longitudinal direction. This small tension revealed that it is limited to put a certain tension on a web during spinning process, and hence, it is limited to straighten and align the CNTs in the web by applying a tension. For this reason, it is almost impossible to make the CNT yarns close-packed by twisting webs and we performed postspin processing on the spun yarns.

14.4.4 Postspin twisting of as-spun yarns
14.4.4.1 Postspin twisting effect

As postprocessing, a postspin twisting has been applied on as-spun yarns in order to enhance mechanical properties [29], which is a similar process used for conventional spun yarns. Since tension in a web is limited by weak van der Waals interactions of small CNT surfaces, it is difficult to apply high tension on a web during a spinning process. However, once a web is spun into a yarn in which CNTs contact each other with large surface area, tensile strength of the yarn is improved and hence higher tension can be applied on that yarn. A CNT yarn was vertically hung by fixing one end of the yarn to a spindle while at the other end weight was attached to apply tension into the yarn. Then, the as-spun CNT yarn was further twisted at different twisting densities.

As a result of the postspin twisting, CNT yarns became thinner and, consequently, the yarn density became higher. The immediate effect of postspin twisting was a drastic increase of the tensile strength and Young's modulus from 418 MPa to 30.6 GPa for as-spun CNT yarn to 772 MPa and 51.1 GPa for post-spin-twisted CNT yarn as Fig. 14.12 shows.

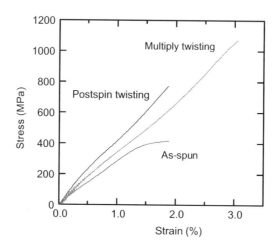

FIGURE 14.12

Stress—strain curves measured on as-spun yarn, post-spin-twisted yarn and multiply-twisted yarn. (For color version of this figure, the reader is referred to the online version of this book.)

Table 14.3 Density, Tensile Strength and Young's Modulus of As-Spun Yarn and Post-spin-Twisted Yarn (Scale Bar: 10 μm)

	As-spun	Post-spin
ρ (g/cm^3)	0.73	1.24
σ (MPa)	418	772
E (GPa)	30.6	51.1

Table 14.3 shows the SEM images of as-spun and post-spin-twisted yarns along with corresponding values for yarns density (ρ), tensile strength (σ), and Young's modulus (E), respectively. After postspin twisting, yarn diameter decreased from 22.8 to 19.2 μm, and density increased from 0.73 to 1.24 g/cm^3. Cross-sectional area of the post-spin-twisted yarn decreased to 70%, but tensile strength and Young's modulus increased to 184% and 166%, respectively. The effective load capacity increased to 130%, from 0.171 N for as-spun yarn to 0.224 N for post-spin-twisted yarn. The reason for the enhancement of mechanical properties is surely attributed to the higher density of the post-spin-twisted yarns. After postspin twisting, the CNT yarns become closely packed as shown in Fig. 14.13, the distance between CNTs into the yarn is smaller and van der Waals interaction is significantly enhanced. However, there is still room for density improvement because the density of MWCNT with a diameter of 40 nm is calculated to be around 2 g/cm^3, assuming the CNT inner diameter of around 5 nm, interlayer spacing of 0.34 nm and C–C bond length of 0.142 nm. Small density of the yarns indicates the presence of voids and interspaces, and a very high porosity of CNT yarns [30].

As described before in the experimental section, we were able to change the postspin twisting density by simply changing the spindle rotation speed. We obtained CNT yarns with twisting densities up to 7/mm and measured their tensile strength as shown in Fig. 14.14. We found that, at a twisting density of 3–4/mm, the tensile strength had a maximum followed by a decrease for higher twisting density. For low twisting densities up to 4/mm, by increasing the twisting density, we improved CNT yarns packing and, as a result, weight density of the yarns became higher. In this region, weight density has a dominant effect on tensile strength enhancement. By further introducing a higher twisting density, because the maximum packing density was achieved, the surface twist angle becomes an important parameter. A large twisting angle gives smaller inward stress during tensile loading of the yarn [31]. As a result, van der Waals interaction between CNTs in the yarn decreases and shear strength degrades, similar to the conventional theory for strength of spun yarns.

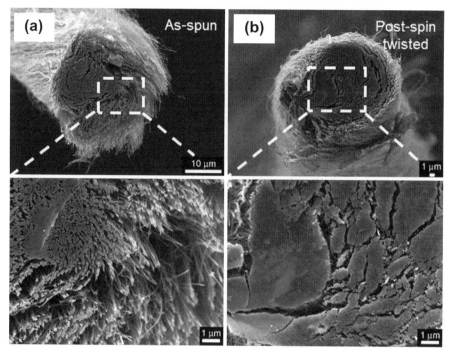

FIGURE 14.13

Cross-sectional observations of (a) as-spun and (b) post-spin-twisted yarns. (For color version of this figure, the reader is referred to the online version of this book.)

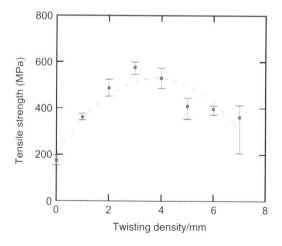

FIGURE 14.14

Twisting density dependence of tensile strength. (For color version of this figure, the reader is referred to the online version of this book.)

14.4.4.2 CNT length dependence

In the theory of conventional spun yarns, it is well known that high aspect ratio of individual fibers improves mechanical properties of the final spun yarns. In order to investigate the similar influence of CNT length on the yarns strength, we used CNTs with lengths ranging from 0.8 to 2.1 mm and results are shown in Fig. 14.15.

The average outer diameter of MWCNTs used in this study was around 40 nm, so the aspect ratio of CNTs was changed from 20,000 for the shortest CNTs to more than 50,000 for the longest ones. With increasing the CNT length, the tensile strength monotonically increased for both as-spun and post-spin-twisted yarns. The same trend has been reported by Fang et al. [32] for yarns obtained from very short CNT arrays. In the case of higher aspect ratio CNTs, contact area per CNT increases leading to more efficient load transfer. According to a conventional theory [33], tensile strength should be saturated in this aspect ratio range, for CNT length higher than several micrometers. This discrepancy arises from the different material properties between conventional small fibers and CNTs which will be discussed in detail later. The linearity of experimental data presented in Fig. 14.15 is perturbed only by the CNT yarns obtained from 1.4-mm-high CNT array. A possible reason might be slightly thicker yarns obtained from that CNT array. Even if the initial CNT web width was kept constant, highly bundled CNT arrays give a denser CNT web and, consequently, thicker CNT yarns. The strength of spun CNT yarns is diameter dependent as discussed in the next paragraph.

14.4.4.3 Yarns' diameter dependence

To study the effect of yarn thickness on tensile property, yarns' diameters were varied from 10 to 70 μm. The diameter of the yarns has been adjusted by changing the initial CNT web width from 1 to 25 mm. A constant web width was obtained by

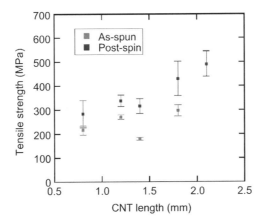

FIGURE 14.15

MWCNT length dependence of tensile strength. (For color version of this figure, the reader is referred to the online version of this book.)

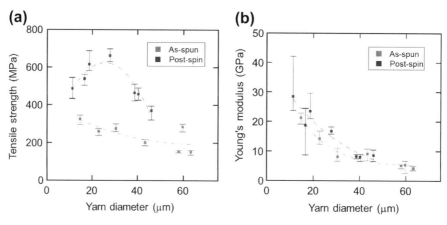

FIGURE 14.16

Yarn diameter dependence of (a) tensile strength and (b) Young's modulus. (For color version of this figure, the reader is referred to the online version of this book.)

engraving cutting lines to arrays along the substrate separated by desired width, allowing us to obtain CNT yarns with uniform diameter for the entire length of the yarn.

Figure 14.16(a) shows the tensile strength variation as a function of CNT yarns diameter for as-spun yarns and post-spin-twisted yarns at a twisting density of 4/mm. For the as-spun CNT yarns, as the yarn diameter decreases, the tensile strength increases as a result of the reduction of weak points in the yarn [34]. On the other hand, for post-spin-twisted yarns, the tensile strength has a maximum at a diameter of 30 μm followed by a reduction for smaller diameters. When the CNT yarn is very thin, it is difficult to apply a high tension and constant twist to the yarns in our spinning system. As a result, the CNTs are not well-aligned parallel with the yarn axis and such thin yarns include a large number of weak points.

As regarding the stiffness of the CNT yarns, we have calculated the Young's modulus from the stress—strain curves and results are shown in Fig. 14.16(b). Young's modulus monotonically decreased with increasing yarns diameter for both as-spun and post-spin-twisted CNT yarns. We can observe a different trend of tensile strength and Young's modulus for very thin yarns. This is because Young's modulus is measured in the elastic region of the stress—strain curves, where the presence of weak points does not influence yet on CNT yarns strain.

For single-spun yarns, CNT packing within the yarn is not uniform and the yarn has a so-called core-sheath structure [35], with high density at the inner part and lower density at the external part. Therefore, the tensile load is mainly supported by the CNTs from the yarn center and very less by the outer CNTs. In order to increase the inward radial force, we considered a multiply twisting process of CNT yarns.

14.4.5 Multiply twisted CNT yarns
14.4.5.1 Multiply twisting of as-spun CNT yarns

As another postspin processing, two or several as-spun or post-spin-twisted CNT yarns were further twisted together. CNT yarns were fixed to a vertically oriented spindle and weights were attached to the free end of each yarn as shown in Fig. 14.17. The spindle can move up by using a motorized slider. The multiply twisted yarns were twisted in the same direction as the original as-spun CNT yarns. Twisting density of the multiply twisted yarns can be controlled by changing the spindle rotation speed and slider speed. Typical rotation speed of the spindle and slider speed were 240/min and 1 mm/s, respectively.

To further improve the close packing of CNT yarns, 2–8 yarns were twisted together. Among these, the strongest proved to be the two-ply twisted yarns, which had a tensile strength of more than 1 GPa as we can see from Fig. 14.12. A simple two-ply twisting can give almost a threefold increase of tensile strength as compared to a single as-spun yarn. Further, for all experiments, we used only two yarns for multiply twisting and we will refer to these as two-ply twisted yarns.

During the two-ply twisting process, as-spun CNT yarns were twisted under a certain tension so that the two-ply twisted yarns were packed densely as compared to the original as spun yarns. Also for the two ply twisted yarns, we observed a significantly larger strain exceeding by almost 1% the strain of the as-spun yarns.

To reduce measurements errors, we assumed that the cross section of a two-ply twisted yarn has an elliptic shape and we measured long and short axis from SEM image. Then we simply calculated the area enclosed by the ellipse as the

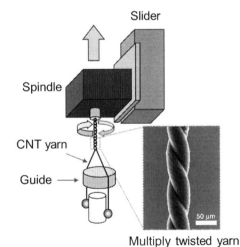

FIGURE 14.17

Multiply twisting of CNT yarns. (For color version of this figure, the reader is referred to the online version of this book.)

cross-sectional area of the two-ply twisted yarn. This elliptic-shaped cross section is due to the low density of as-spun CNT yarns which form distorted circles when twisted together.

14.4.5.2 Twisting tension dependence

As explained before, during the two-ply twisting, a certain tension was applied to the individual as-spun yarns. By changing the weight on the yarns, we modified the applied tension and results are given in Fig. 14.18 and Table 14.4. The twisting tension was calculated by dividing the weight by cross-sectional area of the yarn.

By applying more and more tension during two-ply twisting process, density was monotonically increased beyond the highest density of the post-spin-twisted yarns. Under higher twisting tension, CNTs are well aligned parallel to the yarn axis and

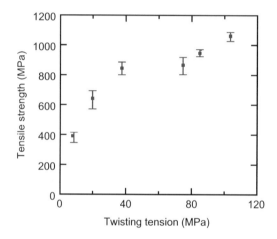

FIGURE 14.18

Tensile strength of two-ply-twisted yarns as a function of twisting tension. (For color version of this figure, the reader is referred to the online version of this book.)

Table 14.4 Twisting Tension Dependence of Mechanical Properties of Two-Ply-Twisted Yarns (Scale Bar: 50 μm)

Twisting tension (MPa)	9	20	38	75	85	104
ρ (g/cm³)	0.83	0.99	1.03	1.25	1.32	1.53
σ (MPa)	383	643	845	867	946	1061
E (GPa)	11	23	28	41	50	51

CNT yarns become denser. Tensile strength and Young's modulus of the two-ply-twisted yarns were significantly improved. At the maximum twisting tension of 104 MPa, we calculated a tensile strength of 1061 MPa and a Young's modulus of 51 GPa. This result indicates that a large twisting tension induced a very tight packing of CNTs into the yarn. The tight packing was also made by further ordering of CNTs produced under a certain tension.

By two-ply twisting, most of interspaces in the yarn were filled with CNTs. In contrast to a single-spun yarn, two-ply twisting induces inward radial force given from outside on contacting yarn surfaces between two yarns. Such an external force is effective to reduce interspaces in the yarn and to increase van der Waals interactions. To further applying inward radial external force, CNT yarns were simply pressed using external weights.

14.4.6 Pressed CNT yarns

We tried to directly apply a strong pressure to two-ply-twisted yarns and compare their performances with as-spun and 2-ply-twisted CNT yarns. All three yarns used for this analysis were spun from the same CNT array at a twisting density of 4/mm. The two-ply-twisted yarn was twisted with an applied tension of 60 MPa and an additional twisting density of 4/mm. Finally, for the pressed yarn, a uniaxial pressure of 2 MPa has been applied to a two-ply-twisted CNT yarn.

As we can see in Table 14.5, the directly pressed yarns had a density as high as 1.89 g/cm^3 and the highest tensile strength of 1068 MPa and Young's modulus of 55 GPa among all the yarns presented in this study. For two-ply-twisted yarns, only CNTs at the interface between the two yarns could be packed with the core of the individual yarns. When an external pressure was applied to the yarns, also the CNTs from the very edge of the two yarns were additionally incorporated to achieve such a high density of the yarns. Consequently, almost all CNTs from the cross section of the yarn effectively support the tensile load and the tensile strength of these yarns had a maximum.

Table 14.5 Comparison of Morphology and Properties of As-Spun, Two-Ply-Twisted and Two-Ply-Pressed Yarns (Scale Bar: 50 μm)

	As-Spun	Two-Ply-Twisted	Two-Ply-Pressed
ρ (g/cm^3)	0.57	1.22	1.89
σ (MPa)	300	676	1068
E (GPa)	36	34	55

14.4.7 Discussions

The main advantage of our dry spinning technology consists in using ultralong and high-quality CNTs. In the case of spun CNT yarns, the load is transferred along the CNTs and discontinuity at both ends acts as defects [36]. In this work, by using millimeter-long CNTs, we could reduce the defect density per unit length. Moreover, when using long CNTs, contacting surface area per CNT was increased and thus van der Waals interaction per CNT was enhanced. These things led to higher tensile strength of the yarns.

If we compare CNT spinning with conventional yarns spinning, there are many similarities. First, in both cases, small fibers are aligned and twisted together to form a much stronger macroscopic yarn. Second, mechanical properties of spun yarns are enhanced when individual fibers become longer and yarn diameter decreases.

However, there are some special features which are characteristics for CNT yarns. Owing to their atomically flat surface, there is almost no friction between CNTs and only van der Waals interaction is responsible for the strength of the CNT yarn. Also, yarns' breaking modes are different. In the case of conventional spun yarns, when the yarn breaks, all short individual fibers are broken. On the other hand, in the case of spun CNT yarn, the tensile strength of each short CNT is huge as compared to the shear force between them [37], which means CNTs in the yarn are not fully locked. Therefore, the reason for CNT yarns' failure can be attributed to relative sliding of CNTs rather than breaking of individual CNTs. One possibility to increase the shear force might be to connect individual nanotubes by direct covalent bonding or bridging, with or without certain functional groups.

In addition to those stated before, we found that the tensile strength of CNT yarns does not obey the conventional theory for mechanical properties of spun yarns. In that theory, the tensile strength of a yarn is given by

$$\sigma_{spun} = \sigma_{fibril} \cos^2 \theta \left(1 - \frac{1}{3L \sin \theta} \sqrt{\frac{dl}{\mu}}\right), \tag{14.2}$$

where σ_{fibril} is the tensile strength of individual CNT, L is the CNT length, d is the CNT diameter, θ is the twisting angle of the yarn, l is the migration length and μ is the friction coefficient [33]. According to this relation, the yarns strength seems to saturate for long CNTs having a diameter of tens of nanometers and length of micrometers or longer, assuming appropriate values for other parameters. The second term of this relation goes to zero under such conditions, and the tensile strength of the spun yarn saturates as $\sigma_{fibril} \cos^2 \theta$. However, in this study, by further using longer CNTs in the millimeter range, the tensile strength will keep raising and saturation point was not reached yet. Moreover, ultimate tensile strength is quite smaller than expected by $\sigma_{fibril} \cos^2 \theta$, assuming the tensile strength of individual CNTs as several tens of GPa [37,38], and a certain twisting angle used in this study. These deviations are caused by the different breaking modes. A better understanding of the tensile properties of the spun CNT yarn requires further studies regarding not only the degree of lock between CNTs but also load transfer from outer walls to inner ones.

For MWCNTs structures, the main channel for load transfer is formed by the outer walls. The strength of the individual CNTs degrades with increasing the number of walls, due to inefficient load transfer from the outer walls to the inner ones [32]. The MWCNTs used in this study have more than 30 walls but are much less deformable as compared to few walled CNTs. Therefore, these CNTs keep their circular cross section and will not be flattened by uniaxial pressure. Owing to this high bending stiffness, CNT bundles become less flexible when array height is shorter than 0.4 mm and CNT arrays lose their spinnability. In contrast, the height of spinnable CNT arrays usually used for CNT yarns spinning is about 0.5 mm [29,34].

The CNT yarn is composed of CNTs separated by air gaps or voids and, from the entire cross section of the yarn, only a small fraction is effective for load transfer. Our CNTs used in this work have an average diameter of 40 nm and, if we suppose a hexagonal close packing of CNTs into the yarns, we calculated that only about 3% represents the effective cross section [14]. Therefore, we believe that close packing of thin and long CNTs is necessary in order to fully utilize the huge potential of CNTs. As for the practical applications of these CNT yarns, it is necessary to obtain a higher effective load capacity. Anyway, at this point, our yarns with diameters of several tens of micrometers can support an effective load of about 0.8 N and can be easily handled, which is another great advantage as compared to dry spun CNT yarns previously reported.

14.5 CONCLUSIONS

In this work, highly drawable MWCNT arrays were produced by chloride-assisted CVD. The drawability was so high that webs could be drawn at high speed from a 2.0-mm-long MWCNT array. By stacking and shrinking the webs, aligned MWCNT paper sheets could be easily fabricated without using a binding material. The MWCNT paper produced was lightweight and flexible, while being mechanically strong and having a high electrical conductivity. The fabrication and mechanical properties of dry spun CNT yarns were reported. As postprocessing, postspin twisting, multiply twisting and direct pressing were applied in order to obtain high mechanical strength and Young's modulus of these CNT yarns. The influence of CNT length, yarn diameter and twisting density was evaluated and discussed. By using millimeter-long CNTs and applying the postspin processes, CNT yarns stronger than 1 GPa and with 50 GPa Young's modulus were obtained. The yarns also have a high specific load capacity. By suppressing the relative sliding of the CNTs in the yarn, and by using thin and long CNTs, the material properties of spun CNT structures will be further improved, which may open the gate toward practical applications of CNT sheets and yarns.

Acknowledgments

The author thanks his former students Y. Suzuki, Y. Minami, J. Muramatsu and Dr A Ghemes for obtaining most of the experimental data presented in this paper.

References

[1] F.H. Gojny, M.G.H. Wichmann, B. Fiedler, I.A. Kinloch, W. Bauhofer, A.H. Windle, et al., Evaluation and identification of electrical and thermal conduction mechanisms in carbon nanotube/epoxy composites, Polymer 47 (2006) 2036–2045.
[2] J.N. Coleman, U. Khan, W.J. Blau, Y.K. Gun'ko, Small but strong: a review of the mechanical properties of carbon nanotube–polymer composites, Carbon 44 (2006) 1624–1652.
[3] S. Yasuda, D.N. Futaba, T. Yamada, M. Yumura, K. Hata, Gas dwell time control for rapid and long lifetime growth of single-walled carbon nanotube forests, Nano Letters 11 (2011) 3617–3623.
[4] K. Hasegawa, S. Noda, Moderating carbon supply and suppressing Ostwald ripening of catalyst particles to produce 4.5-mm-tall single-walled carbon nanotube forests, Carbon 49 (2011) 4497–4504.
[5] G. Zhong, T. Iwasaki, J. Robertson, H. Kawarada, Growth kinetics of 0.5 cm vertically aligned single-walled carbon nanotubes, Journal of Physical Chemistry B 111 (2007) 1907–1910.
[6] Y. Yun, V. Shanov, Y. Tu, S. Subramaniam, M.J. Schulz, Growth mechanism of long aligned multiwall carbon nanotube arrays by water-assisted chemical vapor deposition, Journal of Physical Chemistry B 110 (2006) 23920–23925.
[7] X.S. Li, X.F. Zhang, L.J. Ci, R. Shah, C. Wolfe, S. Kar, et al., Air-assisted growth of ultra-long carbon nanotube bundles, Nanotechnology 19 (2008) 455609.
[8] K. Hata, D.N. Futaba, K. Mizuno, T. Namai, M. Yumura, S. Iijima, Science 306 (2004) 1362.
[9] K. Jiang, Q. Li, S. Fan, Spinning continuous carbon nanotube yarns, Nature 419 (2002) 801.
[10] M. Zhang, K.R. Atkinson, R.H. Baughman, Multifunctional carbon nanotube yarns by downsizing an ancient technology, Science 306 (2004) 1358–1361.
[11] C.D. Tran, W. Humphries, S.M. Smith, C. Huynh, S. Lucas, Improving the tensile strength of carbon nanotube spun yarns using a modified spinning process, Carbon 47 (2009) 2662–2670.
[12] Y. Nakayama, Synthesis, nanoprocessing, and yarn application of carbon nanotubes, Japanese Journal of Applied Physics 47 (2008) 8149–8156.
[13] X. Zhang, Q. Li, T.G. Holesinger, P.N. Arendt, J. Huang, P.D. Kirven, et al., Ultrastrong, stiff, and lightweight carbon-nanotube fibers, Advanced Materials 19 (2007) 4198–4201.
[14] Y. Inoue, Y. Suzuki, Y. Minami, J. Muramatsu, Y. Shimamura, et al., Anisotropic carbon nanotube papers fabricated from multiwalled carbon nanotube webs, Carbon 49 (2011) 2437–2443.

[15] Y. Inoue, K. Kakihata, Y. Hirono, T. Horie, A. Ishida, H. Mimura, One-step grown aligned bulk carbon nanotubes by chloride mediated chemical vapour deposition, Applied Physics Letters 92 (2008) 213113.

[16] A.M. Rao, A. Jorio, M.A. Pimenta, M.S.S. Dantas, R. Saito, G. Dresselhaus, et al., Polarized Raman study of aligned multiwalled carbon nanotubes, Physical Review Letters 84 (2000) 1820–1823.

[17] Y. Zhang, G. Zou, S.K. Doorn, H. Htoon, L. Stan, M.E. Hawley, et al., Tailoring the morphology of carbon nanotube arrays: from spinnable forests to undulating foams, ACS Nano 3 (2009) 2157–2162.

[18] J.G. Park, J. Smithyman, C.Y. Lin, A. Cooke, A.W. Kismarahardja, S. Li, et al., Effects of surfactants and alignment on the physical properties of single walled carbon nanotube buckypaper, Journal of Applied Physics 106 (2009) 104310.

[19] H.E. Boyer, T.L. Gall, Metals Handbook, Desk Edition, American Society for Metals, Ohio, 1985.

[20] K. Liu, Y. Sun, L. Chen, C. Feng, X. Feng, K. Jiang, et al., Controlled growth of super-aligned carbon nanotube arrays for spinning continuous unidirectional sheets with tunable physical properties, Nano Letters 8 (2008) 700–705.

[21] K. Naito, J.M. Yang, Y. Xu, Y. Kagawa, Enhancing the thermal conductivity of polyacrylonitrile- and pitch-based carbon fibers by grafting carbon nanotubes on them, Carbon 48 (2010) 1849–1857.

[22] J. Hone, M.C. Llaguno, M.J. Biercuk, A.T. Johnson, B. Batlogg, Z. Benes, et al., Thermal properties of carbon nanotubes and nanotube-based materials, Applied Physics A 74 (2002) 339–343.

[23] Y. Yue, X. Huang, X. Wang, Thermal transport in multiwall carbon nanotube buckypapers, Physics Letters A 374 (2010) 4144–4151.

[24] E. Pop, D. Mann, Q. Wang, K. Goodson, H. Dai, Thermal conductance of an individual single-wall carbon nanotube above room temperature, Nano Letters 6 (2006) 96–100.

[25] P. Kim, L. Shi, A. Majumdar, P.L. McEuen, Thermal transport measurements of individual multiwalled nanotubes, Physical Review Letters 87 (2001) 215502.

[26] B. Bourlon, C. Miko, L. Forro, D.C. Glattli, A. Bachtold, Determination of the intershell conductance in multiwalled carbon nanotubes, Physical Review Letters 93 (2004) 176806.

[27] D.J. Yang, Q. Zhang, G. Chen, S.F. Yoon, J. Ahn, S.G. Wang, et al., Thermal conductivity of multiwalled carbon nanotubes, Physical Review B 66 (2002) 165440.

[28] A. Ghemes, Y. Minami, J. Muramatsu, M. Okada, H. Mimura, Y. Inoue, Fabrication and mechanical properties of carbon nanotube yarns spun from ultra-long multi-walled carbon nanotube arrays, Carbon 50 (2012) 4579.

[29] X. Zhang, Q. Li, Y. Tu, Y. Li, J.Y. Coulter, L. Zheng, et al., Strong carbon-nanotube fibers spun from long carbon-nanotube arrays, Small 3 (2) (2007) 244–248.

[30] M. Miao, J. McDonnell, L. Vockovic, S.C. Hawkins, Poisson's ratio and porosity of carbon nanotube dry-spun yarns, Carbon 48 (2010) 2802–2811.

[31] J. Zhao, X. Zhang, J. Di, G. Xu, X. Yang, X. Liu, et al., Double-peak mechanical properties of carbon-nanotube fibers, Small 6 (22) (2008) 2612–2617.

[32] S. Fang, M. Zhang, A.A. Zakhidov, R.H. Baughman, Structure and process-dependent properties of solid-state spun carbon nanotube yarns, Journal of Physics: Condensed Matter 22 (2010) 334221.

[33] J.W.S. Hearle, P. Grosberg, S. Backer, Structural Mechanics of Fibers, Yarns and Fabrics, Wiley Interscience, New York, 1969.
[34] K. Liu, Y. Sun, R. Zhou, H. Zhu, J. Wang, L. Liu, et al., Carbon nanotube yarns with high tensile strength made by a twisting and shrinking method, Nanotechnology 21 (4) (2010) 045708.
[35] K. Sears, C. Skourtis, K. Atkinson, N. Finn, W. Humphries, Focused ion beam milling of carbon nanotube yarns to study the relationship between structure and strength, Carbon 48 (2010) 4450–4456.
[36] N. Behabtu, M.J. Green, M. Pasquali, Carbon nanotube-based neat fibers, Nano Today 3 (5–6) (2008) 24–34.
[37] M.F. Yu, O. Lourie, M.J. Dyer, K. Moloni, T.F. Kelly, R.S. Ruoff, Strength and breaking mechanism of multiwalled carbon nanotubes under tensile load, Science 287 (5453) (2000) 637–640.
[38] B.G. Demczyk, Y.M. Wang, J. Cumings, M. Hetman, W. Han, A. Zettl, et al., Direct mechanical measurement of the tensile strength and elastic modulus of multiwalled carbon nanotubes, Material Science and Engineering A 334 (2002) 173–178.

CHAPTER 15

Transport Mechanisms in Metallic and Semiconducting Single-walled Carbon Nanotubes: Cross-over from Weak Localization to Hopping Conduction

Kazuhiro Yanagi

Department of Physics, Tokyo Metropolitan University, Hachioji, Tokyo, Japan

CHAPTER OUTLINE

15.1 Introduction	415
15.2 Relationship between MS Ratio and Conductivity of SWCNT Networks	416
15.3 Summary	422
References	423

15.1 INTRODUCTION

Single-walled carbon nanotubes (SWCNTs) are cylindrical graphitic tubes with diameters of about 1 nm. They exhibit metallic or semiconducting characteristics depending on how the graphitic sheets are rolled (chirality). Because of their remarkable electric characteristics, SWCNTs are expected to be used in a wide range of applications such as field-effect transistors and conducting films. In such applications, networks of SWCNTs are fabricated in the devices. Thus, a fundamental understanding of the electric transport properties in such networks is of great importance for improvement of device performance. A number of studies concerning this issue have been reported; however, most of the studies have been performed on the mixture of the metallic and semiconducting SWCNTs because SWCNTs are produced in such a mixed state [1–6]. In a single bundle of nanotubes, it has been suggested that conductance is affected by two-dimensional weak localization (WL) processes, due to the interference of electron waves, which are classified as quantum transport phenomena [7]. However, the observed temperature dependence of resistance in nanotube networks, which are networks formed by nanotube bundles, has been mainly explained by variable range hopping (VRH) models that assume strong localization of electrons, indicating the presence of strong disorder at contact points between the bundles. These previous studies were performed on networks with an uncontrolled mixture of metallic and semiconducting SWCNTs. Apparently, the presence of semiconducting SWCNTs impedes good conduction and introduces a high degree of disorder. It has been reported that chemical doping of nanotube networks can change the conduction mechanisms [6]; however, for the reported pure SWCNT networks, phonon-assisted hopping processes were found to play a major role in conduction. In the case of metallic nanowires and nanograins, although boundaries exist between the wires and grains, quantum transport has been observed in macroscopic networks of such structures [8]. However, it has not yet been established whether such transport is possible in the absence of intentional chemical treatment on nanotube networks. After the discovery of efficient sorting methods by Prof. Hersam's group in the Northwestern University [9], effective metallic—semiconducting separation has been achieved by various techniques, allowing the formation of high-purity networks [10]. As a result, it is possible to freely adjust the content ratio of semiconducting and metallic types in SWCNT networks. In this chapter, we discuss how the presence of semiconducting types (or content ratio of semiconducting and metallic SWCNTs, MS ratio) affects the conduction mechanisms in SWCNT networks [11].

15.2 RELATIONSHIP BETWEEN MS RATIO AND CONDUCTIVITY OF SWCNT NETWORKS

The correlation between the MS ratio and the conductivity of SWCNT networks [11] is investigated in this section. Several sheets of SWCNT bucky paper were prepared with controlled MS ratios (Figure 15.1). Optical absorption spectra of the samples, which were dispersed in deoxycholate sodium salt 1% solutions from the sheets, are shown in Figure 15.1. These samples were produced by a careful purification process followed by metal—semiconductor separation using density-gradient ultracentrifugation of SWCNTs initially produced by an arc discharge method (Meijo-Nanocarbon Co.) [11,12]. The average diameter of the nanotubes was estimated to be about 1.46 nm from X-ray diffraction profiles.

As seen in the optical absorption spectra of the high-purity samples (Metal and Semi) shown in Figure 15.1, no bands associated with the other conduction type could be identified, indicating a purity of more than 95%. The relative content of semiconducting SWCNTs (Semi%) in the Metal and the Semi samples were defined here as 0% and 100%, respectively. The MS ratio of the remaining three samples was evaluated from the ratio of the intensity of the M_{11} and S_{22} optical absorption bands (here S_{ii} and M_{ii} indicates the ith optical transition of the metallic and semiconducting SWCNTs) as follows. The Semi% of these intermediate samples was estimated by a linear combination of the optical absorption spectra of the Metal and the Semi. (Here we use a term "Semi% = 100%" for the Semi sample; this does not directly mean that the content of metallic SWCNTs in Semi was zero).

15.2 Relationship between MS Ratio and Conductivity of SWCNT Networks

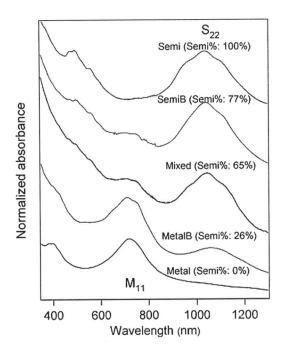

FIGURE 15.1

Optical absorption spectra of SWCNTs with different MS ratio.

Figure 15.2 shows the resistivity of the Metal, Semi, and several other MS ratio samples as a function of temperature. In the Metal sample, the resistivity was almost constant as the temperature decreased; however, in the Semi sample, the resistance rapidly increased and became more than 10 MΩ (beyond our instrumentation limit) at 4 K, indicating the insulating properties of the Semi network at low temperature. The resistivity of the Semi network was almost seven orders of magnitude larger than that of the Metal network at low temperature, clearly indicating that a different transport mechanism is involved. In our samples, dR/dT is always negative between $T = 3$ and 380 K, suggesting a high degree of intertubular coupling.

Figure 15.3 shows the temperature dependence of resistance normalized by the resistance value at $T = 380$ K for the five samples with different MS ratios. It can be seen that all the curves are different from each other. At low temperature, the normalized resistance becomes lower as the content of semiconducting SWCNTs decreases. Contact points between nanotubes or nanotube bundles would be expected to disrupt smooth conduction by acting as scattering centers and potential barriers. Although the number of such disorders depends on the densities and lengths of the nanotubes and nanotube bundles, the effects of these variations can be eliminated by evaluating the normalized resistance. Thus, the observed differences in the normalized resistance indicate that changing the MS ratio affects the intrinsic physical properties of the disorders.

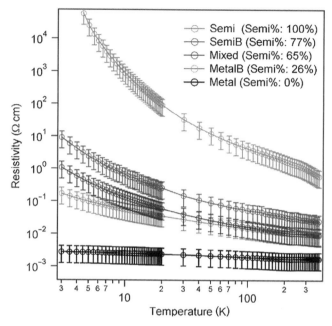

FIGURE 15.2

Temperature dependence of resistivity of SWCNTs with different MS ratio. (For color version of this figure, the reader is referred to the online version of this book.)

Next, we analyzed the temperature dependence of the resistance using VRH models based on the following equation:

$$R(T) = R_0 \exp\left[\left(\frac{T_0}{T}\right)^{\frac{1}{d+1}}\right].$$

Here, d has a value of either 2 or 3 in Mott-VRH, reflecting the dimensionality of a system in which electrons hop (VRH model proposed by Mott, which is referred to simply as VRH) [13] or it has a value of 1 in Coulomb gap VRH by Efros & Shklovskii (ES-VRH), reflecting the existence of a Coulomb gap due to Coulomb interactions between localized electrons [14]. T_0 is the characteristic temperature used in the VRH model. In the Semi sample, the temperature dependence could be reproduced by the model with $d = 1$, which indicates the presence of Coulomb interactions between localized electrons, inducing a Coulomb gap in the density of states of conduction (ES-VRH) [14]. However, in the SemiB and Mixed samples, the appropriate value of d was 2, and in the MetalB sample, it was 3. Remarkably, for the Metal sample, the dependence could not be fitted with any value of d in the range 1−3.

We first consider conduction in pure semiconducting SWCNTs networks, which was found to occur by the ES-VRH mechanism ($d = 1$). This type of

15.2 Relationship between MS Ratio and Conductivity of SWCNT Networks

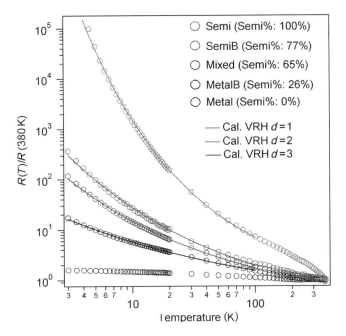

FIGURE 15.3

Normalized resistance of five different samples with different MS ratio as a function of temperature, Semi (red circles), SemiB (brown circles), Mixed (dark-brown circles), MetalB (green circles), and Metal (blue circles). The data for the Semi sample could be reproduced using the VRH model with $d = 1$ (red line), that for the SemiB and Mixed samples using VRH with $d = 2$ (brown lines), and that for MetalB by VRH with $d = 3$ (green line). However, the line shape of Metal cannot be reproduced by any dimension of VRH. (For interpretation of the references to color in this figure legend, the reader is referred to the online version of this book.)

conduction has already been reported in networks with a small volume fraction (less than 8%) of SWCNTs embedded among insulating polymers, or in high-resistivity nanotube networks produced by annealing at a relatively high temperature [6,15]. As the resistance of the Semi sample indicates, semiconducting SWCNTs can be regarded as insulating materials at low temperature. Therefore, on the basis of theoretical studies on the transition between ES-VRH and VRH by Shklovskii [16,17], the conduction process in the Semi sample can be explained in the following scenario. In semiconducting SWCNTs, uncontrolled donors and acceptors lead to random charging of the metallic SWCNTs within the bundles. When the content of metallic SWCNTs is small, Coulomb interactions between electrons of distant metallic SWCNTs induce a soft Coulomb gap in the density of states of conduction. Therefore, ES-VRH conduction occurs. However, when the content of metallic SWCNTs increases, these SWCNTs enhance the screening of the Coulomb potential, which results in a weakening of Coulomb interactions, and the conduction mechanism changes from ES-VRH to the conventional VRH. This theoretical prediction is in good agreement with our observation that an increase in the content of metallic SWCNTs changed the conduction mechanism from ES-VRH ($d = 1$) to VRH ($d = 2$ or 3).

Therefore, conduction in the Semi sample is assumed to be caused by hopping between charged (unintentionally doped) semiconducting SWCNTs and residual metallic SWCNTs that could not be detected in the optical absorption spectra. Benoit et al. proposed that the volume fraction of SWCNTs, which were in a mixed metallic and semiconducting state, can be estimated from the T_{ES} value (defined as the value of T_0 for $d = 1$) [15]. Theoretically, for a concentration n of metallic wires of length L, $T_{ES} \propto 1/(nL^3)^2$ [16]. The T_{ES} value of the Semi sample in this study was determined to be 650 K. From the data reported by Benoit et al. (the ratio of metallic to semiconducting SWCNTs in their samples is assumed to be 1—2), this T_{ES} value indicates that the volume fraction of metallic SWCNTs is less than about 1%, supporting the high purity of the Semi sample.

As the content of metallic SWCNTs increases, the value of d becomes large. Such a systematic change in the d value is also observed in mixed systems of conducting and insulating polymers [18,19]. For example, the relative content of conducting polymers mixed in insulating polymers influences the value of d, and the d values vary from 1 to 3 as the content of the conducting polymer is increased. The physical reasons for this are still a matter of some discussion, but we suggest that the enhancement of conduction channels caused by the increase in metallic SWCNT content is one cause for the systematic change in d. There are some similarities between the conduction mechanisms in conducting polymers and SWCNTs. However, while the transport mechanisms in conducting polymers can be explained by VRH conduction for any volume fraction, conduction in the pure metallic SWCNT network cannot be explained by a simple VRH scenario. To clarify the conduction mechanism in the latter case, we carried out magneto resistance (MR) measurements on the samples.

Figure 15.4 shows MR data for the Metal (Semi%: 0%), MetalB (Semi%: 26%) and Mixed (Semi%: 65%) samples. The MR of the Mixed sample is negative for low magnetic fields, followed by the positive upturn in a high magnetic field region. Such phenomena are consistent with previous studies. Kim et al. studied magnetotransport phenomena using several theoretical models and concluded that the positive and negative MR are caused by VRH and WL, respectively [3]. Remarkably, the positive upturn in the MR decreases with decreasing Semi%, and no such upturn occurs for the Metal sample. This suggests that WL is the dominant factor for conduction in the Metal sample. These MR results clearly indicate that hopping processes were reduced in the Metal sample, which agrees with the temperature dependence of the resistance (Figs 15.2 and 15.5).

15.2 Relationship between MS Ratio and Conductivity of SWCNT Networks

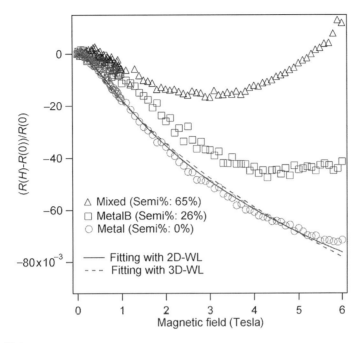

FIGURE 15.4

The magnetoresistance in nanotube networks for three samples with different MS ratios. Blue circles, green squares and black triangles indicate data for Metal (Semi% = 0%), MetalB (Semi% = 26%), and Mixed (Semi% = 65%) samples, respectively. Solid and dotted red lines indicate the fitting results using two-dimensional and three-dimensional weak localization (WL) models, respectively. (For interpretation of the references to color in this figure legend, the reader is referred to the online version of this book.)

The MR of the Metal sample is well reproduced by two-dimensional WL theory using the following equation [20]:

$$\Delta\sigma(H) = \frac{e^2}{2\pi^2\hbar^2}\left[\ln x + \varphi\left(\frac{1}{2} + \frac{1}{x}\right)\right].$$

Here $x = 4L_{th}^2 eH/\hbar c$ and L_{th} is the decoherence length (effective scale size for quantum interference effects). L_{th} was determined to be 26 nm from the fitting. We also tried to fit the data using a three-dimensional WL model, but the two-dimensional model produces a somewhat better fit than the three-dimensional model [11]. The resistance of the Metal sample exhibited a logarithmic dependence on T at low temperature, supporting the two dimensionality of the WL processes [11]. The bundle sizes of metallic nanotubes were estimated to be approximately 10 nm, and this value is less than the estimated decoherence length; therefore, it would lead to two-dimensional rather than three-dimensional electron

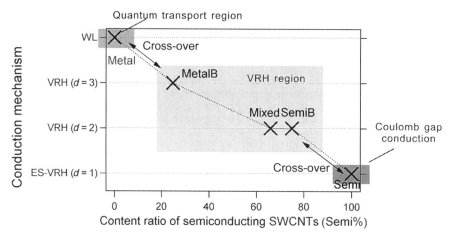

FIGURE 15.5

Phase diagram of conduction mechanisms in SWCNT networks as a function of MS ratio. Relationship among relative content of semiconducting SWCNTs (Semi%), d values in VRH analyses, and conduction mechanisms. WL, weak localization; VRH, variable range hopping; ES-VRH, Coulomb gap VRH by Efros & Shklovskii. (For color version of this figure, the reader is referred to the online version of this book.)

conduction. Thus, we assume that the anisotropy of the nanotube bundles influenced the dimensionality of the WL.

15.3 SUMMARY

In summary, the conduction mechanisms in SWCNT networks were depicted as a function of the MS ratio. The mechanisms are different depending on the MS ratio, showing cross-overs between the conduction mechanisms, from quantum transport, VRH, and ES-VRH, as a function of the MS ratio. Increasing the percentage of semiconducting SWCNTs first causes a transition from WL to VRH and then from VRH to ES-VRH. As mentioned by Anderson [21], a cross-over from the weak to strong localization regimes occurs as the strength of disorders increases. In SWCNT networks, the characteristics of the disorders can be controlled by varying the MS ratio. The presence of semiconducting SWCNTs was found to be the main factor causing localization of the conduction electrons. One of the possible origins for strong disorders is the Schottky barriers between metallic and semiconducting SWCNTs. Another possibility is the potential barriers among semiconducting SWCNTs due to the random distribution of nanotube diameters and the resulting inconsistence of band gap energies. In pure metallic SWCNT networks, boundaries between the nanotubes or bundles act as weak sources of disorder, and as a result, quantum transport was achieved in such macroscopic networks. The phase diagram developed will become the basis of a general concept for understanding the conduction mechanisms in macroscopic networks of not only SWCNTs but also nanoparticles with various electronic structures.

References

[1] V. Skakalova, A.B. Kaiser, Z. Osvath, G. Vertesy, L.P. Biro, S. Roth, Applied Physics A 90 (2008) 597−602.
[2] Y. Yosida, I. Oguro, Journal of Applied Physics 86 (1999) 999−1003.
[3] G.T. Kim, E.S. Choi, D.C. Kim, D.S. Suh, Y.W. Park, K. Liu, G. Duesberg, S. Roth, Physical Review B 58 (1998) 16064−16069.
[4] T. Takano, T. Takenobu, Y. Iwasa, Journal of the Physical Society of Japan 77 (2008) 124709−124713.
[5] L. Langer, V. Bayot, E. Grivei, J. Issi, J.P. Heremans, C.H. Olk, L. Stockman, C. Haesendonck, Y. Bruynseraede, Physical Review Letters 76 (1996) 479−482.
[6] J. Vavro, J.M. Kikkawa, J.E. Fischer, Physical Review B 71 (2005) 155410−155420.
[7] G.C. McIntosh, G.T. Kim, J.G. Park, V. Krstic, M. Burghard, S.H. Jhang, S.W. Lee, S. Roth, Y.W. Park, Thin Solid Films 417 (2002) 67−71.
[8] P. Lee, T. Ramakrishnan, Reviews of Modern Physics 57 (1985) 287−337.
[9] M. Arnold, A. Green, J. Hulvat, S. Stupp, M. Hersam, Nature Nanotechnology 1 (2006) 60−65.
[10] M. Hersam, Nature Nanotechnology 3 (2008) 387−394.
[11] K. Yanagi, H. Udoguchi, S. Sagitani, Y. Oshima, T. Takenobu, H. Kataura, T. Ishida, K. Matsuda, Y. Maniwa, ACS Nano 4 (2010) 4027−4032.
[12] K. Yanagi, Y. Miyata, H. Kataura, Applied Physics Express 1 (2008) 034003−034005.
[13] N. Mott, Conduction in Non-crystalline Materials, Oxford Univ. Press, Oxford, 1987.
[14] B.I. Shklovskii, A. Efros, Electronic Properties of Doped Semiconductors, Springer-Verlag, Berlin, 1984.
[15] J.M. Benoit, B. Corraze, O. Chauvet, Physical Review B 65 (2002) 241405−241408.
[16] T. Hu, B. Shklovskii, Physical Review B 74 (2006) 054205−054209.
[17] J. Zhang, B. Shklovskii, Physical Review B 70 (2004) 115317−115329.
[18] M. Reghu, C. Yoon, Y. Yang, D. Moses, P. Smith, A. Heeger, Y. Cao, Physical Review B 50 (1994) 13931−13941.
[19] J. Planès, A. Wolter, Y. Cheguettine, A. Proñ, F. Genoud, M. Nechtschenin, Physical Review B 58 (1998) 7774−7785.
[20] A. Kawabata, Solid State Communications 34 (1980) 431−432.
[21] P.W. Anderson, E. Abrahams, T.V. Ramakrishnan, Physical Review Letters 43 (1979) 718−720.

Thermal Conductivity of Nanotube Assemblies and Superfiber Materials

16

Michael B. Jakubinek

Division of Emerging Technologies, National Research Council Canada, Ottawa, ON, Canada

CHAPTER OUTLINE

16.1 Introduction	425
16.2 Thermal Conductivity and Measurement Issues for CNT Materials	426
16.3 Individual Carbon Nanotubes	430
16.4 Carbon Nanotube Bundles	432
16.5 Carbon Nanotube Composites	435
16.6 CNT Buckypaper and Thin Films	437
16.7 CNT Superfiber Materials	441
16.7.1 CNT arrays	441
16.7.2 CNT sheets and yarns	445
16.8 Boron Nitride Nanotubes	449
16.9 Challenges and Opportunities	449
Acknowledgments	451
References	451

16.1 INTRODUCTION

Carbon nanotubes (CNTs) have been shown to possess exceptional properties at the individual tube level, including high strength, flexibility, and the highest electrical and thermal conductivities of any known material. These properties, together with a high aspect ratio and low density, have generated much excitement surrounding the potential to develop new engineering materials where the CNT properties are leveraged to create macroscopic multifunctional materials. The majority of this work has centered on CNT-enabled polymer composites, and such materials have found commercial applications, including as battery additives and in composites for sporting goods and electrostatic discharge [1]. A drawback of most composites is the low loading of CNTs. In recent years, nanotube films, paper ("buckypaper"), arrays, sheets, fibers and yarns have been developed and these assemblies, in particular in high-packing-density, aligned configurations, offer a promising route to achieve higher strength and conductivity in macroscopic materials [2,3]. Such superfiber materials, addressed throughout this book, can be expected to figure prominently in applications [3]. While the largest amount of attention has been focused on the production of these materials and on their mechanical and electrical properties, other properties including thermal conductivity are both of fundamental interest and importance for applications such as heat dissipation in composite structures and thermal interface materials. With a focus on experimental results, this chapter addresses the thermal conductivity in individual CNTs, nanotube bundles and macroscopic assemblies in order of increasing hierarchy. CNT assemblies are taken to include CNT bundles, CNT networks in composites, CNT films, buckypapers, and CNT superfiber materials (arrays, sheets and yarns). Significant variability exists both for the reported thermal conductivities of individual CNTs and the thermal conductivities measured for macroscopic CNT assemblies, which range from comparable to metals to aerogel-like.

16.2 THERMAL CONDUCTIVITY AND MEASUREMENT ISSUES FOR CNT MATERIALS

The thermal conductivity, κ, of a solid describes how well it conducts heat and is defined from Fourier's law

$$\mathbf{J} = -\kappa \nabla T, \quad (16.1)$$

as the proportionality constant relating the temperature gradient across the sample, ∇T, to the heat flux per unit area, $\mathbf{J}$. Values for typical bulk solids at room temperature range from around 0.2 W/m K for polymers to 2300 W/m K for diamond [4]. While κ is a tensor property and has directionally dependent values for anisotropic materials, experiments are typically designed to create one-dimensional (1D) heat flow and measurements are performed along directions of interest. This is the case in nanotube assemblies, where measurements are often performed along directions of high alignment (e.g. fibers). κ anisotropy is also well known for graphite, which has a room-temperature thermal conductivity close to that of diamond for the in-plane direction but two orders of magnitude lower in the cross-plane direction [4].

In solids, heat is carried by electrons and by lattice vibrations (phonons). Both carriers can contribute to the measured thermal conductivity but, in practice, one mechanism is often dominant. Electrons are the dominant heat carriers in metals and, because the electron is the carrier of both heat and electricity, their κ and electrical conductivity (σ) are proportional as described by the Wiedemann–Franz law

$$\frac{\kappa}{\sigma T} = L_0, \quad (16.2)$$

where L_0, the Lorentz number, is approximately 2.45×10^{-8} Ω/K^2 for metals. To compare the contributions of electrons and phonons to heat transport, it is common to calculate an effective Lorentz number, $L = \kappa/(\sigma T)$. For CNT materials, including bundles, mats, arrays, sheets and fibers (e.g. Refs. [5–8]), this number is two orders of magnitude larger than its theoretical value for metals (i.e. thermal conductivity is

16.2 Thermal Conductivity and Measurement Issues for CNT Materials

much higher than could be explained by electronic heat conduction). This measurement has not been done for individual metallic single-walled carbon nanotubes (SWCNTs), where κ_e could be significant; however, it is widely accepted that phononic heat conduction is the dominant process in CNTs and CNT assemblies.

Phonon thermal conductivity can be qualitatively understood from the Debye model

$$\kappa = \frac{1}{3} C v \lambda, \qquad (16.3)$$

which applies a phonon gas picture to describe conduction. In Eqn (16.3), C is the heat capacity per unit volume, and v and λ are the phonon velocity and mean free path, respectively. For amorphous materials, λ is short at all temperatures and κ follows the temperature dependence of C and reaches ~ 1 W/m K at room temperature for many materials (lower for polymers due to their low speed of sound). For pure, insulating crystals, λ is usually long at low temperatures leading to much higher κ. Such materials typically have a κ peak below room temperature due to the increase in Umklapp scattering, which reduces λ and causes κ to decrease at higher T. The Debye model representation and Fourier's law, wherein the dependence of heat flux on the temperature gradient indicates that heat transfer is a diffusive process, can reasonably be called into question in the case of individual CNTs. The basic assumption of diffusive heat transport (i.e. sample dimension, $L \gg \lambda$) becomes invalid if $\lambda > L$ and the heat transport becomes ballistic. In this case, κ becomes a function of length, because phonons travel the full length of the sample without scattering. CNTs have large λ resulting from their 1D character and simulations have shown that thermal conductivity does not converge to the diffusive limit even when the CNT length reaches 1 μm [9]. Other calculations suggest that κ does not converge until the SWCNT length reaches the millimeter range [10]. As noted by Shiromi and Maruyama [9], κ is still useful for comparison between reports, even though it may be more accurate to specify only the thermal conductance of CNTs. Despite questions about the ballistic—diffusive character of heat transport in individual CNTs, Fourier's law is still applied to define measurements of κ but, in cases with significant ballistic contribution, κ may no longer be an intrinsic property.

Measurements of κ for CNT materials are complicated by the challenging sample preparation and small heat flux due to low cross-sectional area. Various reports have employed a range of approaches from conventional methods used for bulk materials, to thin film methods, to custom apparatus. In the archetypal steady-state measurement, 1D heat flow is created along a sample mounted between a heater and a heat sink. κ is determined from Eqn (16.1) using the heater power, cross-sectional area, and direct measurement of the steady-state temperature difference across a known distance. This has been employed for CNT arrays [5]. Steady-state methods have the advantage of having the most straightforward interpretation and well-understood experimental errors; however, the need for the sample to support a heater and thermometer in a conventional steady-state measurement poses a challenge for applying this method to CNT materials. Variations including

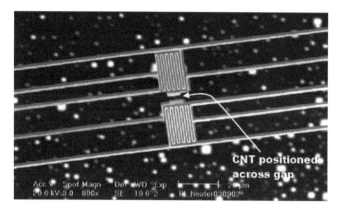

FIGURE 16.1

SEM image of a suspended microdevice used for measuring the thermal conductance of CNTs [14]. A CNT (or CNT bundle) is positioned across the gap between the heater/thermometer islands. The islands are supported by long, narrow silicon nitride arms, which are necessary to minimize thermal conductance of the supports and enable measurement of individual CNTs.

Source: Reprinted with permission from Ref. [14]. Copyright 2005 American Chemical Society. doi:10.1021/nl0519443.

comparative methods [11–13], where the sample is placed in series with a reference of known thermal conductivity, and microfabricated sensors [6,14] (Fig. 16.1) have been applied to CNT bundles/assemblies and individual CNTs/bundles, respectively. For nanotube superfiber materials (buckypaper, fibers/yarns, and array-drawn sheets), the parallel thermal conductance (PTC) method [7,8,13,15,16] provides an effective variation of the steady-state technique. In the PTC method (Fig. 16.2), the sample conductance is in parallel with a post supporting a heater and a differential thermocouple. The background conductance due to the support and the heater/thermometer wires must be minimized and then carefully subtracted. In the case of CNT fiber/yarn, multiple samples have been placed in parallel across the gap to increase the thermal conductance [7]. This also has the beneficial effect of averaging over a longer sample length.

The other major class of measurement method is transient methods, where temperature response is recorded as a function of time. This includes methods based on self-heating, where the sample is used as a resistive heater, and optical heating. Commonly used methods include the 3ω method, which also has been employed for individual CNTs, and laser flash. Advantages of these methods include faster measurement times, as it is not necessary to achieve a steady state, and that it is not necessary to attach heaters/thermometers. However, extraction of the thermal properties is less direct and in some cases samples must be coated. Some transient methods, including the common laser flash method, also measure thermal diffusivity (α)

$$\alpha = \frac{\kappa}{\rho C_p}, \tag{16.4}$$

16.2 Thermal Conductivity and Measurement Issues for CNT Materials

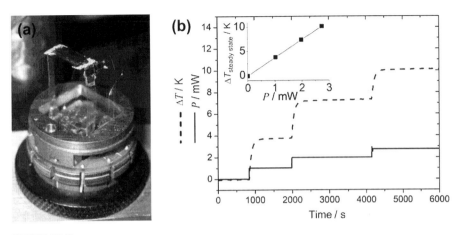

FIGURE 16.2

(a) Measurement platform for PTC measurements shown with a CNT sheet mounted across the sample supports (radiation shield not shown). *Source: Photo courtesy Michel Johnson (Dalhousie University).* (b) Raw data showing a single measurement of thermal conductance, determined from the slope of the $\Delta T_{steady\ state}$ vs heater power (P) sweep at a fixed base temperature. Three such measurements are required to determine the thermal conductances due to the supports, sample, and radiation, and then the sample thermal conductivity is extracted.

Source: Reprinted from Ref. [7]. Copyright (2012), with permission from Elsevier. http://dx.doi.org/10.1016/j.carbon.2011.08.041.

where ρ is the density and C_p is the specific heat capacity, rather than directly measuring κ. This adds an additional source of error from C_p; however, from a number of reports (e.g. [8,17,18]) C_p(CNT) ~ C_p(graphite) ~0.7 J/g K at room temperature. The uncertainty resulting from using C_p(graphite) is likely reasonable compared to κ uncertainties related to sample variability and density estimation. For calculation of κ from α at low temperatures, where C_p can be affected by low-dimensional behavior, or high temperature, where deviation from C_p(graphite) has been reported [17], a direct measurement of heat capacity might become important. General thermal conductivity measurement methods are addressed in more detail by Tritt and Weston, for bulk samples [13], and by Borca-Tascuic and Chen for thin samples [19].

The discussion of thermal conductivity and measurement methods implicitly focused on homogeneous materials. For heterogeneous materials including conventional composites and nanotube assemblies, which can be considered composites of CNTs and air (empty space in the case of vacuum measurements), κ is a function of position within the material. Measurements of heat conduction yield an apparent thermal conductivity, κ_{app}, if the external dimensions of the sample (or the density based on the total mass and external dimensions in the case of α measurements) are used. κ_{app}, also known as effective thermal conductivity or volume-averaged thermal conductivity, is defined here as the thermal conductivity of a homogenous material with equivalent outer dimensions. As noted by Fischer [20], translating the measured thermal conductance of CNT assemblies into thermal conductivity is not straightforward due to the difficulty of defining the cross-sectional area. One option is to consider only the outer dimensions, which yields κ_{app}. However, in CNT bundles and other low-density CNT assemblies, the true cross-sectional area through which heat flows is much smaller than given by the outer dimensions, as the heat is transported almost entirely by the CNTs. A common practice is to scale the κ_{app} by the ratio of theoretical CNT density (ρ_{CNT}) to the apparent density (ρ_{app}) [5,21,22]. In the case of aligned CNTs spanning the full length of the sample, the volume density equals the area density and this scaling is equivalent to using the cross-sectional area of only the CNTs instead of the larger area measured from the external dimensions. The second approach is valuable in order to compare measurements of different assemblies and to compare the performance of a CNT assembly to that of an individual CNT, as well as to study the effects of factors like defects within CNTs. In practice, apparent thermal conductivity is often the relevant quantity for applications.

For thermal conductivity results summarized in this chapter (Tables 16.1–16.4), the κ_{app} is reported wherever it could be determined from the published information. κ, defined here as the thermal conductivity of the CNTs in the assembly, is reported only where it was given in the original publications. These values should be considered as approximations when making comparisons as various studies make different assumptions for density, ranging from 1.3 g/cm^3, a theoretical prediction for SWCNTs, to 2.25 g/cm^3 for graphite. The correction is strictly valid only for the idealized case of straight CNTs along the measurement direction such that the area fraction of CNTs in each slice is equal to their volume fraction. For CNT assemblies, particularly low-density bundles, arrays and sheets, the difference between κ of the CNTs and κ_{app} of the assembly can be large. For well-aligned assemblies, the most accurate comparison might be κ_{app}/SG, where the apparent specific gravity (SG) relative to water is used instead of the apparent density for the convenience of retaining the units of thermal conductivity. This approach is similar to that employed in characterizing CNT fiber mechanical properties [29].

16.3 INDIVIDUAL CARBON NANOTUBES

Based on the observation of defect-free CNTs longer than the crystallite diameter in highly oriented pyrolytic graphite, and comparison to the in-plane properties of graphite, Ruoff and Lorents suggested that the thermal conductivity of CNTs could exceed that of diamond [30]. Molecular dynamics simulations have predicted room-temperature κ values up to 6600 W/m K, much higher than that for nearly isotopically pure diamond, which is associated with long phonon mean free path [31]. Experimental measurements are challenging due to the difficulties in manipulating and contacting individual nanotubes and the small thermal conductance. Several methods have been reported including the microfabricated device (Fig. 16.1), which was the first to measure the thermal conductivity of an individual multiwalled carbon nanotube (MWCNT) to be more than 3000 W/m K at room temperature [6]. This is two to three orders of magnitude larger than typical measured values (κ_{app}) for bulk samples, yet it still represents a lower limit for the MWCNT in that study as the measured thermal conductance included both the sample and the contacts. Independent assessment of the contact thermal resistance suggested that 3000 W/m K may be a significant underestimate [32].

The observation of a peak in $\kappa(T)$ for MWCNTs around 320 K, which is much higher than that observed for diamond and graphite, is indicative of a long λ [6]. The same measurement method was applied to SWCNTs and shows an equal or larger value for SWCNTs (Fig. 16.3(a)) [14]. Measurements of κ for suspended SWCNTs using a self-heating method produced a comparable result, ~3500 W/m K, at room temperature and showed decreasing thermal conductivity from 300 to 800 K (Fig. 16.3(b)) [33]. The different temperature dependences observed in the two reports, $d\kappa/dT$ ($T<300$ K) >0 and $d\kappa/dT$ ($T>300$ K) <0 [14,33], indicate a similar peak near room temperature for SWCNTs as has been observed for MWCNTs.

Other room-temperature experimental results for suspended, individual SWCNTs include 2000 W/m K using a T-type nanosensor [34], and 2400 W/m K

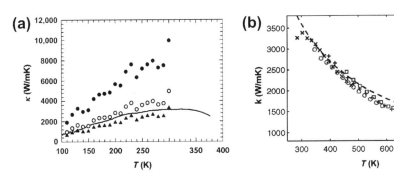

FIGURE 16.3

(a) SWCNT thermal conductivity measured using a suspended microdevice assuming SWCNT diameters of 1 nm (●), 2 nm (○), or 3 nm (▲). For similarly prepared SWCNTs, the diameter was in the range of 1–2 nm for most of the SWCNTs, although 2–3 nm diameters were observed occasionally. The line is the measurement result of an individual MWCNT. *Source: Reprinted with permission from Ref. [14]. Copyright 2005 American Chemical Society.* http://dx.doi.org/10.1021/nl051044e. (b) High-temperature thermal conductivities of an individual SWCNT extracted from I–V data, where the different symbols report data obtained at a range of ambient temperatures (250–400 K).
Source: Reprinted with permission from Ref. [33]. Copyright 2006 American Chemical Society. http://dx.doi.org/10.1021/nl052145f.

by a Raman shift technique [35]. Lower values have been reported, including 300 W/m K and 600 W/m K measured by 3ω methods [28,36]. The latter reports are for chemical vapor deposition (CVD)-grown MWCNTs, and higher defect density in CVD CNTs has been suggested as a factor [28]. In direct measurements of CVD MWCNTs using suspended microthermometers, κ ranged from 40 to 340 W/m K [32,37]. MWCNTs with more walls and defects were found to have lower κ, and κ decreased with electron irradiation [37]. Due to challenges with sample preparation and experimental measurements, defects and other factors expected to affect thermal conductivity (e.g. diameter, length, chirality, defects, and wall number) [9,10,34,38–46] have been explored most extensively in simulations. Simulations have suggested that chirality changes affect κ by up to 20% [46]. However, others found that κ was insensitive to chirality [38,39]. In contrast, nanotube chirality is known to have a dramatic effect on electrical conductivity. The dependence on length can be understood in terms of the transition from diffusive to ballistic heat transport.

Despite significant variability between experimental reports and conflicting simulations, which also vary widely from values equal to or greater than the experimental reports [31,38,40,43,46,47] to values an order of magnitude lower [39,42], the existing literature generally confirms the early expectation that the intrinsic thermal conductivity of CNTs can be very high. For high-quality long CNTs, κ exceeds that of diamond due to a high λ for CNTs, as evidenced by experimental observations of a peak in κ vs T around 300 K, a higher temperature than that observed for diamond and graphite.

16.4 CARBON NANOTUBE BUNDLES

While the first measurements on CNT bundles predate measurements on individual CNTs, they are addressed second in this chapter based on their hierarchy in the understanding of assemblies of CNTs. Experimental κ results for CNT bundles are summarized in Table 16.1. In comparison to individual CNTs, the thermal conductance of bundles and mats is less difficult to measure due to less-demanding sample preparation and larger conductance values (due to larger cross-sectional area). However, the interpretation of the measured results is further complicated, beyond the issues encountered for individual CNTs, by effects of CNT–CNT interaction, interfacial thermal resistances at CNT–CNT contacts, and determination of the cross-sectional area as discussed previously in relation to apparent thermal conductivity.

Experimental results for aligned CNT bundles are one to two orders of magnitude lower than those for individual CNTs, even when the low apparent density of the bundles is taken into account. The apparent thermal conductivity of the low-density bundles is an additional one to two orders of magnitude lower. Similar differences are observed for electrical conductivity, σ. σ can be up to $\sim 10^6$ S/cm for metallic SWCNTs, which also carry very large current densities [48,49], while

Table 16.1 Thermal Conductivity of Carbon Nanotube Mats and Bundles at Room Temperature

Type of Bundle/ Material	κ_{app} W/m K	κ W/m K	ρ_{app} g/cm^3	Measurement Type	Ref
Aligned MWCNT (CVD array)	**0.38**	25	**0.02**	κ; 3ω	[18]
SWCNT mat (arc discharge)	0.7	35	**0.027**	κ; comparative	[11]
Aligned SWCNT bundle				κ; microfabricated Device	[23]
Small bundle ($d \sim 10$ nm)	20				
Large bundle ($d \sim 150$ nm)	2.5				
MWCNT pellet (CVD), sintered	2.8–4.2		1.4	α; laser flash	[24]
MWCNT pellet	1.6		0.51	α; laser flash	[25]
Aligned SWCNT rope (arc discharge)	3.6			κ, α; 3ω	[26]
Aligned MWCNT (CVD array)				κ	[27]
As produced	2.5				
CH$_3$OH treated; annealed 100 °C	8				
Annealed 1250 °C	3				
Annealed 2800 °C	23				
Aligned MWCNT (CVD array)				κ, 3ω	[28]
Over 100 MWCNTs/ bundle		150			
$\sim$50 MWCNTs/ bundle		200			
$\sim$20 MWCNTs/ bundle		300			
Aligned MWCNT rope (CVD)				κ & α	[17]
Bundle 1	5.8		0.116	Electrothermal	
Bundle 2	7.6		0.234		

NB: Plain text values are as reported in the original references. Values in bold were estimated from information in the paper.

SWCNT bundles and mats have $\sigma \sim 10^2$–10^4 S/cm depending on the alignment [12,50]. Lower thermal conductivities for the aligned MWCNT bundles stripped from CVD-grown arrays could be related to the higher defect density of CVD tubes, as described above in the case of individual CNTs. Tall arrays in particular can have higher defect densities and amorphous carbon content, which has been noted as a possible explanation for decreases in κ_{app} with height for tall MWCNT arrays in one study [5]. Another study observed increases in κ_{app} after high-temperature annealing steps [27], which can reduce defects. The highest κ_{app} value in Table 16.1, excluding a sample annealed at 2800 °C, is 20 W/m K for a small bundle of aligned SWCNTs. In the same study, a large SWCNT bundle was reported to have lower thermal conductivity [23]. Theoretical calculations indicate that higher relative contributions from the thermal resistances at the bundle–electrode contacts, due to larger bulk conductance of bundles, could be a major factor in the difference between these large and small SWCNT bundles, and individual CNTs [51]. Simulations have also shown that thermal conductivity decreased by roughly a factor of three for close-packed bundles in comparison to individual SWCNTs [43]. The effect of bundle size was explored by Aliev et al. [28], who measured individual MWCNTs and bundles of increasing size and found that κ decreased by approximately four times as the bundle size increased from an individual MWCNT to over 100 MWCNTs (Fig. 16.4). The decrease in κ is attributed to coupling between

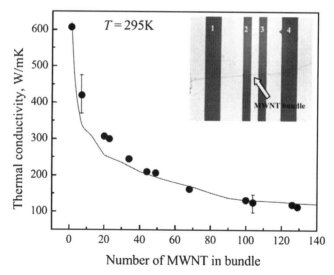

FIGURE 16.4

Thermal conductivity of MWCNT bundles grown by CVD, measured using the 3ω method and normalized to an ideal high-density bundle [28]. For small bundles, κ decreases with increasing bundle size and reaches $\sim 150 \pm 15$ W/m K for large bundles, roughly four times lower than that of the individual MWCNT.

Source: Reprinted with permission from Ref. [28]. http://dx.doi.org/10.1088/0957-4484/21/3/035709.

CNTs in bundles, where bundling restricts out-of-plane vibrations and, therefore, suppresses low-lying optical modes that contribute significantly to heat conduction in individual CNTs at room temperature [28]. When heat transfer between CNTs is involved, such as in unaligned mats or bundles contacted from the side, the interfacial thermal resistance between the nanotubes further reduces the thermal conductivity. In this case, heat transfer is inhibited by small contact areas and high interfacial thermal resistance at CNT−CNT contacts, estimated from simulations to be $>10^{-8}$ m^2 K/W even for short CNT−CNT separation [52]. Measurement of CNT−CNT thermal contact resistance using a device of the type shown in Fig. 16.1 led to an experimental estimate of $\sim 10^{-9}$ m^2 K/W [53]. Such resistances can lead to CNT assemblies with thermally insulating properties. For packed beds composed of 10−20 vol% CNTs, produced by compressing random mats of CNTs, $\kappa_{app} < 0.2$ W/m K has been reported due to the dominant effect of CNT−CNT thermal contact resistance [54].

16.5 CARBON NANOTUBE COMPOSITES

CNT composites, and in particular polymer composites, are the most-studied macroscopic CNT-enabled materials. These materials have been the subject of several detailed reviews (e.g. Refs. [55−58]), including a recent review of the thermal conductivity of CNT−polymer composites [58]. While not falling within the category of superfiber materials, assemblies of CNTs within composites enabled the first CNT engineering materials and the larger body of work on CNT−polymer composites illustrates the challenges in translating the high thermal conductivity of individual CNTs to macroscopic materials.

The electrical conductivity of polymers can be enhanced dramatically, by factors of 10^{10}−10^{15}, following the incorporation of relatively small amounts of CNTs, which form a percolating network [55−57]. Typical polymers are insulators, but CNT−polymer composites have $\sigma \sim 10^{-5}$ to 1 S/cm well above their percolation thresholds of ~ 0.005 to a few wt% [55−57]. Consideration of only the high thermal conductivity and high aspect ratio of individual CNTs suggests that dramatic κ enhancements also would be expected ($\sim 50\times$ for 1 vol% randomly oriented CNTs) [59]. In contrast, experimental reports for CNT−polymer composites all show very modest κ enhancements, usually less than a factor of two for 1 wt% randomly oriented CNTs (Fig. 16.5). These results are generally attributed to a large contribution from interfacial thermal resistances (R_K), which are associated with CNT−polymer and CNT−CNT interfaces. For CNT−polymer interfaces, both experiments on CNTs in liquid and molecular dynamics simulations estimate R_K (CNT−polymer) to be $\sim 10^{-8}$ m^2 K/W [59,81]. Models based on noninteracting CNTs using effective medium theory (EMT) can account for the low experimental results through inclusion of R_K for the CNT−polymer interface [82−84]. A model based on percolation might be required to account for nonlinearity at higher loading [60]. However, simulations indicate that the CNT−CNT thermal resistance is also

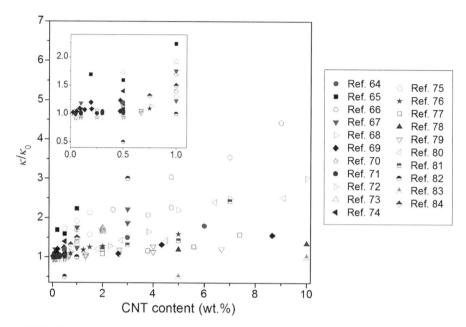

FIGURE 16.5

Representative experimental results [60–80] for thermal conductivity of randomly oriented CNT–polymer composites relative to the matrix thermal conductivity (κ_0) at room temperature. A single symbol is used for each reference, although several datasets include multiple types of composite, e.g. MWCNT, SWCNT, and variations in functionalization or processing method.

Source: Adapted from M.B. Jakubinek, PhD thesis, Dalhousie University, 2009.

high, $\sim 10^{-8}$ m² K/W or greater, and the contact areas at CNT–CNT junctions are small [52,60]. No studies have observed a large jump in κ coinciding with that observed for σ due to percolation.

Several experimental studies have extracted R_K for CNT–polymer interfaces by applying EMT models. The results, $(0.2–80) \times 10^{-8}$ m² K/W [60,65,75], are within an order of magnitude of the simulations and measurements for CNTs in liquids. However, in this approach, a multitude of other factors (e.g. CNT quality, purity, functionalization, straightness, orientation, dispersion, and κ reduction due to bundling) are lumped into the apparent R_K value. The effect of changing R_K independent of these other factors, which vary from sample to sample, has been observed in temperature-dependent measurements. R_K is known to increase at low temperature [85], and one of the first reports for a CNT–epoxy composite showed a κ enhancement that decreased as the temperature was reduced from 300 to 30 K [61]. Measurements have shown that despite an increase in κ at 300 K, κ of the composite can be lower than that of the matrix at low temperatures [75]. Both results are consistent with κ limited by R_K in composites [80]. Given the challenge posed by interfacial thermal resistance, along with issues including limited CNT loading, difficulty in dispersing CNTs, and lack of alignment in many cases, CNT assemblies that can more readily provide for higher CNT content and/or CNT alignment may offer a preferred route to achieve improved thermal conductivity in macroscopic CNT materials. For example, larger thermal conductivity values have been reported for composites prepared by infiltrating preformed vertically oriented CNT arrays [86] and CNT buckypaper [87].

16.6 CNT BUCKYPAPER AND THIN FILMS

Several factors that limit κ in CNT composites can be addressed by assembling densely packed CNTs to create macroscopic materials composed of nominally 100% CNTs. The first such CNT materials, often called buckypapers, were planar networks of randomly oriented CNTs produced by vacuum filtration of CNT suspensions [2]. Sheets with thickness from ~ 10 μm to several hundred microns can be peeled from the filter to produce a free-standing paper with comparable thickness and similar nonwoven structure as paper (Fig. 16.6). Experimental results for the thermal conductivity of these materials are summarized in Table 16.2, along with measurements for similarly structured thin networks (CNT films) and aligned films/buckypapers. Thin CNT films (Fig. 16.6) are also made by filtration, as well as other solution processes and direct growth [88]. Semitransparent films (less than a few hundred nanometers) are usually supported by substrates, which complicates thermal measurements. Novel measurement techniques using films suspended over holes <100 μm or ~ 2 mm in diameter have been reported [89,90]. The measurements of aligned buckypapers and films are mostly of materials prepared by filtration in high magnetic fields. A novel report involved flattening a vertically

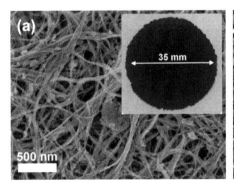

FIGURE 16.6

SEM images and photos of (a) SWCNT buckypaper (thickness ~ 70 μm). *Source: Courtesy of Jingwen Guan (National Research Council Canada)* and (b) a thin SWCNT film (thickness ~ 100 nm) produced by vacuum filtration.

Table 16.2 Thermal Conductivity of In-Plane Oriented Carbon Nanotube Films and Buckypaper at Room Temperature

Sample	κ_{app} W/m K	κ W/m K	ρ_{app} g/cm³	Measurement Type	Ref
SWCNT films (magnetically aligned)				κ; comparative & self-heating	[12]
Magnetically aligned		220			
Random		30			
SWCNT films					[90]
Vacuum filtration ($t = 100$ nm)	83		1.5	κ; bolometer	
Vacuum filtration ($t = 1$ μm)	75		1.5		
Direct growth ($t = 35$ nm)	30				
CNT films ($t = 50-200$ nm)				κ; Raman shift	[89]
SWCNT (arc discharge)	64				
SWCNT (HiPCO)	30				
MWCNT (catalytic CVD)	18				
MWCNT (thermal CVD)	24				
SWCNT buckypaper (magnetically aligned)				κ; comparative	[95]
$H = 7$ T, κ(parallel)	**4.5**	10	0.6		
$H = 7$ T, 11	0.6			κ(perpendicular)	5
$H = 26$ T, κ(parallel)	**41**	60	0.9		
$H = 26$ T, 11	0.9			κ(perpendicular)	**7.4**
SWCNT buckypaper (magnetically aligned)				κ; comparative	[87]
Magnetically aligned, κ(parallel)	42				
Random	19				

Table 16.2 Thermal Conductivity of In-Plane Oriented Carbon Nanotube Films and Buckypaper at Room Temperature—Cont'd

Sample	κ_{app} W/m K	κ W/m K	ρ_{app} g/cm^3	Measurement Type	Ref
MWCNT buckypaper (aligned from array)				κ; self-heating	[91]
Aligned, κ(parallel)	153	331	0.62		
Aligned,				κ(perpendicular)	72
	0.62				
Random	81		0.54		
MWCNT buckypaper (160 μm thick)				κ; electrothermal/ Raman α; electrothermal	[96]
Buckypaper 1	1.19		0.459		
Buckypaper 2	2.92		0.543		
MWCNT buckypaper (stack of 6)				κ; steady state	[97]
As prepared	2		0.60		
Sintered (900 °C)	5		0.95		
Sintered (1200 °C)	8		1.14		
Sintered (1500 °C)	11		1.16		
SWCNT buckypaper ($t = 70$ μm)	2.4		0.42	κ; steady state	[98]
MWCNT buckypaper (through thickness)	0.5–1.7		0.26–0.47	κ; steady state	[94]

NB: Plain text values are as reported in the original references. Values in bold were estimated from information in the paper.

aligned CNT array with a cylinder ("domino pushing") [91]. Aligned CNT networks have also been produced by drawing sheets from arrays (summarized later in Table 16.4). The density of these sheets is often low compared to those of films/buckypapers [8,92]; however, a recent report described a high-density buckypaper (1.39 g/cm^3) with remarkably high apparent thermal conductivity (approaching 800 W/m K) produced by drawing a sheet from an MWCNT array [93].

The experimental results show significant variability, from ~1 to ~200 W/m K. The highest values are reported for aligned CNT cases, and the thermal conductivities reported for thin CNT films are higher than most reports for thicker buckypapers. Most studies have measured in-plane thermal conductivities; however, several buckypapers measured in the cross-plane (through thickness) direction have shown $\kappa_{app} = 0.5-1.7$ W/m K, which is competitive for applications of thermal interface materials [94].

In addition to effects of bundling on κ, the thermal conductivity of CNT networks in buckypaper and CNT films appears to be limited by CNT–CNT interfacial thermal resistance. Volkov and Zhigilei used numerical simulations to evaluate κ_{app} of randomly oriented CNT films composed of straight (10,10) CNTs (Fig. 16.7) with a density of 0.2 g/cm^3 [99,100], somewhat lower than the density of the films and buckypapers in Table 16.2. Under the assumption of high (infinite) κ_{CNT}, the in-plane thermal conductivity of the film increased quadratically with CNT length (ranging from 1 to 200 for CNTs from 100 nm to 1 μm) [99]. The through-thickness thermal conductivity was two to three orders of magnitude lower and increased less dramatically with CNT length. In a subsequent model, the effect of finite κ_{CNT} was found to be quite significant, especially for longer CNTs (Fig. 16.7) [100]. Similar observations have been made for electrical conductivity

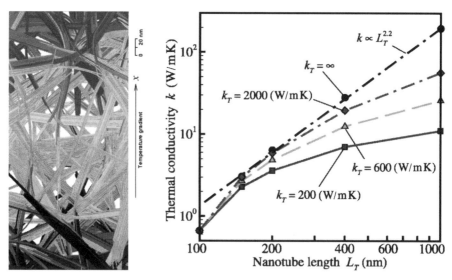

FIGURE 16.7

Example of the CNT film structure used for simulation of thermal conductivity of a film of (10,10) CNTs with an apparent density of 0.2 g/cm^3 and the simulation results showing the effects of CNT length and κ_{CNT}.

Source: Reprinted with permission from Ref. [100]. Copyright 2012, American Institute of Physics. http://dx.doi.org/10.1063/1.4737903.

of CNT films, which is simpler to study as films can be measured on electrically insulating substrates and thermal isolation is much more difficult. Electrical transport is also found to be dominated by junction resistance, but explaining the variation in σ requires variation in SWCNT quality (σ_{CNT}) [101,102]. Bundle size also can be expected to contribute as bundling has been shown to reduce κ relative to isolated CNTs. Further, debundling was found to be critical for improving σ, where decreasing bundle size increases the density of CNT–CNT junctions providing more pathways for electrical transport [101]. The κ simulations and related electrical conductivity work provide strong indications that variations in CNT quality (i.e. κ_{CNT}), length and bundle size are factors in the wide range of experimental results.

16.7 CNT SUPERFIBER MATERIALS
16.7.1 CNT arrays

Films of vertically aligned CNTs, in contrast to the in-plane alignment seen above in buckypapers and CNT films, are produced by CVD growth on substrates. These aligned CNT assemblies, typically called CNT arrays, forests or carpets, can be grown to a range of heights from micron-scale films to centimeter-tall 3D assemblies (Fig. 16.8). In terms of their thermal properties, the arrays themselves are interesting materials to study long CNTs and for applications as thermal interface materials and in composites [22,86,105–107]. CNT arrays are also part of a set of superfiber materials including CNT ribbons/sheets, and fibers/yarns that are produced by drawing from the CNT arrays (Fig. 16.9) [2,3,109,110]. CNT fibers can also be produced by dry spinning of double-walled carbon nanotubes (DWCNTs) from a floating catalyst CVD method [111] or by solution spinning using superacid-dispersed SWCNTs [112].

Thermal conductivity results for CNT arrays at room temperature are summarized in Table 16.3. With a few exceptions where $\kappa_{app} = 75-250$ W/m K, the thermal conductivities of the as-grown arrays are often modest (κ_{app}/SG ~ $10-30$ W/m K) in comparison to those of individual CNTs. κ_{app}/SG is used for comparisons here, instead of the more common κ, as reports make different assumptions about ρ_{CNT} used to calculate κ. Scanning electron microscopic (SEM) images also show that at high magnification the assumption of the MWCNTs as straight cylinders deviates from the actual structure (Fig. 16.8). Waviness of the CNTs also explains why the κ_{app} values measured perpendicular to the CNT direction are only ~20 and ~40 times lower than in the vertical direction [5,113]. Larger anisotropy could be expected based on the MWCNT diameter and the spacing of cylindrical CNTs. Most measurements were performed at room temperature on MWCNT samples. Temperature-dependent measurements over the range from 2 K to 390 K show increasing κ over the full range [5,27], with no evidence of the peak observed for individual CNTs. SWCNT arrays have been much less studied, in comparison to MWCNTs, because growth methods for such arrays are relatively undeveloped. However, one study has reported thermal

Table 16.3 Thermal Conductivity of Vertically Grown CNT Arrays at Room Temperature

Sample	κ_{app} W/m K	κ W/m K	ρ_{app} g/cm^3	κ_{app}/SG W/m K	Measurement Type	Ref
MWCNT array (L = 10–50 μm)	15	200			κ; photothermal	[114]
MWCNT array (L = 30 μm)	0.145	27.3	0.0117	12	κ; photothermal	[21]
MWCNT array (L = 1.6 mm)	8.7		0.234	37	α; 3ω	[115]
MWCNT array (L = 13 μm)	74				κ; 3ω	[105]
MWCNT array (L = 7 μm)	250				κ; thermoreflectance	[116]
MWCNT array (L = 20 μm)	2.85	40	0.185	15	κ; photothermal	[22]
SWCNT/few-walled array (L = 2 mm)					α; flash (in air)	[113]
As grown, κ(parallel)	3–6.4		0.153	24		
As grown, κ(perpendicular)	0.08–0.14					
Annealed 2800 C, κ(parallel)	10–15					
MWCNT array (L = 20 μm)	60	750^1	0.18	330	α; laser flash	[117]
MWCNT array (L = 100 μm)	8.3				α; laser flash	[118]

16.7 CNT Superfiber Materials

Sample							Method	Ref.
Aligned MWCNT (CVD array)							κ	[27]
As produced	2.5							
CH$_3$OH treated; annealed 100 °C	8							
Annealed 1250 °C	3							
Annealed 2800 °C	23							
MWCNT array								
κ(parallel) (L = 2–6 mm)	1		35		0.06	17	κ; steady state	[5]
κ(perpendicular)	0.06							
MWCNT array (L = several mm)	0.3	~2–3		**0.02**	~**0.1–0.3**	16	~5–20	κ; comparative	[119]
MWCNT array (L = 0.7–1.2 mm)	**1.36**		27^2		**0.065**	21	α; laser flash	[120]
MWCNT array (L = 25 µm)	180					30%		κ; infrared pyrometry	[121]

NB: Plain text values are as reported in the original references. Values in bold were estimated from information in the paper.

^{1}The calculation of κ, from Eqn (16.4), was based on an MWCNT density of 2.25 g/cm^3 [122]. This same density was used here to estimate the apparent density. A lower value, 1.65 g/cm^3, is also noted in the paper. Both values are within the range of densities used for individual MWCNTs in other studies.

^{2}A scaled value of 540 W/m K for the individual CNTs, which was obtained by scaling the measured thermal diffusivity (30 mm^2/s) to 100% packing and then calculating κ from Eqn (16.4) [120,123], was also reported in this study. However, this scaling may not accurately represent an individual CNT as, unlike κ, α should not scale with the CNT fraction. The thermal conductivity and heat capacity per unit volume of air can be neglected relative to CNTs; therefore, for a CNT array, both the thermal conductivity and heat capacity per unit volume, ([pC$_p$] = f$_{CNT}$*[pC$_p$]$_{CNT}$ + (1 − f$_{CNT}$)*[pC$_p$]$_{air}$, from rule of mixtures where f is the CNT volume fraction), scale by equal factors with increasing CNT fraction and the thermal diffusivity does not change.

FIGURE 16.8

Tall MWCNT arrays produced by water-assisted CVD [103,104]. These arrays are composed of aligned multiwalled carbon nanotubes with an outer diameter of approximately 10–20 nm and are spaced every 70–100 nm. SEM images show that the MWCNT are well aligned on the 0.1-mm scale but that waviness is apparent at smaller scales.

Source: Images reprinted from Ref. [5]. Copyright (2010), with permission from Elsevier. http://dx.doi.org/10.1016/j.carbon.2010.06.063.

conductivity for an array with a significant content of SWCNTs (a mixture of SWCNTs and few-walled MWCNTs) [113] and the thermal conductivity is comparable to the reported values for MWCNT arrays.

Considering the results for individual CNTs and the discussion of CNT buckypaper and films with in-plane orientation, it is likely that the CNT quality (defect density) is a limitation in the κ of arrays. Arrays are grown by CVD processes, and measurements on individual CNTs discussed above suggest that higher defect densities lead to lower κ_{CNT}. Several measurements on arrays provide indirect evidence of the importance of CNT quality. High-temperature annealing in inert atmosphere has improved κ_{app} between 2 and 10 times [27,113]. Thermal conductivity also has been observed to decrease with increasing Raman D/G band ratio [120], a common measure of CNT quality, and with increasing CNT length (Fig. 16.10) [5]. Electrical conductivity showed the same dependence on array height for the same arrays [5]. While long CNTs are desirable in principle, MWCNT quality in arrays often decreases with the array height and defects and

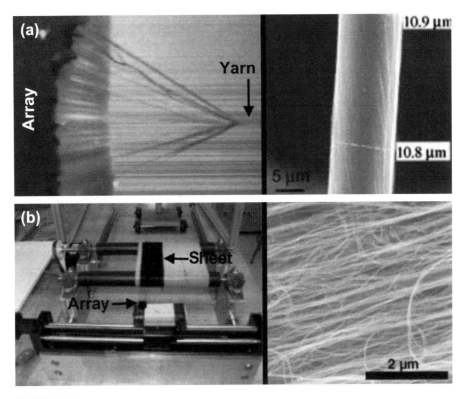

FIGURE 16.9

(a) Drawing of MWCNT yarn [108]. *Source: Reprinted from Ref. [108]. Copyright (2010), with permission from Elsevier.* http://dx.doi.org/10.1016/j.compscitech.2010.02.022 and (b) MWCNT sheets [8].
Source: Reprinted from Ref. [8]. Copyright (2012), with permission from Elsevier. http://dx.doi.org/10.1016/j.carbon.2012.04.067.

nonuniformity limits the length of spinnable arrays. The dependence of κ_{app} on height of the tall (millimeter–centimeter) arrays also indicates that κ_{app} is limited by the CNT properties rather than contacts, which is not unexpected given the measured thermal resistance and thermal contact resistances estimated from results for short arrays [5].

16.7.2 CNT sheets and yarns

As introduced above, drawing from MWCNT arrays has been used to produce both CNT ribbons/sheets and fibers/yarns (Fig. 16.9). Thermal conductivity has been measured for CNT arrays, sheets and yarn produced from the same water-assisted CVD growth system [5,7,8]. The arrays were measured by the longitudinal

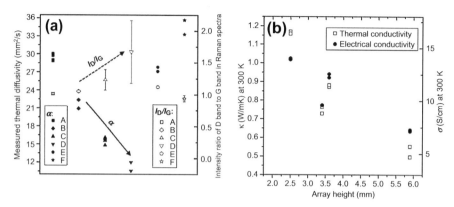

FIGURE 16.10

Evidence of the effect of MWCNT quality on the apparent thermal conductivity of MWCNT arrays from (a) decrease in thermal diffusivity with increasing Raman D/G ratio for samples subjected to increasing microwave irradiation time (samples B, C, D), where arrows were added here. *Source: Reprinted from Ref. [120]. Copyright (2012), with permission from Elsevier. http://dx.doi.org/10.1016/j.carbon.2011.11.038.* (b) Decrease in apparent thermal and electrical conductivities with array height [5].

Source: With data from Ref. [5].

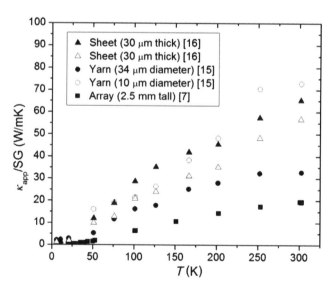

FIGURE 16.11

Comparison of apparent thermal conductivities on a per density basis (κ_{app}/SG) for 2.5-mm-tall MWCNT arrays [5], and MWCNT yarns [7] and sheets [8] drawn from 0.5-mm-tall MWCNT arrays. All arrays were produced by the same water-assisted CVD growth process [103].

steady-state method and sheets and yarns were measured by a steady-state technique based on parallel thermal conductances. Unsurprisingly, the room-temperature thermal conductivity of the yarn ($\kappa_{app} \sim 60$ W/m K for 10 μm diameter) is higher than that of the lower density sheet (~ 2.5 W/m K) and array (~ 1.2 W/m K). On a per density basis, the yarn and sheet both have high κ_{app}/SG compared to the array (Fig. 16.11). This result likely could be explained by improved alignment created during the drawing process, although MWCNT quality may also be a factor as the materials are not from the same production run and the measured array was taller (2.5 mm) than the arrays used for drawing sheets and yarns (0.5 mm). Variation in the curvature of κ_{app} vs T for the sheets and yarns at low temperature, in comparison to the array, might be due to greater difficulty in accurately measuring the low thermal conductances of the small sheet and yarn samples at the lowest temperatures. Comparable values for κ_{app}/SG at room temperature are obtained for the array-drawn sheets and fibers reported by Aliev et al., who report $\kappa = 50$ W/m K based on a densely packed sheet ($\rho = 1.54$ g/cm^3) [92].

Table 16.4 summarizes thermal conductivity reports for CNT fibers/yarns and array-drawn sheets. In comparison to individual CNTs, bundles, and arrays, there is relatively good agreement among reports, with most reports showing ~ 30 W/m K for dry-spun fibers from arrays or aerosol CVD [7,92,124] and somewhat lower values (~ 20 W/m K) for acid-spun fibers [112,125]. Acid spinning is typically done with shorter CNTs (e.g. HiPCO SWCNTs, ~ 1 μm or less) but has the advantage that SWCNTs can be used, including from methods that tend to result in lower defect densities (i.e. arc discharge or laser vaporization). Maximizing the alignment and density of the yarns is important to optimize κ_{app}. Thermal conductivity has been reported for only one yarn with $\rho > 1$ g/cm^3. This yarn is composed of collapsed DWCNTs [124], which is beneficial in terms of density and may also improve CNT–CNT heat transfer due to larger contact area than uncollapsed CNTs, and had the second highest κ_{app} for a CNT yarn. Less intuitively, smaller diameter yarn also was found to be beneficial. The highest κ_{app} reported for a CNT yarn, 60 ± 20 W/m K for a 10-μm-diameter yarn, was almost two times higher than the κ_{app} for a 34-μm-diameter yarn from the same process [7]. Similar observations were made for electrical conductivity in measurements of a larger number of samples (10 yarns of varying diameter) [7]. This trend of increased κ and σ with decreased diameter was attributed primarily to differences in migration length. Migration length, the longitudinal distance over which a CNT shifts from the yarn surface to the interior and back [110], decreases with larger diameter. κ for low-density MWCNT sheets (scaled to account for low density) was also found to decrease with thickness (number of layers), but in this case the difference was attributed to poor contact between the interior sheets and the electrodes in the stacked structure [92]. The highest apparent thermal conductivity for a nanotube superfiber material (κ_{app}/SG approaching 600 W/m K), reported very recently for a high-density, array-drawn MWCNT paper [93], is the first report of a superfiber material with thermal conductivity exceeding that of metals.

Table 16.4 Thermal Conductivity of CNT Fibers, Yarns, and Array-Drawn Sheets at Room Temperature

Sample	κ_{app} W/m K	κ W/m K	ρ_{app} g/cm³	κ_{app}/SG W/m K	Measurement Type	Ref
Aligned MWCNT sheet from array					κ, α; 3ω, laser flash	[28,92]
Parallel	**0.1**	50	**0.003**	33		
Perpendicular		2.1				
Aligned MWCNT sheet (multilayer) from array	2.5		0.039	65	κ; steady state	[8]
Aligned MWCNT sheet/buckypaper	470–770		0.81–1.39	~570	κ; self-heating	[93]
SWCNT fiber (solution spun)						
22% stretch	4				κ; comparative	[126]
58% stretch	10					
SWCNT fiber (acid spun)	21		0.87	24	κ; comparative	[112]
SWCNT fiber (acid spun)					κ; comparative	[125]
220 μm	5					
60 and 110 μm	18					
MWCNT yarn (array spun)	26		0.8	33		[92]
MWCNT yarn (array spun)			0.9		κ; steady state	[7]
10 μm diameter	60			70		
34 μm diameter	30			33		
DWCNT yarn (aerosol spun)	40		1.2	33	α; Angstrom method	[124]

NB: Plain text values are as reported in the original references. Values in bold were estimated from information in the paper.

16.8 BORON NITRIDE NANOTUBES

This chapter has focused on CNT materials, for which there is by far the largest amount of literature and production methods readily provide commercial quantities of nanotube material. However, related materials can offer advantages over CNTs for development of engineering materials. CNT-related materials include graphene and varieties of non-carbon nanotubes including BN, BCN, SiC, ZnS and others. In terms of superfiber materials with desirable thermal properties, boron nitride nanotubes (BNNTs) offer potential advantages over CNTs. The thermal stability of BNNTs in air (>1100 K) is significantly higher than that of CNTs [127]. Based on a higher calculated specific heat capacity for BNNT and the Debye model (Eqn (16.3)), it was suggested that BNNTs could have higher κ than CNTs [128]. There are few experimental measurements for thermal conductivity of BNNTs. Measurements on BNNT bundles led to a prediction of high κ_{BNNT}, although lower than κ_{CNT}, and noted the lower isotopic purity of boron compared to carbon as factor [129]. A thermal conductivity of $\sim$325 W/m K measured on isotopically pure ^{11}B multiwalled BNNTs was equal to that of similar-diameter CNTs, while a natural-abundance BNNT had lower κ [130]. This is consistent with a reduction on the phonon mean free path to isotope scattering.

To date, fundamental experimental research and, in particular, application studies of BNNT have been limited in comparison to CNTs, which can be attributed to the more limited availability of material due to greater difficulties in mass production of BNNTs [131,132]. There are only a few reports of BNNT assemblies, including BNNT films [133,134] and arrays [135]. The first BNNT yarn, a 3-cm-long yarn has been made by finger-twisting cottonlike, as-produced BNNT material [136]. Aside from BNNT bundles, the thermal conductivity of BNNT assemblies has not been reported. A report of thermal conductivity for BNNT–polymer composite films [137] showed κ within the range reported for CNT–polymer composites. At present, production methods for BNNT do not offer sufficient quantities and scalability; however, the activity and advances on BNNT are increasing. Advances in the production and chemistry of BNNTs, including a supplier (BNNT, LLC) aiming to have BNNTs for sale in commercial quantities in early 2013 [136,138], and recent work at the National Research Council Canada, where quantities of $\sim$10 g raw BNNTs have been produced in single 1-h runs using an intrinsically scalable production method, can be expected to provide the availability and quantity of BNNT required to further the characterization and development of BNNT assemblies.

16.9 CHALLENGES AND OPPORTUNITIES

The understanding of thermal conductivity of CNT assemblies and superfiber materials is less developed in comparison to some other properties; however, reports in recent years have begun to uncover the story. Given the challenges in making literature comparisons, due to large variations in the CNT material and measurement methods, systematic studies of key parameters are particularly important. Examples include correlation of increased defect density observed by SEM with reduction of κ for individual CNTs [37], which likely is a factor for the high variability between studies, and experimental observation of a κ reduction with increasing bundle size [28]. The latter is of key importance in translating the properties of CNTs to macroscopic materials. In bulk CNT assemblies and superfiber materials, reports showing trends in κ with Raman D/G ratio [120], length [5], and postheating [27,97,113] all suggest that the CNT quality is a limiting factor. Superfiber materials tend to be composed of long, substrate-grown CVD MWCNTs, which typically have higher defect densities [2]. This suggests that there is significant potential for improvement on the current κ results as methods of synthesis and post-treatment improve. Already, significant increases in κ have been reported for high-temperature-annealed CNT arrays [27,113]. While most theoretical work has been focused on individual CNTs and CNT composites, recent modeling of a CNT network resembling a two-dimensional randomly oriented thin film or buckypaper also indicates that κ_{CNT} is a limiting factor despite a common assumption that thermal resistance is primarily due to CNT–CNT contacts [100].

Other key challenges include optimizing the apparent density and CNT–CNT contacts in CNT assemblies. It is a common practice to report both the apparent thermal conductivity of low-density CNT assemblies and the thermal conductivity of a related, fully dense CNT material (between 1.3 and 2.25 g/cm^3). The latter reflects the thermal conductivity of the CNTs. The apparent thermal conductivities likely are most relevant for applications, in particular for low-density assemblies. However, the thermal conductivity of CNTs is an important quantity to evaluate factors such as CNT quality and length, etc., and assemblies such as papers and fibers can offer higher density. Based on a combination of experimental results and modeling, an optimal MWCNT cable, consisting of long ($\sim$1.5 mm) MWCNTs touching each other over $\sim$2–3% of their length, has been described to minimize both κ reductions due to bundling and CNT–CNT contact resistance [28]. Such a cable could have higher thermal conductivity than copper [28], and a high-density CNT assembly with thermal conductivity significantly higher than copper has now been reported [93]. Conversely, a high-porosity three-dimensional CNT network with superlow thermal conductivity ($\kappa_{app} = 0.035$ W/m K) has also been reported and such low-κ assemblies might be useful for development of new thermoelectric materials [139].

CNT fibers/yarns are particularly promising multifunctional materials [3]. While there are only a handful of reports of their thermal conductivity and their properties remain significantly below those of individual nanotubes (and in some cases below those of traditional high-performance carbon fibers), this is a new material that offers multifunctional capabilities and presents considerable potential for improvement. Such materials already have been produced with densities over 1 g/cm^3 [124] and spun from several millimeter long MWCNTs [140], which are two important requirements for achieving superfiber materials with high thermal conductivity. The largest measured result for a CNT fiber ($\kappa_{app} = 60 \pm 20$ W/m K) for a 10-μm-diameter yarn, which was significantly higher than a larger diameter yarn from the same source, suggests that multiply yarns with finer strands may offer better performance. However, there are presently no reports of systematic studies for κ of CNT fiber/yarn (e.g. as a function of diameter, CNT length, twist, drawing rate, number of strands, annealing and other posttreatments, etc.), which would significantly improve understanding of the potential of CNT yarn superfiber materials.

Other opportunities for superfiber materials with improved thermal properties include developments related to BNNTs (e.g. by Hurst and Misra) and irradiation-induced cross-linking (e.g. by Espinosa and Filleter), which can be expected to influence both κ_{CNT} and CNT—CNT thermal contact resistance. Superfiber materials also provide a starting point for advanced composite materials (e.g. by Song et al.). Composites based on superfiber materials can be expected to enable higher thermal conductivity than composites prepared by dispersion of CNT powders. This has been seen in reports for arrays and buckypaper [86,87,119]. Composites using CNT fibers integrated into a matrix are a relatively recent development and have been the subject of few studies to date [124,141,142]. Recently, an apparent thermal conductivity of over 20 W/m K for an aligned CNT fiber—epoxy composite containing 10 wt% CNT fiber was reported [124]. This enhancement is larger than expected based on the thermal conductivity of the fiber ($\kappa_{app} = 40$ W/m K), which led to the conclusion that infiltration with epoxy improves the effective properties of the fiber [124].

Studies of thermal conductivity and performance of nanotube superfiber materials are still in their early stages. The potential for nanotube superfiber materials to transform engineering design through realization of superior properties based on nanotubes in macroscopic engineering materials is an exciting proposal and many opportunities remain for discoveries and developments related to the thermal properties of these materials.

Acknowledgments

Much of my experience in this field was developed at Dalhousie University under the supervision of Prof. Mary Anne White and with support from an NSERC Canada Graduate Scholarship and Fellowships from the Killam Trusts and the Sumner Foundation. I would like to acknowledge Mike Johnson (Dalhousie U), who I worked with on many measurement challenges, as well as collaborations with the groups of Vesselin Shanov (U Cincinnati), Karen Winey (U Penn), David Carey (U Surrey) and my current group at the National Research Council Canada led by Benoit Simard. I also acknowledge valuable discussions with J.-H. Pöhls (Dalhousie U) and correspondence with W. Lin (Georgia Tech), H. Xie (SSPU, Shanghai), A. Aliev (UT Dallas) and D.G. Cahill (UIUC), and thank M.A. White, M. Johnson, B. Ashrafi, B. Simard and C. Kingston for their comments on sections of this chapter.

References

[1] M. Endo, M.S. Strano, P.M. Ajayan, Topics in Applied Physics 111 (2008) 13.
[2] L. Liu, W. Ma, Z. Zhang, Small 7 (2011) 1504.

[3] C. Jayasinghe, W. Li, Y. Song, J.L. Abot, V.N. Shanov, S. Fialkova, S. Yarmolenko, S. Sundaramurthy, Y. Chen, W. Cho, S. Chakrabarti, G. Li, Y. Yun, M.J. Schulz, MRS Bulletin 35 (9) (2010) 682.
[4] S.R. Phillpot, A.J.H. McGaughey, Materials Today 8 (2005) 18.
[5] M.B. Jakubinek, M.A. White, G. Li, C. Jayasinghe, W. Cho, M.J. Schulz, V. Shanov, Carbon 48 (2010) 3947.
[6] P. Kim, L. Shi, A. Majumdar, P.L. McEuen, Physical Review Letters 87 (2001) 215502.
[7] M.B. Jakubinek, M.B. Johnson, M.A. White, C. Jayasinghe, G. Li, W. Cho, M.J. Schulz, V. Shanov, Carbon 50 (2012) 244.
[8] J.-H. Pöhls, M.B. Johnson, M.A. White, R. Malik, B. Ruff, C. Jayasinghe, M.J. Schulz, V. Shanov, Carbon 50 (2012) 4175.
[9] J. Shiomi, S. Maruyama, International Journal of Thermophysics 31 (2007) 1945.
[10] L. Lindsay, D.A. Broido, N. Mingo, Physical Review B 73 (2009) 125407.
[11] J. Hone, M. Whitney, C. Piskoti, A. Zettl, Physical Review B 59 (1999) R2514. (J. Hone, M. Whitney, A. Zettl, Synthetic Metals 103 (1999) 2498).
[12] J. Hone, M.C. Llaguno, N.M. Nemes, A.T. Johnson, J.E. Fischer, D.A. Walters, M.J. Casavant, J. Schmidt, R.E. Smalley, Applied Physics Letters 77 (2000) 666.
[13] T.M. Tritt, D. Weston, Measurement techniques and considerations for determining thermal conductivity of bulk materials, in: T.M. Tritt (Ed.), Thermal Conductivity: Theory, Properties, and Applications, Kluwer Academic, New York, 2004.
[14] C. Yu, L. Shi, Z. Yao, D. Li, A. Majumdar, Nano Letters 5 (2005) 1842.
[15] B.M. Zawilski, R.T. Littleton IV, T.M. Tritt, Review of Scientific Instruments 72 (2001) 1770.
[16] K. Aaron, Masters Thesis in Physics, Clemson University, 2005.
[17] X. Huang, J. Wang, G. Eres, X. Wang, Carbon 49 (2011) 1680.
[18] W. Yi, L. Lu, Z. Dian-lin, Z.W. Pan, S.S. Xie, Physical Review B 59 (1999) R9015.
[19] T. Borca-Tasciuc, G. Chen, Experimental techniques for thin-film thermal conductivity characterization, in: T.M. Tritt (Ed.), Thermal Conductivity: Theory, Properties, and Applications, Kluwer Academic, New York, 2004.
[20] J.E. Fisher, Carbon nanotubes: structure and properties, in: Y. Gogotsi (Ed.), Nanomaterials Handbook, CRC Press, Boca Raton, FL, 2006.
[21] X. Wang, Z. Zhong, J. Xu, Journal of Applied Physics 97 (2005) 064302.
[22] Y. Xu, Y. Zhang, E. Suhir, Journal of Applied Physics 100 (2006) 074302.
[23] L. Shi, D. Li, C. Yu, W. Jang, D. Kim, Z. Yao, P. Kim, A. Majumdar, Journal of Heat Transfer 125 (2003) 881.
[24] H.L. Zhang, J.-F. Li, K.F. Yao, L.D. Chen, Journal of Applied Physics 97 (2005) 114310. (H.-L. Zhang, J.-F. Li, B.-P. Zhang, K.-F. Yao, W.-S. Lin, H. Wang, Physical Review B 75 (2007) 205407).
[25] H. Xie, Journal of Materials Science 42 (2007) 3695.
[26] J. Hou, X. Wang, P. Vellelacheruvu, J. Guo, C. Liu, H.-M. Cheng, Journal of Applied Physics 100 (2006) 124314.
[27] R. Jin, Z.X. Zhou, D. Mandrus, I.N. Ivanov, G. Eres, J.Y. Howe, A.A. Puretzky, D.B. Geohegan, Physica B 388 (2007) 326.
[28] A.E. Aliev, M.H. Lima, E.M. Silverman, R.H. Baughman, Nanotechnology 21 (2010) 035709.
[29] J.J. Vilatela, A.H. Windle, Advanced Materials 22 (2010) 4959.
[30] R.S. Ruoff, D.C. Lorents, Carbon 33 (1995) 925–930.

[31] S. Berber, Y.-Y. Kwon, D. Tománek, Physical Review Letters 84 (2000) 4613–4616.
[32] J. Yang, Y. Yang, S.W. Waltermire, T. Gutu, A.A. Zinn, T.T. Xu, Y. Chen, D. Li, Small 7 (2011) 2334.
[33] E. Pop, D. Mann, Q. Wang, K. Goodson, H. Dai, Nano Letters 6 (2006) 96–100.
[34] M. Fujii, X. Zhang, H. Xie, H. Ago, K. Takahashi, T. Ikuta, H. Abe, T. Shimizu, Physical Review Letters 95 (2005) 065502.
[35] Q. Li, C. Liu, X. Wang, S. Fan, Nanotechnology 20 (2009) 145702.
[36] T.-Y. Choi, D. Poulikakos, J. Tharian, U. Sennhauser, Nano Letters 6 (2006) 1589.
[37] M.T. Pettes, L. Shi, Advanced Functional Materials 19 (2009) 3918.
[38] M.A. Osman, D. Srivastava, Nanotechnology 12 (2001) 21.
[39] G. Zhang, B. Li, Journal of Chemical Physics 123 (2005) 114714.
[40] J. Wang, J.-S. Wang, Applied Physics Letters 88 (2006) 111909.
[41] Z.L. Wang, D.W. Tang, X.B. Li, X.H. Zheng, W.G. Zhang, L.X. Zheng, Y.T. Zhu, A.Z. Jin, H.F. Yang, C.Z. Gu, Applied Physics Letters 91 (2007) 123119.
[42] M.C.H. Wu, J.-Y. Hsu, Nanotechnology 20 (2009) 145401.
[43] J. Che, T. Cagin, W.A. Goddard III, Nanotechnology 11 (2000) 65.
[44] J.A. Thomas, R.M. Iutzi, A.J.H. McGaughey, Physical Review B 81 (2010) 045413.
[45] A. Cao, J. Qu, Size, Journal of Applied Physics 112 (2012) 013503.
[46] M. Grujicic, G. Cao, W.N. Roy, Journal of Materials Science 40 (2005) 1943.
[47] D. Donadio, G. Galli, Physical Review Letters 99 (2007) 255502.
[48] P. Avouris, Molecular electronics with carbon nanotubes, Accounts of Chemical Research 35 (2002) 1026.
[49] P.L. McEuen, J.-Y. Park, MRS Bulletin 29 (4) (2004) 272.
[50] A.B. Kaiser, G. Düsberg, S. Roth, Physical Review B 57 (1998) 1418.
[51] R. Prasher, Physical Review B 77 (2008) 075424.
[52] H. Zhong, J.R. Lukes, Physical Review B 74 (2006) 125403.
[53] J. Yang, S. Waltermire, Y. Chen, A.A. Zinn, T.T. Xu, D. Li, Applied Physics Letters 96 (2010) 023109.
[54] R.S. Prasher, X.J. Hu, Y. Chalopin, N. Mingo, K. Lofgreen, S. Volz, F. Cleri, P. Keblinski, Physical Review Letters 102 (2009) 105901.
[55] M. Moniruzzaman, K.I. Winey, Macromolecules 39 (2006) 5194.
[56] W. Brauhofer, J.Z. Kovacs, Composites Science and Technology 69 (2009) 1486.
[57] K.I. Winey, T. Kashiwagi, M. Mu, MRS Bulletin 32 (4) (2007) 348.
[58] Z. Han, A. Fina, Progress in Polymer Science 36 (2011) 914.
[59] S.T. Huxtable, D.G. Cahill, S. Shenogin, L. Xue, R. Ozisik, P. Barone, M. Ursey, M.S. Strano, G. Siddons, M. Shim, P. Keblinski, Nature Materials 2 (2003) 731.
[60] R. Haggenmueller, C. Guthy, J.R. Lukes, J.E. Fischer, K.I. Winey, Macromolecules 40 (2007) 2417.
[61] M.J. Biercuk, M.C. Llaguno, M. Radosavljevic, J.K. Hyun, A.T. Johnson, J.E. Fischer, Applied Physics Letters 80 (2002) 2767.
[62] A. Yu, M.E. Itkis, E. Bekyarova, R.C. Haddon, Applied Physics Letters 89 (2006) 133102.
[63] G.-W. Lee, J.I. Lee, S.-S. Lee, M. Park, J. Kim, Journal of Materials Science 40 (2005) 1259.
[64] C.H. Liu, H. Huang, Y. Wu, S.S. Fan, Applied Physics Letters 84 (2004) 4248.
[65] M.B. Bryning, D.E. Milkie, M.F. Islam, J.M. Kikkawa, A.G. Yodh, Applied Physics Letters 87 (2005) 161909.

[66] A. Moisala, Q. Li, L.A. Kinloch, A.H. Windle, Composites Science and Technology 66 (2006) 1285–1288.
[67] F.H. Gojny, M.H.G. Wichmann, B. Fiedler, I.A. Kinloch, W. Bauhofer, A.H. Windle, K. Schulte, Polymer 47 (2006) 2036.
[68] C.H. Liu, S.S. Fan, Applied Physics Letters 86 (2005) 123106.
[69] S. Wang, R. Liang, B. Wang, C. Zhang, Carbon 47 (2009) 53.
[70] Y. Yang, M.C. Gupta, J.N. Zalameda, W.P. Winfree, Micro & Nano Letters 3 (2008) 35.
[71] Y.S. Song, J.R. Youn, Carbon 43 (2005) 1378.
[72] E.T. Thostenson, T.-W. Chou, Carbon 44 (2006) 3022.
[73] P. Bonnet, D. Siriude, B. Garnier, O. Chauvet, Applied Physics Letters 91 (2007) 201910.
[74] Y. Xu, G. Ray, B. Abdel-Magid, Composites A 37 (2006) 114.
[75] M.B. Jakubinek, M.A. White, P.C.P. Watts, D. Carey, Materials Research Society Symposium Proceedings 1022 (2007) II03–06.
[76] C. Guthy, F. Du, S. Brand, K.I. Winey, J.E. Fischer, Journal of Heat Transfer 129 (2007) 1096.
[77] F. Du, C. Guthy, T. Kashiwagi, J.E. Fischer, K.I. Winey, Journal of Polymer Science Part B 44 (2006) 1513.
[78] L.E. Evseeva, S.A. Tanaeva, Mechanics of Composite Materials 44 (2008) 487.
[79] J.E. Peters, D.V. Papavassiliou, B.P. Grady, Macromolecules 41 (2008) 7274.
[80] M.B. Jakubinek, M.A. White, M. Mu, K.I. Winey, Applied Physics Letters 96 (2010) 083105.
[81] S. Shenogin, L. Xue, R. Ozisik, P. Keblinski, D.G. Cahill, Journal of Applied Physics 95 (2004) 8136.
[82] C.-W. Nan, G. Liu, Y. Lin, M. Li, Applied Physics Letters 85 (2004) 3549.
[83] F. Deng, Q.-S. Zheng, L.-F. Wang, C.-W. Nan, Applied Physics Letters 90 (2007) 021914.
[84] F. Deng, Q. Zheng, Acta Mechanica Solida Sinica 22 (2009) 1.
[85] E.T. Swartz, R.O. Pohl, Reviews of Modern Physics 61 (1989) 605.
[86] H. Huang, C. Liu, Y. Wu, S. Fan, Advanced Materials 17 (2005) 1652.
[87] P. Gonnet, Z. Liang, E.S. Choi, R.S. Kadambala, C. Zhang, J.S. Brooks, B. Wang, L. Kramer, Current Applied Physics 6 (2006) 119.
[88] L. Hu, D.S. Hecht, G. Gruner, Chemical Reviews 110 (2010) 5790.
[89] D. Kim, L. Zhu, C.-S. Han, J.-H. Kim, S. Baik, Raman Langmuir 27 (2011) 14533.
[90] M.E. Itkis, F. Borondics, A. Yu, R.C. Haddon, Nano Letters 7 (2007) 900.
[91] D. Wang, P. Song, C. Liu, W. Wu, S. Fan, Nanotechnology 19 (2008) 075609.
[92] A.E. Aliev, C. Guthy, M. Zhang, S. Fang, A.A. Zakhidov, J.E. Fischer, R.H. Baughman, Carbon 45 (2007) 2880.
[93] L. Zhang, G. Zhang, C. Liu, S. Fan (Article ASAP), Nano Letters (2012), http://dx.doi.org/10.1021/nl302374.
[94] H. Chen, M. Chen, J. Di, G. Xu, H. Li, Q. Li, Journal of Physical Chemistry C 116 (2012) 3903.
[95] J.E. Fischer, W. Zhou, J. Vavro, M.C. Llaguno, C. Guthy, R. Haggenmueller, M.J. Casavant, D.E. Walters, R.E. Smalley, Journal of Applied Physics 93 (2003) 2157.
[96] Y. Yue, X. Huang, X. Wang, Physics Letters A 374 (2010) 4414.

[97] K. Yang, J. He, P. Puneet, Z. Su, M.J. Skove, J. Gaillard, T.M. Tritt, A.M. Rao, Journal of Physics: Condensed Matter 22 (2010) 334215.
[98] M.B. Jakubinek, J. Guan, M.B. Johnson (unpublished).
[99] A.N. Volkov, L.V. Zhigilei, Physical Review Letters 104 (2010) 215902.
[100] A.N. Volkov, L.V. Zhigilei, Applied Physics Letters 101 (2012) 043113.
[101] P.N. Nirmalraj, P.E. Lyons, S. De, J.N. Coleman, J.J. Boland, Nano Letters 9 (2009) 3890.
[102] P.E. Lyons, S. De, F. Blighe, V. Nicolosi, L.P.C. Pereira, M.S. Ferreira, J.N. Coleman, Journal of Applied Physics 104 (2008) 044302.
[103] Y.H. Yun, V. Shanov, Y. Tu, S. Subramaniam, M.J. Schulz, Journal of Physical Chemistry B 110 (2006) 23920.
[104] V.N. Shanov, A. Gorton, Y.-H. Yun, M.J. Schulz. US Patent US2008/0095695A1, 2008.
[105] X.J. Hu, A.A. Padilla, J. Xu, T.S. Fisher, K.E. Goodson, Journal of Heat Transfer 128 (2006) 1109.
[106] M.A. Panzer, G. Zhang, D. Mann, X. Hu, E. Pop, H. Dai, K.E. Goodson, Journal of Heat Transfer 130 (2008) 052401.
[107] H. Peng, X. Sun, Chemical Physics Letters 471 (2009) 103−105.
[108] J.L. Abot, Y. Song, M.S. Vatsavaya, S. Medikonda, Z. Kier, C. Jayasinghe, N. Rooy, V.N. Shanov, M.J. Schulz, Composites Science and Technology 70 (2010) 1113.
[109] K. Jiang, Q. Li, S. Fan, Nature 419 (2002) 801.
[110] M. Zhang, K.R. Atkinson, R.H. Baughman, Science 306 (2004) 1358.
[111] K. Koziol, J. Vilatela, A. Moisala, M. Motta, P. Cunniff, M. Sennett, A. Windle, Science 318 (2007) 1892.
[112] L.M. Ericson, H. Fan, H. Peng, V.A. Davis, W. Zhou, J. Sulpizio, Y. Wang, R. Booker, J. Vavro, C. Guthy, A.N.G. Para-Vasquez, M.J. Kim, S. Ramesh, R.K. Saini, C. Kittrell, G. Lavin, H. Schmidt, W.W. Adams, W.E. Billups, M. Pasquali, W.-F. Hwang, R.H. Huage, J.E. Fischer, R.E. Smalley, Science 305 (2004) 1447.
[113] I. Ivanov, A. Puretzky, G. Eres, H. Wang, Z. Pan, H. Cui, R. Jin, J. Howe, D.B. Geohegan, Applied Physics Letters 89 (2006) 223110.
[114] D.J. Yang, S.G. Wang, Q. Zhang, P.J. Sellin, G. Chen, Physics Letters A 329 (2004) 207. (D.J. Yang, Q. Zhang, G. Chen, S.F. Youn, J. Ahn, S.G. Wang, Q. Zhou, Q. Wang, J.Q. Li, Physical Review B 66 (2002) 165440).
[115] T. Borca-Tasciuc, S. Vafaei, D.-A. Borca-Tasciuc, B.Q. Wei, R. Vajtai, P.M. Ajayan, Journal of Applied Physics 98 (2005) 054309.
[116] T. Tong, Y. Zhao, L. Delzeit, A. Kashani, M. Meyyappan, IEEE Transactions Components Packaging Technology 30 (2007) 92.
[117] H. Xie, A. Cai, X. Wang, Physics Letters A 369 (2007) 120.
[118] S. Shaikh, L. Li, K. Lafdi, J. Huie, Carbon 45 (2007) 2608.
[119] A.M. Marconnet, N. Yamamoto, M.A. Panzer, B.L. Wardle, K.E. Goodson, ACS Nano 5 (2011) 4818.
[120] W. Lin, J. Shang, W. Gu, C.P. Wong, Carbon 50 (2012) 1591.
[121] M. Gaillard, H. Mbitsi, A. Petit, E. Amin-Chalhoub, C. Boulmer-Leborgne, N. Semmar, E. Milton, J. Mathias, S. Kouassi, Journal of Vacuum Science and Technology B 29 (2011) 041805. (M. Gaillard, E. Amin-Chalhoub, N. Semmar, A. Petit, A.-L. Thomann, C. Boulmer-Leborgne, Proc. 13th Int. Conference on Plasma Surface Engineering, 2012).

[122] H. Xie, Personal Communication, Shanghai Second Polytechnic University, August 2012.
[123] W. Lin, Personal Communication, Georgia Institute of Technology, August 2012.
[124] J.J. Vilatela, R. Khare, A.H. Windle, Carbon 50 (2012) 1227.
[125] W. Zhou, J. Vavro, C. Guthy, K.I. Winey, J.E. Fischer, L.M. Ericson, S. Ramesh, R. Saini, V.A. Davis, C. Kittrell, M. Pasquali, R.H. Huage, R.E. Smalley, Journal of Applied Physics 95 (2004) 649.
[126] S. Badaire, V. Pichot, C. Zakri, P. Poulin, P. Launois, J. Vavro, C. Guthy, M. Chen, J.E. Fischer, Journal of Applied Physics 96 (2004) 7509.
[127] D. Goldberg, Y. Bando, K. Kurashima, T. Sato, Scripta Materialia 44 (2001) 1561.
[128] Y. Xiao, X.H. Yan, J. Xiang, Y.L. Mao, Y. Zhang, J.X. Cao, J.W. Ding, Applied Physics Letters 84 (2004) 4626.
[129] C.W. Chang, W.-Q. Han, A. Zettl, Applied Physics Letters 86 (2005) 173102.
[130] C.W. Chang, A.M. Mennimore, A. Afanasiev, D. Okawa, T. Ikuno, H. Garcia, D. Li, A. Majumdar, A. Zettl, Physical Review Letters 97 (2006) 085901.
[131] J. Wang, C.H. Lee, Y.K. Yap, Nanoscale 2 (2010) 2028.
[132] A. Pakdel, C. Zhi, Y. Bando, D. Golberg, Materials Today 15 (6) (2012) 256.
[133] L.H. Li, Y. Chen, Langmuir 26 (2010) 5135.
[134] L.H. Li, Y. Chen, A.M. Glushenkov, Journal of Materials Chemistry 20 (2010) 9679.
[135] Y. Wang, Y. Yamamoto, H. Kiyono, S. Shimada, Journal of Nanomaterials 2008 (2008) 606283.
[136] M.W. Smith, K.C. Jordan, C. Park, J.-W. Kim, P.T. Lillehei, R. Crooks, J.S. Harrison, Nanotechnology 20 (2009) 505604.
[137] T. Terao, C. Zhi, Y. Bando, M. Mitome, C. Tang, D. Goldberg, Journal of Physical Chemistry C 114 (2010) 4340.
[138] BNNT, LLC. (http://bnnt.com).
[139] J. Chen, X. Gui, Z. Wang, Z. Li, R. Xiang, K. Wang, D. Wu, X. Xia, Y. Zhou, Q. Wang, Z. Tang, L. Chen, ACS Applied Materials and Interfaces 4 (2012) 81.
[140] C. Jayasinghe, C. Chakrabarti, M.J. Schulz, V. Shanov, Journal of Materials Research 26 (2011) 645.
[141] A.E. Bogdanovich, P.D. Bradford, Composites Part A 41 (2010) 230.
[142] R.J. Mora, J.J. Vilatela, A.H. Windle, Composites Science and Technology 69 (2009) 1558.

CHAPTER 17

Three-dimensional Nanotube Networks and a New Horizon of Applications

Ana Laura Elías[1], Néstor Perea-López[1], Lakshmy Pulickal Rajukumar[1], Amber McCreary[1], Florentino López-Urías[1,2], Humberto Terrones[1], Mauricio Terrones[1,3,4]

[1] *Department of Physics and Center for 2-Dimensional and Layered Materials, The Pennsylvania State University, University Park, PA, USA,* [2] *Advanced Materials Department, IPICYT, San Luis Potosí, SLP, México,* [3] *Department of Materials Science and Engineering and Department of Chemistry, The Pennsylvania State University, University Park, PA, USA,* [4] *Research Center for Exotic Nanocarbons (JST), Shinshu University, Wakasato, Nagano, Japan*

CHAPTER OUTLINE

17.1 Introduction	458
17.2 Nanotube Network Types	459
17.2.1 Covalent CNT junctions	460
17.2.2 van der Waals or noncovalent CNT junctions	461
17.3 Theoretical Studies	461
17.3.1 The simplest nanotube junction: bent nanotube by 5-7 pairs	461
17.3.2 Schwarzites	461
17.3.3 CNT junctions and networks: structure, properties and transport	462
17.4 Synthesis of CNT Networks	468
17.4.1 Template approaches	469
17.4.2 Synthesis of van der Waals or noncovalent CNT networks	470
17.4.3 Electron beam irradiation at high temperatures	472
17.4.4 Secondary growth approaches	473
17.4.5 The role of sulfur during CNT synthesis	475
17.4.6 Carbon nanotube sponges	477
17.5 Applications	479
17.5.1 Electronic applications: memory devices, transparent electrodes, transistors, sensors	479
17.5.2 Bioapplications: biosensors, artificial muscles, cell growth scaffolds	483
17.5.3 Mechanical applications	484
17.5.4 Oil absorption applications	486
17.6 Perspectives	487
Acknowledgments	487
References	487

17.1 INTRODUCTION

Following the discovery of fullerenes and the structural identification of carbon nanotubes (CNTs) [1–3], investigation of these materials and novel carbon structures intensified considerably. A single-walled carbon nanotube (SWNT) can be formed by rolling up a graphene sheet [4–6]. These tubes exhibit high stiffness (large Young's modulus of about 1 TPa) [7] as well as different chiralities (orientations of their hexagonal lattice along the tube axis). The electronic properties of SWNTs vary with the diameter and chirality; armchair SWNTs are metallic, whereas zigzag and chiral SWNTs exhibit either metallic or semiconductive behavior [8]. CNTs can be used as components in electronic devices, as adsorbents and as reinforcement materials [9]. Nowadays, one of the most important challenges in this field is the assembly of CNTs in order to construct nanostructured micro- or macroscopic systems, i.e. bottom-up approach. In this way, parallel and serial circuits made of SWNTs can be important components of electronic devices, or large three-dimensional (3D) networks made of CNTs can be used as adsorbent materials or scaffolds for proliferation of cells, seeking applications in bone tissue regeneration, for instance.

A theoretical study of junctions of SWNTs was first reported by Lambin et al. [10]. They demonstrated that the introduction of pentagon—heptagon pair defects into the hexagonal lattice of a single CNT can change the chirality of the tube and modify its electronic structure. The authors used a tight-binding approach for studying junctions made of semiconductor—metallic CNTs, or vice versa. This type of junction can be seen in Fig. 17.1, where a pentagonal carbon defect is located at the elbow (positive curvature) and the heptagonal ring is located at the interior (negative curvature) of the junction. From local density of states (LDOS) calculations, the pentagonal (heptagonal) ring exhibits additional states in the valence (conduction) band that affect the electronic transport and chemical activity (see below). Immediately after Lambin [10], Chico et al. [11] reported similar theoretical results, and Terrones et al. published the first transmission electron microscopy (TEM) image of a multiwalled carbon nanotube (MWNT) exhibiting a 30° bend and discussed the possibility of having different electronic properties within SWNT by only adding one pentagon—heptagon pair [12] (Fig. 17.1). Since these publications, several models and theoretical calculations of multiple junctions involving CNTs have appeared in the literature. In 1999, Yao et al. [13] reported the first measurements of current—voltage curves in a SWNT intramolecular junction, similar to that reported by Lambin et al. [10]. They found that these metal—semiconductor junctions behave like rectifying diodes with nonlinear transport characteristics. More recently, a method for determining the chirality of CNTs, using aberration-corrected microscopy images, has been reported and bent double-walled CNTs have been indexed [14], thus demonstrating that these tubes indeed could exhibit different electronic properties (Fig. 17.1(c)). Different synthesis methods, such as chemical vapor deposition (CVD), laser irradiation, etc. have emerged in the search for CNT junctions. Several experimental reports have found that the use of a sulfur precursor in the CVD method is a key piece in the synthesis of CNT branching

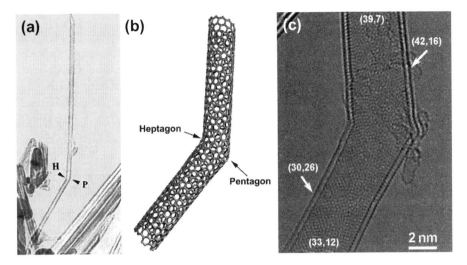

FIGURE 17.1

(a) TEM image of a 30° bent MWNT [12] and (b) a SWNT molecular model of a 30° bent nanotube with a heptagon–pentagon pair. Note that the tube chirality changes by 30° before and after the bend. (c) HRTEM image of a double-walled bent carbon nanotube showing tube chiralities corresponding to a (39,7) tubule inside a (42,16) tube joined to a (33,12) SWNT inside a (30,26) tubule. According to electronic calculations, the (33,12) tube is metallic, whereas the rest are semiconducting [14]. (For color version of this figure, the reader is referred to the online version of this book.)

Source: Reproduced by permission of [12] © 1996 The Royal Society, and [14] © 2012 The Royal Society of Chemistry. Image taken from Refs [12] and [14].

[15–24]. In this chapter, the latest advances related to 3D networks based on CNTs are presented. In addition, different synthesis and characterization techniques, theoretical models and calculations, applications and perspectives of novel CNT networks are discussed.

17.2 NANOTUBE NETWORK TYPES

SWNTs and MWNTs are constructed from sp^2-hybridized C atom sheets; each carbon atom is connected to three other neighbors. However, for MWNTs, the tubules are nested and only interact via van der Waals forces. Since SWNTs and MWNTs possess an almost inert surface, the covalent bonding between CNT surfaces needs to be established via surface chemical functionalization, doping, surface acid treatment, or aggressive methods that could combine electron beam irradiation and high temperatures. In this way, CNTs (SWNTs and MWNTs) could be interconnected covalently. In this section, we will discuss the different ways of joining CNTs.

17.2.1 Covalent CNT junctions

The covalent junctions are formed by establishing covalent bonds between CNTs. The bonded elements could be purely carbon or foreign atoms capable of forming interconnections. For example, covalent junctions could be created by the assistance of metallic particles that, under high temperature and electron irradiation, could result in metal–carbon covalent junctions [25]. Alternatively, this type of metal–carbon link could be obtained via high-temperature CVD or other techniques

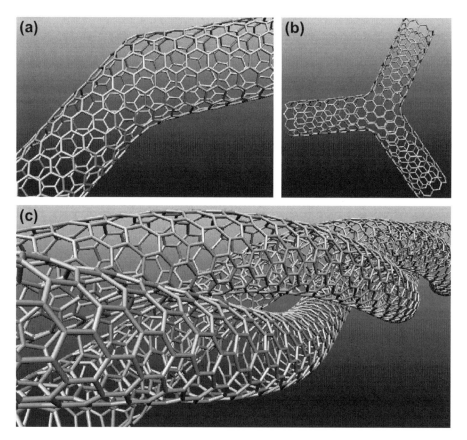

FIGURE 17.2

Different models of carbon nanotube junctions: (a) zigzag–armchair CNTs junction; this structure exhibits a pentagon at the elbow part and a heptagon at the inner part (see also Fig. 17.1). (b) Three-way CNT junction; at the intersection the junction exhibits negative curvature with additional heptagonal rings. These types of structures could be the building blocks of supergraphene CNT networks. (c) Entangled CNTs that are not attached covalently but could percolate. These noncovalent junctions are also known as van der Waals junctions. (For color version of this figure, the reader is referred to the online version of this book.)

(see Section 17.4) [24,26–28]. When the nanotube junctions are made of carbon atoms, the entire structure would keep the sp^2 hybridization. It is noteworthy that the formation of covalent junctions would also involve the presence of heptagonal carbon rings, thus maintaining the sp^2 hybridization (see Figs 17.1 and 17.2).

17.2.2 van der Waals or noncovalent CNT junctions

Two or more CNTs can interact by putting one nanotube close to another nanotube (see Fig. 17.2(c)). The interactions in these types of junctions are of the van der Waals type (or noncovalent). Therefore, percolating networks could be formed with CNTs overlapping each other in order to create a conductive van der Waals CNT junction system. The synthesis of overlapped CNTs could be assisted by chemical functionalization or acid treatment of the CNT surfaces. Although different chemical species can be anchored on the nanotubes by chemical functionalization, acid treatments could remove surface atoms thus producing defects such as vacancies that are usually passivated with OH groups. Examples of van der Waals junctions will be discussed in the following sections [29–33].

17.3 THEORETICAL STUDIES

17.3.1 The simplest nanotube junction: bent nanotube by 5-7 pairs

The simplest SWNT junction is the one consisting of a bent nanotube by the introduction of 5-7 pairs (see Figs 17.1 and 17.2(a)). This junction covalently interconnects nanotubes with different chiralities, thus making metallic–semiconductor devices possible [10–12,14,34]. In this context, the transport in bent SWNTs or one-dimensional (1D) nanotube junctions was studied by Chico et al. [35]. The authors have found that when the defects forming the junction (5-7 pairs) of two metallic SWNTs are arranged asymmetrically, there is conductance; if the defects preserve the rotational symmetry of the two tubes, conductance gaps are generated [35]. Tamura et al. [36] studied several cases of 1D SWNT junctions with 5-7 pairs and found that the conductance depends only on the ratio R_2/R_1, where R_1 and R_2 are the circumferences of the thinner and thicker nanotubes, respectively; thus when this ratio is large ($R_2 \gg R_1$) the conductance decays as the inverse third power of the length of the junction. Confined electronic states have been predicted for 1D junctions with 5-7 pairs in SWNTs and these defects could behave as quantum dots [37]. Menon and Srivastava [38] considered the possibility of producing "T"-junctions (quasi-two-dimensional (2D) junctions) by the introduction of heptagons, producing metal–semiconductor-metal contacts.

17.3.2 Schwarzites

It is important to note that 2D and 3D networks of SWNTs are closely related to 2D and 3D schwarzites. In particular, schwarzites consist of triply periodic minimal surfaces (TPMS) decorated with carbon atoms in which the negative Gaussian curvature is achieved by the introduction of carbon rings containing more than six atoms [39–41]. Since schwarzites can be thought of as compact sp^2-hybridized carbon surfaces [42], they follow the expression: $2N_4 + N_5 - N_7 - 2N_8 = 12(1 - g)$ where N_4, N_5, N_7 and N_8 denote carbon rings with 4, 5, 7 and 8 atoms per primitive unit cell, respectively, and g is the genus or number of handles of the primitive cell. The simplest schwarzite is the P-type (primitive type), which consists of a cubic cell with 12 octagonal rings of carbon or 24 heptagonal rings of carbon, thus exhibiting a genus three per primitive cell (see Figs 17.3(a),(b)). This cubic cell of the P-type schwarzite could be visualized as the linking piece to interconnect six SWNTs (see Figs 17.3(c),(d)). A 3D view of this structure is shown in Fig. 17.4. In a similar way, the diamond TPMS could be used to construct the superdiamond nanotube framework [43] (see Fig. 17.5). While the primitive cell of this diamond-like structure exhibits a genus three (24 heptagons, see Fig. 17.5(a)), its cubic cell has a genus 12 because the cubic cell is a face-centered cubic type structure (Fd3m space group), thus containing four primitive cells (see Fig. 17.5(b)). It is worth mentioning that CNTs do not contribute to the Gaussian curvature of the framework since they are made of only hexagonal rings of carbon that do not exhibit angular excess or deficit.

17.3.3 CNT junctions and networks: structure, properties and transport

The theoretical work of CNT junctions has mainly focused on four aspects: (1) structure of the junctions, (2) mechanical properties, (3) electronic properties and (4) quantum transport. In order to understand the properties of CNT junctions, different classical and quantum calculation models have been applied. For classical models, thousands of atoms could be taken into account. For example, force-field calculations and semi-empirical potentials have been successfully used to relax (structural energy minimization) carbon nanostructures or to study their mechanical properties [43]. When using quantum approaches, simple tight-binding models or density functional theory (DFT) calculations are limited by the number of atoms. For example, in order to study a simple periodic supercubic structure made of CNTs, hundreds of atoms are needed in the unit cell, therefore DFT or tight-binding calculations become a difficult computational task. Fig. 17.6 depicts molecular models of different 3D networks of CNTs. An innovative theoretical study on the formation of 2D and 3D ordered covalent networks was performed by Romo-Herrera et al. [43]. The authors introduced a hierarchical concept for different architectures, which leads to 2D nanotube networks such as the supersquare and supergraphene systems and to 3D networks including the supercubic and superdiamond systems. They found that the 3D architectures support extremely high unidirectional stress (see Fig. 17.6). In this context, it is also noteworthy that Coluci et al. [44] studied supernanotubes; these superstructures are based on a graphene architecture in which the carbon bonds are replaced by SWNTs, and the carbon atoms by Y-like tubular junctions. The authors found that these supertubes could exhibit either metallic or semiconducting behavior. More recently, Novaes and coworkers [45] have

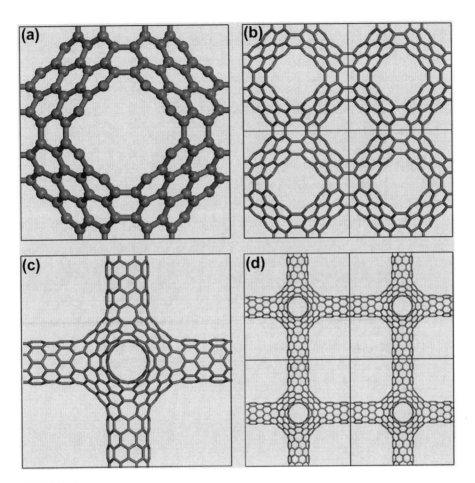

FIGURE 17.3

From schwarzites to 3D nanotube networks. (a) Cubic cell of the P-type schwarzite exhibiting 12 octagonal rings of carbon and genus three per primitive cell. The cell contains 192 atoms and belongs to the *Pm3m* space group. (b) Four cells of the structure shown in (a) exhibiting the octagonal rings of carbon to connect them. (c) Cubic unit cell obtained from (a) by joining the cell to six (8,0) zigzag SWNTs. This cell has 576 atoms and exhibits 24 heptagonal rings (genus three per cubic cell and *Pm3m* space group). (d) Four cells of (c) showing the 3D SWNT network. (For color version of this figure, the reader is referred to the online version of this book.)

used DFT to study junctions of SWNTs and graphene layers for different SWNT lengths. These authors found that in metallic SWNTs, the conductance is independent of the tube length but changes greatly at the graphene–SWNT junction; when using semiconductor SWNTs, the conductance mainly depends on the nanotube length and is independent of the intersections. Xu et al. [46] studied superstructures using CNTs

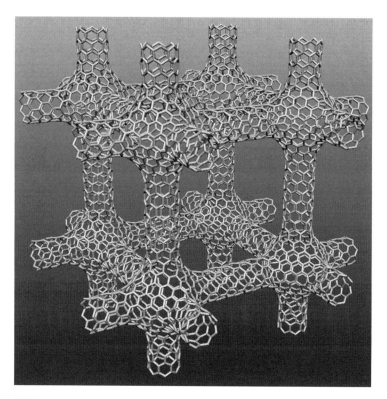

FIGURE 17.4

Molecular model of a 3D view of the schwarzites joined by (8,0) zigzag SWNTs. The unit cell can be seen in Fig. 17.3(c). (For color version of this figure, the reader is referred to the online version of this book.)

and graphene (pillared graphene). It was found that the elongation responses and stress—strain relationships are nearly linear and the calculated strength, fracture strain and Young's moduli become lower than those reported for pristine graphene or SWNTs. The alterations in thermal and mechanical performances are ascribed to the bonds present in the junctions. An algorithm based on diophantine equations for the construction of three CNT junctions was presented by Laszlo [47]. He constructed a CNT junction among (10,0), (10,1) and (10,2) SWNTs by relaxing the structural coordinates, using the conjugate gradient method in conjunction with the Brenner potential [48]. Transport properties of Y-junctions (YJs) were also studied by Meunier et al. [49]. It was observed that a CNT-terminated YJ does not display an asymmetric conductance spectrum, but behaves as a junction between undoped semiconductors. The authors demonstrated that similar behaviors are expected for any three-terminal CNT device. In addition, YJs have been used for ion separation by Beckman et al. [50]. Here, a KCl electrolyte was separated into K^+ and Cl^- ions using a YJ with an (8,8) nanotube

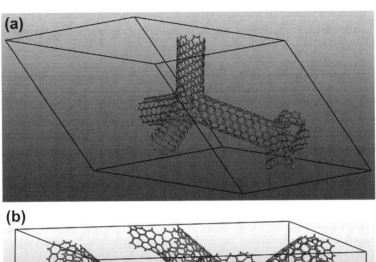
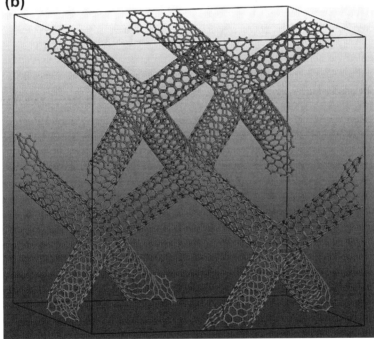

FIGURE 17.5

The 3D superdiamond network using SWNTs. (a) Primitive unit cell with 960 atoms and 24 heptagons (genus 3) exhibiting tetrahedral junctions used to interconnect (6,6) armchair SWNTs and (b) cubic cell with 3840 atoms (genus 12). (For color version of this figure, the reader is referred to the online version of this book.)

466 CHAPTER 17 Three-dimensional Nanotube Networks

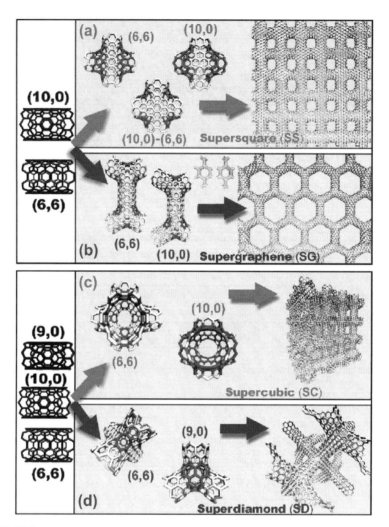

FIGURE 17.6

Atomic models of 2D and 3D SWNT networks, where all nonhexagonal rings are highlighted in red. (a) Supersquare (SS) 2D network composed of (6,6) or (10,0) SWNTs, with a cubic geometry [43]. (b) Supergraphene (SG) 2D network composed of (6,6) or (10,0) SWNTs, with a hexagonal geometry [43]; (c) supercubic (SC) 3D network and (d) superdiamond (SD) 3D network using (6,6), (10,0) or (9,0) SWNT [43]. (For interpretation of the references to color in this figure legend, the reader is referred to the online version of this book.)

Reproduced by permission of [43] © 2007 American Chemical Society

as the main branch and a negatively charged (5,5) and a positively charged (6,6) nanotube as the two smaller branches interconnected to the main branch.

Tight-binding molecular dynamics (TBMD) simulations have been performed in order to understand the covalent interconnection of crossing CNTs [27]. In

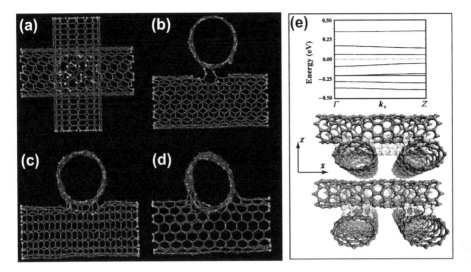

FIGURE 17.7

Different models and calculations of three-dimensional networks based on SWNTs. (a—d) Tight-binding molecular dynamics simulations of the coalescence of two SWNTs [27]. (e) Crossed junctions are modeled by two-dimensional grids of zigzag SWNTs [51]. (For color version of this figure, the reader is referred to the online version of this book.)

Reproduced by permission of [27] © 2002 American Physical Society, and [51] © 2004 American Physical Society.

this context, two overlapped (8,8) SWNTs with a 90° angle between them, and possessing a few vacancies (emulating electron irradiation effects at 1000 °C), were taken into account in the TBMD calculations (see Fig. 17.7(a)). The time step used in the simulations was 0.7 fs, and the total simulation time was 220 ps. Fig. 17.7(b) shows the simulation results after 10 ps, where the two SWNTs start to link through carbon chains. After 100 ps, there is a clear covalent interconnection established between the nanotubes (see Fig. 17.7(c)) and the interconnection still contains dangling bonds and sp^3-bonded C atoms. Mainly sp^2-bonded C atoms could be observed after 200 ps (Fig. 17.7(d)). These calculations were used to explain experimental results of "nanowelding" of SWNTs inside a high-resolution transmission electron microscope (HRTEM), which will be further discussed in the synthesis of CNT networks section. Dag et al. [51] conducted a first-principles study of parallel and crossed junctions of SWNTs (see Fig. 17.7(e)). The van der Waals junctions were modeled by 2D grids of zigzag SWNTs. The atomic and electronic structure, stability, and energetics of the junctions were studied for different magnitudes of contact forces by bringing the tubes toward each other, hence inducing radial deformations. Under relatively weak contact forces, the CNTs are linked with intertube bonds allowing a significant electron conductance passing through the junction [51].

Figure 17.8 depicts the LDOS of zigzag—armchair SWNTs junctions. The LDOS were calculated in four different regions of the structure: pentagon, heptagon, and

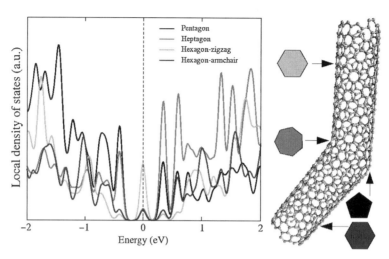

FIGURE 17.8

Local density of states (LDOS) of four different regions of the SWNT junction. Results indicate that the pentagon introduces mostly states in the valence band, whereas the heptagon in the conduction band. In the right-hand side of the figure, the four different regions are indicated. In all cases, the LDOS were averaged over the number of atoms of the respective polygon (pentagon, hexagon and heptagon). Results were obtained using a tight-binding model with an electronic hopping of $t_0 = 2.66$ eV for first nearest neighbors. (For color version of this figure, the reader is referred to the online version of this book.)

two hexagons localized in the zigzag and armchair nanotubes. The LDOS were averaged over the number of atoms of the corresponding polygon. All polygons exhibit electronic states at the Fermi level (see the vertical dashed line in Fig. 17.8) with a large peak for the hexagon localized in the zigzag nanotube. The pentagon (heptagon) exhibits additional electronic states in the valence (conduction) band. In Fig. 17.9, the total electronic density of states for a semiconducting zigzag (8,0) SWNT, a schwarzite structure, and the 3D network formed from these nano-architectures are shown (see the insets in Fig. 17.9). It is noteworthy that all the systems exhibit a semiconducting behavior; results for the schwarzite structures are qualitatively in agreement with previous first-principles calculations [52]. At this point, additional investigations related to CNT networks' mechanical, transport, adsorption, and optical properties are needed to fully understand their behavior and elucidate potential applications.

17.4 SYNTHESIS OF CNT NETWORKS

There are a number of experimental methods used for producing CNT junctions and networks. Several of them will be discussed in this section, including CVD, electron

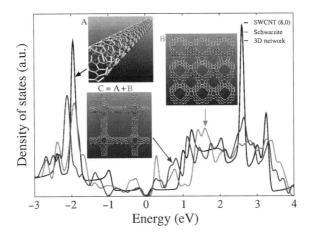

FIGURE 17.9

Electronic density of states (DOS) of an (8,0) SWNT, a schwarzite structure and a schwarzite–nanotube 3D network. For details about the structures, see Fig. 17.3. Note that all the structures exhibit a semiconducting behavior. Results were obtained using a tight-binding model with an electronic hopping t_{ij} between i and j atoms given by $t_{ij} = t_0(R_0/R_{ij})^2$, where $t_0 = 2.66$ eV, $R_0 = 1.42$ Å, and R_{ij} is the distance between i and j atoms, which was taken up to a cutoff distance of 5.6 Å. (For color version of this figure, the reader is referred to the online version of this book.)

beam irradiation at high temperatures, secondary growth approaches and sulfur- or boron-assisted methods.

17.4.1 Template approaches

The first report on the controlled formation of pores within anodized aluminum oxide (AAO) templates was published in 1989 [53]. It was not until 2000 that AAO templates were used for the first time to synthesize branched MWNTs [54]. MWNTs were grown inside the pores of AAO templates at 650 °C using acetylene as a carbon source and Co as a catalyst. TEM images of these branched MWNTs are depicted in Figs 17.10(a),(b). A HRTEM image is shown in Fig. 17.10(c), exhibiting the graphitic planes of the branched MWNTs. In 2005, Meng et al. [26] reported an ingenious generic approach for the synthesis of hierarchically branched pores in AAO, which served as a template for fabricating interconnected MWNTs and nanowires (see Figs 17.10(d)–(g)). It was previously established that by reducing the anodizing voltage by a factor of $1/\sqrt{2}$, one pore in a given AAO template could be divided into two [54]. This method was generalized by Meng et al. [26] since they demonstrated that reducing the anodizing voltage by a factor of $1/\sqrt{n}$ indeed creates n branches within the pores of AAO. The authors reported subsequent reductions of the anodizing voltage by a factor of $1/\sqrt{m}$ to produce a second generation of

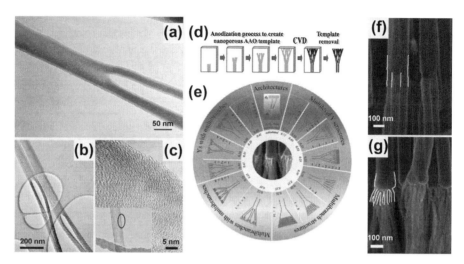

FIGURE 17.10

(a,b) Low-magnification TEM images of MWNT Y-junctions produced by the template method [54]; (c) HRTEM image from the MWNT walls shown in the inset, exhibiting the graphitic 002 planes of the nanostructure [54]; (d) sketch of the experimental approach used to synthesize MWNT junctions using nanoporous ionized aluminum oxide (AAO) templates [26]; (e) diagram exhibiting the variety of architectures that can be synthesized using this templated approach; the scheme includes theoretical and experimental values of the pore diameter ratios, showing an excellent agreement [26]; and (f,g) SEM images of the so-called multibranched structures [26]. (For color version of this figure, the reader is referred to the online version of this book.)

Reproduced by permission of [54] © 1999 Nature Publishing Group, and [26] © 2005 PNAS.

m pores from each of the n pores. A scheme of the generalization of the AAO template method can be seen in Fig. 17.10(e), where experimental and theoretical values for pore diameter showed an excellent agreement. MWNTs were grown inside the pores of this AAO template also by the pyrolysis of acetylene at 650 °C. The template was successfully removed after the MWNT growth by the action of hydrofluoric acid (HF). Scanning electron microscopy (SEM) images of the branched MWNTs are depicted in Figs 17.10(f),(g), where one MWNT was divided into three or n smaller diameter MWNTs.

17.4.2 Synthesis of van der Waals or noncovalent CNT networks

The successful evolution of CNTs toward applications demands the possibility of arranging them with well-defined patterns, either *in situ* during growth or by post-growth processing. Arrays of aligned CNTs have been fabricated; however, very useful electronic applications depend on the interactions of crossing CNTs.

The concept of noncovalent networks originated from the seminal work of Lieber and coworkers [55], who proposed the assembly of 1D structures into

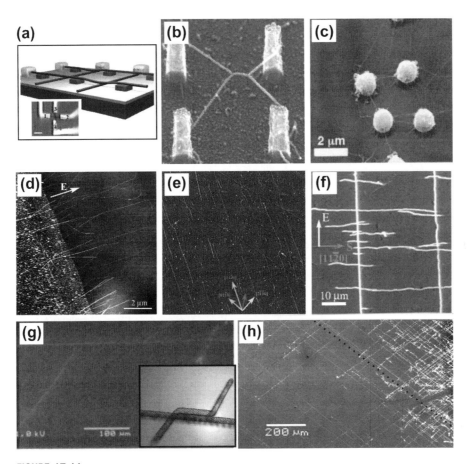

FIGURE 17.11

(a) Schematic representation of a binary digit stored in the crossing of two CNTs and its experimental prototype [55]. (b) Noncovalent overlapped CNT junctions grown on patterned substrates where nanotubes grew horizontally [56]. (c) SWNT network generated by SiO_x microspheres randomly distributed on a substrate [57]. (d) Directional growth of CNT biased by an electric field [58]. (e) Directional growth of CNTs following an atomic step on a crystalline surface [59]. (f) The combination of the approaches shown in (d) and (e) results in a squared CNT network in 2D [29]. (g,h) SEM images of CNTs directionally grown on a substrate and reoriented by rotating the substrate in situ, resulting in nanotube networks. Results in (g) and (h) were reported by Hofmann et al. [60]. (For color version of this figure, the reader is referred to the online version of this book.)

Reproduced by permission of [55] © 2000 American Association for the Advancement of Science, [56] © 2002 American Institute of Physics, [57] © 2003 American Institute of Physics, [58] © 2002 American Chemical Society, [59] © 2004 John Wiley and Sons, [29] © 2006 American Chemical Society, and [60] © 2008 American Chemical Society.

functional networks in order to create random access memory devices. Fig. 17.11(a) depicts the concept and the prototype of the information bit based on two crossing CNTs. Several reports on van der Waals 2D CNT networks on a number of substrates have been published thereafter. Since CNTs usually grow perpendicular to the substrate when using CVD approaches, the use of substrates with SiO_x pillars containing catalyst nanoparticles was one of the first approaches to promote the directed growth of 2D nanotube networks [56,57]. Fig. 17.11(b) shows the growth of nanotubes by squared arrays of silicon oxide pillars, and Fig. 17.11(c) depicts the same concept using randomly arranged protuberances instead of pillars.

A breakthrough related to the directed growth of 2D CNT networks on substrates occurred when CNTs were able to grow horizontally from flat substrates (see Figs 17.11(d)–(h)). Initially, the use of an electric field to promote the directional growth of nanotubes was reported [58]. A few years later, atomic steps in faceted crystalline surfaces were used to horizontally grow SWNTs, following the edges of monoatomic steps on the substrate (see Fig. 17.11(e)) [59]. The combination of both approaches led to the effective formation of orthogonal CNT networks on crystalline surfaces [29]. More recently, it was found that by changing the orientation of the substrate during CNT growth, it is possible to create a CNT network [60]. Figs 17.11(f)–(h) depict SEM micrographs of van der Waals 2D CNT networks created by both approaches.

Dispersed CNTs dried on flat substrates result in the formation of a random 2D network [61], but the van der Waals junctions formed are not periodic, and the interactions among specific CNTs are not measurable. In these systems, the overall film behavior is caused by the collective effect of individual CNTs, just like in transparent electrodes (TEs) and sensors, where electrons percolate randomly between overlapped CNTs within the film. Devices constructed with random CNT networks will be covered in Section 17.5.

It is evident that most of the progress on noncovalent junctions has been achieved on flat substrates. However, the 3D assembly of CNTs has been mainly obtained by templated or regrowth processes [62,63]. In this context, CNT aerogels were first reported by Bryning and collaborators [63], and more recently, Zou and colleagues obtained a hierarchical honeycomb 3D structure based on van der Waals CNT interactions [63]. The key step in this process was the use of block polymers in order to create a gel, which after drying, retains a hierarchical 3D structure (see Figs 17.12(a)–(c)). This material is an aerogel since its density is only 4 mg/cm^2. Very high surface area and good electrical conductivity make this material a very attractive candidate in the fabrication of sensors or pressure transducers.

17.4.3 Electron beam irradiation at high temperatures

Ion or electron beam irradiation at elevated temperatures was proposed to create junctions of SWNTs and MWNTs. Ajayan, Banhart, Terrones and coworkers published a large number of reports along this research line [25,27]. As discussed in

17.4 Synthesis of CNT Networks

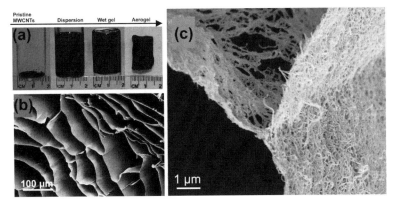

FIGURE 17.12

CNT aerogels. (a) Protocol for the formation of hierarchical CNT network in 3D, where MWNTs are dispersed and gelated and finally form the aerogel, after removal of the solvent. (b) SEM images of the honeycomb structure retained in the aerogel after gelation and drying; and (c) a detail of the walls formed by entangled CNTs. (For color version of this figure, the reader is referred to the online version of this book.)

Source: From [63] and Reproduced by permission of [63] © 2010 American Chemical Society.

the theoretical studies a report related to electron beam welding of SWNTs was published in 2002 [27]. In particular, SWNTs were dispersed in EtOH and placed onto TEM grids and high electron beam irradiation at 800 °C was used to create "YJs" or "T"- or "X"-junctions inside a HRTEM. It was demonstrated that the electron beam manipulation could transform "X"-junctions into "YJ" or "T"-junctions (see Figs 17.13(a)–(d)). The authors demonstrated that these CNT junctions were formed via the formation–reconstruction of vacancies and interstitials, induced by the electron beam (see also Figs 17.7(a)–(d) in the theoretical studies section).

This well-established SWNT welding technique inspired the further creation of MWNT junctions using the catalyst particle inside the nanotube as a soldering material [25]. In particular, thin MWNTs produced in the presence of Co were dispersed in EtOH and placed onto TEM Cu grids. The high-temperature irradiation was conducted inside a HRTEM chamber when two relatively thin MWNTs with few walls overlapped and a Co particle or nanowire was located at the intersecting point, as shown in Fig. 17.13(e). A scanning tunneling microscopy holder was also used inside the HRTEM in order to perform biasing and Joule heating as an alternative method to create such junctions (see Figs 17.13(e)–(h)).

17.4.4 Secondary growth approaches

The secondary growth method could be considered as the most straightforward procedure to synthesize CNT networks. The method uses already grown CNTs as substrates and catalytic particles that cover the tubes. These catalysts might lead

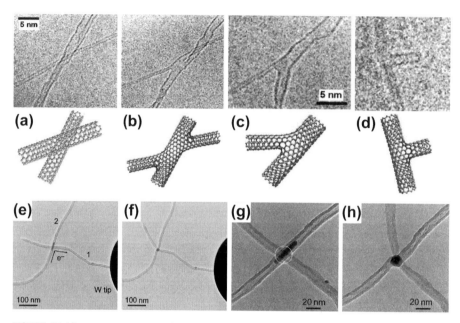

FIGURE 17.13

(a–d) HRTEM images and molecular models of "X"-, "Y"- and "T"-junctions of 2-nm SWNTs created by electron beam irradiation under the microscope. In all the models, the non-hexagonal carbon (heptagons) rings are highlighted in red [27]. (e,f) TEM images of an overlapped MWNT junction turned into a covalent junction, after Joule heating and electron irradiation inside the microscope [25]. The arrows shown in (e) exhibit the current direction. (g,h) Higher magnification TEM images corresponding to (e) and (f), respectively. The circle in (g) indicates the position of the electron beam [25]. (For interpretation of the references to color in this figure legend, the reader is referred to the online version of this book.)

Reproduced by permission of [27] © 2002 American Physical Society, and [25] © 2009 John Wiley and Sons.

to the formation of CNT networks after regrowing tubes on them [64,65]. However, the effectiveness of this approach is limited by the ability to adequately disperse individual CNTs (SWNTs and MWNTs). The degree of difficulty to disperse CNTs is proportional to their aspect ratio; SWNT bundles are usually very long and narrow and are hence extremely difficult to disperse. Since MWNTs grow in dense forests, it is necessary to separate them in order to provide space between them and allow the growth of a second generation of nanotube branches depicted in Fig. 17.14.

Another aspect that has to be considered for the secondary growth of CNTs is the size of catalytic nanoparticles. The ability of the primary CNT to support a catalytic particle with dimensions ranging between 5 and 50 nm requires the presence of large-diameter CNTs. For this reason, secondary CNT growth has been successfully reported on large-diameter MWNTs and carbon nanofibers rather than on individual SWNTs [66]. Three-dimensional CNT networks were synthesized by Lepró et al. [28]. The authors created branched CNTs and networks based on heterojunctions

FIGURE 17.14

Different carbon nanostructures grown on already synthesized MWNTs by the CVD method. (a) Aligned MWNTs (carpets) synthesized by the CVD method. (b) Dispersion in solvents by an ultrasonic bath. (c) Catalyst nanoparticles deposited by physical or chemical methods; (d,e) second-generation CNTs grown by CVD using methane and toluene as carbon source, respectively. Results in (d) were reported by Sun et al. [66] and those in (e) by Lepro et al. [28]. (For color version of this figure, the reader is referred to the online version of this book.)

Reproduced by permission of [66] © 2004 Elsevier, and [28] © 2007 American Chemical Society.

consisting of undoped and nitrogen-doped MWNTs and were able to obtain "YJ" and "T"-junction morphologies (see Fig. 17.14(e)).

17.4.5 The role of sulfur during CNT synthesis

Sulfur has been an important element for the synthesis of carbon fibers over several decades [15−24]. However, it has also been shown that sulfur was considered a strong agent for inhibiting certain catalytic processes, and therefore it has been avoided in some industrial processes related to carbon materials [15].

The formation of the so-called "sea urchin" nanoparticles was reported in 1983 [16] (see Fig. 17.15(g)). Several routes were used to achieve the synthesis of these interesting carbon structures, all of them involving sulfur in the synthesis process. In 1993, the interplay of sulfur adsorption and carbon deposition on Co catalysts was investigated [17]. In two separate reports in 1994, sulfur was found to increase the production of carbon fibers or filaments [18,19], although it was also reported that certain features of the fibers could be modified (length, for instance). In 2000, the formation of YJ was reported through the pyrolysis of thiophene and nickelocene at 1000 °C. The authors reported a stacked cone morphology within these nanofibers

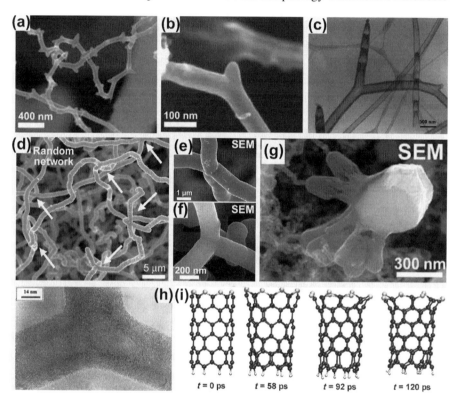

FIGURE 17.15

(a,b) SEM images of MWNT branching [23]. (c) TEM image of a YJ induced by the presence of sulfur [21]. (d,f) Hollow carbon fiber junctions synthesized by the thermal decomposition of nickelocene in the presence of sulfur, hydrogen and argon [24]. (g) Higher sulfur/carbon ratio in the nickelocene decomposition experiment led to the formation of sea urchin-like structures entangled with carbon fibers [24]. (h) HRTEM image of a YJ [20]. (i) Theoretical calculations explaining the role of sulfur at the atomic level [23]. (For color version of this figure, the reader is referred to the online version of this book.)

Reproduced by permission of [23] © 2003 John Wiley and Sons, and [21] © 2005 Elsevier, [24] © 2009 John Wiley and Sons, and [20] © 2000 American Institute of Physics.

(see Fig. 17.15(h)) [20]. The same research group reported the synthesis of CNT junctions in the presence of sulfur in 2005 [21] (see Fig. 17.15(c)). Helicoidal CNTs have also been synthesized in the presence of sulfur in 2006 [22]. In these experiments, thiophene vapor was used and, alternatively, a sulfonated sol—gel metal catalyst. The authors also reported the presence of YJ and they could neither find an explanation for this behavior nor for the appearance of helically coiled nanotubes.

In 2008, a report on the sulfur effect in CNTs, including both experimental and theoretical simulation data, was published [23]. Several YJs were produced by the authors following similar experimental conditions established in previous publications [20], involving the pyrolysis of nickelocene and thiophene in an inert atmosphere (see Figs 17.15(a),(b)). MWNT branching was extensively studied in this report, as well as the formation of elbows and protuberances within MWNTs. Energy-dispersive X-ray spectroscopy indicated the presence of sulfur within the positive curvature of the tubular structures (protuberances and elbows). Furthermore, molecular dynamics simulations were carried out on an open zigzag SWNT saturated with S atoms at the edge [23], and it was noted that sulfur induced a widening in the open end of the SWNT indicating the introduction of negative curvature within the graphitic lattice (see Fig. 17.15(i)). The stability of sulfur atoms was also studied in different positions within a bent SWNT. These simulations confirmed the fact that S atoms are more energetically stable when located at negatively curved regions; positive curvature regions in the graphitic lattice were also found to be energetically favorable to allocate S atoms.

Subsequently, the same research group reported the formation of 3D CNT networks, bound together via covalent YJs (with diameters of a few micrometers), that were grown in the presence of S (see Figs 17.15(d)—(f)) [24]. Sea urchin-like carbon nanostructures were also reported (see Figs 17.15(g)). A sulfur—carbon ratio was established in order to obtain all the different kinds of carbon structures and networks by this CVD approach, thus confirming the key role that sulfur has on the CNT growth (see Fig. 17.16).

17.4.6 Carbon nanotube sponges

The bulk synthesis of a 3D CNT sponge was reported for the first time in 2010 by Cao and collaborators [30]. The sponges were synthesized by a single-step CVD process, involving the decomposition of a mixture of ferrocene and 1,2-dichlorobenzene at 860 °C under an H_2/Ar atmosphere (see Figs 17.17(a),(b)). The precursor solution was fed into the reaction zone by a continuous automatic injection at a rate of 0.13 ml/min. The CNT sponge material exhibited remarkable structural stability. Mechanical properties of this material were studied through compressive strain cyclic measurements and it was found that this material can experience large strain deformations and recover to approximately its original shape. The CNT sponge was reported to have a high absorption capability (from 80 to 180 times its own weight). The most striking reported properties of these CNT sponges were the extreme hydrophobicity and oleophilicity. A more detailed study in the

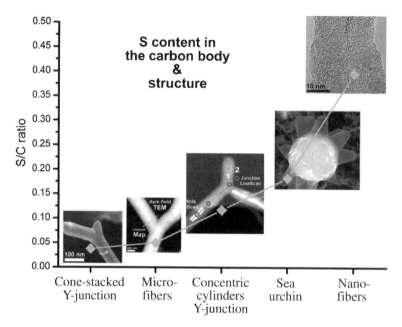

FIGURE 17.16

A variety of carbon nanostructures synthesized with different sulfur/carbon ratios [24]. This plot summarizes the role of sulfur in the synthesis of CNTs, which has been studied by many researchers for several decades [15–24].

Reproduced by permission of [24] © 2009 John Wiley and Sons.

oleophilicity of the CNT sponges was reported later [67] (see section oil absorption applications for more details).

Not long after the appearance of these reports, a report related to a new type of CNT sponge was released [68]. The material was composed of boron-doped MWNTs with elbow-like junctions (see Figs 17.17(c)–(g)). Theoretical simulations were performed in order to establish the most favorable position of B, N and S atom within a bent SWNT, as it was reported for the case of sulfur [23]. A boron atom located in a negative curvature region was found to be energetically more stable when compared to planar or positively curved regions. This theoretical study was in perfect agreement with experimental electron energy loss spectroscopy, indicating the presence of B atoms within the negatively curved regions of the elbow-like junctions. Mechanical tests on this material were also performed under compressive strain. Similar to the work of Cao et al., the B-doped MWNT sponges were extremely hydrophobic and oleophilic.

Further experiments in order to optimize the growth of the boron-doped CNT sponges are currently underway. Preliminary results indicate that by increasing the amount of boron in the precursor solution, several covalent multi-junctions of thicker MWNTs appear. The mechanical properties of this material are still under

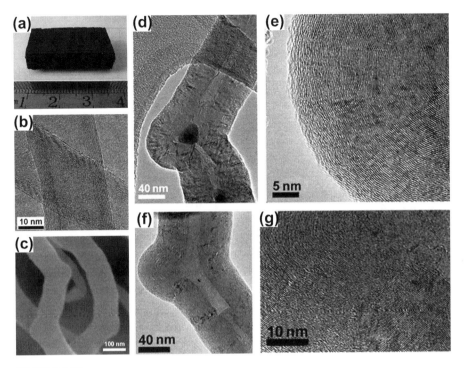

FIGURE 17.17

Light, porous, flexible carbon nanotube sponges. (a) A monolithic sponge with a size of 4 cm × 3 cm × 0.8 cm and a bulk density of 7.5 mg/cm^3 [30]. (b) TEM image of large-cavity, thin-walled CNTs [30]. (c) SEM image of boron-doped MWNT 3D material, and (d–g) TEM and HRTEM images of elbow-like junctions present in the boron-doped MWNT sponge.

Reproduced by permission of [30] © 2009 John Wiley and Sons.

intense investigation and the bulk production of this material needs to be developed in the near future.

The synthesis temperature and reaction times are also under investigation, since they could have important effects on the overall morphology of the boron-doped CNT sponge. Preliminary data showed that by doubling the synthesis time (1 h) and by increasing the temperature, a much more dense material can be obtained. The synthesis of CNT sponges by these two routes [30,68] represents the greatest achievement to date in the formation of bulk and stable 3D architectures.

17.5 APPLICATIONS

17.5.1 Electronic applications: memory devices, transparent electrodes, transistors, sensors

Because of their good electrical conductivity and flexibility, CNT networks could be used in the fabrication of different electronic devices. There is a major interest in creating CNT networks for electronic applications, including memory devices that are able to 10^{12} bits/cm^2. The state of the art for CNT manipulation is moving forward and alternative methods [15–24,30,68] for producing such high-density CNT networks, using bottom-up or top-down approaches, have appeared. A bottom-up approach to create these CNT networks was first proposed by Lieber et al. [69]. This methodology is based on the directional growth of CNTs. Here, two families of directionally growing nanotubes are allowed to extend orthogonally to each other creating a squared noncovalent (van der Waals) network. A number of challenges are yet to be addressed before the mentioned bit density is achieved. For example, defects and impurities contained in CNTs are responsible for creating non-perfectly straight CNTs that are needed to create such a high-density crossing array. Fig. 17.18(a) shows an array of bits produced by this approach [70]. It is noteworthy that the CNT crossing density is several orders of magnitude smaller than the hypothetical value (10^{12} bits/cm^2) and that the family of CNTs directed by the gas flow is not as straight as those following the crystalline facets of the substrate. On the other hand, one top-down approach consists of the orthogonal transfer of aligned CNT films [31]. This has resulted in well-ordered orthogonal CNT networks with higher densities than those shown in Fig. 17.18(b). For both approaches, the addition of ohmic contacts to each CNT still needs to be solved.

Orthogonal van der Waals networks could be potentially advantageous; however, random van der Waals networks are easier to fabricate and hence their applications might be broader. Among the most prominent are TEs, thin-film field-effect transistors and chemical sensors. Thin and random CNT networks could be used as TEs, which are the key in lighting and display industry. Up to now, indium tin oxide (ITO) is the TE of choice, but industry is eagerly looking for a substituting because indium is a very uncommon metal and its price is increasing rapidly. Random van der Waals CNT networks are in the group of advanced materials that are closer to substituting the ITO as TEs [71]. Nevertheless, the implementation of very high transparency and high electrical conductivities is still a great challenge. In CNT devices, such as TEs, there is a compromise between thickness and transparency, and it could be solved either by using highly conductive CNTs or by promoting enhanced electrical connections established between adjacent tubes, in such a way that electrons travel longer distances with less energy loss. In this context, various research groups are attempting novel functionalization routes and alternative chemical treatments in order to improve the electrical conductivity of CNTs networks while preserving a high transparency [72–74].

In order to fabricate thin-film transistors (TFTs) with random CNT networks, a variety of methods have been reported [75,76]. An additional advantage of the CNT network transistors is that they are usually transparent and it is possible to mount them on flexible and transparent substrates. This is very important for display technologies, in which the formation of smaller pixels is required. Fig. 17.18(c) shows a prototype CNT TFT array fabricated on Poly(ethylene-terephthalate) PET [75]. Ink-jet printing has allowed high-throughput fabrication of these SWNT TFTs [76].

17.5 Applications

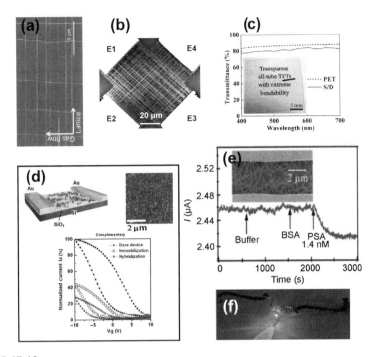

FIGURE 17.18

High-density memory devices based on arrays of crossed CNT consist of orthogonal CNT networks that can be fabricated by (a) bottom-up [70] or (b) top-down methodologies [31]. (c) Thin-film transistors deposited on transparent and flexible substrates [75], (d) biochemical DNA sensors based on an FET device structure [77]. (e) CNT biochemical sensors for specific prostate specific antigen (PSA) developed using SWNT networks in a chemiresistor device [78]. and (f) LED energized using electrical conductors made of cotton yarns embedded with CNTs [32]. (For color version of this figure, the reader is referred to the online version of this book.)

Reproduced by permission of [70] © 2009 American Chemical Society, [31] © 2007 American Chemical Society, [75] © 2006 John Wiley and Sons, [77] © 2007 American Chemical Society, [78] © 2005 American Chemical Society, and [32] © 2008 American Chemical Society.

Using a similar device configuration to a TFT, CNT networks connected to source and drain electrodes could be used to detect minute amounts of chemicals either in gas or liquid forms (see Fig. 17.18(d),(e)) [77,78]. In principle, CNTs could be functionalized in such a way that certain molecules present in the medium modify the source to drain conductance by changing the mean free path of the electrons traveling along the network. As an example of the impressive sensing ability of CNTs, we should mention the work of Gui and collaborators who used CNT van der Waals networks to detect deoxyribonucleic acid (DNA). Fig. 17.18(d) shows a schematic layout of a TFT they used for the electrical detection of DNA. This detector uses a SWNT network as the Field Effect Transistor (FET) channel where DNA molecules attach and cause

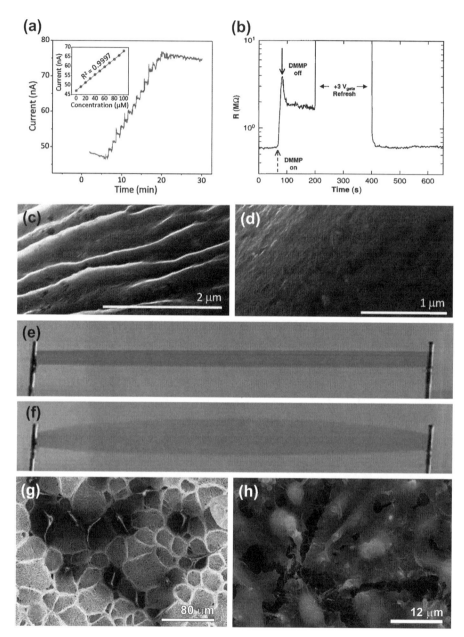

FIGURE 17.19

(a) Results from amperometric H_2O_2-sensing measurements carried out in phosphate buffered saline (PBS, 0.1M pH 7.4) PBS (0.1 M pH 7.4) using a SWNT-based biosensor [82]. It is possible to detect 10 nM increments of H_2O_2 concentration by observing the a conductance modification [77]. A variety of devices such as this one have been fabricated for sensing other chemicals or biomolecules (see Fig. 17.18(e)) [78].

The volumetric component of the 3D networks, along with their increased surface area and good electrical conductivity, make them suitable for applications where the storage of gaseous species or ions is important. For example, it has been found that CNT networks in conjunction with metal oxides constitute efficient electrodes for electrochemical batteries and supercapacitors, in which the electrical conductivity and the large surface area of the CNT network are key parameters in the battery performance [79,80]. Another interesting aspect of the 3D nanotube networks with high electrical conductivity is their use as electrical conductors; CNT aerogels with densities of a few milligrams per cubic centimeter could provide excellent substitutes for metal wires [81]. Shim and collaborators used percolating CNT networks embedded in cotton yarns as fairly good conductors ($\rho \sim 20 \ \Omega$/cm) [32]. In addition, the authors proposed to exploit the good sensitivity of CNTs to fabricate wearable biosensors (see below). Fig. 17.18(f) shows the use of CNT-embedded cotton yarns to energize a light-emitting diode (LED) [32].

17.5.2 Bioapplications: biosensors, artificial muscles, cell growth scaffolds

CNT networks have opened a new area in biosensing. It is noteworthy that all these applications are only possible due to the intrinsic low resistivity of CNTs. Electrochemical biosensors made from nanomaterials have attracted a great deal of attention because of the high sensitivity that they can offer. Researchers at the Purdue University [82] have developed nanocube-augmented CNT networks that can detect glucose and hydrogen peroxide. The material is fabricated by initially growing SWNTs in a porous anodic alumina template followed by electrodepositing metallic Au/Pd nanocubes onto its surface. The biosensing characteristics were found to be enhanced with SWNT pretreatments in H_2SO_4 and NaOH. Fig. 17.19(a) shows the current (nA) vs time (min) response of Au/Pd nanocube when exposed to an increasing concentration of hydrogen peroxide.

corresponding increase in current (inset scatter plot). (b) Plot of resistance vs time for a thin-film transistor made of SWNT networks when dimethyl methylphosphonate (DMMP) is introduced [33]. (c,d) A suggested method for using SWNT-adsorbed cotton yarn as a biosensor, showing the surface of the modified yarn before (c) and after (d) an antibody/antigen reaction. The extensive change of the surface structure when exposed to a target protein could be used in biosensing [32]. (e) Photograph of a 55 mm × 2 mm carbon nanotube sheet before and (f) after actuation when applying a 5 kV voltage [83]. (g,h) SEM image of a honeycomb-shaped 3D MWNT-based network used as a scaffold to grow L929 mouse fibroblasts (g) after a day of growth and (h) after 7 days of growth [86]. (For color version of this figure, the reader is referred to the online version of this book.)

Reproduced by permission of [82] © 2009 American Chemical Society, [33] © 2003 American Institute of Physics, [32] © 2008 American Chemical Society, [83] © 2009 American Association for the Advancement of Science, and [86] © 2004 American Chemical Society.

Another interesting biosensing application of CNT networks is in the detection of nerve agents [33]. It has been found that the resistance of semiconducting SWNT devices changes when exposed to certain types of gaseous analytes. When CNTs are exposed to dimethyl methylphosphonate (DMMP), which is a nerve agent with electron donor (Lewis base) properties, a charge transfer is observed. It is possible to detect up to part per billion levels of DMMP using this technique. Fig. 17.19(b) shows how the resistivity of the SWNT device increases while exposing SWNTs to DMMP. Upon removal of DMMP, the resistivity does not fall back to its initial value, but after applying a positive bias of 3 V, it is possible to recover the pre-exposure resistance.

As mentioned in Section 17.5.1, Shim and coworkers have developed CNT composite yarns with high electrical conductivities [32]. Multifunctional fabrics were produced using SWNT-coated cotton yarns. These yarns were fabricated by a simple process involving dispersing the nanotubes in diluted Nafion—ethanol or poly(sodium 4-styrene sulfonate)—water solutions followed by dipping cotton threads in them and eventually drying them. Very low resistivities (as low as 20 Ω/cm) have been observed in these cotton yarns after repeating this process a few times. This is demonstrated by using the threads as connectors between an LED device and a battery (see Fig. 17.18(f)). These SWNT—cotton yarns were tested as biosensors for albumin, which is a protein present in the blood. SEM images of such SWNT—cotton yarns are depicted in Figs 17.19(c),(d).

Baughman and coworkers [83] have fabricated CNT aerogel muscles, which are essentially long aerogel sheets drawn from MWNT forests, and have a density of ~1.5 mg/cm^3. The sheets are actuated by applying a voltage between a counterelectrode and the nanotube sheet electrode. These long aerogel sheets can achieve up to ~220% width actuation thus making them suitable materials for fabricating artificial muscles. Photographs of the sheet before and after applying a voltage for actuation are shown in Figs 17.19(e),(f). The actuation can be permanently frozen by inducing a van der Waals bonding between a substrate and the sheet. This is very useful for device applications such as TEs.

CNTs have also been used as scaffolds for biological applications. Two different research groups [84,85] reported the formation of CNT scaffolds from vertically aligned MWNTs. A honeycomb-shaped 3D structure was obtained by wetting aligned arrays of MWNTs with water or organic solvents and allowing them to dry. The formation of these so-called carbon nanofoams was driven by the action of capillarity forces. Following a similar approach, Correa-Duarte et al. [86] functionalized MWNTs arrays in an acid solution containing 1:3 nitric/sulfuric acid for 12 h and obtained similar nanofoam arrays. They reported the use of these 3D structures as scaffolds for cell seeding and tissue growth. Figs 17.19(g),(h) depict SEM images of L929 mouse fibroblasts grown on a MWNT scaffold.

17.5.3 Mechanical applications

CNTs are mechanically robust and good electrical conductors. In functional composites, for instance, it is possible to produce conducting polymer nanocomposites by dispersing small amounts of CNTs in insulating polymers. This new type of electrically conductive CNT/polymer nanocomposite can be applied as piezoresistive or

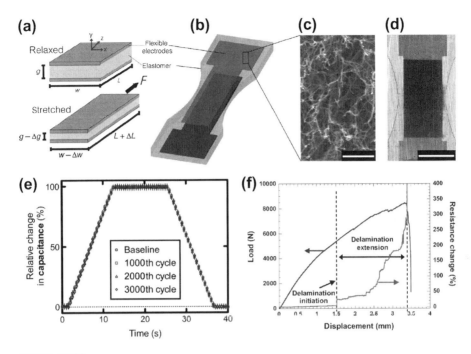

FIGURE 17.20

(a–c) Poisson capacitor containing a MWNT-overlapped networks as electrodes that can be used as a strain gauge [87]. In (a), the distance between electrodes diminishes as the capacitor is strained, which causes an increase in capacitance [87]. (b) Schematic of the device, exhibiting flexible MWNT-based electrodes and the elastomer portion [87]. (c) SEM image of the MWNT network used as an electrode (scale bar 500 nm) [87]. (d) Photograph of mounted MWNT electrodes, exhibiting a darker region where the electrodes overlap (scale bar 0.75 cm) [87]. (e) Relative change in capacitance plotted for 1000th, 2000th and 3000th cycles, exhibiting the high stability of the capacitor over many cycles [87]. (f) Load and resistance change for CNTs that form a conductive network throughout a polymer matrix [88]. It can be seen that the nanotube network is highly sensitive to damage progression. (For color version of this figure, the reader is referred to the online version of this book.)

Reproduced by permission of [87] © 2012 American Chemical Society, and [88] © 2006 John Wiley and Sons.

resistance-type strain sensors of high sensitivity. In this context, van der Waals junctions of MWNTs have been tested for application in state of the art strain sensors based on the Poisson effect. Cohen and coworkers [87] have reported an inexpensive and relatively fast fabrication method of a strain gauge using SWNTs and hydrophobic patterned Si, as shown in Fig. 17.20(a)–(d). A percolation MWNT-based network has also been used to prepare electrodes, in which mechanical deformations result in capacitance changes (piezocapacitance). The device showed an outstanding stability, even after 3000 cycles of 100% strain, only exhibiting a 3% variation (Fig. 17.20(e)). According to the authors, this sensor has the highest gauge factor reported for similar devices, 0.99.

Thostenson et al. [88] showed that conductive percolating CNT networks in traditional fiber composites can accurately detect the onset, nature, and progression

FIGURE 17.21

(a) Photograph of a droplet of water on the surface of boron-doped MWNT 3D material exhibiting an extraordinary hydrophobicity. (b) A 15 mm × 15 mm × 15 mm CNT sponge block [67] under an oil absorption cycle. The oil is easily burnt off and the sponge retains its original shape, and it is ready to start another oil absorption cycle. (For color version of this figure, the reader is referred to the online version of this book.)

Reproduced by permission of [67] © 2011 Elsevier.

of damage (see Fig. 17.20(f)). The authors claim that the sensitivity of the technique for damage sensing may have broad applications, including the assessment of self-healing strategies. An extensive review on electrical conductivity and piezoresistivity of CNT/polymer nanocomposites was published by Alamusi et al. [89].

17.5.4 Oil absorption applications

A few of the striking features of the CNT sponges discussed above are their hydrophobicity, oleophilicity and large surface area. These characteristics make these sponges an ideal material to absorb oil and remediate oil spill disasters.

In this context, the boron-doped MWNT sponge material [68] has shown outstanding performance in oil absorption applications. Fig. 17.21(a) depicts a droplet of water on the surface of a sponge, confirming its superhydrophobicity. B-doped MWNT sponges were tested for absorption capacity of several common solvents such as hexanes, ethanol, toluene and chloroform. For chloroform, an outstanding weight-to-weight absorption capacity (W) of 123 was found. W values were also obtained for oils, such as kerosene and engine oil. W values of around 80 and 60 were found for the cases of engine oil and kerosene, respectively.

Undoped CNT sponges [67] synthesized from CVD (pyrolysis of 1,2-dichlorobenzene) have been also tested for oil spill absorption [30]. The absorption capabilities of this undoped sponge were found to be greater than 100 g/g (grams of pollutant absorbed per gram of sponge) when tested with oils and organic solvents having viscosities in the range 3–200 cP. The cyclic performance of oil absorption and desorption was studied and the CNT sponges maintained capacities of about

40 g/g even after 10 absorption—desorption cycles. The sponge retains its original shape after the oil is burnt off, as shown in Fig. 17.21(b).

Ever since CNTs have been synthesized, there have been many successful attempts at using them as hosts for hydrogen storage [90]. It has been reported that doping of CNTs would improve their hydrogen storage capacity because of the improved binding energy of hydrogen molecules by means of charge-induced dipole interactions on doping. Porous carbon has also been successfully tested for its use in this regard [91]. CNT sponges, particularly B-doped sponges [68], encompass features of both these materials and we envisage they could perform better than pure SWNTs or MWNTs; however, further research is still needed along this research line.

17.6 PERSPECTIVES

Synthetic routes for producing CNT networks have been successful, and it has been possible to grow a wide variety of 2D and 3D CNT networks. Therefore, the next step should be focused on the morphology control and size of CNT branches. Self-assembly and CVD methods will play a key role in the multiscale production of CNT networks. Similar to welding and plumbing nanotube connections (elbow, T-connection, Y-connection, double Y-connection, etc.), CNTs could be assembled/grown so as to produce 3D CNT networks. Obtaining periodic CNT networks still remains a challenge. The use of porous templates for producing 3D CNT networks also needs to be exploited further. Studies on adsorbent capacity, mechanical properties, sensing efficiency, and toxicity of 3D networks based on CNTs are still needed. More investigations on composite materials based on biological systems and CNT 3D networks are waiting to be unveiled. For example, CNT networks could improve bone tissue regeneration. Experiments involving the synthesis of SWNT networks (covalent or van der Waals) are still needed since most of the published work deals with MWNT networks.

Acknowledgments

This work is supported by the US Air Force Office of Scientific Research MURI grant FA9550-12-1-0035. MT thanks JST, Japan, for funding the Research Center for Exotic NanoCarbons, under the Japanese regional Innovation Strategy Program by the Excellence. We are grateful to J. M. Romo-Herrera, R. Lv and M. Crespo-Ribadeneira for technical assistance and useful discussions.

References

[1] H.W. Kroto, J.R. Heath, S.C. Obrien, R.F. Curl, R.E. Smalley, C-60-buckminsterfullerene, Nature 318 (6042) (1985) 162—163.

[2] A. Oberlin, M. Endo, T. Koyama, Filamentous growth of carbon through benzene decomposition, Journal of Crystal Growth 32 (3) (1976) 335—349.

[3] S. Iijima, Helical microtubules of graphitic carbon, Nature 354 (6348) (1991) 56–58.
[4] R. Saito, G. Dresselhaus, M.S. Dresselhaus, Topological defects in large fullerenes, Chemical Physics Letters 195 (5–6) (1992) 537–542.
[5] G.D.R. Saito, M.S. Dresselhaus, in: Physical Properties of Carbon Nanotubes, Imperial College Press, London, 1998.
[6] R.H. Baughman, A.A. Zakhidov, W.A. de Heer, Carbon nanotubes—the route toward applications, Science 297 (5582) (2002) 787–792.
[7] M.M.J. Treacy, T.W. Ebbesen, J.M. Gibson, Exceptionally high Young's modulus observed for individual carbon nanotubes, Nature 381 (6584) (1996) 678–680.
[8] R. Saito, M. Fujita, G. Dresselhaus, M.S. Dresselhaus, Electronic-structure of chiral graphene tubules, Applied Physics Letters 60 (18) (1992) 2204–2206.
[9] M. Terrones, Science and technology of the twenty-first century: synthesis, properties and applications of carbon nanotubes, Annual Review of Materials Research 33 (2003) 419–501.
[10] P. Lambin, A. Fonseca, J.P. Vigneron, J.B. Nagy, A.A. Lucas, Structural and electronic-properties of bent carbon nanotubes, Chemical Physics Letters 245 (1) (1995) 85–89.
[11] L. Chico, V.H. Crespi, L.X. Benedict, S.G. Louie, M.L. Cohen, Pure carbon nanoscale devices: nanotube heterojunctions, Physical Review Letters 76 (6) (1996) 971–974.
[12] M. Terrones, W.K. Hsu, J.P. Hare, H.W. Kroto, H. Terrones, D.R.M. Walton, Graphitic structures: from planar to spheres, toroids and helices, Philosophical Transactions of the Royal Society A: Mathematical Physical and Engineering Sciences 354 (1715) (1996) 2025–2054.
[13] Z. Yao, H.W.C. Postma, L. Balents, C. Dekker, Carbon nanotube intramolecular junctions, Nature 402 (6759) (1999) 273–276.
[14] T. Hayashi, H. Muramatsu, D. Shimamoto, K. Fujisawa, T. Tojo, Y. Muramoto, T. Yokomae, T. Asaoka, Y.A. Kim, M. Terrones, M. Endo, Determination of the stacking order of curved few-layered graphene systems, Nanoscale 4 (20) (2012) 6419–6424.
[15] J. Oudar, Sulfur adsorption and poisoning of metallic catalysts, Catalysis Reviews: Science and Engineering 22 (2) (1980) 171–195.
[16] M. Egashira, H. Katsuki, Y. Ogawa, S. Kawasumi, Whiskerization of carbon beads by vapor-phase growth of carbon-fibers to obtain sea-urchin type particles, Carbon 21 (1) (1983) 89–92.
[17] M.S. Kim, N.M. Rodriguez, R.T.K. Baker, The interplay between sulfur adsorption and carbon deposition on cobalt catalysts, Journal of Catalysis 143 (2) (1993) 449–463.
[18] G.G. Tibbetts, C.A. Bernardo, D.W. Gorkiewicz, R.L. Alig, Role of sulfur in the production of carbon-fibers in the vapor-phase, Carbon 32 (4) (1994) 569–576.
[19] T. Kato, K. Kusakabe, S. Morooka, Effect of sulfur on formation of vapor-grown carbon-fiber, Journal of Materials Science Letters 13 (5) (1994) 374–377.
[20] B.C. Satishkumar, P.J. Thomas, A. Govindaraj, C.N.R. Rao, Y-junction carbon nanotubes, Applied Physics Letters 77 (16) (2000) 2530–2532.
[21] F.L. Deepak, N.S. John, A. Govindaraj, G.U. Kulkarni, C.N.R. Rao, Nature and electronic properties of Y-junctions in CNTs and N-doped CNTs obtained by the pyrolysis of organometallic precursors, Chemical Physics Letters 411 (4–6) (2005) 468–473.

[22] C. Valles, M. Perez-Mendoza, P. Castell, M.T. Martinez, W.K. Maser, A.M. Benito, Towards helical and Y-shaped carbon nanotubes: the role of sulfur in CVD processes, Nanotechnology 17 (17) (2006) 4292–4299.

[23] J.M. Romo-Herrera, B.G. Sumpter, D.A. Cullen, H. Terrones, E. Cruz-Silva, D.J. Smith, V. Meunier, M. Terrones, An atomistic branching mechanism for carbon nanotubes: sulfur as the triggering agent, Angewandte Chemie, International Edition 47 (16) (2008) 2948–2953.

[24] J.M. Romo-Herrera, D.A. Cullen, E. Cruz-Silva, D. Ramirez, B.G. Sumpter, V. Meunier, H. Terrones, D.J. Smith, M. Terrones, The role of sulfur in the synthesis of novel carbon morphologies: from covalent Y-junctions to sea-urchin-like structures, Advanced Functional Materials 19 (8) (2009) 1193–1199.

[25] J.A. Rodriguez-Manzo, M.S. Wang, F. Banhart, Y. Bando, D. Golberg, Multibranched junctions of carbon nanotubes via cobalt particles, Advanced Materials 21 (44) (2009) 4477.

[26] G.W. Meng, Y.J. Jung, A.Y. Cao, R. Vajtai, P.M. Ajayan, Controlled fabrication of hierarchically branched nanopores, nanotubes, and nanowires, Proceedings of the National Academy of Sciences of the United States of America 102 (20) (2005) 7074–7078.

[27] M. Terrones, F. Banhart, N. Grobert, J.C. Charlier, H. Terrones, P.M. Ajayan, Molecular junctions by joining single-walled carbon nanotubes, Physical Review Letters 89 (7) (2002).

[28] X. Lepro, Y. Vega-Cantu, F.J. Rodriguez-Macias, Y. Bando, D. Golberg, M. Terrones, Production and characterization of coaxial nanotube junctions and networks of CNx/CNT, Nano Letters 7 (8) (2007) 2220–2226.

[29] A. Ismach, E. Joselevich, Orthogonal self-assembly of carbon nanotube crossbar architectures by simultaneous graphoepitaxy and field-directed growth, Nano Letters 6 (8) (2006) 1706–1710.

[30] X.C. Gui, J.Q. Wei, K.L. Wang, A.Y. Cao, H.W. Zhu, Y. Jia, Q.K. Shu, D.H. Wu, Carbon nanotube sponges, Advanced Materials 22 (5) (2010) 617.

[31] S.J. Kang, C. Kocabas, H.S. Kim, Q. Cao, M.A. Meitl, D.Y. Khang, J.A. Rogers, Printed multilayer superstructures of aligned single-walled carbon nanotubes for electronic, applications, Nano Letters 7 (11) (2007) 3343–3348.

[32] B.S. Shim, W. Chen, C. Doty, C.L. Xu, N.A. Kotov, Smart electronic yarns and wearable fabrics for human biomonitoring made by carbon nanotube coating with polyelectrolytes, Nano Letters 8 (12) (2008) 4151–4157.

[33] J.P. Novak, E.S. Snow, E.J. Houser, D. Park, J.L. Stepnowski, R.A. McGill, Nerve agent detection using networks of single-walled carbon nanotubes, Applied Physics Letters 83 (19) (2003) 4026–4028.

[34] P. Lambin, V. Meunier, Structural properties of junctions between two carbon nanotubes, Applied Physics A: Materials Science and Processing 68 (3) (1999) 263–266.

[35] L. Chico, L.X. Benedict, S.G. Louie, M.L. Cohen, Quantum conductance of carbon nanotubes with defects, Physical Review B 54 (4) (1996) 2600–2606.

[36] R. Tamura, M. Tsukada, Conductance of nanotube junctions and its scaling law, Physical Review B 55 (8) (1997) 4991–4998.

[37] L. Chico, M.P.L. Sancho, M.C. Munoz, Carbon-nanotube-based quantum dot, Physical Review Letters 81 (6) (1998) 1278–1281.

[38] M. Menon, D. Srivastava, Carbon nanotube 'T junctions': nanoscale metal-semiconductor-metal contact devices, Physical Review Letters 79 (22) (1997) 4453–4456.

[39] A.L. Mackay, H. Terrones, Diamond from graphite, Nature 352 (6338) (1991) 762.

[40] H. Terrones, A.L. Mackay, The geometry of hypothetical curved graphite structures, Carbon 30 (8) (1992) 1251–1260.

[41] H. Terrones, A.L. Mackay, Triply periodic minimal-surfaces decorated with curved graphite, Chemical Physics Letters 207 (1) (1993) 45–50.

[42] H. Terrones, M. Terrones, W.K. Hsu, Beyond c-60-graphite structures for the future, Chemical Society Reviews 24 (5) (1995) 341.

[43] J.M. Romo-Herrera, M. Terrones, H. Terrones, S. Dag, V. Meunier, Covalent 2D and 3D networks from 1D nanostructures: designing new materials, Nano Letters 7 (3) (2007) 570–576.

[44] V.R. Coluci, D.S. Galvao, A. Jorio, Geometric and electronic structure of carbon nanotube networks: 'super'-carbon nanotubes, Nanotechnology 17 (3) (2006) 617–621.

[45] F.D. Novaes, R. Rurali, P. Ordejon, Electronic transport between graphene layers covalently connected by carbon nanotubes, ACS Nano 4 (12) (2010) 7596–7602.

[46] L.Q. Xu, N. Wei, Y.P. Zheng, Z.Y. Fan, H.Q. Wang, J.C. Zheng, Graphene-nanotube 3D networks: intriguing thermal and mechanical properties, Journal of Materials Chemistry 22 (4) (2012) 1435–1444.

[47] I. Laslo, Construction of atomic arrangement for carbon nanotube junctions, Physica Status Solidi B: Basic Solid State Physics 244 (11) (2007) 4265–4268.

[48] D.W. Brenner, Empirical potential for hydrocarbons for use in simulating the chemical vapor-deposition of diamond films, Physical Review B 42 (15) (1990) 9458–9471.

[49] V. Meunier, M.B. Nardelli, J. Bernholc, T. Zacharia, J.C. Charlier, Intrinsic electron transport properties of carbon nanotube Y-junctions, Applied Physics Letters 81 (27) (2002) 5234–5236.

[50] J.H. Park, S.B. Sinnott, N.R. Aluru, Ion separation using a Y-junction carbon nanotube, Nanotechnology 17 (3) (2006) 895–900.

[51] S. Dag, R.T. Senger, S. Ciraci, Theoretical study of crossed and parallel carbon nanotube junctions and three-dimensional grid structures, Physical Review B 70 (20) (2004).

[52] M.Z. Huang, W.Y. Ching, T. Lenosky, Electronic-properties of negative-curvature periodic graphitic carbon surfaces, Physical Review B 47 (3) (1993) 1593–1606.

[53] R.C. Furneaux, W.R. Rigby, A.P. Davidson, The formation of controlled-porosity membranes from anodically oxidized aluminum, Nature 337 (6203) (1989) 147–149.

[54] J. Li, C. Papadopoulos, J. Xu, Nanoelectronics—growing Y-junction carbon nanotubes, Nature 402 (6759) (1999) 253–254.

[55] T. Rueckes, K. Kim, E. Joselevich, G.Y. Tseng, C.L. Cheung, C.M. Lieber, Carbon nanotube-based nonvolatile random access memory for molecular computing, Science 289 (5476) (2000) 94–97.

[56] Y. Homma, Y. Kobayashi, T. Ogino, T. Yamashita, Growth of suspended carbon nanotube networks on 100-nm-scale silicon pillars, Applied Physics Letters 81 (12) (2002) 2261–2263.

[57] A.M. Cassell, G.C. McCool, H.T. Ng, J.E. Koehne, B. Chen, J. Li, J. Han, M. Meyyappan, Carbon nanotube networks by chemical vapor deposition, Applied Physics Letters 82 (5) (2003) 817–819.

[58] E. Joselevich, C.M. Lieber, Vectorial growth of metallic and semiconducting single-wall carbon nanotubes, Nano Letters 2 (10) (2002) 1137–1141.

[59] A. Ismach, L. Segev, E. Wachtel, E. Joselevich, Atomic-step-templated formation of single wall carbon nanotube patterns, Angewandte Chemie, International Edition 43 (45) (2004) 6140–6143.

[60] M. Hofmann, D. Nezich, A. Reina, J. Kong, In-situ sample rotation as a tool to understand chemical vapor deposition growth of long aligned carbon nanotubes, Nano Letters 8 (12) (2008) 4122–4127.

[61] M. Endo, H. Muramatsu, T. Hayashi, Y.A. Kim, M. Terrones, N.S. Dresselhaus, 'Buckypaper' from coaxial nanotubes, Nature 433 (7025) (2005) 476.

[62] M.B. Bryning, D.E. Milkie, M.F. Islam, L.A. Hough, J.M. Kikkawa, A.G. Yodh, Carbon nanotube aerogels, Advanced Materials 19 (5) (2007) 661–664.

[63] J.H. Zou, J.H. Liu, A.S. Karakoti, A. Kumar, D. Joung, Q.A. Li, S.I. Khondaker, S. Seal, L. Zhai, Ultralight multiwalled carbon nanotube aerogel, ACS Nano 4 (12) (2010) 7293–7302.

[64] H. Yu, Z.F. Li, G.H. Luo, F. Wei, Growth of branch carbon nanotubes on carbon nanotubes as support, Diamond and Related Materials 15 (9) (2006) 1447–1451.

[65] A.A. El Mel, A. Achour, W. Xu, C.H. Choi, E. Gautron, B. Angleraud, A. Granier, L. Le Brizoual, M.A. Djouadi, P.Y. Tessier, Hierarchical carbon nanostructure design: ultra-long carbon nanofibers decorated with carbon nanotubes, Nanotechnology 22 (43) (2011).

[66] X. Sun, R. Li, B. Stansfield, J.P. Dodelet, S. Desilets, 3D carbon nanotube network based on a hierarchical structure grown on carbon paper backing, Chemical Physics Letters 394 (4–6) (2004) 266–270.

[67] X.C. Gui, H.B. Li, K.L. Wang, J.Q. Wei, Y. Jia, Z. Li, L.L. Fan, A.Y. Cao, H.W. Zhu, D.H. Wu, Recyclable carbon nanotube sponges for oil absorption, Acta Materialia 59 (12) (2011) 4798–4804.

[68] D.P. Hashim, N.T. Narayanan, J.M. Romo-Herrera, D.A. Cullen, M.G. Hahm, P. Lezzi, J.R. Suttle, D. Kelkhoff, E. Munoz-Sandoval, S. Ganguli, A.K. Roy, D.J. Smith, R. Vajtai, B.G. Sumpter, V. Meunier, H. Terrones, M. Terrones, P.M. Ajayan, Covalently bonded three-dimensional carbon nanotube solids via boron induced nanojunctions, Scientific Reports 2 (2012).

[69] Y. Huang, X.F. Duan, Y. Cui, L.J. Lauhon, K.H. Kim, C.M. Lieber, Logic gates and computation from assembled nanowire building blocks, Science 294 (5545) (2001) 1313–1317.

[70] B. Zhang, G. Hong, B. Peng, J. Zhang, W. Choi, J.M. Kim, J.Y. Choi, Z. Liu, Grow single-walled carbon nanotubes cross-bar in one batch, Journal of Physical Chemistry C 113 (14) (2009) 5341–5344.

[71] A. Kumar, C.W. Zhou, The race to replace tin-doped indium oxide: which material will win? ACS Nano 4 (1) (2010) 11–14.

[72] J.W. Jo, J.W. Jung, J.U. Lee, W.H. Jo, Fabrication of highly conductive and transparent thin films from single-walled carbon nanotubes using a new non-ionic surfactant via spin coating, ACS Nano 4 (9) (2010) 5382–5388.

[73] Y. Oh, D. Suh, Y.J. Kim, C.S. Han, S. Baik, Transparent conductive film fabrication using intercalating silver nanoparticles within carbon nanotube layers, Journal of Nanoscience and Nanotechnology 11 (1) (2011) 489−493.

[74] M. Pyo, E.G. Bae, Y. Cho, Y.S. Jung, K. Zong, Composites of low bandgap conducting polymer-wrapped MWNT and poly(methyl methacrylate) for low percolation and high transparency, Synthetic Metals 160 (19−20) (2010) 2224−2227.

[75] Q. Cao, S.H. Hur, Z.T. Zhu, Y.G. Sun, C.J. Wang, M.A. Meitl, M. Shim, J.A. Rogers, Highly bendable, transparent thin-film transistors that use carbon-nanotube-based conductors and semiconductors with elastomeric dielectrics, Advanced Materials 18 (3) (2006) 304.

[76] E. Gracia-Espino, G. Sala, F. Pino, N. Halonen, J. Luomahaara, J. Maklin, G. Toth, K. Kordas, H. Jantunen, M. Terrones, P. Helisto, H. Seppa, P.M. Ajayan, R. Vajtai, Electrical transport and field-effect transistors using inkjet-printed SWCNT films having different functional side groups, ACS Nano 4 (6) (2010) 3318−3324.

[77] E.L. Gui, L.J. Li, K.K. Zhang, Y.P. Xu, X.C. Dong, X.N. Ho, P.S. Lee, J. Kasim, Z.X. Shen, J.A. Rogers, S.G. Mhaisalkar, DNA sensing by field-effect transistors based on networks of carbon nanotubes, Journal of the American Chemical Society 129 (46) (2007) 14427−14432.

[78] C. Li, M. Curreli, H. Lin, B. Lei, F.N. Ishikawa, R. Datar, R.J. Cote, M.E. Thompson, C.W. Zhou, Complementary detection of prostate-specific antigen using In_2O_3 nanowires and carbon nanotubes, Journal of the American Chemical Society 127 (36) (2005) 12484−12485.

[79] C.Y. Lee, H.M. Tsai, H.J. Chuang, S.Y. Li, P. Lin, T.Y. Tseng, Characteristics and electrochemical performance of supercapacitors with manganese oxide-carbon nanotube nanocomposite electrodes, Journal of the Electrochemical Society 152 (4) (2005) A716−A720.

[80] J.Y. Lee, K. Liang, K.H. An, Y.H. Lee, Nickel oxide/carbon nanotubes nanocomposite for electrochemical capacitance, Synthetic Metals 150 (2) (2005) 153−157.

[81] T. Bordjiba, M. Mohamedi, L.H. Dao, New class of carbon-nanotube aerogel electrodes for electrochemical power sources, Advanced Materials 20 (4) (2008) 815.

[82] J.C. Claussen, A.D. Franklin, A. ul Haque, D.M. Porterfield, T.S. Fisher, Electrochemical biosensor of nanocube-augmented carbon nanotube networks, ACS Nano 3 (1) (2009) 37−44.

[83] A.E. Aliev, J.Y. Oh, M.E. Kozlov, A.A. Kuznetsov, S.L. Fang, A.F. Fonseca, R. Ovalle, M.D. Lima, M.H. Haque, Y.N. Gartstein, M. Zhang, A.A. Zakhidov, R.H. Baughman, Giant-stroke, superelastic carbon nanotube aerogel muscles, Science 323 (5921) (2009) 1575−1578.

[84] N. Chakrapani, B.Q. Wei, A. Carrillo, P.M. Ajayan, R.S. Kane, Capillarity-driven assembly of two-dimensional cellular carbon nanotube foams, Proceedings of the National Academy of Sciences of the United States of America 101 (12) (2004) 4009−4012.

[85] H. Liu, S.H. Li, J. Zhai, H.J. Li, Q.S. Zheng, L. Jiang, D.B. Zhu, Self-assembly of large-scale micropatterns on aligned carbon nanotube films, Angewandte Chemie, International Edition 43 (9) (2004) 1146−1149.

[86] M.A. Correa-Duarte, N. Wagner, J. Rojas-Chapana, C. Morsczeck, M. Thie, M. Giersig, Fabrication and biocompatibility of carbon nanotube-based 3D networks as scaffolds for cell seeding and growth, Nano Letters 4 (11) (2004) 2233−2236.

[87] D.J. Cohen, D. Mitra, K. Peterson, M.M. Maharbiz, A highly elastic, capacitive strain gauge based on percolating nanotube networks, Nano Letters 12 (4) (2012) 1821–1825.

[88] E.T. Thostenson, T.W. Chou, Carbon nanotube networks: sensing of distributed strain and damage for life prediction and self healing, Advanced Materials 18 (21) (2006) 2837.

[89] N. Alamusi, H. Hu, S. Fukunaga, Y.L. Atobe, J.H. Liu Li, Piezoresistive strain sensors made from carbon nanotubes based polymer nanocomposites, Sensors 11 (11) (2011) 10691–10723.

[90] G.E. Froudakis, Why alkali-metal-doped carbon nanotubes possess high hydrogen uptake, Nano Letters 1 (10) (2001) 531–533.

[91] K.M. Thomas, Hydrogen adsorption and storage on porous materials, Catalysis Today 120 (3–4) (2007) 389–398.

CHAPTER 18

A Review on the Design of Superstrong Carbon Nanotube or Graphene Fibers and Composites

Nicola Pugno

Laboratory of Bio-Inspired and Graphene Nanomechanisms, Center for Materials and Microsystems, Fondation Bruno Kessler, Povo, Italy

CHAPTER OUTLINE

18.1 Introduction	495
18.2 Hierarchical Simulations and Size Effects	497
18.3 Brittle Fracture	499
18.4 Elastic-Plasticity, Fractal Cracks and Finite Domains	503
18.5 Fatigue	504
18.6 Elasticity	505
18.7 Atomistic Simulations	506
18.8 Nanotensile Tests	507
18.9 Thermodynamic Limit	508
18.10 Sliding Failure	513
18.11 Conclusions	515
References	516

18.1 INTRODUCTION

A space elevator basically consists of a cable attached to the Earth surface for carrying payloads into space [1]. If the cable is long enough, i.e. around 150 Mm (a value that can be reduced by a counterweight), the centrifugal forces exceed the gravity of the cable that will work under tension [2]. The elevator would stay fixed geosynchronously; once sent far enough, climbers would be accelerated by the Earth's rotational energy. A space elevator would revolutionize the methodology for carrying payloads into space at low cost, but its design is very challenging. The most critical component in the space elevator design is undoubtedly the cable [3–5] that requires a material with very high strength and low density.

Considering a cable with constant cross section and a vanishing tension at the planet surface, the maximum stress–density ratio, reached at the geosynchronous orbit, is for the Earth equal to 63 GPa/(1300 kg/m^3), corresponding to 63 GPa if the low carbon density is assumed for the cable. Only recently, after the rediscovery of carbon nanotubes [6], such a large failure stress has been experimentally measured, during tensile tests of ropes composed of single-walled carbon nanotubes (SWCNTs) [7] and multiwalled carbon nanotubes consist of graphene [8], both expected to have an ideal strength of ~100 GPa. Note that for steel (density of 7900 kg/m^3 and maximum strength of 1.5 GPa), the maximum stress expected in the cable would be of 383 GPa, whereas for Kevlar (density of 1440 kg/m^3 and strength of 3.6 GPa) of 70 GPa, both much higher than their strengths [3].

However, an optimized cable design must consider a uniform tensile stress profile rather than a constant cross-sectional area [2]. Accordingly, the cable could be built of any material by simply using a large enough taper ratio, that is the ratio between the maximum (at the geosynchronous orbit) and minimum (at the Earth's surface) cross-sectional area. For example, for steel or Kevlar, a giant and unrealistic taper ratio would be required, 10^{33} or 10^8, respectively, whereas for carbon nanotubes and graphene, it must theoretically be only 2^9. Thus, the feasibility of the space elevator seems to become only currently plausible [9,10] thanks to the discovery of carbon nanotubes. The cable would represent the largest engineering structure, hierarchically designed from the nano- (single nanotube or graphene ribbon with length of the order of 100 nm) to the megascale (space elevator cable with a length of the order of 100 Mm). Pushed by this problem we have worked on the design of nanotube or graphene fibers and composites during the last ten years.

In this chapter, the asymptotic analysis on the role of defects for the nanotubes or graphene fibers and composites, based on new theoretical deterministic and statistical approaches of quantized fracture mechanics (QFM) proposed by the author [11–14], and their extension to nonasymptotic regimes, elastic-plasticity, rough cracks and finite domains are reviewed. The role of thermodynamically unavoidable atomistic defects with different size and shape is thus quantified on brittle fracture, fatigue and elasticity for nanotubes, graphene and related bundles and composites. The results are compared with atomistic and continuum simulations and nanotensile tests Specific key simple formulas for the design of a flaw-tolerant space elevator megacable are reported, suggesting that it would need a taper ratio (for uniform stress) of about 1-2 orders of magnitude larger than as today erroneously proposed.

The chapter is organized in 10 short sections, as follows. After this introduction, reported as the first section, we start calculating the strength of nanotube and graphene bundles and composites by using ad hoc hierarchical simulations, discussing the related size effect. In Section 3, the strength reduction of a single nanotube or graphene ribbon and of a nanotube bundle containing defects with given size and shape is calculated; the taper ratio for a flaw-tolerant space elevator cable is accordingly derived. In Section 4, elastic-plastic (or hyperelastic) materials, rough cracks and finite domains are discussed. In Section 5, the fatigue life time is evaluated for a single nanotube or graphene and for a related bundle. In Section 6, the related Young's modulus degradations are quantified. In Sections 7 and 8, we compare our results on strength and elasticity with atomistic simulations and tensile tests of carbon nanotubes. In Section 9, we demonstrate that defects are thermodynamically unavoidable, evaluating the minimum defect size and corresponding maximum achievable strength. In Section 10, the fiber or composite design against sliding failure is presented. The last section presents our concluding remarks.

18.2 HIERARCHICAL SIMULATIONS AND SIZE EFFECTS

To evaluate the strength of carbon nanotube cables, the hierarchical fiber bundle model, formerly proposed [3], has been adopted [15]. Multiscale simulations are necessary in order to tackle the size scales involved, spanning over ~10 orders of magnitude from nanotube/graphene length (~100 nm) to kilometer-long cables, and also to provide useful information about cable scaling properties with length.

The cable is modeled as an ensemble of stochastic "springs", arranged in parallel sections. Linearly increasing strains are applied to the fiber bundle, and at each algorithm iteration, the number of fractured springs is computed (fracture occurs when local stress exceeds the nanotube graphene failure strength) and the strain is uniformly redistributed among the remaining intact springs in each section.

In silico stress–strain experiments have been carried out according to the following hierarchical architecture. Level 1: the nanotubes graphene (single springs, Level 0) are considered with a given elastic modulus and failure strength distribution and composed a 40×1000 lattice or fiber. Level 2: again a 40×1000 lattice composed by second level "springs," each of them identical to the entire fiber analyzed at the first level, is analyzed with the elastic modulus as input and stochastic strength distribution derived as the output of the numerous simulations to be carried out at the first level. And so on. Five hierarchical levels are sufficient to reach the size scale of the megameter from that of the nanometer (Fig. 18.1).

The Level 1 simulation is carried out with springs $L_0 = 10^{-7}$ m in length, $w_0 = 10^{-9}$ m in width, with Young's modulus $E_0 = 10^{12}$ Pa and strength σ_f randomly distributed according to the nanoscale Weibull statistics [16] $P(\sigma_f) = 1 - \exp[-(\sigma_f/\sigma_0)^m]$, where P is the cumulative probability. Fitting to the experiments on carbon nanotubes [7,8], we have derived for carbon nanotubes $\sigma_0 = 34$ GPa and $m = 2.7$ [16]. Then, Level 2 is computed, and so on. The results are summarized in Fig. 18.2, in which a strong size effect is observed, up to length of ~1 m.

Given the decaying σ_f vs cable length L obtained from simulations, it is interesting to fit the behavior with simple analytical scaling laws. Various laws exist in the literature, and one of the most used is the multifractal scaling law (MFSL [17,18]). This law has been recently extended toward the nanoscale [19]:

$$\frac{\sigma_f}{\sigma_{macro}} = \sqrt{1 + \frac{l_{ch}}{L + l_0}} \tag{18.1}$$

where σ_f is the failure stress, σ_{macro} is the macrostrength, L is the structural characteristic size, l_{ch} is the characteristic internal length and l_0 is defined via $\sigma_f(l=0) = \sigma_{macro}\sqrt{1 + \frac{l_{ch}}{l_0}} \equiv \sigma_{nano}$, where σ_{nano} is the nanostrength. Note that

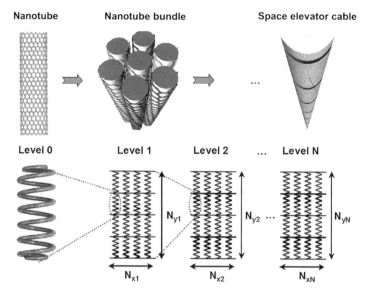

FIGURE 18.1

Schematization of the adopted multiscale simulation procedure to determine the space elevator cable strength. Here, $N=5$, $N_{x1} = N_{x2} = \ldots N_{x5} = 40$ and $N_{y1} = N_{y2} = \ldots N_{y5} = 1000$, so that the total number of nanotubes or graphene ribbons in the space elevator cable is $N_{tot} = (1000 \times 40)^5 \approx 10^{23}$ [15] From: Multiscale stochastic simulations as in-silico tensile testing of nanotube-based megacables, Small 4, 2008. (For color version of this figure, the reader is referred to the online version of this book.)

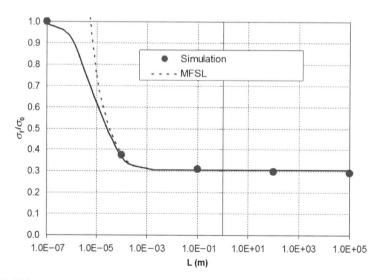

FIGURE 18.2

Comparison between simulations (dots) and analytical scaling law (Eqn (18.1)) for the failure strength of the nanotube bundle as a function of its length; the asymptote is at 10.20 GPa [15]. The dotted line corresponds to Co=o, thus to [17]. (For color version of this figure, the reader is referred to the online version of this book.)

for $l_0 = 0$, this law is that reported in [17]. Here, we can choose σ_{nano} as the nanotube stochastic strength, i.e. $\sigma_{nano} = 34$ GPa. The computed macrostrength is $\sigma_{macro} = 10.20$ GPa. The fit with Eqn (18.1) is shown in Fig. 18.1 for the various L considered at the different hierarchical levels. The best fit is obtained for $l_{ch} = 5 \times 10^{-5}$ m, where the analytical law is practically coincident with the simulated results. Thus, for a carbon nanotube or graphene megacable, we have numerically derived a plausible strength $\sigma_C = \sigma_{macro} \approx 10$ GPa.

18.3 BRITTLE FRACTURE

By considering QFM [11–14], the failure stress σ_N for a nanotube or graphene having atomic size q (the "fracture quantum") and containing an elliptical hole of half-axes a perpendicular to the applied load (or nanotube or graphene axis) and b can be determined including in the asymptotic solution [12] the contribution of the far-field stress. We accordingly derive

$$\frac{\sigma_N(a,b)}{\sigma_N^{(theo)}} = \sqrt{\frac{1 + 2a/q(1 + 2a/b)^{-2}}{1 + 2a/q}}, \quad \sigma_N^{(theo)} = \frac{K_{IC}}{\sqrt{q\pi/2}} \quad (18.2)$$

where $\sigma_N^{(theo)}$ is the theoretical (defect-free) nanotube or graphene strength (~100 GPa, see Table 18.1) and K_{IC} is the material fracture toughness. The self-interaction between the tips has been neglected here (i.e. $a \ll \pi R$, W with R the nanotube radius or W, the graphene width) and would further reduce the failure stress. For atomistic defects (having characteristic length of few Ångstrom) in nanotubes or graphene (having characteristic diameter of several nanometers), this hypothesis is fully verified. However, QFM can easily treat also the self-tip interaction starting from the corresponding value of the stress-intensity factor (reported in the related handbooks). The validity of QFM has been recently confirmed by atomistic simulations [3–5,12,13,20], but also at larger size scales [12,13,21] and for fatigue crack growth [14,22,23].

Regarding the defect shape, for a sharp crack perpendicular to the applied load, $a/q = $ const and $b/q \rightarrow 0$; thus, $\sigma_N \approx \sigma_N^{(theo)}/\sqrt{1 + 2a/q}$, and for $a/q \gg 1$, i.e. large cracks, $\sigma_N \approx K_{IC}/\sqrt{\pi a}$ in agreement with linear elastic fracture mechanics (LEFM); note that LEFM can (1) only treat sharp cracks and (2) unreasonably predict an infinite defect-free strength. On the other hand, for a crack parallel to the applied load, $b/q = $ const and $a/q \rightarrow 0$, and thus, $\sigma_N = \sigma_N^{(theo)}$, as it must be. In addition, regarding the defect size, for self-similar and small holes, $a/b = $ const & $a/q \rightarrow 0$ and coherently $\sigma_N = \sigma_N^{(theo)}$; furthermore, for self-similar and large holes, $a/b = $ const & $a/q \rightarrow \infty$ and we deduce $\sigma_N \approx \sigma_N^{(theo)}/(1 + 2a/b)$ in agreement with the stress concentration posed by elasticity; but elasticity (coupled with a maximum stress criterion) unreasonably predicts (3) a strength independent from the hole size and (4) tending to zero for cracks. Note the extreme consistency of Eqn (18.2) that, removing all the

Table 18.1 Atomistic Simulations [30–34] vs QFM Predictions, for Nanocracks of Size n or Nanoholes of Size m in carbon nanotubes or graphene ribbons. (From: The role of defects in the design of the space elevator cable: from nanotube to megatube, Acta Materialia 55 2007) [4]

Graphene/Nanotube Type	Nanocrack (n) and Nanohole (m) sizes	Strength [GPa] by QM (MTB-G2) and MM (PM3; M) QM/MM Atomistic or QFM Calculations
[5,5]	Defect-free	105 (MTB-G2); 135 (PM3)
[5,5]	$n=1$ (Sym.+H)	85 (MTB-G2), 79 (QFM); 106 (PM3), 101 (QFM)
[5,5]	$n=1$ (Asym. +H)	71 (MTB-G2), 79 (QFM); 99 (PM3), 101 (QFM)
[5,5]	$n=1$ (Asym.)	70 (MTB-G2), 79 (QFM); 100 (PM3), 101 (QFM)
[5,5]	$n=2$ (Sym.)	71 (MTB-G2), 63 (QFM); 105 (PM3), 81 (QFM)
[5,5]	$n=2$ (Asym.)	73 (MTB-G2), 63 (QFM); 111 (PM3), 81 (QFM)
[5,5]	$m=1$ (+H)	70 (MTB-G2), 68 for long tube, 79 (QFM); 101 (PM3), 101 (QFM)
[5,5]	$m=2$ (+H)	53 (MTB-G2), 50 for long tube, 67 (QFM); 78 (PM3), 86 (QFM)
[10,10]	Defect-free	88 (MTB-G2); 124 (PM3)
[10,10]	$n=1$ (Sym.+H)	65 (MTB-G2), 66 (QFM)
[10,10]	$n=1$ (Asym. +H)	68 (MTB-G2), 66 (QFM)
[10,10]	$n=1$ (Sym.)	65 (MTB-G2), 66 (QFM); 101 (PM3), 93 (QFM)
[10,10]	$n=2$ (Sym.)	64 (MTB-G2), 53 (QFM); 107 (PM3), 74 (QFM)
[10,10]	$n=2$ (Asym.)	65 (MTB-G2), 53 (QFM); 92 (PM3), 74 (QFM)
[10,10]	$m=1$ (+H)	56 (MTB-G2), 52 for long tube, 66 (QFM); 89 (PM3), 93 (QFM)
[10,10]	$m=2$ (+H)	42 (MTB-G2), 36 for long tube, 56 (QFM); 67 (PM3), 79 (QFM)
[50,0]	Defect-free	89 (MTB-G2)
[50,0]	$m=1$ (+H)	58 (MTB-G2); 67 (QFM)
[50,0]	$m=2$ (+H)	46 (MTB-G2); 57 (QFM)
[50,0]	$m=3$ (+H)	40 (MTB-G2); 44 (QFM)
[50,0]	$m=4$ (+H)	36 (MTB-G2); 41 (QFM)
[50,0]	$m=5$ (+H)	33 (MTB-G2); 39 (QFM)
[50,0]	$m=6$ (+H)	31 (MTB-G2); 37 (QFM)
[100,0]	Defect-free	89 (MTB-G2)
[100,0]	$m=1$ (+H)	58 (MTB-G2); 67 (QFM)

Table 18.1 Atomistic Simulations [30–34] vs QFM Predictions, for Nanocracks of Size n or Nanoholes of Size m in carbon nanotubes or graphene ribbons. (From: The role of defects in the design of the space elevator cable: from nanotube to megatube, Acta Materialia 55 2007) [4]—Cont'd

Graphene/ Nanotube Type	Nanocrack (n) and Nanohole (m) sizes	Strength [GPa] by QM (MTB-G2) and MM (PM3; M) QM/MM Atomistic or QFM Calculations
[100,0]	$m = 2$ (+H)	47 (MTB-G2); 57 (QFM)
[100,0]	$m = 3$ (+H)	42 (MTB-G2); 44 (QFM)
[100,0]	$m = 4$ (+H)	39 (MTB-G2); 41 (QFM)
[100,0]	$m = 5$ (+H)	37 (MTB-G2); 39 (QFM)
[100,0]	$m = 6$ (+H)	35 (MTB-G2); 37 (QFM)
[29,29]	Defect-free	101 (MTB-G2)
[29,29]	$m = 1$ (+H)	77 (MTB-G2); 76 (QFM)
[29,29]	$m = 2$ (+H)	62 (MTB-G2); 65 (QFM)
[29,29]	$m = 3$ (+H)	54 (MTB-G2); 50 (QFM)
[29,29]	$m = 4$ (+H)	48 (MTB-G2); 46 (QFM)
[29,29]	$m = 5$ (+H)	45 (MTB-G2); 44 (QFM)
[29,29]	$m = 6$ (+H)	42 (MTB-G2); 42 (QFM)
[47,5]	Defect-free	89 (MTB-G2)
[47,5]	$m = 1$ (+H)	57 (MTB-G2); 67 (QFM)
[44,10]	Defect-free	89 (MTB-G2)
[44,10]	$m = 1$ (+H)	58 (MTB-G2); 67 (QFM)
[40,16]	Defect-free	92 (MTB-G2)
[40,16]	$m = 1$ (+H)	59 (MTB-G2); 69 (QFM)
[36,21]	Defect-free	96 (MTB-G2)
[36,21]	$m = 1$ (+H)	63 (MTB-G2); 72 (QFM)
[33,24]	Defect-free	99 (MTB-G2)
[33,24]	$m = 1$ (+H)	67 (MTB-G2); 74 (QFM)
[80, 0]	Defect-free	93 (M)
[80, 0]	$n = 2$	64 (M); 56 (QFM)
[80, 0]	$n = 4$	50 (M); 43 (QFM)
[80, 0]	$n = 6$	42 (M); 35 (QFM)
[80, 0]	$n = 8$	37 (M); 32 (QFM)
[40, 0] (nested by a [32, 0])	Defect-free	99 (M)
[40, 0] (nested by a [32, 0])	$n = 2$	73 (M); 69 (QFM + vdW interaction ~10 GPa)
[40, 0] (nested by a [32, 0])	$n = 4$	57 (M); 56 (QFM + vdW interaction ~10 GPa)
[40, 0] (nested by a [32, 0])	$n = 6$	50 (M); 48 (QFM + vdW interaction ~10 GPa)
	$n = 8$	

(continued)

Table 18.1 Atomistic Simulations [30–34] vs QFM Predictions, for Nanocracks of Size n or Nanoholes of Size m in carbon nanotubes or graphene ribbons. (From: The role of defects in the design of the space elevator cable: from nanotube to megatube, Acta Materialia 55 2007) [4]—Cont'd

Graphene/ Nanotube Type	Nanocrack (n) and Nanohole (m) sizes	Strength [GPa] by QM (MTB-G2) and MM (PM3; M) QM/MM Atomistic or QFM Calculations
[40, 0] (nested by a [32, 0])		44 (M); 44 (QFM + vdW interaction ~10 GPa)
[100,0]	Defect-free	89 (MTB-G2)
[100,0]	$n = 4$	50 (M); 41 (QFM)
[10,0]	Defect free	124 (QM); 88 (MM);
[10,0]	$n = 1$	101 (QM) 95 (QM/MM) 93 (QFM); 65 (MM) 66 (QFM)

The QFM predictions are here obtained simply considering in Eqn (18.2) $2a/q = n$, $2b/q = 1$ for cracks of size n or $a/q = b/q = (2m - 1)/\sqrt{3}$ for holes of size m. Quantum mechanics (QM) semiempirical calculations (PM3 method), molecular mechanics (MM) calculations (modified Tersoff–Brenner potential of second generation (MTB-G2), modified Morse potential (M)) and coupled QM/MM calculations. The symbol (+H) means that the defect was saturated with hydrogen. Symmetric and asymmetric bond reconstructions were also considered; the tubes are "short", if not otherwise specified. We have roughly ignored in the QFM predictions the difference between symmetric and asymmetric bond reconstruction, hydrogen saturation and length effect (for shorter tubes, an increment in the strength is always observed as an intrinsic size effect), noting that the main differences in the atomistic simulations are imputable to the used potential. For nested nanotubes, a strength increment of ~10 GPa is here assumed to roughly take into account the van der Walls (vdW) interaction between the walls.

limitations (1–4) represents the first law capable of describing in a unified manner all the size and shape effects for the elliptical holes, including cracks as limit case. In other words, Eqn (18.2) shows that the two classical strength predictions based on stress intensifications (LEFM) or stress concentrations (elasticity) are only reasonable for "large" defects; Eqn (18.2) unifies their results and extends its validity to "small" defects ("large" and "small" are here with respect to the fracture quantum). Equation (18.2) shows that even a small defect can dramatically reduce the mechanical strength.

An upper bound of the cable strength can be derived assuming the simultaneous failure of all the defective nanotubes or graphene layers present in the bundle. Accordingly, imposing the critical force equilibrium (mean-field approach) for a cable composed of nanotubes or graphene layers in numerical fractions f_{ab} containing holes of half-axes a and b, we find the cable strength σ_C (ideal if $\sigma_C^{(\text{theo})}$) in the following form:

$$\frac{\sigma_C}{\sigma_C^{(\text{theo})}} = \sum_{a,b} f_{ab} \frac{\sigma_N(a,b)}{\sigma_N^{(\text{theo})}} \qquad (18.3)$$

The summation is extended to all the different holes; the numerical fraction f_{00} of nanotubes/graphene is defect-free and $\sum_{a,b} f_{ab} = 1$. If all the defective nanotubes/graphene in the bundle contain identical holes $f_{ab} = f = 1 - f_{00}$, and the following simple relation between the strength reductions holds: $1 - \sigma_C/\sigma_C^{(theo)} = f(1 - \sigma_N/\sigma_N^{(theo)})$.

Thus, the taper ratio λ needed to have a uniform stress in a space elevator cable [2], under the centrifugal and gravitational forces, must be larger than its theoretical value, in order to design a flaw-tolerant megacable. In fact, according to our analysis, we deduce ($\lambda = e^{const. \rho_C/\sigma_C} \geq \lambda^{(theo)} \approx 1.9$ for carbon nanotubes/graphene; ρ_C denotes the material density) the following:

$$\frac{\lambda}{\lambda^{(theo)}} = \lambda^{(theo)} \left(\frac{\sigma_C^{(theo)}}{\sigma_C} - 1 \right) \tag{18.4}$$

Equation (18.4) shows that a small defect can dramatically increase the taper ratio required for a flaw-tolerant megacable and thus, nearly proportionally, its mass.

18.4 ELASTIC-PLASTICITY, FRACTAL CRACKS AND FINITE DOMAINS

The previous equations are based on linear elasticity, i.e. on a linear relationship $\sigma \propto \varepsilon$ between stress σ and strain ε. In contrast, let us assume $\sigma \propto \varepsilon^\kappa$, where $\kappa > 1$ denotes hyperelasticity, as well as $\kappa < 1$ denotes elastic-plasticity. The power of the stress singularity will accordingly be modified [24] from the classical value 1/2 to $\alpha = \kappa/(\kappa + 1)$. Thus, the problem is mathematically equivalent to that of a reentrant corner [25], and consequently, we predict

$$\frac{\sigma_N(a, b, \alpha)}{\sigma_N^{(theo)}} = \left(\frac{\sigma_N(a, b)}{\sigma_N^{(theo)}} \right)^{2\alpha}, \quad \alpha = \frac{\kappa}{\kappa + 1} \tag{18.5}$$

A crack with a self-similar roughness, mathematically described by a fractal with noninteger dimension $1 < D < 2$, would similarly modify the stress singularity, according to $\alpha = (2 - D)/2$ [18,26]; thus, with Eqn (18.5), we can also estimate the role of the crack roughness. Both plasticity and roughness reduce the severity of the defect, whereas hyperelasticity enlarges its effect. For example, for a crack composed by n adjacent vacancies, we found $\sigma_N/\sigma_N^{(theo)} \approx (1 + n)^{-\alpha}$.

However, note that among these three effects, only elastic-plasticity may have a significant role in carbon nanotubes/graphene; in spite of this, fractal cracks could play an important role in nanotube/graphene bundles as a consequence of their larger size scale, which would allow the development of a crack surface roughness. Hyperelasticity is not expected to be relevant in this context (but important for biological materials such as spider silk).

According to LEFM and assuming the classical hypothesis of self-similarity ($a_{max} \propto L$), i.e. the largest crack size is proportional to the characteristic structural size L, we expect a size effect on the strength in the form of the power law $\sigma_C \propto L^{-\alpha}$. For linear elastic materials, $\alpha = 1/2$ as classically considered, but for elastic-plastic materials or fractal cracks, $0 \leq \alpha \leq 1/2$ [24], whereas for hyperelastic materials, $1/2 \leq \alpha \leq 1$, suggesting an unusual and stronger size effect. This parameter would represent the maximum slope (in a bi-log plot) of the scaling as reported in Fig. 18.1.

Equation (18.2) does not consider the defect—boundary interaction. The finite width $2W$ can be treated by applying QFM starting from the related expression of the stress-intensity factor (reported in handbooks). However, to have an idea of the defect—boundary interaction, we apply an approximated method [27], deriving the following correction $\sigma_N(a,b,W) \approx C(W)\sigma_N(a,b)$, $C(W) \approx (1-a/W)/(\sigma_N(a,b)|_{q\to W-a}/\sigma_N^{(theo)})$ (note that such a correction is valid also for $W \approx a$, whereas for $W \gg a$, it becomes $C(W \gg a) \approx 1 - a/W$. Similarly, the role of the defect orientation β could be treated by QFM considering the related stress-intensity factor; roughly, one could use the self-consistent approximation $\sigma_N(a,b,\beta) \approx \sigma_N(a,b)\cos^2\beta + \sigma_N(b,a)\sin^2\beta$.

18.5 FATIGUE

The superstrong fibers can be cyclically loaded, e.g. in a space elevator cable by the climbers carrying the payloads, thus fatigue could play a role on their design. By integrating the quantized Paris' law, that is an extension of the classical Paris' law recently proposed especially for nanostructure or nanomaterial applications [14,22,23], we derive the following number of cycles to failure (or life time):

$$\frac{C_N(a)}{C_N^{(theo)}} = \frac{(1+q/W)^{1-m/2} - (a/W+q/W)^{1-m/2}}{(1+q/W)^{1-m/2} - (q/W)^{1-m/2}}, \quad m \neq 2 \quad (18.6a)$$

$$\frac{C_N(a)}{C_N^{(theo)}} = \frac{\ln[(1+q/W)/(a/W+q/W)]}{\ln[(1+q/W)/(q/W)]}, \quad m = 2 \quad (18.6b)$$

where $m > 0$ is the material Paris' exponent. Note that according to Wöhler, $C_N^{(theo)} = K\Delta\sigma^{-k}$, where K and k are material constants and $\Delta\sigma$ is the amplitude of the stress range during the oscillations. Even if fatigue experiments in nanotubes/graphene are still to be performed, their behavior is expected to be intermediate between those of Wöhler and Paris, as displayed by all the known materials, and the quantized Paris' law basically represents their asymptotic matching (as QFM basically represents the asymptotic matching between the strength and toughness approaches).

Only defects remaining self-similar during fatigue growth have to be considered, thus only a crack (of half-length a) is of interest in this context. By means of Eqn (18.6), the time to failure reduction can be estimated, similarly to the brittle fracture treated by Eqn (18.2).

For a bundle, considering a mean-field approach (similarly to Eqn (18.3)) yields

$$\frac{C_C}{C_C^{(theo)}} = \sum_a f_a \frac{C_N(a)}{C_N^{(theo)}} \qquad (18.7)$$

Better predictions could be derived integrating the quantized Paris' law for a finite width strip. However, we note that the role of the finite width is already included in Eqn (18.6), even if these are rigorously valid in the limit of W tending to infinity.

18.6 ELASTICITY

Consider a nanotube/graphene of lateral surface A under tension and containing a transversal crack of half-length a. Interpreting the incremental compliance, due to the presence of the crack, as a Young's modulus (here denoted by E) degradation, we find $\frac{E(a)}{E^{(theo)}} = 1 - 2\pi \frac{a^2}{A}$ [28]. Thus, recursively, considering Q cracks (in the megacable, 10^{12}–10^{20} defects are expected, see Section 2) having sizes a_i or, equivalently, M different cracks with multiplicity Q_i ($Q = \sum_{i=1}^{M} Q_i$), noting that $n_i = \frac{2a_i}{q}$ represents the number of adjacent vacancies in a crack of half-length a_i, with q atomic size, and $v_i = \frac{Q_i n_i}{A/q^2}$ its related numerical (or volumetric) vacancy fraction, we find [28]

$$\frac{E}{E^{(theo)}} = \prod_{i=1}^{Q} \frac{E(a_i)}{E^{(theo)}} \approx 1 - \xi \sum_{i=1}^{M} v_i n_i \qquad (18.8)$$

with $\xi \geq \pi/2$, where the equality holds for isolated cracks. Equation (18.8) can be applied to nanotubes or graphene and related bundles containing defects in volumetric percentages v_i.

Forcing the interpretation of our formalism, we note that $n_i = 1$ would describe a single vacancy, i.e. a small hole. Thus, as a first approximation, different defect geometries from cracks to circular holes, e.g. elliptical holes, could in principle be treated by Eqn (18.8); we have to interpret n_i as the ratio between the transversal and longitudinal (parallel to the load) defect sizes ($n_i = a_i/b_i$). Introducing the i-th defect eccentricity e_i as the ratio between the lengths of the longer and shorter axes, as a first approximation, $n_i(\beta_i) \approx e_i \cos^2\beta_i + 1/e_i \sin^2\beta_i$, where β_i is the defect orientation. For a single-defect typology, $\frac{E}{E^{(theo)}} \approx 1 - \xi vn$, in contrast to the common assumption $\frac{E}{E^{(theo)}} \approx 1 - v$, rigorously valid only for the cable density, for which $\frac{\rho_C}{\rho_C^{(theo)}} \equiv 1 - v$. Note that the failure strain for a defective nanotube, graphene or related bundle can also be predicted by $\varepsilon_{N,C}/\varepsilon_{N,C}^{(theo)} = (\sigma_{N,C}/\sigma_{N,C}^{(theo)})/(E/E^{(theo)})$.

In contrast to what happens for the strength, large defectiveness is required to have a considerable elastic degradation, even if we have shown that sharp transversal defects could have a role. For example, too-soft space elevator cables would become dynamically unstable [29].

18.7 ATOMISTIC SIMULATIONS

Let us study the influence on the strength of nanocracks and circular nanoholes. n atomic adjacent vacancies perpendicular to the load correspond to a blunt nanocrack of length $2a \approx nq$ and thickness $2b \approx q$ (or $2a \approx nq$ with a radius at tips of $b^2/a \approx q/2$). Similarly, nanoholes of size m can be considered: the index $m = 1$ corresponds to the removal of an entire hexagonal ring, $m = 2$ to the additional removal of the six hexagons around the former one (i.e. the adjacent perimeter of 18 atoms), $m = 3$ to the additional removal of the neighboring 12 hexagonal rings (next adjacent perimeter), and so on (thus, $a = b \approx q(2m-1)/\sqrt{3}$). Quantum mechanics (QM), semiempirical (PM3 method), molecular mechanics (MM; with a modified Tersoff—Brenner potential of second generation (MTB-G2) or a modified Morse potential (M)) and coupled QM/MM calculations [30–34] are reported and extensively compared in Table 18.1 with the QFM nonasymptotic predictions of Eqn (18.2) (differently from the asymptotic comparison reported in Refs [3,12]). The comparison shows a relevant agreement, confirming and demonstrating that just a few vacancies can dramatically reduce the strength of a single nanotube/graphene, or of a nanotube/graphene bundle as described by Eqn (18.3) that predicts for $f \approx 1$, $\sigma_C/\sigma_C^{(\text{theo})} \approx \sigma_N/\sigma_N^{(\text{theo})}$. Assuming large holes ($m \to \infty$) and applying QFM to a defective bundle ($f \approx 1$), we predict $1 - \sigma_C/\sigma_C^{(\text{theo})} \approx 1 - \sigma_N/\sigma_N^{(\text{theo})} \approx 67\%$; but nanocracks surely would be even more critical, especially if interacting with each other or with the boundary. Thus, the expectation for a nanotube/graphene megacable of strength larger than ~ 33 GPa is unlikely.

Note that an elastic ($\kappa \approx 1$), nearly perfectly plastic ($\kappa \approx 0$) behavior with a flow stress at ~ 30–35 GPa for strains larger than ~ 3–5% has been recently observed in tensile tests of carbon nanotubes [35], globally suggesting $\kappa \approx 0.6$–0.7; similarly, numerically computed stress–strain curves [36] reveal for an armchair (5,5) carbon nanotube $\kappa \approx 0.8$, whereas for a zigzag (9,0) nanotube $\kappa \approx 0.7$, suggesting that the plastic correction reported in Section 4 could have a role in nanotubes or graphene.

Regarding elasticity, we note that Eqn (18.8) can be viewed as a generalization of the approach proposed in Ref. [37], being able to quantify the constants k_i fitted by atomistic simulations for three different types of defect [28]. In particular, rearranging Eqn (18.8) and in the limit of three small cracks, we deduce $\frac{E_{\text{th}}}{E} \approx 1 + k_1 c_1 + k_2 c_2 + k_3 c_3$, identical to their law (Eqn (18.15)), in which $c_i = Q_i/L$ is the linear defect concentration in a nanotube of length L and radius R and $k_i = \frac{\xi c_i n_i^2 q^2}{\pi^2 R}$. These authors consider 1, 2 and 3 atoms missing, with and without reconstructed bonds; for nonreconstructed bonds two alternative defect orientations were investigated for 2 and 3 atoms missing. Even if their defect geometries are much more complex than the nanocracks that we consider here, the comparison between our approach and their atomistic simulations, which does not involve best-fit parameters, shows a good agreement [28].

18.8 NANOTENSILE TESTS

The discussed tremendous defect sensitivity, described by Eqn (18.2), is confirmed by a statistical analysis based on Nanoscale Weibull Statistics (NWS) [16] applied to nanotensile tests. According to this treatment, the probability of failure P for a nearly defect-free nanotube/graphene under a tensile stress σ_N is independent from its volume (or surface), in contrast to classical Weibull Statistics [38], namely,

$$P = 1 - \exp - N_N \left(\frac{\sigma_N}{\sigma_0}\right)^w \qquad (18.9)$$

where w is the nanoscale Weibull modulus, σ_0 is the nominal failure stress (i.e. corresponding to a probability of failure of 63%) and $N_N \equiv 1$. In classical Weibull statistics, $N_N \equiv V/V_0$ for volume dominating defects (or $N_N = A/A_0$ for surface dominating defects), i.e. N_N is the ratio between the volume (or surface) of the structure and a reference volume (or surface). The experimental data on carbon nanotubes [7,8] were treated [16] according to nanoscale and classical Weibull statistics: the coefficients of correlation were found to be much higher for the nanoscale statistics than for the classical one (0.93 against 0.67, $w \approx 2.7$ and $\sigma_0 \approx 31-34$ GPa). The data set on MWCNT tensile experiments [39] has also been statistically treated [3]. The very large, highest measured strengths denote interactions between the external and internal walls, as pointed out by the same authors [39] and recently quantified [14]. Thus, the measured strengths cannot be considered plausible for describing the strength of a SWCNT. Such experiments were best-fitted with $\sigma_0 \approx 108$ GPa (but not significant for the strength of a single nanotube) and $w \approx 1.8$ (coefficient of correlation 0.94). In Fig. 18.3, the new data set [40] is treated [4] by applying NWS ($N_N \equiv 1$, $w \approx 2.2$, $\sigma_0 \approx 25$ GPa) and compared with the other nanoscale statistics [3,4] deduced from the other data sets [7,8,39]. Note that volume- or surface-based Weibull statistics are identical in treating the external wall of the tested nanotubes, just an atomic layer thick. We have found a poor coefficient of correlation also treating this new data set with classical Weibull statistics, namely 0.51 (against 0.88 for NWS, see Fig. 18.3).

All these experimental data [7,8,39,40] are treated in Table 18.2, by applying QFM in the form of Eqn (18.2): nonlinear multiple solutions for identifying the defects corresponding to the measured strength clearly emerge; however, these are quantifiable, showing that a small defect is sufficient to rationalize the majority of the observed strong strength reductions.

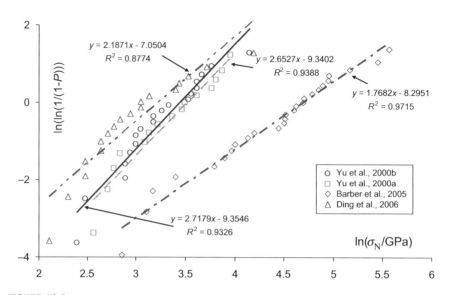

FIGURE 18.3

Nanoscale Weibull statistics, straight lines, applied to nanotensile experiments on carbon nanotubes. From: The role of defects in the design of the space elevator cable: from nanotube to megatube, Acta Materialia 55 2007 [4]. (For color version of this figure, the reader is referred to the online version of this book.)

Finally, the new experimental results [39] are differently treated in Table 18.3, with respect to both strength and elasticity, assuming the presence of transversal nanocracks. The ideal strength is assumed to be of 100 GPa and the theoretical Young's modulus of 1 TPa; by Eqn (18.2), the crack length n is calculated and introduced in Eqn (18.8) to derive the related vacancy fraction v ($\xi = \pi/2$).

18.9 THERMODYNAMIC LIMIT

Defects are thermodynamically unavoidable, especially at the megascale. At the thermal equilibrium, the vacancy fraction $f = n/N \ll 1$ (n is the number of vacancies and N is the total number of atoms) is estimated as follows [41]:

$$f \approx e^{-E_1/(k_B T_a)} \qquad (18.10)$$

where $E_1 \approx 7$ eV is the energy required to remove one carbon atom and T_a is the absolute temperature at which the carbon is assembled, typically in the range between 2000 and 4000 K. Thus, $f \approx 2.4 \times 10^{-18} - 1.6 \times 10^{-9}$. For example for a space elevator megacable, having a carbon weight of $\sim$5000 Kg, the total number of atoms is $N \approx 2.5 \times 10^{29}$, thus a huge number of equilibrium defects in the range $n \approx 0.6 \times 10^{12} - 3.9 \times 10^{20}$ is expected in agreement with a recent discussion [42] and observations [43].

Table 18.2 Experiments vs QFM Predictions; Strength Reduction $\sigma_N(a,b)/\sigma_N^{(theo)}$ Derived According to Eqn (18.2) for carbon nanotubes and graphene. From: The role of defects in the design of the space elevator cable: from nanotube to megatube, Acta Materialia 55 2007 [4]

$\sigma_N/\sigma_N^{(theo)}$	2a/q	2b/q	0	1	2	3	4	5	6	7	8	9	10	∞
0			1.00*	1.00*	1.00*	1.00*	1.00*	1.00*	1.00*	1.00*	1.00*	1.00*	1.00*	1.00
1			0.71*	0.75	0.79	0.82	0.85	0.87*	0.88*	0.90	0.91	0.91	0.92	1.00
2			0.58	0.60*	0.64*	0.68	0.71*	0.73	0.76	0.78*	0.79	0.81	0.82	1.00
3			0.50	0.52	0.54*	0.58	0.61	0.64*	0.66*	0.68	0.70*	0.72	0.74	1.00
4			0.45	0.46	0.48	0.51*	0.54*	0.56	0.59	0.61	0.63	0.65	0.67	1.00
5			0.41	0.42	0.44*	0.46	0.48	0.51*	0.53*	0.55*	0.58	0.59	0.61	1.00
6			0.38	0.38	0.40	0.42	0.44*	0.47	0.49*	0.51*	0.53*	0.55*	0.57	1.00
7			0.35	0.36	0.37	0.39	0.41	0.43	0.45	0.47	0.49*	0.51*	0.53*	1.00
8			0.33	0.34	0.35	0.37	0.38	0.40	0.42	0.44*	0.46	0.48	0.49*	1.00
9			0.32	0.32	0.33	0.34	0.36	0.38	0.40	0.41	0.43	0.45	0.46	1.00
10			0.30*	0.30*	0.31	0.33	0.34	0.36	0.37	0.39	0.41	0.42	0.44*	1.00
11			0.29	0.29	0.30*	0.31	0.32	0.34	0.35	0.37	0.39	0.40	0.42	1.00
12			0.28	0.28	0.29	0.30*	0.31	0.32	0.34	0.35	0.37	0.38	0.40	1.00
13			0.27	0.27	0.28	0.29	0.30*	0.31	0.32	0.34	0.35	0.36	0.38	1.00
14			0.26	0.26	0.27	0.27	0.29	0.30*	0.31	0.32	0.34	0.35	0.36	1.00
15			0.25	0.25	0.26	0.27	0.27	0.29	0.30*	0.31	0.32	0.34	0.35	1.00
16			0.24*	0.24*	0.25	0.26	0.27	0.28	0.29	0.30*	0.31	0.32	0.33	1.00
17			0.24*	0.24*	0.24*	0.25	0.26	0.27	0.28	0.29	0.30*	0.31	0.32	1.00
18			0.23	0.23	0.24*	0.24*	0.25	0.26	0.27	0.28	0.30*	0.30*	0.31	1.00
19			0.22*	0.22*	0.22*	0.23	0.24*	0.25	0.26	0.27	0.28	0.29	0.30*	1.00
20			0.22*	0.22*	0.22*	0.23	0.24*	0.24*	0.25	0.26	0.27	0.28	0.29	1.00
21			0.21	0.21	0.22*	0.22*	0.23	0.24*	0.25	0.25	0.26	0.27	0.28	1.00

(Continued)

Table 18.2 Experiments vs QFM Predictions; Strength Reduction $\sigma_N(a,b)/\sigma_N^{(theo)}$ Derived According to Eqn (18.2) for carbon nanotubes and graphene. From: The role of defects in the design of the space elevator cable: from nanotube to megatube, Acta Materialia 55 2007 [4]—Cont'd

$\sigma_N/\sigma_N^{(theo)}$	2a/q	2b/q	0	1	2	3	4	5	6	7	8	9	10	∞
22			0.21	0.21	0.21	0.22*	0.22*	0.23	0.24*	0.25	0.26	0.27	0.28	1.00
23			0.20	0.21	0.21	0.21	0.22*	0.23	0.23	0.24*	0.25	0.26	0.27	1.00
24			0.20	0.20	0.20	0.21	0.21	0.22*	0.23	0.24*	0.24*	0.25	0.26	1.00
25			0.20	0.20	0.20	0.20	0.21	0.22*	0.22*	0.23	0.24*	0.25	0.26	1.00
26			0.19	0.19	0.20	0.20	0.20	0.21	0.22*	0.22*	0.23	0.24*	0.25	1.00
27			0.19	0.19	0.19	0.20	0.20	0.21	0.21	0.22*	0.23	0.24*	0.24*	1.00
28			0.19	0.19	0.19	0.19	0.20	0.20	0.21	0.22*	0.22*	0.23	0.24*	1.00
29			0.18	0.18	0.19	0.19	0.19	0.20	0.20	0.21	0.22*	0.23	0.23	1.00
30			0.18	0.18	0.18	0.19	0.19	0.19	0.20	0.21	0.21	0.22*	0.23	1.00
31			0.18	0.18	0.18	0.18	0.19	0.19	0.20	0.20	0.21	0.22*	0.22*	1.00
32			0.17*	0.17*	0.18	0.18	0.18	0.19	0.19	0.20	0.21	0.21	0.22*	1.00
33			0.17*	0.17*	0.17*	0.18	0.18	0.19	0.19	0.20	0.20	0.21	0.21	1.00
34			0.17*	0.17*	0.17*	0.17*	0.18	0.19	0.19	0.20	0.20	0.21	0.21	1.00
35			0.17*	0.17*	0.17*	0.17*	0.17*	0.18	0.19	0.19	0.20	0.20	0.21	1.00
36			0.16	0.16	0.17*	0.17*	0.17*	0.18	0.18	0.19	0.19	0.20	0.20	1.00
37			0.16	0.16	0.16	0.17*	0.17*	0.17*	0.18	0.18	0.19	0.19	0.20	1.00
38			0.16	0.16	0.16	0.16	0.17*	0.17*	0.18	0.18	0.19	0.19	0.20	1.00
39			0.16	0.16	0.16	0.16	0.17*	0.17*	0.17*	0.18	0.18	0.19	0.19	1.00

18.9 Thermodynamic Limit

n													
40	**0.16**	**0.16**	**0.16**	**0.16**	**0.16**	**0.16**	0.17*	0.18	0.18	0.19			1.00
41	**0.15**	**0.15**	**0.16**	**0.16**	**0.16**	**0.16**	0.17*	0.17*	0.18	0.18	0.19		1.00
42	**0.15**	**0.15**	**0.15**	**0.16**	**0.16**	**0.16**	0.17*	0.17*	0.18	0.18	0.19		1.00
43	**0.15**	**0.15**	**0.15**	**0.15**	**0.16**	**0.16**	0.16	0.17*	0.17*	0.18	0.18		1.00
44	**0.15**	**0.15**	**0.15**	**0.15**	**0.15**	**0.16**	0.16	0.17*	0.17*	0.18	0.18		1.00
45	**0.15**	**0.15**	**0.15**	**0.15**	**0.15**	**0.16**	0.16	0.16	0.17*	0.17*	0.18		1.00
46	**0.15**	**0.15**	**0.15**	**0.15**	**0.15**	**0.15**	0.16	0.16	0.17*	0.17*	0.18		1.00
47	0.14	**0.15**	**0.15**	**0.15**	**0.15**	**0.15**	0.16	0.16	0.16	0.17*	**0.17***		1.00
48	0.14	0.14	0.14	**0.15**	**0.15**	**0.15**	0.16	0.16	0.16	**0.17***	**0.17***		1.00
49	0.14	0.14	0.14	0.14	**0.15**	**0.15**	0.15	0.16	0.16	**0.17***	**0.17***		1.00
50	0.14	0.14	0.14	0.14	0.14	**0.15**	0.15	0.15	0.16	**0.16**	**0.17***		1.00
∞	0.00	0.00	0.00	0.00	0.00	0.00	0.00	0.00	0.00	0.00	0.00		$(1+2a/b)^{-1}$

In **bold** type are represented the 15 different nanostrengths measured on single-walled carbon nanotubes in bundle [7]; whereas in *italic*, we report the 19 nanostrengths measured on multiwalled carbon nanotubes [8], and in underlined type, the most recent 18 observations [35]. All the data are reported with the exception of the five smallest values of 0.08, 0.10 [35], 0.11 [8], 0.12 [8,35] and 0.13 [7], for which we would need for example adjacent vacancies (2b/q ∼ 1) in number n = 2a/q = 138–176, 90–109, 75–89, 64–74 and 55–63 respectively. The 26 strengths measured in Ref. [39] are also treated (asterisks), simply assuming two interacting walls for 100 < $\sigma_N^{(exp)}$ ≤ 200 Gpa (thus, $\sigma_N = \sigma_N^{(exp)}/2$) or three interacting walls for 200 < $\sigma_N^{(exp)}$ ≤ 300 Gpa ($\sigma_N = \sigma_N^{(exp)}/3$). All the experiments are referred to $\sigma_N^{(theo)} = 100$ GPa (q ∼ 0.25 nm). If all the nanotubes in the cable contain identical holes, $\sigma_C/\sigma_C^{(theo)} = \sigma_N/\sigma_N^{(theo)}$.

Table 18.3 Fracture strength, Young's modulus, non-linear elasticity, crack size and defect content estimations from Modulus nanotensile tests. From: Modulus, fracture strength, and brittle vs plastic response of the outer shell of arc-grown multiwalled carbon nanotubes, Experimental Mechanics 47 2006 [35]

MWCNT Number and Fracture Typology	Strength [GPa]	Young's Modulus [GPa]	κ	n	v (%)
1 (multiple load A)	8.2	1100	1.01	148	0.07
2 (clamp failed)	10	840	0.98	100	0.23
3	12	680	1.00	69	0.44
4 (failure at the clamp)	12	730	0.98	69	0.40
5 (multiple load B)	14	1150	1.02	51	0.14
6 (multiple load a)	14	650	0.97	51	0.62
7	15	1200	1.05	44	0.11
8	16	1200	1.02	39	0.13
9	17	960	1.00	34	0.49
10	19	890	0.97	27	0.74
11 (multiple load b)	21	620	0.99	22	1.51
12 (multiple load I)	21	1200	0.99	22	0.22
13 (multiple load II)	23	1250	0.99	18	0.17
14	30	870	1.00	11	1.92
15 (plasticity observed)	31	1200	0.59 (0.99)	10	0.49
16 (plasticity observed)	34	680	0.69 (1.02)	8	3.80
17 (multiple load III)	41	1230	1.03	5	0.69
18 (failure at the clamp)	66	1100	0.98	2	4.90

The new results [35] are here treated with respect to both strength and elasticity, assuming the presence of transversal nanocracks composed by n adjacent vacancies [4]. The constitutive parameter κ has been estimated as $\kappa \approx \ln(\varepsilon_N)/\ln(\sigma_N/E)$ for all the tests: note the low values for the two nanotubes that revealed plasticity (in brackets, the values calculated up to the incipient plastic flow are also reported). The ideal strength is assumed to be of 100 GPa and the theoretical Young's modulus of 1300 GPa; by Eqn (18.2), the crack length n is calculated and introduced in Eqn (18.8) to derive the related vacancy fraction v ($\xi = \pi/2$). Fracture in two cases was observed at the clamp; in one case, the clamp itself failed, thus the deduced strength represents a lower bound of the nanotube strength. Three nanotubes were multiple loaded (in two a,b and A,B or in three I,II, and III steps), i.e. after the breaking in two pieces of a nanotube, one of the two pieces was again tested and fractured at a higher stress. Two nanotubes displayed a plastic flow. A vacancy fraction of the order of few ‰ is estimated, suggesting that such nanotubes are much more defective than as imposed by the thermodynamic equilibrium, even if the defects are small and isolated. However, note that other interpretations are still possible, e.g. assuming the nanotubes coated by an oxide layer and rationalizing the ratio between the observed Young's modulus and its theoretical value as the volumetric fraction (for softer coating layers) of carbon in the composite structure.

The strength of the cable will be dictated by the largest transversal crack on it, according to the weakest link concept. The probability of finding a nanocrack of size m in a bundle with vacancy fraction f is $P(m) = (1-f)f^m$, and thus, the number M of such nanocracks in a bundle composed by N atoms is $M(m) = P(m)N$. The size of the largest nanocrack, which typically occurs once, is found from the solution to the equation $M(m) \approx 1$, which implies [44]:

$$m \approx -\ln[(1-f)N]/\ln f \approx -\ln N/\ln f \qquad (18.11)$$

For example, we deduce a size $m \approx 2-4$ for the largest thermodynamically unavoidable defect in the considered megacable. Inserting Eqns (18.11) and (18.10) into Eqn (18.2) evaluated for a transversal crack ($b \approx 0$ and $2a/q \approx m$), we deduce the statistical counterpart of Eqn (18.2) and thus, the following thermodynamical maximum achievable strength:

$$\frac{\sigma_N(N)}{\sigma_N^{(theo)}} \leq \frac{\sigma_N^{(max)}(N)}{\sigma_N^{(theo)}} = \frac{1}{\sqrt{1 + \frac{k_B T_a}{E_1} \ln N}} \qquad (18.12)$$

Then, inserting Eqn (18.12) into Eqns (18.3) and (18.4), the maximum cable strength and minimum taper ratio can be statistically deduced. The corresponding maximum achievable strength, at unavoidable limit (at least at the thermodynamic equilibrium), is ~45 GPa and the corresponding flaw-tolerant taper ratio is ~4.6. But the larger taper ratio implies a large cable mass and thus a large number N of atoms. Updating N in our statistical calculation yields the same, thus self-consistent, predictions. Statistically, we expect an even smaller strength, as previously discussed.

18.10 SLIDING FAILURE

The fracture mechanics approach could be of interest to evaluate the strength of nanotube/graphene bundles assuming a sliding failure mode [45]. This hypothesis is complementary to that of intrinsic nanotube/graphene fracture, already treated in the previous sessions. Thus we assume the interactions between adjacent nanotubes/graphene as the weakest links, i.e. that the fracture of the bundle composite is caused by nanotube sliding rather than by their intrinsic fracture.

Accordingly, the energy balance during a longitudinal delamination (here "delamination" has the meaning of Mode II crack propagation at the interface between adjacent nanotubes/graphene ribbons) dz under the applied force F is

$$d\Phi - F du - 2\gamma(P_C + P_{vdW})dz = 0 \qquad (18.13)$$

where dΦ and du are the strain energy and elastic displacement variation, respectively, due to the infinitesimal increment in the compliance caused by the delamination dz; P_{vdW} describes the still existing van der Waals attraction (e.g. attractive part of the Lennard-Jones potential) for vanishing nominal contact nanotube perimeter $P_C = 6a$ (the shear force between two nanotube/graphene layers becomes zero for nominally negative contact area); γ is the surface energy of the interface.

Elasticity poses $\frac{d\Phi}{dz} = -\frac{F^2}{2ES}$, where S is the cross-sectional surface area of the nanotube, whereas according to Clapeyron's theorem $Fdu = 2d\Phi$. Thus, the following simple expression for the bundle composite strength ($\sigma_C = F_C/S$, effective stress and cross-sectional surface area are here considered; F_C is the force at fracture) is predicted:

$$\sigma_C^{(\text{theo})} = 2\sqrt{E\gamma \frac{P}{S}} \qquad (18.14)$$

in which it appears as the ratio between the effective perimeter ($P = P_C + P_{vdW}$) in contact and the cross-sectional surface area of the nanotubes.

Equation (18.14) can be considered valid also for the entire bundle composite, since we are assuming here the same value P/S for all the nanotubes/graphene ribbons; the fact that the strength is not a function of the numbers or bundle composite size is for the same reason, i.e. because we are not assuming here a "defect" size distribution for S/P—that basically represents a characteristic defect size—but a constant value; of course, assuming a statistical distribution for the characteristic defect size S/P with the upper limit proportional to the structural size (the larger the structure the larger the largest defect) would imply a size effect, thus a dependence on the nanotube/graphene numbers or bundle composite size. Nevertheless, here we are interested in the simplest model (1) and in the upperbound strength predictions (2), thus we do not consider statistics into Eqn (18.14).

Note that Eqn (18.14) is basically the asymptotic elastic limit for sufficiently long overlapping length, that is, the length along with two adjacent nanotubes/graphene ribbons are nominally in contact; for overlapping length smaller than a critical value, the strength increases by increasing the overlapping length; for a single nanotube/graphene, this overlapping length is of the order of 10 μm, whereas it is expected to be larger for bundles/composites, (e.g. of the order of several millimeters, as confirmed experimentally). This critical length is $\ell_C \approx 6\sqrt{\frac{hES}{PG}}$ where h and G are the thickness and shear modulus of the interface, respectively. It suggests that increasing the size scale $L \propto \sqrt{S} \propto P \propto h$, this critical length increases too, namely $\ell \propto L$, thus the strength increases by increasing the overlapping length in a wider range; however, note that the achievable strength is reduced since, $\sigma_C^{(\text{theo})} \propto \sqrt{h}\ell^{-1} \propto \sqrt{P/S} \propto L^{-1/2}$, if $L \propto \ell \propto h$: increasing the overlapping length ad infinitum is not a way to indefinitely increase the strength. The real strength could be significantly smaller, not only because $\ell < \ell_C$ but also as a consequence of the misalignment of the nanotubes/graphene ribbons with respect to the bundle axis.

Assuming a nonperfect alignment in the bundle, described by a non-zero angle β (here assumed identical for all the nanotubes/graphene ribbons, even if—also in this case, as for the characteristic defect size S/P—a proper statistics could be invoked for this parameter), the longitudinal force carried by the nanotubes/graphene ribbons will be $F/\cos\beta$, thus the equivalent Young' modulus of the bundle/composite will be $E\cos^2\beta$, as can be evinced by the corresponding modification of the energy balance during delamination; accordingly,

$$\sigma_C = 2\cos\beta\sqrt{E\gamma\frac{P}{S}} \tag{18.15}$$

Applying eq. 18.15 to nanotubes, The maximal achievable strength is predicted for collapsed [45] perfectly aligned (sufficiently overlapped) nanotubes in a bundle or nanotubes even not collapsed in a composite, i.e. $\frac{P}{S} \approx \frac{1}{Nt}$, $\beta = 0$:

$$\sigma_C^{(\text{theo},N)} = 2\sqrt{\frac{E\gamma}{Nt}} \tag{18.16}$$

Taking $E = 1$ TPa (Young's modulus of graphene), $\gamma = 0.2$ N/m (surface energy of graphene; however, note that in reality, γ could be also larger as a consequence of additional dissipative mechanisms, e.g. fracture and friction in addition to adhesion, or presence of a matrix in a nanotube/graphene composites), the predicted maximum strength for single walled nanotube ($N=1$) fiber/composite is $\sigma_C^{(\max,\text{CNT})} = \sigma_C^{(\text{theo},1)} = 48.5$ GPa, whereas for double- or triple-walled nanotubes, $\sigma_C^{(\text{theo},2)} = 34.3$ GPa or $\sigma_C^{(\text{theo},3)} = 28.0$ GPa. For graphene, the surface area for load transfer is doubled with respect to the case of the single carbon nanotube (the inner surface area does not contribute) and thus from eqn (18.15) we predict [46],

$$\sigma_C^{(\max,G)} = \sqrt{2}\sigma_C^{(\max,\text{CNT})} \tag{18.17}$$

This result is important and suggests that graphene is superior even to nanotubes for designing superstrong fibers and composites. These maximum stress predictions assume a vanishing matrix content in the fibers/composites; in general we expect:

$$\sigma_C^{(\text{Composite})} \cong \sigma_C^{(G,\text{CNT})}f + \sigma_C^{(\text{matrix})}(r - P)$$

where P is the volumetric content of graphene or carbon nanotubes in the composite.

18.11 CONCLUSIONS

The strength of a real, thus defective, carbon nanotube or graphene superstrong fibers is expected to be greatly reduced with respect to the theoretical strength of carbon nanotubes or graphene. Accordingly, in this chapter, key simple formulas for the design of superstrong carbon nanotube and graphene fibers and composites have been reviewed based on the analysis reported in previous papers by the same author. Graphene has been demonstrated to be superior even with respect to carbon nanotubes.

References

[1] Y.V. Artsutanov, Kosmos na elektrovoze, komsomol-skaya pravda. July 31 (1960); contents described in Lvov, V, Science 158 (1967) 946−947.

[2] J. Pearson, The orbital tower: a spacecraft launcher using the Earth's rotational energy, Acta Astronautica 2 (1975) 785−799.

[3] N. Pugno, On the strength of the nanotube-based space elevator cable: from nanomechanics to megamechanics, Journal of Physics—Condensed Matter 18 (2006) S1971−S1990.

[4] N. Pugno, The role of defects in the design of the space elevator cable: from nanotube to megatube, Acta Materialia 55 (2007) 5269−5279.

[5] N. Pugno, Space elevator: out of order? Nano Today 2 (2007) 44−47.

[6] S. Iijima, Helical microtubules of graphitic carbon, Nature 354 (1991) 56−58.

[7] M.F. Yu, B.S. Files, S. Arepalli, R. Ruoff, Tensile loading of ropes of single wall carbon nanotubes and their mechanical properties, Physical Review Letters 84 (2000) 5552−5555.

[8] M.F. Yu, O. Lourie, M.J. Dyer, K. Moloni, T.F. Kelly, R. Ruoff., Strength and breaking mechanism of multiwalled carbon nanotubes under tensile load, Science 287 (2000) 637−640.

[9] B.C. Edwards, Design and deployment of a space elevator, Acta Astronautica 10 (2000) 735−744.

[10] B.C. Edwards, E.A. Westling, The Space Elevator: A Revolutionary Earth-to-Space Transportation System, Spageo Inc, 2003.

[11] N. Pugno, A Quantized Griffith's Criterion, Fracture Nanomechanics, Meeting of the Italian Group of Fracture, Vigevano, Italy, September 25−26, 2002.

[12] N. Pugno, R. Ruoff, Quantized fracture mechanics, Philosophical Magazine 84 (2004) 2829−2845.

[13] N. Pugno, Dynamic quantized fracture mechanics, International Journal of Fracture 140 (2006) 158−168.

[14] N. Pugno, New quantized failure criteria: application to nanotubes and nanowires, International Journal of Fracture 141 (2006) 311−328.

[15] N. Pugno, F. Bosia, A. Carpinteri, Multiscale stochastic simulations as in-silico tensile testing of nanotube-based megacables, Small 4 (2008) 1044-1052.

[16] N. Pugno, R. Ruoff, Nanoscale Weibull statistics, Journal of Applied Physics 99 (2006) 1−4.

[17] A. Carpinteri, Scaling laws and renormalization groups for strength and toughness of disordered materials, International Journal of Solid and Structures 31 (1994) 291−302.

[18] A. Carpinteri, N. Pugno, Are the scaling laws on strength of solids related to mechanics or to geometry? Nature Materials 4 (2005) 421−423.

[19] N. Pugno, A general shape/size-effect law for nanoindentation, Acta Materialia 55 (2007) 1947−1953.

[20] M. Ippolito, A. Mattoni, L. Colombo, N. Pugno, The role of lattice discreteness on brittle fracture: how to reconcile atomistic simulations to continuum mechanics, Physical Review B 73 (2006) 104111−1/6.

[21] D. Taylor, P. Cornetti, N. Pugno, The fracture mechanics of finite crack extensions, Engineering Fracture Mechanics 72 (2005) 1021−1028.

[22] N. Pugno, M. Ciavarella, P. Cornetti, A. Carpinteri, A unified law for fatigue crack growth, Journal of the Mechanics and Physics of Solids 54 (2006) 1333–1349.

[23] N. Pugno, P. Cornetti, A. Carpinteri, New unified laws in fatigue: from the Wöhler's to the Paris' regime, Engineering Fracture Mechanics 74 (2007) 595–601.

[24] J.R. Rice, G.F. Rosengren, Plane strain deformation near a crack tip in a power-law hardening material, Journal of the Mechanics and Physics of Solids 16 (1968) 1–12.

[25] A. Carpinteri, N. Pugno, Fracture instability and limit strength condition in structures with re-entrant corners, Engineering Fracture Mechanics 72 (2005) 1254–1267.

[26] A. Carpinteri, B. Chiaia, Crack-resistance behavior as a consequence of self-similar fracture topologies, International Journal of Fracture 76 (1996) 327–340.

[27] Q.Z. Wang, Simple formulae for the stress-concentration factor for two- and three-dimensional holes in finite domains, Journal of Strain Analysis 73 (2002) 259–264.

[28] N. Pugno, Young's modulus reduction of defective nanotubes, Applied Physics Letters 90 (2007) 043106.

[29] N. Pugno, H. Troger, A. Steindl, M. Schwarzbart, On the stability of the track of the space elevator, Proc. of the 57th International Astronautical Congress, October 2–6, 2007b, (Valencia, Spain).

[30] S.L. Mielke, D. Troya, S. Zhang, J.-L. Li, S. Xiao, R. Car, R.S. Ruoff, G.C. Schatz, T. Belytschko, The role of vacancy defects and holes in the fracture of carbon nanotubes, Chemical Physics Letters 390 (2004) 413–420.

[31] N. Pugno. Mimicking nacre with super-nanotubes for producing optimized super-composites. Nanotechnology 17 (2006) 5480-5484.

[32] T. Belytschko, S.P. Xiao, R. Ruoff, Effects of Defects on the Strength of Nanotubes: Experimental-Computational Comparisons, Los Alamos National Laboratory, 2002. Preprint Archive, Physics, arXiv:physics/0205090.

[33] S. Zhang, S.L. Mielke, R. Khare, D. Troya, R.S. Ruoff, G.C. Schatz, T. Belytschko, Mechanics of defects in carbon nanotubes: atomistic and multiscale simulations, Physical Review B 71 (2005) 115403 1–12.

[34] R. Khare, S.L. Mielke, J.T. Paci, S. Zhang, R. Ballarini, G.C. Schatz, T. Belytschko, Coupled quantum mechanical/molecular mechanical modelling of the fracture of defective carbon nanotubes and grapheme sheets, Physical Review B 75 (2007) 075412.

[35] W. Ding, L. Calabri, K.M. Kohlhaas, X. Chen, D.A. Dikin, R.S. Ruoff, Modulus, fracture strength, and brittle vs plastic response of the outer shell of arc-grown multiwalled carbon nanotubes, Experimental Mechanics 47 (2006) 25–36.

[36] M. Meo, M. Rossi, Tensile failure prediction of single wall carbon nanotube, Engineering Fracture Mechanics 73 (2006) 2589–2599.

[37] M. Sammalkorpi, A. Krasheninnikov, A. Kuronen, K. Nordlund, K. Kaski, Mechanical properties of carbon nanotubes with vacancies and related defects, Physical Review B 70 (2004) 245416–245421/8.

[38] W. Weibull, A statistical theory of the strength of materials, Ingeniörsvetens kapsakademiens, Handlingar 151 (1939).

[39] A.H. Barber, I. Kaplan-Ashiri, S.R. Cohen, R. Tenne, H.D. Wagner, Stochastic strength of nanotubes: an appraisal of available data, Composite Science and Technology 65 (2005) 2380–2386.

[40] I. Kaplan-Ashiri, S.R. Cohen, K. Gartsman, V. Ivanovskaya, T. Heine, G. Seifert, I. Wiesel, H.D. Wagner, R. Tenne, On the mechanical behavior of WS_2 nanotubes under axial tension and compression, Proceeding of the National Academy of Science USA 103 (2006) 523–528.

[41] C. Kittel, Introduction to Solid State Physics, John Wiley & Sons, New York, 1966.

[42] H.K.D.H. Bhadeshia, 52nd Hatfield memorial lecture—large chunks of very strong steel, Materials Science and Technology 21 (2005) 1293–1302.

[43] Y. Fan, B.R. Goldsmith, P.G. Collins, Identifying and counting point defects in carbon nanotubes, Nature Materials 4 (2005) 906–911.

[44] P.D. Beale, D.J. Srolovitz, Elastic fracture in random materials, Physical Review B 37 (1988) 5500–5507.

[45] N. Pugno, The design of self-collapsed super-strong nanotube bundles, Journal of the Mechanics and Physics of Solids 58 (2010) 1397–1410.

[46] N. Pugno, Towards the Artsutanov's dream of the space elevator: the ultimate design of a 35 GPa stronger tether thanks to graphene, Acta Astronautica (2012). (Available on line).

CHAPTER 19

Transition from Tubes to Sheets—A Comparison of the Properties and Applications of Carbon Nanotubes and Graphene

Xiaogan Liang

Department of Mechanical Engineering, University of Michigan, Ann Arbor, MI, USA

CHAPTER OUTLINE

19.1 Overview	520
19.2 Electronic Band Structures of Monolayer Graphene and Carbon Nanotubes	520
19.2.1 Electronic band structure of monolayer graphene	520
19.2.2 Band structure of SWNTs derived from graphenes	522
19.2.3 Band structures of graphene nanoribbons	524
19.3 Comparison of Physical Properties and Device Applications between Graphenes and Carbon Nanotubes	526
19.3.1 Electrical properties	527
19.3.1.1 Mobility characteristics and high-frequency electronic applications	527
19.3.1.2 Bandgap engineering and ON/OFF switching characteristics	532
19.3.2 Mechanical properties	541
19.3.2.1 Elastic properties	541
19.3.2.2 Applications in bendable electronics and photovoltaics	543
19.3.3 Optical and optoelectronic properties	547
19.3.3.1 Linear optical response in terahertz and infrared spectral ranges	547
19.3.3.2 Active optoelectronic materials	553
19.3.3.3 Plasmon and electromagnetic excitations	555
19.4 Summary	556
References	558

19.1 OVERVIEW

Although the discovery of carbon nanotubes (CNTs) occurred more than 10 years earlier than the first exfoliation of a monolayer of graphene in experiments [1,2], for a long time, graphene has been regarded as the most basic building block for graphitic materials of all dimensionalities, including fullerenes [3], carbon fibers [4], and graphite [5,6]. Therefore, it is naturally easy to discuss molecular structure, electronic band structure, and physical properties of CNTs in terms of its counterpart graphene. In particular, a single-walled carbon nanotube (SWNT) can be conceptually formed by cutting and wrapping a graphene sheet into a seamless cylinder along specific and discrete "chiral" angles, and capping the two ends of the tube with hemispheres of the buckyball structure [7]. The combination of the rolling angle and radius of this piece of graphene determines the final electronic band structures as well as physical properties of the SWNT; for example, whether the nanotube shell is a metal or semiconductor [7–11]. Therefore, the band structures and physical properties of graphene and CNTs are intrinsically correlated with each other but exhibit significant differences due to variation in morphology. The understanding of such relationships and differences is important for the practical application of carbon-based nanomaterials and nanostructures.

In this chapter, we will first summarize the basic characteristics of electronic band structures of two-dimensional (2D) graphene sheets and SWNTs. Afterward, we will review the comparison of electrical, mechanical, and optical properties as well as device applications between graphene sheets and nanotubes. Here, many examples of device prototype demonstrations and applications are cited from the most recently published works (e.g. works since 2010). Therefore, the content presented in this chapter is expected to serve as an introductory level guideline for researchers and engineers to choose appropriate materials for specific new applications as well as a brief overview of current progress of the research field related to graphene and nanotube-based functional devices.

19.2 ELECTRONIC BAND STRUCTURES OF MONOLAYER GRAPHENE AND CARBON NANOTUBES

The energy band structures of monolayer graphene and CNTs in combination with their dimensionalities are responsible for their unique electronic properties. This section summarizes the essential quantum mechanics associated with the formation of electronic band structures of graphene and CNTs.

19.2.1 Electronic band structure of monolayer graphene

The lattice structure of graphene consists of a hexagonal arrangement of carbon atoms as shown in Fig. 19.1(a). Three atomic orbitals originated from carbon atoms, 2s, $2p_x$, and $2p_y$, are hybridized into three sp^2 orbitals in the base plane of the

19.2 Electronic Band Structures of Monolayer Graphene and Carbon Nanotubes

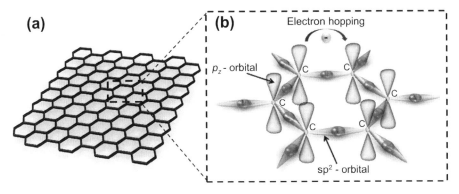

FIGURE 19.1

(a) Lattice structure of graphene. (b) sp^2 hybridization of carbon atoms to form the 2D crystal structure of graphene as well as delocalized π orbitals.

graphene sheet, leaving the $2p_z$ orbital perpendicular to the base plane, as illustrated in Fig. 19.2(b). The hybridized sp^2 orbitals are responsible for covalent σ bonds between the adjacent carbon atoms that are responsible for the robust 2D crystal structure of graphene, while the $2p_z$ orbital out of the plane forms π bonds. The energy levels of π orbitals in the graphene lie near the Fermi energy level and therefore provide delocalized states for transporting electrons, which are responsible for the electrical conductivity of graphene [11].

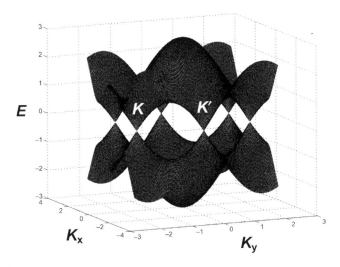

FIGURE 19.2

Energy band structure of a monolayer graphene sheet plotted within the first Brillouin zone.

The electronic bands of a monolayer graphene are derived from delocalized π orbitals, as discussed above. The specific band structure can be approximately calculated by using tight-binding models [11,12]. The detailed theoretical discussion and derivation can be found in previous publications [11,12]. Here, we review and summarize the main conclusions. The energy dispersion (or the energy spectrum) in a monolayer graphene can be calculated based on tight-binding models and approximately expressed by Eqn (19.1), where $\mathbf{k}_x$, $\mathbf{k}_y$, and $\mathbf{k}_z$ are the wave vector components; γ_0 is the tight-binding integral that measures the strength of exchange interaction between nearest-neighbor carbon atoms [11,12].

$$E = E_0 \pm \gamma_0 \sqrt{1 + 4\cos\left(\frac{\sqrt{3}\mathbf{k}_x a}{2}\right)\cos\left(\frac{\mathbf{k}_y a}{2}\right) + 4\cos^2\left(\frac{\mathbf{k}_y a}{2}\right)} \quad (19.1)$$

In Eqn (19.1), the negative sign branch represents the valence bands of graphene originated from bonding π orbitals, and the positive sign branch represents the conduction bands originated from antibonding π^* orbitals [11]. The energy band structure described by Eqn (19.1) is illustrated in Fig. 19.2 [12]. Graphene's unique physical properties are mainly attributed to this distinctive band electronic spectrum. As illustrated in Fig. 19.2, the band structure of a graphene monolayer features conical shape valleys located at the corners of the Brillouin zone, which are categorized into two sets of inequivalent Dirac points (K and K') within the Brillouin zone [12]. Intervalley scattering between K and K' needs a large momentum transfer (on the order of the inverse lattice spacing of graphene) and hence is strongly suppressed [13,14]. At the low energy limit (i.e. the states near the Dirac point), graphene has a linear energy spectrum [15]. This indicates that the electrons in such honeycomb lattice completely lose their effective mass, resulting in massless Dirac fermions [16,17], which are described by a Dirac-like equation $H = v_F \sigma \cdot p$ with Fermi velocity $v_F \sim 10^6$ m/s and a 2D pseudospin matrix σ describing the contribution from two nonequivalent sublattices in graphene (here p is the momentum operator) [15]. Owning to such underlying physics, graphene exhibits exotic electronic properties such as a pronounced ambipolar effect, high carrier mobility (up to 20,000 cm^2/V s), even at a high carrier concentration ($n > 10^{12}$/cm^2), a stable 2D crystal structure, and potential to realize ballistic transport at room temperature (the phase coherent length L is $\sim$300 nm at 300 K) [15].

19.2.2 Band structure of SWNTs derived from graphenes

An SWNT is indeed a rolled-up cylinder of a monolayer graphene sheet. The rolling fashion is uniquely designated by using a chiral vector $\vec{C} = n_1\vec{a_1} + n_2\vec{a_2}$, as illustrated in Fig. 19.3 [11]. An SWNT can be formed by wrapping a graphene sheet in a way that the carbon atoms located at two ends of $\vec{C}$ are in connection with each other [11]. Here, the numbers n_1 and n_2 are wrapping indices and they can be used to uniquely define and denote various SWNTs with different chirality [11]. For

19.2 Electronic Band Structures of Monolayer Graphene and Carbon Nanotubes

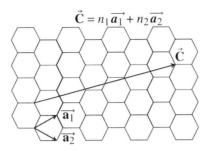

FIGURE 19.3

SWNT defined by a chiral vector $\vec{C} = n_1 \vec{a_1} + n_2 \vec{a_2}$.

example, indices (n, n) and $(n, 0)$ are referred to armchair and zigzag SWNTs, respectively [11]. The absolute values of n_1 and n_2 also determine the ultimate diameters of SWNTs [11].

The electronic band structure of an SWNT with a seamless hollow cylinder structure can be calculated from that of 2D graphene sheets by taking into account the boundary condition expressed in Eqn (19.2), where $\mathbf{k}_\perp$ is the wave vector along the circumference direction of a tube and m is an integer [11]. This boundary condition can be regarded as a periodic condition for the wave function of electrons moving along an SWNT. It means that the original wave function derived from the 2D graphene model should repeat itself as it encircles around a tube and leads to quantized values of $\mathbf{k}_\perp$ for the conductive states in the tube [11]. Based on this scenario, the specific band structure of SWNTs can be deduced by cutting the cross-sectional profiles of the energy spectrum of 2D graphene along a set of discrete $\mathbf{k}_\perp$ values given by the periodic boundary condition (Eqn (19.2)), as illustrated in Fig. 19.4, which is called the *zone-folding scheme* [11,18]. Each cross-sectional cutting generates a one-dimensional (1D) subband, as exemplified in Fig. 19.4(b) and (c). The specific band structures of various SWNTs are therefore determined by the spacing and direction angles of discrete $\mathbf{k}_\perp$ states in the Brillouin zone (i.e. momentum space), which are dependent on the chirality of SWNTs (or wrapping indices) [11].

$$\mathbf{k}_\perp \cdot \mathbf{C} = 2\pi m \quad (19.2)$$

The transport properties of SWNTs, for example, whether they are semiconducting or metallic materials, are mainly determined by the delocalized electronic states near the Fermi level that is related to the allowed $\mathbf{k}_\perp$ states in the proximity of K (or K′) points [11,19]. In particular, if the allowed $\mathbf{k}_\perp$ states exactly pass through K and K′ points as illustrated in Fig. 19.4(b), two linear bands crossing at the Fermi level are created and SWNTs are metallic without a bandgap (such a linear energy spectrum is similar to that of 2D graphene). However, if the allowed $\mathbf{k}_\perp$ states have a misalignment with the K points as illustrated in Fig. 19.4(c), two parabolic bands with a nonzero bandgap are formed in SWNTs and the tube exhibits semiconducting transport properties (e.g., a large modulation range of the tube conductivity). Based on such

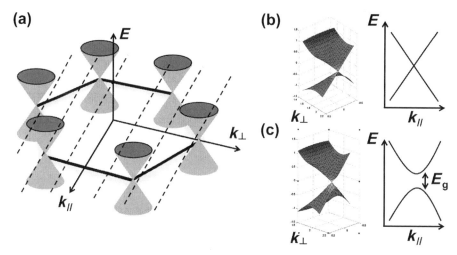

FIGURE 19.4

(a) Illustration of the zone-folding scheme to obtain 1D subbands of SWNTs based on the 2D energy dispersion of graphene. 1D energy dispersions of (b) a metallic CNT and (c) a semiconducting SWNT.

Source: Adapted from Ref. [18].

a zone-folding scheme near the K points, all SWNTs with wrapping indices (n_1, n_2) can be classified into three different species by using an important index p, as listed below [20]. Here, p is defined as the remainder when the difference between n_1 and n_2 is divided by 3 (i.e. $3q + p = |n_1 - n_2|$). Readers can find more detailed derivation of this conclusion in other relevant literature [11,18,20–22].

(a) If $p = 0$, SWNTs are metallic with linear 1D subbands crossing at the K points. Therefore, armchair SWNTs with wrapping indices of (n, n) are always metallic tubes for any integer number n.
(b) If $p = 1$ or 2, SWNTs are semiconducting with a nonzero bandgap that can be estimated by using the empirical equation $E_g \sim 0.7/d$ (nm). Here, d is the diameter of SWNTs [20,21].

19.2.3 Band structures of graphene nanoribbons

Recently, a great effort has been invested to produce graphene nanoribbons (GNRs) that are another group of 1D (or quasi-1D) nanostructures derived from 2D graphene sheets, as illustrated in Fig. 19.5. GNRs have been experimentally demonstrated to be able to open an energy bandgap and hence enable semiconductor-related applications [23–28]. For a GNR, the formation of an energy bandgap E_g is attributed to the quantum confinement [24]. Given the first-order approximation without considering the types of the edge termination and crystallinity, E_g inversely scales with the ribbon width w (i.e. $E_g = \alpha/w$), where α is the fitting parameter and measured to

19.2 Electronic Band Structures of Monolayer Graphene and Carbon Nanotubes

FIGURE 19.5

Graphene nanoribbons, another group of 1D nanostructures derived from 2D graphene.

be 0.8−1.0 nm eV [24,25,27]. For a field effect transistor (FET) made from a GNR (Fig. 19.6), the gate-modulated ON/OFF current ratio exponentially scales with the bandgap (i.e. $I_{ON}/I_{OFF} \propto \exp(E_g/kT)$) [29]. Therefore, the increase of bandgap can significantly improve the ON/OFF current ratio (I_{ON}/I_{OFF}), which is the basis for constructing graphene-based integrated circuits (ICs). To produce a bandgap in GNRs that is comparable to the bandgaps of conventional semiconductors (e.g. Si), a sub-10 nm ribbon width is required [25,27].

For nanostructures of graphene and other emerging 2D crystals, the edge morphology (i.e. roughness and crystallographic orientation) plays an important role in determining their transport properties. The theoretical works have indicated that GNRs with different types of edge terminations (zigzag or armchair) have distinct band structures as shown in Fig. 19.7 [12]. Readers are referred to previous literature for a more detailed derivation of the band structures of GNRs with specific edge

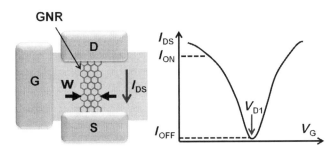

FIGURE 19.6

Schematic of a field-effect transistor made from a single graphene nanoribbon and sketch of its characteristic transport curve (I_{ds}−V_g). Bandgap of GNRs: $E_g = \alpha/w$ ($\alpha \sim 0.8–1$ nm eV); ON/OFF current ratio of the FET: $I_{ON}/I_{OFF} \propto \exp(E_g/kT)$.

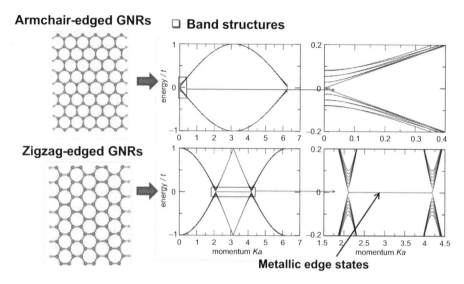

FIGURE 19.7

Graphene nanoribbons (GNRs) with different edge morphologies, resulting in distinct band structures.

Source: Reprinted figure with permission from Ref. [12]. Copyright (2009) by the American Physical Society. http://rmp.aps.org/abstract/RMP/v81/i1/p109_1

terminations [12]. In short, the 1D subbands of a GNR can be derived from electronic states of 2D graphene by imposing a boundary condition different from the periodic one for SWNTs [12]. Figure 19.7(a) shows that armchair-edged GNRs have parabolic 1D subbands with a nonzero bandgap near the Fermi level and thus may have potential applications for semiconductor-related applications (e.g. switch transistors) [12]. Figure 19.7(b) shows that zigzag-edged GNRs do not have a bandgap but always have metallic edge states near the Fermi level, which results in metallic transport characters. Such metallic edge states have become a very interesting and important topic in the areas of graphene-based nanoelectronics, quantum mechanics, and condensed matter physics [30–33]. Especially, zigzag-edged GNRs with metallic edge states have been theoretically and experimentally studied as quantum transport channels for generating and controlling "spin"- and "valley"-polarized currents in the ballistic regime for future low-dissipation electronics [12,34,35].

19.3 COMPARISON OF PHYSICAL PROPERTIES AND DEVICE APPLICATIONS BETWEEN GRAPHENES AND CARBON NANOTUBES

Graphene and SWNTs have relevant but significantly different physical properties due to the variation in the molecular structures of sp^2 carbon atoms and morphology of quantum confinement. In this section, we systematically compare the electrical, mechanical, and optical properties of these two carbon-based nanomaterials. Here, the discussion concentrates on the properties and characters that have been experimentally investigated and are directly associated with practical and well-identified applications. Readers who are interested in theoretically proposed characters and newly identified properties of graphene and CNTs for potential future applications are referred to the literature with more specific focus on such topics [15].

19.3.1 Electrical properties

19.3.1.1 Mobility characteristics and high-frequency electronic applications

The electron mobility characterizes how quickly an electron (or charged carrier) can move through a solid material (e.g. metals or semiconductors), when pulled by an electric field [29]. In semiconductors, there is an analogous mobility for holes called the *hole mobility*. Generally, the term *carrier mobility* refers to both electron and hole mobilities in semiconductors and semimetals. The carrier mobility is an important parameter and determines the operation speed and high-frequency response character of electronic devices (e.g. switch transistors, logic gates, and demodulators) made from solid materials [29].

As shown in Fig. 19.2, graphene has a linear energy dispersion near the Fermi level. One of the direct consequences of such a linear energy spectrum is that the charged carriers in graphene behavior like Dirac fermions with extremely high electrical mobility [36]. Experimentally measured values of the carrier mobility in graphene strongly depend on the supporting substrates that bear the graphene films and other medium materials [37—41]. Table 19.1 lists experimentally measured electron and hole mobilities of graphenes coated on various dielectric substrates, suspended graphene, and graphene epitaxially grown on SiC substrates [37—41]. Here, the suspended graphene and turbostratic graphene grown on carbon-terminated SiC substrates exhibit exceptionally high carrier mobilities ($\mu > 200{,}000$ cm^2/V s at room temperature) [36,42—45]. This feature is attributed to the weak electronic coupling between the measured graphene layer and the substrate (or other medium materials) [36,42—45].

Although suspended graphene films exhibit high carrier mobility, such a suspending geometry imposes severe limitations on device architecture and practical applications. Therefore, there is a growing need to find suitable dielectric substrates for hosting graphene devices with a substrate-supported geometry. Thermally grown SiO$_2$ has been widely used as convenient substrates in the fundamental research to support graphene devices, which is simply adopted from conventional silicon-based technology and process. However, as shown in Table 19.1, the graphene features or devices on SiO$_2$ substrates exhibit mobility characteristics (hundreds to thousands of cm^2/V s) far inferior to the intrinsic mobility of graphene [37—40,47—50]. This is attributed to the relatively large density of dangling bonds and charge traps on SiO$_2$ surface [37—40,47—50]. Recently, hexagonal boron nitride

Table 19.1 The Carrier Mobility of Graphene Prepared by Different Processing Methods and Coated on Various Substrates

Preparation Methods and Substrates	Electron Mobility (cm^2/V s)	Hole Mobility (cm^2/V s)	Measurement Methods
Multilayer Graphene Grown on C-terminated SiC [42–45]	>250,000	NA	Far infrared transmission analysis of inter-landau level transition
Monolayer Graphene Grown on Si-Terminated SiC [46]	15–1100	NA	Four-point magneto-transport measurements
CVD-Grown Multilayer Graphene on Polycrystalline Ni films and Transferred onto Plastic and Si Wafers [38]	~3750	NA	Transport characterization of Hall bar devices
UHV-Grown Monolayer Graphene on SiC [21]	470	NA	Transport characterization of Hall bar devices
Ar-Grown Monolayer Graphene on SiC [21]	900	NA	Transport characterization of Hall bar devices
Monolayer Graphene Grown on Copper and Transferred to SiO$_2$ Substrates [47]	~4000	~4000	FET characterization
Electrically Exfoliated Graphene Flakes [48]	NA	1050	FET characterization
Mechanically Exfoliated Graphene Flakes [49]	795	3735	FET characterization
Suspended Graphene [36]	230,000	230,000	Transport characterization of Hall bar devices
Mechanically Exfoliated Graphene on Hexagonal Boron Nitride (h-BN) Substrates [41]	~60,000	~60,000	Electronic transport measurement

(h-BN) has been identified as an excellent substrate for supporting graphene devices because of its atomically smooth surface, extremely low density of dangling bonds and surface charge traps [41]. Mechanically exfoliated graphene on top of h-BN substrates have mobility values almost an order of magnitude higher than those of graphene on SiO$_2$ [41]. The current challenge is to produce single-crystal h-BN substrates over large areas for large-scale applications in electronics.

To exploit outstanding mobility characteristics of graphene, high-speed transistors with graphene-based channels have been demonstrated for radio frequency (RF) applications [51–54]. For example, Avouris' group at IBM characterized FETs fabricated on epitaxially grown graphene on Si-terminated SiC wafer [51]. The FETs with 240 nm gate length exhibited 100 GHz cutoff frequency (f_T) that exceeds that of the fastest Si metal oxide semiconductor FETs (~40 GHz) [51]. More

19.3 Comparison of Physical Properties and Device Applications

recently, the same group demonstrated the faster FETs ($f_T = 155$ GHz) that were fabricated on chemical vapor deposition (CVD)-grown graphene with 40 nm gate length (Fig. 19.8(a)) [54,55]. This work also provided an important insight for realizing future terahertz-frequency transistors, that is, although the substrate-limited mobility is a key factor affecting the operation frequency of long-channel FETs, the contact resistance between graphene channels and metal contacts becomes more critical as the gate length decreases [54,55]. In addition, Duan's group at University of California, Los Angeles reported the fabrication of graphene FETs with self-aligned gates that exhibited 300 GHz operation frequency (Fig. 19.8(b)) [52]. More recently, Duan's group has pushed this record to 427 GHz [53].

For CNTs, mobility is also an important parameter for determining the performance of carbon nanotubes (CNT)-based electronic devices. For example, mobility determines the high-frequency characteristics of CNT-based FETs and the sensitivity of chemical sensors to chemical species [11,56]. Similar to graphene, SWNTs typically have very high intrinsic carrier mobility ($\mu > 100,000$ cm^2/V s) [57]. However, SWNTs exhibit much more complicated mobility characteristics that strongly depend on their chirality, morphology, and device configuration [57].

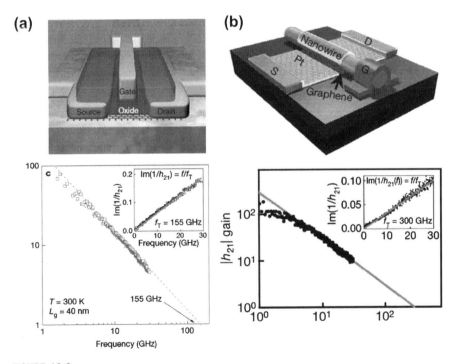

FIGURE 19.8

High-frequency field-effect transistors (FETs) based on single-layer graphene made by (a) Avouris' group at IBM and (b) Duan's group at University of California, los Angeles.

Source: Reprinted by permission from Macmillan Publishers Ltd: [Nature] (Ref. [54]), copyright (2011).
Reprinted by permission from Macmillan Publishers Ltd: [Nature] (Ref. [52]), copyright (2010).

It is hard to define a clear mobility for metallic SWNTs, since the existence of 1D subbands makes it unclear whether the density of charge carriers should be measured from the subband bottom or the band bottom [45,57]. However, metallic SWNTs are expected to have a long mean free path ($L > 10$ μm) in ballistic transport regimes, which is due to the high symmetry of the band structure of armchair SWNTs [45,57]. As illustrated in Fig. 19.4(b), two degenerate 1D subbands in metallic SWNTs are orthogonal to each other and crossing at the Fermi level (i.e. they do not mix). As a result, the backscattering process, in which left-going subband electrons are scattered to the right-going subband, is significantly depressed [45]. For the sake of fundamental research, metallic SWNTs can serve as important test beds for investigating quantum transport in ballistic regimes. Readers who are interested in such a topic are referred to previous publications [58–60]. Because of the high electrical conductivity of metallic SWNTs, they are suitable for practical applications such as transparent electrodes for optoelectronics and chemical sensors [56].

Semiconducting SWNTs have outstanding ballistic transport properties featuring high carrier mobilities comparable to those of single-layer graphenes [57,61]. Several theoretical models were developed for evaluating the mobility in semiconducting SWNTs and identifying its dependence on the tube diameter [62–64]. Pennington and Goldsman developed a model to investigate the mobility characteristics of semiconducting SWNTs in terms of electron–phonon coupling [62]. Their model gives out the electron drift velocity (v_d) versus applied electric field (E) as shown in Fig. 19.9 [62]. The different curves in Fig. 19.9 are for various diameters of zigzag

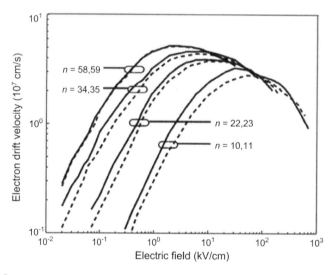

FIGURE 19.9

Calculated drift velocity in semiconducting SWNTs with different indices (n, 0). In the calculation, electron–phonon scattering is regarded as the only scattering scheme.

Source: Reprinted figure with permission from Ref. [62]. Copyright (2003) by the American Physical Society.
http://rmp.aps.org/abstract/RMP/v81/i1/p109_1]

19.3 Comparison of Physical Properties and Device Applications

SWNTs. The carrier mobility can be evaluated by using equation $\mu = v_d/E$. It is clear that the mobility increases with increasing the tube diameter. The maximum mobility is estimated to be ~120,000 cm^2/V s for (59, 0) SWNTs [62]. This work also indicates that at low-field limit, the scattering time of carriers (τ) is approximately proportional to the tube diameter (d), and the effective mass (m^*) at the band bottom (as illustrated in Fig. 19.4(c), the parabolic energy spectra of semiconducting SWNTs have well-defined band bottoms) is inversely proportional to diameter (d) [62]. Therefore, the carrier mobility ($\mu = e\tau/m^*$) is expected to be proportional to the square of diameter (d^2). This analysis suggests that SWNTs with the larger diameter are more suitable for high-speed electronic applications (e.g. RF applications) in comparison with the narrower tubes [57,62,63].

A great amount of effort has been invested to experimentally measure the mobility of CNTs and compare the experimental data with theoretical modeling results, aiming to fully understand the intrinsic conductance behaviors and scattering mechanisms in semiconducting CNTs [61,65–68]. The most common method for measuring the mobility in semiconductors is to measure the Hall mobility, which is applicable for measuring the mobility in 2D graphene, but cannot be used for 1D CNTs [2,12,16]. Here, one has to extract the mobility parameters through characterizing the transport properties of CNT-based FETs [9,57,61]. A detailed description of this method can be found in previous publications [9,57]. For CNT-based FETs with channel length much larger than dielectric thickness and negligible Schottky barriers at CNT–metal contacts (or Ohmic contacts), the effective mobility (μ) and field-effect mobility (μ_{EF}) can be extracted from channel conductance (G)– gate voltage (V_g) characteristic curves of FETs in the linear response region using Eqns (19.3) and (19.4), respectively [57,61]. Here, L is the channel length; c_g is the capacitance per channel length between the CNT and the gate; and V_{th} is the threshold voltage. In 2004, Durkop et al. fabricated semiconducting nanotube FETs with very long channel ($L > 300$ μm) [61]. The intrinsic mobility according to Eqn (19.3) is measured to be over 100,000 cm^2/V s for small gate voltage (Fig. 19.10(a)), and the field-effect mobility according to Eqn (19.4) has a peak at 79,000 cm^2/V s at room temperature (Fig. 19.10(b)) [61]. Finally, the readers need to keep in mind that such calculation methods according to Eqns (19.3) and (19.4) cannot be applied for FETs with significantly large Schottky contact barriers or short channel length (i.e. $L < 10$ s μm) in the ballistic transport regime, in which contact resistance, ballistic transport behavior, or tunneling through the Schottky barriers may dominate the conductance properties of the devices [11,57].

$$\mu = \frac{LG}{C_g(V_g - V_{th})} \quad (19.3)$$

$$\mu = \frac{L}{C_g}\frac{\partial G}{\partial V_g} \quad (19.4)$$

Owning to the high charge mobility and small geometric capacitance of SWNTs, the carriers on the tubes could respond to the external electrical signals at picosecond

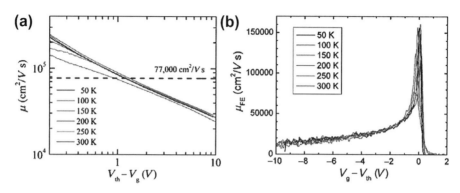

FIGURE 19.10

Intrinsic mobility (a) and field-effect mobility (b) of an FET bearing a ∼300-μm-long CNT channel acquired at various gate voltages. Here, all the mobilities are hole mobilities.

Source: Reprinted with permission from Ref. [61]. Copyright (2004) American Chemical Society.

timescales, corresponding to frequencies close to terahertz regimes [11,69]. Such fast response time opens new opportunities for high-frequency electronic applications. A series of CNT-based RF transistors [69,70], oscillators [71–75], and analog electronic circuits [76,77] has been fabricated and exhibited superior high-frequency performance. The previous experimental and theoretical works indicate that FETs bearing single SWNTs are impractical for RF applications because the electrostatic capacitance of SWNTs is much smaller than the parasitic fringing capacitance between the gates and drain/source electrodes, which can result in large impedance mismatches between the instrumentation and the devices [69,77]. The desirable device configuration for solving this problem is to incorporate densely aligned parallel SWNTs into the device channel [69,77]. For example, Rogers' group demonstrated FETs bearing multiple SWNT channels (the integration density is 3–5 SWNTs/μm; channel length ranges from 900 to 300 nm), as shown in Fig. 19.11(a) [69]. Figure 19.11(b) shows amplitude plots for current gain (H_{21}), unilateral power gain (U), and maximum available gain for frequencies between 0.5 and 50 GHz. The unity current gain frequency (f_T) and unity power gain frequency (f_{max}) are estimated to be around 5 and 9 GHz, respectively [69].

19.3.1.2 Bandgap engineering and ON/OFF switching characteristics

One of the critical drawbacks of single-layer graphene sheets for transistor-related applications is the poor ON/OFF current ratio (I_{ON}/I_{OFF}) [12,15,78]. This is attributed to the zero bandgap of graphene that leads to significantly high OFF current, corresponding to the nonzero conductivity on the order of $4e^2/h$ at Dirac point [12,15,78]. As mentioned in Section 19.2.1, I_{ON}/I_{OFF} is exponentially proportional to E_g/kT. To achieve an ON/OFF ratio in graphene-based FETs comparable to that of Si FETs, one needs to open a bandgap > 1 eV [29]. Several methods have been tested to open a bandgap in graphene films or nanostructures [79–85]. For

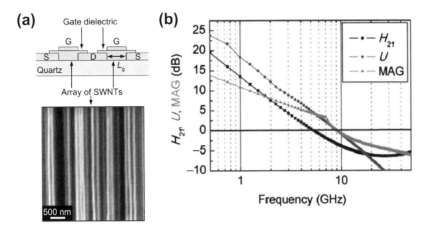

FIGURE 19.11

(a) Top: schematic view of multichannel FETs bearing densely aligned SWNTs; bottom: SEM image of densely aligned SWNTs with a density of 3–5 tubes/μm. (b) Amplitude plots for current gain (H_{21}), unilateral power gain (U), and maximum available gain for frequencies between 0.5 and 50 GHz.

Source: Reprinted with permission from Ref. [69]. Copyright (2009) American Chemical Society.

example, several works have suggested that bilayer graphenes have a nonzero bandgap when subjected to external perturbations (e.g. electric field) [79,86–88]. This bandgap is induced by the coupling of two bonded graphene layers, which fundamentally alters the symmetry of moving electrons [79]. Figure 19.12(a) illustrates the work done by Wang et al., in which a bilayer graphene is gated by using a dual-gate configuration. The gating of the bilayer graphene induces top (D_t) and bottom (D_b) electrical displacement fields [79]. Figure 19.12(b) illustrates that the energy spectrum of a pristine bilayer graphene exhibits zero bandgap (left inset), whereas the asymmetric field can induce a bandgap (Δ) (right inset) [79]. Figure 19.12(c) displays the optically measured bandgap data in comparison with the theoretical predictions based on various models [79]. Wang's work shows that the bandgap in bilayer graphenes can be tuned by the external field over a broad range (0–250 meV) [79].

Fabrication of GNRs also serves as an important solution to open energy bandgap for transistor-related applications [24,25,33,89]. Kim's group systematically investigated the bandgap (E_g) of lithographically patterned GNRs as a function of the ribbon width (w), as shown in Fig. 19.13(a) [24]. This is a pioneer work to demonstrate that the energy gap in lithographically patterned GNRs can be tuned through controlling the average ribbon width [24]. Figure 19.13(b) shows that to obtain a bandgap ($\sim$1 eV) comparable to those of conventional semiconductors, GNRs with sub-10 nm ribbon width are required. Kim's work also indicates that the conventional plasma etching processes can cause structural disorder and localized edge states along GNR edges [24].

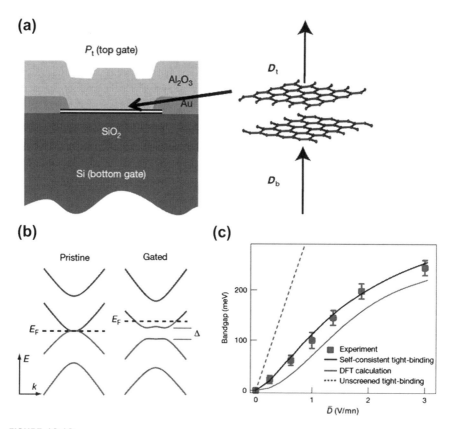

FIGURE 19.12

(a) Illustration of a dual-gated bilayer graphene. (b) Gating of the bilayer graphene induces a nonzero bandgap Δ. (c) Experimental data of bandgap tuned by external field in comparison with the theoretical calculations based on various models.

Source: Reprinted by permission from Macmillan Publishers Ltd: [Nature] (Ref. [79]), copyright (2009).

To minimize the structural disorder induced by the etching process, several plasma-free methods have been demonstrated to create GNRs with ultrasmooth edges. In 2008, Dai's group chemically exfoliated sub-10-nm-wide GNRs from commercially available graphite samples [25]. These GNRs have much smoother edges compared to lithographically patterned ones [25]. FETs bearing sub-10-nm-wide GNRs exhibit a high ON/OFF current ratio ($I_{ON}/I_{OFF} > 10^6$) [25]. This work also gives out the dependence of ON/OFF ratio on the ribbon width [25]. It is strongly indicated that a ribbon width narrower than 10 nm is needed for making FETs with $I_{ON}/I_{OFF} > 10^2$ [25]. The same research group also developed another nanofabrication method, in which CNTs are chemically unzipped into GNRs along specific crystallographic orientations (Fig. 19.14(a)) [47]. High-resolution

19.3 Comparison of Physical Properties and Device Applications

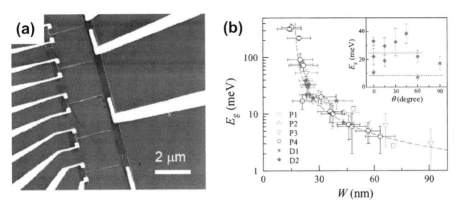

FIGURE 19.13

(a) SEMs of lithographically patterned graphene nanoribbons (GNRs). (b) Measured bandgap data plotted as a function of ribbon width.

Source: Reprinted figure with permission from Ref. [24]. Copyright (2007) by the American Physical Society.
http://prl.aps.org/abstract/PRL/v98/i20/e206805

transmission electron microscopic (HR-TEM) images reveal that such unzipping reaction results in sub-10-nm-wide GNRs with ultrasmooth edges (Fig. 19.14(b)) [47]. FETs made from unzipped GNRs exhibit greatly improved ON/OFF switching characteristics compared to FETs based on 2D graphene films (Fig. 19.14(c)) [47].

Transistors and relevant electronic devices based on a single (or well isolated) ultrasmooth GNR(s) have been extensively studied and exhibit superior electronic characteristics (e.g. high I_{ON}/I_{OFF} ratios and stable quantum transports in ballistic regimes) [25,27,33]. These superior properties are attributed to the high-quality edges of GNRs used in the devices [25,33]. In spite of these progresses, realistic applications in electronics demand large-scale arrays of such high-quality GNRs and devices [90]. Therefore, one needs multiplexing processes to create densely arranged arrays of sub-10 nm GNRs or other relevant nanostructures over large-area substrates. Such spatially multiplexed graphene nanostructures can sustain a much higher driving current for electronic applications or result in a high sensitivity for biological and chemical sensors. Although the nanofabrication approaches mentioned above have been demonstrated to be able to create single (or dispersed) GNRs with ultrasmooth (or even atomically smooth) edges for the purpose of laboratory research [47,91,92], they can hardly be further developed into scalable nanomanufacturing technologies for producing ordered nanostructures over large areas since they either do not have the addressable control of device locations or cannot operate in a high-throughput way.

Recently, block copolymers (BCPs) lithography techniques in combination with plasma etching have been developed to pattern graphene sheets into various densely arranged nanostructures, aiming to obtain nonzero bandgap in patterned graphene, improved ON/OFF switching characteristics of graphene-based FETs, and high

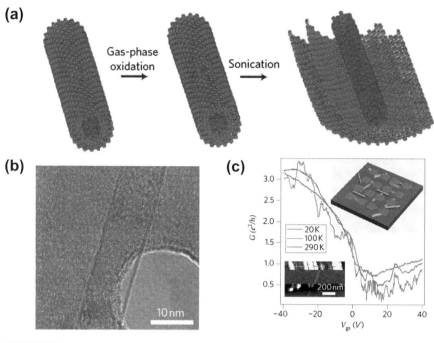

FIGURE 19.14

(a) Schematic illustration of chemical unzipping of a carbon nanotube (CNT) into a graphene nanoribbon (GNR). (b) HR-TEM image of a sub-10-nm-wide GNR that exhibits atomically smooth edges. (c) Temperature-dependent transport characteristic curves of an FET made from a GNR.

Source: Reprinted by permission from Macmillan Publishers Ltd: [Nature Nanotechnology] (Ref. [47]), copyright (2010).

uniformity of graphene nanostructures for large-scale application [85,89,90,93–95]. For example, hexagonally packed graphene nanomeshes (GNMs) [85,93,94] and linear arrays of GNRs [89,90,95] have been fabricated and incorporated into multichannel transistors [85,89,90,93–95]. Figure 19.15 displays three independent works done by Duan et al. [93], Kim et al. [94], and Liang (the author) et al. [85], respectively. All these works are about utilizing BCP-related lithographic methods to pattern GNMs with sub-10-nm-scale interhole spacing [85,93,94]. Such GNM structures have been experimentally demonstrated to be able to function as a network of sub-10 nm GNRs. When incorporated into FETs as conduction channels, they can result in high ON/OFF current ratio ($I_{ON}/I_{OFF} > 100$) and upscalable driving current [85,93,94].

However, to date, these multichannel transistors exhibit noticeably worse transport characteristics in comparison with that of single-GNR transistors mentioned above [85,89,93–95]. For example, the ON/OFF current ratio (I_{ON}/I_{OFF}) data of

19.3 Comparison of Physical Properties and Device Applications

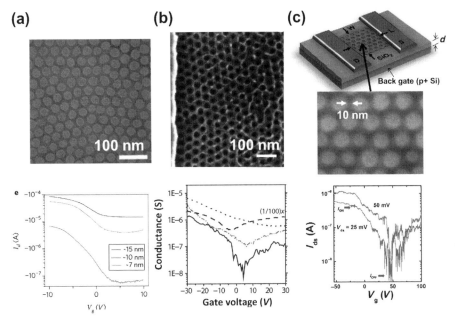

FIGURE 19.15

Graphene nanomeshes (GNMs) fabricated by (a) Duan et al. (Reprinted by permission from Macmillan Publishers Ltd: [Nature Nanotechnology] (Ref. [93]), copyright (2010).); (b) Kim et al. (Reprinted (adapted) with permission from Ref. [94]. Copyright (2010) American Chemical Society); and (c) Liang et al. (Reprinted (adapted) with permission from Ref. [85]. Copyright (2010) American Chemical Society.) and associated transport characteristic curves. GNM-based FETs exhibit much higher ON/OFF current ratio compared to those made from unpatterned graphene.

all the multichannel FETs made from densely arranged GNMs or GNRs are <200 [85,89,90,93,94,96,97], which are far below the typical I_{ON}/I_{OFF} values obtained from the FETs consisting of single or well-isolated GNRs (e.g. Wang et al. reported $I_{ON}/I_{OFF} \sim 10^6$ for a single-GNR transistor) [27]. A recent work done by Liang et al. (the author) indicates that a large ribbon-to-ribbon width variation (RWV) in a multichannel FET bearing densely aligned GNRs can lead to nonsynchronized switching characters of multiple-GNR channels and thus, a poor ON/OFF current ratio [90]. Through process optimization, Liang et al. have fabricated 8 nm half-pitch GNRs with the minimal RWV value of ~ 2.4 nm (3σ value), as shown in Fig. 19.16(a) [90]. The corresponding FETs exhibit relatively high ON/OFF current ratio values well above 10 (Fig. 19.16(b)) [90]. Although such ON/OFF ratio data are among the highest values ever reported for transistors consisting of densely arranged GNRs [89,97], a much higher ON/OFF ratio (>1000) is still needed for making multi-GNR FETs suitable for practical applications. Therefore, new nanomanufacturing processes based on different physical and chemical

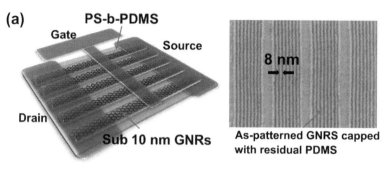

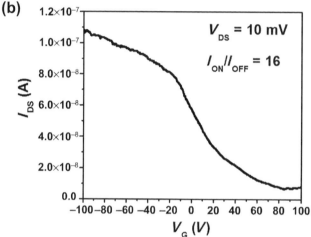

FIGURE 19.16

(a) Multichannel field-effect transistors made from densely aligned sub-10 nm half-pitch graphene nanoribbons that are patterned by using directed self-assembly of cylindrical-morphology BCPs followed with plasma etching. (b) Transport characteristic curve of a transistor bearing nanoribbons with 8 nm half-pitch and 3σ RWV of 2.4 nm [90].

schemes are needed to manufacture densely aligned graphene nanostructures with ultrasmooth (or ideally, atomically smooth) edges for reliable large-scale applications [90].

In comparison with 2D graphene, 1D semiconducting SWNTs have a nonzero bandgap (E_g) that is inversely proportional to the tube diameter ($E_g \sim 0.7/d$ (nm)) [11]. FETs consisting of a single semiconducting tube usually exhibit a high ON/OFF current ratio [98]. Figure 19.17(a) shows the transport characteristics of an FET consisting of a semiconducting SWNT made by Avouris et al. that exhibits an ON/OFF ratio $>10^4$ [98]. As a comparison, Fig. 19.17(b) displays the gate-dependent conductance curve of a typical metallic SWNT that exhibits a very weak gating effect due to the zero bandgap of the metallic tube [57].

19.3 Comparison of Physical Properties and Device Applications

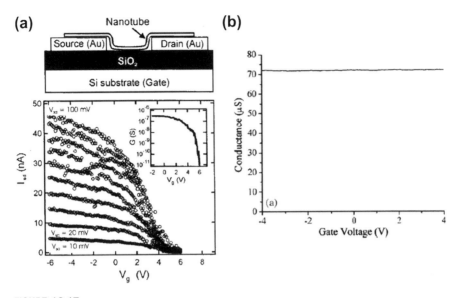

FIGURE 19.17

Gate-dependent transport characteristics of (a) a semiconducting SWNT and (b) a metallic SWNT.

Sources: (a) Reprinted from Ref. [98], Copyright (1999), with permission from Elsevier.
(b) Reprinted from Ref. [57], copyright (2004), with permission from IOP publishing.
(http://iopscience.iop.org/0953-8984/16/18/R01/).

To make CNT-based FETs able to provide a high ON/OFF current ratio, a large (or upscalable) driving current and a high integration density, densely arranged multiple CNTs are usually incorporated into FET arrays to form multiple conduction channels [11,72,99,100]. The nanofabrication processes for creating such multichannel devices need a well-defined control over the tube density (D) that is measured in the number of tubes per unit area for random network films of CNTs or tubes per length for aligned linear arrays as well as the purity of semiconducting SWNTs [11,100,101]. Current synthetic routes to CNTs typically yield mixture of metallic and semiconducting SWNTs [102–104]. This feature results in the final transport properties of CNT networks that highly depend on the tube density and device geometry [102–104]. Rogers' group systematically studied the transport characteristics of multi-CNT thin-film transistors (TFTs) with different channel lengths, tube densities, and degrees of tube alignment (Fig. 19.18(a) schematically illustrates the device structure and Fig. 19.18(b) shows the SEM image of an exemplary FET consisting of aligned SWNTs with ~25-μm-long channel length) [105]. As indicated by the transport characteristic curves in Fig. 19.18(c), given the coverage density and the alignment degree of SWNTs, TFTs with relatively longer channel lengths exhibit a relatively higher ON/OFF current ratio [105]. This experimental observation is well interpreted by using a stick-percolation transport model [105]. This model implies that for TFTs

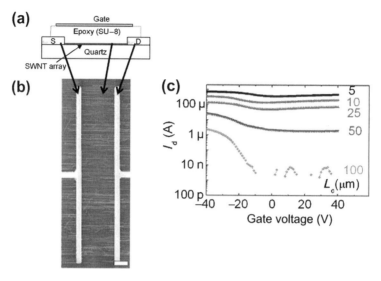

FIGURE 19.18

(a) Illustration of multi-CNT FETs. (b) Top-view SEM image of an exemplary FET with aligned SWNTs ($L_c \sim 25$ μm). (c) I_{ds}–V_g characteristics of FETs comprising coexisting metallic and semiconducting SWNTs plotted as a function of total channel length. The percolation transport mechanism becomes dominant in long-channel FETs ($L_c > L_s$), which makes high ON/OFF ratios possible. However, such network films of coexisting metallic and semiconducting tubes are not suitable for making short-channel transistors.

Source: Reprinted (adapted) with permission from Ref. [105]. Copyright (2007) American Chemical Society.

with relatively long channel length (i.e. channel length (L_c) > length of SWNTs (L_s)), slightly nonaligned networks of SWNTs could make percolation transport mechanism dominant and result in high ON/OFF current ratios, whereas the short channel TFTs (i.e. $L_c < L_s$) typically have relatively lower ON/OFF ratios that approximately reflect the relative number ratio of metallic and semiconducting SWNTs incorporated in the FETs [105]. To ultimately solve such an issue about low ON/OFF ratios due to the heterogeneity of SWNTs, additional methods have been studied, aiming to thoroughly eliminate or disenable the electrical conduction of metallic SWNTs in FETs [76,100]. Zhou's and Wong's groups developed an automated electrical breakdown process to remove metallic and highly conductive semiconducting SWNTs without destroying other semiconducting tubes [76]. Figure 19.19 shows that this process is performed in a stepwise manner and the ON/OFF current ratios of multi-CNT FETs can be enhanced by about three orders of magnitude after a couple of cycles of electrical breakdown processes [76]. Here, most of the semiconducting SWNTs remain intact during the process because they can be selectively turned into OFF state by applying an appropriate gate voltage during a breakdown process cycle [76]. Meanwhile, there are other methods for separating metallic tubes from semiconducting ones [106–110].

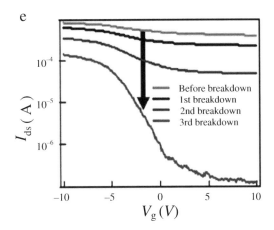

FIGURE 19.19

Stepwise evolution of I_{ds}–V_g characteristics of an FET comprising coexisting metallic and semiconducting SWNTs when subjected to electrical breakdown processes that can selectively remove metallic tubes and leave semiconducting ones intact.

Source: Reprinted (adapted) with permission from Ref. [76]. Copyright (2009) American Chemical Society.

Readers who are interested in this topic are referred to the literature that focus on the selective synthesis and fabrication processes of CNT-based devices [106–110].

19.3.2 Mechanical properties
19.3.2.1 Elastic properties

Both graphene sheets and CNTs have very stable molecular structures and exhibit superior mechanical properties [111–115]. Recent research works have indicated that graphene may be the hardest material known today [116]. It is believed to be even harder than diamond and about 300 times harder than steel [116]. A CNT is conceptually formed by cutting and wrapping a graphene sheet into a seamless cylinder along specific and discrete "chiral" angles. So its mechanical property may also depend on tube diameter and chirality [117,118].

Graphene, as an atomically thin crystal material, exhibits nonlinear elastic behaviors when subjected to external loading [116]. The elastic stress response induced by uniaxial strains can be expressed by Eqn (19.5), where σ^{2D} is the second Piola–Kirchhoff stress in 2D films, ε is the uniaxial strain, E^{2D} is the Young's modulus for 2D graphene (note that here the unit of 2D elastic modulus is [N/m]), and D^{2D} is the third-order elastic modulus [116].

$$\sigma^{2D} = E^{2D}\varepsilon + D^{2D}\varepsilon^2 \tag{19.5}$$

$$\sigma_i^{2D} = \sqrt{\frac{FE^{2D}}{4\pi R}} \tag{19.6}$$

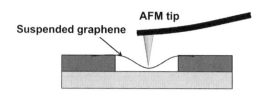

FIGURE 19.20

Schematic illustration of the AFM-based nanoindentation method for evaluating the mechanical properties of suspended graphene over lithographically patterned holes [116].

Hone et al. used an atomic force microscope (AFM) equipped with diamond tips to perform nanoindentation and measure critical mechanical properties of graphene monolayers suspended over open holes, as illustrated in Fig. 19.20 [116]. In this experiment, E^{2D} is measured to be 340 ± 50 N/m; D^{2D} is -690 ± 120 N/m; the intrinsic strength, σ_i^{2D}, is estimated to be 42 ± 4 N/m [116]. In particular, σ_i^{2D} is deduced from the values of measured breaking force (F) by using Eqn (19.6), where R is the AFM tip radius [116]. By assuming an effective graphene thickness of 0.335 nm, these parameters for 2D films can be converted to conventional (or effective) Young's modulus of $E = 1.0 \pm 0.1$ TPa; third-order elastic stiffness of $D = -2.0 \pm 0.4$ TPa; intrinsic breaking stress of $\sigma_i = 130 \pm 10$ GPa that is corresponding to a maximum strain of $\varepsilon_i = 0.25$ [116]. Based on these experimentally determined mechanical properties, graphene is believed to be the hardest material known today [116]. However, it should be pointed out that the intrinsic strength mentioned above is acquired from nanoscopic graphene films free of defects [116]. In principle, this quantity is the maximum stress that can be supported by pristine and defect-free graphene films prior to failure [116]. In reality, for the macroscopic applications, the breaking strength of graphene and other graphitic materials is still limited by the presence and areal density of crystal defects and boundaries of crystal domains [119].

Ruoff's group investigated the mechanical response of SWNT ropes by using a tensile-loading setup [117], in which an AFM tip was manipulated to quantitatively measure tensile stress–strain curves of SWNT ropes, as shown in Fig. 19.21 [117]. In Ruoff et al.'s work, Young's modulus values of SWNTs measured range from 320 to 1470 GPa and the average breaking strength ranges from 13 to 52 GPa [117].

Although CNTs exhibit a very high rigidity in the axial direction similar to the in-plane rigidity of 2D graphene sheets, they are experimentally observed to be much more compliant in the radial direction [120]. Hence, a CNT has an effective modulus that refers to the elastic response to the deformation caused by an anisotropic indentation load applied in radial directions [121]. Based on Hertzian contact model, this effective modulus of CNTs is estimated to be from 0.3 to 4 GPa [121]. A series of theoretical and experimental works have been performed to identify various elastic response behaviors in the radial directions of CNTs that include elastic buckling under bending and torsional loads (Fig. 19.22(a)) [120], cross-sectional deformation of tubes due to van de Waals forces (Fig. 19.22(b)) [122], and collapsed MWCNTs with mechanical twists (Fig. 19.23) [123]. In the previous

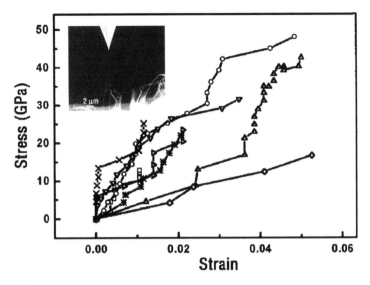

FIGURE 19.21

Stress versus strain curves of SWNT ropes obtained from tensile-loading experiments. The inset SEM image shows the experimental setup, in which an AFM tip is used to apply controllable tensile stress onto SWNT ropes and measure resultant strain.

Source: Reprinted figure with permission from Ref. [117]. Copyright (2000) by the American Physical Society.
http://prl.aps.org/abstract/PRL/v84/i24/p5552_1

works, the modeling works were mainly based on molecular dynamics (MD), [120,124,125], and the experimental observations of the mechanical response behaviors at nanoscopic scales were performed using HR-TEM [120,122,123]. The theoretical modeling works done by Yakobson et al. indicate that the first buckling pattern of CNTs is initiated at a nominal compressive strain of 0.05 [124,125]. The energetics analysis further shows the dependence of CNT geometry on the tube diameter and has been used for evaluating the effect of the tube diameter on the cross-sectional deformation caused by van der Waals forces, bending, or other mechanical twists [120,122,123]. To be more specific, CNTs with sub-2 nm diameter have a stable form with circular cross section. If the tube diameter is between 2 and 4 nm, both circular and collapsed forms are possible. SWNTs with diameter >6 nm tend to collapse and form twisted nanoribbons [120]. However, these arguments are only applicable to single or well-isolated SWNTs. SWNTs in a tightly packed bundle tend to stabilize each other against the complete collapse [120].

19.3.2.2 Applications in bendable electronics and photovoltaics

As discussed above, owing to the robust C—C bonds in the 2D honey-comb lattice structure, graphene and CNTs exhibit very high in-plane stiffness and breaking strength, whereas they are compliant to the mechanical load applied along out-plane directions [116,120,122]. Such special mechanical properties make graphene,

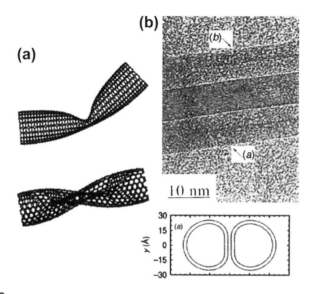

FIGURE 19.22

(a) Simulated buckling state of SWNTs under bending (top) and torsional (bottom) loads. (b) HRTEM image of two adjacent multiwalled carbon nanotubes (MWCNTs) a and b. Tube a has 10 fringes and tube b has 12 fringes. The average interlayer spacing for inner layers and outer layers of the tube "a" are measured to be 0.338 nm and 0.345 nm, respectively. For tube b, the according spacing parameters are 0.343 nm and 0.351 nm, respectively. These 0.07 and 0.08 nm differences are attributed to the compressive force acting in the contact region of two tubes. The bottom inset shows the simulated cross-sectional deformation from perfectly cylindrical shells caused by van der Waals force between two tubes.

Sources: (a) Reprinted by permission from Ref. [120], copyright (2002); (b) Reprinted by permission from Macmillan Publishers Ltd: [Nature] (Ref. [122]), copyright (1993).

few-layer graphene films, and network films of CNTs able to serve as excellent functional as well as structural materials for making lightweight, bendable, and low-cost electronic and energy devices [99,126,127].

Bae et al. developed a roll-to-roll method for producing 30-inch-wide graphene films that can serve as flexible transparent electrodes (Fig. 19.24(a)) [126]. Figure 19.24(b) shows the dependence of the sheet resistance of a graphene film and a conventional indium tin oxide (ITO) film on the tensile strain [126]. Both conductive films are deposited on 188-μm-thick polyethylene terephthalate (PET) substrates. Compared to the ITO electrode, the graphene electrode exhibits much higher electromechanical stability under both tensile and compressive strains up to 6% [126]. The research team, in collaboration with Samsung, also demonstrated a working graphene-based touch-screen panel connected to a computer (Fig. 19.24(c)) [126].

Zhou et al. demonstrated highly flexible photovoltaic (PV) cells with a structure of PET substrate/CVD-grown graphene or ITO/PEDOT:PSS/copper phthalocyanine

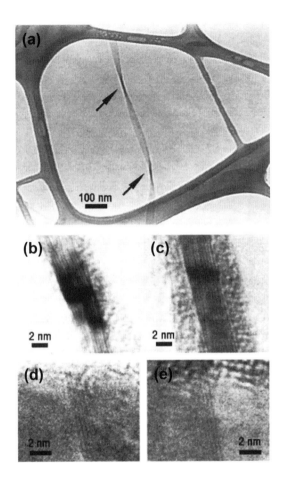

FIGURE 19.23

(a) TEM of a twisted MWCNT with two indicated twists. (b) and (c) HRTEM images of exactly the fringes at twists 1 and 2, respectively. (d) and (e) HRTEM images of fringes along the edges of untwisted segments (five along the left edge; five along the right edge). TEM images in (b)–(e) indicate that this MWCNT has collapsed into a ribbon-like structure with five collapsed nested shells.

Source: Reprinted figures with permission from Ref. [123]. Copyright (2001) by the American Physical Society.

(CuPc)/fullerene (C60)/bathocuproine (BCP)/aluminum (Al), in which graphene and ITO serve as transparent anodes [127]. Figure 19.25(a) shows the PV characteristic curves of a graphene-based cell plotted as a function of the bending angles (2θ) (the inset photograph shows the setup of the bending experiment). It is shown that the graphene-based PV cell remains a good performance even under a very large bending angle ($2\theta = 138°$), whereas the ITO-based PV cell exhibits a seriously degraded performance under a bending angle of $2\theta = 60°$ (Fig. 19.25(b) and (c)). Such a dramatic

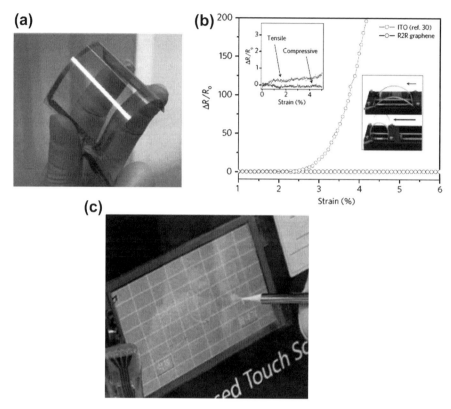

FIGURE 19.24

(a) A graphene-based transparent electrode that exhibits a good flexibility. (b) Sheet resistance of a graphene-based electrode and an ITO electrode as a functional elastic strain. (c) A working graphene-based touch-screen panel.

Source: Reprinted by permission from Macmillan Publishers Ltd: [Nature Nanotechnology] (Ref. [126]), copyright (2010).

difference in electromechanical stability is attributed to the superior mechanical properties of graphene and brittle nature of ITO, as further demonstrated by the SEM images of these two films after subjection to the bending test (Fig. 19.25(d)) [127]. The ITO film shows observable microcracks after the bending test, while no signs of microcracks or fissures were observed on the bended graphene film [127].

In comparison with graphene, the network films of semiconducting CNTs are more suitable for semiconductor-related applications such as analog electronic components and logic gates. For example, Javey et al. demonstrated bendable ICs made from semiconducting CNT networks for digital, analog, and RF applications [99]. Figure 19.26(a) shows the schematic view of the flexible nanotube thin film transistor (TFT); and Fig. 19.26(b) shows the AFM image of the channel of the TFT, which consists of random networks of semiconducting CNTs [99]. Figure 19.26(c)

19.3 Comparison of Physical Properties and Device Applications

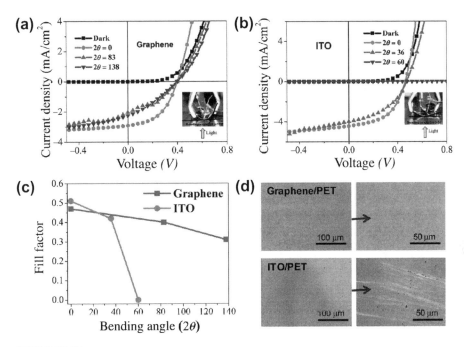

FIGURE 19.25

Current density versus voltage characteristics of graphene-based (a) and ITO-based (b) PV cells subjected to different bending angles (2θ). Insets show the setup for the bending test. (c) Fill factor of graphene-based and ITO-based cells plotted as a function of the bending angles. (d) SEM images showing the surface structure of graphene-based (top) and ITO-based (bottom) cells after being subjected to the bending angles.

Source: Reprinted (adapted) with permission from Ref. [127]. Copyright (2010) American Chemical Society.

displays a photograph showing the extreme bendability of the flexible nanotube circuits, where the samples are being rolled onto a metal rod with a diameter of only 2.5 mm. Besides the excellent flexibility, TFTs with different channel lengths exhibit reasonably high ON/OFF current ratios (Fig. 19.26(d)) [99]. Graphene has also been implemented for making bendable TFTs and exhibited superior mechanical properties [128–131]. However, graphene-based TFTs usually have much lower ON/OFF ratios due to the zero bandgap of intrinsic graphene [128,129].

19.3.3 Optical and optoelectronic properties

19.3.3.1 Linear optical response in terahertz and infrared spectral ranges

The optical properties of graphene are related to its electrodynamic properties, the frequency dependence of optical conductivity, and the collective excitations. In graphene, the linear dispersion behavior (or energy spectrum) of electrons near the Fermi level can result in very interesting linear and nonlinear electrodynamic response characters in microwave and terahertz regimes [132]. The frequency-dependent optical

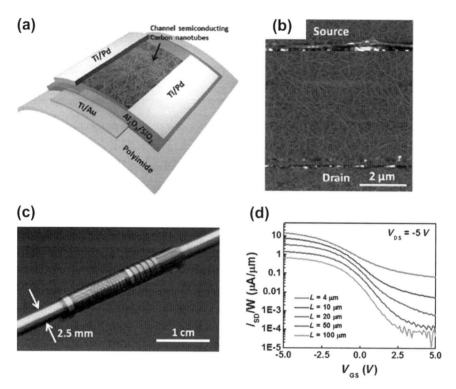

FIGURE 19.26

(a) Schematic view of a bendable TFT made from carbon nanotube networks. (b) AFM image showing the TFT channel bearing randomly arranged tubes. (c) Photograph showing the extreme bendability of the flexible nanotube circuits, where the nanotube device samples are being rolled onto a metal rod with a diameter of only 2.5 mm. (d) $I_{ds}-V_{gs}$ transport characteristics of carbon nanotube TFTs with different channel lengths.

Source: Reprinted (adapted) with permission from Ref. [99]. Copyright (2012) American Chemical Society.

conductivity of graphene can be written as $\sigma(\omega) = \sigma_{intra}(\omega) + \sigma_{inter}(\omega)$, where $\sigma_{intra}(\omega)$ and $\sigma_{inter}(\omega)$ are conductivity contributions associated with intraband and interband transitions, respectively [133–135]. The intraband conductivity $\sigma_{intra}(\omega)$, as expressed in Eqn (19.7), takes a Drude-model form, whereas the interband conductivity $\sigma_{inter}(\omega)$ has the form expressed in Eqn (19.8).

$$\sigma_{intra}(\omega) = \frac{i2\pi q^2 g_s g_v T}{h^2(\omega + i\gamma)} \ln\left(e^{\mu/2T} + e^{-\mu/2T}\right) \qquad (19.7)$$

$$\sigma_{inter}(\omega) = \frac{iq^2 g_s g_v}{4h} \int_0^\infty \left(\frac{\sinh(s)}{\cosh(\mu/T) + \cosh(s)} \times \frac{\hbar(\omega + i\gamma)/2T}{s^2 - [\hbar(\omega + i\gamma)/2T]^2}\right) ds \qquad (19.8)$$

where q is the electron charge; $g_s = 2$ is the spin degeneracy; $g_v = 2$ is the valley degeneracy factor; T is the temperature; h is the Planck's constant (or $\hbar = h/2\pi$); γ is the momentum scattering rate; and μ is the chemical potential[133–135].

Figure 19.27(a) plots the total optical conductivity $\sigma(\omega)$ [132]. At the high-frequency limit, $\sigma(\omega)$ approaches a universal conductivity value (i.e. $\sigma(\omega \to \infty) \sim \frac{q^2}{4\hbar}$) [132]. Based on $\sigma(\omega)$, the measurable transmission T can be evaluated by using Eqn (19.9), where t is the transmission amplitude and c is the speed of light. Figure 19.27(b) shows the corresponding T as a function of ω/μ. At the high-frequency limit, T is close to a universal value $T = 1 - \frac{\pi q^2}{\hbar c}$.

$$T = |t|^2, \quad t = \left[1 + \frac{2\pi\sigma(\omega)}{c}\right]^{-1} \tag{19.9}$$

These theoretically predicted behaviors of $\sigma(\omega)$ and $T(\omega)$ have been experimentally validated [136]. Figure 19.27(c) shows the real (upper inset) and imaginary (bottom inset) parts of $\sigma(\omega)$, experimentally determined by using infrared spectroscopy, which are highly consistent with the calculated $\sigma(\omega)$ function [136]. Geim et al. also reported the experimental result that the transparency of single-layer graphene features a universal transmission value of T = (1+ $0.5\pi q^2/\hbar c$) over the range of visible frequencies that is expected for 2D Dirac fermions [137].

The terahertz and infrared applications (e.g., graphene-based optical modulators) of graphene require methods able to modulate the transmission of light of these frequencies. Based on the electronic structures of single-layer graphene, the optical conductivity as well as transmission can be modulated in a continuous fashion by adjusting the Fermi energy level (E_F) [133–135]. The change of E_F can induce the change of the Drude-like conductivity and thus, the optical response associated with intraband transitions in the terahertz, as indicated by Eqn (19.7) [133–135]. Moreover, the interband transitions in graphene exhibit a "$2E_F$ onset" in the mid-infrared (i.e. in case of $T \sim 0$ K, interband transition in graphene is possible only when the photon energy is larger than $2E_F$). So the interband optical conductivity is also strongly correlated to E_F [133]. Recently, Ren et al. demonstrated the modulation of terahertz and infrared transmission spectra of gated single-layer graphene (Fig. 19.28(a)), in which a back gate was used to control the values of E_F [138]. Figure 19.28(b) displays the terahertz transmission spectra as a function of gate voltages, which demonstrates that the applied gate voltage can cause a modulation of transmitted terahertz wave power by ~20% [138]. In addition, the bottom/right inset in Fig. 19.28(b) plots the transmitted terahertz wave power and the DC resistance of the gated graphene layer together, which exhibit very similar gate (or E_F) dependence [138]. This also indicates that the intraband conductivity of graphene in the terahertz range exhibits Drude-like frequency dependence with Fermi-energy-dependent magnitude [138].

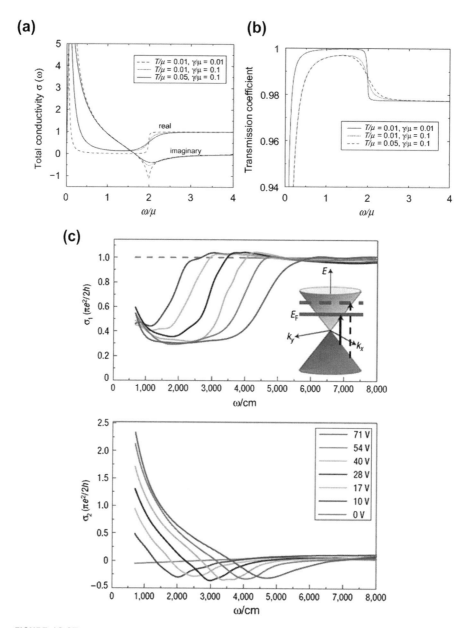

FIGURE 19.27

(a) The theoretically calculated total optical conductivity $\sigma(\omega)$ of graphene. [132]. (b) The corresponding transmission coefficient $T(\omega)$. (c) Experimentally determined $\sigma(\omega)$ [136].

Sources: (b) Reprinted with permission from Ref. [132]. Copyright [2011], American Institute of Physics.
(c) Reprinted (adapted) by permission from Macmillan Publishers Ltd: [Nature Physics] (Ref. [136]), copyright (2008).

19.3 Comparison of Physical Properties and Device Applications

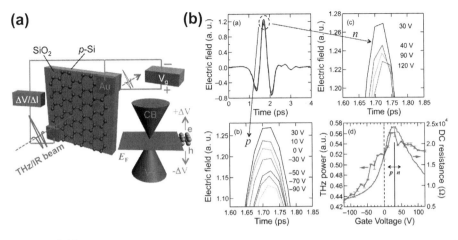

FIGURE 19.28

(a) Schematic illustration of the experimental setup for measuring the terahertz and mid-infrared transmission spectra of a single-layer graphene biased with a gate voltage [138]. (b) Gate–voltage-dependent terahertz wave transmission as a function of applied gate voltages. The zoomed inset views show that the terahertz wave power exhibits an observable dependence on the gate voltages ranging from −90 to 120 V [138]. Such dependence is similar to the gate dependence of the DC resistance.

Source: Reprinted (adapted) with permission from Ref. [138]. Copyright (2012) American Chemical Society.

Similar to single-layer graphene, SWNTs also exhibit a number of interesting response behaviors in the frequency ranges of terahertz and infrared [19,139–142]. Although the terahertz/infrared spectroscopic studies of SWNTs started significantly earlier than those on graphene, the researchers in this field are still facing a large number of conflicting results and contradicting interpretations [139]. Such a situation is partly attributed to the broad variety of growth methods for preparing SWNT samples, such as CVD, Arc discharge, and laser ablation [139]. One common spectral feature that many researchers have observed is a broad absorption peak around 135/cm or 4 THz that was firstly reported by Ugawa et al. (Fig. 19.29) [19,141]. The researchers have not reached a consensus about the interpretation of this feature [139,143]. Some people attributed it to the interband absorption in mod-3 nonarmchair nanotubes with bandgaps induced by tube curvatures [36,37,144], whereas some believed that it is due to the absorption associated with plasma oscillations along the tube [139,145,146].

In comparison with 2D graphene, 1D SWNTs exhibit a strong anisotropy in optical properties [7]. For example, Ren et al. recently reported the polarization-dependent terahertz transmission measurements on films of aligned SWNTs, as illustrated in Fig. 19.30(a) [147]. The fabrication of such SWNT-based polarizers includes the upright growth of nanotubes on patterned catalyst substrates followed with a mechanical transfer onto sapphire substrates [147]. Figure 19.30(b) displays

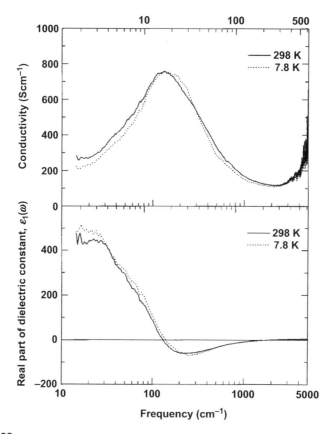

FIGURE 19.29

Real parts of the conductivity (top) and dielectric function (bottom) of a film of SWNTs acquired through a Kramers–Kronig (K–K) analysis of the reflectance.

Source: Reprinted figures with permission from Ref. [141]. Copyright (1999) by the American Physical Society.
http://prb.aps.org/abstract/PRB/v60/i16/pR11305_1

the optical and SEM images of aligned nanotubes, which exhibit macroscopic alignment of nanotubes with excellent uniformity [147]. Such polarizers exhibit an extremely high degree of anisotropy in the transmission of terahertz waves (0.1–2.2 THz). When the light polarization is perpendicular to the nanotubes, the films of SWNTs are highly transparent despite the total thickness of the films (the total thickness is determined by the total number of stacked SWNT films). However, when the polarization is parallel to the nanotubes, there is strong absorption of the light [147]. As shown in Fig. 19.30(c), such SWNT-based terahertz polarizers remain a high degree of polarization and extinction ratio throughout the entire frequency range of the experiment (i.e. 0.1–2.2 THz) [147].

19.3 Comparison of Physical Properties and Device Applications

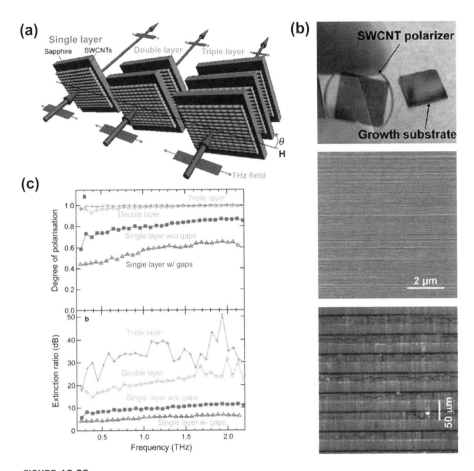

FIGURE 19.30

(a) Illustration showing the use of multiple SWNT films to build up terahertz polarizers. (b) Optical and SEM images of macroscopically aligned SWNTs over sapphire substrates. (c) Degrees of polarization and extinction of SWNT-based terahertz polarizers with different numbers of SWNT films.

Source: Reprinted (adapted) with permission from Ref. [147]. Copyright (2012) American Chemical Society.

19.3.3.2 Active optoelectronic materials

Semiconducting SWNTs are direct-bandgap materials that hold significant potentials to be implemented as active materials for optical and optoelectronic applications [148]. Especially, the unique band structures of semiconducting SWNTs result in several attractive properties suitable for PV applications, such as high absorption due to high-efficiency carrier multiplication effects [149], high carrier mobility [61], diameter-dependent bandgaps [150], and the availability of Ohmic contacts with metallic electrodes [101,151]. However, one of the critical drawbacks for

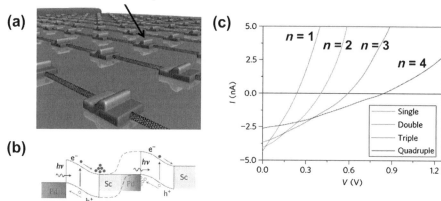

FIGURE 19.31

(a) Illustration showing multiple SWNT photovoltaic cells connected in series. (b) Band diagram of two photovoltaic cells connected in series, which illustrates the excitation and accumulation of electron–hole pairs under illumination. (c) I–V characteristics for the photovoltaic module with various numbers (n) of SWNT cells connected in series.

Source: Reprinted by permission from Macmillan Publishers Ltd: [Nature Photonics] (Ref. [152]), copyright (2011).

practical PV applications is that the typical photogenerated voltage of a semiconducting SWNT is too small ($V \sim 0.2$) [152]. Recently, Yang et al. reported the generation of multiplied photovoltage by introducing a virtual contact into the nanotube (Fig. 19.31) [152]. As illustrated in Fig. 19.31(a), the SWNT-based PV module with multiple virtual contacts can be regarded as multiple SWNT cells connected in series. The band diagram of two of the connected cells illustrated in Fig. 19.31(b) indicates that the introduction of the virtual scandium/palladium contact results in the cascaded structures and introduces a potential valley in the conduction band near the scandium pad as well as a hill in the valence band near the palladium pad, which can induce an effective open-circuit voltage almost twice that created in a single cell [152]. Based on this principle, Yang et al. has demonstrated SWNT PV modules with an improved photovoltage close to 1 V (Fig. 19.31(c)) [152].

In addition to PV applications, semiconducting SWNTs have been studied as active materials for other photonic devices, such as microcavity-controlled nanotube-based emitters at infrared wavelengths [150], FET-based ambipolar infrared emitters [153,154], and infrared photodetectors made from densely aligned nanotube arrays [155]. All these photonic applications are based on the direct-bandgap character of semiconducting SWNTs. Therefore, metallic nanotubes and graphene can be hardly used for these applications without additional functioning processes to tailor their band structures. However, metallic SWNTs, graphene, and few-layer-graphene have been tested for making electrode or electroabsorption components of photonic devices [126,156,157]. As mentioned above (Section 19.3.2.2), large-area graphene and graphene/CNT composite films have been used as transparent and flexible electrodes for PV or display applications [126,157]. Furthermore, recently, Xiang Zhang research group demonstrated a graphene-based broadband optical modulator, in which a gated graphene film is implemented as an electroabsorption modulator [156]. This modulator can create the modulation of guided light at frequencies over 1 GHz, provide a broad operation spectrum ranging from 1.35 to 1.6 μm, and have a small footprint (~25 μm^2) for future device integration [156].

19.3.3.3 Plasmon and electromagnetic excitations

All the optoelectronic and photonic properties of graphene and CNTs discussed above, such as universal absorption at terahertz/infrared frequencies and tunable interband transitions in optical response, can be well described by the model of single-particle excitation of electrons [136,148,158]. Plasmons in these nanomaterials are associated with the collective excitations of 2D massless electrons. The plasmon excitations in graphene have been studied [159–161]. In the long-wavelength limit (i.e. $k \ll k_F$, where k is the wave vector of 2D plasmon waves and k_F is the Fermi wave vector), 2D plasmons in graphene have a square-root $\omega(k)$ dispersion, as expressed by Eqn (19.10). Here, the plasmon frequency (ω_p) exhibits a ¼ power dependence on the 2D electron density (n_s), which is different from $\omega_p = n_s^{1/2}$ dependence observed for the 2D plasmons in conventional 2D electron gases (2DEG) [132].

$$\omega_p(k) = \left(\frac{q^2 g_s g_v \mu k}{\hbar^2}\right)^{1/2} \propto n_s^{1/4} k^{1/2} \quad (19.10)$$

Although the plasmon excitations and resonances in graphene have been investigated using inelastic electron scattering spectroscopy and inelastic scanning tunneling microscope (STM) [162,163], there are much fewer works specifically studying the light-plasmon coupling in graphene [164]. This is because it is very hard to couple electromagnetic radiation with plasmon excitations in a large-area graphene film [164]. However, in a recent breakthrough work done by Wang group, a strong light-plasmon coupling and plasmon resonance were observed in lithographically patterned graphene microribbon arrays (Fig. 19.32(a)) [164]. Such a strong light-plasmon coupling in graphene indicates that the integrated oscillator strength of plasmons in graphene is one order of magnitude stronger than that in conventional 2DEG, and the plasmon resonance results in prominent room-temperature absorption peaks at terahertz frequencies [164]. In addition, this work demonstrates that the plasmon resonance frequency (ω_p) can be tuned over a broad range of terahertz frequencies by electrostatic doping (Fig. 19.32(b)) [164]. As shown in Fig. 19.32(b), experimentally measured ω_p data exhibit a ¼ power dependence on the electron density (n), which is consistent with Eqn (19.10) and serves as an important signature of the plasmon excitations in 2D massless Dirac electrons

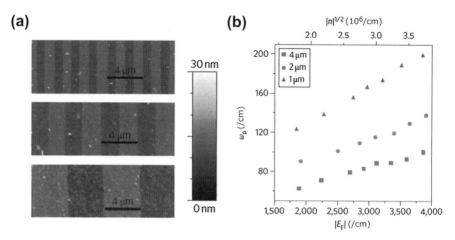

FIGURE 19.32

(a) AFM images of lithographically patterned graphene microribbons with ribbon width of 1 μm, 2 μm, and 4 μm. (b) Experimentally measured plasmon resonance frequency (ω_p) plotted as a function Fermi energy level E_F or equivalent square-roots of electron densities ($n^{1/2}$).

Source: Reprinted by permission from Macmillan Publishers Ltd: [Nature Nanotechnology] (Ref. [164]), copyright (2011).

present in graphene [132,164]. This work provides knowledge that could serve as the foundation for making graphene-based terahertz metamaterials.

CNTs also exhibit a rich spectrum of collective electronic excitation modes (plasmons) in the infrared/visible and ultraviolet (UV) frequency ranges, which has been revealed by using electron energy loss spectroscopy (EELS) [165,166]. The plasmon modes in CNTs and graphene are related because they have highly correlated electronic structures, which can be described as an analogy for excitations in a 2D sheet and along a tube axis [167]. However, SWNTs exhibit two distinct UV absorption peaks for the light with polarization along and perpendicular to the nanotube [167]. For vertically aligned SWNTs, the on-axis polarization peak is at ~4.5 eV and the cross-polarized peak is at ~5.2 eV [168,169]. In addition, EELS works also identified a linear dispersion behavior of on-axis π plasmons, and the linearity was observed to extend over one-third of the Brillouin zone (the magnitude of corresponding wave vectors are up to ~$1/\text{Å}^1$) [167]. A more detailed discussion about the plasmon modes in SWNTs is beyond the scope of this chapter. The readers who are interested in such discussions are referred to previously published works [170–173].

19.4 SUMMARY

In summary, this chapter provides readers an introduction to the main features of the electronic band structure and transport behavior in graphene and SWNTs as well as associated applications in electronics, mechanics, photonics, etc. These two kinds of carbon-based nanomaterials share some common electronic structure associated with the hexagonally packed carbon atoms based on sp^2 hybridization. Their difference in band structures and transport properties stems from the morphological variation in quantum confinement. The electronic structure of 2D graphene features two sets of unequal energy valleys at the boundary of the Brillouin zone. The transport states near the Dirac point exhibit a linear energy spectrum (or dispersion) that makes the carriers in graphene behave like massless Dirac fermions. These unique quantum transport properties make graphene a platform for exploring new mesoscopic science and also an excellent material for making low-dissipation electronic units, ultrafast transistors operating at hundreds of GHz to 1 THz frequencies, bendable transparent electrodes for PV and display devices, or terahertz/infrared plasmonic metamaterials for photonic devices. Intrinsically, graphene is a gapless semimetal, which may be a drawback of graphene for electronic applications requiring a high ON/OFF ratio. Various graphene nanostructures have been studied, aiming to open a large bandgap comparable to that of conventional semiconductors. This field is still active, and people are looking for new nanofabrication or nanomanufacturing methods to create graphene nanostructures with well-defined edge morphologies and crystallinity. Compared to graphene, CNTs apparently exhibit more complicated band structures and transport characteristics, which are highly dependent on the wrapping manner (or "chiral" angles), tube diameters, and number of walls. Here, our discussion focuses on the main properties of SWNTs to fit to the scope of this chapter and capture the most critical physical properties of CNTs. Individual or well-isolated semiconducting SWNTs can be used for making high-speed switch transistors with a much higher ON/OFF ratio than that of graphene-based transistors because of their nonzero bandgap. The scale-up application requires multichannel transistors made of bundles or network films of SWNTs. Because metallic and semiconducting nanotubes are usually synthesized together in the bundles or films, a great deal of effort has been invested to selectively eliminate metallic nanotubes. Similar to graphene, SWNTs exhibit very high in-plane stiffness and breaking strength, whereas they are compliant to the mechanical load applied along out-plane directions. Such unique mechanical properties make SWNTs an ideal structural material for making bendable electronic and energy devices. SWNTs also exhibit a rich spectrum of excitation states at terahertz/infrared and UV frequencies and have a strong anisotropy in optical responses. These characteristics can be implemented for polarization-sensitive optical devices operating at terahertz/infrared frequencies. In addition, semiconducting SWNTs are direct-bandgap semiconductors that have been demonstrated as active materials for PV cells, optical emitters, and infrared detectors. This chapter aims to provide readers basic knowledge about choosing appropriate carbon-based nanomaterials for specific applications in electronics, mechanical composites, optics, and optoelectronics.

References

[1] S. Iijima, Helical microtubules of graphitic carbon, Nature 354 (Nov 7 1991) 56–58.
[2] K.S. Novoselov, A.K. Geim, S.V. Morozov, D. Jiang, Y. Zhang, S.V. Dubonos, I.V. Grigorieva, A.A. Firsov, Electric field effect in atomically thin carbon films, Science 306 (Oct 22 2004) 666–669.
[3] H.W. Kroto, J.R. Heath, S.C. Obrien, R.F. Curl, R.E. Smalley, C-60-buckminsterfullerene, Nature 318 (1985) 162–163.
[4] A. Oberlin, M. Endo, T. Koyama, Filamentous growth of carbon through benzene decomposition, Journal of Crystal Growth 32 (1976) 335–349.
[5] W. Bollmann, J. Spreadborough, Action of graphite as a lubricant, Nature 186 (1960) 29–30.
[6] M. Monthioux, V.L. Kuznetsov, Who should be given the credit for the discovery of carbon nanotubes? Carbon 44 (Aug 2006) 1621–1623.
[7] M.S. Dresselhaus, G. Dresselhaus, P. Avouris, Carbon Nanotubes: Synthesis, Structure, Properties, and Applications. No. 18 in Topics in Applied Physics, Springer, Berlin, 2001.
[8] W. Park, Y.S. Min, Properties, synthesis, purification, and integration of carbon nanotubes for the electronic device applications, Encyclopedia of Nanotechnology 1–2 (2009) 845–883.
[9] H.J. Dai, Carbon nanotubes: synthesis, integration, and properties, Accounts of Chemical Research 35 (Dec 2002) 1035–1044.
[10] J. Kong, C. Zhou, A. Morpurgo, H.T. Soh, C.F. Quate, C. Marcus, H. Dai, Synthesis, integration, and electrical properties of individual single-walled carbon nanotubes, Applied Physics A: Materials Science & Processing 69 (Sep 1999) 305–308.
[11] J. Kong, A. Javey, Carbon nanotube electronics, Integrated Circuits and Systems (2009), http://dx.doi.org/10.1007/978-0-387-69285-2.
[12] A.H. Castro Neto, F. Guinea, N.M.R. Peres, K.S. Novoselov, A.K. Geim, The electronic properties of graphene, Reviews of Modern Physics 81 (Jan–Mar 2009) 109–162.
[13] A.F. Morpurgo, F. Guinea, Intervalley scattering, long-range disorder, and effective time-reversal symmetry breaking in graphene, Physical Review Letters 97 (Nov 10 2006).
[14] S.V. Morozov, K.S. Novoselov, M.I. Katsnelson, F. Schedin, L.A. Ponomarenko, D. Jiang, A.K. Geim, Strong suppression of weak localization in graphene, Physical Review Letters 97 (Jul 7 2006).
[15] A.K. Geim, Graphene: status and prospects, Science 324 (Jun 2009) 1530–1534.
[16] K.S. Novoselov, A.K. Geim, S.V. Morozov, D. Jiang, M.I. Katsnelson, I.V. Grigorieva, S.V. Dubonos, A.A. Firsov, Two-dimensional gas of massless Dirac fermions in graphene, Nature 438 (Nov 10 2005) 197–200.
[17] Y.B. Zhang, Y.W. Tan, H.L. Stormer, P. Kim, Experimental observation of the quantum Hall effect and Berry's phase in graphene, Nature 438 (Nov 2005) 201–204.
[18] E. Minot, Tuning the Band Structure of Carbon Nanotubes, Ph.D. thesis, Cornell University, 2004.
[19] A. Ugawa, A.G. Rinzler, D.B. Tanner, Far-infrared gaps in single-wall carbon nanotubes, Ferroelectrics 249 (2001) 145–154.
[20] R. Saito, G. Dresselhaus, M.S. Dresselhaus, Physical Properties of Carbon Nanotubes, Imperial College Press, 1998.

[21] R. Saito, M. Fujita, G. Dresselhaus, M.S. Dresselhaus, Electronic-structure of graphene tubules based on C-60, Physical Review B 46 (Jul 15 1992) 1804–1811.
[22] R. Saito, M. Fujita, G. Dresselhaus, M.S. Dresselhaus, Electronic-structure of chiral graphene tubules, Applied Physics Letters 60 (May 4 1992) 2204–2206.
[23] Y.W. Son, M.L. Cohen, S.G. Louie, Energy gaps in graphene nanoribbons, Physical Review Letters 97 (Nov 24 2006).
[24] M.Y. Han, B. Özyilmaz, Y.B. Zhang, P. Kim, Energy band-gap engineering of graphene nanoribbons, Physical Review Letters 98 (May 18 2007) 206805.
[25] X.L. Li, X.R. Wang, L. Zhang, S. Lee, H.J. Dai, Chemically derived ultrasmooth graphene nanoribbon semiconductors, Science 319 (2008) 1229–1232.
[26] Y.M. Lin, V. Perebeinos, Z.H. Chen, P. Avouris, Electrical observation of subband formation in graphene nanoribbons, Physical Review B 78 (Oct 2008) 4.
[27] X.R. Wang, Y.J. Ouyang, X.L. Li, H.L. Wang, J. Guo, H.J. Dai, Room-temperature all-semiconducting sub-10 nm graphene nanoribbon field-effect transistors, Physical Review Letters 100 (May 2008) 23.
[28] X. Liang, A.S.P. Chang, Y. Zhang, B.D. Harteneck, H. Choo, D.L. Olynick, S. Cabrini, Electrostatic force assisted exfoliation of prepatterned few-layer graphenes into device sites, Nano Letters 9 (Jan 2009) 467–472.
[29] R.S. Muller, T.I. Kamins, M. Chan, Device Electronics for Integrated Circuits, third ed., John Wiley & Sons, 2002. p.431.
[30] W. Yao, S.A. Yang, Q. Niu, Edge states in graphene: from gapped flat-band to gapless chiral modes, Physical Review Letters 102 (Mar 6 2009).
[31] T. Wassmann, A.P. Seitsonen, A.M. Saitta, M. Lazzeri, F. Mauri, Structure, stability, edge states, and aromaticity of graphene ribbons, Physical Review Letters 101 (Aug 29 2008).
[32] D.A. Abanin, P.A. Lee, L.S. Levitov, Spin-filtered edge states and quantum hall effect in graphene, Physical Review Letters 96 (May 5 2006).
[33] X.R. Wang, Y.J. Ouyang, L.Y. Jiao, H.L. Wang, L.M. Xie, J. Wu, J. Guo, H.J. Dai, Graphene nanoribbons with smooth edges behave as quantum wires, Nature Nanotechnology 6 (Sep 2011) 563–567.
[34] J.L. Garcia-Pomar, A. Cortijo, M. Nieto-Vesperinas, Fully valley-polarized electron beams in graphene, Physical Review Letters 100 (Jun 13 2008).
[35] A. Rycerz, J. Tworzydlo, C.W.J. Beenakker, Valley filter and valley valve in graphene, Nature Physics 3 (Mar 2007) 172–175.
[36] K.I. Bolotin, K.J. Sikes, Z. Jiang, M. Klima, G. Fudenberg, J. Hone, P. Kim, H.L. Stormer, Ultrahigh electron mobility in suspended graphene, Solid State Communications 146 (Jun 2008) 351–355.
[37] C. Berger, Z.M. Song, T.B. Li, X.B. Li, A.Y. Ogbazghi, R. Feng, Z.T. Dai, A.N. Marchenkov, E.H. Conrad, P.N. First, W.A. de Heer, Ultrathin epitaxial graphite: 2D electron gas properties and a route toward graphene-based nanoelectronics, Journal of Physical Chemistry B 108 (Dec 30 2004) 19912–19916.
[38] K.S. Kim, Y. Zhao, H. Jang, S.Y. Lee, J.M. Kim, K.S. Kim, J.H. Ahn, P. Kim, J.Y. Choi, B.H. Hong, Large-scale pattern growth of graphene films for stretchable transparent electrodes, Nature 457 (Feb 5 2009) 706–710.
[39] K.V. Emtsev, A. Bostwick, K. Horn, J. Jobst, G.L. Kellogg, L. Ley, J.L. McChesney, T. Ohta, S.A. Reshanov, J. Rohrl, E. Rotenberg, A.K. Schmid, D. Waldmann, H.B. Weber, T. Seyller, Towards wafer-size graphene layers by atmospheric pressure graphitization of silicon carbide, Nature Materials 8 (Mar 2009) 203–207.

[40] X.S. Li, W.W. Cai, J.H. An, S. Kim, J. Nah, D.X. Yang, R. Piner, A. Velamakanni, I. Jung, E. Tutuc, S.K. Banerjee, L. Colombo, R.S. Ruoff, Large-area synthesis of high-quality and uniform graphene films on copper foils, Science 324 (Jun 5 2009) 1312–1314.

[41] C.R. Dean, A.F. Young, I. Meric, C. Lee, L. Wang, S. Sorgenfrei, K. Watanabe, T. Taniguchi, P. Kim, K.L. Shepard, J. Hone, Boron nitride substrates for high-quality graphene electronics, Nature Nanotechnology 5 (Oct 2010) 722–726.

[42] Y. Shibuta, J.A. Elliott, Interaction between two graphene sheets with a turbostratic orientational relationship, Chemical Physics Letters 512 (Aug 25 2011) 146–150.

[43] S. Shallcross, S. Sharma, E. Kandelaki, O.A. Pankratov, Electronic structure of turbostratic graphene, Physical Review B 81 (Jun 29 2010) 165105.

[44] T. Jayasekera, K.W. Kim, M. Buongiorno Nardelli, Electronic and structural properties of turbostratic epitaxial graphene on the 6H-SiC (000-1) surface, Materials Science Forum 717-720 (2012) 595–600.

[45] M. Orlita, C. Faugeras, P. Plochocka, P. Neugebauer, G. Martinez, D.K. Maude, A.L. Barra, M. Sprinkle, C. Berger, W.A. de Heer, M. Potemski, Approaching the Dirac point in high-mobility multilayer epitaxial graphene, Physical Review Letters 101 (Dec 31 2008).

[46] C. Berger, Z. Song, T. Li, X. Li, A.Y. Ogbazghi, R. Feng, Z. Dai, A.N. Marchenkov, E.H. Conrad, P.N. First, W.A. de Heer, Ultrathin epitaxial graphite: 2D electron gas properties and a route toward graphene-based nanoelectronics, Journal of Physical Chemistry 108 (2004) 19912–19916.

[47] L.Y. Jiao, X.R. Wang, G. Diankov, H.L. Wang, H.J. Dai, Facile synthesis of high-quality graphene nanoribbons, Nature Nanotechnology 5 (May 2010) 321–325.

[48] X. Liang, Electrostatic Force-Assisted Printing of Pre-Patterned Few-Layer-Graphenes into Device Sites, US Patent, 2008, 8057863.

[49] T.J. Echtermeyer, M.C. Lemme, J. Bolten, M. Baus, M. Ramsteiner, H. Kurz, Graphene field-effect devices, European Physical Journal-Special Topics 148 (Sep 2007) 19–26.

[50] X. Liang, Z. Fu, S.Y. Chou, Graphene transistors fabricated via transfer-printing in device active-areas on large wafer, Nano Letters 7 (Dec 2007) 3840–3844.

[51] Y.M. Lin, C. Dimitrakopoulos, K.A. Jenkins, D.B. Farmer, H.Y. Chiu, A. Grill, P. Avouris, 100-GHz transistors from wafer-scale epitaxial graphene, Science 327 (Feb 5 2010) 662.

[52] L. Liao, Y.C. Lin, M.Q. Bao, R. Cheng, J.W. Bai, Y.A. Liu, Y.Q. Qu, K.L. Wang, Y. Huang, X.F. Duan, High-speed graphene transistors with a self-aligned nanowire gate, Nature 467 (Sep 16 2010) 305–308.

[53] R. Cheng, J.W. Bai, L. Liao, H.L. Zhou, Y. Chen, L.X. Liu, Y.C. Lin, S. Jiang, Y. Huang, X.F. Duan, High-frequency self-aligned graphene transistors with transferred gate stacks, Proceedings of the National Academy of Sciences of the United States of America 109 (Jul 17 2012) 11588–11592.

[54] Y.Q. Wu, Y.M. Lin, A.A. Bol, K.A. Jenkins, F.N. Xia, D.B. Farmer, Y. Zhu, P. Avouris, High-frequency, scaled graphene transistors on diamond-like carbon, Nature 472 (Apr 7 2011) 74–78.

[55] Y.Q. Wu, D.B. Farmer, A. Valdes-Garcia, W.J. Zhu, K.A. Jenkins, C. Dimitrakopoulos, P. Avouris, and Y.M. Lin, Record High RF Performance for Epitaxial Graphene Transistors, IEEE International Electron Devices Meeting (Iedm), 2011.

[56] J. Kong, N.R. Franklin, C.W. Zhou, M.G. Chapline, S. Peng, K.J. Cho, H.J. Dai, Nanotube molecular wires as chemical sensors, Science 287 (Jan 28 2000) 622−625.

[57] T. Durkop, B.M. Kim, M.S. Fuhrer, Properties and applications of high-mobility semiconducting nanotubes, Journal of Physics: Condensed Matter 16 (May 12 2004) R553−R580.

[58] S. Datta, Electron Transport in Mesoscopic Systems, Cambridge University Press, Cambridge, 1995.

[59] C.T. White, T.N. Todorov, Carbon nanotubes as long ballistic conductors, Nature 393 (May 21 1998) 240−242.

[60] D. Mann, A. Javey, J. Kong, Q. Wang, H.J. Dai, Ballistic transport in metallic nanotubes with reliable Pd ohmic contacts, Nano Letters 3 (Nov 2003) 1541−1544.

[61] T. Durkop, S.A. Getty, E. Cobas, M.S. Fuhrer, Extraordinary mobility in semiconducting carbon nanotubes, Nano Letters 4 (Jan 2004) 35−39.

[62] G. Pennington, N. Goldsman, Semiclassical transport and phonon scattering of electrons in semiconducting carbon nanotubes, Physical Review B 68 (Jul 15 2003).

[63] R.J. Nicholas, A. Mainwood, L. Eaves, Introduction. Carbon-based electronics: fundamentals and device applications, Philosophical Transactions of the Royal Society A: Mathematical Physical and Engineering Sciences 366 (Jan 28 2008) 189−193.

[64] H.N. Nguyen, H.C. d'Honincthun, C. Chapus, A. Bournel, S. Galdin-Retailleau, P. Dollfus, N. Locatelli, Monte Carlo modeling of Schottky contacts on semiconducting carbon nanotubes, Simulation of Semiconductor Processes and Devices 2007 (2007) 313−316.

[65] Y. Miyata, K. Shiozawa, Y. Asada, Y. Ohno, R. Kitaura, T. Mizutani, H. Shinohara, Length-sorted semiconducting carbon nanotubes for high-mobility thin film transistors, Nano Research 4 (Oct 2011) 963−970.

[66] B. Xu, Y.D. Xia, J. Yin, X.G. Wan, K. Jiang, A.D. Li, D. Wu, Z.G. Liu, The effect of acoustic phonon scattering on the carrier mobility in the semiconducting zigzag single wall carbon nanotubes, Applied Physics Letters 96 (May 3 2010).

[67] Y. Zhao, A. Liao, E. Pop, Multiband mobility in semiconducting carbon nanotubes, IEEE Electron Device Letters 30 (Oct 2009) 1078−1080.

[68] V. Perebeinos, J. Tersoff, P. Avouris, Mobility in semiconducting carbon nanotubes at finite carrier density, Nano Letters 6 (Feb 2006) 205−208.

[69] C. Kocabas, S. Dunham, Q. Cao, K. Cimino, X.N. Ho, H.S. Kim, D. Dawson, J. Payne, M. Stuenkel, H. Zhang, T. Banks, M. Feng, S.V. Rotkin, J.A. Rogers, High-frequency performance of submicrometer transistors that use aligned arrays of single-walled carbon nanotubes, Nano Letters 9 (May 2009) 1937−1943.

[70] S. Rosenblatt, H. Lin, V. Sazonova, S. Tiwari, P.L. McEuen, Mixing at 50 GHz using a single-walled carbon nanotube transistor, Applied Physics Letters 87 (Oct 10 2005).

[71] A. Srivastava, Y. Xu, Y. Liu, A.K. Sharma, C. Mayberry, CMOS LC voltage controlled oscillator design using multiwalled and single-walled carbon nanotube wire inductors, ACM Journal on Emerging Technologies in Computing Systems 8 (Aug 2012).

[72] A.A. Pesetski, J.E. Baumgardner, S.V. Krishnaswamy, H. Zhang, J.D. Adam, C. Kocabas, T. Banks, J.A. Rogers, A 500 MHz carbon nanotube transistor oscillator, Applied Physics Letters 93 (Sep 22 2008).

[73] J.W. Kang, H.J. Hwang, Operating frequency in a triple-walled carbon-nanotube oscillator, Journal of the Korean Physical Society 49 (Oct 2006) 1488–1492.

[74] J.W. Kang, K.O. Song, O.K. Kwon, H.J. Hwang, Carbon nanotube oscillator operated by thermal expansion of encapsulated gases, Nanotechnology 16 (Nov 2005) 2670–2676.

[75] V. Sazonova, Y. Yaish, H. Ustunel, D. Roundy, T.A. Arias, P.L. McEuen, A tunable carbon nanotube electromechanical oscillator, Nature 431 (Sep 16 2004) 284–287.

[76] K. Ryu, A. Badmaev, C. Wang, A. Lin, N. Patil, L. Gomez, A. Kumar, S. Mitra, H.S.P. Wong, C.W. Zhou, CMOS-analogous wafer-scale nanotube-on-insulator approach for submicrometer devices and integrated circuits using aligned nanotubes, Nano Letters 9 (Jan 2009) 189–197.

[77] C. Kocabas, H.S. Kim, T. Banks, J.A. Rogers, A.A. Pesetski, J.E. Baumgardner, S.V. Krishnaswamy, H. Zhang, Radio frequency analog electronics based on carbon nanotube transistors, Proceedings of the National Academy of Sciences of the United States of America 105 (Feb 5 2008) 1405–1409.

[78] A.K. Geim, K.S. Novoselov, The rise of graphene, Nature Materials 6 (March 2007) 183.

[79] Y.B. Zhang, T.T. Tang, C. Girit, Z. Hao, M.C. Martin, A. Zettl, M.F. Crommie, Y.R. Shen, F. Wang, Direct observation of a widely tunable bandgap in bilayer graphene, Nature 459 (Jun 11 2009) 820–823.

[80] X.L. Li, X.R. Wang, L. Zhang, S.W. Lee, H.J. Dai, Chemically derived, ultrasmooth graphene nanoribbon semiconductors, Science 319 (Feb 29 2008) 1229–1232.

[81] I.I. Naumov, A.M. Bratkovsky, Semiconducting graphene nanomeshes, Physical Review B 85 (May 25 2012).

[82] W. Oswald, Z.G. Wu, Energy gaps in graphene nanomeshes, Physical Review B 85 (Mar 22 2012).

[83] H. Sahin, S. Ciraci, Structural, mechanical, and electronic properties of defect-patterned graphene nanomeshes from first principles, Physical Review B 84 (Jul 29 2011).

[84] K. Lopata, R. Thorpe, S. Pistinner, X.F. Duan, D. Neuhauser, Graphene nanomeshes: onset of conduction band gaps, Chemical Physics Letters 498 (Oct 8 2010) 334–337.

[85] X.G. Liang, Y.S. Jung, S.W. Wu, A. Ismach, D.L. Olynick, S. Cabrini, J. Bokor, Formation of bandgap and subbands in graphene nanomeshes with sub-10 nm ribbon width fabricated via nanoimprint lithography, Nano Letters 10 (Jul 2010) 2454–2460.

[86] L.A. Falkovsky, Gate-tunable bandgap in bilayer graphene, Journal of Experimental and Theoretical Physics 110 (Feb 2010) 319–324.

[87] Bandgap controlled in bilayer graphene, MRS Bulletin 34 (Sep 2009) 630–631.

[88] E. Sano, T. Otsuji, Bandgap engineering of bilayer graphene for field-effect transistor channels, Japanese Journal of Applied Physics 48 (Sep 2009).

[89] L.Y. Jiao, L.M. Xie, H.J. Dai, Densely aligned graphene nanoribbons at ∼35 nm pitch, Nano Research 5 (Apr 2012) 292–296.

[90] X.G. Liang, S. Wi, Transport characteristics of multichannel transistors made from densely aligned sub-10 nm half-pitch graphene nanoribbons, ACS Nano (October 13 2012). Publication Date (Web).

[91] L. Tapasztó, G. Dobrik, P. Lambin, L.P. Biró, Tailoring the atomic structure of graphene nanoribbons by scanning tunneling microscope lithography, Nature Nanotechnology 3 (2008) 397–401.

[92] J.M. Cai, P. Ruffieux, R. Jaafar, M. Bieri, T. Braun, S. Blankenburg, M. Muoth, A.P. Seitsonen, M. Saleh, X.L. Feng, K. Mullen, R. Fasel, Atomically precise bottom-up fabrication of graphene nanoribbons, Nature 466 (Jul 22 2010) 470–473.

[93] J. Bai, X. Zhong, S. Jiang, Y. Huang, X. Duan, Graphene nanomesh, Nature Nanotechnology 5 (2010) 190–194.

[94] M. Kim, N.S. Safron, E. Han, M.S. Arnold, P. Gopalan, Fabrication and characterization of large-area, semiconducting nanoperforated graphene materials, Nano Letters 10 (2010) 1125–1131.

[95] X. Liang, S. Wi, Patterning densely arranged sub-10 nm graphene nanoribbons for making multi-channel transistors. Accepted, ACS Nano (2012).

[96] X. Liang, Y.-S. Jung, S. Wu, A. Ismach, D. L. Olynick, S. Cabrini, and J. Bokor, Graphene Nanomeshes with Sub-10 nm Ribbon Width Fabricated via Nanoimprint Lithography in Combination with Block Copolymer Self-Assembly, Proceeding of International Conference on Electron, Ion, and Photon Beam Technology and Nanofabrication, 2010.

[97] Z.H. Pan, N. Liu, L. Fu, Z.F. Liu, Wrinkle engineering: a new approach to massive graphene nanoribbon arrays, Journal of the American Chemical Society 133 (Nov 9 2011) 17578–17581.

[98] H.R. Shea, R. Martel, T. Hertel, T. Schmidt, P. Avouris, Manipulation of carbon nanotubes and properties of nanotube field-effect transistors and rings, Microelectronic Engineering 46 (May 1999) 101–104.

[99] C. Wang, J.C. Chien, K. Takei, T. Takahashi, J. Nah, A.M. Niknejad, A. Javey, Extremely bendable, high-performance integrated circuits using semiconducting carbon nanotube networks for digital, analog, and radio-frequency applications, Nano Letters 12 (Mar 2012) 1527–1533.

[100] Y.C. Che, C. Wang, J. Liu, B.L. Liu, X. Lin, J. Parker, C. Beasley, H.S.P. Wong, C.W. Zhou, Selective synthesis and device applications of semiconducting single-walled carbon nanotubes using isopropyl alcohol as feedstock, ACS Nano 6 (Aug 2012) 7454–7462.

[101] Z.Y. Zhang, S. Wang, L. Ding, X.L. Liang, T. Pei, J. Shen, H.L. Xu, O. Chen, R.L. Cui, Y. Li, L.M. Peng, Self-aligned ballistic n-type single-walled carbon nanotube field-effect transistors with adjustable threshold voltage, Nano Letters 8 (Nov 2008) 3696–3701.

[102] Q. Cao, J.A. Rogers, Random networks and aligned arrays of single-walled carbon nanotubes for electronic device applications, Nano Research 1 (Oct 2008) 259–272.

[103] C. Kocabas, M. Shim, J.A. Rogers, Spatially selective guided growth of high-coverage arrays and random networks of single-walled carbon nanotubes and their integration into electronic devices, Journal of the American Chemical Society 128 (Apr 12 2006) 4540–4541.

[104] C. Kocabas, M.A. Meitl, A. Gaur, M. Shim, J.A. Rogers, Aligned arrays of single-walled carbon nanotubes generated from random networks by orientationally selective laser ablation, Nano Letters 4 (Dec 2004) 2421–2426.

[105] C. Kocabas, N. Pimparkar, O. Yesilyurt, S.J. Kang, M.A. Alam, J.A. Rogers, Experimental and theoretical studies of transport through large scale, partially aligned arrays of single-walled carbon nanotubes in thin film type transistors, Nano Letters 7 (May 2007) 1195–1202.

[106] W.W. Zhou, S.T. Zhan, L. Ding, J. Liu, General rules for selective growth of enriched semiconducting single walled carbon nanotubes with water vapor as in situ etchant, Journal of the American Chemical Society 134 (Aug 29 2012) 14019−14026.

[107] L. Ding, A. Tselev, J.Y. Wang, D.N. Yuan, H.B. Chu, T.P. McNicholas, Y. Li, J. Liu, Selective growth of well-aligned semiconducting single-walled carbon nanotubes, Nano Letters 9 (Feb 2009) 800−805.

[108] R. Voggu, A. Govindaraj, C.N.R. Rao, Selective synthesis of metallic and semiconducting single-walled carbon nanotubes, Indian Journal of Chemistry Section A: Inorganic, Bio-Inorganic Physical Theoretical & Analytical Chemistry 51 (Jan−Feb 2012) 32−46.

[109] C.Z. Loebick, R. Podila, J. Reppert, J. Chudow, F. Ren, G.L. Haller, A.M. Rao, L.D. Pfefferle, Selective synthesis of subnanometer diameter semiconducting single-walled carbon nanotubes, Journal of the American Chemical Society 132 (Aug 18 2010) 11125−11131.

[110] W.H. Chiang, M. Sakr, X.P.A. Gao, R.M. Sankaran, Nanoengineering Ni_xFe_{1-x} catalysts for gas-phase, selective synthesis of semiconducting single-walled carbon nanotubes, ACS Nano 3 (Dec 2009) 4023−4032.

[111] K. Milowska, M. Birowska, J.A. Majewski, Mechanical and electrical properties of carbon nanotubes and graphene layers functionalized with amines, Diamond and Related Materials 23 (Mar 2012) 167−171.

[112] S.H. Tzeng, J.L. Tsai, Characterizing the mechanical properties of graphene and single walled carbon nanotubes, Journal of Mechanics 27 (Dec 2011) 461−467.

[113] D. Askari, M.N. Ghasemi-Nejhad, Effects of vacancy defects on mechanical properties of graphene/carbon nanotubes: a numerical modeling, Journal of Computational and Theoretical Nanoscience 8 (Apr 2011) 783−794.

[114] S.Y. Yang, W.N. Lin, Y.L. Huang, H.W. Tien, J.Y. Wang, C.C.M. Ma, S.M. Li, Y.S. Wang, Synergetic effects of graphene platelets and carbon nanotubes on the mechanical and thermal properties of epoxy composites, Carbon 49 (Mar 2011) 793−803.

[115] A.V. Herrera-Herrera, M.A. Gonzalez-Curbelo, J. Hernandez-Borges, M.A. Rodriguez-Delgado, Carbon nanotubes applications in separation science: a review, Analytica Chimica Acta 734 (Jul 13 2012) 1−30.

[116] C. Lee, X.D. Wei, J.W. Kysar, J. Hone, Measurement of the elastic properties and intrinsic strength of monolayer graphene, Science 321 (Jul 18 2008) 385−388.

[117] M.F. Yu, B.S. Files, S. Arepalli, R.S. Ruoff, Tensile loading of ropes of single wall carbon nanotubes and their mechanical properties, Physical Review Letters 84 (Jun 12 2000) 5552−5555.

[118] V.M. Harik, Mechanics of carbon nanotubes: applicability of the continuum-beam models, Computational Materials Science 24 (Jun 2002) 328−342.

[119] S.L. Zhang, S.L. Mielke, R. Khare, D. Troya, R.S. Ruoff, G.C. Schatz, T. Belytschko, Mechanics of defects in carbon nanotubes: atomistic and multiscale simulations, Physical Review B 71 (Mar 2005).

[120] D. Qian, G.J. Wagner, W.K. Liu, M.-F. Yu, R.S. Ruoff, Mechanics of carbon nanotubes, Applied Mechanics Reviews 55 (2002) 495−534.

[121] M.F. Yu, T. Kowalewski, R.S. Ruoff, Investigation of the radial deformability of individual carbon nanotubes under controlled indentation force, Physical Review Letters 85 (Aug 14 2000) 1456−1459.

[122] R.S. Ruoff, J. Tersoff, D.C. Lorents, S. Subramoney, B. Chan, Radial deformation of carbon nanotubes by van-der-Waals forces, Nature 364 (Aug 5 1993) 514–516.

[123] M.F. Yu, M.J. Dyer, J. Chen, D. Qian, W.K. Liu, R.S. Ruoff, Locked twist in multiwalled carbon-nanotube ribbons, Physical Review B 64 (Dec 15 2001).

[124] B.I. Yakobson, C.J. Brabec, J. Bernholc, Nanomechanics of carbon tubes: instabilities beyond linear response, Physical Review Letters 76 (Apr 1 1996) 2511–2514.

[125] B.I. Yakobson, R.E. Smalley, Fullerene nanotubes: C-1000000 and beyond, American Scientist 85 (Jul–Aug 1997) 324–337.

[126] S. Bae, H. Kim, Y. Lee, X.F. Xu, J.S. Park, Y. Zheng, J. Balakrishnan, T. Lei, H.R. Kim, Y.I. Song, Y.J. Kim, K.S. Kim, B. Ozyilmaz, J.H. Ahn, B.H. Hong, S. Iijima, Roll-to-roll production of 30-inch graphene films for transparent electrodes, Nature Nanotechnology 5 (Aug 2010) 574–578.

[127] L.G. De Arco, Y. Zhang, C.W. Schlenker, K. Ryu, M.E. Thompson, C.W. Zhou, Continuous, highly flexible, and transparent graphene films by chemical vapor deposition for organic photovoltaics, ACS Nano 4 (May 2010) 2865–2873.

[128] S.K. Lee, H.Y. Jang, S. Jang, E. Choi, B.H. Hong, J. Lee, S. Park, J.H. Ahn, All graphene-based thin film transistors on flexible plastic substrates, Nano Letters 12 (Jul 2012) 3472–3476.

[129] C. Yan, J.H. Cho, J.H. Ahn, Graphene-based flexible and stretchable thin film transistors, Nanoscale 4 (2012) 4870–4882.

[130] Q.Y. He, S.X. Wu, S. Gao, X.H. Cao, Z.Y. Yin, H. Li, P. Chen, H. Zhang, Transparent, flexible, all-reduced graphene oxide thin film transistors, ACS Nano 5 (Jun 2011) 5038–5044.

[131] M. Jin, H.K. Jeong, W.J. Yu, D.J. Bae, B.R. Kang, Y.H. Lee, Graphene oxide thin film field effect transistors without reduction, Journal of Physics D: Applied Physics 42 (Jul 7 2009).

[132] S.A. Mikhailov, Nonlinear Electrodynamics And Optics Of Graphene, Physics of Semiconductors: 30th International Conference on the Physics of Semiconductors, 1399, (2011).

[133] T. Ando, Y.S. Zheng, H. Suzuura, Dynamical conductivity and zero-mode anomaly in honeycomb lattices, Journal of the Physical Society of Japan 71 (May 2002) 1318–1324.

[134] L.A. Falkovsky, S.S. Pershoguba, Optical far-infrared properties of a graphene monolayer and multilayer, Physical Review B 76 (Oct 2007).

[135] V.P. Gusynin, S.G. Sharapov, Transport of Dirac quasiparticles in graphene: Hall and optical conductivities, Physical Review B 73 (Jun 2006).

[136] Z.Q. Li, E.A. Henriksen, Z. Jiang, Z. Hao, M.C. Martin, P. Kim, H.L. Stormer, D.N. Basov, Dirac charge dynamics in graphene by infrared spectroscopy, Nature Physics 4 (Jul 2008) 532–535.

[137] R.R. Nair, P. Blake, A.N. Grigorenko, K.S. Novoselov, T.J. Booth, T. Stauber, N.M.R. Peres, A.K. Geim, Fine structure constant defines visual transparency of graphene, Science 320 (Jun 6 2008) 1308.

[138] L. Ren, Q. Zhang, J. Yao, Z.Z. Sun, R. Kaneko, Z. Yan, S. Nanot, Z. Jin, I. Kawayama, M. Tonouchi, J.M. Tour, J. Kono, Terahertz and infrared spectroscopy of gated large-area graphene, Nano Letters 12 (Jul 2012) 3711–3715.

[139] L. Ren, Q. Zhang, S. Nanot, I. Kawayama, M. Tonouchi, J. Kono, Terahertz dynamics of quantum-confined electrons in carbon nanomaterials, Journal of Infrared Millimeter and Terahertz Waves 33 (Aug 2012) 846–860.

[140] F. Bommeli, L. Degiorgi, P. Wachter, W.S. Bacsa, W.A. deHeer, L. Forro, Evidence of anisotropic metallic behaviour in the optical properties of carbon nanotubes, Solid State Communications 99 (Aug 1996) 513–517.

[141] A. Ugawa, A.G. Rinzler, D.B. Tanner, Far-infrared gaps in single-wall carbon nanotubes, Physical Review B 60 (Oct 15 1999) R11305–R11308.

[142] T. Kampfrath, K. von Volkmann, C.M. Aguirre, P. Desjardins, R. Martel, M. Krenz, C. Frischkorn, M. Wolf, L. Perfetti, Mechanism of the far-infrared absorption of carbon-nanotube films, Physical Review Letters 101 (Dec 31 2008).

[143] G.Y. Slepyan, M.V. Shuba, S.A. Maksimenko, C. Thomsen, A. Lakhtakia, Terahertz conductivity peak in composite materials containing carbon nanotubes: theory and interpretation of experiment, Physical Review B 81 (May 15 2010).

[144] C. Frischkorn, T. Kampfrath, K. von Volkmann, L. Perfetti, and M. Wolf, Ultrafast changes in the far-infrared conductivity of carbon nanotubes, 33rd International Conference on Infrared, Millimeter and Terahertz Waves, 2008, vols 1–2, pp. 185–185.

[145] N. Akima, Y. Iwasa, S. Brown, A.M. Barbour, J.B. Cao, J.L. Musfeldt, H. Matsui, N. Toyota, M. Shiraishi, H. Shimoda, O. Zhou, Strong anisotropy in the far-infrared absorption spectra of stretch-aligned single-walled carbon nanotubes, Advanced Materials 18 (May 2 2006) 1166.

[146] T. Nakanishi, T. Ando, Optical response of finite-length carbon nanotubes, Journal of the Physical Society of Japan 78 (Nov 2009).

[147] L. Ren, C.L. Pint, T. Arikawa, K. Takeya, I. Kawayama, M. Tonouchi, R.H. Hauge, J. Kono, Broadband terahertz polarizers with ideal performance based on aligned carbon nanotube stacks, Nano Letters 12 (Feb 2012) 787–790.

[148] F. Triozon, P. Lambin, S. Roche, Electronic transport properties of carbon nanotube based metal/semiconductor/metal intramolecular junctions, Nanotechnology 16 (Feb 2005) 230–233.

[149] N.M. Gabor, Z.H. Zhong, K. Bosnick, J. Park, P.L. McEuen, Extremely efficient multiple electron-hole pair generation in carbon nanotube photodiodes, Science 325 (Sep 11 2009) 1367–1371.

[150] F.N. Xia, M. Steiner, Y.M. Lin, P. Avouris, A microcavity-controlled, current-driven, on-chip nanotube emitter at infrared wavelengths, Nature Nanotechnology 3 (Oct 2008) 609–613.

[151] A. Javey, J. Guo, D.B. Farmer, Q. Wang, D.W. Wang, R.G. Gordon, M. Lundstrom, H.J. Dai, Carbon nanotube field-effect transistors with integrated ohmic contacts and high-k gate dielectrics, Nano Letters 4 (Mar 2004) 447–450.

[152] L.J. Yang, S. Wang, Q.S. Zeng, Z.Y. Zhang, T. Pei, Y. Li, L.M. Peng, Efficient photovoltage multiplication in carbon nanotubes, Nature Photonics 5 (Nov 2011) 673–677.

[153] J.A. Misewich, R. Martel, P. Avouris, J.C. Tsang, S. Heinze, J. Tersoff, Electrically induced optical emission from a carbon nanotube FET, Science 300 (May 2 2003) 783–786.

[154] M. Freitag, J. Chen, J. Tersoff, J.C. Tsang, Q. Fu, J. Liu, P. Avouris, Mobile ambipolar domain in carbon-nanotube infrared emitters, Physical Review Letters 93 (Aug 13 2004).

[155] Q.S. Zeng, S. Wang, L.J. Yang, Z.X. Wang, T. Pei, Z.Y. Zhang, L.M. Peng, W.W. Zhou, J. Liu, W.Y. Zhou, S.S. Xie, Carbon nanotube arrays based

high-performance infrared photodetector [Invited], Optical Materials Express 2 (Jun 1 2012) 839–848.

[156] M. Liu, X.B. Yin, E. Ulin-Avila, B.S. Geng, T. Zentgraf, L. Ju, F. Wang, X. Zhang, A graphene-based broadband optical modulator, Nature 474 (Jun 2 2011) 64–67.

[157] T. Battumur, S.H. Mujawar, Q.T. Truong, S.B. Ambade, D.S. Lee, W. Lee, S.H. Han, S.H. Lee, Graphene/carbon nanotubes composites as a counter electrode for dye-sensitized solar cells, Current Applied Physics 12 (Sep 2012) E49–E53.

[158] K.F. Mak, M.Y. Sfeir, Y. Wu, C.H. Lui, J.A. Misewich, T.F. Heinz, Measurement of the optical conductivity of graphene, Physical Review Letters 101 (Nov 7 2008).

[159] M.K. Kinyanjui, C. Kramberger, T. Pichler, J.C. Meyer, P. Wachsmuth, G. Benner, U. Kaiser, Direct probe of linearly dispersing 2D interband plasmons in a free-standing graphene monolayer, Europhysics Letters 97 (Mar 2012).

[160] B. Wunsch, T. Stauber, F. Sols, F. Guinea, Dynamical polarization of graphene at finite doping, New Journal of Physics 8 (Dec 13 2006).

[161] O. Vafek, Thermoplasma polariton within scaling theory of single-layer graphene, Physical Review Letters 97 (Dec 31 2006).

[162] Y. Liu, R.F. Willis, K.V. Emtsev, T. Seyller, Plasmon dispersion and damping in electrically isolated two-dimensional charge sheets, Physical Review B 78 (Nov 2008).

[163] V.W. Brar, S. Wickenburg, M. Panlasigui, C.H. Park, T.O. Wehling, Y.B. Zhang, R. Decker, C. Girit, A.V. Balatsky, S.G. Louie, A. Zettl, M.F. Crommie, Observation of carrier-density-dependent many-body effects in graphene via tunneling spectroscopy, Physical Review Letters 104 (Jan 22 2010).

[164] L. Ju, B.S. Geng, J. Horng, C. Girit, M. Martin, Z. Hao, H.A. Bechtel, X.G. Liang, A. Zettl, Y.R. Shen, F. Wang, Graphene plasmonics for tunable terahertz metamaterials, Nature Nanotechnology 6 (Oct 2011) 630–634.

[165] M. Sing, V.G. Grigoryan, G. Paasch, M. Knupfer, J. Fink, H. Berger, F. Levy, Unusual plasmon dispersion in the quasi-one-dimensional conductor $(TaSe_4)_2I$: experiment and theory, Physical Review B 57 (May 15 1998) 12768–12771.

[166] L.A. Bursill, P.A. Stadelmann, J.L. Peng, S. Prawer, Surface-plasmon observed for carbon nanotubes, Physical Review B 49 (Jan 15 1994) 2882–2887.

[167] C. Kramberger, R. Hambach, C. Giorgetti, M.H. Rummeli, M. Knupfer, J. Fink, B. Buchner, L. Reining, E. Einarsson, S. Maruyama, F. Sottile, K. Hannewald, V. Olevano, A.G. Marinopoulos, T. Pichler, Linear plasmon dispersion in single-wall carbon nanotubes and the collective excitation spectrum of graphene, Physical Review Letters 100 (May 16 2008).

[168] S. Kazaoui, N. Minami, R. Jacquemin, H. Kataura, Y. Achiba, Amphoteric doping of single-wall carbon-nanotube thin films as probed by optical absorption spectroscopy, Physical Review B 60 (Nov 15 1999) 13339–13342.

[169] Y. Murakami, S. Chiashi, E. Einarsson, S. Maruyama, Polarization dependence of resonant Raman scattering from vertically aligned single-walled carbon nanotube films, Physical Review B 71 (Feb 2005).

[170] M.Z. Herrera, J.L. Gervasoni, Surface and bulk plasmon excitations in carbon nanotubes. comparison with the hydrodynamic model, Nuclear Instruments & Methods in Physics Research Section B: Beam Interactions with Materials and Atoms 267 (Jan 2009) 415–418.

[171] I.V. Bondarev, K. Tatur, and L.M. Woods, Strongly Coupled Surface Plasmon-Exciton Excitations in Small-Diameter Carbon Nanotubes, Conference on Lasers and Electro-Optics & Quantum Electronics and Laser Science Conference, 2008, vols 1—9, pp. 3459—3460.

[172] T. Stockli, Z.L. Wang, J.M. Bonard, P. Sadelmann, A. Chatelain, Plasmon excitations in carbon nanotubes, Philosophical Magazine B: Physics of Condensed Matter: Statistical Mechanics Electronic Optical and Magnetic Properties 79 (Oct 1999) 1531—1548.

[173] T. Stockli, J.M. Bonard, P.A. Stadelmann, A. Chatelain, EELS investigation of plasmon excitations in aluminum nanospheres and carbon nanotubes, Zeitschrift Fur Physik D: Atoms Molecules and Clusters 40 (1997) 425—428.

CHAPTER 20

Multiscale Modeling of CNT Composites using Molecular Dynamics and the Boundary Element Method

Y.J. Liu[1], D. Qian[2], P. He[3], N. Nishimura[4]

[1] *Mechanical Engineering, PO Box 210072, University of Cincinnati, Cincinnati, Ohio 45221-0072, USA,* [2] *Department of Mechanical Engineering, University of Texas at Dallas, ECSN 3.206, 800 West Campbell Rd Mailstop: EC-38, Richardson, TX 75080-3021, USA,* [3] *UES, Inc., Dayton, Ohio, USA,* [4] *Academic Center for Computing and Media Studies, Kyoto University, Kyoto 606-8501, Japan*

CHAPTER OUTLINE

20.1 Introduction	570
20.1.1 Nanocomposites and the challenges in the modeling and simulations	570
20.1.2 Literature review	571
20.1.3 A hierarchical multiscale approach to modeling CNT composites	573
20.2 Nanoscale Simulations Using Molecular Dynamics	574
20.2.1 Basics of molecular dynamics	574
20.2.2 Developing a cohesive interface model for CNT composites using MD	575
20.3 Microscale Simulations Using the Boundary Element Method	578
20.3.1 The boundary element method	578
20.3.2 The fast multipole method	580
20.4 Numerical Examples	581
20.4.1 Further study of CNT pull-out tests using MD	581
20.4.2 Fracture analysis of CNT/polymer composites using MD	584
20.4.3 Large-scale BEM models for CNT composites	584
20.4.4 Effect of the cohesive interface in modeling CNT/polymer composites	586
20.5 Discussions	589
Acknowledgments	590
References	591

20.1 INTRODUCTION

20.1.1 Nanocomposites and the challenges in the modeling and simulations

Nanocomposites, reinforced with carbon nanotubes (CNTs), nanoplatelets or nanoparticles (Fig. 20.1), are very attractive modern materials that have great potentials in engineering, such as for mechanical, civil, aerospace, electronic and biomedical engineering applications. For example, CNTs [1–9] have been found to possess extremely high stiffness, strength and resilience, and are considered to be the best reinforcing fibers so far to make lightweight, strong and tough composites for modern structures.

There has been tremendous interest in the modeling and simulations of the CNT composites in order to characterize their mechanical properties for potential engineering applications. Both molecular dynamics (MD) and continuum mechanics, and combinations of both, have been attempted for this purpose. MD approach is necessary in the study of nanocomposites, especially for investigations of the local interactions of CNTs with matrix materials, interface properties, and failure modes. However, MD simulations at present are limited to small length and time scales, due to the limitations of the current computing power. For example, all current MD simulation results of CNT composites have been limited to models with one or a few CNTs in a matrix. CNT fibers in a real composite are likely to have different shapes and sizes. They can be straight or curved, short or long, aligned or oriented arbitrarily, and distributed randomly. All these factors make the estimates of the mechanical properties of CNT composites very difficult using only the MD approach, if not impossible. Large-scale representative volume elements (RVEs) with hundreds or thousands of CNT fibers at the microscale may be deemed necessary in characterizing CNT composites.

In addition to the scale and size of adequate models for modeling CNT composites, interface models have been found to be another key factor. Interfaces between CNT fibers and the matrix are critical for load transfer in CNT composites, besides alignment and dispersion of the CNTs in a matrix. It is observed in experiments that the load transfer mechanism in CNT composites is weak due to the difficulties in making strong bond between CNT fibers and

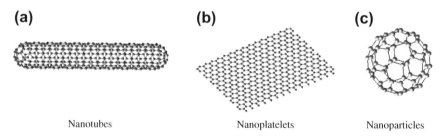

FIGURE 20.1

Possible nanostructures that can be used as fillers in developing nanocomposites.

polymer matrix [7,10–12]. CNTs often interact with the molecules of the polymer matrix through mainly van der Waals forces. Therefore, perfect bonding or any other strong bond interface conditions in various MD or continuum models of CNT composites will result in overestimated effective stiffness for the composites. New interface models that can account for the weak interactions between CNTs and their host matrix need to be developed. These improved models of the CNT composites will improve the correlation of the simulation results with the experimental ones. The improved interface models can also be employed to study the effects of different interface parameters and hence help to design better CNT composites.

20.1.2 Literature review

Atomistic models, such as MD simulations [13–15], have been applied widely in nanoscale research. For example, MD approaches have provided abundant simulation results for understanding the behaviors of individual or bundled CNTs [1,16–23] at nanoscale. Atomistic simulations on evaluating the interfacial and mechanical properties of CNT-based composites have also been presented in Refs [24–31] during the past few years. These studies have covered issues such as elastic properties, interfacial strength based on pull-out tests, chemical functionalization and load transfer mechanisms. Owing to the difficulties in validating the assumptions using experiments, some assumptions used in these simulations remain to be verified. Most importantly, bridging the gap between small (atomistic) and large (continuum) length scales in modeling CNT composites remains an important challenge due to limitations of current computing algorithms and resources.

Recently, Griebel and Hamaekers [27] carried out a detailed study on evaluating the effective properties of CNT–polymer composites using the MD approach. Two MD models of the CNT composites were considered, one with a short single-walled CNT (of 6 nm in length) in the polymer, and the other with an infinitely long one. To reduce the computational complexity, a united-atom potential is also proposed for modeling the polymer, besides the full Brenner potential [27]. Their MD models predict that along the CNT direction, the effective elastic modulus is about two times the polymer modulus for the short CNT case (with a CNT volume fraction of 2.8%), and about 30 times higher for the long CNT case (with a CNT volume fraction of 6.5%). They also compared their MD results with those using the rules of mixtures. Surprisingly, the MD predictions are found to be within the range of those using the rules of mixtures, with the largest decrease being 15% for the long CNT case and the largest increase being 25% for the short CNT case, compared with the results using the rules of mixtures [27].

Continuum mechanics approaches have been applied successfully for simulations of the mechanical responses of individual CNTs which are treated as beams, thin shells or solids in cylindrical shapes [9,32–36]. Although efficient in computing and able to handle models at larger length scales, simulation results obtained using the continuum mechanics approaches should be interpreted correctly. Attention should be paid to the overall deformations or load transfer mechanisms, rather than to local properties, such as those at the interface between CNTs and a matrix, where the physics should be addressed by MD simulations. Characterization of a CNT composite requires only the knowledge of its global responses, such as the displacement and stress fields at the boundaries of an RVE. Thus, the continuum mechanics approaches may be adequate and sufficient in modeling CNT composites in this regard. Some research results along this line have demonstrated the usefulness of the continuum approaches.

Pipes and Hubert's work [37] seems to be the first among others in characterizing CNT composites using a continuum mechanics approach. Applying the traditional textile-mechanics approach and anisotropic elasticity theory, they studied the behavior of a CNT composite wherein the fibers are layered cylinders with layers containing arrays of CNTs arranged to form a hexagonal cross section and following a helical curve. Stress distributions and effective elastic properties are evaluated in Ref. [37] using the two continuum mechanics approaches. Liu and Chen [38–40] applied the finite element and boundary element methods (FEM/BEM) for the study of CNT composite models, where RVEs containing one or multiple CNTs are modeled as thin elastic layer in the shape of a capsule (for short CNT) or an open cylinder (for long CNT). Effective elastic properties of the CNT composites are evaluated and compared with the rules of mixtures (including an extended rule of mixture derived for short-fiber composites). The detailed FEM models in Refs [38–40] reveal that the "stress" gradient across the interface of the CNT and matrix is very high and the number of elements can become prohibitively large for the FEM in large-scale modeling of CNT composites, if the continuum models can be used. Fisher, Bradshaw and Brinson [41,42] employed the FEM to study the effect of the waviness of CNT fibers in a CNT composite. Their FEM model predicts that even slight curvature of the CNTs can result in significant decrease of the effective stiffness of the CNT composites, which is consistent with experimental observations. Owing to the large size of the FEM model for detailed three dimensional (3-D) studies, the RVE used in Refs [41,42] contains only a segment of a CNT with the surrounding matrix material. All the aforementioned investigations based on the continuum approaches suggest that micromechanics is still relevant and useful in the study of CNT composites for characterizations based on the global responses of CNT composites.

Multiscale approaches have also been attempted for modeling CNT composites, albeit with very small models that only include one or two CNT fibers. Odegard, Gates, and others published the most comprehensive results in Ref. [43] on evaluations of effective elastic properties of CNT composites using an MD-based multiscale approach. They combined the MD simulation with the micromechanics approach using the equivalent-continuum model they developed earlier [44]. In their approach, the MD is employed to calculate the properties of an effective fiber that contains a CNT surrounded by the polymer in a cylindrical shape. This effective fiber is then used in the micromechanics homogenization model for evaluations of the effective properties of the CNT composites. Results of computed effective Young's moduli and shear moduli using this approach are presented for different CNT lengths and volume fractions. Direct comparison of the computed results with the available experimental data is also provided and good agreement is shown in Ref. [43]. Interestingly, this multiscale approach is also compared with the continuum approach developed by Pipes and Hubert in Ref. [37] in a recent paper [45]. Overlap of the results of the two different models is shown for a large range of CNT volume fractions for a CNT–polymer composite with the CNT fiber length equal to 500 nm. It is also stated in Ref. [45] that results obtained by the two methods are in general about 15% lower than those predicted by the classical rule of mixtures.

A fast multipole BEM (FMBEM) for large-scale modeling of CNT composites using a rigid-inclusion approach was developed recently [46,47]. In this BEM approach to modeling CNT composites, a rigid-inclusion model was employed to represent the CNTs in an elastic polymer matrix due to the high stiffness of the CNTs. The perfect bonding conditions between the CNT fibers and matrix were used in these BEM models. Very large 3-D RVEs containing up to 16,000 CNT fibers and with the total degrees of freedom (DOFs) above 28 million were successfully solved by using this FMBEM. Although the rigid-inclusion model may be sufficient for modeling the CNTs, the perfect bonding conditions for the interfaces may not be adequate. A new cohesive interface model is developed recently in Ref. [48] to improve the BEM models in Refs [46,47] for modeling CNT composites. This cohesive interface model is developed with the help of MD simulations of CNT pullouts from a polymer matrix under several load conditions. Numerical results show that the cohesive interface model has a significant impact on the estimated effective Young's moduli of the CNT/polymer composites. Lower effective Young's moduli are obtained with the new interface model, as compared with those based on perfect bonding conditions. These new results are closer to those observed in experiments and are believed to be more realistic.

20.1.3 A hierarchical multiscale approach to modeling CNT composites

A hierarchical multiscale approach has been developed for modeling and simulations of nanocomposites using MD simulations at the nanoscale and the BEM at the microscale. The developed methodology can be applied to systematically characterize microscale and macroscale mechanical properties of nanocomposites from their nanoscale structures and physics.

At the nanoscale, MD is employed to study the nanostructures, mainly the CNTs in this work, that can be used as fillers in nanocomposites. The effective mechanical properties of these nanostructures and the interface models for nanocomposites can be extracted from these nanoscale studies. Both the interface and effective CNT properties will be used in the continuum models at the microscale (Fig. 20.2).

At the microscale, large-scale continuum models of nanocomposites are studied using an FMBEM. In the BEM, only the boundary and interfaces of an RVE of a composite need to be discretized. The interface models will be applied in the microscale BEM models that will be used to extract the effective mechanical properties of the nanocomposites at the microscale. Owing to the modeling and

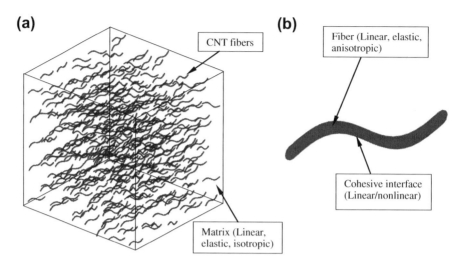

(a) An RVE with curved CNT fibers

(b) Models for the CNT and interface

FIGURE 20.2

A microscale model for CNT composites. (For color version of this figure, the reader is referred to the online version of this book.)

computing efficiencies of the fast BEM, tens of thousands of nanoscale fibers can be modeled directly in the RVE by the BEM. Furthermore, the fibers can be arbitrarily shaped, distributed or aligned in the RVE (Fig. 20.2), to represent the realistic configurations in the current nanocomposite samples.

This chapter is organized as follows. In Section 20.2, the basics of MD are reviewed and the approach to determine cohesive interface models using the MD simulations is described. In Section 20.3, the boundary integral equation (BIE) formulations are presented and the FMBEM used to solve the BIE equations is described. In Section 20.4, numerical results are presented to show the effects of the new interface model and the potentials of the multiscale approach using MD and FMBEM. Sections 20.5 are discussions and conclusions, respectively. Related references are provided at the end of this chapter.

20.2 NANOSCALE SIMULATIONS USING MOLECULAR DYNAMICS

20.2.1 Basics of molecular dynamics

The method of MD [13–15] is a classical approach of simulating atomic systems with large numbers of atoms. In this technical approach, the interaction among the atoms is governed by certain interatomic potential. Correspondingly, the spatial coordinate $\mathbf{r}_i$ of the i-th atom is governed by the Newton's second law, that is

$$m_i \frac{d^2 \mathbf{r}_i}{dt^2} = -\nabla V + \mathbf{F}_i^{ext} \tag{20.1}$$

in which m_i and $\mathbf{r}_i$ are the mass and spatial coordinates of the i-th atom, respectively, V is the empirical potential for the system, $\mathbf{F}_i^{ext}$ is the external force and ∇ denotes the spatial gradient. In general, the length and time scales that are treated in MD are still relatively small when compared with the continuum scale. In order to extend the properties that are obtained from MD to macroscopic scale and systems, thermodynamic ensembles and periodical boundary conditions are usually implemented. These approaches will be combined with time integration algorithms such as the Verlet method [49,50] and other high-order methods to evolve the system. The key aspect to ensure the accuracy of the simulation is the formulation of the interatomic potentials. Based on the nature of the interatomic bond, these can be categorized as bonded and nonbonded interaction. The Lennard-Jones potential, for instance, is a widely used nonbonded potential for many systems. The specific choices of the interatomic potentials for the nanocomposites systems considered in this chapter are discussed in the following sections.

20.2.2 Developing a cohesive interface model for CNT composites using MD

MD simulations can be employed to extract the interfacial properties of the CNT/polymer composites. From these MD results, interface models can be established that can be employed in continuum models of CNT composites at the microscale for large-scale analysis of such composites. In this context, we propose a cohesive interface model based on the MD results for CNT pull-out simulations. In these MD simulations, a CNT with surrounding polymer chains is modeled using MD. The CNT is pulled in the axial direction, or compressed in the radial direction, or twisted in the circumferential direction so that the relative displacement and force relations can be determined (Fig. 20.3).

For example, for a CNT in polymer and under the three load conditions, the compliance matrix that relate the traction $\mathbf{t}$ and displacement $\mathbf{u}$ at the CNT/polymer interface can be written in the following form in the cylindrical coordinate system (Fig. 20.3):

$$\begin{Bmatrix} \Delta u_z \\ \Delta u_r \\ \Delta u_\theta \end{Bmatrix} = \begin{bmatrix} C_z & 0 & 0 \\ 0 & C_r & 0 \\ 0 & 0 & C_\theta \end{bmatrix} \begin{Bmatrix} t_z \\ t_r \\ t_\theta \end{Bmatrix}, \quad (20.2)$$

where Δu is the relative displacement of the CNT to the polymer, and the coupling effects (off-diagonal terms in $\mathbf{C}$) are ignored in this study. The compliance matrix $\mathbf{C}$ in Eqn (20.2) will be applied at each node on the interface in the BEM models at the microscale.

To determine the compliance matrix $\mathbf{C}$ in Eqn (20.2) for the cohesive interface model, detailed MD simulations have been carried out using models of a CNT with surrounding polymer chains. For load conditions, the CNT is pulled out of the polymer in the axial z-direction or rotated relative to the polymer in the circumferential direction, or the entire model is compressed in the radial r-direction

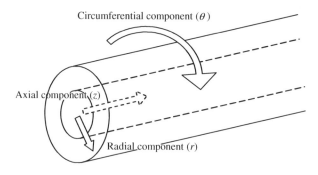

FIGURE 20.3

The local cylindrical coordinates for determining the compliance coefficients.

(Fig. 20.3). With these three load conditions, the components of the compliance matrix in the local (z, r, θ) coordinates can be determined.

The models of polyethylene (PE) systems with both amorphous and crystalline polymer chains are used in the MD simulations. In both cases, neither cross-link chemical bonds nor coatings are assumed to be present. The CNT is modeled by the Tersoff–Brenner potential [51,52]. The interactions between the CNT and polymer matrix are governed by the Lennard-Jones pair potential with the 6–12 functional dependence. The PE system is modeled by the united-atom potential [53], which considers bond stretching, dihedral and torsional effects. In the case of amorphous polymer chains, the CNT chosen is a (10, 0) single-walled CNT with 40 Å length, 3.85 Å radius and 380 atoms. The system is located in a cell of $40 \times 40 \times 40$ Å^3 and has 4100–4300 atoms, depending on the CNT–polymer spacing. The system is simulated using the MD approach. A representative configuration with amorphous polymer chains is shown in Fig. 20.4(a). A uniform velocity

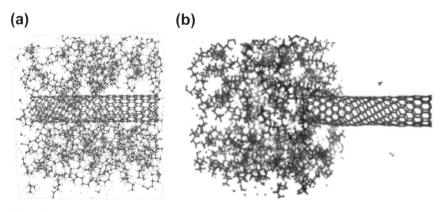

FIGURE 20.4

CNT pull-out test with *amorphous* polymer chains: (a) initial and (b) final configurations of the CNT pull-out test.

20.2 Nanoscale Simulations Using Molecular Dynamics

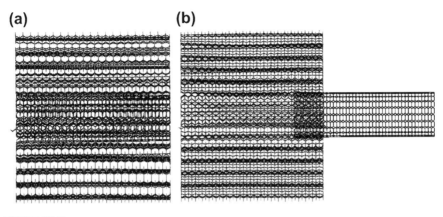

FIGURE 20.5

CNT pull-out test with *crystalline* polymer chains: (a) initial and (b) final configurations of the CNT pull-out test.

of 5 Å/ps is applied to the boundary atoms of the CNT, so that it pulls out the CNT from the matrix (Fig. 20.4(b)).

Figure 20.5(a) shows the CNT embedded in a crystalline polymer matrix. After the initial configuration is relaxed, the cohesive forces between the CNT and polymer by either pulling it out along the axial direction (Fig. 20.5(b)) or rotating it with respect to its own axis are examined. Plotted in Fig. 20.6(a), (b) are the cohesive forces per unit length vs the axial displacement and rotation angle, respectively. It is found that the nature of these curves show periodicities that match those of the

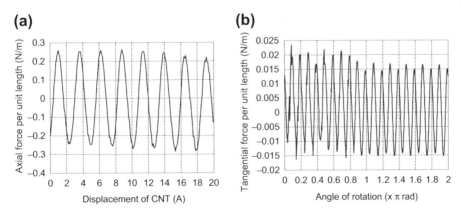

FIGURE 20.6

(a) The axial cohesive force between the CNT and polymer as a function of the displacement. (b) The circumferential cohesive force as a function of the rotation angle. (For color version of this figure, the reader is referred to the online version of this book.)

CNTs in the axial and circumferential directions. This indicates that such cohesive force can be well characterized in a form similar to the Mohr—Coulomb model used in continuum mechanics, although additional considerations of the periodicity are needed. Finally, the variations in the slopes of the cohesive force curves are found to be extremely small which are subsequently used to determine the coefficients in the compliance matrix **C** in Eqn (20.2) and used in the microscale BEM models.

20.3 MICROSCALE SIMULATIONS USING THE BOUNDARY ELEMENT METHOD

In this subsection, we first review the BIE formulation for multidomain elasticity problems for determining the displacements and tractions on the boundary and interfaces in a composite material model. Then, we discuss the challenges in using the conventional BEM for solving these BIEs. Finally, we present the fast multipole method (FMM) used to accelerate the solutions of the BEM systems of equations.

20.3.1 The boundary element method

Consider a 3-D elastic domain V (matrix) embedded with n elastic inclusions V_α, such as fibers, particles or other type of inclusions (Fig. 20.7). The boundary of the model is S_0 and the interfaces between the inclusions and the matrix are S_α ($\alpha = 1, 2, ..., n$).

For the matrix domain V, the BIE relating the displacement and traction fields on the boundary and interfaces can be written as (see, e.g. Refs [54–58]) follows:

$$\frac{1}{2}\mathbf{u}(\mathbf{x}) = \int_S [\mathbf{U}(\mathbf{x},\mathbf{y})\,\mathbf{t}(\mathbf{y}) - \mathbf{T}(\mathbf{x},\mathbf{y})\,\mathbf{u}(\mathbf{y})]dS(\mathbf{y}), \quad \forall \mathbf{x} \in S, \qquad (20.3)$$

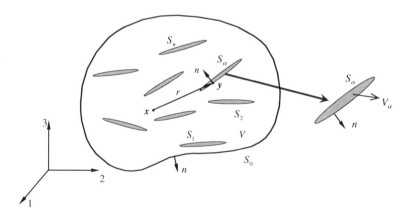

FIGURE 20.7

A material domain V filled with other constitutes occupying V_α. (For color version of this figure, the reader is referred to the online version of this book.)

20.3 Microscale Simulations Using the Boundary Element Method

in which $S = \cup_{\alpha=0}^{n} S_\alpha$ is assumed to be smooth at the source point **x**; **u** and **t** are the displacement and traction vectors on S, respectively. The two kernel functions **U(x,y)** and **T(x,y)** in Eqn (20.3) are the displacement and traction components of the fundamental solution (Kelvin's solution), respectively, which can be found in any references on the BEM (see, e.g. Refs [54−58]). The first integral with the **U** kernel is weakly singular ($O(1/r)$, with r being the distance between **x** and **y**), whereas the second with the **T** kernel is strongly singular ($O(1/r^2)$, a Cauchy principal value integral), when the field point **y** approaches the source point **x**. Regularized forms of the BIEs can be used to avoid the singular integrals in the BEM solutions of the BIEs (see, e.g. Refs [59−61]).

For the inclusions, we have the following set of BIEs similar to BIE (3):

$$\frac{1}{2}\mathbf{u}^{(\alpha)}(\mathbf{x}) = \int_{S_\alpha} [\mathbf{U}^{(\alpha)}(\mathbf{x},\mathbf{y})\,\mathbf{t}^{(\alpha)}(\mathbf{y}) - \mathbf{T}^{(\alpha)}(\mathbf{x},\mathbf{y})\,\mathbf{u}^{(\alpha)}(\mathbf{y})]dS(\mathbf{y}), \quad \forall \mathbf{x} \in S_\alpha, \quad (20.4)$$

for $\alpha = 1, 2, ..., n$, in which $\mathbf{u}^{(\alpha)}$ and $\mathbf{t}^{(\alpha)}$ are the displacement and traction of inclusion V_α on its boundary S_α, and $\mathbf{U}^{(\alpha)}(\mathbf{x},\mathbf{y})$ and $\mathbf{T}^{(\alpha)}(\mathbf{x},\mathbf{y})$ are the kernels determined with the material properties of inclusion V_α.

On boundary S_0, one component of **u** or **t** at any point must be given and the other component will be the unknown. On the interfaces, we apply the interface model for the cohesive interfaces between the fibers (CNTs) and matrix. In this cohesive interface model, the difference between the displacement fields $\mathbf{u}^{(\alpha)}$ in the fiber (CNT) and **u** in the matrix is related to the traction **t** at the interface (from the matrix domain) in the following form:

$$\mathbf{u}^{(\alpha)}(\mathbf{y}) - \mathbf{u}(\mathbf{y}) = \mathbf{C}(\mathbf{y})\mathbf{t}(\mathbf{y}), \quad \forall \mathbf{y} \in S_\alpha, \quad (20.5)$$

where **C** is the compliance matrix determined by the MD simulations (Section 20.2). For perfect bonding interface conditions, we have $\mathbf{C} = \mathbf{0}$.

BIE (3) and BIEs in (4) (with $\alpha = 1, 2, ..., n$), together with the interface conditions in Eqn (20.5), provide sufficient equations for solving the displacements and tractions on the entire boundary S_0 and interfaces S_α (with $\alpha = 1, 2, ..., n$).

The CNTs are very stiff with Young's moduli around 1 TPa as compared to polymers which have their Young's moduli around only a few GPa. Thus, in some cases, for example, when CNTs are straight, a rigid-inclusion or rigid-fiber model can be employed in the microscale models to simplify the analysis. In such models, the displacement vector for a rigid inclusion or fiber is given by

$$\mathbf{u}^{(\alpha)}(\mathbf{y}) = \mathbf{d} + \boldsymbol{\omega} \times \mathbf{p}(\mathbf{y}), \quad (20.6)$$

where **d** is the rigid-body translational displacement vector, $\boldsymbol{\omega}$, the rotation vector and **p**, a position vector for point **y** measured from a reference point (such as the center of the fiber). Thus, BIE (3) is reduced to

$$\frac{1}{2}\mathbf{u}(\mathbf{x}) = \int_S \mathbf{U}(\mathbf{x},\mathbf{y})\,\mathbf{t}(\mathbf{y})dS(\mathbf{y}), \quad \forall \mathbf{x} \in S, \quad (20.7)$$

which is the BIE used in Refs [46,47] for modeling CNT composites. More details in deriving Eqn (20.7) can be found in Refs [46,47].

In the rigid inclusion case, BIE (4) is no longer needed. To determine **d** and **ω** for each fiber, additional equations can be obtained by considering the equilibrium of each fiber in the form of the following two equations [46,47]:

$$\int_{S_\alpha} \mathbf{t}(\mathbf{y}) dS(\mathbf{y}) = \mathbf{0}; \qquad (20.8)$$

$$\int_{S_\alpha} \mathbf{p}(\mathbf{y}) \times \mathbf{t}(\mathbf{y}) dS(\mathbf{y}) = \mathbf{0}; \qquad (20.9)$$

for $\alpha = 1, 2, \ldots, n$. Equation (20.8) represents the equilibrium of the forces, whereas Eqn (20.9) that of the moments, for each fiber. BIE (7) and Eqns (20.6), (20.8) and (20.9) can be solved simultaneously to obtain the unknown traction **t**, and rigid-body motions **d** and **ω** for all the fibers in the rigid-fiber models for CNT composites [46,47].

In the BEM for solving the BIEs, the boundary and interfaces are discretized using surface elements and the resulting system of linear equations from the BIEs will be solved to obtain **u** and **t** on all the interfaces. The stresses and strains on the boundary of a model (e.g. an RVE) can be calculated and then the effective properties of the material can be determined from these stress and strain fields [46,47,62].

20.3.2 The fast multipole method

In the conventional BEM for solving the BIEs, such as BIE (3), a linear system of equations in the form

$$\mathbf{A}\boldsymbol{\lambda} = \mathbf{b}, \qquad (20.10)$$

will need to be formed explicitly and solved using either direct or iterative solvers. In Eqn (20.10), **A** is the coefficient matrix, $\boldsymbol{\lambda}$, the unknown vector containing the unknown nodal displacement and traction on the boundary (and nodal displacement and traction on the interfaces, for multidomain problems), and **b**, the known vector from the given boundary conditions. The dimensions of matrix **A** is $N \times N$, where $N = 3m$ with m being the number of nodes on the boundary (and all the interfaces, for multidomain problems) for 3-D problems. In general, matrix **A** is dense and nonsymmetrical. Thus, the conventional BEM approach can only solve small BEM models with a few thousands of nodes on a desktop PC, due to the computing cost that is $O(N^2)$ for computing matrix **A** (for **x** at every node, integrations need to be done over all the elements/nodes) and $O(N^3)$ for solving the system of equations with direct solvers.

The FMM [63–68] can be employed to accelerate the solutions of the BEM systems of equations (Eqn (20.10)) dramatically. The main idea of the FMBEM is to employ iterative solvers (such as GMRES (generalised minimal residuals)) to

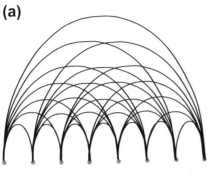 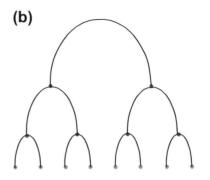

(a) Conventional BEM approach ($O(N^2)$))   (b) FMBEM approach ($O(N)$ for large N)

FIGURE 20.8

A graphical illustration of the conventional BEM and the fast multipole BEM. (For color version of this figure, the reader is referred to the online version of this book.)

solve Eqn (20.10) and employ the FMM to accelerate the matrix–vector multiplication ($\mathbf{A}\lambda$), without ever forming the matrix $\mathbf{A}$ explicitly. In the FMBEM, the node-to-node interactions in the conventional BEM are replaced by cell-to-cell interactions by using a hierarchical tree structure of cells containing groups of elements (Fig. 20.8, in which dots indicate nodes/cells and lines indicate the interactions needed). This is possible by using the multipole and local expansions of the integrals and some translations (termed M2M, M2L, and L2L translations). With all these expansions and translations, an order $O(N)$ efficiency in computing and memory requirement can be achieved by the FMBEM [46,47,66,69–71].

20.4 NUMERICAL EXAMPLES

20.4.1 Further study of CNT pull-out tests using MD

A further study of the CNT pull-out tests using MD for a different system is reported here. We consider a virtual pull-out test of CNT embedded in polystyrene matrix. A similar test conducted for CNT embedded in PE has been discussed in Section 20.2 (see also Ref. [53]). In this case, the force field for polystyrene includes three interatomic potentials: bond potential, angle potential and dihedral potential. Because polystyrene is a nonpolarized material, electrostatic forces are ignored in this research. The only intermolecular potential considered is the short ranged (van der Waals) potential. The total energy of the polystyrene matrix, sum of the interatomic energy as well as energy from intermolecular, can be expressed as follows:

$$E_{polystyrene} = E_{bond} + E_{angle} + E_{dihedral} + E_{vdw} \quad (20.11)$$

For bonding interaction, Morse's formulation was used; for angle interaction, harmonic cosine formulation was used; for dihedral interaction, cosine formulation

Table 20.1 Parameters for Morse Potential

Parameter	C–C Bond	C–H Bond	C=C Bond	Units
E_0	70.000	70.000	140.000	kcal/mol
r_0	1.53	1.09	1.42	Å
k	2.236	2.236	2.236	(Å)$^{-1}$

Table 20.2 Parameters for Harmonic Cosine Potential

Parameter	C–C–C	H–C–H	C–C–H/H–C–C	Units
k	112.5	112.5	112.5	kcal/mol
θ_0	109.47	109.47	109.47	Deg

Table 20.3 Parameters for Cosine Potential

Parameter	H–C–C–H/H–C–C–C	C–C–C–H/C–C–C–C	Units
A	0.1667	0.1111	kcal/mol
δ	0	0	Deg
m	3	3	–

was used. For intermolecular interaction, 6–12 formation (Lennard-Jones potential) was used. The parameters for each potential are listed in Tables 20.1–20.3. The details of the functional forms of the potential can be found in Ref. [53].

The intermolecular interactions in the composite systems studied were modeled with Lennard-Jones potential as well, using Girifalco's parameters for representing the C–C interactions. Van der Waals forces acting between atoms of CNT were not accounted for. The interactions between CNT atoms and polymer atoms were defined based on a cutoff radius of 10 times the C–C bond length, since the contribution of atoms beyond this distance is very small. The interactions were updated periodically in the simulation. The period of updating will be determined to satisfy both the accuracy and efficiency requirements of the simulation procedure. The composite systems studied for interfacial properties were based on a through, dangling bonding free, open-ended (10, 10) armchair CNT inside an amorphous polystyrene matrix (Fig. 20.9). Displacement boundary conditions are imposed on the right end of the tube at 0.005 Å/step.

In Fig. 20.9, we show the images of nanotube pulling out process. These correspond to the initial and final steps of the load. The interfacial shear stress acting on

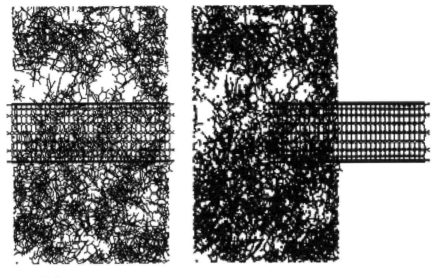

FIGURE 20.9

Images of nanotube pulling-out process.

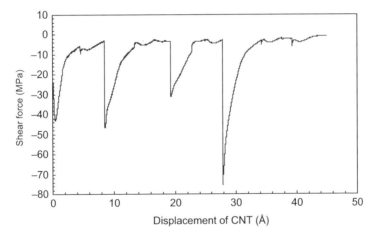

FIGURE 20.10

Shear stress acted on entire carbon nanotube during the pull-out test. (For color version of this figure, the reader is referred to the online version of this book.)

the entire nanotube and two layers of carbon atoms at the middle of CNT are evaluated at each step of the load. A typical shear stress vs pull-out displacement curve is shown in Fig. 20.10. Here, the shear stress is defined as the total load experienced by the CNT in the axial direction divided by its surface area. One can first see that the shear stress is oscillatory in nature and shows certain pattern of periodicity. This is mainly due to periodical structure of the CNT chosen for this study. The peak force observed here gives the threshold value for the interlayer sliding. This pattern is very similar to the friction force at the continuum scale. Each peak in the curve represents a barrier in order for the stick-to-slip to take place. The information obtained from this MD study will then be provided to the subsequent BEM studies for evaluations of systems involved large number of tubes embedded in the polymer.

20.4.2 Fracture analysis of CNT/polymer composites using MD

MD simulations can also be applied to study the strength of CNT composites at the nanoscale. In this example, we use the same MD method to study the fracture behavior of the CNT/polymer composite at the large strain range, as illustrated in Fig. 20.11. We considered an uncoated (10, 10) CNT embedded in amorphous PE matrix. To evaluate the effect of volume fraction, the model size ranges from 80,000 to 600,000 atoms. We found that the effect of CNT reinforcement is strongly dependent on the volume fractions and external pressures. For the system studied here, a critical volume fraction of about 8% is observed. Further addition of CNT beyond this point leads to degradation of the properties. The Young's modulus and tensile strength of the composite is significantly less than that of pure PE. This indicates poor load transfer from the matrix to CNT. This is justified since the length of the nanotube used is many orders of magnitude less than the critical length for good load transfer, 70–80 μm, obtained from our calculations. On the contrary, the strength and elastic modulus increases with increasing volume fraction of nanotubes, but the values are lower as compared to pure PE. The upper bound of our computation is limited by the account of nonbonded interactions, which contributes to the interfacial property and consumes a large amount of memory.

20.4.3 Large-scale BEM models for CNT composites

Large-scale analysis of fiber-reinforced composites, with the CNT composites as special cases, has been conducted with the developed FMBEM code to extract the

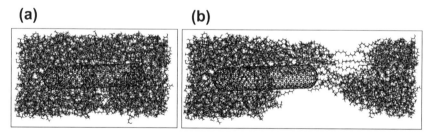

FIGURE 20.11

Fracture at nanoscale due to tensile loading in a CNT/polymer composite (volume fraction = 7%): (a) the original nanocomposite structure and (b) the fracture pattern at 45% engineering strain [53].

20.4 Numerical Examples

effective mechanical properties of the composites, such as the effective Young's modulus of a composite. Both perfect bonding and cohesive interface models have been applied in these BEM models. Up to now, the CNTs in the BEM composite models have been regarded as rigid inclusions. So, these models can only handle composites with straight CNT fibers. A detailed discussion on using this approach to model CNT composites is given in Refs [46,47]. FMBEM code for handling elastic CNT fibers and cohesive interface models is being developed and the results will be reported elsewhere.

A typical BEM model, or RVE, for modeling CNT composites is shown in Fig. 20.12. The RVE is stretched in the x-direction and the averaged strains and stresses on the two data-collection surfaces are obtained from the BEM solutions to determine the effective Young's modulus. Figure 20.13 shows an RVE with 2197 short randomly distributed fibers and the boundary element mesh used to discretize each fiber on the interfaces. Figure 20.14 shows another RVE with 2000 long CNT fibers and the total number of BEM equations (degrees of freedom) equal to 3,612,000. The largest BEM model studied so far is eight times larger than the one shown in Fig. 20.14, which has total DOFs reaching 28.8 millions. These very large BEM models of CNT composites were solved successfully with the FMBEM on a supercomputer at the Kyoto University [46,47].

Based on the models shown in Fig. 20.14 and perfect bonding condition, the estimated effective longitudinal Young's moduli (E_{eff}) of a CNT composite against the

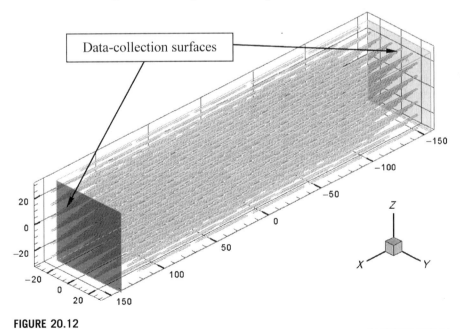

FIGURE 20.12

A microscale BEM model (RVE) containing 720 CNT fibers. (For color version of this figure, the reader is referred to the online version of this book.)

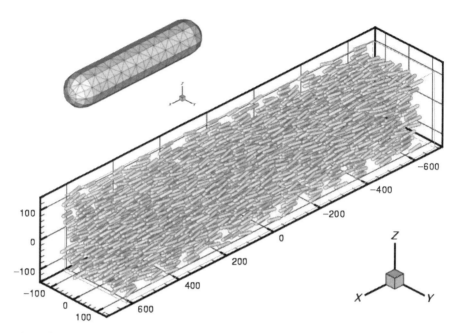

FIGURE 20.13

An RVE containing 2197 randomly distributed short fibers using the fast BEM with the total DOFs = 3,018,678 (inset shows the boundary mesh on the fibers). (For color version of this figure, the reader is referred to the online version of this book.)

CNT volume fractions are plotted in Fig. 20.15, and compared with the data in Ref. [43]. The length of the CNT fibers considered is 50 nm, the radius is 0.7 nm and the thickness is 0.34 nm. The volume of the CNT is calculated by considering it as a hollow cylinder with the outer radius equal to 0.7 nm. For the matrix material, Young's modulus of 1 GPa and Poisson's ratio of 0.4 are used here. The BEM results were obtained for CNT volume fractions up to 10.48%. All the results obtained by using the developed BEM approach based on the rigid-inclusion models are close to the results in Ref. [43] which are based on the MD and a multiscale approach. For the small RVE cases, the results of the aligned random and uniform cases are very close, possibly due to the fact that the random cases are only slight perturbations of the corresponding uniform ones. The results in the large RVE show significant improvement (half way closer to the curve from Ref. [43]). More details of this study can be found in Refs [46,47].

20.4.4 Effect of the cohesive interface in modeling CNT/polymer composites

The same BEM models are applied with cohesive interface conditions to estimate the effective Young's moduli of CNT/polymer composites as those in Refs [46,47], where perfect bonding interface conditions are used.

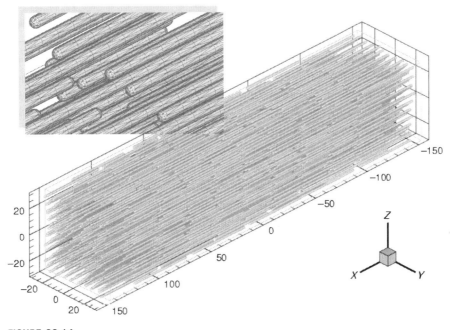

FIGURE 20.14

An RVE containing 2000 CNT fibers with the total DOF = 3,612,000 (with CNT length = 50 nm, volume fraction = 10.48%. The inset shows a close-up of the tubes with the boundary element mesh). (For color version of this figure, the reader is referred to the online version of this book.)

From the MD simulations of CNT pull-out tests, the axial component in the local z-direction of the **C** matrix is found to be $C_z = 3.506$, and the radial component $C_r = 0.02157$ (in the nondimensionalized system). There is a large difference between the values of C_z and C_r (more than 160 times). In this investigation, we set the coefficients of **C** to be $C_{ij} = C\delta_{ij}$ for simplicity, and carry out computation with various choices of C. In the determination of the effective moduli, the effect of the axial slip of the CNT is considered important. Therefore, we have to pay attention to the values of C_z used in the analysis.

The following three cases of the interface properties are considered in determining the effective moduli:

Case 1: $C = 0$ (no interfacial effects, perfect bonding);
Case 2: $C = C_r = 0.02157$ (strong interfacial cohesive force);
Case 3: $C = C_z = 3.506$ (weak interfacial cohesive force).

The estimated effective longitudinal Young's moduli (E_{eff}) of the CNT composites against the CNT volume fractions are plotted in Fig. 20.16 for the three cases of C. Case 1 is based on the perfect bonding interface model used earlier [46,47], which

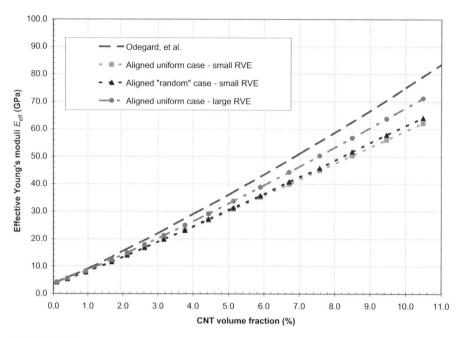

FIGURE 20.15

Estimated effective moduli with perfect bonding interface model by the BEM as compared with the result by Odegard et al. [43]. (For color version of this figure, the reader is referred to the online version of this book.)

has been shown to be very close to the results using the MD and continuum mechanics model in Ref. [43] (Fig. 7 in Ref. [43], with CNT length = 50 nm). Case 2 represents a strong interfacial cohesive force case and the estimated effective moduli are very close to those in Case 1 as shown in Fig. 20.16. This suggests again that any interface models based on a strong cohesive force model will yield results that are close to those based on perfect bonding interface models. In other words, perfect bonding models would be sufficient if strong interfaces do exist. Case 3 in Fig. 20.16 represents a weak interface between the CNTs and the matrix, since $C_z = 3.506$ is the compliance tangent to the interface and in the axial direction of the CNTs. In experiments of CNT composites, slip has been observed in CNT/polymer composite samples [7,10–12]. The estimated effective Young's moduli in Case 3 are indeed much smaller than those in Cases 1 and 2, as expected. About 50% decrease of the values is observed for all the CNT volume fractions. These lower values for the effective moduli are believed to be closer to those in experimental studies (see Ref. [43] for a few limited experimental data points that are lower than the MD simulation results with the strong interface model). However, direct comparisons cannot be made at this time due to the lack of data from the experiments with the same parameters. More studies can be made when the experimental data are available.

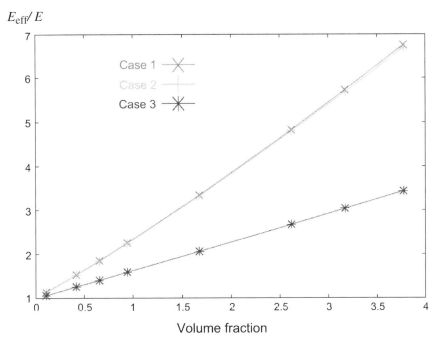

FIGURE 20.16

Effects of the cohesive interface parameter C on the effective moduli of the composites: Case 1: $C=0$ (perfect bonding, the earlier results in Ref. [46,47]); Case 2: $C=C_r=0.02157$ (large interface stiffness); Case 3: $C=C_z=3.506$ (small interface stiffness). (For color version of this figure, the reader is referred to the online version of this book.)

20.5 DISCUSSIONS

A hierarchical multiscale approach for modeling CNT/polymer composites is presented and the preliminary numerical results are reported in this chapter. In this multiscale approach, the MD approach is applied at the nanoscale to characterize the CNT properties and the interface behaviors. Cohesive interface models can be extracted from these nanoscale MD simulations that can be applied to large BEM models of the CNT composites at the microscale. FMBEM has been developed for large-scale simulations of the CNT composites at the microscale. Preliminary results clearly show the potential of the MD-BEM approach and the efficiencies of the MD method at the nanoscale and the FMBEM at the microscale. The FMBEM code that can handle models with elastic CNT fibers and other more general behaviors of the CNT and matrix materials (as shown in Fig. 20.2) is being developed and the results will be reported later elsewhere.

The cohesive interface model based on MD simulation results is only the first step in developing multiscale computational models for such composites. The MD models need to be further investigated and verified against experimental results when they are available. More MD simulations with different parameters are definitely needed to find the intricate characteristics of the CNT/polymer interfaces and their theoretical models. The interface models should be verified with real physical tests, such as nanoscale CNT pull-out tests. Although these tests are still extremely difficult to perform with the current technologies, these experiments will be very important in validations of the MD simulation results and the developed interface models.

From the BEM results obtained so far, it is clear that the interface plays an important role in load transfers and thus the effective properties of the CNT composites. Variations in the interface stiffness can have a significant impact to the results of these estimated properties of the CNT composites. Establishing a valid and effective interface model is a crucial step in the modeling and simulations of the CNT composites. The MD approach employed in this study seems to be a fundamental and effective method for establishing such interface models. The BEM results using the cohesive interface model based on the MD simulations show marked decreases of the effective Young's moduli for the CNT/polymer composites, which are closer to the current experimental observations. More studies are needed, for example, to verify if the cohesive interface model is scale dependent, and to develop other more sophisticated interface models (e.g. with nonlinear properties).

The developed BEM can be extended readily to account for more complicated physics or interactions of the CNT fibers in a composite. Elasticity of the CNTs can be introduced readily in this model to replace the assumption of rigidity for the CNT fibers. Curved CNT fibers, which are often the case in current CNT/polymer composite samples, can also be considered once the rigid-body assumption for the CNT fibers is replaced. An effective-fiber model based on the MD and similar to the one developed in Ref. [43] can also be combined with the BEM models. With the development in computing hardware, it is quite possible in the near future to directly combine the MD model for the CNTs and the BEM model for the matrix to develop a multiscale method that can handle larger models and account for more intricate physics in CNT/polymer composites, such as debonding or other failure modes, besides evaluations of their effective properties.

Acknowledgments

The first author (Y.J.L.) would like to acknowledge the National Science Foundation support through the grant (CMS-0508232). The second author (D.Q.) would like to acknowledge the support by the NSF (Grants CMMI-0600583, -0700107 and -1120382). Any opinions, findings, conclusions, or recommendations expressed in these documents are those of the author(s) and do not necessarily reflect the views of the National Science Foundation. This work was also supported by the start-up fund from the University of Texas at Dallas and in part by an allocation of computing time from the Ohio Supercomputer Center.

References

[1] M. Buongiorno Nardelli, J.-L. Fattebert, D. Orlikowski, C. Roland, Q. Zhao, J. Bernholc, Mechanical properties, defects and electronic behavior of carbon nanotubes, Carbon 38 (11−12) (2000) 1703−1711.

[2] C. Li, T.-W. Chou, Elastic moduli of multi-walled carbon nanotubes and the effect of van der Waals forces, Composites Science and Technology 63 (11) (2003) 1517−1524.

[3] J.P. Lu, Elastic properties of carbon nanotubes and nanoropes, Physical Review Letters 79 (7) (1997) 1297−1300.

[4] J.P. Lu, Elastic properties of single and multilayered nanotubes, Journal of Physics and Chemistry of Solids 58 (11) (1997) 1649−1652.

[5] D. Qian, G.J. Wagner, W.K. Liu, M.-F. Yu, R.S. Ruoff, Mechanics of carbon nanotubes, Applied Mechanics Reviews 55 (6) (2002) 495−533.

[6] R.S. Ruoff, D.C. Lorents, Mechanical and thermal properties of carbon nanotubes, Carbon 33 (7) (1995) 925−930.

[7] E.T. Thostenson, Z.F. Ren, T.-W. Chou, Advances in the science and technology of carbon nanotubes and their composites: a review, Composites Science and Technology 61 (2001) 1899−1912.

[8] M.M.J. Treacy, T.W. Ebbesen, J.M. Gibson, Exceptionally high Young's modulus observed for individual carbon nanotubes, Nature 381 (6584) (1996) 678−680.

[9] E.W. Wong, P.E. Sheehan, C.M. Lieber, Nanobeam mechanics: elasticity, strength, and toughness of nanorods and nanotubes, Science 277 (5334) (1997) 1971−1975.

[10] P.M. Ajayan, L.S. Schadler, C. Giannaris, A. Rubio, Single-walled carbon nanotube-polymer composites: strength and weakness, Advanced Materials 12 (10) (2000) 750−753.

[11] D. Qian, E.C. Dickey, R. Andrews, T. Rantell, Load transfer and deformation mechanisms in carbon nanotube-polystyrene composites, Applied Physics Letters 76 (20) (2000) 2868−2870.

[12] L.S. Schadler, S.C. Giannaris, P.M. Ajayan, Load transfer in carbon nanotube epoxy composites, Applied Physics Letters 73 (26) (1998) 3842−3844.

[13] D. Frenkel, B. Smit, Understanding Molecular Simulation: From Algorithms to Applications, Academic Press, London, 1996.

[14] J.M. Haile, Molecular Dynamics Simulation, Wiley Interscience, New York, 1992.

[15] D.C. Rapaport, The Art of Molecular Dynamics Simulation, Cambridge University Press, Cambridge, 1995.

[16] C.F. Cornwell, L.T. Wille, Elastic properties of single-walled carbon nanotubes in compression, Solid State Communications 101 (8) (1997) 555−558.

[17] G.H. Gao, T. Cagin, W.A. Goddard, Energetics, structure, mechanical and vibrational properties of single-walled carbon nanotubes, Nanotechnology 9 (3) (1998) 184−191.

[18] T. Halicioglu, Stress calculations for carbon nanotubes, Thin Solid Films 312 (1−2) (1998) 11−14.

[19] J. Han, A. Globus, R. Jaffe, G. Deardorff, Molecular dynamics simulations of carbon nanotube-based gears, Nanotechnology 8 (3) (1997) 95−102.

[20] J.-W. Kang, H.-J. Hwang, Mechanical deformation study of copper nanowire using atomistic simulation, Nanotechnology 12 (3) (2001) 295−300.

[21] M. Macucci, G. Iannaccone, J. Greer, J. Martorell, D.W.L. Sprung, A. Schenk, I.I. Yakimenko, K.-F. Berggren, K. Stokbro, N. Gippius, Status and perspectives of nanoscale device modelling, Nanotechnology 12 (2) (2001) 136–142.

[22] S.B. Sinnott, O.A. Shenderova, C.T. White, D.W. Brenner, Mechanical properties of nanotubule fibers and composites determined from theoretical calculations and simulations, Carbon 36 (1–2) (1998) 1–9.

[23] D. Srivastava, M. Menon, K. Cho, Computational nanotechnology with carbon nanotubes and fullerenes, Computing in Science and Engineering 3 (4) (2001) 42–55.

[24] S.J.V. Frankland, A. Caglar, D.W. Brenner, M. Griebel, Molecular simulation of the influence of chemical cross-links on the shear strength of carbon nanotube-polymer interfaces, Journal of Physical Chemistry B 106 (12) (2002) 3046–3048.

[25] S.J.V. Frankland, V.M. Harik, Analysis of carbon nanotube pull-out from a polymer matrix, Surface Science 525 (1–3) (2003) L103–L108.

[26] S.J.V. Frankland, V.M. Harik, G.M. Odegard, D.W. Brenner, T.S. Gates, The stress-strain behavior of polymer-nanotube composites from molecular dynamics simulation, Composites Science and Technology 63 (11) (2003) 1655–1661.

[27] M. Griebel, J. Hamaekers, Molecular dynamics simulations of the elastic moduli of polymer-carbon nanotube composites, Computer Methods in Applied Mechanics and Engineering 193 (17–20) (2004) 1773–1788.

[28] K. Liao, S. Li, Interfacial characteristics of a carbon nanotube polystyrene composite system, Applied Physics Letters 79 (25) (2001) 4225–4227.

[29] K. Mylvaganam, L.C. Zhang, Chemical bonding in polyethylene-nanotube composites: a quantum mechanics prediction, Journal of Physical Chemistry B 108 (17) (2004) 5217–5220.

[30] S. Namilae, N. Chandra, C. Shet, Mechanical behavior of functionalized nanotubes, Chemical Physics Letters 387 (4–6) (2004) 247–252.

[31] M. Wong, M. Paramsothy, X.J. Xu, Y. Ren, S. Li, K. Liao, Physical interactions at carbon nanotube-polymer interface, Polymer 44 (25) (2003) 7757–7764.

[32] S. Govindjee, J.L. Sackman, On the use of continuum mechanics to estimate the properties of nanotubes, Solid State Communications 110 (4) (1999) 227–230.

[33] D. Qian, W.K. Liu, R.S. Ruoff, Mechanics of C_{60} in nanotubes, The Journal of Physical Chemistry B 105 (44) (2001) 10753–10758.

[34] C.Q. Ru, Column buckling of multiwalled carbon nanotubes with interlayer radial displacements, Physical Review B 62 (24) (2000) 16962–16967.

[35] C.Q. Ru, Axially compressed buckling of a double walled carbon nanotube embedded in an elastic medium, Journal of the Mechanics and Physics of Solids 49 (6) (2001) 1265–1279.

[36] K. Sohlberg, B.G. Sumpter, R.E. Tuzun, D.W. Noid, Continuum methods of mechanics as a simplified approach to structural engineering of nanostructures, Nanotechnology 9 (1) (1998) 30–36.

[37] R.B. Pipes, P. Hubert, Helical carbon nanotube arrays: mechanical properties, Composites Science and Technology 62 (3) (2002) 419–428.

[38] X.L. Chen, Y.J. Liu, Square representative volume elements for evaluating the effective material properties of carbon nanotube-based composites, Computational Materials Science 29 (1) (2004) 1–11.

[39] Y.J. Liu, X.L. Chen, Continuum models of carbon nanotube-based composites using the boundary element method, Electronic Journal of Boundary Elements 1 (2) (2003) 316–335.

[40] Y.J. Liu, X.L. Chen, Evaluations of the effective materials properties of carbon nanotube-based composites using a nanoscale representative volume element, Mechanics of Materials 35 (1–2) (2003) 69–81.

[41] F.T. Fisher, R.D. Bradshaw, L.C. Brinson, Effects of nanotube waviness on the modulus of nanotube-reinforced polymers, Applied Physics Letters 80 (24) (2002) 4647–4649.

[42] F.T. Fisher, R.D. Bradshaw, L.C. Brinson, Fiber waviness in nanotube-reinforced polymer composites—I: modulus predictions using effective nanotube properties, Composites Science and Technology 63 (11) (2003) 1689–1703.

[43] G.M. Odegard, T.S. Gates, K.E. Wise, C. Park, E.J. Siochi, Constitutive modeling of nanotube-reinforced polymer composites, Composites Science and Technology 63 (11) (2003) 1671–1687.

[44] G.M. Odegard, T.S. Gates, L.M. Nicholson, K.E. Wise, Equivalent-continuum modeling of nano-structured materials, Composites Science and Technology 62 (14) (2002) 1869–1880.

[45] G.M. Odegard, R.B. Pipes, P. Hubert, Comparison of two models of SWCN polymer composites, Composites Science and Technology 64 (7–8) (2004) 1011–1020.

[46] Y.J. Liu, N. Nishimura, Y. Otani, Large scale modeling of carbon-nanotube composites by the boundary element method based on a rigid-inclusion model, Computational Materials Science 34 (2) (2005) 173–187.

[47] Y.J. Liu, N. Nishimura, Y. Otani, T. Takahashi, X.L. Chen, H. Munakata, A fast boundary element method for the analysis of fiber-reinforced composites based on a rigid-inclusion model, Journal of Applied Mechanics 72 (1) (2005) 115–128.

[48] Y.J. Liu, N. Nishimura, D. Qian, N. Adachi, Y. Otani, V. Mokashi, A boundary element method for the analysis of CNT/polymer composites with a cohesive interface model based on molecular dynamics, Engineering Analysis with Boundary Elements 32, No. 4 (2008) 299–308.

[49] H.J.C. Berendsen, W.F. van Gunsteren, Dynamics Simulation of Statistical Mechanical Systems, North Holland, Amsterdam, 1986.

[50] L. Verlet, Computer experiments on classical fluids. I. Thermodynamical properties of Lennard-Jones molecules, Physical Review 159 (1) (1967) 98.

[51] D.W. Brenner, Empirical potential for hydrocarbons for use in simulating the chemical vapor-deposition of diamond films, Physical Review B 42 (15) (1990) 9458–9471.

[52] J. Tersoff, Empirical interatomic potential for carbon, with applications to amorphous-carbon, Physical Review Letters 61 (25) (1988) 2879–2882.

[53] V. Mokashi, D. Qian, Y.J. Liu, A study on the tensile response and fracture in carbon nanotube-based composites using molecular mechanics, Composites Science and Technology 67 (3–4) (2007) 530–540.

[54] P.K. Banerjee, The Boundary Element Methods in Engineering, McGraw-Hill, New York, 1994.

[55] C.A. Brebbia, J. Dominguez, Boundary Elements—An Introductory Course, McGraw-Hill, New York, 1989.

[56] T.A. Cruse, Boundary Element Analysis in Computational Fracture Mechanics, Kluwer Academic Publishers, Dordrecht, The Netherlands, 1988.

[57] J.H. Kane, Boundary Element Analysis in Engineering Continuum Mechanics, Prentice Hall, Englewood Cliffs, NJ, 1994.

[58] S. Mukherjee, Boundary Element Methods in Creep and Fracture, Applied Science Publishers, New York, 1982.

[59] Y.J. Liu, T.J. Rudolphi, Some identities for fundamental solutions and their applications to weakly-singular boundary element formulations, Engineering Analysis with Boundary Elements 8 (6) (1991) 301–311.

[60] Y.J. Liu, T.J. Rudolphi, New identities for fundamental solutions and their applications to non-singular boundary element formulations, Computational Mechanics 24 (4) (1999) 286–292.

[61] S. Mukherjee, CPV and HFP integrals and their applications in the boundary element method, International Journal of Solids and Structures 37 (45) (2000) 6623–6634.

[62] X.L. Chen, Y.J. Liu, An advanced 3-D boundary element method for characterizations of composite materials, Engineering Analysis with Boundary Elements 29 (6) (2005) 513–523.

[63] Y. Fu, K.J. Klimkowski, G.J. Rodin, E. Berger, J.C. Browne, J.K. Singer, R.A.V.D. Geijn, K.S. Vemaganti, A fast solution method for three-dimensional many-particle problems of linear elasticity, International Journal for Numerical Methods in Engineering 42 (1998) 1215–1229.

[64] J.E. Gomez, H. Power, A multipole direct and indirect BEM for 2D cavity flow at low Reynolds number, Engineering Analysis with Boundary Elements 19 (1997) 17–31.

[65] A.A. Mammoli, M.S. Ingber, Stokes flow around cylinders in a bounded two-dimensional domain using multipole-accelerated boundary element methods, International Journal for Numerical Methods in Engineering 44 (1999) 897–917.

[66] N. Nishimura, Fast multipole accelerated boundary integral equation methods, Applied Mechanics Reviews 55 (2002) 299–324 (4 July).

[67] N. Nishimura, K. Yoshida, S. Kobayashi, A fast multipole boundary integral equation method for crack problems in 3D, Engineering Analysis with Boundary Elements 23 (1999) 97–105.

[68] A.P. Peirce, J.A.L. Napier, A spectral multipole method for efficient solution of large-scale boundary element models in elastostatics, International Journal for Numerical Methods in Engineering 38 (1995) 4009–4034.

[69] Y.J. Liu, A new fast multipole boundary element method for solving large-scale two-dimensional elastostatic problems, International Journal for Numerical Methods in Engineering 65 (6) (2005) 863–881.

[70] Y.J. Liu, N. Nishimura, The fast multipole boundary element method for potential problems: a tutorial, Engineering Analysis with Boundary Elements 30 (5) (2006) 371–381.

[71] N. Nishimura, Y.J. Liu, Thermal analysis of carbon-nanotube composites using a rigid-line inclusion model by the boundary integral equation method, Computational Mechanics 35 (1) (2004) 1–10.

CHAPTER

Development of Lightweight Sustainable Electric Motors

21

Brad Ruff[1], Weifeng Li[1], Rajiv Venkatasubramanian[1], David Mast[1], Anshuman Sowani[1], Mark Schulz[1], Timothy J. Harned[2]

[1] *University of Cincinnati, Nanoworld Smart Materials and Devices Laboratory, Cincinnati, OH, USA,* [2] *Parker Motor Design Center, Parker Hannifin Corporation—Parker Aerospace, Portsmouth, NH, USA*

CHAPTER OUTLINE

21.1 Electromagnetic Devices with Nanoscale Materials	595
21.2 Electric Motor Development	598
21.2.1 Rationale for developing electric motors using nanoscale materials	598
21.2.2 Material preparation and technical design	601
21.2.2.1 CNT materials development and characterization	601
21.2.2.2 Electromagnetic materials development and characterization	606
21.2.2.3 Motor components design and testing	612
21.3 Conclusions	620
References	620

21.1 ELECTROMAGNETIC DEVICES WITH NANOSCALE MATERIALS

Electromagnetic devices (transformers, motors, generators, and solenoids) are the main energy prime movers in our society. Motors and generators that are lightweight and built without relying on diminishing permanent magnet materials and rare earth and copper will be critical for developing a strong economy and a sustainable environment. Multifunctional electromagnetic materials including carbon nanotube (CNT) wire and superparamagnetic nanoparticle magnetic core material are new materials that can be used to replace copper and iron on conventional motors and to design a lightweight high-performance electric motor that is an example of a new generation of energy conversion devices that will power up society into the future. CNT wire will be uniquely manufactured using new procedures to synthesize long, high-quality nanotubes and scale up manufacturing of the nanotubes. Superparamagnetic nanoparticles in a polymer matrix will replace iron/permanent magnets in electric motors. Long CNTs will also be used to form a multifunctional nanocomposite structural material to form the frames of motors. The novelty of this approach comes from the synergy that these three new materials offer thus serving the goal for a next generation (nex-gen) power efficient and lightweight motor. The carbon motor will have another advantage over conventional motors for select applications; the nanomaterials motor is not susceptible to corrosion because the motor is metal-free. Only the metallic nanoparticles are metal, and they are encapsulated in a polymer matrix. The selected materials for the carbon motor design [1–28] will represent a commercial opportunity and will open up the door to CNT materials design for many advanced applications [3–5,18,24].

Moreover, there is an urgency to develop new materials for electric motors because the demand for electric motors is increasing and, as of April 2011, it became well known among direct drive manufacturers, the continuity of the supply of rare-earth metals is now a threat to the industry [7]. Herein, nanoscale-based materials that will not rely on rare-earth metals and that will have wide application in the electric motor industry are developed. Key technology drivers will be magnetic materials to replace the iron core in motors, CNT electromagnets to replace rare-earth magnets, CNT wire to replace copper, and nanocomposite material for the motor structure. Also a new motor design using the new materials is discussed. Conventional linear and rotary electric motors, generators, and actuators have been too bulky and heavy to be used in developing advanced lightweight vehicles and machines. However, nanomaterials-based (CNT and magnetic nanoparticles) electric motors could have applications in all types of transportation systems, lightweight structures, biomedical microdevices, and smart structures. Carbon motors and actuators may be integrated onto or within laminated composite structures to make composites smart by adding actuation. The motor could be used on concept future turboelectric aircraft or electric automobiles where lightweight and high performance are important. Fig. 21.1 shows different applications that could benefit from a carbon motor. Currently, 15 new and upcoming electric automobiles on the market [1] might benefit from lightweight and high-power carbon motor technology.

FIGURE 21.1

Applications where carbon electric motors and generators may reduce mass and improve performance: (a) NASA composite aircraft concept with turboelectric generators at the wing tips and electric fans across the rear of the fuselage; (b) electric automobiles (GM, Audi); (c) wind turbines that now have a heavy generator; and (d) biomedical microdevices and the da Vinci surgical robot [25–28]. (For color version of this figure, the reader is referred to the online version of this book.)

The principle of operation of the electric motor is based on Faraday's law of induction which describes the force produced by the interaction between a current carrying conductor and a magnetic field. The law cannot change, but what can be improved are the materials used to construct the motor and that define the performance characteristics of the motor. Materials used to construct motors have not changed much between 1910 and 2010. Copper and iron are always used, and rare-earth magnets are used to obtain the highest performance. Our vision is to use nanoscale materials to develop nex-gen electric motors and generators that are sustainable and efficient. This will be a new sustainable energy pathway because rare-earth materials which are expensive, and copper and iron which are heavy materials used in motor manufacturing, will be replaced with abundant, low-cost, and lightweight materials. The use of nanoscale materials matched to the electromagnetic design of the motor will produce a new motor that is lightweight and sustainable to manufacture. Furthermore, the nex-gen motor will not be a disruptive technology to industry. New materials used in the motor are based on nanocarbon and nanoiron and can be readily manufactured. Components of the nex-gen motor can be individually integrated into existing motors to improve their performance. Also, the entire new motor design can be easily integrated into machinery and applications. There are no adverse environmental or societal impacts of widespread adoption of the carbon and magnetic nanoparticle-based bulk materials because they are coated or encapsulated into polymer matrix materials and form macroscale hybrid materials with no loose nanoparticles. Development of the new motor will require integration of different technologies. The nex-gen motor will require integration of fundamental research related to nanoscale materials development with reengineering of the electromagnetic design of the motor. The electric motor application is perfect for a systems level integration of nanoscale materials as it demands a multi-functional material to meet the performance requirements.

The importance of the research to develop a new motor is to enable a new technology in the area of nanoscale materials. This chapter provides the initial experimental and theoretical basis to put these exciting new nanoscale materials into electromagnetics applications, which consider structural, electrical, and thermal requirements. The nex-gen motor is designed based on the synergy of integrating different materials together that can overcome limitations of existing materials, and designing a motor to take advantage of the properties of the new materials. CNT wire or carbon wire is formed by drawing nanotubes and coating them with a thin dielectric to form nanolitz wire thus eliminating eddy current losses that limit the performance of copper wire. Nanotube materials produced using other methods allow eddy currents and have a skin effect due to lateral conduction. Superparamagnetic nanoparticles will also be used in a polymer matrix to produce an assembly of particles that have low hysteresis and high saturation allowing replacement of iron core and rare-earth metals in motors. Our modeling showed that an optimal electromagnetic design of the motor to use these materials involves a universal motor design that can operate at high electrical frequency and where the motor rotational speed is independent of the electrical frequency. Coupling the materials design to the electromagnetics application is intellectually novel and important and will provide information that can be used by the general research community for developing many types of power electronic devices.

Providing the new longer nanotube material and the superparamagnetic material to industry will be the first step in the motor development. Industry will then integrate the new materials into unlimited applications. Related to the motor, the objective is to reduce the weight of existing motors, generators, and actuators by 50% or more by integrating nanomaterials. Nanomaterials being developed are (1) CNT wire—multifunctional, strong, lightweight, electrically and thermally conductive; (2) superparamagnetic nanoparticles—electrical insulators, exchange coupled, and little hysteresis; and (3) nanotube composites—structural material to replace metals, tough, lightweight, and high thermal conductivity. The electromagnetics design provides new concepts that can be used to develop many types of power control/conversion devices (switches, transformers, and antennas) besides motors.

21.2 ELECTRIC MOTOR DEVELOPMENT

There are several advantages of nanoscale materials for electromagnetics design. Carbon materials reduce the weight and the inertia of components, and allow higher speeds of rotating machinery. Weight savings is most important on spacecraft and aircraft although automotive and electric drive applications could also benefit in terms of cost of ownership and performance. Other advantages of carbon motors are fast acceleration, high torque, and the ability to operate at higher temperatures and at higher speeds. Future turboelectric aircraft could benefit from lightweight carbon electric motors but the efficiency of the motor must be close to copper which is why long CNTs are sought in this project. Small-diameter motors will have ultrahigh magnetic field intensity and produce large force/size because the force is proportional to the electrical current/distance between conductors. Eliminating iron cores and the losses due to eddy currents in the iron is a big factor. Carbon materials and electric motor design are discussed next to provide rationale for choosing the first carbon motor to be developed as a test bed for the new carbon wire, structural material, and nanoparticle superparamagnetic materials.

21.2.1 Rationale for developing electric motors using nanoscale materials

Almost the entire electric motor can eventually be manufactured from nanoscale materials. The housing and shaft of the motor can be made of nanotube composite. The carbon components will be lighter than the metal components. With the use of nanotubes, heat conduction will be improved. Silicon carbide-sealed full-ceramic ball bearings will be used, which are lighter and have lower friction than steel, generate less heat, are wear and corrosion resistant, are lubrication and maintenance free, and have low thermal expansion. Ceramic bearings [2] also do not conduct electricity and are not magnetic. CNT wire [3] can replace copper wire in the motor.

Nanotubes have a large maximum current density and can conduct huge current to generate strong magnetic fields. Electrical power conduction using lightweight nanotube material is limited by the temperature range of the application. Thus, thermal design and cooling of the motor are important design aspects. The resistivity of nanotube yarn decreases with increasing temperature, which is an advantage compared to copper. Reducing the electrical resistance of nanotube yarn will be a goal to increase the power and improve the efficiency of carbon motors [4–28]. The long nanotubes to be manufactured will exceed the electrical conductivity of copper on a per weight basis, and potentially match copper on a per area basis depending if we dope, functionalize, and postprocess anneal the nanotubes. Measurements indicate annealing improves CNT conductivity by a factor of 5. Functionalization with metal particles also increases conductivity. Doping has been shown to increase the electrical conductivity of nanotubes [29]. All these improvements may eventually allow the conductivity of CNT thread to approach the electrical conductivity of copper (the electrical conductivity of short, perfect armchair nanotubes is similar to copper). Then large currents in the motor windings in the carbon motor will produce ultrahigh magnetic flux [4], which translates into high performance [24]. Recently, the thermal conductivity of our annealed nanotubes has also been measured to be very large, about 400 W/(m K). The nanotubes had a covalently bonded metal coating at the ends to achieve this high thermal conductivity. Thus, interfacing the nanotubes with metals is an important part of the design.

Carbon electric motors might also form electromechanical actuators located inside of open structures, e.g. within a wing, or on the outer surfaces of composite structures that could strain structures using mechanical advantage through cables, linkages, screw drives, or local bender elements. Both rotary and linear motors can be built using carbon. In the future, the frame of the motor could be part of the composite structure. Using long CNTs minimizes the need for spinning [5] to provide high strength to allow CNT yarn to replace carbon fiber in composites which would toughen composite materials and improve heat conduction. There will be many applications of carbon motors, actuators, switches, tiny devices, and composite materials across many industries. The importance of improving electromagnetic devices is highlighted by the importance of energy to our society. Most of the energy generated and used in civilization is related to electromagnetic devices including motors, generators, solenoids, actuators, transformers, wires, shielding, cables, and antennas. Thus, improvement in the performance of electromagnetic devices will significantly impact our economy. An example is energy recovery in automobiles using a nex-gen motor/generator system. The motor can operate as a generator and convert kinetic energy to electrical energy as a means of braking the automobile. The harvested electrical energy would be stored in the battery and used later to power the automobile. Factors that are critical in motor design include dissipating heat from high current densities; minimizing i^2R loss; avoiding core material saturation (e.g. silicon steel limit is 1.8 T) by high flux density; increasing the pole count to get more torque but the magnetic frequency increases and the magnetic losses in iron may exceed the copper losses; and the design must be balanced or the motor can be overheated. A new core material that could provide the same saturation but increase electrical conductivity, or a core material with a higher saturation with the same or lower electrical conductivity would be important improvements. Also, permeability is in general less important than conductivity. Recoil permeability should be reduced to allow increasing the magnetic frequency, e.g. by increasing the number of poles. Eliminating rare-earth magnets would be beneficial. A high-strength housing to reduce motor weight is also beneficial. These are described below.

1. *A magnetic core material that is light, carries equal or greater flux than iron laminations, and minimizes eddy current losses* is possible to fabricate by using magnetic nanoparticles coated with an electrical insulator and the insulated magnetic nanoparticles distributed in a polymer. When the diameter of magnetic nanoparticles goes below about 20 nm, the particles become superparamagnetic which means the particles consist of a single domain where their magnetization on average is zero. But an external magnetic field is able to magnetize the nanosize magnetic particles and their magnetic susceptibility is much larger than that of paramagnets. When an external magnetic field is applied to an assembly of superparamagnetic nanoparticles, their magnetic moments tend to align along the applied field leading to a net magnetization. The magnetization curve (magnetization as a function of the applied field) is in S shape, which means the particles behave as paramagnets with large magnetic moments. The magnetic field can be switched back and forth at high frequency which thus opens the possibility for the operation of high-frequency motors. Coating the particles will provide electrical insulation and maintain the magnetic properties to serve as a core material. A company developed a CNT composite coated with Ni/Co nanoparticles that showed superparamagnetism [18]. General Electric Company just started a large project to develop nanoscale materials to replace rare-earth materials.

2. *A carbon wire that is light and carries large current with low impedance* is possible using long CNTs that are postprocessed and operate at high frequency where the impedance is low. Annealing, templating, coating, and doping are being developed to lower the resistivity and electrical impedance of CNT yarn. Presently in the literature, the resistivity of CNT thread on a per-weight basis is lower than that of copper. At high frequency, the advantage may be greater. Thus, based on our initial results and the literature (especially from China), the chance of success to develop improved wire material for motors is high.

3. *A motor structure that is light and thermally conductive* will become possible as the properties of CNT materials improve. CNT thread is much tougher than carbon fiber and CNT arrays have recently been shown to have high thermal conductivity. Composite materials reinforced with CNT threads will become commercially available as CNT materials production is increased. Our initial results and the literature indicate the chance of success to develop high-strength

structural material that is lightweight and has high thermal conductivity for use in motors is high. In summary, advances are likely in all three areas to enable development of the "*Next Generation of Electric Motors*" which will have broad applications in industry.

21.2.2 Material preparation and technical design

Specific steps to further develop CNT materials and to design and demonstrate select components of the carbon electric motor [30–69], [70–95] are described in this chapter. Any type of electric motor could be developed using CNT thread and metal nanoparticle core materials. One consideration in carbon motor design is CNT wire resistance is greater than copper at present, but the electrical impedance decreases at high frequency, and decreases slightly with increased temperature (opposite to copper). Thus, alternating current (AC) motors operating at high-frequency excitation, if possible, may be preferred for use with CNT thread. Coils in CNT motors can be in parallel to reduce the driving voltage to compensate for the higher resistance of CNT thread, although the power loss will not change because the same current will be used. The small diameter of CNT thread (10 μm) and our processing method also allow development of lightweight electromagnets to replace permanent magnets. Properties of the CNT and magnetic material must be evaluated and modeled to determine the optimal configuration for the carbon motor. The carbon motor will be capable of higher speed than metal motors. The performance of this design is being evaluated in concert with the Parker Hannifin Motor Design Group [70–72,76]. The geometry of the motor is not finalized as modeling the magnetic field is complex. Overall, an efficient magnetic design using Halbach arrays and CNT wire has been outlined. Research efforts are outlined to design and build the nex-gen carbon motor. The organizational plan is shown in Fig. 21.2. The three major motor components that will be "attacked" to replace with carbon-based materials are (1) copper wire, (2) soft iron core, and (3) the motor housing. The copper wire will be replaced by CNT thread, which will be posttreated by annealing, doping, and coated with dielectric polymer, thus forming conductive wire [3,29,33,96–117]. The next component, soft iron, will be replaced with a magnetic composite consisting of nanoscale superparamagnetic particles. The last component is the motor housing and our plan is to make it of CNT-based polymer composite, which will provide electrical insulation and will be able to effectively manage the heat released by the motor. For this, the composite will be thermally conductive in preferred directions, thanks to the highly conductive and oriented CNTs incorporated into the polymer. Materials mentioned in the organization plan have already been manufactured at small scale.

21.2.2.1 CNT materials development and characterization

This important step is to grow long CNT that can be twisted into multistrand yarn or wire. Longer CNT will reduce the number of junctions in the wire and increase its conductivity and strength.

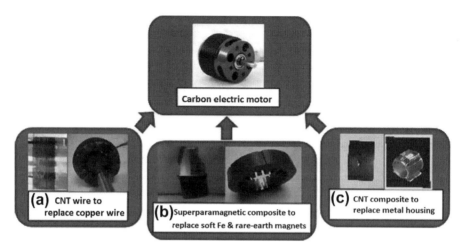

FIGURE 21.2

Organizational plan for developing the nanoscale materials and electric motor. Prototypes of CNT wire, magnetic material, and CNT composite have already been manufactured in the UC Nanoworld Lab. (For color version of this figure, the reader is referred to the online version of this book.)

21.2.2.1.1 Synthesis and postprocessing of long CNT and characterization

The main goal of the synthesis research efforts is to produce a CNT-based wire that will replace copper wire in the nex gen carbon motor. The materials research includes the following procedures:

1. Synthesis of high-quality, dense and vertically aligned CNT arrays.
2. Postprocessing yarn by high-temperature annealing, doping, and metal decoration.
3. Coating of the CNT thread and yarn with dielectric polymer.
4. Future work will explore synthesis of superlong and continuous nanotubes.

21.2.2.1.1.1 The first procedure is synthesis of high-quality, dense and vertically aligned CNT arrays. Our team is routinely producing high-quality, highly dense CNT arrays on 4 inch Si substrates using CVD processes. The CVD process is run at 750 °C and is based on a gas mixture containing ethylene, hydrogen, argon and water vapor. The arrays have been characterized in terms of their properties [110]. Lately, our industrial partner General Nano [30] started manufacturing CNT arrays on 12″ × 16″ stainless-steel foil as a fast route to scaling up the process. The length of the CNT arrays ranges from 300 to 500 μm. The average diameter of the CNTs is 10–15 nm, and the spacing between CNTs is 70–100 nm. The samples have a rather high CNT coverage density ($\sim 10^{10}$ CNT/cm^2). Our team operates a state-of-the-art CVD system for growing CNT arrays.

21.2.2.1.1.2 The second procedure is postprocessing yarn by annealing, doping, and metal decoration. Two major tools that will be used for posttreatment are a high-temperature furnace and a thermal system for doping the CNT thread. The high-temperature annealing will be conducted at a temperature of 2500 °C which was proven to improve the quality of the CNT tested by high-resolution transmission electron microscopy and Raman spectroscopy as reported in our published results [33]. An initial external annealing study was performed for our CNT thread. Our CNT arrays and dry spun yarn contain almost no impurities and high-temperature annealing greatly improved their quality and made the nanotubes stiffer and straighter. Before annealing, CNTs were wavy and bent easily. Annealed CNT are straighter and do not bend as readily and are more difficult to coil. The Raman G/D ratio increased from 1.78 (typical for as-produced thread) to 9.57 for thread annealed at 2500 °C. The G/D ratio is the ratio between the G and D peaks in the Raman Spectroscopy. The G and D peaks are generally associated with graphitic structure and defects respectively. The G/D ratio is inversely related to the relative amount of defects in the CNT structure. Annealing made the nanotubes in the array cleaner but the nanotubes bundled at the ends.

Doping is also important to improve electrical conductivity. The doping efforts are inspired by a recently published paper but we will modify the approach to consider a higher temperature dopant material [29] for use in electric motors. Optimizing the reactor pressure, exposure time, temperature and reactor design, and use of nitrogen or other dopants will be the focus of the effort. Decoration of our CNT thread/yarn with gold nanoclusters has been explored through a partnership with the Air Force Research Laboratories in Dayton, OH. Some preliminary results are shown in Fig. 21.3. The high-resolution Scanning Electron Microscopic (SEM) images in Fig. 21.3 demonstrate that tiny nanoclusters of gold have been successfully deposited on the tubes. This technique is capable of creating a very thin film of gold on the tips

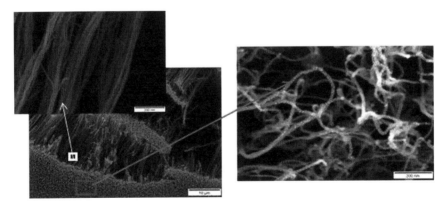

FIGURE 21.3

High-resolution SEM images of CNT array metalized with gold using a proprietary vacuum technique developed by Christopher Muratore at the AFRL. The conductivity of CNT wire is increased. (For color version of this figure, the reader is referred to the online version of this book.)

of the CNT incorporated into the polymer matrix, which is expected to decrease the resistivity of the CNT thread/yarn. It is anticipated that electrical conductivity will be improved in a much greater proportion than the mass of the gold added. The rationale for this is the combination of the two types of conduction mechanisms (CNT and metal) which are synergistic especially related to defects (junctions) in the CNT thread. The decoration will be followed by coating the fiber with dielectric polymer.

21.2.2.1.1.3 The third procedure is coating CNT thread and yarn with dielectric polymer. The coating approach to apply dielectric polymer on the CNT yarn surface includes cleaning the CNT fiber followed by electrostatic coating using a Terronics coating applicator. Our team has preliminary experience using this approach. The monomer cure will be performed either at 140 °C in a conventional furnace or by using an ultraviolet lamp. Both facilities are available at University of Cincinnati (UC) and have been successfully tested. The cleaning operation prior to coating with polymer will be performed using plasma etching. An atmospheric pressure microwave plasma torch has been built at the UC Nanoworld lab and is available to be employed for this processing. Tensile strength, modulus and strain will be evaluated using a state-of-the-art Instron Micro Testing Machine Model 5948 with pneumatic, foot-activated grippers and resolution of millinewton. The resistivity will be measured by a four-probe method with Labview software and hardware.

21.2.2.1.2 Forming CNT yarn and wire from long CNT and characterization

Capabilities to spin CNT array into thread and yarn and their characterization are shown in Fig. 21.4. The results have been published elsewhere [111–113]. Continuous improvement of the array spinnability is achieved by better substrate design, efficient catalyst system, strict control of the water vapor, and optimized growth process. An improved spinning process that allows greater productivity is based on a new spinning machine designed and made at the UC Nanoworld. In addition, we plan to increase the length of the spinnable CNT arrays and improve their quality [3]. The properties of the CNT thread and yarn will be further improved by densification with solvents, thermal annealing, plasma treatment, doping and coating.

Thread is currently spun by drawing directly from the nanotube array. The smaller diameter produces higher strength based on our results in Fig. 21.5(a). But CNTs longer than 0.5 mm cannot be easily dry spun into yarn [4]. In contrast,

FIGURE 21.4

Spinning CNT arrays into CNT wire: (a) spinning machine; (b) spun CNT thread on a Teflon drum; (c) SEM of CNT thread; and (d) SEM of 2-ply yarn. (For color version of this figure, the reader is referred to the online version of this book.) Scale bars are: (c) 20 microns and (d) 50 microns.

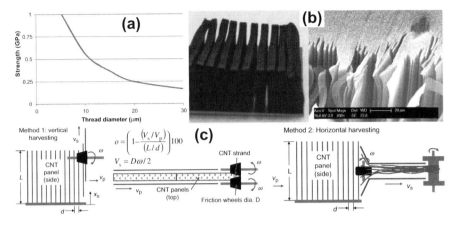

FIGURE 21.5

Developing strong and conductive yarn: (a) strength of as-produced yarn increases with decreasing diameter; (b) long CNT panels (grown in Nanoworld at UC) that will be used to spin yarn; (c) microspinning to be developed and investigated in the research facility, $o = \%$ overlap of CNT which must be carefully controlled. (For color version of this figure, the reader is referred to the online version of this book.) Scale bar is 20 microns.

cotton fibers are about 1.5 cm long (e.g. Pima long staple cotton). We plan to achieve a 20× increase in the length of CNT that can be dry spun into strong thin yarn. Our approach is to grow CNT in the shape of thin panels (Fig. 21.5(b)) and develop a mechanical system to draw and combine the CNT from the panels (Fig. 21.5(c)) into a sliver that can be spun into yarn. Vertical spinning is shown first but the friction wheels can be turned 90° and horizontal spinning can also be tried at any position along the length of the CNT panel. A miniaturized spinning machine to spin long CNT panels into yarn is being developed with collaborators [77–79]. UC can produce 2.2-cm-long CNT that potentially can be spun into strong yarn.

A new approach to be developed is to form cable by spinning multiple threads and combining them. Individual threads are coated to allow many individual conductors to be combined into a rope. A second coating step puts a dielectric coating on the outside of the cable. This approach of coating small-diameter threads (5–10 μm in diameter) and combining into a cable cannot be done using the floating catalyst method of nanotube thread manufacturing, which is a competitive method of nanotube synthesis. Only a single thread can be produced from the floating catalyst furnace and this thread is much larger in diameter than the thread that can be spun from a nanotube forest. Dry spinning of thread has the following advantages for electronics and electromagnetics applications:

1. CNT strands can be decorated with metal nanoparticles and then twisted into yarn to tailor the electrical properties of the material for the motor.

2. Eventually very long CNT will be used which will improve the strength and conductivity of the yarn.
3. The strands can be functionalized with a dielectric to prevent lateral conduction which will produce a nanolitz wire and reduce the skin effect, which will still occur in CNT thread.
4. The electrically insulated strands will prevent lateral conduction in thread and sheet which will prevent eddy currents and hysteresis losses when using the material for magnetics and supercapacitors.
5. It produces the purest material (almost catalyst free).
6. The array or strands can be conveniently doped to increase electrical conductivity at low frequency.
7. CNT can be annealed/templated to heal defects and improve quality.

An electrostatic method of coating allows coating individual bundles or strands of nanotubes with magnetic nanoparticles and a dielectric coating. A micrometer-thin coating improves the thread mechanical and electrical properties. Coating causes CNTs to bond to each other, which transfers a greater shear load between CNTs as compared to load transfer due to friction and van der Waals forces. The properties also improve because solvent in the polymer evaporates and surface tension causes the thread to shrink in diameter, which increases properties because the cross-sectional area decreases. Coating CNT thread is an important step in developing the motor, to provide an electrical insulation and to improve the mechanical and electrical properties.

21.2.2.1.3 Fabrication and characterization of the nanocomposite structural material

CNTs are ideal structural reinforcing materials for composites due to their excellent mechanical properties. We propose a novel tough-ply fabricated from long spinnable CNT forest to achieve high in-plane tensile strength. The fabrication process is shown in Fig. 21.6. A long spinnable CNT forest is electrically annealed for 4–6 h at 2500 °C. The annealed CNT forest is drawn and twisted into a CNT thread, which is subjected to polymer coating process for electrical insulation. Multiple CNT threads are placed together and unidirectionally aligned for a preimpregnation process with polymer so that a unidirectional CNT thread prepreg ply could be accomplished. This procedure could be repeated and multiple unidirectional CNT thread prepreg plies are stacked and consolidated with a hot press. This produces nanocomposites with high tensile strength.

21.2.2.2 Electromagnetic materials development and characterization

A replacement for the iron core and the permanent magnets in motors will be developed. Several carbon materials were considered and tested. But the magnetic properties were not satisfactory. Finally, superparamagnetic particles were selected and a composite was formed as described below.

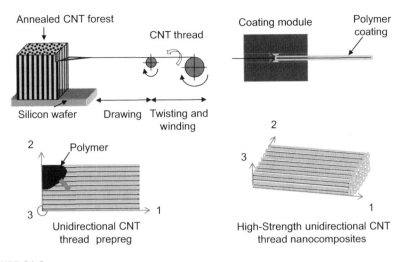

FIGURE 21.6

Forming CNT composite materials for the motor frame. (For color version of this figure, the reader is referred to the online version of this book.)

21.2.2.2.1 Superparamagnetic particle core design and testing

One approach for manufacturing of nanomagnetic materials is by powder synthesis using insulated coated magnetic nanoparticles and consolidation of the powder into exchange-coupled cores [102]. When reducing the particle size and the separation between neighboring particles to the nanometer scale, it was found that a Co- or Fe-based nanocomposite can possess permeability similar to bulk Co or Fe. The increased permeability is due to the exchange coupling effect which leads to magnetic ordering within a grain and extends to neighboring particles within a characteristic distance called the *exchange length* L_{ex}. Thus, neighboring grains separated by distances shorter than L_{ex} can be magnetically coupled by exchange interaction. The permeability of an exchange-coupled nanocomposite can be higher than the permeability of its bulk counterpart. Nanocomposite soft magnetic materials possess higher initial permeability and a higher cut-off frequency than iron core materials. This advance greatly opens up the parameter space for motor design. The nanocomposite material is beginning to be offered commercially [102]. Eddy currents will be minimized in CNT materials, as explained in Fig. 21.7. We anticipate nanomagnetics is going to provide a new generation of core material for motor design.

A simple model of the magnetization can be used for initial design of the new core material. If all the magnetic nanoparticles are similar with the same energy barrier and same magnetic moment and their easy axes are all oriented parallel to the applied field, the temperature is low enough, and the particles are closely spaced, then the material behaves like a paramagnet and the magnetization of the assembly

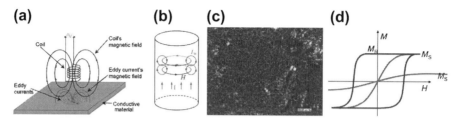

FIGURE 21.7

Eddy currents and magnetic losses are minimized in CNT materials with magnetic nanoparticle coatings. (a) Eddy currents are circular currents formed in a material. When a composite of superparamagnetic nanoparticles and a polymer are used, it does not conduct electrically very well. Thus, eddy currents will be small and will reduce heating of the motor and power loss (Wikipedia, [118]). (b) Also, the skin effect in CNTs is small (Wikipedia, [119]) as each CNT shell is so thin, eddy currents are negligible. (c) Ni particles coated onto a CNT array, top view. (d) Superparamagnetic material (red) has high saturation magnetization (M_S) and no remanence ($M_R = 0$) and no hysteresis loop, whereas ferromagnetic material (blue) has high remanence and paramagnetic material (green) has low saturation. (For interpretation of the references to color in this figure legend, the reader is referred to the online version of this book.)

of particles is approximated by $M = \mu n \tanh\left(\dfrac{\mu \mu_0 H}{k_B T}\right)$, where n is the density of nanoparticles in the sample, μ_0 is the magnetic permeability of vacuum, μ is the relevant magnetic moment or permeability of a nanoparticle [106], M is the magnetization of the material (the magnetic dipole moment per unit volume measured in amperes per meter), k_B is the Boltzmann constant, T is the temperature, and H is the magnetic field strength also measured in amperes per meter. The magnetic induction B (or flux density in Tesla units) is related to H by the relationship $B = \mu_0(H + M) = \mu_0(H + \chi_v H) = \mu_0(1 + \chi_v)H = \mu H$, where χ_v is the volume magnetic susceptibility (degree of magnetization of a volume of material due to an applied magnetic field), and $\mu = (1 + \chi_v)$ is the magnetic permeability of the material. Note that SI units should be used for the above derivations, as opposed to electromagnetic and centimeter–gram–second units.

A recent sample made in our lab is a superparamagnetic composite (Fig. 21.8). The mass of the sample is 0.6638 g. The dimensions of the cylindrical sample are height = 1.28 cm and diameter = 0.64 cm. Volume of the sample = 0.41 cc. Density of the sample = 1.62 g/cc. Particle type: gamma iron oxide nanoparticles (size: 20 nm). Percentage by weight of nanoparticles: 44.7% w/w. Binder used: Toolfusion epoxy (10 parts of Toolfusion 1 A resin + 2 parts of Toolfusion 1 B hardener). Percentage by weight of epoxy: 55.3% w/w. Density of nanoparticles from MSDS: 5.24 g/cc at 20 °C. Mass of the nanoparticles in sample = 0.2967 g. Volume of nanoparticles in sample = 0.0565 cc. Volume fraction of nanoparticles in sample = 0.138. The electrical and thermal conductivity, strength, B–H and magnetization curves of

FIGURE 21.8

Superparamagnetic composite to replace iron. (For color version of this figure, the reader is referred to the online version of this book.)

the material are being measured and will be compared to theory. High-frequency operation of the material will be tested.

21.2.2.2.2 Magnetic nanotubes and nanowires

Electric machines such as electric motors, transformers, and solenoids operate based on Faraday's law, which describes the force produced by the interaction between a current carrying conductor and a magnetic field. The materials that are used to construct the motor define its performance. Copper and iron materials have always been used in the fabrication of electric motors. Rare-earth magnets are utilized to obtain the highest magnetic performance. In this section, nanoscale materials are to be used to develop new electric motors, transformers, and actuators. The use of nanoscale materials will produce new electric devices that are lightweight with high energy density. The new materials are based on nanoscale carbon and magnetic materials, which can be readily manufactured. It is expected the weight of existing electromagnetic devices can be reduced by half, by use of nanomaterials. Nanomaterials that will be used are (1) CNT wire that is strong, lightweight, electrically and thermally conductive and (2) superparamagnetic nanoparticles which are electrical insulators that are exchange coupled, with little hysteresis. Electromagnetic designs using nanomaterials will allow development of many types of biomedical microdevices and tiny machines that will change engineering design. Copper wire may be replaced by doped CNT thread or superfiber. Iron core in motors must be replaced by a lighter material. Different magnetic nanomaterials are being investigated to

FIGURE 21.9

Magnetic nanoscale materials: (a) magnetite (Fe_3O_4) nanotubes, from Vijay Varadan, U. Arkansas. *Nanomedicine: Design and Applications of Magnetic Nanomaterials, Nanosensors and Nanosystems*, Vijay K. Varadan, Linfeng Chen, Jining Xie, 2008 John Wiley & Sons [120]; (b) Ni nanowires manufactured at UC; (c) 25 μm nickel wire cut into small pieces with lengths around 5 mm and aligned under a strong external magnetic field, all wires are standing up; the distance between the magnet and the wires is about 3 cm, the wires are widely distributed; and (d) nickel wires are put into an elastomer and a magnet at the bottom is used to attract the wires. (For color version of this figure, the reader is referred to the online version of this book.)

replace iron for electromagnetic applications. Such promising materials are magnetite nanotubes as shown in Fig. 21.9(a) and Ni nanowires as shown in Fig. 21.9(b,c). These materials can be used to build motors, transformers and solenoids. The approach suggested is to align the magnetic nanotubes/nanowires in a polymer matrix. The alignment direction of the nanotubes/nanowires would be in the direction of the magnetic flux in the motor or transformer. There are several important advantages from making a magnetic material using nanomaterials. First, eddy currents, which cause heating and reduce the efficiency of motors, will be very small because magnetic composites will not have electrical conduction laterally. Second, if the nanotubes/nanowires are small enough in diameter, the magnetic flux in the composite may be linked by exchange coupling which means the magnetic flux density of the composite may be close to the strength of the magnetic flux in iron material. Moreover, if the diameter of the nanotube/nanowire is small enough, the nanotube/nanowire is superparamagnetic. There are several methods to synthesize magnetic nanotubes. One common method is to use anodic aluminum oxide templates with pore diameter of several hundreds of nanometers [121–124]. Another method is synthesis of single crystalline nanotubes by wet etching [125]. Different magnetic nanotubes have different magnetic properties. One way to increase the saturation magnetization is to fill the large-diameter magnetic nanotubes with nanoparticles. Accordingly, the density of the core will increase.

Superparamagnetic materials can operate at higher frequency with little hysteresis loss. The properties of iron nanotubes and nanowires are reported in Ref. [120]. On the down side, there are difficulties fabricating magnetic nanocomposites with a high volume fraction of the magnetic material and the polymer composite will not operate at high temperature as an iron core. One particular hypothesis is the iron nanotubes must be aligned in a magnetic field and well dispersed in the polymer during curing in order to achieve the desired magnetic effects. In the past, there have been efforts to form a magnetic composite material by integrating micrometer-size iron filings into the polymer. The large-size iron particles are multidomain and are not expected to have superparamagnetic properties. Recently, effort is focusing on integrating magnetic nanoparticles into polymer. The theory is the nanoparticles will provide the superparamagnetic effect and exchange coupling. But using nanoparticles is different than previous work using micrometer-sized iron filings. The magnetic nanoparticles are new materials which have potential to produce a new type of magnetic material with properties that exceed those of iron or iron-filled polymers. Besides lightweight, another advantage of using magnetic nanotube and nanowire is the anisotropic property. That is in only the axis direction, the external magnetic field will be magnified. In the cross-direction, the change is less because of high magnetic reluctance. This property is important to ensure sensitive devices operate in complicated electromagnetic fields.

21.2.2.2.3 Carbon wire design and testing

Carbon wire is designed by drawing fine CNT strands, coating each strand with a dielectric, and twisting the strands to form thread, and twisting the threads to form yarn. The yarn will be a nanolitz wire that eliminates the skin effect. Testing is performed using a Solartron impedance analyzer. At high frequency, the impedance of CNT thread is lower than copper due to the capacitance of thread and absence of the skin effect. As the CNT resistance is reduced, the cross-over frequency is moved to the left. It may be possible to operate the motor in the tens of kilohertz frequency which may reduce the resistance of the CNT thread. Coating the thread with metal particles, doping, and using superlong CNT will make the thread competitive with copper.

21.2.2.2.4 Thermal testing of components and the motor

A limitation of electric motor design is the dissipation of heat within the motor. Metals are good thermal conductors and CNTs are an excellent thermal conductor along their axis. An experiment was performed using an infrared (IR) camera FLI T640, where a copper wire and CNT thread were connected to a heat sink at their ends, as shown in Fig. 21.10. Even though the ends of the two conductors were at a similar temperature, the temperature in the center of the copper wire was higher than the CNT. This indicates CNT thread will allow a larger amount of current to be used without the conductor reaching a critical temperature. When the power source is cycled on and off, the temperature of the CNT drops to ambient temperature immediately after the current is removed. In contrast, the copper wire takes several seconds to return to ambient temperature. CNT thread would allow the heat from the system to be dissipated quicker. It could allow the motor to run at very high loads for short periods of time without building up excessive heat in the system. The thermal conductivity of CNT is greater than copper and the specific heat of CNT is lower than copper. If the resistivity of CNT is reduced to approach copper, then CNT will be a superior conductor for electromagnetic devices to reduce the buildup of heat in the system.

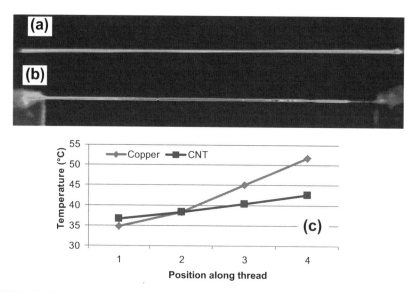

FIGURE 21.10

Comparison of relative thermal conductivity in copper wire and CNT thread heated by electrical current: (a) thermal image of CNT thread; (b) and of copper wire (red is hotter); (c) surface temperature obtained by an IR camera FLIR T640 of copper wire and CNT thread at similar points along their half-length. (For interpretation of the references to color in this figure legend, the reader is referred to the online version of this book.)

21.2.2.3 Motor components design and testing

Depending on the unique properties of carbon nanomaterials, the design and working principle may have some difference between this nex-gen motor and traditional motor.

21.2.2.3.1 Motor design and analysis

The main idea in the motor design is to take advantage of lightweight and high maximum current density of CNT materials. The properties of CNT and conventional materials are listed in Table 21.1. Besides measuring the properties of thread, the properties for different form factors of nanotube thread must be measured. This includes measuring the impedance of thread and sheet along with metal coating. CNTs are drawn from arrays and are twisted to form thread. Also, in the future, using long CNT to form thread will further increase the electrical conductivity and other properties of the thread (strength, thermal conductivity, and toughness).

Advantages of different motors [6–24] were considered to help develop a design of a carbon motor. A universal motor design is described below. The universal motor is also called a *series-wound motor*. The universal motor is the most used electrical motor in the world. It has many advantages including low cost, easy to use, lightweight, small size, high speed, high starting torque, and good performance at

Table 21.1 Electrical Properties of CNT and Conventional Materials

	Density (g/cm^3)	Resistivity (Ω cm)	Current Density (A/cm^2)	Cross-over Frequency where Impedance of CNT Material = Copper (Equal CS Areas)	Thermal Conductivity (W/mK)
Copper	8.9	1.7×10^{-6}	10^3	—	400
Aluminum	2.7	2.8×10^{-6}	—	NA	237
Iron	7.9	9.6×10^{-6}	—	NA	80
SWCNT (no defects)	1.8	1×10^{-6}	$\sim 10^5$	DC	3000
CNT (superlong, triple-wall, postprocessed)	1.4	$2-3 \times 10^{-6}$ (projected)	$\sim 10^5$	Estimated to be low frequency	600 (projected)
CNT thread	1.0	$\sim 1.7 \times 10^{-4}$	$\sim 10^4$	~ 100 MHz	~ 100

overload. It also has disadvantages such as low efficiency, higher cost, and short life. The universal motor operates similar to a typical direct motor (DC) motor, without a permanent magnet, except the stator and rotor coils are connected in series. This allows the motor to operate on DC as well as any frequency AC current. It unique in that the rotation speed of the motor is independent of the input frequency. Operating at high frequency will reduce the impedance of the nanotube thread. Universal motors generally run at high speeds, making them useful where high speed and lightweight are desirable. Universal motors have the highest horsepower-to-weight ratio of all the types of electric motors. The goal is to develop a motor that has no bulk iron, no copper, and no rare-earth metals. A carbon universal motor with CNT and magnetic coated particles may eliminate the limitation on high-frequency operation due to the iron core and thus allows a motor to use CNT thread at high frequency where the impedance is low. A carbon universal motor would also reduce the magnetic losses from high-frequency operation. The stator magnets, too, are electromagnets. The two stators are wound in the same direction so as to give a field in the same direction and the rotor has a field which reverses twice per cycle because it is connected to brushes. When you drive such a motor with AC, the current in the coil changes twice in each cycle (in addition to changes from the brushes), but the polarity of the stators changes at the same time, so these changes cancel out.

In the universal motor, the reactance of the series field and armature windings must be reduced as much as practicable. The reactance voltage drop due to the armature winding can be practically eliminated by use of a compensating winding. The compensating winding is connected in series with the armature winding (conductive compensation) and arranged such that the ampere-turns of the compensating winding oppose and neutralize the ampere-turns of the armature [23]. To realize this compensation, the compensating winding is displaced by 90 electrical degrees from the field winding. The compensating winding also improves commutation considerably. This is an advantage since the magnetic field of a universal motor is weakened by lowering the reactance of the series field winding. If the compensating winding is short circuited (inductive compensation), the ACs in the armature are induced by transformer action into the shorted compensating winding, thus, effectively cancelling the reactive armature currents.

This initial analysis done on noncompensated AC motor is to show that the rotational speed of the universal motor depends on the magnitude of the applied voltage, and not on the source frequency. For a single-phase universal motor, the principle of operation is the same as the DC electrical motor. Considering a single wire, the electrical potential generated in a magnetic field is $e = Blv$. Here, e is the electrical potential with units of volts, B is the magnetic field density with units of Tesla, l is the effective length of the wire with units of meters, and v is the linear velocity of the wire with units of m/s. For a rotor with N wires and b parallel circuits, the electrical potential generated, E, is given by $E = Bl \cdot \frac{N}{2b} \cdot v$. E has units of volts; the speed of the wire is $v = \pi D n$, where D is the diameter of the rotor with units of meters and

n is the rotational speed of the rotor with units of revolution per second. Suppose the number of pole pairs is p, the arc length of each pole is τ, then $2p\tau = \pi D$. Now, $E = Bl \cdot \frac{N}{2b} \cdot v = Bl \frac{N}{A} \cdot p\tau n$. For a single-phase motor powered by an AC source, if there is only one parallel circuit in the rotor ($b = 1$), we can simplify the equation above; $\sqrt{2}E = \Phi Nn$. Here $\Phi = Bl\tau$ is the magnetic flux with unit of webers, and $\sqrt{2}$ changes the E from peak voltage to RMS voltage. The rotational speed is n with units of revolution per second; $n = \frac{\sqrt{2}E}{\Phi N}$. The rotational speed of the motor does not depend on the frequency of the electrical source. It depends on the electrical potential E.

The MATLAB model of the electrical and torque characteristics of a universal (or series) motor is given next. The Universal Motor block represents the electrical and torque characteristics of a universal (or series) motor using the equivalent circuit model shown in Fig. 21.11. In the model, R_a is the armature resistance; L_a is the armature inductance; R_f is the field winding resistance; and L_f is the field winding inductance. The universal motor is also modeled in MATLAB SimElectronics. The L_{af} is calculated using the $R_a + R_f$, rated voltage, power, and the rotational frequency. The MATLAB algorithm computes the motor torque as follows. The magnetic field in the motor induces the following back electromotive force (emf) v_b in the armature: $v_b = L_{af} i_f \omega$ where L_{af} is the constant of proportionality and ω is the angular velocity of the rotor in radians per second. The mechanical power is equal to the power reacted by the back emf: $P = v_b i_f = L_{af} i_f^2 \omega$. The motor torque is $T = P/\omega = L_{af} i_f^2$. The torque–speed characteristic for the Universal Motor block model is related to the parameters in Fig. 21.11.

For the steady-state torque–speed relationship when using a DC supply, L has no effect. Sum the voltages around the loop: $V = (R_f + R_a) i_f + v_b = (R_f + R_a + L_{af}\omega) i_f$. Solving the preceding equation for i_f and substituting this value into the equation for torque gives $T = L_{af}(V/(R_f + R_a + L_{af}\omega))^2$, which is for the DC supply only. The block uses the rated speed and power to calculate the rated torque. The block uses the rated torque and rated speed values in the preceding equation plus the corresponding electrical power to determine values for $R_f + R_a$ and L_{af}.

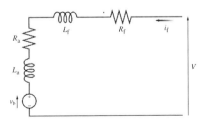

FIGURE 21.11

Universal motor equivalent circuit model.

For AC analysis, the inductive terms L_a and L_f are used in the model. The motor inertia J and damping B can also be used. The output torque is $T = \dfrac{V^2 L_{af}}{(R_f + R_a + L_{af}\omega_R)^2} - J\dot{\omega}_R - B\omega_R$ for an AC supply. The power output is $P = T\omega_R$. In the parameter study, the inertial and damping terms in the torque equation were neglected. Different motors were studied to investigate the power output. They are the conventional copper motor, a CNT motor with existing CNT material wires, and a CNT motor with future CNT wires. The parameters of the copper motor are from an MATLAB sample. Supposing the motor rotation per minute is directly related to the current passing through the coils, we obtain results as listed in Table 21.2. In the table, we first calculate the L_{af} using the suggested value from MATLAB. L_{af} is constant for a defined motor. Then the power output (mechanical power) was calculated using voltage input, L_{af}, resistance, and rotational speed. The power input is calculated using voltage input and current input which are based on the power output, rotational speed, and L_{af}. The efficiency of universal motor is lower than other electric motors but the universal motor is the highest power motor for a fixed volume. For a CNT motor, a larger current can pass through the wires and heat can be conducted out of the motor. Thus, a larger voltage might be applied on the CNT motor. For CNT motor 1, suppose the resistivity of CNT wire is eventually close to copper, a doubled input voltage will improve the rotational speed for the same load. When considering the weight factor, the advantage of the CNT motor is even greater as shown in the "power-to-weight ratio" column in Table 21.2. We set the weight factor of the copper motor as 1 and the CNT motor as 0.5 to represent the approximate weight difference of these two motors. With the CNT resistivity, where we are now and which is about 10 times larger than copper, the efficiency is low. Table 21.2 shows that reducing the wire resistance is the primary goal because it greatly affects motor performance. Higher frequency of the electrical source will reduce the electrical impedance of the carbon wire and improve the performance of the motor. Otherwise, a larger diameter wire could be used which would partially offset the weight reduction of using CNT and magnetic nanoparticles. We also have to evaluate the performance of the composite magnetic material. Table 21.2 represents the optimal performance that we might expect. Motor 3 is also having a good design because the power/weight is doubled.

21.2.2.3.2 Universal motor prototype construction and performance evaluation

The motor design could take advantage of the variety of CNTs and other carbon materials that are available. The design would use circular power rails that could be made from CNT composite materials. These rails would be on either side of the rotor and act as both the power supply and the structure of the rotor. The AC Halbach arrays would be arranged between the power rails. The power rails would then be supported on the rotor shaft by carbon composite spokes. The design of the motor is illustrated in Figs 21.12 and 21.13. The stator would have very similar construction to the rotor. The number of poles would be different than in the rotor to ensure

Table 21.2 Study of Motor Parameters for the Carbon-Wound Universal Electric Motor

Material	$R_a + R_f$ (Ω)	Rated Speed (rpm)	Voltage (Volts)	L_{af} (mH)	Power Output (W)	Power Input (W)	Efficiency %	Power to Weight Ratio
Copper	133	6500	200	200	75	111	68	75
CNT motor1	133	13,000	400	200	266	390	68	531
CNT motor2	266	6500	400	200	135	396	34	269
CNT motor3	133	6500	200	200	75	111	68	150

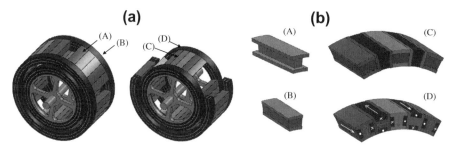

FIGURE 21.12

Concept AC commutated carbon motor: (a) *Motor*: (A) stator-field coils; (B) stator power rail; (C) rotor armature coils; (D) rotor power rail. (b) *Construction of a carbon-based Halbach array*: (A) the first step is to have high permeability, low coercively cores made from nano-magnetic material; (B) the next step is to wind the core with CNT thread or long CNT conductors; (C) finally, assemble the coils in a rotating manner to produce a Halbach array; (D) cross-sectional diagram of the current direction in each coil (white arrows), and the resulting local magnetic flux in each coil (red arrows). (For interpretation of the references to color in this figure legend, the reader is referred to the online version of this book.)

adequate starting torque and efficient commutation. The stator would also be composed of a set of power rails with sets of alternating Halbach arrays between them. The structure would be self-supporting and fixed directly to the frame of the device that it powers. The materials and methods that will be required to construct a carbon motor are all practical. The commutation or the switching of the poles is extremely important in the operation of the motor. The simplest method is to use brushes and a segmented contact point on the rotor. It is also possible to use

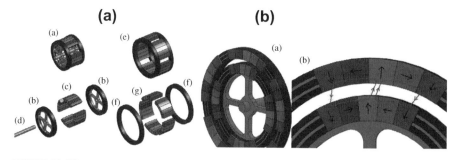

FIGURE 21.13

Concept AC commutated carbon motor: (a), (A) rotor assembly; (B) rotor power rails; (C) rotor armature coils; (D) carbon rotor axle; (E) stator assembly; (F) stator power rails; and (G) stator-field coils. Cross section detail of rotor and stator coil configuration: (b), (A) motor cross section showing the location of each pole and (B) detailed schematic of magnetic flux interaction of the rotor and stator coils. (For color version of this figure, the reader is referred to the online version of this book.)

a magnetic or an optical encoder to determine the position of the rotor relative to the stator and then use electronics to do the commutation. The use of electronics allows for complex control of the motor to optimize torque at low and high rotational speeds. Such motor controles are avaiable for Brushless DC motors and would have to be modified to work with AC.

High-temperature dielectric coatings are needed on top of the CNT thread. Materials such as boron nitride, aluminum oxide and glass could be used. Heat transfer in the motor will be modeled including the thermal properties of the nanoparticle core material and CNT thread. Magnetic and eddy current losses are expected to be reduced using superparamagnetic materials but high-frequency operation will generate more heat. A thermal analysis will help to produce a candidate motor design.

21.2.2.3.3 Electromagnetic Halbach array fabrication and bench test

The concept of using small coils of CNT thread arranged in a Halbach array is a novel idea to improve motor torque or to eliminate rare-earth magnets [104]. Five small coils were made using copper wire. The core material was machined in to "I" shapes from part of a laminate silicon steel transformer core. Each core is a 0.44″ cube with a 0.125″ wide and 0.125″ deep slot cut in two opposing sides. Each coil was then wound with 100 turns of copper magnet wire. The coils were then wired in series (Fig. 21.14).

First, the coils were arranged all in the same direction and placed in a series next to each other. This is a typical design that could be used in any electric motor. The other configuration that was tested is a Halbach array. The coils were oriented in a rotating pattern. The same coils were used in each configuration. There was some difference in the performance of each coil so they were kept in the same order for a better comparison. In each test, the current was kept constant at 0.5 A. The hall sensor can only measure magnetic flux in one direction. So for simplicity, we only measured the vertical component of the field. Each coil was tested on the top and then flipped over to measure the field on the bottom. The maximum magnetic flux density for the series coils was 20 G. The maximum magnetic flux density for the Halbach Array coils was 40 G. The series of similarly oriented coils produced a series of plateaus that were larger in the middle. This magnitude varies slightly between coils which could easily be due to variations in the steel core material or slight air gaps between the coils. The field lines all come out of the top of the coils

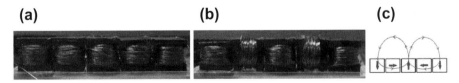

FIGURE 21.14

Component testing: (a) standard series of coils all oriented in the same direction; (b) Halbach array of electromagnetic coils; and (c) theoretical magnetic flux lines for the Halbach array. (For color version of this figure, the reader is referred to the online version of this book.)

then reenter on the bottom. The polarity of the magnetic field is opposite between the top and bottom. This is due to the rotation, relative to the hall sensor, of the coils to measure the field on the bottom. Relative to the coils themselves, the field is in a constant direction through the core. The Halbach array of coils had more complicated results. The top showed the desired effect of increased flux density. As predicted, the field is in a different direction in the middle than on the end of the coils. This is because the magnetic field lines come out of the array in the middle and then reenter the array on the ends. On the bottom of the Halbach array, the field is near zero except for the large positive and negative spikes that appear to happen right over the center of the second and fourth coils. The Halbach concept was proved but we found that greater precision is needed in manufacturing and joining the coils.

21.2.2.3.4 Composite motor structure design and testing
The motor structural design will be based on strength and thermal considerations. Thermal modeling is normally done using specialized software. The composite material testing is mentioned in another chapter. A small prototype motor will be built and tested.

21.3 CONCLUSIONS
This chapter outlines a design for an electric motor based on the use of nanoscale materials. Significant development work is needed to up-scale the properties of the nanomaterials and to validate a prototype design. If the nanoscale materials motor is successful, the payoff would be large.

References

[1] List of Electric Automobiles, http://www.hybridcars.com/electric-car.
[2] Ceramic Ball Bearings, http://www.vxb.com/Merchant2/merchant.mvc?Screen=CTGY&Store_Code=bearings&Category_Code=FullCeramicBearings.
[3] C. Jayasinghe, S. Chakrabarti, M.J. Schulz, V. Shanov, Spinning yarn from long carbon nanotube arrays, Journal of Materials Research 26 (5) (2011).
[4] M. Schulz, W. Li, Nanomedicine Devices "Tiny Machines" to Change the Outcome for Patients. Nanotechnology Materials and Devices Workshop, University of Cincinnati, 4 October 2010.
[5] UC Nanoworld and Smart Materials and Devices Laboratory, www.min.uc.edu/nanoworldsmart.
[6] http://www.femm.info/wiki/inductionmotorexample; http://www.mathworks.com/help/toolbox/physmod/elec/ref/inductionmotor.html.
[7] http://motionsystemdesign.com/news/rare-earth-magnet-shortage-0411/.
[8] S.E. Lyshevski, Electromechanical Systems, Electric Machines, and Applied Mechatronics, CRC, 1999.
[9] W. Bolton, Mechatronics: Electronic Control Systems in Mechanical and Electrical Engineering, third ed., Pearson Education, 2004.

References

[10] http://www.designnews.com/document.asp?doc_id=214685, Slotless, Brushless DC Motor.

[11] http://www.servo-motors-controls.com/sep2004slotlessmotors.htm, Slotless Motors.

[12] http://www.motionshop.com/pr/Koford-25mm-Frame-Slotless-BlessMtr.shtml, Slotless Motor.

[13] http://www.directindustry.com/prod/thingap/slotless-brushless-dc-electric-motors-55987-365802.html, Coreless DC Motor.

[14] http://www.servo2go.com/support/files/REASONS%20FOR%20TURNING%20TO%20SLOTLESS%20MOTOR%20TECHNOLOGY.pdf, Advantages of Brushless, Slotless Motors.

[15] http://www.citizen-micro.com/tec/corelessmotor.html, Coreless Motor.

[16] http://www.thingap.com/; http://www.thingap.com/videos.htm.

[17] http://www.grc.nasa.gov/WWW/RT/2006/RX/RX54S-siebert.html, NASA Glenn Ironless Motor.

[18] D. Shi, P. He, P. Zhao, F.F. Guo, F. Wang, C. Huth, X. Chaud, S.L. Bud'ko, J. Lian, Magnetic alignment of Ni/Co-coated carbon nanotubes in polystyrene composites, Composites: Part B 42 (2011) 1532–1538.

[19] http://www.animations.physics.unsw.edu.au/jw/electricmotors.html#universal.

[20] http://www.launchpnt.com/portfolio/aerospace/uav-electric-propulsion/.

[21] http://www.launchpnt.com/.

[22] http://en.wikipedia.org/wiki/Halbach_array.

[23] R.H. Engelmann, W.H. Middendort, Handbook of Electric Motors, Marcel Dekker, New York, 1994.

[24] UC Invention Disclosure No. 112-016, Carbon Electromagnetic Materials to Replace Copper, Iron and Rare Earth Metals in Electric Motors, August 2011, M.J. Schulz, W. Li, B. Ruff, D. Mast, V.N. Shanov.

[25] http://www.boston.com/bigpicture/2009/03/robots.html.

[26] http://www.festo.com/cms/en_corp/11369.htm.

[27] http://en.wikipedia.org/wiki/Electricity_generation.

[28] http://en.wikipedia.org/wiki/Unmanned_aerial_vehicle.

[29] Y. Zhao, J. Wei, R. Vajtai, P.M. Ajayan, E.V. Barrera, Iodine doped carbon nanotube cables exceeding specific electrical conductivity of metals, Science Reports 1 (2011) 83, http://dx.doi.org/10.1038/srep00083. Online.

[30] General Nano, Nanotube Materials Producer, www.generalnanollc.com.

[31] Atkins & Pearce, Braiding and Nanotube Fiber Formation, www.atkinsandpearce.com.

[32] R. Saito, G. Dresselhaus, M.S. Dresselhaus, Physical Properties of Carbon Nanotubes, Imperial College Press, London, 1998.

[33] V. Shanov, W. Li, D. Mast, C. Jayasinghe, N. Mallik, W. Cho, P. Salunke, G. Li, Y. Yun, S. Yarmolenko, Synthesis of carbon nanotube materials for biomedical applications (Chapter 2), in: M. Schulz, V. Shanov, Y. YeoHeung (Eds.), Nanomedicine, Artech House Publishers, August 2009.

[34] V.N. Shanov, R.S. Kukreja, G. Maheshwari, G. Li, P. Salunke, Y. Yun, V.D. Krstic, M.J. Schulz, Effect of Thermal Treatment on Carbon Nanosphere Chains, NIST Nanotech, Santa Clara, CA, 20–24 May 2007.

[35] V. Shanov, A. Gorton, Y.-H. Yun, M. J. Schulz, (patent pending), Univ. Cincinnati 107-044 Catalyst and Method for Manufacturing Carbon Nanostructured Materials, 10.17.06.

[36] Y. Yun, V. Shanov, Y. Tu, S. Subramaniam, M.J. Schulz, Growth mechanism of long aligned multiwall carbon nanotube arrays by water-assisted chemical vapor deposition, Journal of Physical Chemistry B 110 (2006) 23920.

[37] V. Shanov, Y.-H. Yun, M.J. Schulz, Synthesis and characterization of carbon nanotube materials (review), Journal of the University of Chemical Technology and Metallurgy 41 (4) (2006) 377–390.

[38] V.N. Shanov, Y. Yun, S. Yarmolenko, S. Neralla, J. Sankar, Y. Tu, A. Gorton, M.J. Schulz, Substrate design for long multi-wall carbon nanotube arrays grown by chemical vapor deposition, International Conf. on Composites Engineering (ICCE), Boulder Colorado, 2–7 July 2006.

[39] S.S. Xiea, W.Z. Li, Z.W. Pan, B.H. Chang, L.F. Sun, Carbon nanotube arrays, European Physical Journal, D 9 (1999) 85–89.

[40] S. Yarmolenko, S. Neralla, J. Sankar, V. Shanov, Y. Yun, M.J. Schulz, The Effect of Substrate and Catalyst Properties on the Growth of Multi-wall Carbon Nanotube Arrays, Materials Research Society, Fall Meeting, 2005.

[41] Nanoworld Laboratory, Unpublished Results on Research, www.min.uc.edu/nanoworldsmart.

[42] M.J. Schulz, A. Kelkar, M. Sundaresan, Nanoengineering of Structural, Functional, and Smart Materials, CRC Press, Boca Raton, 2006.

[43] Y. Shiratori, H. Hiraoka, M. Yamamoto, Vertically aligned carbon nanotubes produced by radio-frequency plasma-enhanced chemical vapor deposition at low temperature and their growth mechanism, Materials Chemistry and Physics 87 (2004) 31–38.

[44] W. Huang, Y. Wang, G. Luo, F. Wei, 99.9% Purity multi-walled carbon nanotubes by vacuum high-temperature annealing, Carbon 41 (2003) 2585–2590.

[45] H. Wang, Z. Xu, G. Eres, Order in vertically aligned carbon nanotube arrays, Applied Physics Letters 88 (2006).

[46] D.B. Geohegan, A.A. Puretzky, I.N. Ivanov, S. Jesse, G. Eres, In situ growth rate measurements and length control during chemical vapor deposition of vertically aligned multiwall carbon nanotubes, Applied Physics Letters 83 (9) (2003).

[47] D.-H. Kim, H.-S. Jang, C.-D. Kim, D.-S. Cho, H.-S. Yang, H.-D. Kang, B.-K. Min, H.-R. Lee, Dynamic growth rate behavior of a carbon nanotube forest characterized by in situ optical growth monitoring, Nano Letters 3 (6) (2003).

[48] L.A. Zheng, M.J. Oconnell, S.K. Doorn, X.Z. Liao, Y.H. Zhao, E.A. Akhadov, M.A. Hoffbauer, B.J. Roop, Q.X. Jia, R.C. Dye, D.E. Peterson, S.M. Huang, J. Liu, Y.T. Zhu, Ultra-long single-wall carbon nanotubes, Nature Materials 3 (October 2004).

[49] M. Hiramatsu, H. Nagao, M. Taniguchi, H. Amano, Y. Ando, M. Hori, High-rate growth of films of dense, aligned double-walled carbon nanotubes using microwave plasma-enhanced chemical vapor deposition, Japanese Journal of Applied Physics, Express Letter 44 (22) (2005).

[50] S. Huang, X. Cai, J. Liu, Growth of millimeter-long and horizontally aligned single-wall carbon nanotubes on flat substrates, Journal of American Chemical Society 125 (2003).

[51] S. Huang, M. Woodson, R. Smalley, J. Liu, Growth mechanism of oriented long single wall carbon nanotubes using "fast-heating" chemical vapor deposition process, Nano Letters 4 (6) (2004).

[52] S. Chakrabarti, T. Nagasaka, Y. Yoshikawa, L. Pan, Y. Nakayma, Growth of super long aligned brush-like carbon nanotubes, Japanese Journal of Applied Physics, Express Letter 45 (28) (2006).

[53] Q. Li, X. Zhang, R.F. DePaula, L. Zheng, Y. Zhao, L. Stan, T.G. Holesinger, P.N. Arendt, D.E. Peterson, Y.T. Zhu, Sustained growth of ultralong carbon nanotube arrays for fiber spinning, Advanced Materials 18 (2006) 3160−3163.

[54] X.Z. Liao, A. Serquis, Q.X. Jia, D.E. Peterson, Y.T. Zhu, H.F. Zu, Effect of catalyst composition on carbon nanotube growth, Applied Physics Letters 82 (16) (2003).

[55] J.S.C. Kim, The role of hydrogen in the growth of carbon nanotubes: a study of the catalyst state and morphology, Submitted to the Dept. of Mat. Science and Engineering for the Degree of Bachelor of Science at MIT, June 2006.

[56] P. Salunke, S. Yarmolenko, S. Neralla, J. Sankar, K. Fischbach, G. Li, Yun, M. Schulz, V. Shanov, Substrate preparation by magnetron sputtering and CVD growth of carbon nanotube arrays, Proceedings 14th International Mechanical Engineering Congress, IMECE2007-43190, 11−15 November 2007, Seattle, WA.

[57] K. Hasegawa, S. Noda, Y. Yamaguchi, Growth window and possible mechanism of millimeter-thick single-walled carbon nanotube forests, Journal of Nanoscience and Nanotechnology 8 (11) (2008).

[58] J.Y. Raty, F. Gygi, G. Galli, Growth of carbon nanotubes on metal nanoparticles: a microscopic mechanism form Ab Initio molecular dynamics simulations, Physical Review Letters (2005) 95.

[59] O.A. Louchev, T. Laude, H. Kanda, Diffusion controlled kinetics of carbon nanotube forest growth by chemical vapor deposition, Journal of Chemical Physics 118 (16) (2003) 7622.

[60] A special section on SWCNT growth mechanisms, Journal of Nanoscale Science and Technology 8 (11) (November 2008). (American Scientific Publishers).

[61] Cleantech Technologies, Unpublished Technical Reports, www.cleantechnano.com/research.html.

[62] D. Pech, M. Brunet, H. Durou, P. Huang, V. Mochalin, Y. Gogotsi, P.-L. Taberna, P. Simon, Ultrahigh-power micrometre-sized supercapacitors based on onion-like carbon, Nature Nanotechnology (2010), http://dx.doi.org/10.1038/nnano.2010.162.

[63] N. Sano, H. Wang, I. Alexandrou, M. Chhowalla, K.B.K. Teo, G.A.J. Amaratunga, Properties of carbon onions produced by an arc discharge in water, Journal of Applied Physics 92 (5) (1 September 2002).

[64] T. Cabioc'h, J.C. Girard, M. Jaouen, M.F. Denanot, G. Hug, Carbon onions thin film formation and characterization, Europhysics Letters 38 (6) (1997) 471−475.

[65] Animations of Electric Motor Operation, www.animations.physics.unsw.edu.au/jw/electricmotors.html.

[66] Information on Electric Motors, not available in other forms, en.wikipedia.org/wiki/Electric_motor.

[67] Information on Magnetism, only available on site, www.coolmagnetman.com/magacmot.htm.

[68] Information on Motor Design, www.designworldonline.com/ArticleDetails.aspx?id=1973.

[69] R. Raffaelle, B. Landai, J. Harribs, S. Bailey, A. Hepp, Carbon nanotubes for power applications, Materials Science and Engineering B, 116 (205) 233−243.

[70] UC Invention Disclosure, Carbon Electric Motor, in process.

[71] UC Invention Disclosure, Hybrid Electrical Fiber, in process.
[72] Potential Opportunities for Parker Hannifin in Carbon Nanotube Materials, Discussion with Pete Buca, Newry Report, 9 July 2010.
[73] NASA Glenn presentation by G. Brown, Materials Aspects of Turboelectric Aircraft Propulsion, 2009 Annual Meeting, Fundamental Aeronautics Program, Subsonic Fixed Wing Project.
[74] J.L. Abot, et al., Self-Sensing Composite Materials, Provisional Patent UC109−064, 2009.
[75] C. Jayasinghe, W. Li, Y. Song, J.L. Abot, V.N. Shanov, S. Fialkova, S. Yarmolenko, S. Sundaramurthy, Y. Chen, W. Cho, S. Chakrabarti, G. Li, Y. Yun, M.J. Schulz, Nanotube responsive materials, MRS Bulletin 35 (9) (2010).
[76] Parker Hannifin Corp., www.parker.com.
[77] Material Properties, en.wikipedia.org.
[78] K. Jiang, Q. Li, S. Fan, Spinning continuous carbon nanotube yarns, Nature 419 (2002).
[79] K. Atkinson, S. Hawkins, C. Huynh, C. Skourtis, J. Dai, M. Zhang, S. Fang, A. Zakhidov, S. Lee, A. Aliev, C. Williams, R. Baughman, Multifunctional carbon nanotube yarns and transparent sheets: fabrication, properties and applications, Physics B 394 (2007) 339−343.
[80] K. Jiang, S. Fan, Q. Li, Method for Fabricating CNT Yarn, US Patent 7,045,108 B2, 2006.
[81] K. Liu, Y. Sun, L. Chen, C. Feng, X. Feng, K. Jiang, Y. Zhao, S. Fan, Controlled growth of super-aligned carbon nanotube arrays for spinning continuous unidirectional sheets with tunable physical properties, Nano Letters 8 (2) (2008) 700.
[82] C. Feng, K. Liu, J.-S. Wu, L. Liu, J.-S. Cheng, Y. Zhang, Y. Sun, Q. Li, S. Fan, K. Jiang, Flexible, stretchable, transparent conducting films made from superaligned carbon nanotubes, materials views, Advanced Functional Materials 20 (2010) 885−891.
[83] P.B. Amama, C.L. Pint, L. McJilton, S.M. Kim, E.A. Stach, P.T. Murray, R.H. Hauge, B. Maruyama, Role of water in super growth of single-walled carbon nanotube carpets, Nano Letters 9 (1) (2009).
[84] D. LóPez, I. Cendoya, F. Torres, J. Tejada, C. Mijangos, Preparation and characterization of poly(vinyl alcohol)-based magnetic nanocomposites. 1. Thermal and mechanical properties, Journal of Applied Polymer Science 82 (2001) 3215−3222.
[85] L.A. Garcıa-Cerda, M.U. Escareño-Castro, M. Salazar-Zertuche, Preparation and characterization of polyvinyl alcohol−cobalt ferrite nanocomposites, Journal of Non-Crystalline Solids 353 (2007) 808.
[86] M.J. Schulz, V.N. Shanov, Y. YeoHeung, Nanomedicine Design of Particles, Sensors, Motors, Implants, Robots, and Devices, with a Supplementary Materials and Solutions Manual, In: Engineering in Medicine and Biology, Artech House Publishers, October 2009.
[87] Y.K. Ko, B.S. Seo, D.S. Park, H.J. Yang, W.H. Lee, P.J. Reucroft, J.G. Lee, Additive vapour effect on the conformal coverage of a high aspect ratio trench using MOCVD copper metallization from (hfac)Cu(DMB) precursor, Semiconductor Science and Technology 17 (9) (September 2002) 978−982.
[88] M.F. Bain, Y.H. Low, D.C.S. Bien, J.H. Montgomery, B.M. Armstrong, H.S. Gamble, Investigation into the selectivity of CVD iron from $Fe(CO)_5$ precursor on various metal and dielectric patterned substrates, Surface and Coatings Technology 201 (22−23) (25 September 2007) 8998−9002.

[89] Nanonex, 1 Deer Park Drive, Suite O, Monmouth Junction, NJ 08852, USA, Nanonex 2600 NanoImprint Lithography System, for thermoplastic and UV curable resists plus photolithography.

[90] W. Li, H. Zhang, C. Wang, Y. Zhang, L. Xu, K. Zhu, Raman characterization of aligned carbon nanotubes produced by thermal decomposition of hydrocarbon vapor, Applied Physics Letters 70 (20) (1997).

[91] M.B. Nardelli, J-L. Fattenbert, D. Orlikowski, C. Rolan, Q. Zhao, J. Bernholc, Mechanical Properties, defects and electronic behavior of carbon nanotubes, Carbon 38 (2000) 1703–1711.

[92] R.A. EiLeo, B.J. Landi, R.P. Raffaelle, Purity assessment of multiwalled carbon nanotubes by Raman spectroscopy, Journal of Applied Physics 101 (2007).

[93] J. Chen, J.Y. Shan, T. Tsukada, F. Munekane, A. Kuno, M. Matsuo, T. Haysahi, Y.A. Kim, M. Endo, The structural evolution of thin multiwalled carbon nanotubes during isothermal annealing, Carbon 45 (2007).

[94] T. Telytschko, S.P. Xiao, G.C. Schatz, R.S. Ruoff, Atomistic simulations of nanotube fracture, Physical Review B 65 (2002).

[95] Market Report on Electric Motors in the Integral hp Range in the World for 2015, www.streetinsider.com/Press+Releases/Global+Integral+Horsepower+Motors+Market+to+Reach+$12.1+Billion+by+2015,+According+to+New+Report+by+Global+Industry+Analysts,+Inc./5888727.html.

[96] http://www.journalamme.org/papers_ammc06/192.pdf, Nanocrystalline Iron Based Powder Cores for High Frequency Applications.

[97] http://www.imego.com/Expertise/Electromagnetic-sensors/Magnetic-nanoparticles/index.aspx.

[98] D. Hasegawa, H. Yang, T. Ogawad, M. Takahashia, Challenge of ultra high frequency limit of permeability for magnetic nanoparticle assembly with organic polymer—application of superparamagnetism, Journal of Magnetism and Magnetic Materials 321 (7) (April 2009) 746–749. Proceedings of the Fourth Moscow International Symposium on Magnetism.

[99] A.-H. Lu, E.L. Salabas, F. Schuth, Magnetic nanoparticles: synthesis, protection, functionalization, and application, Angewandte Chemie International Edition 46 (2007) 1222–1244.

[100] G. Reiss, A. Hütten, Magnetic nanoparticles, applications beyond data storage, Nature Materials 4 (October 2005). www.nature.com/naturematerials.

[101] P.M. Raj, H. Sharma, G.P. Reddy, D. Reid, N. Altunyurt, M. Swaminathan, R. Tummala, V. Nair, Novel nanomagnetic materials for high-frequency RF applications, Electronic Components and Technology Conference, May 31 2011–June 3 2011, pp. 1244–1249, IEEE, Lake Buena Vista, FL.

[102] http://www.inframat.com/magnetic.htm, Magnetic Nanocomposite Research.

[103] T. Dey, Polymer-coated magnetic nanoparticles: surface modification and end-functionalization, Journal of Nanoscience and Nanotechnology 6 (8) (August 2006) 2479–2483.

[104] http://www.popsci.com/science/article/2011-01/new-nanocomposite-magnets-could-reduce-demand-rare-earth-elements.

[105] J.-W. Park, I.S. Yoo, W.-S. Chang, E.-C. Lee, H. Ju, B.H. Chung, B.S. Kim, Magnetic moment measurement of magnetic nanoparticles using atomic force microscopy, Measurement Science and Technology 19 (1) (2008).

[106] M.D. Lima, S. Fang, X. Lepró, C. Lewis, R. Ovalle-Robles, J. Carretero-González, E. Castillo-Martínez, M.E. Kozlov, J. Oh, N. Rawat, C.S. Haines, M.H. Haque, V. Aare, S. Stoughton, A.A. Zakhidov, R.H. Baughman, Biscrolling nanotube sheets and functional guests into yarns, Science 331 (7 January 2011).
[107] http://www-als.lbl.gov/index.php/news-and-publications/alsnews/580-large-magnetization-at-carbon-surfaces.html, Large Magnetization at Carbon Surfaces.
[108] http://hptdrivesystems.thomasnet.com/item/units-high-frequency-hf-motorized-spindle-units-ac/hf-spindle-units-with-ac-motor/mf-100-120-2-2?.
[109] http://www.mathworks.com/help/toolbox/physmod/elec/ref/universalmotor.html.
[110] M. Jakubinek, M. White, G. Li, C. Jayasinghe, W. Cho, M. Schulz, V. Shanov, Thermal and electrical conductivity of tall, vertically aligned carbon nanotube arrays, Carbon 48 (2010) 3947–3952.
[111] M.B. Jakubinek, M.B. Johnson, M. Anne White, C. Jayasinghe, G. Li, W. Cho, M.J. Schulz, V. Shanov, Thermal and electrical conductivity of array-spun multi-walled carbon nanotube yarns, Carbon 50 (2012) 244–248.
[112] J.H. Pöhls, M.B. Johnson, M. Anne White, R. Malik, B. Ruff, C. Jayasinghe, M.J. Schulz, V. Shanov, Physical properties of array-spun multiwalled carbon nanotube sheets physical properties of sheet drawn from arrays of multiwalled carbon nanotubes, Carbon (2012).
[113] J. Kluener, MS Thesis, Characterization of CNT Thread and Application for Firefighters Garment, University of Cincinnati, 11 July 2011.
[114] W. Qian, Q. Weizhong, N. Jingqi, C. Anyuan, N. Guoqing, W. Yao, H. Ling, Z. Qiang, H. Jiaqi, W. Fei, 100 mm Long semiconducting triple-walled carbon nanotubes, Advanced Materials (2010) 1867–1871.
[115] X. Wang, Q. Li, J. Xie, Z. Jin, J. Wang, Y. Li, K. Jiang, S. Fan, Fabrication of ultralong and electrically uniform single-walled carbon nanotubes on clean substrates, Nano Letters 9 (9) (2009).
[116] Ultra-Long Carbon Nanotubes, Columbia University NSEC, http://www.cise.columbia.edu/nsec/research/nuggets.php?subsection=long.
[117] W. Fei, Structure Control and Mass Production of CNTs. Nanotechnology Materials and Devices Workshop, University of Cincinnati, 3 October 2011.
[118] http://www.eurondt.com/EDDY%20CURRENT.html.
[119] http://en.wikipedia.org/wiki/Skin_effect.
[120] V.K. Varadan, L. Chen, J. Xie, Nanomedicine: Design and Applications of Magnetic Nanomaterials, Nanosensors and Nanosystems, Wiley, ISBN 978-0-470-03351-7.
[121] Y. Xu, Y. Liang, L. Jiang, H. Wu, H. Zhao, D. Xue, Preparation and magnetic properties of ZnFe2O4 nanotubes, Journal of Nanomaterials (2011) 5. Article ID 525967.
[122] Y.C. Sui, R. Skomski, K.D. Sorge, D.J. Sellmyer, Magnetic nanotubes produced by hydrogen reduction, Journal of Applied Physics 95 (11) (1 June 2004).
[123] L. Chen, J. Xie, K.R. Aatre, J Yancey, S Chetan, M Srivatsan, V.K. Varadan, Synthesis of hematite and maghemite nanotubes and study of their applications in neuroscience and drug delivery, Proceedings of SPIE 7980 798008–1.
[124] L. Chen, J. Xie, J. Yancey, M. Srivatsan, V.K. Varadan, Magnetic nanoparticles and nanotubes for biomedical applications, Proceedings of SPIE 7291 729108–1.
[125] Z. Liu, D. Zhang, S. Han, C. Li, B. Lei, W. Lu, J. Fang, C. Zhou, Single crystalline magnetite nanotubes, Journal of American Chemical Society 127 (2005) 6–7.

CHAPTER 22

Multiscale Laminated Composite Materials

Yi Song, Bolaji Suberu, Vesselin Shanov, Mark Schulz

Nanoworld Smart Materials and Devices Laboratory, University of Cincinnati, Cincinnati, OH, USA

CHAPTER OUTLINE

22.1 Introduction	627
22.2 Fabrication and Characterization of MWCNT Array-Reinforced Laminated Composites	628
22.2.1 CNT array synthesis	629
22.2.2 Transfer of CNT arrays onto prepreg laminae	629
22.2.3 Consolidation of IM7-CNT array layup	630
22.2.4 Short beam shear test setup	631
22.2.5 Three-point bending test setup	632
22.2.6 Iosipescu interlaminar shear test setup	633
22.3 Results and Discussion	634
22.3.1 Short beam shear test	634
22.3.2 Three-point bending test	637
22.3.3 Iosipescu interlaminar shear test	641
22.4 Conclusions	645
References	645

22.1 INTRODUCTION

Aluminum is a multifunctional material; it is strong, stiff, lightweight, withstands moderate temperatures, and has good thermal and electrical conductivity. It also has limitations, i.e. it will fatigue and crack, it has a large coefficient of thermal expansion, and it corrodes in a chloride environment. Carbon fiber polymeric matrix composite materials are replacing aluminum in aircraft and other applications where lightweight, high strength, and stiffness are important. After developing numerous military systems using composites, along with gaining the advantage of being lightweight, several disadvantages are becoming apparent in the aerospace industry. Laminated composites are poor thermal conductors transversely and cannot get the heat out of the structure; the composite cannot withstand high temperature, absorbs moisture, is not electrically conductive enough to discharge static electricity or lightning strikes, and is brittle and susceptible to impact damage and delamination. These limitations are restricting the flight envelope of aircraft or are requiring retrofit of materials to improve some property of composites; however, this offsets the main benefit of weight reduction gained from using the composite material in the first place [1–34]. A multiscale composite material may overcome some of the limitations of carbon microfiber composites. A structural composite material that is multifunctional is most desired for many applications including aircraft and automotive body and frame components and any application where impact and high loads might be experienced. If CNT can be integrated into a matrix material, the resulting laminated composite could have improved properties. There are various ways CNT can be integrated into composites and this is a continuing area of investigation. A nanothermosetting composite material would be stiff and strong. A nanothermoplastic composite material could deform under load but would not fracture and could be restored to its original shape. A nanotube array core material could form a sandwich composite using nanotube face sheets. The resulting all-nanotube composite sandwich would be multiscale—having many exceptional properties—including high strength, high stiffness, lightweight, electrical and thermal conductivity, and toughness.

Existing high-performance composite materials are fabricated using epoxy and carbon or glass microfibers that are about 7 μm in diameter. This material is not only extremely strong and stiff but also very brittle. What is needed is a material that can take an impact without fracturing and failing [13–18,20–39]. The objective of the work in this chapter is to develop a CNT multiscale material that will absorb a large amount of energy from an impact and not fracture. The multiscale material will have a suite of key properties. Development of the material is possible because of CNT arrays that can be tailored to have different geometry, and different forms of CNT materials can be used to reinforce composites. The array between plies of composites is investigated in this chapter to reinforce composites. Other forms of CNT used for reinforcing composites are under development and results will be reported in the future.

22.2 FABRICATION AND CHARACTERIZATION OF MWCNT ARRAY-REINFORCED LAMINATED COMPOSITES

The procedure to fabricate the multiscale composite material is as follows: (1) synthesize CNT as a vertical forest, (2) transfer the vertically aligned CNT (VACNT) arrays to the prepreg laminate, and (3) consolidate and cure the CNT arrays and carbon fabric plies with epoxy into a laminate. Automobile panels and aircraft frames constructed this way would be strong and lightweight. A lightweight, tough, low-cost material could be transformative for transportation systems. Composite materials with CNT inside may replace existing composites that are too brittle and metals that are too heavy. The new tough lightweight material would also be

22.2 Fabrication and Characterization of MWCNT Array

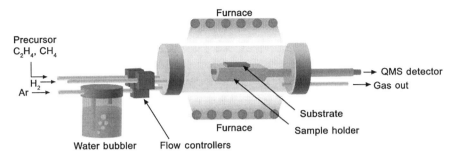

FIGURE 22.1

Schematic illustration of chemical vapor deposition (CVD) synthesis. (For color version of this figure, the reader is referred to the online version of this book.)

more electrically and thermally conductive than traditional composites and would have broad applications.

22.2.1 CNT array synthesis

The CNTs grow on a silicon wafer through chemical vapor deposition (CVD), as shown in Fig. 22.1. CVD is a common method for the fabrication of carbon nanotubes. A substrate is prepared with a layer of metallic catalyst particles. The diameters of the CNTs depend on the size of the metal particles. The substrate is then heated to approximately 700 °C. A process gas (nitrogen or hydrogen) and a carbon-based gas (methane or ethylene) are bled into the reactor in order to produce CNTs. CNTs grow at sites of the metal catalyst particles. The carbon-based gas is broken apart at the surface of the catalyst particle, and the carbon is transported to the edges of the particle, where it forms the nanotubes [34].

22.2.2 Transfer of CNT arrays onto prepreg laminae

Transfer of VACNT arrays on the prepreg is a critical step of the process since the feasibility and scale-up of this process is highly dependent on the height of the VACNT array. Unidirectional IM7/977-3 prepregs are selected for the fabrication of multiscale composites. Long CNT arrays could result in imperfect resin infiltration on a prepreg laminate and reduce overall performance of the hybrid composite. Based on previous research work done in this field [34], a 20-μm-high VACNT array was used to enable a satisfactory resin infiltration.

The transfer process of the VACNT array onto 19 plies of carbon fiber prepreg was performed using hot iron and dry ice. Here, the VACNT array was inverted on the prepreg surface in such a way that a moderately hot iron can be placed on the substrate and moderate pressure is applied while moving the hot iron in a forward and backward manner until the VACNT arrays are fully transferred to the prepreg surface. While the hot press iron is applied, dry ice is also applied on the back of

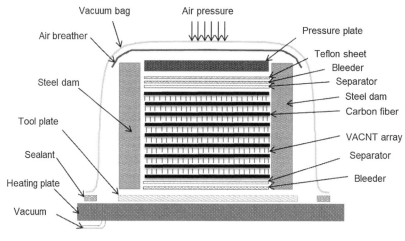

FIGURE 22.2

Schematic diagram of multiscale composite layup in a vacuum bag [24]. (For color version of this figure, the reader is referred to the online version of this book.)

the substrate to create a cooling effect during the transfer process, which enables the resin to cool fast and in turn keeps the 977-3 epoxy in a B-stage cure regime where the epoxy remains tacky. The advantage of tacky surface of the prepreg is being utilized for effective transfer printing. The VACNT arrays are subsequently transferred on each of the 19 plies of prepreg and the individual plies are stacked up based on the ply stacking sequence and are ready for vacuum bagging process [24].

22.2.3 Consolidation of IM7-CNT array layup

The surface dimension of IM7 prepreg is 5×2 inch^2 and the ply orientation is followed by $(0°, 90°, +45°, -45°, 0°, 90°, +45°, -45°, 0°, 90°)_s$. The IM7-CNT array layup was placed in a vacuum bag shown in Fig. 22.2 and then subjected to a curing cycle of 977-3 resin. This layup was heated with a temperature increase rate of 5 to 355 °F for 6 h and then cooled down to room temperature with a temperature decrease rate of 5 °F/min. The laminate manufacturing is shown in Fig. 22.2.

During the fabrication process, a multiscale composite panel reinforced with 20-μm CNT arrays and a baseline composite panel without CNT arrays were fabricated. Then these two composite panels were cut into test samples for quasi-static tests including short beam shear test, three-point bending test, and Iosipescu interlaminar shear test. Five samples were tested for short beam shear test and three samples were tested for three-point bending test and Iosipescu interlaminar shear test. An Instron 4465 tensile testing system with a 2-kN load cell was used, which is calibrated to measure quasi-static loads ranging from 5 N to 2 kN. The resolution is 1 mN for load and 1 lm for displacement and a data sampling point of 800 points/s [24].

22.2.4 Short beam shear test setup

In the short beam shear test, the determination of interlaminar shear stress is based on classical (Euler–Bernoulli) beam theory. For a beam of rectangular cross-section loaded in short beam shear, the maximum interlaminar shear stress occurs at the mid-thickness of the beam between the center and end supports and is calculated to be

$$F^{sbs} = 0.75 \cdot \frac{P_m}{bh} \qquad (22.1)$$

where the F^{sbs} is the short beam strength, P_m is the maximum load observed during the test, b is the sample width and h is the sample thickness.

The experimental setup for short beam shear test followed ASTM standard D2344/D2344M. The thickness of the sample is critical since it dictates the span-to-thickness ratio of the sample and also the recommended mode of failure. The samples fail in different modes depending on the span-to-thickness ratio [40]. For a state of pure shear to exist between the laminate in a standard short beam shear test, the span-to-thickness ratio should be at least 4. Hence, the thickness of the laminate was designed to achieve the nominal thickness required for this test, as shown in Fig. 22.3 [24].

The span length for the short beam shear test is equal to four times the thickness. The sample fiber direction contains greater than 10% volume fraction of 0° plies in the laminate/test axis of the specimen. The laminate is subjected to a short beam shear test at a constant crosshead speed of 0.05 inch/min. Each sample was placed on two support pins and the load was applied via a 6-mm loading nose. The compressive load is applied until fracture occurs and maximum failure load is used in the calculation to determine the apparent short beam shear strength of the laminate ply interface. Based on visual inspection of the failed sample, it is seen that the sample failed by delamination, which occurred at the midplane of the laminate, as shown in Fig. 22.3. The test was automatically

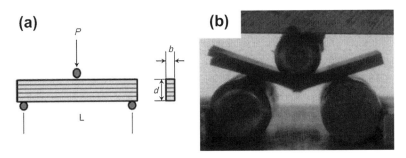

FIGURE 22.3

Short beam shear (SBS) test for (a) a multiscale composite sample and (b) shear failure mode at the midplane where the maximum shear stress occurs. (For color version of this figure, the reader is referred to the online version of this book.)

stopped when the applied load dropped by 30% or the crosshead travel exceeded the specimen nominal thickness. Five specimens were tested for each specimen type (baseline without MWCNTs and multiscale composite with MWCNTs). The equation for computing strength of the specimen is based on classical Euler–Bernoulli beam theory, which assumes that shear stresses vary parabolically through the thickness [40], which is only approximately correct for a short beam and for the experimental three-point bending loading conditions. The quasi-isotropic laminate configurations of fibers make the interlaminar shear stress state complex to analyze; however, the short beam shear test allows for a quantitative and relative comparison of the short beam shear strength of composites with and without CNTs [24].

22.2.5 Three-point bending test setup

For a beam with rectangular cross-section subjected to three-point bending, the maximum flexural strength of the sample usually occurs at the bottom of the beam, which is subjected to maximum tensile stresses, and is calculated using the following formula as recommended by ASTM 790:

$$\sigma_f = 3PL/2bd^2 \tag{22.2}$$

where P is the maximum load observed during the test, L is the span length between the two supports, b is the width of sample, and d is the sample thickness. Also, the flexural strain was calculated by using the formula below as recommended. This is based on the fractional change in length of an element at the outer surface of the specimen where the maximum strain occurs.

$$\varepsilon_f = 6Dd/L^2 \tag{22.3}$$

where D is the maximum deflection of the center of the beam, d is the thickness of the sample and L is the support span length of the sample.

Based on ASTM specification, the basic difference between the short beam shear test and three-point bending test is the span-to-thickness ratio. ASTM D790 specification with a span-to-thickness ratio of 10:1 was used for cutting the samples into the desired dimension. Figure 22.4 shows an experimental setup for the three-point bending test.

All the test conditions in the three-point bending test are the same as those for short beam shear test except for the span length. A span length of 1 inch was selected for the three-point bending test. The span-to-depth ratio is 10. The thickness of the sample used in this test is also the same as that used for short beam shear test since it dictates the span-to-thickness ratio of the sample and also the mode of failure of the sample as recommended by ASTM D790. For a flexural failure to exist in a standard three-point bending test [41], the span-to-thickness ratio should be large enough to develop a high bending moment at the central loading nose. The experiments measure the effectiveness of the CNTs to improve the flexural strength and stiffness of the laminated composite.

FIGURE 22.4

Three-point bending test setup. (For color version of this figure, the reader is referred to the online version of this book.)

22.2.6 Iosipescu interlaminar shear test setup

The Iosipescu-type shear test can produce a state of pure interlaminar shear stress and guarantee interlaminar composite failure instead of composite–adhesive interface failure [42]. The specimen with two 90° symmetrically and centrally located V-notches is clamped in the Iosipescu-type fixture and tightened with eight threaded bolts as shown in Fig. 22.5. The Iosipescu-type shear fixture consists of one pair of steel blocks (A), another pair of steel blocks (B), eight threaded bolts and two dowel pins. The compressive load is applied through the two dowel pins. The shear stress can be calculated according to the following equation [34,42]:

$$\tau_{31} = \frac{P}{t_n w} \tag{22.4}$$

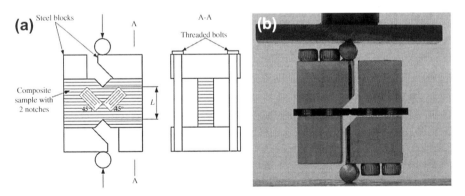

FIGURE 22.5

Iosipescu interlaminar shear fixture. (a) Loading configuration with strain gages shown and (b) shear fixture mounted in the Instron machine. (For color version of this figure, the reader is referred to the online version of this book.)

where τ_{31} is shear stress in the cross-sectional area of the specimen, P is the out-of-plane load that is measured by the load cell, t_n is the thickness between the two tips of V-notches (the thickness in the out-of-plane direction), and w is the width of the cross-sectional area (the width in the in-plane direction). Two strain gages as shown in Fig. 22.5(a) are mounted on the specimen at 45° and −45° orientations so that the shear strain can be recorded. The Iosipescu-type shear fixture assembly with strain gages mounted on the specimen is shown in Fig. 22.5(b). The interlaminar shear strain, γ_{31}, can be calculated with the following equation [34,42]:

$$\gamma_{31} = \varepsilon_{45°} - \varepsilon_{-45°} \qquad (22.5)$$

The algebraic difference of the two strain gage readings, $\varepsilon_{45°}$ and $\varepsilon_{-45°}$ gives the interlaminar shear strain [34,42]. The shear stress versus shear strain curve can then be produced.

22.3 RESULTS AND DISCUSSION

Composite laminates were prepared using a hot press. Samples for three types of testing were cut from the laminates using a diamond wheel. Results of the testing are described for the three types of testing.

22.3.1 Short beam shear test

Five samples from both 20-ply baseline laminated composite specimen and 20-ply multiscale composite specimen were tested using a short beam shear test fixture with a span-to-depth ratio of 4. All tests are carried out in the same conditions under room temperature. The load−displacement results acquired from the Instron software are used to estimate the shear strength−displacement data. This data is plotted as shown in Figs 22.6 and 22.7. Strain gages were not used in the test, thus the shear strain is not reported. The mean shear strength is computed for both samples for comparison.

Figures 22.6 and 22.7 show the shear strength−deflection plots obtained during the test. The slopes of the initial parts of the curves were very small. This is due to the slight uneven thickness of the samples. The loading nose gently touched one side of the sample when the test started and the touch line between the loading nose and the upper surface of the sample extended to the other side of the sample across the sample width as the test continued. When the curves began to increase substantially in Fig. 22.7, the loading nose fully touched the sample through the sample width. The results for the tested specimens demonstrate good repeatability. From Table 22.1 it can be seen that the mean short beam shear strength of the 20-ply baseline laminated composite is 40.5 MPa with a standard deviation of about 1.8 MPa, while that of the 20-ply 20-μm CNT array-reinforced multiscale composite is 61.5 MPa with a standard deviation of 6.8 MPa, which is an acceptable data correlation. The shear strength in the baseline and reinforced samples is fairly low and depends on several factors including the curing cycle, the composite layup, the thickness of the sample and the bearing stress. The improvement in strength depends on several

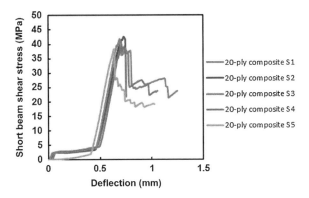

FIGURE 22.6

Short beam shear stress versus deflection test curves for IM7 20-ply baseline laminated composites samples. (For color version of this figure, the reader is referred to the online version of this book.)

factors also including the geometry of the reinforcement and how well the reinforcement is wetted by the epoxy.

The introduction of the 20-μm-long CNT array between each ply leads to about 50% increase of the short beam shear strength while increasing the sample thickness from a mean value of 0.089 to 0.097 inch (the increased thickness is used in computing the stress in the reinforced sample). The 9% increase in thickness is undesirable and can be reduced by increasing the temperature/time and pressure in the hot press curing process. From visual inspection of the failed sample, it

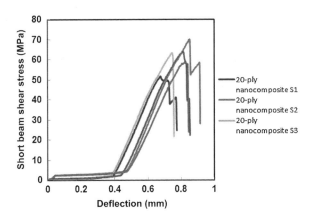

FIGURE 22.7

Short beam shear stress versus deflection test curves for IM7 20-ply multiscale composites samples. (For color version of this figure, the reader is referred to the online version of this book.)

Table 22.1 Short Beam Shear Test Results

		Width (inch)	Thickness (inch)	Max Load (N)	Short Beam Shear Strength (MPa)	Thickness/Ply (inch)
20-ply baseline laminated composite	S1	0.277	0.088	855.5	40.8	0.0044
	S2	0.278	0.091	926.2	42.6	0.00455
	S3	0.282	0.091	923.3	41.8	0.00455
	S4	0.261	0.089	783.6	39.2	0.00445
	S5	0.242	0.088	700.5	38.2	0.0044
	Mean	0.268	0.089	837.8	40.5	0.00447
	Standard deviation	0.017	0.002	–	1.8	0.000076
20-ply 20-μm CNT array multiscale composite	S1	0.266	0.099	1170.7	51.7	0.00495
	S2	0.26	0.097	1518.9	70.0	0.00485
	S3	0.265	0.097	1400.0	63.3	0.00485
	S4	0.268	0.098	1444.4	63.9	0.0049
	S5	0.267	0.095	1282.5	58.8	0.00475
	Mean	0.265	0.097	1363.3	61.5	0.00486
	Standard deviation	0.003	0.001	–	6.8	0.000074

was observed that the samples failed in shear at the midplane, which is a matrix-dominated failure. This failure is believed to have occurred in the samples when the shear stress applied exceeded the maximum shear strength of the fiber matrix interface of the composite plies.

The composite baseline and multiscale composite samples all failed in a typical brittle failure mode. At the maximum shear stress applied, there is a fiber/matrix debonding at the interface of the center plies, which might result from interface slipping or matrix microcracking. The CNT-reinforced multiscale composite failed at higher shear stress due to efficient load transfer through the matrix and between the transversely aligned CNT arrays and the adjacent plies (Fig. 22.8).

22.3.2 Three-point bending test

Three samples each were tested from both the 20-ply baseline laminated composite specimen and the 20-ply CNT array-reinforced multiscale composite specimen. The samples were tested using a three-point bending test fixture with a span-to-depth ratio of 10. All tests were carried out under the same conditions at room temperature. The stress versus deflection curves for the baseline samples are shown in Fig. 22.9. The stress versus deflection curves for the nanotube array-reinforced samples are shown in Fig. 22.10.

It can be seen from Table 22.2 that the mean flexural strength of the 20-ply baseline laminated composite is 829 MPa with a standard deviation of about 58 MPa, while that of the 20-ply 20-µm CNT array-reinforced composite is 921 MPa with a standard deviation of 78 MPa, which is an acceptable data correlation. The introduction of the

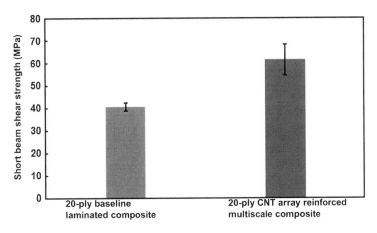

FIGURE 22.8

Bar chart showing a comparison between the shear strength of 20-ply baseline composite and 20-µm-high CNT array-reinforced multiscale composite. (For color version of this figure, the reader is referred to the online version of this book.)

CHAPTER 22 Multiscale Laminated Composite Materials

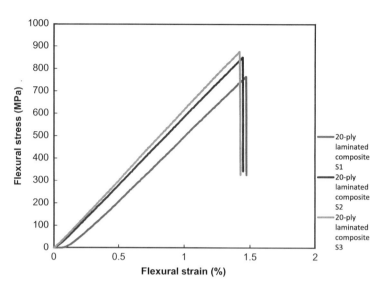

FIGURE 22.9

Three-point bending test curves for IM7 20-ply baseline laminated composites samples. Flexural strain is calculated and is approximate. (For color version of this figure, the reader is referred to the online version of this book.)

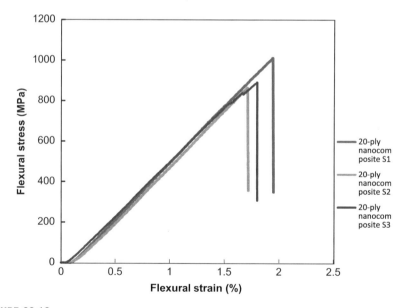

FIGURE 22.10

Three-point bending test curves for IM7 20-ply CNT array-reinforced multiscale composites samples. Flexural strain is calculated and is approximate. (For color version of this figure, the reader is referred to the online version of this book.)

Table 22.2 Three-Point Bending Test Results Including Approximate Strain

		Width (inch)	Thickness (inch)	Flexural Strength (MPa)	Max Flexural Strain (%)	Thickness/Ply (inch)
20-ply baseline laminated composite	S1	0.284	0.09	763	1.47	0.0045
	S2	0.277	0.09	850	1.45	0.0045
	S3	0.269	0.088	874	1.43	0.0044
	Mean	0.277	0.089	829	1.45	0.0045
	Standard deviation	0.008	0.001	58	0.02	0.00006
20-ply 20-μm CNT array-reinforced multiscale composite	S1	0.25	0.096	1010.49	1.95	0.0048
	S2	0.268	0.098	862.8	1.72	0.0049
	S3	0.27	0.1	891.36	1.8	0.005
	Mean	0.263	0.098	922	1.82	0.0049
	Standard deviation	0.011	0.002	78.33	0.12	0.0001

20-μm-high CNT array between each ply leads to 11% increase in flexural strength. It can be seen from Figs 22.9 and 22.10 that both baseline and CNT-reinforced composites failed by brittle failure since there is no ductile—brittle transition phase in the stress versus displacement curve. Here, there was a sharp drop in strength when the samples were loaded to reach their yield strength. Also, it is believed that the mode of reinforcement of the VACNT array played a less-significant effect on the flexural strength of the hybrid composite since the arrays are transversely aligned in between the laminae. There will be a maximum effect of CNT reinforcement in bending if the arrays are aligned longitudinally to the carbon fiber axis, which will enable the fibers to carry more load and improve the stiffness of the composite. The flexural strain was also computed in order to analyze the toughness and stiffness of both the baseline and the array-reinforced composites. The flexural strain was calculated based on the displacement of the beam and the geometry of the sample and is approximate. Figure 22.11 shows a bar chart with error bar of the flexural strength of both baseline composite and the CNT array composite. From Table 22.2 it can also be seen that the baseline composite sample has a mean ultimate flexural strain of about 1.5%, while the CNT array-reinforced composite has a mean ultimate flexural strain of about 1.8%. This shows an approximately 20% increase in ultimate flexural strain when compared to the baseline samples. Therefore, the CNT array reinforcement imparted a reasonable increase in the strength of the hybrid composite, which allowed the multiscale composite to withstand higher bending stresses and prolonged the crack initiation and propagation process. The CNT array modified the interphase and better interlocked the adjacent carbon fiber plies and thus improved the fracture toughness of the multiscale composite when compared to the baseline sample [24].

The CNT array reinforcement was not expected to increase the flexural strength. The three-point bending strength increased 11% by CNT array reinforcement. The failure mode is shear but compression/tension failure of the surface plies was

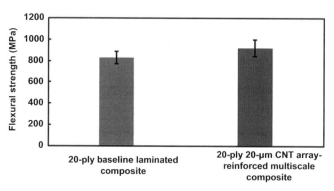

FIGURE 22.11

Bar chart showing a comparison between the flexural strength of 20-ply baseline composite and 20-μm CNT array-reinforced multiscale composite. (For color version of this figure, the reader is referred to the online version of this book.)

expected. The beam L/t ratio is 10:1, where L and t are the length and thickness of the specimen respectively, the failure is still due to interlaminar shear, and shear strength is very low. The CNT perpendicular array between plies does not directly reinforce bending because the failure is on the surface of the beam. The composite with CNT array reinforcement is thicker than the composite without reinforcement. The stress in the CNT-reinforced beam is

$$\sigma_{CNT} = (h/h_{CNT})^2 \sigma \qquad (22.6)$$

Thus a small increase in thickness of the beam with CNT can increase the strength. Thus the increase in strength is attributed mostly to the increased thickness of the beam. When the thickness is controlled to not increase, the flexural strength should not increase.

22.3.3 Iosipescu interlaminar shear test

The interlaminar shear stress in Iosipescu interlaminar shear test is calculated as recommended by ASTM D5379, where the maximum load prior to failure is divided by the cross-section of the specimen. Three samples from both the 20-ply baseline laminated composite specimen and the 20-ply CNT array-reinforced multiscale composite specimen were tested with an Iosipescu interlaminar shear test fixture in which all tests are carried out in the same conditions under room temperature. The load—displacement results acquired from the Instron software are used to estimate the shear strength—displacement data and a plot is acquired for this test. In the Iosipescu interlaminar shear test, accurate interlaminar shear strain can be obtained by using two strain gages mounted between the two V-notch tips in the sample. The interlaminar shear strain is equal to the sum of the strains from the two strain gages. Strain gages were not used in this particular testing, so shear stress is graphed against displacement. Plots from each test are averaged to determine the shear strength of both samples for comparison. Figures 22.12 and 22.13 show the shear strength

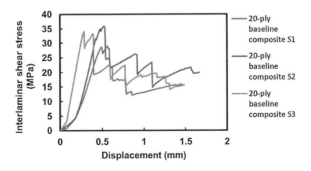

FIGURE 22.12

Iosipescu interlaminar shear test curves for 20-ply baseline laminated composites samples. (For color version of this figure, the reader is referred to the online version of this book.)

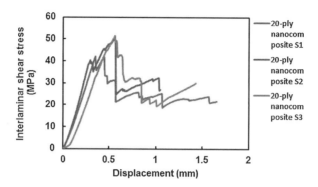

FIGURE 22.13

Iosipescu interlaminar shear test curves for 20-ply CNT array-reinforced multiscale composites samples. (For color version of this figure, the reader is referred to the online version of this book.)

versus deflection plots obtained during the test. The results are shown in bar chart form in Fig. 22.14. The results obtained for each of the tested specimens demonstrate good repeatability. It can also be seen from Table 22.3 that the standard deviation of sample dimensions as regards the width and tip-to-tip distance from the mean value is within a close interval. The average interlaminar shear strength is 32.9 MPa with a standard deviation of 3.5 MPa, while the CNT array-reinforced multiscale composite shows an interlaminar shear strength value of 47.8 MPa with a standard deviation of 5.1 MPa. This represents acceptable data correlation.

It can be seen that influence of CNT reinforcement on the hybrid composite brought about 47% increase in interlaminar shear strength. From Figs 22.12 and

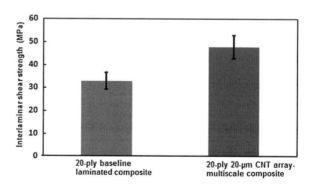

FIGURE 22.14

Bar chart showing a comparison of the Iosipescu interlaminar shear strength of 20-ply baseline composite and 20-μm CNT array-reinforced multiscale composite. (For color version of this figure, the reader is referred to the online version of this book.)

Table 22.3 Interlaminar Shear Test Results

		Width (inch)	Tip-Tip Thickness (inch)	Interlaminar Shear Strength (psi)	Interlaminar Shear Strength (MPa)
Baseline composite	S1	0.266	0.066	4211.7	29
	S2	0.294	0.06	5210.9	35.9
	S3	0.259	0.056	4932.4	34
	Mean	0.273	0.061	4785	33
	Standard deviation	0.019	0.005	515.7	3.6
Nanocomposite	S1	0.257	0.07	6103.2	42.1
	S2	0.273	0.065	7267.7	50.1
	S3	0.31	0.069	7457.5	51.4
	Mean	0.280	0.068	6942.8	47.9
	Standard deviation	0.027	0.003	733.2	5.1

22.13, the shear strength—displacement test plots for both types of samples follow the same shear failure process where there is a steep increase in the shear strength value with corresponding increase in displacement until a point is reached where there is a failure that precipitates a slight drop in shear strength. This drop is believed to be a result of the onset of unstable shearing in the matrix, failure of the fiber—matrix bond, matrix microcracking or a combination of these. This point denotes the ultimate shear failure of the specimen after which an alternate drop and increase in load far below the peak value of the shear strength is observed. Based on specimen observations, both notch roots in the samples were devoid of fiber crushing, which is sometimes observed in Iosipescu test samples [24,42] due to stress concentration at the tips and also at the failure initiation points. The VACNT arrays help in reinforcing the matrix since shear failure is mostly matrix dominated and may be a combination of other failure modes.

The mechanism observed for interlaminar shear strength increase as a result of CNT reinforcement is based on an energy dissipation mechanism in which energy is absorbed when the horizontal shear failure at the fiber/matrix interface tries to propagate through the transversely aligned CNT array, which in turn increases the load that is needed to propagate the crack through the strong and stiff CNT arrays, thereby increasing the interfacial shear strength of the hybrid composite. A second mechanism involves the mechanical interlocking of transversely aligned CNT arrays between the fiber/matrix interface of the composite. This interlocking mechanism increased the resistance of the hybrid composite to shear failure, thus enhancing the fracture toughness and interlaminar shear strength. Thus the presence of CNTs aligned perpendicular to the surface of a quasi-isotropic laminate (i.e. through-the-thickness direction) considerably contributes to a more efficient shear stress load transfer and enhances the interface properties of the laminated composites. It is believed that the CNTs create a crack bridging effect to delay crack propagation and it is also assumed that the CNTs are also pulled in tension while undergoing an in-plane shear effect at the midplane interface. Figure 22.15 is an

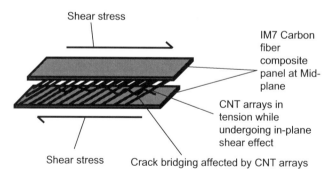

FIGURE 22.15

Illustration how CNTs might create a crack bridging effect to delay crack propagation. (For color version of this figure, the reader is referred to the online version of this book.)

illustration of how CNTs might create a crack bridging effect to delay crack propagation.

22.4 CONCLUSIONS

Aligned CNT arrays can interlock the adjacent carbon fabric prepreg plies and increase the load transfer between the CNT array-matrix-carbon fiber interphase. The CNTs create a crack bridging effect to delay crack propagation and the CNTs are also pulled in tension while undergoing an in-plane shear effect at the midplane interface. A combination of these effects can improve both in-plane and interlaminar shear properties of the CNT array-reinforced laminated multiscale composite [24]. Other architectures for reinforcing composites using CNT materials are under development and are expected to improve the strength of composites in all the failure modes. Thus multiscale composites are predicted to exceed the properties of conventional composites.

References

[1] M.J. Sundaresan, M.J. Schulz, A. Ghoshal, Structural Health Monitoring Static Test of a Wind Turbine Blade, performed August 1999, National Renewable Energy Laboratory Technical Report, NREL/SR-500—28719, published March 2002. (paper available at NREL's web site).

[2] M.J. Schulz, A. Kelkar, M. Sundaresan, Nanoengineering of Structural, Functional, and Smart Materials, CRC Press, Boca Raton, 2006.

[3] J.L. Abot, Zachary Kier, Vesselin N. Shanov, et al., Self-Sensing Composite Materials. Provisional Patent UC 109-064, (2009).

[4] Mark J. Schulz, Vesselin N. Shanov, Dong Qian, et al., *Carbon Nanotube Superfiber Development*, 5th WSEAS Int. Conf. on NANOTECHNOLOGY (NANOTECHNOLOGY'13) Cambridge, UK, February 20—22, 2013.

[5] Yi Song, et al., SHM applications of nanotechnologies for sensor development and communication, Nanotechnologies and smart materials for SHM, Ed. A. Catalano, G. Fabbrocino and C. Rainieri, ISBN 978-88-88102-47-4, Campobasso, Italy, 2012.

[6] Chaminda Jayasinghe, Weifeng Li, Yi Song, et al., Nanotube responsive materials, MRS Bulletin 35 (September 2010).

[7] J.L. Abot, Y. Song, M.J. Schulz and V.N. Shanov, Novel carbon nanotube array-reinforced laminated composite materials with higher interlaminar elastic properties, Composites Science and Technology, http://dx.doi.org/10.1016/j.compscitech.2008.05.023.

[8] Mark J. Schulz, Gunjan Maheshwari, Jandro Abot, et al., Responsive nanomaterials for engineering asset evaluation and condition monitoring, BNDT, Insight Journal (June 2008).

[9] Y. Song, Joe Kluener, Gary Conroy, et al., Carbon Nanotube Materials for Smart Composites, 242nd ACS National Meeting, Denver, Colorado, 28 August—1 September 2011.

[10] D.B. Mast, Chaminda Jayasinghe, Supriya Chakobarty, et al., Carbon Nanotube Threads, Yarns and Ribbons: Making the Transition from Materials to High

Performance Devices for Defense Applications, 2009 Nano Technology for Defense Conference, Burlingame, CA, April 2009.

[11] N. Mallik, Chaminda Jayasinghe, Supriya Chakraborty, et al., Fabrication of Threads from Carbon Nanotube Arrays, Proceedings of Processing and Fabrication of Advanced Materials-XVII, VII, 15—17 December 2008, New Delhi, India.

[12] G. Maheshwari, J. Abot, A. Song, E. Head, V. Shanov, M. Schulz, C. Jayasinghe, P. Salunke, Y.H. Yun, Powering up nanoparticles: versatile carbon materials for use in engineering and medicine, SPIE Newsroom, 10.1117/2.1200809.1305.

[13] J.L. Abot, Y. Song, M.J. Schulz, V.N. Shanov, Novel Carbon Nanotube Array-Reinforced Laminated Composite Material, SAMPE Conference, Cincinnati, Ohio, 2007.

[14] V.N. Shanov, Gyeongrak Choi, Gunjan Maheshwari, et al., Structural Nanoskin Based on Carbon Nanosphere Chains, SPIE Smart Structures Conference, San Diego, CA, March 2007.

[15] Nilanjan Mallik, Jandro Abot, Y. Song, et al., Carbon nanotube array based smart materials (Chapter 16), in: M. Umeno, Prakash R. Somani (Eds.), Carbon Nanotubes: Synthesis, Properties and Applications, Applied Science Innovations Pvt. Ltd., India, 2009.

[16] Inpil Kang, Gunjan Maheshwari, YeoHeung Yun, et al., (published), in: Christian Boller, Fu-Kuo Chang, Yozo Fujino (Eds.), Nanoengineering of Sensory Materials, Chapter in the Encyclopedia of Structural Health Monitoring, Wiley, 2008, www.wiley.co.uk/eshm.

[17] UC 109-076, 04/09/09, De-Icing of Solid Structures by Surface Heating Through CNT Thread and Ribbon, V. Shanov, J. Abot, Y. Song, M. Schulz.

[18] UC 109-077, 04/10/09, Remote, Non-Destructive Approach for Detecting Defects and Cracks in Solid Structures, V. Shanov, C. Jayasinghe, J. Abot, Y. Song, M. Schulz.

[19] Haibo Zhao, Fuh-Gwo Yuan, in: H. Felix Wu (Ed.), Carbon Nanotube Yarn Sensors for Structural Health Monitoring of Composites, Nondestructive Characterization for Composite Materials, Aerospace Engineering, Civil Infrastructure, and Homeland Security 2011, Vol. 7983, SPIE, 2011, p. 79830.

[20] UC 111-007, Carbon Nanotube Thread for Distributed Sensing, W. Li, S. Sundaramurthy, V. Shanov, M. Schulz, 15 July 2010.

[21] UC Invention Disclosure, 112-088, High Volume Fraction Composites (HVFC), B. Suberu, Y. (Albert) Song, V.N. Shanov, M.J. Schulz, et al., 28 March 2012.

[22] M.E. Orazem, B. Tribollet, Electrochemical Impedance Spectroscopy, Wiley, Hoboken, 2008.

[23] www.generalnanollc.com/; www.min.uc.edu/nanoworldsmart; www.atkinsandpearce.com/.

[24] MS Thesis, Next Generation Composite Materials Reinforced with Carbon Nanotubes, B. Suberu, University of Cincinnati, 2012.

[25] www.nsf.gov/news/news_summ.jsp?cntn_id=108992&;org=NSF&from=news.

[26] Press release by UC, Spinning Carbon Nanotubes Spawns New Wireless Applications, http://www.uc.edu/News/NR.aspx?id=9743.

[27] Shafi Ullah Khan, Jang-Kyo Kim, Impact and delamination failure of multiscale carbon nanotube-fiber reinforced polymer composites: a review, International Journal of Aeronautical and Space Sciences 12 (2) (2011) 115—133.

[28] I.M. Daniel, O. Ishai, Engineering Mechanics of Composite Materials, Oxford, New York, 2006.

[29] B.N. Cox, G. Flanagan, Handbook of Analytical Methods for Textile Composites, NASA Contractor Report 4750, 1997.
[30] B.N. Cox, Delamination and buckling in 3D composites, Journal of Composite Materials 28 (1994) 114–126.
[31] J.L. Abot, I.M. Daniel, Through-thickness mechanical characterization of woven fabric composites, Journal of Composite Materials 38 (2004) 543–553.
[32] Weifeng Li, Chaminda Jayasinghe, Vesselin Shanov, Mark Schulz, Spinning carbon nanotube nanothread under a scanning electron microscope, Materials 4 (2011) 1519–1527, http://dx.doi.org/10.3390/ma4091519.
[33] http://www.microstrain.com/; http://www.data-linc.com/dd1000.htm.
[34] PhD Thesis, Multifunctional Composites Using Carbon Nanotube Fiber Materials, Y. Song, University of Cincinnati, 2011.
[35] www.f22raptor.com/af_airframe.php; www.boeing.com/defense-space/military/f22/f22facts.html.
[36] Seminar M. Schulz, B. Suberu, Y. Song, V. Shanov, A. Hehr, Multifunctional Nanocomposite Materials to Revolutionize Aerospace Systems, GE Global Research Center Noise and Vibration Symposium 2012, GE Niskayuna, New York, June 19.
[37] UC Invention Disclosure 112-071, Patterned Carbon Nanotube Arrays for Interlaminar Reinforcement of Composite Materials, M.J. Schulz, B. Ruff, W. Li, V.N. Shanov, A. Krishnaswamy, Y. Song, B. Suberu, G. (Lucy) Li, S.N. Yarmolenko, S. Fialkova, 3 March 2012.
[38] UC Invention Disclosure, 112-088, High Volume Fraction Composites (HVFC), B. Suberu, Y. Song, V.N. Shanov, M.J. Schulz, G. (Lucy) Li, C. Jayasinghe, J. Kim, 28 March 2012.
[39] UC Invention Disclosure, Multi-Scale Composite Materials with Increased Design Limits.
[40] ASTM D3039/D3039M-06. Standard Test Method for Short-Beam Strength of Polymer Matrix Composite Materials and Their Laminates. ASTM International.
[41] ASTM D3039/D3039M-10. Standard Test Methods for Flexural Properties of Unreinforced and Reinforced Plastics and Electrical Insulating Materials. ASTM International.
[42] ASTM D5379/D5379M-05. Standard Test Methods for Constituent Content of Composite Materials, ASTM International.

CHAPTER 23

Aligned Carbon Nanotube Composite Prepregs

Xin Wang[1], Philip D. Bradford[1], Qingwen Li[2], Yuntian Zhu[1]
[1] *North Carolina State University, Raleigh, NC, USA,* [2] *Suzhou Institute of Nano-Tech and Nano-Bionics, Suzhou, China*

CHAPTER OUTLINE

23.1 Introduction	649
23.2 Recent Advances in the Fabrication of Aligned Composite Prepregs	652
23.2.1 Fabricating aligned prepregs from drawable superaligned CNT sheets	652
23.2.2 Fabricating aligned prepregs directly from aligned CNT arrays by shear pressing	657
23.3 Mechanical and Physical Properties of CNT Composite Prepregs	660
23.3.1 Effect of CNT alignment and straightness	660
23.3.2 Effect of CNT volume fraction	662
23.3.3 Effect of CNT types	662
23.3.4 Effect of matrix types	663
23.4 Opportunities and Challenges	663
23.5 Conclusions and Outlook	664
References	665

23.1 INTRODUCTION

Carbon nanotubes (CNTs), a new member of the fullerene family, have generated intense scientific and technological interest since their identification in 1991 [1]. The combination of remarkable mechanical and physical properties allows CNTs to be envisioned as a revolutionary candidate for a wide spectrum of applications. Owing to the strong carbon–carbon covalent bonding and the potential for a seamless defect-free structure, CNTs have demonstrated an experimentally measured tensile strength and Young's modulus of ~50–60 GPa [2,3] and ~1 TPa [4,5], respectively. In addition, the room temperature thermal conductivities of CNTs have been measured as high as 3500 W/m K [6] for single-walled nanotubes (SWNTs) and 3000 W/m K for multiwalled nanotubes (MWNTs) [7]. These values are superior even to monocrystalline diamond (2000 W/m K) [8]. Metallic SWNTs and MWNTs exhibit electrical conductivities on the order of 10^5–10^6 S/cm [9,10].

Despite the superior properties of individual CNTs, it remains a challenge to translate the properties to macroscopic composites with any preferred alignment. Aligned CNT composites that have a combination of superb mechanical and physical properties would find a large range of applications in the aerospace, automotive, electronics and sporting industries. Currently, there are four main challenges in the field of high-performance CNT composite fabrication: (1) utilizing the highest possible aspect ratio CNTs; (2) obtaining the highest volume fractions of CNTs possible while maintaining CNT dispersion in the matrix; (3) achieving a high degree of CNT alignment and straightness; and (4) developing simple, efficient, continuous and scalable technologies for industrial production.

The dispersion of CNTs in polymer matrices has been extensively explored for preparing CNT/polymer composites. In this method, CNT composites are usually formed by dispersing nanotubes in a polymer with a suitable solvent and energetically agitating the solution before solvent evaporation. There are three major characteristics for the dispersion method. First, this method relies on the efficient dispersion of CNTs, and thus usually produces composites with low CNT volume factions and can only utilize short nanotubes. This is due to the fact that long CNTs tend to agglomerate and tangle in polymer solution, making dispersion very difficult. Therefore, short nanotubes are generally utilized in the dispersion method. If longer CNTs are used, they are generally broken down into smaller segments during extended sonication. Low volume fraction random dispersions are very suitable for some electrical or thermal applications such as static charge dissipation and sensors [11] or for modest improvement in matrix materials used in traditional composites. For example, Sandler et al. [12] dispersed nanotubes with a length of a few micrometers in epoxy, and fabricated composites with only 0.1–0.5 vol.% CNTs to achieve percolation. The second defining characteristic for the dispersion method is the use of surfactants or functionalization agents to aid in CNT dispersion. Many of the covalent and noncovalent functionalization techniques also have the function of improving the CNT–polymer bonding [13,14]. Sidewall functionalization, however, imparts defects to initially seamless tubes, lowering their mechanical potential. Finally, nanocomposites produced by the dispersion method are usually composed of nanotubes with random orientations, which minimize the potential composite properties. Electrostatic force has been utilized to improve the nanotube alignment of dispersed CNTs through a self-assembly process [15]. However, in general, short-CNT composites made by solution dispersion and melt processing show a relatively low tensile strength (<500 MPa).

Paper-like CNT films, known as buckypapers, have attracted much attention for the development of CNT composites because they can contain high volume fractions of CNTs. They have also shown potential for use in applications such as catalysis, filtration, sensors, actuators and electrodes for supercapacitors or lithium-ion batteries [16]. Generally, the fabrication of buckypaper involves dispersion and filtration of a suspension of CNTs. Recently, there have been newly developed approaches in which CNT arrays were directly pushed or printed into buckypapers, which feature highly aligned CNTs in planar form. Wang et al. [17] presented a "domino pushing" method for preparing large areas of aligned thick buckypapers from vertically aligned CNT arrays. Similarly, a "transfer-printing" [18] approach was demonstrated to roll across CNT arrays grown on a silicon substrate. Buckypapers, with alignment induced by stretching the random CNT structures, have demonstrated excellent properties when infiltrated with a polymer matrix and cured. Cheng et al. [19] demonstrated a mechanical stretching strategy for producing high-mass fraction (~60 wt.%) MWNT composites with the tensile strength and modulus reaching 2 GPa and 169 GPa, respectively. After functionalizing the buckypaper and infiltrating it with aerospace-grade bismaleimide (BMI) resin, they further improved the composite strength and modulus to 3 and 350 GPa, respectively [20]. This was the first report to demonstrate that the mechanical properties of CNT composites can exceed the state-of-the-art unidirectional carbon fiber composites. Nanocomposites made with buckypaper and polycarbonate solution containing 40–60 wt.% CNTs exhibited promising electrical conductivities (>200 S/cm), which are 1.7–3 times higher than that of the neat buckypaper [21]. The electrical properties of the conductive composites would make them useful for electromagnetic interference shielding, lightning strike protection and electrostatic discharge protection. Pristine aligned buckypaper has shown a thermal conductivity of ~42 W/m K [22]. However, high thermal conductivities in buckypaper/polymer composites have not been realized, which is mainly due to limited tube–tube coupling, high-interfacial thermal resistance and defects that scatter phonons.

CNT fibers (also often called CNT yarns when twist is present) containing axially aligned and densely packed CNTs have exhibited remarkable specific tensile strength and elastic modulus values. CNT fibers can be produced from aerogels [23,24], solutions [25–27] and drawable arrays [28–35]. Since the development of a dry-drawing method in 2002 [34], much effort has been spent on producing fibers (or yarns) from superaligned CNT arrays. For example, Zhang et al. added subsequent twisting to the drawing process, and synthesized CNT fibers with stiffness up to 330 GPa [31] and strength up to 3.3 GPa [28]. Tran et al. [36] used a modified dry-spinning method, which added several rollers to improve CNT alignment and polyurethane resin as matrix, to obtain a high tensile strength of approximately 2 GPa. Fang et al. [37] improved interfacial shear strength between CNTs by curing with polyimide, and enhanced the fiber strength by 30% from 1.58 to 2.06 GPa. For electrical properties, Zhao et al. [38] reported the fabrication of iodine-doped, double-walled nanotube fibers having electrical resistivity reaching $\sim 10^{-7}$ $\Omega \cdot$m, whose specific electrical conductivity surpasses copper. Such fibers are extremely promising for fabrication of CNT fiber-based composites featuring high strength or excellent conductivity.

CNT fibers have a similar diameter range as traditional carbon fibers. Making prepreg materials out of these CNT fibers is a natural extension for making composites. However, the production rate of CNT fibers has progressed slowly and thus only a few studies have used hundreds or thousands of such fibers to make small composite samples. Bogdanovich and Bradford [39–41] conducted pioneering research in textile assemblies of CNTs. They fabricated 3-D braids from 25-ply CNT fibers, infused them with epoxy resin systems and examined the morphology and mechanical properties of the composites. It was found that epoxy resin penetrated not only around the CNT fibers but also within the fibers when the viscosity of the resin system was low. Their final composite samples reached strength of 325 MPa and modulus of 24 GPa. Mora et al. [42] produced small unidirectional composites with 100 m of CNT fiber. They optimized their composite processing to reach a strength of 250 MPa with a modulus of 18 GPa. The major limitation for developing CNT fiber composites is that fiber production has been limited mainly to research labs, and thereby limiting the availability of fibers for prepreg and composite fabrication.

CNT sheets drawn from superaligned arrays can be directly used for fabricating CNT composites. Cheng et al. [43,44] was the first group to utilize solid-state drawn CNT sheets for producing aligned CNT composites. By dry-drawing the CNT sheets by hand, stacking multiple layers together and then infiltrating the aligned stack with epoxy resin, they produced composites with 16.5 wt.% CNTs that were homogenously dispersed and aligned in the epoxy matrix. The tensile strength and Young's modulus of the composites were 231.5 MPa and 20.4 GPa, respectively. However, a simple and rapid fabrication technique to simultaneously meet the critical structural features and realize the reinforcing potential of CNTs is still lacking. In the following sections, we describe our latest efforts to develop new approaches for fabricating CNT composites with critical structural characteristics and excellent mechanical, thermal and electrical properties, which are also promising for future industrial scale-up.

23.2 RECENT ADVANCES IN THE FABRICATION OF ALIGNED COMPOSITE PREPREGS

A number of techniques exist for making CNT/polymer composites, including both dry (solution-sate) and wet (solid-state) processing methods. Compared with wet methods, the advantages of dry methods include utilizing long nanotubes, better control of CNT orientations and easier processing. Among dry CNT materials, vertically aligned CNT arrays grown on a substrate by chemical vapor deposition, with high purity, high quality and unidirectional alignment, provide an exciting and promising opportunity for fabricating aligned and continuous CNT composites. The key to synthesizing CNT composites from vertically aligned arrays is to convert the vertically aligned nanotubes to the horizontal direction, and preserve the good alignment and high aspect ratio of CNTs in the resulting composite structure. In this section, we will discuss two major approaches—synthesis from drawable superaligned CNT sheets and synthesis by directly "shear-pressing" CNT arrays, and their key technical issues and mechanisms.

23.2.1 Fabricating aligned prepregs from drawable superaligned CNT sheets

Drawable superaligned CNT arrays were first developed by Fan et al. in 2002 [34]. Drawable arrays can be viewed as a special type of vertically aligned CNT array from which CNTs with lengths up to millimeters can be drawn continuously from the array. There are a few research groups capable of growing highly drawable CNT arrays [29,34,35,45–49]. Synthesis of drawable CNT arrays is determined by a narrow size distribution of catalyst particles, high nucleation density, and clean CNT surfaces [50]. These characteristics are enabled only when precisely controlled and finely adjusted growth conditions are used [51,52]. Despite the advances in practical exploration of drawable CNT arrays, the fundamental mechanism governing the drawing process is not yet fully revealed. Kuznetsov et al. [53] believed that a network of individual CNTs or small bundles serve as interconnections between large-diameter MWNT bundles in the array. Proper densities of interconnections allow pulling out of CNT sheets through preferentially peeling off and self-densification at the top and bottom of the array. In another study, Zhu et al. [54] reported the formation of entangled structures at the ends of CNT bundles at the bottom or top of the arrays. They concluded that the self-entanglement effect of CNTs is responsible for maintaining the continuity during the CNT drawing process.

Through utilizing drawable superaligned CNT sheets, we have developed "spray-winding" process [55], which has produced CNT/polyvinyl alcohol (PVA) composites with a tensile strength of 1.8 GPa, Young's modulus of 40–96 GPa, toughness of 38–100 J/g and electrical conductivity as high as 780 S/cm. These composites are stronger than other reported CNT/PVA composites (100–600 MPa) [56–59] and CNT/PVA composite fibers (1.5 GPa) [60], and comparable to the strongest CNT/PVA fiber (2.05 GPa) [61] reported very recently. The high performance arises from the long CNTs, high level of CNT alignment, high CNT fraction, and good CNT dispersion in PVA matrix, which are obtained simultaneously.

During the spray-winding process, CNT sheets consisting of CNTs with a length of ∼300–1000 μm (aspect ratio ∼30,000–50,000) are pulled out from superaligned CNT arrays. The CNT sheets have a thickness of a few micrometers and density of <0.01 g/cm. The sheets are continuously wound onto a rotating mandrel in a layer-by-layer fashion while a fine atomized spray of a polymer solution is directed onto each layer in situ. A schematic of the process is shown in Fig. 23.1. Compared to regular infiltration methods that cannot avoid disturbance to CNT alignment due to migration of polymer molecules and flow of the liquid, the "spray-winding" process deposits polymer onto each layer of well-aligned CNT sheet and maintains the aligned structure. Additionally, the spray-winding process allows the polymer matrix to penetrate between individual or very small bundles of CNTs instead of large bundles found in most buckypapers because the polymer is introduced before large bundles can be formed. Therefore, the CNTs or their bundles are integrated with the matrix at molecular level, which is critical to ensuring the effectiveness of load transfer, and thereby enhancing the mechanical properties of the resulting composites. Furthermore, the size and thickness of the unidirectional composite prepregs are tunable by choosing an appropriate mandrel diameter, sheet width, and number of revolutions. The weight fraction of the CNT composite prepregs can also be tuned by varying the concentration of the polymer solution, and high weight fractions are achieved by using very dilute solutions. Most importantly, this fabrication approach directly produces composite prepregs

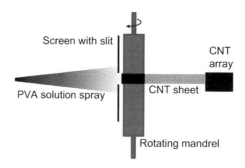

FIGURE 23.1

Schematic view of spray winding. A CNT sheet is drawn out of a drawable array and continuously wound onto a rotating mandrel, on which atomized droplets of PVA solution are deposited. A screen with a slit is used to control the spray area [55]. (For color version of this figure, the reader is referred to the online version of this book.)

within one step, and the production rate depends only on the rotation speed of the mandrel. The process is much faster than other existing approaches such as resin transfer molding [43] and layer-by-layer assembly [15,57,62]. Thus, the described approach is conducive to large-scale productions and suitable for both thermoplastic and thermosetting polymers.

One critical issue that prevents the full utilization of the reinforcing potential of the CNTs is their waviness. Micrographs suggest that individual nanotubes are wavy microscopically [63]. CNTs with different degrees of waviness do not carry an applied load uniformly, cannot be packed densely, and have poor intertube connections, all of which adversely affect the strength, stiffness, and conductivity of the resulting CNT composites. The adverse effect of CNT waviness on composite properties has been clearly demonstrated through modeling [64–66]. To solve this issue, a stretching system was incorporated to the spray-winding process to straighten the wavy CNTs before embedding them into a polymer matrix [67], as illustrated in Fig. 23.2(a). The modified process is called "stretch winding", which used a pair of stretching rods to apply tension to the CNT sheets and stretch them before taking up onto the mandrel. Figure 23.2(b) shows the stretch-winding process, where the CNT sheet travels horizontally and passes through a tensioning system. As the CNT sheet passes around a pair of stationary rods, the increased tension stretches the CNT sheets. The CNT sheet was stretched according to a stretch ratio $(L_S - L_0)/L_0$, where L_S and L_0 are the length of the CNT sheet after and before stretching, respectively. The stretch ratio can be controlled by adjusting the height of the stretching rods so as to adjust the contact angle between the CNT sheet and the rods.

CNT/BMI composites, with a volume fraction of 46%, were fabricated by the stretch-winding approach and exhibited a combined strength and Young's modulus exceeding current carbon fiber-reinforced polymer (CFRP) composites. The CNT composite that was stretched by 12% achieved a tensile strength of 3.8 GPa and

23.2 Recent Advances in the Fabrication of Aligned Composite Prepregs

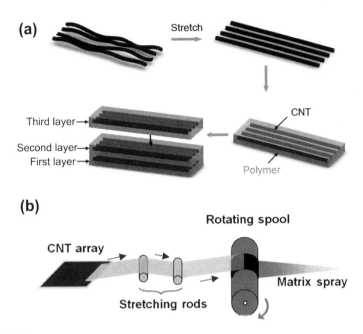

FIGURE 23.2

(a) Schematic illustration of the concept of straightening CNTs before embedding them in polymer matrix in a layer-by-layer fashion. (b) Schematic illustration of the experimental setup for the stretch-winding process [67]. (For color version of this figure, the reader is referred to the online version of this book.)

Young's modulus of 293 GPa, as shown in Fig. 23.3(a). The average strength and stiffness of four CNT composite samples were 3.5 and 266 GPa, respectively. Figure 23.3(b) shows that the CNT/BMI nanocomposites have a density of 1.25 g/cm^3 which is significantly lower than that of traditional CFRP (1.6 g/cm^3). As a result, the specific strength of the composites stretched by 12% reached 3.0×10^7 cm, which is at least 30% higher than that for the best unidirectional CFRP composites. These nanocomposites also showed a high thermal conductivity of 41 W/m K and electrical conductivity of 1230 S/cm. The thermal conductivity is higher than that of unidirectional polyacrylonitrile-based CFRP (4.5 W/m K) [68] and other reported high-volume fraction CNT composites (0.4–1.3 W/m K) [69]. These results indicate that stretching improves the alignment and straightness of CNTs, increases their load carrying efficiency, and promotes a more efficient transfer of phonons and electrons along the CNT length direction of the composites. Figure 23.4 shows the flexible CNT/BMI composites prepared by the stretch-winding approach.

Postfabrication stretching of prepregs also proved effective in reducing CNT waviness and enhancing the properties of the composites. In another effort [70], high volume fraction and aligned CNT/nylon composites fabricated by rotary winding and layer-by-layer polymer infusion were poststretched via a simple local

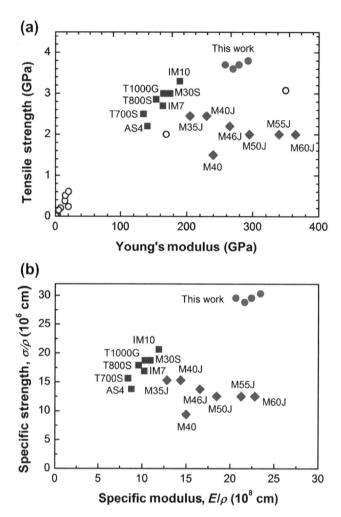

FIGURE 23.3

(a) Comparison of the tensile strength and Young's modulus of currently available engineering carbon fiber-reinforced polymer (CFRP) composites as well as CNT composites reported previously. Note that all data are from composite samples. The blue-filled squares are from existing high-strength CFRP and the green-filled diamonds are from those of currently existing high-modulus CFRP. The unfilled circles are best CNT composites reported previously. (b) Comparison of the specific tensile strength and specific modulus of the stretch-wound composites with the best engineering CFRP. The data were calculated based on a density of 1.6 g/cm^3 for CFRP (60 vol.%) and 1.25 g/cm^3 for the stretch-wound composites (46 vol.%) [67]. (For interpretation of the references to color in this figure legend, the reader is referred to the online version of this book.)

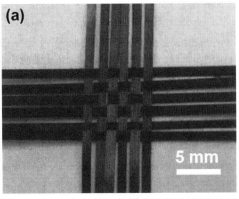
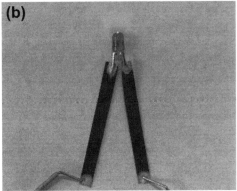

FIGURE 23.4

Optical photographs of CNT/BMI composite sheets synthesized by stretch winding. (a) A textile woven from CNT composite films. (b) The CNT composite films as segments of conductive media loaded with an LED light bulb [67]. (For color version of this figure, the reader is referred to the online version of this book.)

heating and stretching method that uniformly stretched the nanotubes together with the nylon matrix. The tensile strength, Young's modulus and electrical conductivity showed a 191%, 294% and 207% improvement, respectively, for composites that were poststretched by 7%. Through stretching the CNT sheet and/or stretching the prepreg before curing, it was concluded that macroscopically aligned CNTs and microscopically reduced waviness are critical for improving the mechanical and physical properties of CNT prepreg composites.

23.2.2 Fabricating aligned prepregs directly from aligned CNT arrays by shear pressing

Aligned CNT arrays can be directly "pushed down" to form CNT buckypapers or films for composite fabrication. There have been newly developed approaches to directly convert CNT arrays into buckypapers, comprising aligned CNTs in planar form. For example, Wang et al. [17] presented a "domino pushing" method for preparing aligned thick buckypapers from vertically aligned CNT arrays. In our work [71], a "shear-pressing" method was developed to produce aligned CNT buckypapers from CNT arrays. Figure 23.5(a) shows a schematic of the shear-pressing setup and Fig. 23.5(b) shows the first generation shear-pressing device. The pressing process shears the CNTs uniformly from their vertical orientation to a horizontal orientation and consolidates the structure by removing a large amount of empty space within the array, while preserving the alignment of the nanotubes.

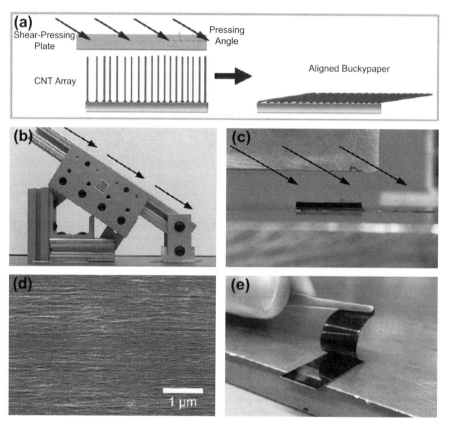

FIGURE 23.5

Overview of the process of shear-pressing CNT arrays showing (a) a schematic of the morphology of the array before and after shear pressing; (b) the shear-pressing device; (c) close-up of the parallel plates right before the array is shear pressed; (d) SEM image of the end of the preform showing the alignment of CNTs; and (e) the aligned CNT preform is easily removed from the substrate by hand and ready for resin infusion [71]. (For color version of this figure, the reader is referred to the online version of this book.)

23.2 Recent Advances in the Fabrication of Aligned Composite Prepregs

This alignment was confirmed by SEM imaging, with an example shown in Fig. 23.5(d). Domino pushing CNT arrays should theoretically produce a similar morphology to the new shear-pressing method. It is our experience that the shear-pressing process works for a wider range of array types. This is due to the fact that during the shear-pressing operation, the array is consolidated into the aligned sheet all in one motion. During the domino pushing operation, loose attachment of the array to the substrate can prevent the material from pressed in a uniform manner. After shear pressing, the aligned preforms can be easily removed from the growth substrate (Fig. 23.5(e)) and then soaked in an epoxy resin solution to form prepregged sheets.

The dry CNT preforms exhibited a failure strength of 16 MPa, and the cured prepregs showed a tensile strength reaching 300 MPa and a modulus of 15 GPa. To further improve the mechanical properties of the composites by reducing CNT waviness, the resin-infused prepregs were exposed to a strain of 5% before curing, followed by cutting the sheet into test strips and then curing. Such pre-cure-stretching of the prepreg increased the maximum tensile strength values of the composite by 33% to 402 MPa and elastic modulus by 50% to 22.3 GPa. These results demonstrate once again that straightening wavy CNTs can effectively improve the strength and stiffness of the composites. One of the challenges of this technology is to provide the best resin infusion possible. Owing to the fact that much of the empty space is removed during the pressing, many of the CNTs form much larger bundles. This bundling reduces the effective reinforcement of the aligned CNTs in the composite due to less effective load transfer. Current work is being completed to add resin to the array before pressing to increase uniformity of CNT dispersion within the prepregs.

The manual machine seen in Fig. 23.5 is now upgraded to a fully functioning computer-controlled shear-pressing machine, built by 3TEX, Inc. The device, shown in Fig. 23.6(a), has two individual pressing plates that are 200 × 200 mm. One plate moves only in the vertical direction and the other moves only in a single horizontal direction. Both plates are controlled independently by high-precision stepper motors. Through the programmable computer controls, the angle of pressing can be changed. It is also possible to press the arrays using a nonlinear, elliptical trajectory to minimize any induced waviness during the shear-pressing operation. The lower pressing plate contains force sensors to collect data on how the different pressing trajectories apply vertical force to the CNT arrays during the pressing. Computer control and highly precise movements ensure repeatability of the operation. Figure 23.6(b) shows multiple large CNT arrays that were each shear pressed in a few seconds.

Advantages of the shear-pressing approach are the following. (1) Aligned CNT arrays containing long nanotubes can be converted into aligned CNT preforms in seconds. (2) The preforms meet the desired characteristics of millimeter-long CNTs, high volume fraction, high degree of CNT alignment, and fast processing speed simultaneously. (3) The parallel alignment and through-the-thickness morphology of CNTs in the preforms facilitate resin infusion of the preform and

 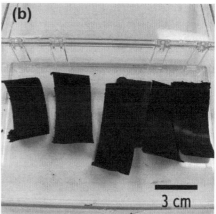

FIGURE 23.6

(a) Automated computer-controlled shear-pressing device. (b) Large shear-pressed aligned CNT sheets ready for prepregging and lamination. (For color version of this figure, the reader is referred to the online version of this book.)

thus create composites with both good unidirectional and through-the-thickness properties.

23.3 MECHANICAL AND PHYSICAL PROPERTIES OF CNT COMPOSITE PREPREGS

The superb mechanical and physical properties of individual CNTs are the major motivation for the research and development of high-performance composites. So far the stiffest CNT composites (350 GPa) were achieved through mechanical alignment, functionalization and subsequent infiltration of CNT buckypaper with BMI matrix [20]. Without functionalization, these nanocomposites demonstrated unprecedentedly high electrical conductivity of 5500 S/cm along the alignment direction [19]. In our work, CNT/BMI composites with record-high tensile strength of 3.8 GPa were achieved by the stretch-winding approach [67], using drawable superaligned CNTs as the starting materials. When assembling the CNTs into macroscopic composites, many factors, such as nanotube alignment, waviness, volume fraction, and interaction with the matrix, affect the final properties of the composites.

23.3.1 Effect of CNT alignment and straightness

CNT alignment refers to the mean nanotube orientation direction that is parallel to the long axis of the material. CNT waviness can be considered as local misalignment along the individual nanotube. As reported by many studies [19,44,72–74], high degree of CNT alignment directly leads to improved unidirectional mechanical,

23.3 Mechanical and Physical Properties of CNT Composite Prepregs

thermal and electrical properties. In this work, both the spray-/stretch-winding and shear-pressing approaches utilized aligned CNT arrays and successfully converted the nanotubes into horizontally aligned CNT films. The reported high strength, stiffness, and thermal and electrical conductivities of the composites, which are superior to those of their counterparts fabricated by dispersion method, are direct results of the aligned building blocks.

The CNT alignment and straightness within composites can be characterized by polarized Raman spectroscopy [75,76]. Specifically, the shift of the intensity ratio of G-band peaks is measured. Theoretically, Raman intensity change is proportional to $\cos^4\theta$ versus the CNT orientation angle (θ) [75]. The ratio of the intensity of the G-band in the parallel configuration to the perpendicular configuration ($I_{G\parallel}/I_{G\perp}$) can be used to describe CNT alignment degree; higher CNT alignment degree results in a higher intensity ratio. For the unstretched CNT/BMI composite film in our work, the intensity ratio ($I_{G\parallel}/I_{G\perp}$) is approximately 1.6. When the composite was stretched by 10% and 12%, the intensity ratio increased to 6.2 and 7.6, respectively, suggesting that the stretch-winding process significantly improved the alignment of CNTs in the nanocomposites. Owing to the improved CNT alignment, the mechanical, thermal and electrical properties of the composites were significantly enhanced. As shown in Fig. 23.7, the pristine CNT sheet has a strength of 300 MPa and Young's modulus of 21 GPa. The unstretched composites exhibited strength of 2.0 GPa and Young's modulus of 130 GPa. The stretching raised the strength and Young's modulus to as high as 3.5 GPa and 266 GPa, respectively, at the stretch ratio of 12%. Similarly, a higher degree of CNT alignment showed higher thermal conductance and electrical conductivity along the aligned direction. Stretching by 12% increased the average thermal conductivity values by 186% to 40 W/m K and average electrical conductivity by 43% to 1180 S/cm for the composites,

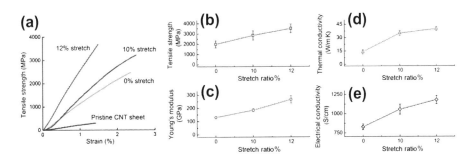

FIGURE 23.7

(a) Typical stress—strain curves of pristine CNT sheet, unstretched and stretched composites, demonstrating a significant improvement of the mechanical properties through aligning and straightening of CNTs. Effect of stretching on (b) tensile strength, (c) Young's modulus, (d) thermal conductivity, and (e) electrical conductivity of the composites [67]. (For color version of this figure, the reader is referred to the online version of this book.)

compared to unstretched composites. Similar increases in mechanical properties were observed after stretching shear-pressed composite prepreg samples before curing.

23.3.2 Effect of CNT volume fraction

Higher CNT volume fraction results in better mechanical performance of the composites, according to the rule of mixtures. High CNT fraction (~40–65 wt.%) composites with good dispersion can be made by the spray-winding approach. To investigate the effect of CNT volume fraction on properties of CNT composites, we produced CNT/PVA and CNT/BMI [77] composites with different CNT fractions by adjusting the solution concentration. The mechanical properties of both CNT/PVA and CNT/BMI composites showed a strong dependence on the CNT fractions. As the weight fraction increased from 30 to 65 wt.%, the tensile strength of the CNT/PVA composites increased from approximately 400 MPa to 1.8 GPa. The highest strength was found for the composite containing 65 wt.% of CNTs, by using the 1.0 g/l PVA solution, which was the optimal polymer concentration. The electrical conductivity of the CNT/PVA composites is also strongly dependent on the CNT fraction, with higher weight fractions consistently resulting in higher electrical conductivities of the composites, until reaching the optimal polymer concentration. CNT/BMI composites demonstrated similar CNT fraction dependency in both tensile strength and electrical conductivity to those for CNT/PVA composites. The optimal concentration of the BMI solution was also found to be 1.0 g/l, although the corresponding weight fraction varied as a result of the different polymer types. When the polymer solution was further diluted, insufficient polymer resulted in polymer-deficient areas throughout the composite, which reduced load transfer between CNTs and thus negatively impacted the strength and stiffness.

23.3.3 Effect of CNT types

Mechanical properties of CNT composites strongly depend on the CNT morphology such as tube diameter, wall thickness and tube length (aspect ratio). Generally, the as-grown drawable MWNT arrays have a height from a few hundred micrometers to millimeters, and aspect ratios can be as large as 50,000 [29,50,60,78]. Drawable MWNT arrays with different CNT morphologies were used to fabricate CNT composites using the spray-winding method. When using a type of thick CNTs that have 50 walls and a diameter of 45 nm, the rotary wound CNT/nylon composites exhibited a tensile strength of ~200–600 MPa and an elastic modulus of ~15–60 GPa, and the spray-wound CNT/PVA composites exhibited a tensile strength ~350–1000 MPa and an elastic modulus of ~20–70 GPa [79]. In comparison, when using CNTs with a small diameter of ~7–10 nm and ~4–6 walls, the spray-wound CNT/PVA composites showed a tensile strength of ~1.5–1.8 GPa and an elastic modulus of ~40–60 GPa, and the spray-wound CNT/BMI composites showed a tensile strength of ~1.5–2.0 GPa and an elastic modulus of ~60–130 GPa. Both types of CNTs demonstrated a high aspect ratio of >15,000, much larger than the minimum that was required to effectively transfer load across the CNT—polymer interface. We found that CNTs with larger diameters and thicker walls resulted in composites with lower strength and stiffness. This is because CNTs with larger diameters and thicker walls often contain more defects (imperfection in crystallization) in structure and have slippage between CNT walls, thereby reducing the mechanical properties of the composites. Many theoretical and experimental results also indicate that the elastic modulus of CNTs or CNT bundles becomes lower with the increase of the diameter of the CNTs or the CNT bundles [80,81].

23.3.4 Effect of matrix types

The spray-winding approach and its derivatives work for both thermoplastic and thermosetting polymers, given that a suitable solvent can be found to dissolve the polymer. In order to produce high-toughness composites, thermoplastic polymers such as PVA, nylon and thermoplastic polyimides should be used; they have flexible long-chain structures and can undergo large strains before failure. As an example, the spray-wound CNT/PVA composites exhibited a high toughness of 100 J/g. If high strength and stiffness or high-temperature resistance is the target property for the composites, thermosetting polymers such as BMI, epoxy and thermosetting polyimide should be used. They have cross-linked structures, which are stiff, providing the maximum load transfer and also have high glass transition temperatures. The CNT/BMI composites showed a much higher tensile strength and Young's modulus than those for CNT/PVA composites fabricated by the same method. The CNT/BMI composites also exhibited a high thermal decomposition temperature >400 °C.

CNT—matrix interfacial properties, which are matrix-dependent, are vital in determining the stress-transfer efficiency across the interface. Molecular dynamics studies have demonstrated that different types of polymers have different conformations along the CNT surface, and also result in different interaction energy between the polymer and the CNT [82,83]. Some other studies [84—86] have suggested that a combination of backbone stiffness and aromaticity can result in more stabilized polymer adsorption to the CNT surface and thus lead to enhanced CNT—matrix interaction.

23.4 OPPORTUNITIES AND CHALLENGES

Approaches of fabricating drawable CNT-based prepregs and shear-pressed CNT array prepregs are promising for making future high-performance CNT composites. Prepregs fabricated by spray-winding and stretch-winding approaches demonstrate many advantages in terms of exceptional properties, easy fabrication and the potential to be scaled up for industrial production. We have demonstrated stretch-wound CNT/BMI composite samples with combined strength and stiffness superior to current best engineering carbon fiber composites, when the samples are small in size. We believe that with improved CNT quality and further engineering of the fabrication technique, the excellent mechanical, electrical and thermal properties can be translated to larger composites.

The prerequisite for producing high-strength CNT composites is to synthesize high-quality CNT arrays comprising long and strong individual CNTs that can be converted into continuous and uniform CNT sheets. However, it has been a big challenge for the research community to produce CNT arrays with high and consistent quality, a high level of drawability and high growth throughput. These challenges must be addressed in order to produce CNT composites that are superior to high-performance carbon fiber composites on a commercial scale.

Another challenge is how to further transfer the intrinsic properties of individual CNTs to macroscopic structures when scaling up the composite fabrication process. While the initial recorded properties of some aligned CNT prepreg composites are impressive, they still have not reached the full potential. Further increased level of CNT alignment and straightness, improvement of bonding between CNTs and the matrix, and minimization of defects in the composite structures are undoubtedly crucial for further breakthrough.

23.5 CONCLUSIONS AND OUTLOOK

Aligned CNT composites fabricated by the spray-/stretch-winding approach possess all the critical structural characteristics found in traditional unidirectional composites. As a result, the composites produced using this technique have ultrahigh tensile strength, high elastic modulus, excellent thermal and electrical conductivities, low density and high thermal degradation resistance. The combined ultrahigh strength and stiffness makes the composites promising as next generation composite materials. The exceptional mechanical properties together with multifunctionality also make them promising as building blocks for future smart textiles, unidirectional thermal interface materials, flexible conductors, sensors, etc. Aligned CNT composites fabricated by shear-pressing approach also possess desired structural features and excellent properties. Such simple and direct techniques open the door for developing large-scale fabrication of aligned CNT composites.

Although aligned CNT composites have been successfully developed through using drawable CNT arrays or aligned buckypapers, the research efforts on aligned CNT composites are still in their infancy. Better understanding of multiscale structure—property relationship for CNT composites needs to be established. This includes further study of the effect of CNT type, morphology and waviness at the nanoscale, the effect of CNT arrangement, alignment, intertube coupling and stress transfer at the microscale, and the effect of prepreg thickness and lay-up orientation at the macroscale on final composite properties. Moreover, some challenges remain for synthesis of high and consistent quality of CNT arrays and transfer of unique properties of individual CNTs to large composites. Further sustained effort needs to be made in order to realize the full potential of CNTs in aligned prepregs and composites, to enable future aerospace technologies and inspire new devices that can capitalize on the unique properties.

References

[1] S. Iijima, Helical microtubules of graphitic carbon, Nature 354 (6348) (1991) 56–58.
[2] M.F. Yu, O. Lourie, M.J. Dyer, K. Moloni, T.F. Kelly, R.S. Ruoff, Strength and breaking mechanism of multiwalled carbon nanotubes under tensile load, Science 287 (5453) (2000) 637–640.
[3] M.F. Yu, B.S. Files, S. Arepalli, R.S. Ruoff, Tensile loading of ropes of single wall carbon nanotubes and their mechanical properties, Physical Review Letters 84 (24) (2000) 5552–5555.
[4] M. Treacy, T. Ebbesen, J. Gibson. Exceptionally high Young's modulus observed for individual carbon nanotubes. 381 (6584) (1996) 678–680.
[5] E.W. Wong, P.E. Sheehan, C.M. Lieber, Nanobeam mechanics: elasticity, strength, and toughness of nanorods and nanotubes, Science 277 (5334) (1997) 1971.
[6] E. Pop, D. Mann, Q. Wang, K. Goodson, H. Dai, Thermal conductance of an individual single-wall carbon nanotube above room temperature, Nano Letters 6 (1) (2006) 96–100.
[7] P. Kim, L. Shi, A. Majumdar, P. McEuen, Thermal transport measurements of individual multiwalled nanotubes, Physical Review Letters 87 (21) (2001) 215502.
[8] L. Wei, P. Kuo, R. Thomas, T. Anthony, W. Banholzer, Thermal conductivity of isotopically modified single crystal diamond, Physical Review Letters 70 (24) (1993) 3764–3767.
[9] P.L. McEuen, M.S. Fuhrer, H. Park, Single-walled carbon nanotube electronics, Nanotechnology, IEEE Transactions 1 (1) (2002) 78–85.
[10] A. Bachtold, M. Henny, C. Terrier, C. Strunk, C. Schönenberger, J.P. Salvetat, J.M. Bonard, L. Forro, Contacting carbon nanotubes selectively with low-ohmic contacts for four-probe electric measurements, Applied Physics Letters 73 (2) (1998) 274.
[11] K.I. Winey, T. Kashiwagi, M. Mu, Improving electrical conductivity and thermal properties of polymers by the addition of carbon nanotubes as fillers, MRS Bulletin 32 (4) (2007) 348–353.
[12] J. Sandler, M. Shaffer, T. Prasse, W. Bauhofer, K. Schulte, A. Windle, Development of a dispersion process for carbon nanotubes in an epoxy matrix and the resulting electrical properties, Polymer 40 (21) (1999) 5967–5971.
[13] M.U. Khan, V.G. Gomes, I.S. Altarawneh, Synthesizing polystyrene/carbon nanotube composites by emulsion polymerization with non-covalent and covalent functionalization, Carbon 48 (10) (2010) 2925–2933.
[14] T.J. Simmons, J. Bult, D.P. Hashim, R.J. Linhardt, P.M. Ajayan, Noncovalent functionalization as an alternative to oxidative acid treatment of single wall carbon nanotubes with applications for polymer composites, ACS Nano 3 (4) (2009) 865–870.
[15] M. Olek, J. Ostrander, S. Jurga, H. Mohwald, N. Kotov, K. Kempa, M. Giersig, Layer-by-layer assembled composites from multiwall carbon nanotubes with different morphologies, Nano Letters 4 (10) (2004) 1889–1895.
[16] T. Chou, L. Gao, E.T. Thostenson, Z. Zhang, J. Byun, An assessment of the science and technology of carbon nanotube-based fibers and composites, Composites Science and Technology 70 (1) (2010) 1–19.

[17] D. Wang, P.C. Song, C.H. Liu, W. Wu, S.S. Fan, Highly oriented carbon nanotube papers made of aligned carbon nanotubes, Nanotechnology 19 (7) (2008) 075609.
[18] E.J. Garcia, B.L. Wardle, A. John Hart, Joining prepreg composite interfaces with aligned carbon nanotubes, Composites Part A 39 (6) (2008) 1065–1070.
[19] Q. Cheng, J. Bao, J.G. Park, Z. Liang, C. Zhang, B. Wang, High mechanical performance composite conductor: multi-walled carbon nanotube sheet/bismaleimide nanocomposites, Advanced Functional Materials 19 (20) (2009) 3219–3225.
[20] Q. Cheng, B. Wang, C. Zhang, Z. Liang, Functionalized carbon-nanotube sheet/bismaleimide nanocomposites: mechanical and electrical performance beyond carbon-fiber composites, Small 6 (6) (2010) 763–767.
[21] G.T. Pham, Y.B. Park, S.R. Wang, Z.Y. Liang, B. Wang, C. Zhang, P. Funchess, L. Kramer, Mechanical and electrical properties of polycarbonate nanotube buckypaper composite sheets, Nanotechnology 19 (32) (2008).
[22] P. Gonnet, S.Y. Liang, E.S. Choi, R.S. Kadambala, C. Zhang, J.S. Brooks, B. Wang, L. Kramer, Thermal conductivity of magnetically aligned carbon nanotube buckypapers and nanocomposites, Current Applied Physics 6 (1) (2006) 119–122.
[23] Y.L. Li, I.A. Kinloch, A.H. Windle, Direct spinning of carbon nanotube fibers from chemical vapor deposition synthesis, Science 304 (5668) (2004) 276.
[24] K. Koziol, J. Vilatela, A. Moisala, M. Motta, P. Cunniff, M. Sennett, A. Windle, High-performance carbon nanotube fiber, Science 318 (5858) (2007) 1892–1895.
[25] B. Vigolo, A. Pénicaud, C. Coulon, C. Sauder, R. Pailler, C. Journet, P. Bernier, P. Poulin, Macroscopic fibers and ribbons of oriented carbon nanotubes, Science 290 (5495) (2000) 1331–1334.
[26] A.B. Dalton, S. Collins, E. Munoz, J.M. Razal, V.H. Ebron, J.P. Ferraris, J.N. Coleman, B.G. Kim, R.H. Baughman, Super-tough carbon-nanotube fibres, Nature 423 (6941) (2003) 703.
[27] L.M. Ericson, H. Fan, H.Q. Peng, V.A. Davis, W. Zhou, J. Sulpizio, Y.H. Wang, R. Booker, J. Vavro, C. Guthy, A.N.G. Parra-Vasquez, M.J. Kim, S. Ramesh, R.K. Saini, C. Kittrell, G. Lavin, H. Schmidt, W.W. Adams, W.E. Billups, M. Pasquali, W.F. Hwang, R.H. Hauge, J.E. Fischer, R.E. Smalley, Macroscopic, neat, single-walled carbon nanotube fibers, Science 305 (5689) (2004) 1447–1450.
[28] X.F. Zhang, Q. Li, T.G. Holesinger, P.N. Arendt, J.Y. Huang, P.D. Kirven, T.G. Clapp, R.F. DePaula, X.Z. Liao, Y.H. Zhao, L.X. Zheng, D.E. Peterson, Y.T. Zhu, Ultrastrong, stiff, and lightweight carbon-nanotube fibers, Advanced Materials 19 (23) (2007) 4198–4201.
[29] Q.W. Li, X.F. Zhang, R.F. DePaula, L.X. Zheng, Y.H. Zhao, L. Stan, T.G. Holesinger, P.N. Arendt, D.E. Peterson, Y.T. Zhu, Sustained growth of ultralong carbon nanotube arrays for fiber spinning, Advanced Materials 18 (23) (2006) 3160–3163.
[30] Q.W. Li, Y. Li, X.F. Zhang, S.B. Chikkannanavar, Y.H. Zhao, A.M. Dangelewicz, L.X. Zheng, S.K. Doorn, Q.X. Jia, D.E. Peterson, P.N. Arendt, Y.T. Zhu, Structure-dependent electrical properties of carbon nanotube fibers, Advanced Materials 19 (20) (2007) 3358–3363.
[31] X.F. Zhang, Q.W. Li, Y. Tu, Y.A. Li, J.Y. Coulter, L.X. Zheng, Y.H. Zhao, Q.X. Jia, D.E. Peterson, Y.T. Zhu, Strong carbon-nanotube fibers spun from long carbon-nanotube arrays, Small 3 (2) (2007) 244–248.

[32] L.X. Zheng, X.F. Zhang, Q.W. Li, S.B. Chikkannanavar, Y. Li, Y.H. Zhao, X.Z. Liao, Q.X. Jia, S.K. Doorn, D.E. Peterson, Y.T. Zhu, Carbon-nanotube cotton for large-scale fibers, Advanced Materials 19 (18) (2007) 2567–2570.

[33] X. Zhang, K.L. Jiang, C. Feng, P. Liu, L. Zhang, J. Kong, T.H. Zhang, Q.Q. Li, S.S. Fan, Spinning and processing continuous yarns from 4-inchwafer scale super-aligned carbon nanotube arrays, Advanced Materials 18 (12) (2006) 1505–1510.

[34] K.L. Jiang, Q.Q. Li, S.S. Fan, Nanotechnology: spinning continuous carbon nanotube yarns—carbon nanotubes weave their way into a range of imaginative macroscopic applications, Nature 419 (6909) (2002) 801.

[35] M. Zhang, K.R. Atkinson, R.H. Baughman, Multifunctional carbon nanotube yarns by downsizing an ancient technology, Science 306 (5700) (2004) 1358–1361.

[36] C.D. Tran, S. Lucas, D.G. Phillips, L.K. Randeniya, R. Baughman, T. Tran-Cong, Manufacturing polymer/carbon nanotube composite using a novel direct process, Nanotechnology 22 (14) (2011) 145302–145310.

[37] C. Fang, J. Zhao, J. Jia, Z. Zhang, X. Zhang, Q. Li, Enhanced carbon nanotube fibers by polyimide, Applied Physics Letters 97 (18) (2010) 181906.

[38] Y. Zhao, J. Wei, R. Vajtai, P.M. Ajayan, E.V. Barrera, Iodine doped carbon nanotube cables exceeding specific electrical conductivity of metals, Scientific Reports 1 (2011). (Article number: 83).

[39] A.E. Bogdanovich, P.D. Bradford, Carbon nanotube yarn and 3-D braid composites. Part I: tensile testing and mechanical properties analysis, Composites Part A 41 (2) (2010) 230–237.

[40] P.D. Bradford, A.E. Bogdanovich, Carbon nanotube yarn and 3-D braid composites. Part II: dynamic mechanical analysis, Composites Part A 41 (2) (2010) 238–246.

[41] A. Bogdanovich, P. Bradford, F. Shaoli, M. Zhang, R.H. Baughman, S. Hudson, Fabrication and mechanical characterization of carbon nanotube yarns, 3-D braids, and their composites, Sampe Journal 43 (1) (2007) 6–19.

[42] R.J. Mora, J.J. Vilatela, A.H. Windle, Properties of composites of carbon nanotube fibres, Composites Science and Technology 69 (10) (2009) 1558–1563.

[43] Q.F. Cheng, J.P. Wang, J.J. Wen, C.H. Liu, K.L. Jiang, Q.Q. Li, S.S. Fan, Carbon nanotube/epoxy composites fabricated by resin transfer molding, Carbon 48 (1) (2010) 260–266.

[44] Q.F. Cheng, J.P. Wang, K.L. Jiang, Q.Q. Li, S.S. Fan, Fabrication and properties of aligned multiwalled carbon nanotube-reinforced epoxy composites, Journal of Materials Research 23 (11) (2008) 2975–2983.

[45] M. Zhang, S. Fang, A.A. Zakhidov, S.B. Lee, A.E. Aliev, C.D. Williams, K.R. Atkinson, R.H. Baughman, Strong, transparent, multifunctional, carbon nanotube sheets, Science 309 (5738) (2005) 1215.

[46] K.R. Atkinson, S.C. Hawkins, C. Huynh, C. Skourtis, J. Dai, M. Zhang, S. Fang, A.A. Zakhidov, S.B. Lee, A.E. Aliev, Multifunctional carbon nanotube yarns and transparent sheets: fabrication, properties, and applications, Physica B 394 (2) (2007) 339–343.

[47] K. Liu, Y. Sun, L. Chen, C. Feng, X. Feng, K. Jiang, Y. Zhao, S. Fan, Controlled growth of super-aligned carbon nanotube arrays for spinning continuous unidirectional sheets with tunable physical properties, Nano Letters 8 (2) (2008) 700–705.

[48] S. Zhang, L. Zhu, M.L. Minus, H.G. Chae, S. Jagannathan, C.P. Wong, J. Kowalik, L.B. Roberson, S. Kumar, Solid-state spun fibers and yarns from 1-mm long carbon

nanotube forests synthesized by water-assisted chemical vapor deposition, Journal of Materials Science 43 (13) (2008) 4356–4362.

[49] Y. Inoue, K. Kakihata, Y. Hirono, T. Horie, A. Ishida, H. Mimura, One-step grown aligned bulk carbon nanotubes by chloride mediated chemical vapor deposition, Applied Physics Letters 92 (21) (2008) 213113.

[50] K. Jiang, J. Wang, Q. Li, L. Liu, C. Liu, S. Fan, Superaligned carbon nanotube arrays, films and yarns: a road to applications, Advanced Materials 23 (9) (2011) 1154–1161.

[51] C.P. Huynh, S.C. Hawkins, Understanding the synthesis of directly spinnable carbon nanotube forests, Carbon 48 (4) (2010) 1105–1115.

[52] Y. Zhang, G. Zou, S.K. Doorn, H. Htoon, L. Stan, M.E. Hawley, C.J. Sheehan, Y. Zhu, Q. Jia, Tailoring the morphology of carbon nanotube arrays: from spinnable forests to undulating foams, ACS Nano 3 (8) (2009) 2157–2162.

[53] A.A. Kuznetsov, A.F. Fonseca, R.H. Baughman, A.A. Zakhidov, Structural model for dry-drawing of sheets and yarns from carbon nanotube forests, ACS Nano 5 (2) (2011) 985–993.

[54] C. Zhu, C. Cheng, Y. He, L. Wang, T. Wong, K. Fung, N. Wang, A self-entanglement mechanism for continuous pulling of carbon nanotube yarns, Carbon 49 (15) (2011) 4996–5001.

[55] W. Liu, X. Zhang, G. Xu, P.D. Bradford, X. Wang, H. Zhao, Y. Zhang, Q. Jia, F.G. Yuan, Q. Li, Producing superior composites by winding carbon nanotubes onto a mandrel under a poly (vinyl alcohol) spray, Carbon 49 (14) (2011) 4786–4791.

[56] J.N. Coleman, M. Cadek, R. Blake, V. Nicolosi, K.P. Ryan, C. Belton, A. Fonseca, J.B. Nagy, Y.K. Gun'ko, W.J. Blau, High-performance nanotube-reinforced plastics: understanding the mechanism of strength increase, Advanced Functional Materials 14 (8) (2004) 791–798.

[57] B.S. Shim, J. Zhu, E. Jan, K. Critchley, S. Ho, P. Podsiadlo, K. Sun, N.A. Kotov, Multiparameter structural optimization of single-walled carbon nanotube composites: toward record strength, stiffness, and toughness, ACS Nano 3 (7) (2009) 1711–1722.

[58] M.S.P. Shaffer, A.H. Windle, Fabrication and characterization of carbon nanotube/poly (vinyl alcohol) composites, Advanced Materials 11 (11) (1999) 937–941.

[59] Y. Hou, J. Tang, H. Zhang, C. Qian, Y. Feng, J. Liu, Functionalized few-walled carbon nanotubes for mechanical reinforcement of polymeric composites, ACS Nano 3 (5) (2009) 1057–1062.

[60] J. Jia, J. Zhao, G. Xu, J. Di, Z. Yong, Y. Tao, C. Fang, Z. Zhang, X. Zhang, L. Zheng, A comparison of the mechanical properties of fibers spun from different carbon nanotubes, Carbon 49 (4) (2011) 1333–1339.

[61] K. Liu, Y. Sun, X. Lin, R. Zhou, J. Wang, S. Fan, K. Jiang, Scratch-resistant, highly conductive, and high-strength carbon nanotube-based composite yarns, ACS Nano 4 (10) (2010) 5827–5834.

[62] A.A. Mamedov, N.A. Kotov, M. Prato, D.M. Guldi, J.P. Wicksted, A. Hirsch, Molecular design of strong single-wall carbon nanotube/polyelectrolyte multilayer composites, Nature Materials 1 (3) (2002) 190–194.

[63] Y. Zhang, C.J. Sheehan, J. Zhai, G. Zou, H. Luo, J. Xiong, Y.T. Zhu, Q.X. Jia, Polymer-embedded carbon nanotube ribbons for stretchable conductors, Advanced Materials 22 (28) (2010) 3027–3031.

[64] R.D. Bradshaw, F.T. Fisher, L.C. Brinson, Fiber waviness in nanotube-reinforced polymer composites-II: modeling via numerical approximation of the dilute strain concentration tensor, Composites Science and Technology 63 (11) (2003) 1705–1722.

[65] F.T. Fisher, R.D. Bradshaw, L.C. Brinson, Fiber waviness in nanotube-reinforced polymer composites-1: modulus predictions using effective nanotube properties, Composites Science and Technology 63 (11) (2003) 1689–1703.

[66] F. Deng, Q.S. Zheng, L.F. Wang, C.W. Nan, Effects of anisotropy, aspect ratio, and nonstraightness of carbon nanotubes on thermal conductivity of carbon nanotube composites, Applied Physics Letters 90 (2) (2007) 021914.

[67] X. Wang, Z. Yong, Q. Li, P.D. Bradforda, W. Liu, D.S. Tucker, W. Cai, H. Wang, F.G. Yuan, Y.T. Zhu, Ultrastrong, stiff and multifunctional carbon nanotube composites, Materials Research Letters 1 (1) (2013) 19–25.

[68] C.T. Pan, H. Hocheng, Evaluation of anisotropic thermal conductivity for unidirectional FRP in laser machining, Composites Part A 32 (11) (2001) 1657–1667.

[69] K. Yang, M. Gu, Y. Guo, X. Pan, G. Mu, Effects of carbon nanotube functionalization on the mechanical and thermal properties of epoxy composites, Carbon 47 (7) (2009) 1723–1737.

[70] X. Wang, P.D. Bradford, W. Liu, H. Zhao, Y. Inoue, J.P. Maria, Q. Li, F.G. Yuan, Y. Zhu, Mechanical and electrical property improvement in CNT/Nylon composites through drawing and stretching, Composites Science and Technology 71 (14) (2011) 1677–1683.

[71] P.D. Bradford, X. Wang, H. Zhao, J. Maria, Q. Jia, Y.T. Zhu, A novel approach to fabricate high volume fraction nanocomposites with long aligned carbon nanotubes, Composites Science and Technology 70 (13) (2010) 1980–1985.

[72] J.G. Park, Q. Cheng, J. Lu, J. Bao, S. Li, Y. Tian, Z. Liang, C. Zhang, B. Wang, Thermal conductivity of MWCNT/epoxy composites: the effects of length, alignment and functionalization, Carbon 50 (6) (2012) 2083–2090.

[73] B. Pradhan, R.R. Kohlmeyer, J. Chen, Fabrication of in-plane aligned carbon nanotube–polymer composite thin films, Carbon 48 (1) (2010) 217–222.

[74] E.S. Choi, J.S. Brooks, D.L. Eaton, M.S. Al-Haik, M.Y. Hussaini, H. Garmestani, D. Li, K. Dahmen, Enhancement of thermal and electrical properties of carbon nanotube polymer composites by magnetic field processing, Journal of Applied Physics 94 (9) (2003) 6034–6039.

[75] H. Gommans, J. Alldredge, H. Tashiro, J. Park, J. Magnuson, A. Rinzler, Fibers of aligned single-walled carbon nanotubes: polarized Raman spectroscopy, Journal of Applied Physics 88 (5) (2000) 2509–2514.

[76] J.E. Fischer, W. Zhou, J. Vavro, M.C. Llaguno, C. Guthy, R. Haggenmueller, M.J. Casavant, D.E. Walters, R.E. Smalley, Magnetically aligned single wall carbon nanotube films: preferred orientation and anisotropic transport properties, Journal of Applied Physics 93 (4) (2003) 2157–2163.

[77] X. Wang, W. Liu, Q. Jiang, M.S. Harb, Q. Li, Y. Zhu, Multifunctional Nanoprepregs Based on Aligned Carbon Nanotube Sheets. MRS Proceedings, Cambridge Univ. Press, 2012.

[78] A. Ghemes, Y. Minami, J. Muramatsu, M. Okada, H. Mimura, Y. Inoue, Fabrication and mechanical properties of carbon nanotube yarns spun from ultra-long multi-walled carbon nanotube arrays, Carbon 50 (12) (2012) 4579–4587.

[79] W. Liu, H. Zhao, Y. Inoue, X. Wang, P.D. Bradford, H. Kim, Y. Qiu, Y. Zhu, Poly(vinyl alcohol) reinforced with large-diameter carbon nanotubes via spray winding, Composites Part A 43 (4) (2012) 587–592.

[80] V. Popov, V. Van Doren, M. Balkanski, Elastic properties of single-walled carbon nanotubes, Physical Review B 61 (4) (2000) 3078–3084.

[81] J.P. Salvetat, G.A.D. Briggs, J.M. Bonard, R.R. Bacsa, A.J. Kulik, T. Stöckli, N.A. Burnham, L. Forró, Elastic and shear moduli of single-walled carbon nanotube ropes, Physical Review Letters 82 (5) (1999) 944–947.

[82] S.S. Tallury, M.A. Pasquinelli, Molecular dynamics simulations of flexible polymer chains wrapping single-walled carbon nanotubes, Journal of Physical Chemistry B 114 (12) (2010) 4122–4129.

[83] S.S. Tallury, M.A. Pasquinelli, Molecular dynamics simulations of polymers with stiff backbones interacting with single-walled carbon nanotubes, Journal of Physical Chemistry B 114 (29) (2010) 9349–9355.

[84] A. Nish, J.Y. Hwang, J. Doig, R.J. Nicholas, Highly selective dispersion of single-walled carbon nanotubes using aromatic polymers, Nature Nanotechnology 2 (10) (2007) 640–646.

[85] D.W. Steuerman, A. Star, R. Narizzano, H. Choi, R.S. Ries, C. Nicolini, J.F. Stoddart, J.R. Heath, Interactions between conjugated polymers and single-walled carbon nanotubes, Journal of Physical Chemistry B 106 (12) (2002) 3124–3130.

[86] W.Z. Yuan, J.Z. Sun, Y. Dong, M. Häussler, F. Yang, H.P. Xu, A. Qin, J.W.Y. Lam, Q. Zheng, B.Z. Tang, Wrapping carbon nanotubes in pyrene-containing poly(phenylacetylene) chains: solubility, stability, light emission, and surface photovoltaic properties, Macromolecules 39 (23) (2006) 8011–8020.

CHAPTER 24

Embedded Carbon Nanotube Sensor Thread for Structural Health Monitoring and Strain Sensing of Composite Materials

Adam Hehr, Yi Song, Bolaji Suberu, Joe Sullivan, Vesselin Shanov, Mark Schulz

University of Cincinnati, College of Engineering and Applied Science, Cincinnati, Ohio USA

CHAPTER OUTLINE

24.1	Introduction	671
24.2	Embedded Sensing Proof of Concept	676
24.3	CNT Sensor Thread Performance	680
	24.3.1 Sensitivity (gauge factor)	682
	24.3.2 Hysteresis	688
	24.3.3 Consistency	691
	24.3.4 Stability	692
	24.3.5 Bandwidth	695
24.4	Carbon Nanotube Thread SHM Architectures	699
24.5	Areas of Strong Multifunctional Potential	704
24.6	Future Work	708
	Acknowledgments	709
	References	709

24.1 INTRODUCTION

The need for speed—i.e. higher performance and efficiency—is driving the increased use of advanced composite materials in high-cost defense and commercial applications (see Fig. 24.1). These materials provide high specific strength and high performance, but ensuring material safety and reliability is a challenge. Currently, over 80% of structural commercial aircraft inspections are done visually [1] (Federal Aviation Administration of the USA, 1997), and visual inspection continues to be the leading technique for composite aircraft [2]. However, these external visual

672 CHAPTER 24 Embedded Carbon Nanotube Sensor Thread

FIGURE 24.1

SHM technology may increase aircraft reliability, safety, and efficiency and reduce operating costs of aircraft. (For color version of this figure, the reader is referred to the online version of this book.)

Sources: F-22 photo taken by Rob Shenk [4], Apache photo courtesy of U.S. Army and taken by Tech. Sgt. Andy Dunaway [5], 787 photo taken by wiki user, Spaceaero2 [6].

inspections do not work reliably well with laminated composite materials due to many failure modes manifesting and progressing from within the material, such as matrix cracking and ply delamination [3].

An example of unseen ply delamination in a laminated composite from an impact can be seen in Fig. 24.2. It should be pointed out that the external indent is very small compared to the internal damage area and that there is no obvious evidence of this internal damage from the exterior. As a consequence of poor failure mode detection

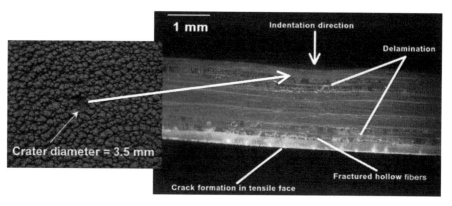

FIGURE 24.2

Ply delamination from an external impact. The composite in the figure would have a similar small indent where there is no obvious evidence of internal damage from the exterior.

Sources: These photos taken and combined from [7] and [8] with permission. Exterior impact photo reprinted from Springer and JOM's The development of multifunctional composite material for use in human space exploration beyond low-earth orbit *is given to the publication in which the material was originally published with kind permission from Springer Science and Business Media. Composite figure reprinted from* A hollow fibre reinforced polymer composite encompassing self-healing and enhanced damage visibility *with permission from Elsevier.*

24.1 Introduction 673

in these materials, material reliability has limited or slowed use in certain applications, such as in commercial air flight, where reliability and safety are of utmost importance.

In order to inspect composite materials in a more effective manner, costly nondestructive techniques such as ultrasonic inspection and somewhat qualitative tap tests have been employed [3]. Besides being costly in capital and accuracy, these techniques are time consuming and the aircraft must be taken out of service for inspection. As a result, planes will be inspected less frequently and thorough inspections can only take place when the plane is out of commission for an extended period. However, by building sensors into the composite material, assessment can be done in real time, can be more accurate, and would be less costly in time, labor, and operator/passenger safety.

The idea of building damage sensing and detection into composite materials is not a new one. Researchers began building traditional strain gauges into composites several decades ago; however, these embedded strain sensors create large stress concentrations in the material, which degrades the fatigue strength and longevity of the material [9]. Additionally, many of the embedded sensor schemes were not dense enough to monitor critical complex geometries like in the F-22 Raptor wings (Fig. 24.3), and the sensor's design life is much shorter than the structure that is being monitored [10]. To mitigate these large stress concentrations, increased sensor density, and increased sensor longevity, small robust continuous sensors have begun to attract attention. Currently, the leading technology that meets these characteristics of small, robust, and easily

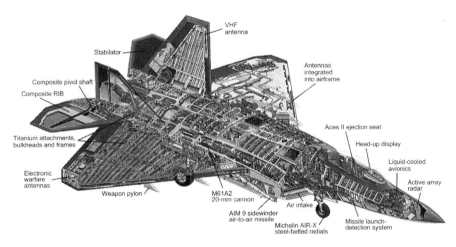

FIGURE 24.3

Internal rib structure of F-22 Raptor. Monitoring the composite ribs in the wings and external composite panels becomes difficult with their complex geometries. (For color version of this figure, the reader is referred to the online version of this book.)

Source: This image is approved by US DoD for use.

distributed is fiber-optic cabling. However, fiber-optic cabling interrogation equipment is currently very expensive, i.e. on the order of 20,000 USD per interrogator. Thus, fiber-optic cabling only becomes feasible if the sensing system is large [10]. Despite this high cost, fiber-optic cable is extremely robust to environmental conditions, can undergo frequent loading, and is immune to electromagnetic radiation [10]. However, this high cost of operation has led researchers to explore other small, robust, and fiber/tape-like options, such as carbon nanotube (CNT) film/sheet [11], fiber [12], thread [13–15] and the carbon fiber itself [16]. In the past 4 years, the University of Cincinnati (UC) Nanoworld Laboratories has been investigating miniature sensors using CNT thread and tape (tape is flat adhesive-backed CNT sheet or thread). As a result, the sensing element is much smaller than conventional sensors [17–29], would be minimally invasive if built into the material, can be used in complex geometries, would withstand the life of the structure, and could improve the material besides damage sensing.

CNT thread spun from a CNT forest has a piezoimpedance property (electrical impedance based on R, L, C properties changes with strain/stress). Thus, CNT thread can be used to measure internal material strain along with detecting damage in composite materials. This change in impedance comes from the lateral overlap and end gaps between the individual nanotubes in the thread due to strain. As a result, the resistance increases with strain and if the thread breaks due to damage in the composite, the impedance becomes large. The first use of CNT sensor thread to detect damage in composite materials was performed at the UC [14,30,31] by Dr Jandro Abot working with the Nanoworld Group at UC. Some of the CNT materials produced in Nanoworld Laboratories can be seen in Fig. 24.4. It should be noted that CNT sheet has similar properties, but its use as a structural health monitoring (SHM) sensor will not be discussed in detail in this chapter. Instead, the primary focus will be on CNT thread.

As mentioned before, these embedded CNT materials have the potential to improve standard composite materials in many ways other than self-sensing, often called multifunctional features. The following are the potential multifunctional features if CNT thread is added to traditional polymeric composite materials:

1. *Strain sensing:* CNT thread is piezoresistive so it can be used to measure strain besides detecting damage. CNT thread also has capacitance, so its impedance decreases significantly (less than copper wire) at high frequencies, which, in turn could allow increased sensitivity to damage/strain [32].
2. *Thermal sensing:* CNT thread can be used as a temperature sensor [13].
3. *Moisture sensing:* CNTs have been used previously as capacitive moisture sensors [33], which would provide valuable information about added water weight and icing risk to the composite.
4. *Oxidation sensing:* Carbon materials easily oxidize or degrade at high temperatures [34], therefore, these materials are limited in their application.

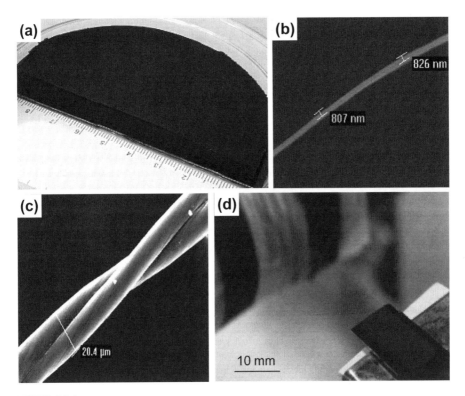

FIGURE 24.4

CNT materials: (a) vertically aligned CNT array, (b) nanothread produced under the microscope, (c) two-ply yarn, and (d) sheet produced by the University of Cincinnati. These materials can be pure CNT or infiltrated with polymer. (For color version of this figure, the reader is referred to the online version of this book.)

Yet, by utilizing the electrochemical properties of CNTs, corrosion or oxidation could be measured and tracked similar to metals by measuring the electrical resistivity [35].

5. *Increased strength and stiffness:* Composite strength will increase based on the volume fraction of CNT thread in comparison to epoxy [36].
6. *Increased damping:* CNT threads and materials exhibit damping on the same or larger order of magnitude of epoxy damping. Thus, CNT integration will enhance or maintain current material damping. Others have used CNTs to improve damping in composite materials already [37,38].
7. *Damage limiting:* Carbon fibers break at small strain ($\sim 1\%$) [39]. CNT thread has higher strain to failure ($\sim 3-5\%$) and will self-limit damage by absorbing strain energy, which holds the composite together and makes it resilient. Others have proven composite toughening with CNTs [40,41].

8. *Improved transport:* CNT thread has high thermal [42] and electrical conductivity in plane, which will moderately improve the properties of the composite.
9. *No significant added weight or size:* There is almost no added mass or volume of the sensor thread and simple lightweight instrumentation is used to interrogate/measure the sensor material.
10. *Other multifunctional areas:* Electrical grounding [40], Electro-magnetic interference (EMI) shielding [43], hydrophobic material [44], energy harvesting [45,46], and self-healing [47].

Enhancements of strength, stiffness toughness, and damping will be made by replacing unneeded epoxy between neighboring fibers with small-diameter CNT thread. This idea is called *High Volume Fraction of CNT Thread* and an invention disclosure was filed for an approach to integrate CNT thread into carbon fibers [36].

The remainder of this chapter focuses on work done with CNT threads for composite material SHM for embedded sensing proof-of-concept work, an assessment of CNT thread strain sensor performance, potential designs for using embedded CNT thread on current aircraft, some explored multifunctional areas, and future work.

24.2 EMBEDDED SENSING PROOF OF CONCEPT

The UC began using CNT thread as a damage and strain sensing element in the 2008–2010 time frame. Several proof-of-concept experiments were done with the CNT thread, and one of the first experiments was embedding the thread into an epoxy beam, which was strained in a four-point bending test rig [48]. This epoxy beam and embedded thread can be seen in Fig. 24.5 along with the schematic of the test in Fig. 24.6.

A four-point bending test was done over a three-point bending test because the strain field should theoretically be constant between each load point. Also, this test assumed that the thread lies in the same plane with respect to the neutral axis. This measurement approach provided a basis for calculating a reasonable estimate of the gauge factor for the embedded thread. In this test, strain and stress within the beam were estimated from Bernoulli beam theory.

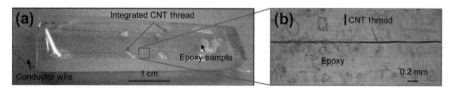

FIGURE 24.5

Epoxy beam with embedded carbon nanotube: (a) thread in epoxy beam; (b) zoomed-in region from (a) showing thread. (For color version of this figure, the reader is referred to the online version of this book.)

Source: From [48] with permission from the publisher.

24.2 Embedded Sensing Proof of Concept

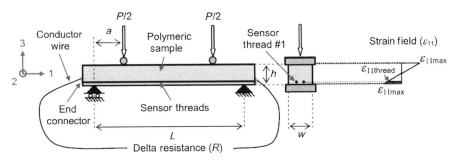

FIGURE 24.6

Schematic of epoxy brick with beam loading conditions, i.e. a four-point bending test. This loading condition creates a constant strain condition between load points. (For color version of this figure, the reader is referred to the online version of this book.)

Source: From [48] with permission from the publisher.

During the experiment, the epoxy was loaded up to about 1% strain at the top surface of the beam to avoid cracking and early failure. However, after several loading cycles, the sample did begin to crack/break. Nonetheless, this test provided a baseline estimate of resistance response of CNT thread to strain while studying consecutive loading cycles of strain. The resistance response from strain can be seen in Fig. 24.7 for a single cycle and for six cycles. Note that the resistance response tracks the stress response fairly well in shape and trends.

The gauge factor can be estimated from this test setup. Thus, the calculated gauge factors for the loading cycles seen above can be seen below in Fig. 24.8. It appeared that the thread consistently had two zones of sensitivity and that the second zone was fairly consistent in response. This experiment estimated a gauge factor of

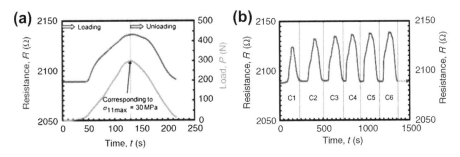

FIGURE 24.7

Resistance response of embedded CNT thread to an applied stress: (a) single cycle and (b) several strain cycles. (For color version of this figure, the reader is referred to the online version of this book.)

Source: From [48] with permission from the publisher.

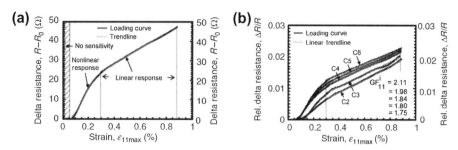

FIGURE 24.8

Analysis of embedded thread resistance response to strain. (a) Sensitivity or gauge factor for a single loading cycle and (b) consistency of the gauge factor to multiple loading cycles. (For color version of this figure, the reader is referred to the online version of this book.)

Source: From [48] with permission from the publisher.

around 1.8–2, which is on the order of foil-type strain gauge performance. However, recent result of thread, which is not embedded in any composite or epoxy, has a far smaller response, as will be seen later in the chapter. This difference between experiments is not well understood and may be due to the effect of epoxy reducing the strain of the thread and is still being investigated.

In addition to embedding CNT thread into epoxy beams, CNT thread sensors were also embedded into composite materials themselves to measure the onset of damage, specifically mode I (crack opening) and II (in-plane shear/sliding) delamination [49]. In order to initiate and measure mode II delamination, a multi-ply composite beam was fabricated with a CNT thread stitched transversely through the ply layers and down the beam, as seen in Fig. 24.9. This composite beam was then loaded to failure in a three-point bending test fixture while the resistance response of the CNT thread was measured. The resistance response from this mode II delamination initiation test can be seen in Fig. 24.9, and it should be highlighted that the thread breakage is directly correlated to the onset of delamination. This test was done with both E-glass fiber composites and IM7/977-3 carbon fiber composites, and both materials had similar CNT thread resistance responses. However, it should be noted that the CNT thread sensors in the carbon fiber composite were coated with a polyurethane coating prior to stitching. This coating was used for electrical isolation of the CNT thread from the carbon fibers, but the interface between the fiber and coating was not well understood prior to the experiment. Consequently, confidence in electrical isolation was low in the carbon fiber composite test.

Mode I delamination was also done on a composite beam with transversely stitched CNT threads. To simulate crack opening, the composite beam's plies were pulled apart; specifically, a crack was built into the beam near the center and a load was applied on the edge of the material to force a crack to propagate along the ply layer interface. While this crack propagated along the interface,

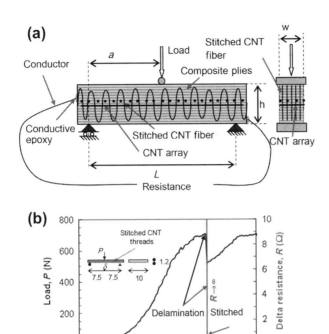

FIGURE 24.9

Three-point bend test for detecting in-plane damage/delamination in laminated composites. (a) Schematic of longitudinal and transverse cross-sections of a laminated composite beam sample instrumented with in-plane sensor arrays and transversely stitched sensor tapes. When the tapes are broken, their resistance increases. (b) Load vs deflection and thread delta resistance (difference between actual and initial resistance) vs deflection curves for an IM7/977-3 10-ply unidirectional composite subjected to three-point bending. It can be seen that one of the sensor threads captures the delamination exactly at its onset since the discontinuity in both curves occurs simultaneously. (For color version of this figure, the reader is referred to the online version of this book.)

Source: Reprinted from Delamination with carbon nanotube thread in self-sensing composite materials with permission from Elsevier.

the resistance responses of the embedded threads were monitored to identify crack location. Besides monitoring the CNT threads, a reference foil-type strain gauge was also mounted to the plies as a reference for comparison. The resistance response from these embedded threads and reference strain gauge can be seen in Fig. 24.10 along with a schematic of the experimental design. As seen from the resistance response of the embedded threads, each thread identifies, in a separable manner, when the crack reaches and passes through the material.

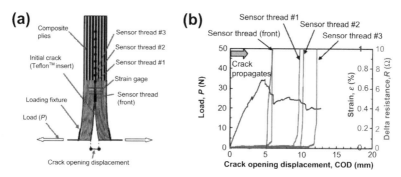

FIGURE 24.10

Detection of mode I delamination in laminated composite materials. (a) Schematic of double-cantilevered beam with a precrack between the central plies. The beam is mounted with a strain gauge to measure crack growth while the plies are pulled apart. (b) Resistance response of the strain gauge and embedded CNT thread from the induced crack along with the applied crack opening load. Note that the foil strain gauge and the closest CNT thread sensor to the strain gauge identify the crack's propagation within a close proximity. Additionally, all the subsequent CNT thread sensors identify when the crack reaches that part of the composite. (For color version of this figure, the reader is referred to the online version of this book.)

Source: Reprinted from Delamination with carbon nanotube thread in self-sensing composite materials with permission from Elsevier.

In addition to showing potential as a piezoresistive strain and damage sensor, the CNT thread can also act as an infrared (IR) damage sensor [50]. If a polymer is reinforced with CNT thread, and the thread is powered by a DC or an AC voltage, it will start Joule heating. This heat can then be measured with an IR camera to see if the thread is intact. If no heat signature is detected, then the thread has broken internal to the structure, which may imply damage. This concept is demonstrated in Fig. 24.11 for two embedded parallel threads. One thread is hooked up to a DC voltage, while the other has no electrical load applied. This approach could act as a stand-alone sensing method or a parallel method to confirm piezoresistive measurements. The schematic in Fig. 24.11 illustrates how this technique could be scaled up to an actual composite in use. It should also be noted that the temperature can remain below 30 °C for measurement purposes, which is safe for composite materials.

24.3 CNT SENSOR THREAD PERFORMANCE

Since Nanoworld Laboratories began sensing strain and damage in polymeric composites with CNT thread, a few other researchers have begun to perform work along the same lines [13,15]. However, no CNT thread strain and damage sensing literature today rigorously examines "how good" the CNT thread strain sensor is

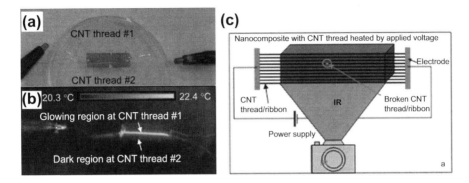

FIGURE 24.11

Remote damage detection with IR thermography. (a) Ambient light photo of two embedded CNT threads with one thread powered with a DC signal; (b) IR image of the same embedded threads, yet it is obvious only one thread is still "intact"; and (c) schematic of SHM of a polymer nanocomposite instrumented with CNT thread using the IR camera for ease in inspection. (For color version of this figure, the reader is referred to the online version of this book.)

Source: From [49] and [50] with permission from the publisher.

performancewise. Consequently, recent efforts at Nanoworld Laboratories have begun to examine the performance of these sensors when they are embedded and not embedded in a polymeric material. The specific areas of performance, which would typically be found on a strain sensor calibration sheet, are listed and defined below.

- Sensitivity: Due to the parameter of strain, sensitivity is essentially the gauge factor for piezoresistive strain sensors.
- Hysteresis: Hysteresis is when there is a dynamic or changing lag or lead in a response from the input. Consequently, things do not behave in a linear manner. Typically, a good sensor is one that demonstrates little to no hysteresis.
- Consistency: Consistency or repeatability here implies accurately measuring the same induced strain consecutively. This can be done by examining what the resistance of the strain gauge is at a particular strain value or by measuring the net change in resistance per strain cycle. In other words, consistency can be analyzed by studying the consistency of the gauge factor for several strain cycles. This gauge factor analysis was done to separate out stability problems, as will be seen and discussed later.
- Stability: As mentioned in the above bullet, stability is an area that requires much attention itself. Stability here implies measurement of drift and variance in the measurement at a particular strain vs time.
- Bandwidth: Bandwidth here implies the maximum (or minimum) frequency of strain that can be confidently measured by the sensor.

Currently, these performance areas have been explored when the CNT thread has not been embedded into any polymeric material. This was primarily done as a benchmark for comparison when the thread was embedded into a polymeric material or any other material for that matter. As a result, all the following performance data is exclusively for nonembedded CNT thread. However, it should be noted that embedded experiments are still in the works and will be carried out. The following subsections will discuss each of these areas in more detail.

24.3.1 Sensitivity (gauge factor)

As mentioned in the above section, sensitivity here implies the gauge factor due to the sensor being used to measure strain. Gauge factor is defined as shown in Eqn (24.1). Thus, the gauge factor can be found by rearranging Eqn (24.1) into a linear equation and fitting a line against the normalized change in resistance ($\Delta R/R_0$) and strain (ε). Consequently, a confident measurement of strain, resistance, and change in resistance needs to be made in order to obtain an accurate gauge factor estimate.

$$\text{Gauge factor (GF)} = \frac{\Delta R}{R_0}\frac{1}{\varepsilon} \qquad (24.1)$$

In order to confidently measure the normalized change in resistance and strain, a custom-designed tensile test coupon combined with a custom-designed Wheatstone bridge circuit were utilized. This custom-designed coupon with detailed components can be seen in Fig. 24.12. This coupon was designed within the limits

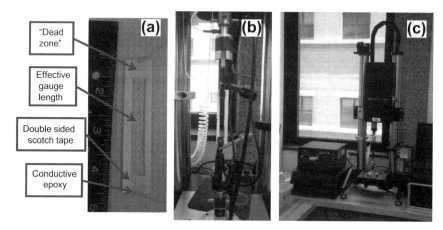

FIGURE 24.12

Sensitivity analysis approach. (a) Sensing sample for making resistance measurements, (b) sensing sample loaded into the grips of the Instron 5948 tensile testing machine, and (c) the Instron 5948 used in these experiments located at Nanoworld Laboratories. (For color version of this figure, the reader is referred to the online version of this book.)

of a high-precision tensile tester at Nanoworld Laboratories, the Instron 5948 (see Fig. 24.12), to maximize the effective gauge length compared to the length of the thread under the machine's grip, called the dead zone in the figure. This approach was taken because (1) strain is calculated by utilizing the displacement output of the moving head of the tensile testing machine and (2) this longer length of thread allowed more thread to be strained, which, in turn, allowed more thread to undergo a change in resistance. In other words, a more confident gauge factor can be measured with a longer effective gauge length. However, from current coupon geometries, an underestimate in the gauge factor of nearly 50% is possible. However, it is assumed that this estimate error is much lower due to the fact that the thread does not strain far into the grip. It should also be noted that the thread is insignificantly stiff relative to the test machine fixture, thus minimal deformation takes place within the gripping structure of the tensile testing machine. Instead, most deformation will take place within the thread. It should also be noted that this strain estimate technique is utilized because using a noncontact video extensiometer to noninvasively measure the thread's strain would be very difficult. To ensure that these assumptions were indeed valid, and they are for the most part, a benchmark gauge factor measurement was done with small-diameter copper wire. Despite the potential for error, this strain measurement approach provides a ballpark estimate in order to draw correlations and perform analysis.

In order to confidently measure the change in resistance, a Wheatstone bridge circuit was utilized as seen in Fig. 24.13. This approach confidently measures the change in resistance because the measurement is a null measurement, or a measurement referenced to zero. Consequently, it is easier to detect small perturbations because the measurement essentially has zero mean. This resistance measurement approach is not as precise as a four-wire measurement, yet the signal to noise ratio is much larger, which, in turn, allows a more accurate resistance change

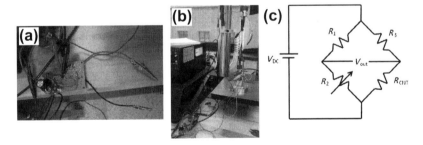

FIGURE 24.13

Wheatstone bridge circuit for resistance measurement. (a) Custom-built Wheatstone bridge circuit for balancing CNT thread against known resistor behavior, (b) custom circuit hooked up to strained specimen in Instron 5948, and (c) schematic of Wheatstone bridge circuit with R_{CNT} representing the CNT thread. This circuit is a null measurement approach to measuring resistance. (For color version of this figure, the reader is referred to the online version of this book.)

measurement to be made [51]. A custom circuit was built with interchangeable balance resistors because (1) CNT thread resistance is not very consistent from thread to thread and this circuit was to be used on more than one thread and (2) a CNT thread can be used as a balance resistor to separate out thermal fluctuations in the testing room. More details on thermal fluctuations and using a CNT thread as a balance resistor will be discussed in the Section 24.3.4. It should be noted, however, that a four-wire measurement approach was done initially with a high-precision data acquisition (DAQ) device, and many of the results presented here were made with a four-wire measurement. This measurement approach was able to fully capture the resistance change due to the high resolution of the DAQ, but this technique does not provide temperature compensation. Thus, short test time measurements could be easily and confidently made with a four-wire technique, such as estimating the gauge factor.

As mentioned before, to validate the assumptions of the tensile test coupon used to estimate the gauge factor, a 90-µm-diameter copper wire's gauge factor was estimated using the same tensile test coupon design as seen in Fig. 24.12. The typical response of copper wire during these tests can be seen in Fig. 24.14. As seen in the figure, the copper wire's gauge factor is correlated with the elastic or linear range of the material. Four samples of copper were used to obtain a mean gauge factor estimate of 2.2 with a standard deviation of 0.66. It is hypothesized that the error here was large because the wire's resistance is incredibly small ($\approx 0.5\ \Omega$) and the change is even smaller. Thus, a precise estimate is difficult to make. Nonetheless, an average of 2.2 is within reasonable bounds of the theoretical response of around 2.1, so it appeared that the tensile test coupon had the potential to obtain a reasonable gauge factor estimate. This theoretical gauge factor response for a cylindrical homogenous material can be seen below in Eqn (24.2), and this equation is animated in Fig. 24.15 to obtain this theoretical value. It should be highlighted that a value of close to 2 was expected because most commercial foil-type strain gauges are made of copper alloys, which have gauge factors on the order of 2.

$$GF_l = 1 + 2\gamma + \pi_l E \cdot (1 - \gamma) \qquad (24.2)$$

where
- γ is the Poisson ratio
- E is the modulus of elasticity
- π is the piezoresistivity constant

The gauge factor for as-spun CNT thread can comparatively be seen in Fig. 24.16. From the figure, two linear zones can be seen that appear to be correlated to the knee in the stress—strain curve of the thread. It should also be highlighted that the elastic region for the stress—strain curve and resistance—strain of the thread is around 1% strain, which is much larger than the elastic region of standard engineering materials and around the breaking strain of many polymeric composites. Thus, this sensor has the potential to measure strains close to material failure.

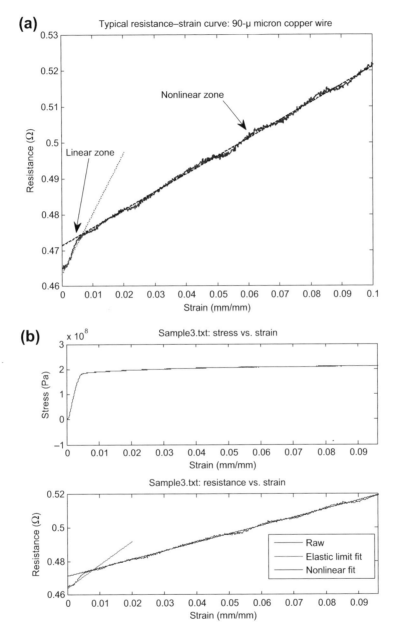

FIGURE 24.14

The resistance response from strain of a 90-μm copper wire. (a) Representative resistance response of copper wire while strained to failure and (b) correlation of linear zone on stress–strain plot with the linear piezoresistive response of the copper. (For color version of this figure, the reader is referred to the online version of this book.)

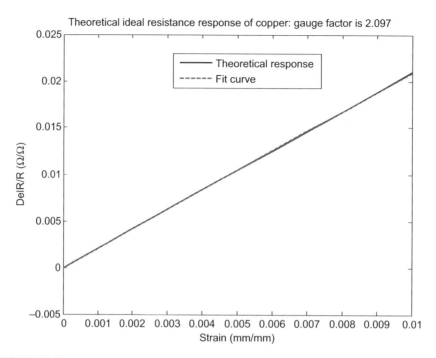

FIGURE 24.15

Theoretical response of 90-μm copper wire to applied strain, assuming that the material remains linear up to 1% strain. A line fit was done because the theoretical relationship is slightly nonlinear. (For color version of this figure, the reader is referred to the online version of this book.)

Although as-spun CNT thread has a large linear range, the gauge factor is typically around 0.2–0.4. This is roughly 10–20% of the sensitivity of copper. It should also be pointed out that CNT thread does not obey the piezoresistive model for a homogenous cylinder as seen in Eqn (24.2). Instead, a given CNT thread is made up of thousands of individual tubes. Thus the piezoresistive effect is suspected to be dominated by these interfaces between the CNTs as seen in CNT film [52]. It is suspected that these interfaces dominate the sensitivity because high Poisson ratios have been reported in CNT thread [53] combined with the gauge factor being less than 1. In other words, CNT thread does not agree with Eqn (24.2). Thus, it could be possible that the piezoresistive coefficient for the radial and lengthwise directions may be counteracting each other. More investigation should be done to prove this hypothesis and to improve fundamental understanding behind this response behavior.

Despite an incomplete understanding of the physical mechanisms behind CNT thread's piezoresistive response, the gauge factor was investigated as a function of the thread's geometry, i.e. the twist angle. This was done to attempt to improve

24.3 CNT Sensor Thread Performance 687

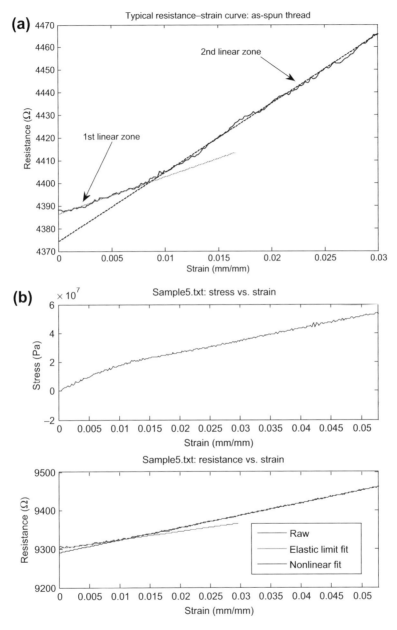

FIGURE 24.16

Resistance response of CNT thread from strain. (a) Illustration of typical resistance response of CNT thread to applied strain and (b) resistance response with synchronized stress–strain curve to illustrate the correlation between the linear sensing zones and linear stress–strain zones of the thread. (For color version of this figure, the reader is referred to the online version of this book.)

on the thread's sensitivity. In order to examine the gauge factor as a function of twist angle, an analysis of variance (ANOVA) was performed on several threads of various twist angles with similar diameters as seen in Fig. 24.17. From the figure, it can be seen that lower twist angles have higher sensitivities, and it is believed that this higher sensitivity comes from the thread's overall resistivity being lower at lower twist angles (see Fig. 24.17). This lower resistivity in turn lowers the resistance, which then increases the gauge factor as defined in Eqn (24.1). Finally, it should be noted that this ANOVA analysis for resistivity vs twist angle was performed on the same thread's used for the ANOVA analysis for gauge factor vs twist angle [54]. It is believed that this lower resistivity at lower twist angles predominately comes from less tube interfacial interaction due to the higher tube alignment against the primary axis at lower twist angles. In summary, it appears that the sensitivity of the CNT thread can be enhanced by lowering the overall resistance of the thread. Thus, it is hypothesized that postspinning densification and annealing may also increase the gauge factor by lowering the overall resistivity of the thread.

24.3.2 Hysteresis

The next sensor performance area examined was hysteresis. Like the abovementioned sensitivity experiments, the Instron 5948 was utilized to provide a low-frequency triangle wave to the thread for some specified amount of cycles. For this particular analysis, a 0.1-Hz signal with just less than 1% strain amplitude was applied to an as-spun thread for 100 cycles. Typical time history responses for this test along with hysteresis plots for stress—strain and resistance—strain can be seen in Fig. 24.18. From the figures, it can be seen that the stress—strain curve demonstrates consistent hysteresis, yet the resistance—strain curve does not appear to have any hysteresis characteristics. Instead, the resistance time history has a significant amount of drift, which will be discussed later in Section 24.3.4. This observed hysteresis behavior in the stress—strain curve is damping, and more analysis on this topic can be seen later in the Section 24.5.

From these 100 cycles, a representative single resistance—strain cycle from a particular test can be seen in Fig. 24.19 along with some statistics for the given test. As mentioned earlier, the thread went through 100 strain cycles. Each of these cycles was examined individually. Thus 100 averages were done for the analysis below.

From the figure, it can be seen that the loading and unloading curves generally follow the same pathway with minimal disagreement. However, this disagreement was quantified by integrating each loading and unloading curve at zero mean and then subtracting the instantaneous difference. The average and standard deviation of this loading, unloading, and difference integration can be seen in the Figure 24.19b. As seen in the table, the loading and unloading integrations were very close in magnitude and error, but this error was fairly high. The difference in loading was small, about 2—4% of the loading and unloading integral. Yet, the variance in this calculation was also high. Consequently, it appears that the CNT thread has the potential to have very low hysteresis (2—4%), but the high variance must be

24.3 CNT Sensor Thread Performance 689

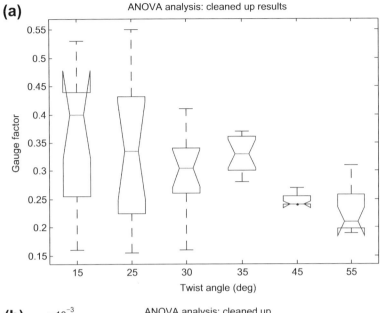

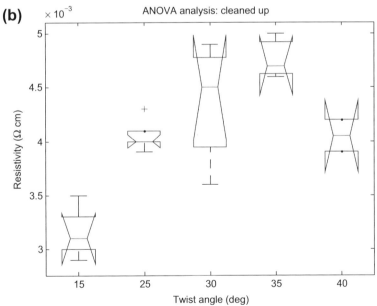

FIGURE 24.17

ANOVA analysis for CNT thread gauge factor optimization with twist angle. (a) ANOVA for CNT thread gauge factor as a function of twist angle. (b) Similar ANOVA was done for CNT thread resistivity as a function of twist angle. Note how the lower resistivity at lower twist angles contributes to a larger gauge factor at lower twist angles. These two analyses were done with 11 CNT threads of various twist angles made in Nanoworld Laboratories [54]. (For color version of this figure, the reader is referred to the online version of this book.)

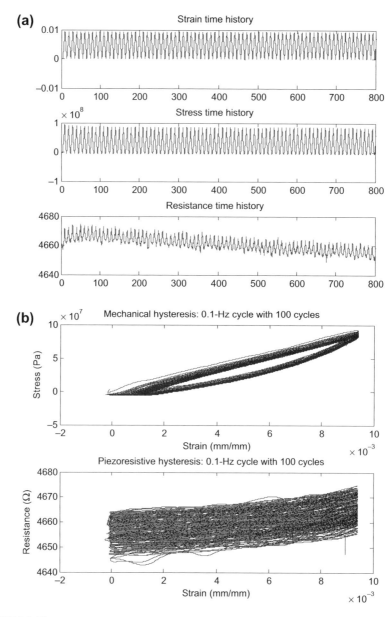

FIGURE 24.18

Hysteresis analysis of CNT thread. (a) Time histories of strain, stress, and resistance of the CNT thread during a typical hysteresis test and (b) time histories of stress and resistance plotted against strain to show consistency and correlations with consecutive strain cycles. Note that the hysteresis loop seen in the mechanical hysteresis plot implies damping. This will be discussed in more detail later. (For color version of this figure, the reader is referred to the online version of this book.)

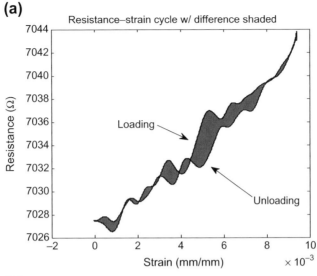

FIGURE 24.19

Hysteresis behavior with statistics. (a) A single resistance–strain cycle for a given hysteresis test. As seen in the plot, the loading and unloading curves generally follow the same given path. (b) Statistics for 100 cycles, or 100 averages. The statistics represent how close the loading and unloading curves are by integrating under the curve without any DC offset (starting value of zero). The two curves are very similar in their average area, showing low hysteresis. (For color version of this figure, the reader is referred to the online version of this book.)

addressed. It is believed that this high variance is predominately from a poor signal to noise ratio (this measurement was done with four-wire approach) combined with temperature variations in the room. More will be discussed on these temperature variations in the Section 24.3.4.

24.3.3 Consistency

Consistency of the CNT thread strain sensor was analyzed by examining the consistency of the gauge factor of the thread after consecutive cycles. Thus, the gauge

Table 24.1 Gauge Factor Consistency

Gauge Factor Consistency			
	GF Loading	GF Unloading	GF Cycle
Average	0.226	0.219	0.222
Standard Deviation	0.0337	0.0373	0.0258
% Standard Deviation	14.91	17.03	11.62

This table shows how close 100 loading, unloading, and cycle point clouds are in their linear fit for the gauge factor. All the line fits are very close, which implies that the thread consistently measures the same change in resistance.

factor was calculated for each cycle of hysteresis curve above. In order to see how consistent the sensor was in loading, unloading, and the full cycle, a line was fit to each of these datasets. In other words, the loading and unloading curves seen in Fig. 24.19 each had a line fit to their trend. Then, a line was fit to the entire cycle (both curves' data points). The results for these line fits are tabulated in Table 24.1, and it should be noted that all line fits are very consistent and that the error is not excessively high. However, similar to the hysteresis analysis, this error is believed to be caused by a poor signal to noise ratio along with slight changes in room temperature. Nonetheless, it appears that CNT thread is indeed consistent in its sensitivity.

24.3.4 Stability

As mentioned earlier, drift in the CNT thread resistance measurement is fairly significant. To quantify the severity of this drift, the resistance values at 0%, 0.47%, and 0.94% strain were analyzed for a typical hysteresis test. Figure 24.20 below illustrates the points analyzed on the hysteresis curve with corresponding resistance values in (b). From the table, a net change of around 13 Ω is consistently measured, but the standard deviation is near 3 Ω. Additionally, this test takes about 15 min to complete, thus resistance values were not measured for a long period. As a result, two resistance measurements of the CNT thread at zero strain over a long period were made. This measurement was made with the bridge circuit seen earlier in Fig. 24.13 with the CNT thread balanced against standard resistors. It was found that the resistance did not stay very stable, in fact the standard deviation increased with time. These measurements can be seen in Fig. 24.21 with the blue and red curves.

These large variations in the resistance values seen in Fig. 24.21 are suspected to come from temperature variations in the testing room. From the two curves, sharp responses can be observed that may correlate to when the air-conditioning was turned on in the laboratory. As a consequence of the thread appearing to be

24.3 CNT Sensor Thread Performance

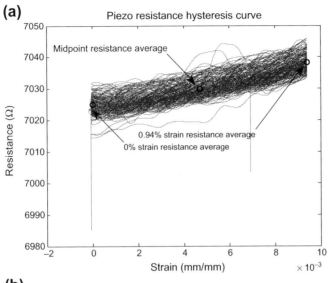

FIGURE 24.20

Resistance stability during strain cycles with statistics. (a) Illustration of CNT thread resistance response drift seen during a typical hysteresis test along with points of concern for analysis. These points are 0%, 0.47%, and 0.94% strain. (b) Statistics for a 100 cycle hysteresis test can be seen in the table. As seen from the table, the standard deviation at a particular strain is low relative to the average, yet the net change is not large enough to say that this deviation is insignificant. (For color version of this figure, the reader is referred to the online version of this book.)

extremely sensitive to temperature, another resistance measurement was made, yet with the CNT thread balanced against another CNT thread to separate out thermal effects. This measurement can also be seen in Fig. 24.21 as the black curve. This approach works because each thread should undergo a very similar change in resistance from temperature, which, in turn, would make the resistance measurement much more stable.

As a consequence of CNT thread being extremely susceptible to temperature variations, a reference thread may need to be embedded into the composite material, which does not undergo strain. This architecture would be difficult to accomplish, but may be possible. However, it should also be noted that once the CNT thread

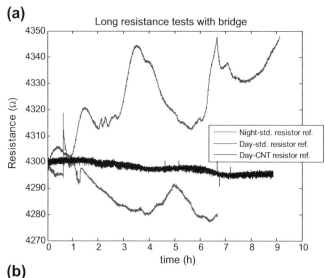

FIGURE 24.21

Resistance stability under no strain for long periods with statistics. (a) CNT thread resistance from a Wheatstone bridge circuit for a long time. The blue and red curves (standard ceramic resistance balance resistor) tend to drift and undergo low-frequency oscillations due to suspected temperature changes in the room (air-conditioned system). The black curve on the other hand does not undergo these dramatic drifts, but remains fairly stable due to the CNT thread being balanced another CNT thread that undergoes similar thermal effects. (b) Key comparison statistics from these three curves. (For interpretation of the references to color in this figure legend, the reader is referred to the online version of this book.)

is embedded into the composite material, the thermal mass of the material will be much larger, which will minimize temperature variations that the thread would experience compared to the thread in open air. Thus, embedded thread may be much more stable.

24.3.5 Bandwidth

The final performance criterion under investigation is the strain bandwidth of the thread. In order to approximate the thread's bandwidth, a calibration measurement was made against a reference strain gauge calibrated from 0 to 1600 Hz. This particular range was chosen because the thread is planned to be embedded into large composite structures. Also, a high bandwidth is needed to detect impact events. These large composite structures typically have many fundamental modes (natural frequencies) of vibration below 10 Hz, yet an impact could be much higher in frequency (100–10,000 Hz). Thus, to be an effective dynamic structural health and strain sensor, these frequency ranges must be measured.

In order to perform this calibration, (1) the sensor must undergo significant strain to cause a measurable resistance response in the thread and (2) the various frequencies of concern must undergo measurable strain. In other words, the structure should have a somewhat broadband response and not many antimodes. In order to achieve these goals, a custom double-cantilever calibration rig made out of aluminum was designed and built. This calibration rig was then mounted to an electromagnetic shaker for excitation. The initial design geometries of this structure were chosen so that there would be large strains and many modes of vibration within the spectrum of concern. This calibration rig can be seen in Fig. 24.22 in its naked form along with sensors mounted to the structure for the calibration measurement. It should be noted

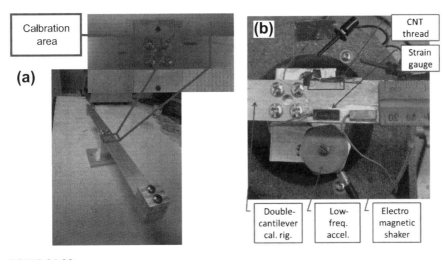

FIGURE 24.22

Custom double-cantilever beam strain calibration rig for CNT thread strain sensor. (a) Custom-built strain calibration rig in its naked form with calibration area highlighted. (b) Calibration area with the thread mounted adjacent to the reference strain gauge. It should be noted that this thread is close to 20 μm in diameter and about 12.5 mm in length. (For color version of this figure, the reader is referred to the online version of this book.)

that the sensors that were used to make this calibration were piezoelectric strain gauges (PCB Piezotronics Inc 740A02 and 740M04), which are calibrated from 0.5 Hz–100 KHz, and a piezoelectric accelerometer (PCB Piezotronics Inc 301A11), which is calibrated from 0.5 Hz–10 KHz. The electromagnetic shaker is an MB-50 and an HP 45670A frequency analyzer was used for the presented measurements.

The thread in the above photograph is mounted close to the surface of the aluminum beam with two epoxy mounting points. It should be highlighted that the thread was not entirely encased in epoxy in order to obtain a baseline measurement prior to infiltrating the entire thread with epoxy. This thread mount is not a true strain measurement of the beam because the thread does not conform to the surface of the beam. Instead, it is a displacement measurement of two points. This technique was acceptable here because the thread was nearly the same size as the strain gauge, so this mounting technique proved not to be a problem. After the epoxy mount points, conductive epoxy was utilized to hook up small wires. The thread could then be hooked up with the Wheatstone bridge circuit seen in Fig. 24.13. A detailed schematic of this mounting configuration can be seen in Fig. 24.23. Finally, it should be noted that this mounting technique may prove to be beneficial in certain sensing applications for strain amplification or separation.

Prior to performing any frequency domain analysis, the CNT thread response time history quality was analyzed. This was done by exciting the calibration rig near resonance with a 35 Hz sine wave while measuring the voltage output from the Wheatstone bridge. This voltage or thread response along with the response of the other sensors used in this experiment can be seen in Fig. 24.24. As seen below,

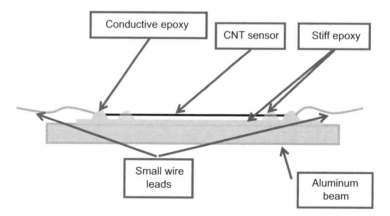

FIGURE 24.23

CNT thread attachment to aluminum calibration rig. As seen in the schematic, drops of epoxy are utilized to keep the thread in tension once mounted, conductive epoxy is used to hook up lead wires, and the thread is mounted above a surface of thin epoxy to ensure electrical insulation and to avoid uneven surfaces. (For color version of this figure, the reader is referred to the online version of this book.)

the CNT thread has a "clean" sine wave output, which is desired for optimal frequency response function (FRF) estimation. This clean signal was obtained by pretensioning the CNT thread during mounting. Thus, when the beam goes into compression, the thread does not go slack. In summary, this test was performed to ensure that the thread would not go slack during testing so that FRF estimation could take place.

After a clean sine wave was obtained near peak excitation, FRFs were generated to (1) determine the bandwidth of the thread and (2) calibrate the thread against the reference strain gauge. An FRF is defined as the output parameter over the input parameter across several frequencies of concern as defined below in Eqn (24.3) for the frequency domain. This function provides a description of how the output's response varies over various frequencies relative to an input, thus system behavior can quickly and easily be understood by measuring this function.

$$\text{FRF}(\omega) = \frac{\text{Output}(\omega)}{\text{Input}(\omega)} \qquad (24.3)$$

To generate FRFs, an excitation technique must be utilized, which puts known frequencies into the system. In its simplest form, this would be a periodic sine wave. However, sine wave excitation is very time consuming, frequency resolution would not be the best, averaging could not be done as easily, and a somewhat larger amplitude signal is needed to visually see the signal. To handle these disadvantages, a chirp excitation (quick periodic sine wave sweep) was utilized combined with 50–100 root mean square averages of various frequency resolutions to estimate FRFs in this chapter. This estimation process for the calibration rig's base acceleration and strain can be seen in Fig. 24.25 as an example. It should be pointed out that the FRFs in this chapter are also plotted with the ordinary coherence function to quantify the quality of the FRF estimate. A coherence of 1 implies good correlation or measurement quality.

One of the first analyses of this system can be seen in Fig. 24.26 comparing FRFs of CNT thread to base acceleration and strain gauge to base acceleration. As seen in the figure, this analysis provided a conservative bandwidth estimate of the thread, which ranged from 0 to 350 Hz. It should be noted that this bandwidth measurement is conservative because the mounting epoxy damps the strain that the CNT thread experiences and limits the received energy. Thus, it is expected that higher bandwidths could be achieved with stiffer epoxy. Nonetheless, stiffer epoxies will still limit the bandwidth of the thread and ultimately the sensing once embedded into composites. Despite this low bandwidth, this frequency range meets the prior requirements of measuring structural modes and detecting lower frequency impacts.

After the CNT thread bandwidth was determined for this setup, a 0–100 Hz FRF was made between the CNT thread and adjacent strain gauge originally seen in Fig. 24.22. This lower frequency range was selected because (1) predetermined rig dynamics showed that this range was best for calibrating the thread and (2) thread response cuts off shortly thereafter. This measurement can be seen in Fig. 24.27. As seen below, coherence is decent in the range of interest except at lower frequencies. It

CHAPTER 24 Embedded Carbon Nanotube Sensor Thread

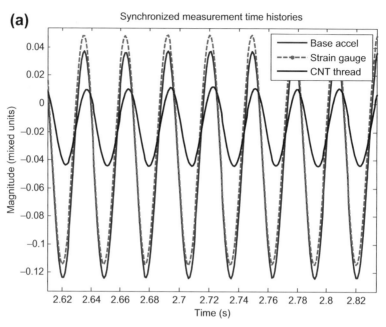

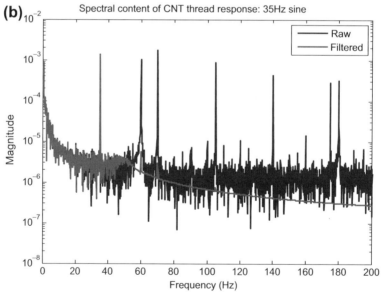

FIGURE 24.24

Sine wave of CNT thread to 35-Hz resonance excitation. Prior to performing frequency domain analysis, the CNT thread was confirmed to have a "clean" sine wave appearance. In order to obtain this clean signal, the thread's output was filtered with a tight low-pass filter.

24.4 Carbon Nanotube Thread SHM Architectures

is believed that this drop comes from low input energy combined with potential standing waves setting up on the thread. Yet, when the coherence improves, the FRF appears "flat" and the phase angle is near zero, as expected. This flatness and near-zero response is expected because the thread is nearly the same size as the reference strain gauge and measures nearly the same strain state on the opposite side of the beam. Thus, calibration can be done to determine the gauge factor within this range, or the gauge factor FRF of the thread can be calculated. This calibration is still being performed.

In summary, this analysis shows that (1) CNT thread can measure frequencies of interest, specifically structural modes and impacts, and (2) the CNT thread can be calibrated against a reference strain gauge to estimate sensitivity or gauge factor over multiple frequencies. This work would not have been possible without the advisement and generosity of the Structural Dynamics Research Lab here at UC.

24.4 CARBON NANOTUBE THREAD SHM ARCHITECTURES

Developing an SHM sensor for composite materials or a structure is only part of the SHM solution. How this sensor will be incorporated into the structure, how many sensors should be used, and how the sensor data will be used are all other important areas that also need to be addressed for a holistic SHM approach. This section will briefly address some of these questions.

On a few previous occasions, it has been mentioned that the CNT thread sensor can detect specific damage modes such as ply delamination and matrix cracking. Matrix cracking can occur from several different mechanisms itself, and it would be advantageous to know which mechanism is the culprit of the cracking. It is believed that the CNT sensor thread can shed light on some of these cracking mechanisms by (1) confidently identifying a crack in the structure and (2) utilizing neighboring thread sensors to help assess what particular mechanism is at play for better health and repair assessments. The sensor thread can be seen in Fig. 24.28 identifying these various cracking mechanisms of concern.

In order to assess the mechanism behind cracking or identifying other damage modes, the sensing thread needs to have a minimum density and a clever incorporation design in order to identify damage in its nascent stages. This incorporation design should provide information about damage location, severity, and potential prognosis in addition to damage mode characterization. One design that would meet these criteria and be easy to incorporate into current composite materials is a cross-weave or grid design. This design can be seen in Fig. 24.29

(a) Clean sine wave time history along with outputs from the strain gauge and accelerometer and (b) frequency domain of CNT thread signal before and after filtering. It should be noted that the thread's output was on the same order as 60-Hz noise. Other peaks in the spectrum are harmonics of this 35-Hz excitation and of the 60-Hz noise. (For color version of this figure, the reader is referred to the online version of this book.)

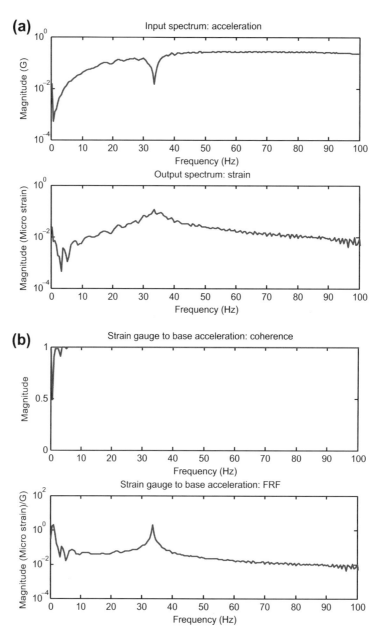

FIGURE 24.25

Frequency domain analysis and frequency response function (FRF) estimation. (a) Linear input spectrum from the base acceleration and output spectrum from the strain gauge. As seen from the plot, the electromagnetic shaker struggles to put low-frequency energy into the system, while the strain gauge spectrum shows a rise near the beam's first resonance. These spectrums can then be used to estimate the FRF of the system. (b) FRF of the system with ordinary coherence. Ordinary coherence is a statistical tool used to ensure/determine FRF quality. (For color version of this figure, the reader is referred to the online version of this book.)

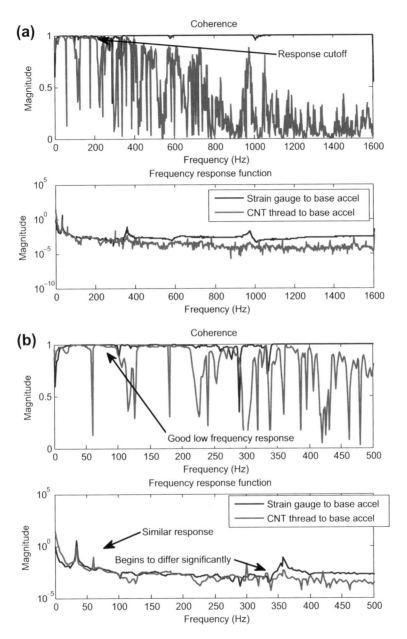

FIGURE 24.26

The 0–1600 Hz FRF of strain gauge and CNT thread to base acceleration. (a) Strain gauge and CNT thread from 0 to 1600 Hz. As seen in the coherence, the output response begins to diminish at higher frequencies. (b) Zoomed in region of the coherence and FRF where this poor response begins to happen. As pointed out in the figure, it seems like the FRFs between the two sensors diverge near 300–350 Hz. (For color version of this figure, the reader is referred to the online version of this book.)

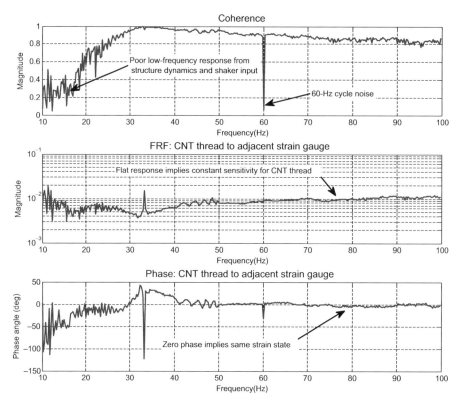

FIGURE 24.27

The 10- to100-Hz FRF of CNT thread to strain gauge. This figure illustrates how the CNT thread can be used as a broadband sensor due to its near-constant sensitivity (flatness) and near-zero phase angle. This FRF can be scaled with calibration values from the strain gauge and Wheatstone bridge circuit to construct a gauge factor FRF. This calibration is still being carried out. (For color version of this figure, the reader is referred to the online version of this book.)

with a potential sensor thread incorporation scheme during the fabrication of the material. Additionally, this fabrication method would easily allow various thread densities for optimal damage identification. As seen in the figure, this design would also utilize multiplexing switches for data channel reduction combined with an algorithm to analyze CNT sensor signals for damage location, severity, and characterization.

Multiplexing will be done in the time domain by pulsing and assessing each sensing thread in close vicinity. Close vicinity monitoring will be implemented to simplify data handling by utilizing sensing nodes to do localized data processing. These sensing nodes will then utilize wireless transmitters to send limited yet meaningful health assessment information to the operator of the aircraft. A simplified

24.4 Carbon Nanotube Thread SHM Architectures

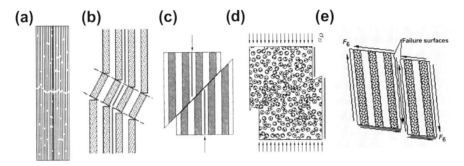

FIGURE 24.28

Micromechanical failure modes in laminated composites. (a) Coalescing of microcracks in a composite subjected to tensile loading in the fiber direction. (b) Failure of a composite with brittle fibers subjected to compression in the direction of the fibers. (c) Shear failure in a composite with a high fiber volume fraction loading in the transverse direction to the fiber. (d) Shear failure in a composite subjected to compression loading in the transverse direction to the fiber. (e) Shear failure in a composite subjected to in-plane shear loading. (For color version of this figure, the reader is referred to the online version of this book.)

Source: All images edited from [55] by Jandro Abot and M. Schulz.

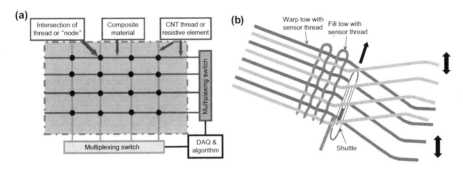

FIGURE 24.29

Array configuration of sensors. (a) Spatial grid design for detailed failure information and (b) industrial scale-up manufacturing process for prepreg plain weave fabrics. (For color version of this figure, the reader is referred to the online version of this book.)

representation of this proposed multiplexed and distributed sensing node design can be seen in Fig. 24.30.

To illustrate how this SHM architecture would be used in a specific application, the F-22 Raptor [57] is used as a design example, as seen in Fig. 24.31, with detailed areas of concern. It should be noted that the F-22 is built by several manufacturers. Boeing takes care of building the aft fuselage, main wings, power supplies, auxiliary power units, auxiliary power generation systems, airframe-mounted accessory drives and the fire-protection system. The aft fuselage is 67% titanium, 22% aluminum and

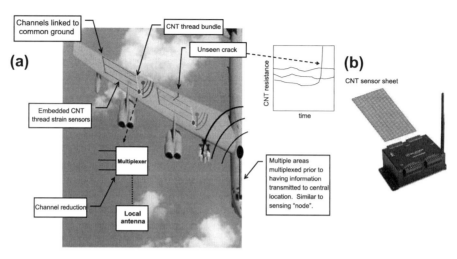

FIGURE 24.30

SHM system communication. (a) CNT thread strain/damage sensor wired multiplexer, local antenna. (b) Off-the-shelf multiplexer and wireless transmitter [56]. (For color version of this figure, the reader is referred to the online version of this book.)

11% composite by weight. The Boeing-built portion of the wing is 42% titanium, 35% composite and 23% aluminum; steel and other materials are present in the form of fasteners, clips and other miscellaneous parts, also by weight. Each wing weighs about 2000 pounds. Titanium reinforcement was added for strength against impact during combat. Nonetheless, most ribs and spars in the wing are composites and it would be beneficial to actively monitor their health.

24.5 AREAS OF STRONG MULTIFUNCTIONAL POTENTIAL

Besides adding strain and damage sensing to a polymeric composite material, CNT thread has the potential to improve the material in other ways as well. In other words, the CNT thread would be multifunctional in how it would improve the overall performance and quality of the material. Some of the potential areas where CNT thread could have an impact on composite material improvement are listed in the introduction of this chapter, but only the areas of current investigation will be covered in this section, specifically damping and toughness.

As observed in earlier sensing hysteresis analysis (see Fig. 24.18), CNT thread exhibits significant hysteresis in the stress—strain curve. This hysteresis can be considered to be mechanical damping or mechanical energy loss since mechanical energy is not being conserved in the loading cycle. This damping can be quantified by finding the hysteresis loop area normalized by the maximum mechanical energy (U_{max}) as defined below in Eqn (24.4). This quantity, u, is called the specific

24.5 Areas of Strong Multifunctional Potential

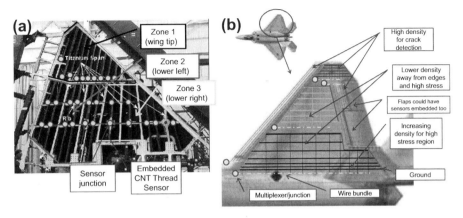

FIGURE 24.31

Example design of the SHM system for retrofit to an aircraft. (a) Sensor thread design for the F-22 interior wing with 29 junctions where sensor connects with multiplexed network; three multiplexers will be used to manage information from specific zones of wing, 39 spar sensor threads and 20 rib sensor threads, Close to 60 total sensors are needed. Each zone could transmit via antenna to central location. (b) Sensor thread design for the F-22 exterior wing with five multiplexers to manage information from specific zones of wing, four to five sensors in front wing edge (close spacing, 1–2″), four to five sensors around flaps (close spacing, 1–2″), 15 to 20 sensors in wing tip grid (dense, 3–5″ spacing), 15–20 sensors in midwing (low density, 8–10″), and 15–20 sensors in lower wing (high and low density, 3–5″ and 10–12″), around 100 total sensors. Each zone transmits wirelessly to the central location. (For color version of this figure, the reader is referred to the online version of this book.)

Source: Inner wing photo courtesy of DOT&E FY 2001 Report.

damping capacity [58], but it can also be thought of as an energy loss percentage. In order to implement this equation numerically, the damping capacity can be calculated by finding the difference of the integrated loading and unloading curves. Then, this quantity can be normalized by the loading curve. These curves are labeled in Fig. 24.32 for both as-spun CNT thread and copper wire. As seen in the figure, the materials were both strained in their elastic regions to approximate values of specific damping capacity. Copper wire was used not only for comparison purposes but also to ensure that the method of generating this hysteresis curve was valid and not an artifact of the Instron 5948's sensors/structure.

$$u = \frac{(\oint \sigma d\varepsilon)}{U_{max}} = \frac{(\int_0^{\varepsilon_{max}} \sigma_{loading} d\varepsilon - \int_0^{\varepsilon_{max}} \sigma_{unloading} d\varepsilon)}{\int_0^{\varepsilon_{max}} \sigma_{loading} d\varepsilon} \qquad (24.4)$$

Once the specific damping capacity is found, the loss factor, or specific damping capacity per unit radian can be found as defined below in Eqn (24.5) [58]. Typically,

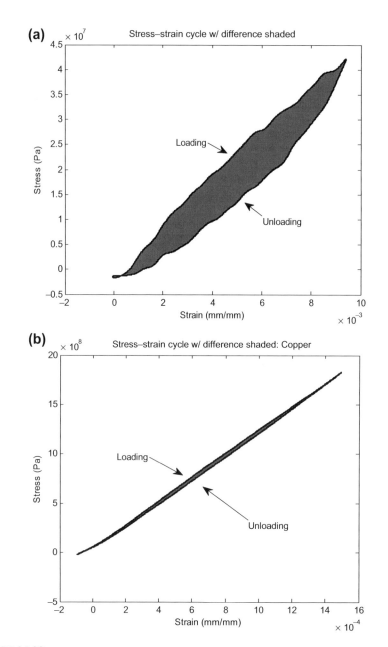

FIGURE 24.32

Energy dissipation through damping generated with Instron 5948. (a) As-spun CNT thread energy loss and (b) copper wire energy loss. The shaded region for the CNT thread is around 30% energy loss and around 3% for the copper wire. (For color version of this figure, the reader is referred to the online version of this book.)

24.5 Areas of Strong Multifunctional Potential

materials are compared with their loss factor or damping ratio. If the loss factor is small, the damping ratio can be approximated by multiplying the loss factor by two. Figure 24.33 compares the loss factors of CNT thread, epoxy, rubber, aluminum, and glass.

$$\text{Loss Factor} = \frac{u}{2\pi} \quad (24.5)$$

As seen in the figure above, the CNT thread is close in magnitude to epoxy resin. It can then be inferred that if CNT thread were to be embedded into a polymeric composite material with an epoxy matrix, the overall material damping will not degrade. Instead, the composite material will maintain or better the effective damping of the material. As a result, CNT thread has the potential to improve composite material damping.

Another strong multifunctional area of CNT thread is toughness or damage limiting. Toughness is defined as the area under the stress—strain curve as defined in Eqn (24.6) and seen in Fig. 24.34 for as-spun CNT thread. Current as-spun CNT thread produced in Nanoworld Laboratories has 100—200 MPa strength and 3—5% elongation. The epoxy resin has 55—130 MPa strength with 2—4% elongation [39]. Thus, the toughness of as-spun CNT thread is on the order of that of epoxy

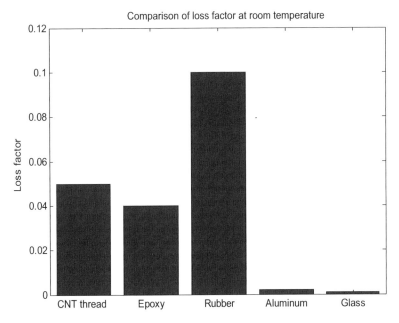

FIGURE 24.33

Loss factor comparison. Calculated CNT thread loss factor is compared against published values for epoxy [59], rubber, aluminum, and glass [58]. (For color version of this figure, the reader is referred to the online version of this book.)

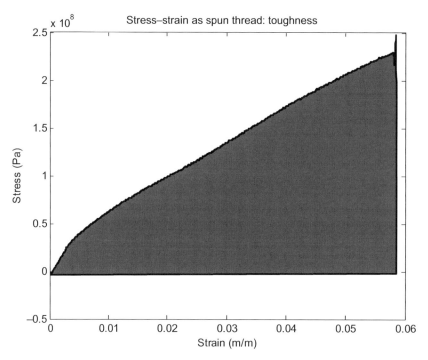

FIGURE 24.34

Stress–strain curve of CNT thread with area under the curve shaded, which represents material toughness. (For color version of this figure, the reader is referred to the online version of this book.)

resin. It can then be inferred that CNT thread has the potential to improve material toughness, which increases the material's resilience to damage.

$$\text{Toughness} = \int_0^{\varepsilon_{break}} \sigma d\varepsilon \qquad (24.6)$$

24.6 FUTURE WORK

Future CNT thread sensor work at Nanoworld Laboratories entails several key areas of investigation to prove sensor feasibility for a composite SHM sensor. These key areas of investigation are listed below with descriptive significance.

- Embedded sensor performance: Although CNT thread appears to perform well outside the composite material, it must have similar if not improved sensor performance once embedded into the material. This will be similar to earlier concept work presented in Section 24.2.

- Sensor invasiveness: It is suspected that the CNT thread sensor will be minimally invasive to the composite material due to its small size. However, this hypothesis will be tested by examining composite breaking strength, stiffness, elongation, and fatigue strength.
- Failure mode identification: Composite materials will be fabricated with voids and cracks. Then, the composite material will be excited with broadband excitation to identify these impregnated damage modes with the strain response of the thread. Additionally, impact testing will be done on the composite material to (1) identify the impact through the strain response and (2) potentially assess the damage through CNT thread breakage strain response.

Acknowledgments

First and foremost, the principal author would like to thank Dr Schulz and Dr Shanov, directors of Nanoworld Labs, for their support, suggestions, and excellent advisement. Next, the author would like to thank the National Science Foundation for their support, and this material is based on work supported by the National Science Foundation under Grant No. 1102690. The author would also like to thank Doug Hurd, manager of the machine shop at UC, for his advisement in test rig design and manufacturing. The author would also like to thank Wenyu Zhao, of the University of Cincinnati Center for Intelligence Maintenance Systems (http://www.imscenter.net/), for his generosity in borrowing an NI 9205 DAQ system long term. Finally, the author would like to thank Dr Allemang and Dr Phillips of the University of Cincinnati Structural Dynamics and Research Lab (SDRL, http://sdrl.uc.edu/) for their generosity for borrowing a shaker, sensors, an FRF analyzer, and laboratory space to perform the strain bandwidth work. These individuals also provided suggestions in design of the calibration rig, analysis techniques, and general suggestions. Their advisement is greatly appreciated.

References

[1] Federal Aviation Administration of the USA. AC 43-204 Visual Inspection of Aircraft, 1997.
[2] Federal Aviation Administration of the USA. AC 20-107B Composite Aircraft Structure, 2009.
[3] A. Kapadia, Non-Destructive Testing of Composite Materials, National Composites Network. http://www.compositesuk.co.uk/LinkClick.aspx?fileticket=14Rxzdzdkjw=&.
[4] File:LockheedMartin F-22A Raptor JSOH.jpg. Wikipedia. (Online) (Cited: November 15, 2012)] http://en.wikipedia.org/wiki/File:Lockheed_Martin_F-22A_Raptor_JSOH.jpg.
[5] File:AH-64D Apache Longbow.jpg. Wikipedia. (Online) (Cited: November 15, 2012) http://en.wikipedia.org/wiki/File:AH-64D_Apache_Longbow.jpg
[6] File:All Nippon Airways Boeing 787-8 Dreamliner JA801A OKJ.jpg. Wikipedia. (Online) (Cited, 2012 (Online) (Cited: November 15, 2012). http://commons.wikimedia.org/wiki/File:All_Nippon_Airways_Boeing_787−8_Dreamliner_JA801A_OKJ.jpg.
[7] J.W.C. Pang, I.P. Bond, A hollow fibre reinforced polymer composite encompassing self-healing and enhanced damage visibility, Composite Science and Technology 65 (11-12) (2005) 1791−1799.

[8] S. Sen, E. Schofield, J.S. O'Dell, L. Deka, S. Pillay, The development of a multifunctional composite material for use in human space exploration beyond low-earth orbit, JOM 61 (1) (2009) 23–31.

[9] K.L. Schaaf, Composite Materials with Integrated Embedded Sensing Networks, PhD Dissertation, UC San Diego (2008).

[10] J.R. Zayas, D.P. Roach, M.A. Rumsey, W.R. Allan, D.A. Horsley, Low-Cost Fiber Bragg Grating Interrogation System for in situ Assessment of Structures, SPIE, Sensors and Smart Structures Technologies for Civil, Mechanical, and Aerospace Systems, San Diego, CA, March 18th 2007.

[11] B.R. Loyola, K.J. Loh, V.L. Saponara, Static and dynamic strain monitoring of GFRP composites using carbon nanotube thin films, SPIE, Sensors and Smart Structures Technologies for Civil, Mechanical, and Aerospace Systems, San Diego, CA, March 2011.

[12] N.D. Alexopoulos, HYPERLINK " C. Bartholome, HYPERLINK "http://www.sciencedirect.com/science/article/pii/S0266353809003807" P. Poulin, HYPERLINK "http://www.sciencedirect.com/science/article/pii/S0266353809003807" Z. Marioli-Riga, Structural health monitoring of glass fiber reinforced composites using embedded carbon nanotube (CNT) fibers, Composites Science and Technology 70(2) (2010) 260–271.

[13] NASA, Multifunctional carbon nanotube yarn sensors, Advanced Sensing and Optical Measurement Branch. (Online) April 2008. (Cited: 2.1.2012.) http://asomb.larc.nasa.gov/research/nano_technology/yarn_sensors.htm.

[14] J.L. Abot, Z. Kier, V.N. Shanov, Y. Song, C. Jayasinghe, S. Sundaramurthy, M.J. Schulz, Self-sensing Composite Materials, UC 109–064 Provisional Patent (2009).

[15] H. Zhao, F. Yuan, Carbon nanotube yarn sensors for structural health monitoring of composites, SPIE, Nondestructive Characterization for Composite Materials, Aerospace Engineering, Civil Infrastructure, and Homeland Security, San Diego, CA, March 2011.

[16] X. Wang, D.D.L. Chung, Continuous carbon fibre epoxy-matrix composite as a sensor of its own strain, Smart Materials & Structures 5 (1996) 796–800.

[17] M.J. Schulz, A. Kelkar, M. Sundaresan, Nanoengineering of Structural, Functional, and Smart Materials, CRC Press, Boca Raton, 2006.

[18] Y. Song, Basics of carbon nanotube materials for structural applications, Nanotechnologies and smart materials for SHM, Ed. A. Catalano, G. Fabbrocino and C. Rainieri, ISBN 978-88-88102-47-4, Campobasso, Italy, 2012

[19] Y. Song, SHM applications of nanotechnologies for sensor development and communication, Nanotechnologies and smart materials for SHM, Ed. A. Catalano, G. Fabbrocino and C. Rainieri, ISBN 978-88-88102-47-4, Campobasso, Italy, 2012.

[20] C. Jayasinghe, W. Li, Y. Song, J.L. Abot, V.N. Shanov, S. Fialkova, S. Yarmolenko, S. Sundaramurthy, Y. Chen, W. Cho, S. Chakrabarti, G. Li, Y. Yun, M.J. Schulz, Nanotube responsive materials, MRS Bulletin 35 (9) (2010) 682–692.

[21] M.J. Schulz, G. Maheshwari, J. Abot, Y. Song, C. Jayasinghe, N. Mallik, V. Shanov, M. Dadhania, Y. Yun, S. Yarmolenko, J. Sankar, Responsive nanomaterials for engineering asset evaluation and condition monitoring, BNDT: Insight Journal (June 2008).

[22] Yi Song, Carbon Nanotube Materials for Smart Composites, 242nd ACS National Meeting, Denver, Colorado, August 28-September 1, 2011.

[23] D.B. Mast, C. Jayasinghe, S. Chakobarty, L. Li, Y. Chen, W. Cho, M.J. Schulz, V. Shanov, J. Abot, S. Pixley, Carbon Nanotube Threads, Yarns and Ribbons: Making the Transition from Materials to High Performance Devices for Defense Applications, Nano Technology for Defense Conference, Burlinghame, CA, April, 2009.

[24] N. Mallik, C. Jayasinghe, S. Chakraborty, P. Salunke, W. Li, W. Cho, L. Lee, E. Head, D. Hurd, J. Abot, Y. Song, V. Shanov, M.J. Schulz, Fabrication of threads from carbon nanotube arrays. New Delhi, India, Proceedings of Processing and Fabrication of Advanced Materials — XVII (December 2008).

[25] V.N. Shanov, G. Choi, G. Maheshwari, G. Seth, S. Chopra, G. Li, Y. Yun, J. Abot, M.J. Schulz, Structural Nanoskin Based on Carbon Nanosphere Chains, SPIE, Sensors and Smart Structures Technologies for Civil, Mechanical, and Aerospace Systems San Diego, CA, March 2007.

[26] N. Mallik, Carbon nanotube array based smart materials, in: M. Umerno, Prakash R. Somani (Eds.), Carbon Nanotubes: Synthesis, Properties and Applications, Applied Science Innovations Pvt. Ltd., India, 2009. (Chapter 16).

[27] Inpil Kang, Nanoengineering of sensory materials, in: Fu-Kuo Chang, Yozo Fujino, Christian Boller (Eds.), Encyclopedia of Structural Health Monitoring, Wiley, 2008.

[28] V. Shanov, J. Abot, Y. Song, M. Schulz, De-Icing of Solid Structures by Surface Heating Through CNT Thread and Ribbon, UC Invention disclosure 109—076 USA (9.4.2009).

[29] G. Maheshwari, J. Abot, A. Song, E. Head, V. Shanov, M. Schulz, C. Jayasinghe, P. Salunke, Y. Yun, Powering up nanoparticles: versatile carbon materials for use in engineering and medicine, SPIE Newsroom, http://dx.doi.org/10.1117/2.1200809.1305.

[30] J.L. Abot, Y. Song, M.J. Schulz, V.N. Shanov, Novel carbon nanotube array-reinforced laminated composite materials with higher interlaminar elastic properties, Composite Science and Technology 68 (13) (2008) 2755—2760.

[31] G. Maheshwari, J. Abot, A. Song, E. Head, V. Shanov, M. Schulz, C. Jayasinghe, P. Salunke, Y. Yun, Novel Carbon Nanotube Array-Reinforced Laminated Composite Material, SAMPE Conference, Cincinnati, OH, 2007.

[32] W. Li, J. Bulmer, B. Ruff, Y. Song, P. Salunke, V. Shanov, M.J. Schulz, Modeling the electrical impedance of carbon nanotube ribbon, Nano LIFE (May 2013).

[33] J.P.M. She, J.T.W. Yeow, Carbon nanotube-enhanced capillary condensation for a capacitive humidity sensor, Nanotechnology 17 (21) (2006).

[34] K. Chawla, Composite Materials, Second Ed., Springer, 1998.

[35] GamryInstruments. Application Note: Getting Started with Electrochemical corrosion measurement, http://www.gamry.com/assets/Application-Notes/Getting-Started-with-Electrochemical-Corrosion-Measurement.pdf.

[36] B. Suberu, Y. Song, V.N. Shanov, M.J. Schulz, High Volume Fraction Composite (HVFC), UC UC Invention Disclosure (28.4.2012) 112—088.

[37] J. Suhr, N. Koratkar, P. Keblinski, P. Ajayan, Viscoelasticity in carbon nanotube composites, Nature Materials 4 (2005) 134—137.

[38] X. Li, C. Levy, A. Agarwal, A. Datye, L. Elaadil, A.K. Keshri, M. Li, Multifunctional carbon nanotube film composite for structure health monitoring and damping, Open Construction and Building Technology Journal 3 (2009) 146—152.

[39] B.D. Agarwal, L.J. Broutman, K. Chandrashekhara, Analysis and Performance of Fiber Composites, Wiley, Hoboken NJ, 2006.

[40] R. Hollertz, S. Chatterjee, H. Gutmann, T. Geiger, F.A. Nüesch, B.T.T. Chu, Improvement of toughness and electrical properties of epoxy composites with carbon nanotubes prepared by industrially relevant processes, Nanotechnology 22 (12) (2011).

[41] R. Weisenberger, M.C. Andrews, Carbon nanotube polymer composites, Current Opinion in Solid State and Materials Science 8 (1) (2004) 31−37.

[42] S. Shaikh, K. Lafdi, A carbon nanotube-based composite for the thermal control of heat loads, Carbon 50 (2) (2012) 542−550.

[43] S.C. Jun, J.H. Choi, S.N. Cha, C.W. Baik, S. Lee, H.J. Kim, J. Hone, J.M. Kim, Radio-frequency transmission characteristics of a multi-walled carbon nanotube, Nanotechnology 18 (25) (2007).

[44] Y. Tzeng, T.S. Huang, Y.C. Chen, C. Y.K.

[45] R. Hu, B.A. Cola, N. Haram, J.N. Barisci, S. Lee, S. Stoughton, G. Wallace, C. Too, M. Thomas, A. Gestos, M.E.D. Cruz, J.P. Ferraris, A.A. Zakhidov, R.H. Baughman, Harvesting waste thermal energy using a carbon-nanotube-basedthermo-electrochemical cell, Nano Letters 10 (3) (2010) 838−846.

[46] Prashant V. Kamat, Harvesting photons with carbon nanotubes, Nano Today 1 (4) (2006) 20−27.

[47] NASA, DRAFT: Nanotechnology Road MapdTA10 (November 2010). http://www.nasa.gov/pdf/501325main_TA10-Nanotech-DRAFT-Nov2010-A.pdf.

[48] J.L. Abot, M.J. Schulz, Y. Song, S. Medikonda, N. Rooy, Novel distributed strain sensing in polymeric materials, Smart Materials & Structures 19 (8) (2010).

[49] J.L. Abot, Y. Song, M.S. Vatsavaya, S. Medikonda, Z. Kier, C. Jayasinghe, N. Rooy, V.N. Shanov, M.J. Schulz, Delamination detection with carbon nanotube thread in self-sensing composite materials, Composites Science and Technology 70 (7) (2010) 1113−1119.

[50] V. Shanov, C. Jayasinghe, J. Abot, Yi Song, M. Schulz, Remote, Non-Destructive Approach for Detecting Defects and Cracks in Solid Structures, UC Invention Disclosure (10.4.2009) 109−077.

[51] E.F. Northrup, Methods of Measuring Electrical Resistance, McGraw Hill, NY, 1912.

[52] W. Obitayo, T. Liu, A review: carbon nanotube-based piezoresistive strain sensors, Journal of Sensors (2012).

[53] M. Miao, J. McDonnell, L. Vuckovic, S.C. Hawkins, Poisson's ratio and porosity of carbon nanotube dry spun yarns, Carbon 48 (10) (2010) 2802−2811.

[54] J.J. Sullivan, A.J. Hehr, Effects of Thread Twist Angle on the Thread Properties: A Preliminary Study, Internal Report of Nanoworld Labs, University of Cincinnati, 2012.

[55] I.M. Daniel, O. Ishai, Engineering Mechanics of Composite Materials, New York, Oxford, 2006.

[56] Data Link. Wireless Multiplexer Product Info. (Online) (Cited: 16.11.2012) http://www.data-linc.com/dd1000.htm.

[57] F22 Specification Facts. (Online) (Cited: 6.10.2013) http://www.f22fighter.com/Specs.htm

[58] C.W.D. Silva, Damping. Vibration: Fundamentals and Practice, CRC Press, Boca Raton, 2000 Chapter 7.

[59] L.E. Nielsen, R.F. Landel, Mechanical Properties of Polymers and Composites, Marcel Dekker, Inc, NY, 1994.

CHAPTER 25

Tiny Medicine

Weifeng Li[1], Brad Ruff[1], John Yin[1], Rajiv Venkatasubramanian[1], David Mast[1], Anshuman Sowani[1], Arvind Krishnaswamy[1], Vesselin Shanov[1], Noe Alvarez[1], Rachit Malik[1], Mark Haase[1], Madhura Patwardhan[1], Mark Schulz[1], Sergey Yarmolenko[2], Svitlana Fialkova[2], Salil Desai[2], Ge Li[3]

[1] *Nanoworld Laboratory, University of Cincinnati, Cincinnati, OH, USA,* [2] *Center for Advanced Materials and Smart Structures, North Carolina A&T State University, Greensboro, NC, USA,* [3] *General Nano, LLC, Cincinnati, OH, USA*

CHAPTER OUTLINE

25.1	The History of Tiny Machines	714
25.2	Nanoscale Materials	718
	25.2.1 Carbon nanotube materials	719
	25.2.2 Magnetic nanotube and nanowire materials	721
	25.2.3 Magnetic nanoparticle materials	724
	25.2.4 Superparamagnetic core design and testing	725
25.3	A Pilot Microfactory for Nanomedicine Devices	727
25.4	Tiny Machines Concepts and Prototype Fabrication	730
	25.4.1 Electromagnetic devices using CNT yarn	731
	25.4.2 Failure mechanisms for in vivo devices	734
	25.4.3 Nanomaterial electric motor	735
	25.4.4 Thermal analysis of nanomaterial components	737
	25.4.5 Composite electromagnetic material	738
	25.4.6 Biodegradability	739
	25.4.7 Carbon nanotube wire	740
	25.4.8 Telescoping nanotubes	742
	25.4.9 Communication with tiny machines	743
25.5	Summary and Conclusions	743
Acknowledgments		744
References		744

Nanotube Superfiber Materials. http://dx.doi.org/10.1016/B978-1-4557-7863-8.00025-6
Copyright © 2014 Elsevier Inc. All rights reserved.

25.1 THE HISTORY OF TINY MACHINES

Nanomaterial robotic devices or nanorobots and other tiny machines enabled by nanotechnology will be used in medicine (and engineering) because they will have greater precision, force, energy density, and computational capability than biological materials and existing microelectromechanical systems. The history of "tiny machines" goes back to 1959 when Richard Feynman, Fig. 25.1, said "there's plenty of room at the bottom" and he talked about tiny machines [1,2]. In a 1984 talk, he said "Development of tiny machines cannot be avoided." Since then there have been advances in scanning electron microscopy (SEM), electron-beam patterning, the discovery of nanotubes, commercial Kleindiek robots for use in SEMs, the emergence of nanotechnology, nanoimprint lithography (NIL), and others. With all these advances, now in 2013, half a century later, an obvious question is: How many different types of nanorobot devices or "tiny machines" have been built? The answer is nano (none). Why? Because the nanoscale materials needed to build the devices have only recently become available and the processing technology needed to build tiny machines is still being developed and put to use for specific applications. The first comprehensive book on the design of nanorobot devices was written by Robert Freitas, Jr, Fig. 25.1. Furthermore, the cost of the instrumentation to build microrobots is very high, which restricts research. This chapter describes several devices such as biomedical micromanipulators that we believe are feasible to build, that are in the process of being built and that may be the first nanorobot devices in the market.

It is important to have an appreciation of the size of the nanorobot devices in order to gage the difficulty in building the parts and developing the metrology methods that will be needed to measure the size of the parts. We can start by saying that there is no yellow submarine yet that can swim around the vasculature tree. The scale of nanorobot devices is illustrated in Fig. 25.2. As a reference, the smallest size particle that can be seen with the naked eye is about a few microns in size. Now consider that nanoparticles are thousand times smaller than what can be seen with

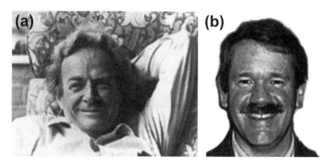

FIGURE 25.1

Richard Feynman (a) and Robert Freitas, Jr (b).

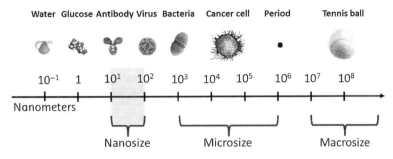

FIGURE 25.2

Practical nanorobot devices are expected to be in the microsize range. (For color version of this figure, the reader is referred to the online version of this book.)

the naked eye. Some nanomaterials are handled in bundles of tubes or larger particles are used, and the size can then be tens of nanometers. Considering the development of tiny machines and robotic devices, there are three size ranges that could be considered that require different levels of complexity to work at. The smallest is the nanosize range where nanoparticles are being intensively developed for drug delivery/therapy. But nanosize electromechanical devices would be too difficult to build and control using existing processing technology. Next is the microsize range, which is a less-investigated intermediate range where nanorobot electromechanical devices can have unique applications and promise to produce advances. Nanoscale materials can be manipulated using existing instrumentation to build devices that are microscale in size. The largest devices are in the macrosize range, which is at the scale of existing technology such as the da Vinci robot manipulators, which use centimeter-sized manipulators and tools.

Devices that could be built in each of the three size ranges are described next to serve as reference points on what might be practical. An example of a nanosize device is a drug delivery shuttle. Development of nanosize devices relies on nanoparticles, chemical attachment of linkers to the particle, and binding of proteins or drugs to the linkers. This nanoparticle can be used for cancer therapy [3]. It would be difficult to build, control, and communicate with electromechanical devices that are in the nanoscale size range. A bioelectronic device called a smart stent is shown in Fig. 25.3. The stent can degrade and disappear on command and has additional possible functions under investigation. The following are the 16 possible applications of stent: angioplasty stent, biliary stent, cerebral stent, colonic stent, or colorectal stent, coronary stent, carotid stent, duodenal stent, gastroduodenal stent, microstents to treat glaucoma, pancreatic stent, prostatic stent, esophageal stent, stent grafts, ureteral stent (ureters), and a vascular stent (vessels other than coronary). There is also a medley of eight possible general applications for intelligent implantable devices: tissue diagnostics, corrosion control, drug delivery, cell delivery, vascular repair, circulating sensors, cancer screening, and implants (stents, pins, plates, and others).

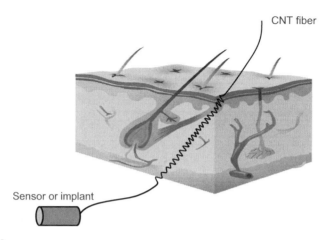

FIGURE 25.3

Example of a microsize nanomedicine device. Carbon nanotube fiber is used for transdermal communication, which is simple and reliable compared to wireless devices. (For color version of this figure, the reader is referred to the online version of this book.)

An area where tiny machines may improve health is the circulatory system. Blood is a liquid with cells and proteins suspended in it. Blood constantly circulates throughout the vascular tree providing the body with oxygen and nutrients, and removing waste. About 5 l of blood are contained in the body. Blood cells are called hematocytes and there are three main types. Red blood cells called erythrocytes carry oxygen to the tissues. Hemoglobin is an iron-containing protein and the main component of red blood cells. Hemoglobin helps supply oxygen and other respiratory gases to tissues. White blood cells called leukocytes fight infections. Platelets called thrombocytes are smaller cells that help blood to clot. These three types of blood cells comprise 45% of the blood, while 55% of the blood volume is plasma, which is the liquid constituent of blood. Plasma contains proteins, glucose, and other dissolved nutrients. Plasma helps blood to clot, transports substances through the blood, and has other functions. Blood is conducted through blood vessels (arteries and veins) and is prevented from clotting by the smoothness of the blood vessels and the finely tuned balance of clotting factors.

There are concepts of nanosize or microsize devices to sort particles in blood. A nanosize device uses impellors in a housing with a small clearance to sort small particles or impurity molecules. This device is too small to be manufactured at this time. But a larger size machine to sort larger particles such as cells may be feasible to build now [4,5]. The larger machine may use antibodies on plungers to capture select cells and move the cell to another chamber, and a cam mechanism releases the cells. This device is complex to build and may be difficult to maintain the antibody effectiveness. Another approach is to develop artificial cells called sensorcytes that can bind to specific cells or particles and the sensorcytes can be collected magnetically. The biochemistry and conjugation is still difficult, and small particles below about 200 nm may be cleared from the body by macrophages. Thus, cell-size sensorcytes may be desirable, also because they have a larger magnetic response to perform different functions on captured cells or tissue or for removal from the body. The sensorcytes, Fig. 25.4, may be injected into the blood stream and be removed from the body magnetically.

Another example of a microsize device is the artificial red blood cell, or reciprocyte. The first respirocyte, an artificial mechanical red cell, was designed by Robert Freitas from the Nanomedicine website, Foresight Institute [6]. The micron-sized device is at the scale that could be manufactured at this time, but some features of the reciprocyte may still be too small to fabricate [6–11]. The need is to deliver oxygen to the blood for people who are injured or who are under heat stress such as soldiers, first responders and firefighters. Nanotubes are reported to be nonthrombogenic (will not cause coagulation of the blood), are small and flexible and may be minimally invasive to tissue. Long nanotubes that are large in diameter would be used in a bundle and the interior of the nanotubes is open and hollow, Fig. 25.5(a,b). This artificial vein is also a nanocatheter that is flexible, strong, and electrically conductive and may be more practical to fabricate in the near future than the respirocyte. The large-diameter long carbon nanotubes (CNTs) are still under development. Another material form is a tube of nanotubes, Fig. 25.5(c), that might be infiltrated with a polymer to form a micron-sized catheter. The needlelike catheter could be a few centimeters long and then connect to tubing just outside the body.

An example of macrosize devices are the manipulators shown in Fig. 25.6. These devices are used for surgical procedures and must be rugged and reliable. However, the centimeter size range of the grippers or tweezers is too large for come applications and smaller tools would allow the surgeon to work with greater precision. Also,

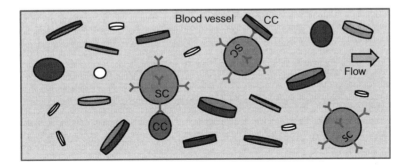

FIGURE 25.4

Sensorcyte (SC) artificial cells as molecular sorters shown capturing cancer cells (CC). Sensorcyte devices could be built at the microscale for sorting larger particles such as bacteria and cells. Sensorcytes are superparamagnetic nanocomposite materials. (For color version of this figure, the reader is referred to the online version of this book.)

718 CHAPTER 25 Tiny Medicine

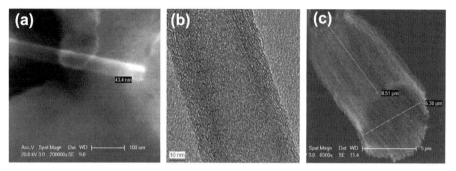

FIGURE 25.5

Nanotube artificial vein concept to deliver oxygen to the blood at time of injury. Nanotube artificial veins could deliver oxygen gas automatically at the time of injury or need from a pressurized canister worn outside the body. (a) Large-diameter CNT, (b) large-diameter nanofiber, and (c) tube of nanotubes. Arrows show the terminations of the inner walls of the carbon nanofiber. Ellipse shows the parallel wall structure. (For color version of this figure, the reader is referred to the online version of this book.)

FIGURE 25.6

Example of macrosize medical devices: (a) jaw tissue sealer, (b) endocutters, (c) 5-mm instruments, (d) hollow tip electrode, and (e) laparoscopic bipolar forceps. (For color version of this figure, the reader is referred to the online version of this book.)

Source: Images of surgical tools from Ethicon Endosurgery Inc. [12].

sensors should be integrated into some tools to guide the surgeon. Microsize nanorobot devices, it is expected, may provide improved surgical outcomes.

Small surgical devices could improve the precision of the da Vinci robot, Fig. 25.7. Nanorobots tethered on flexible catheters and bronchoscopes are other possible applications.

The use of nanoscale materials to build tiny machines and nanorobot devices from the nano- to the macroscale size was introduced in this section. Different types of nanoscale materials available for building the devices are described in the next section [13–58].

25.2 NANOSCALE MATERIALS

Nanoscale materials are enabling building nanorobot devices that are microscale in size. Processing and characterization of the materials is discussed in this section.

25.2 Nanoscale Materials   719

FIGURE 25.7

The da Vinci robot for minimally invasive surgery. (For color version of this figure, the reader is referred to the online version of this book.)

Source: Image from UC College of Medicine [13].

Nanoscale materials can be considered to be molecules or single crystals because of their small size and include carbon and noncarbon materials. The different material forms that are needed to construct the robots are discussed in the following subsections.

25.2.1 Carbon nanotube materials

The fundamental carbon materials include buckyballs, graphene, and nanotubes. Nanotubes are the most important material. Nanotubes may be processed into a family of larger material forms including thread, yarn and sheet, as shown in Fig. 25.8. The synthesis of carbon materials and their different material forms are described well in Chapters 1, 3, 8, etc. of this book and thus are not described in detail here.

CNT thread properties improve as the diameter of the thread decreases. Submicron-diameter thread is shown in Fig. 25.9 along with micron-sized coated

FIGURE 25.8

CNT materials. (a) Forest, (b) winding ribbon, (c) thread, (d) nanonet, and (e) two-ply yarn. An Office of Naval Research (ONR) program manager has suggested that a family of forms of nanomaterials be developed and commercialized for different applications. (For color version of this figure, the reader is referred to the online version of this book.)

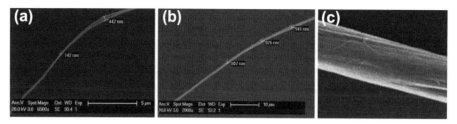

FIGURE 25.9

CNT thread. (a) Nanothread (about 300 nm diameter) twisted using a rotational tool on the Kleindiek manipulator, (b) nanothread (about 800 nm diameter), and (c) microthread (20 μm diameter) with a thin polymer coating.

thread. The coating densifies the thread and can also provide electrical insulation for electrical wiring applications.

Carbon is being investigated to replace copper for certain electronics applications. The electrical properties of annealed CNTs are being measured and the thread is being modified by doping to improve electrical conductivity. Making electrical measurements on nanotubes is shown in Fig. 25.10. The maximum current density of annealed CNT is measured using Kleindiek robots and tips. The current is increased until the CNT fails. The maximum current density in vacuum is greater than that in air.

High-temperature annealing (2800 °C) increases the electrical conductivity of CNT. The resistivity of CNT also decreases with increasing temperature, as shown

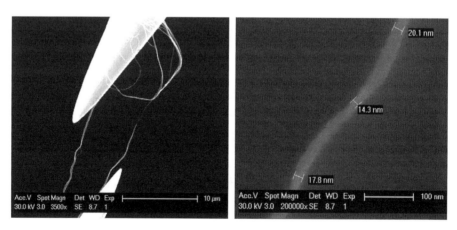

FIGURE 25.10

Using two probes to pass current through a CNT bundle. The right image is a magnification of the red square in the left image. The average cross section was calculated based on the three measurements in the right image. The measured CNT maximum current density at failure in vacuum is 2.6E+07 A/cm^2. Images are at different magnification.

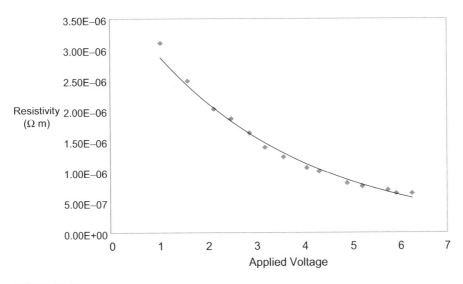

FIGURE 25.11

Measured electrical properties of annealed CNTs in vacuum. Increasing voltage increases the temperature of the CNT (annealing performed by John Bulmer and Larry Christy at the AFRL [14]). (For color version of this figure, the reader is referred to the online version of this book.)

in Fig. 25.11. Note that increasing voltage increases the temperature of the CNT. This test was conducted in vacuum. CNTs (annealed, but not doped or coated) are about 40 times more resistive than copper.

Improving the electrical conduction of CNTs is possible by increasing the number of walls in a CNT of a given diameter. This assumes that the length of the nanotube is much greater than the diameter and that all walls conduct electricity. The increase in cross-sectional area of the nanotube versus the number of walls is shown in Fig. 25.12(a). Increasing from three to seven walls doubles the conductivity of a two-wall CNT of 10 nm diameter. Synthesis of large-diameter CNT is also being investigated. A large-diameter (44 nm) CNT is shown in Fig. 25.12(b).

NIL is being used to pattern nanotubes on substrates. The imprinting can be done on flat substrates. Wells for the catalyst can also be put in the substrate. Fig. 25.13 shows patterning of 40-nm-diameter holes in a polymer resist. NIL for patterning catalyst is under development and may provide the best control over nanotube and forest geometry.

25.2.2 Magnetic nanotube and nanowire materials

Electric machines such as electric motors, transformers, and solenoids operate based on Faraday's law, which describes the force produced by the interaction

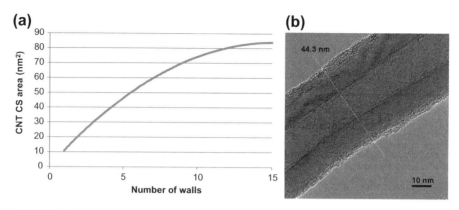

FIGURE 25.12

Increasing the number of walls of nanotubes. (a) CNT cross-sectional (CS) area versus number of walls of a 10-nm-outer-diameter nanotube and (b) a large 44-nm-diameter CNT with about 30 walls. (For color version of this figure, the reader is referred to the online version of this book.)

between a current-carrying conductor and a magnetic field. The materials that are used to construct the motor define its performance. Copper and iron materials have always been used in the fabrication of electric motors. Rare earth magnets are utilized to obtain the highest magnetic performance. In this section, nanoscale materials are proposed to be used to develop new electric motors, transformers, actuators, and devices that are lightweight with high energy density. The new materials are based on nanoscale carbon and magnetic materials, which can be readily manufactured. It is expected that the weight of existing electromagnetic devices can be reduced by half, by use of nanomaterials. Nanomaterials that will be used are (1) CNT wire that is strong, lightweight, and electrically and thermally conductive and

FIGURE 25.13

Patterning 40-nm-diameter holes in a polymer resist using nanoimprint lithography (NIL). Images at different magnifications are shown.

(2) superparamagnetic nanoparticles that are electrical insulators that are exchange coupled, with little hysteresis. Electromagnetic designs using nanomaterials will allow development of many types of biomedical microdevices and tiny machines that will change engineering design. Copper wire may be replaced by doped CNT superfiber. Iron core in motors must be replaced by a lighter material. Different magnetic nanomaterials are being investigated to replace iron for electromagnetic applications. Promising materials are magnetite nanotubes, as shown in Fig. 25.14(a), and Ni nanowires, as shown in Fig. 25.14(b,c). These materials can be used to build motors, transformers and solenoids. The approach suggested is to align the magnetic nanotubes/nanowires in a polymer matrix. The alignment direction of the nanotubes/nanowires would be the direction of the magnetic flux in the motor or transformer. There are several important advantages of making a magnetic material using nanomaterials. First, eddy currents, which cause heating and reduce the efficiency of motors, will be very small because magnetic composites will not have electrical conduction laterally. Second, if the nanotubes/nanowires are small enough in diameter, the magnetic flux in the composite might be linked by exchange coupling, which means the magnetic flux density of the composite may be close to the strength of the magnetic flux in iron material. Moreover, if the diameter of the nanotube/nanowire is small enough, about 10 nm, the nanotube/nanowire might have superparamagnetic properties. Superparamagnetic materials can operate at higher frequency with little hysteresis loss. The properties of iron nanotubes and nanowires are reported in [15—21]. But the properties of nanocomposite magnetic materials and the possible superparamagnetic effect must still be proven for replacing iron core. On the down side, there are difficulties fabricating magnetic nanocomposites with a high volume fraction of the magnetic material and the polymer composite will not operate at a high temperature as iron core. One particular hypothesis is that the iron nanotubes must be aligned in a magnetic field and be well dispersed in the polymer during curing in order to achieve the desired magnetic effects. In the past, there have been efforts to form

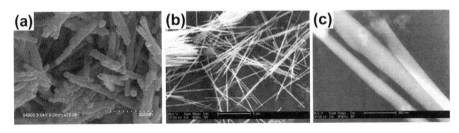

FIGURE 25.14

Magnetic nanoscale materials. (a) Magnetite (Fe_3O_4) nanotubes and (b,c) Ni nanowires at different magnifications manufactured at the University of Cincinnati.

Source: (a) From Wang and Geng, 2010 [59].

a magnetic composite material by integrating micron-sized iron filings into the polymer. The large-size iron particles are multidomain and are not expected to have superparamagnetic properties. Recently, efforts are on to integrate magnetic nanoparticles into polymer. The theory is that the nanoparticles will provide the superparamagnetic effect and exchange coupling. Thus, using nanoparticles is different than previous work using micron-sized iron filings. The magnetic nanoparticles are new materials that have the potential to produce a new type of magnetic material with properties that exceed those of iron or iron-filled polymers.

25.2.3 Magnetic nanoparticle materials

Magnetic nanoparticle materials are nanometer-diameter spherical particles of iron, cobalt, nickel, or other combinations of magnetic materials. As compared to solid materials, and the magnetic nanotubes/nanowires discussed above, nanoparticles are the smallest size magnetic material available. Nanoparticles are about 10 nm in diameter with a single magnetic domain. There are several advantages that make nanoscale materials attractive for electromagnetic design. Superparamagnetic nanoparticles combined with nanotube yarn will be used to enable high-frequency communication and the operation of electromagnetic devices and nanorobots. Carbon nanomaterials reduce the weight and the inertia of components and thus allow higher mechanical rotational velocities. Other advantages of nanomotors are rapid acceleration, high torque, and the ability to operate at higher speeds. However, the efficiency of the motor or device must approach that of copper, which is why long CNTs that have low electrical resistance are sought. Small-diameter motors will have ultrahigh magnetic field intensities and produce large forces, since the force is proportional to the electrical current/distance between conductors. Eliminating iron cores and losses due to eddy currents within the iron is a significant factor for nanorobots.

Magnetic core materials that are light and carry a similar flux as iron laminations minimize eddy current losses. This core may be possible to fabricate by using magnetic nanoparticles that are coated with an electrical insulator and subsequently distributed within a polymer. When the diameters of magnetic nanoparticles are below ~ 20 nm, they become superparamagnetic, which means that the nanoparticles consist of a single domain where their magnetization, on an average, is equivalent to zero. However, an external magnetic field is able to magnetize the magnetic nanoparticles since their magnetic susceptibility is much larger than that of paramagnets. When an external magnetic field is applied to an assembly of superparamagnetic nanoparticles, their magnetic moments tend to align along the applied field, which leads to a net magnetization. The magnetization curve (magnetization as a function of the applied field) is S shaped, which means that the particles behave as paramagnets that possess large magnetic moments. The magnetic field can be switched back and forth at high frequency, which thus opens the possibility for the operation of high-frequency motors and a transformer to power the nanorobot. The coating of the nanoparticles will provide electrical insulation and maintain their magnetic properties to serve as a core material.

25.2.4 Superparamagnetic core design and testing

Manufacturing nanoscale magnetic materials can be done via powder synthesis using coated magnetic nanoparticles. The powder is consolidated into exchange-coupled cores [22–31]. When reducing the particle size and the separation between neighboring particles to the nanometer scale, Co- or Fe-based nanocomposites can possess permeability similar to that of bulk Co or Fe metals. The high permeability is due to the exchange coupling effect, which leads to magnetic ordering within a grain and extends to neighboring particles within a characteristic distance called the exchange length L_{ex}. Thus, neighboring grains separated by distances shorter than L_{ex} may be magnetically coupled by exchange interaction. The permeability of an exchange-coupled nanocomposite might be as high as the permeability of its bulk counterpart. Soft nanocomposite magnetic materials possess higher initial permeability and a higher cutoff frequency than iron core materials. This advance opens up greatly the parameter space for motor and transformer design. This nanocomposite material is beginning to be offered commercially [22,27]. Eddy currents will be minimized in CNT materials, as explained in Fig. 25.15. We anticipate that

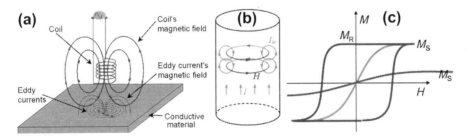

FIGURE 25.15

Eddy currents and magnetic losses are minimized in CNT materials. (a) Eddy currents are circular currents that are formed in a material. When a composite of superparamagnetic nanoparticles and a polymer is used, it does not conduct electricity very well. Thus eddy currents will be small and one of the major problems when using soft iron cores in transformers and motors (eddy currents), which cause heating and power loss, will be eliminated [32]. (b) Also, the skin effect in CNTs is minimized [32–35] as each CNT shell is so thin that eddy currents are correspondingly small if lateral conduction in the thread is prevented. (c) Superparamagnetic material (red) has high saturation magnetization (M_S), no remanence ($M_R = 0$), and no hysteresis loop, whereas ferromagnetic material (blue) has high remanence and paramagnetic material (green) has low saturation. (For interpretation of the references to color in this figure legend, the reader is referred to the online version of this book.)

Source: Figures and caption from Wikipedia [32–34].

nanomagnetics is going to provide a new generation of core materials for the enhancement of electromagnetic design.

Magnetization can be modeled simply for the initial design of the new core material for the tiny machine. If all the magnetic nanoparticles are similar with the same energy barriers and same magnetic moments with their easy axes all oriented parallel to the applied field, the temperature is low enough and the particles are closely spaced, then the material behaves like a paramagnet. The magnetization of the assembly of particles is approximated by $M = \mu n \tanh[(\mu \mu_0 H)/(k_B T)]$ where n in the density of nanoparticles within the sample, μ_0 is the magnetic permeability of vacuum, μ is the relevant magnetic moment or permeability of a nanoparticle [32,34], M is the magnetization of the material (the magnetic dipole moment per unit volume measured in amperes per meter), k_B is the Boltzmann constant, T is temperature, and H is the magnetic field strength, also measured in amperes per meter. The magnetic induction B (or flux density in Tesla) is related to H by the relationship $B = \mu_0(H + M) = \mu_0(H + \chi_v H) = \mu_0(1 + \chi_v)H = \mu H$ where χ_v is the volume magnetic susceptibility (e.g. degree of magnetization of a volume of material due to an applied magnetic field), and $\mu = (1 + \chi_v)$ is the magnetic permeability of the material. SI units should be used for the above calculations.

Magnetic nanoparticles are shown in Fig. 25.16(a). A typical sample made in our laboratory is a magnetic nanoparticle composite (Fig. 25.16(b)). The mass of the sample is 0.66 g. The dimensions of the cylindrical sample are as follows: height = 1.28 cm, diameter = 0.64 cm, volume = 0.41 cc, and density = 1.62 g/cc; particle type, gamma iron oxide nanoparticles (size: 20 nm); percentage by weight of nanoparticles, 44.7% w/w; binder used, Toolfusion epoxy (10 parts of Toolfusion 1 A resin + 2 parts of Toolfusion 1 B hardener); percentage by weight of epoxy, 55.3%; density of nanoparticles from MSDS, 5.24 g/cc at 20 °C; mass of nanoparticles in sample, 0.30 g; volume of nanoparticles in sample, 0.056 cc; and volume fraction of nanoparticles in sample, 0.138.

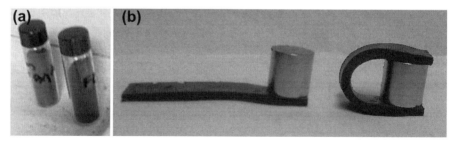

FIGURE 25.16

Development of superparamagnetic composites to replace iron. (a) Vial of iron oxide nanoparticles (left) and ferrousoferric oxide powder (right) and (b) hybrid material iron oxide nanoparticles cast in an elastomer and attracted by a magnet [36].

Testing showed that the specific relative permeability of a nanocomposite using iron oxide nanoparticles is about the same as that of a soft iron material. Continuing work is focusing on increasing the volume fraction of iron particles in the polymer to increase the permeability of the nanocomposite [36]. A tactic to increase the volume fraction of the nanoparticles is to apply a large magnetic field during curing, which moves the nanoparticles to one end of the sample. The lower density section of the sample is removed. The nanocomposite is electrically insulating and lightweight.

25.3 A PILOT MICROFACTORY FOR NANOMEDICINE DEVICES

A research facility is being developed to fabricate nanomaterial-based tiny medical devices that can change the outcome for patients. The goal is to use recent breakthroughs in nanoscale materials in microelectromechanical devices and medical applications. Transitioning engineering breakthroughs into applications requires special facilities to develop miniaturized devices. These facilities must be developed as they are not generally available. A small fabrication facility is required where engineering researchers can manipulate nanostructured materials to reproducibly build prototype engineering and medical devices that physicians and medical researchers need and can evaluate under clinical conditions. This section describes development of a research facility to fabricate prototypes of new nanomaterial devices using nano- and microscale materials. The research facility is to enable scientists, engineers, physicians, designers, students, and industrial partners to develop miniature engineering devices and medical devices that operate in the body. The facility will process nanoscale raw materials such as CNTs and magnetic nanoparticles, and possibly other materials such as boron nitride nanotubes (BNT) and semiconductor nanorods, into intermediate and derivative materials such as functionalized CNT/BNT thread and then use these materials to fabricate electrical and mechanical components for nanorobot devices. Developing the prototyping facility is feasible because of recent advances in robotic manipulators, microscopy, and nanotechnology and formation of a multidisciplinary research team. The research facility physically is a large integrated single instrument built with four connected modules: M1, the doping, coating, spinning & magnetic nanoparticles module; M2, the machining & joining module; M3, the characterization & quality control module; and M4, the assembly & packaging module. Over 20 components are integrated into one large instrument in the research facility including an electrostatic coating component, nanomill machining component, miniature spinning component, electronic and microscopy component, and robots and fixtures.

The pilot microfactory will produce advances in the science and engineering of microfabrication technology and utilize nanoscale materials to build microdevices. The microfactory provides engineers with prototyping tools and nanoscale materials needed to reproducibly build miniature devices. The devices will allow investigating new approaches to medical science and enable critical in vivo measurements never made before. The facility offers researchers a means to economically prototype next-generation devices and to incorporate nanotechnology into engineering, biology and medicine. The first nanomaterials used in the facility are CNT arrays, ribbon, yarn and commercially available magnetic nanoparticles. CNT thread is being made using long nanotubes and is lightweight, stronger than steel, pliable, inert, nontoxic, and electrically and thermally conductive. Initial devices being fabricated will utilize these materials for miniature carbon electronics, electric motors, actuators, carbon wires, sensors, antennae, and actuators for inside and outside the body. These microdevices will monitor and repair the body in ways that were not possible before. The pilot microfactory will enable replacing conventional engineering with exploratory microengineering to produce nanostructured biomedical microdevices. With these devices, physicians will have tools small enough to go inside the body and do what they want to do. This new capability will produce inventions, patents, licensing, income, and jobs.

The microfactory uses commercial Kleindiek robots to grip, apply force, twist, and measure for (1) in-vacuum processing and (2) in-air processing under an optical microscope. There are currently no standard tools or commercial instrumentation available for the construction of biomedical microdevices. The microfactory is an integrated suite of instrumentation employed in the construction of tiny machines and other medical devices. Novel micro- and nanoscale devices produced in the factory will answer important scientific, engineering, and medical questions. It will take 13 steps of scale reduction by factors of two to go from millimeter-scale features in current devices to the precision of submicron-scale features in microscale devices. Considerations in the design of nanorobots and tiny machines in general are as follows: the performance is related to a strong magnetic field B, the efficiency is related to power loss due to electrical resistance of the coil, and the size should be small and depends on the size of the wire and force needed. This instrument consists of several modules and submodules, each of which has specific functionality and when assembled will comprise a single, large, integrated system. The term "pilot microfactory" is used because "pilot" refers to experimental and "factory" is from the analogy of molecular manufacturing. Robert Freitas, Jr, proposed to develop new machines in his book *Nanomedicine* [4]. The "pilot microfactory" is currently under development funded by the National Science Foundation and the University of Cincinnati. The "pilot microfactory" will manipulate raw nanomaterials into actual functional microdevices.

The "microfactory" can be used to answer the following research questions. What practical capabilities might be developed by micro/nanofabrication? Since the practical working distances are short, will micro-/nanoscale machinery enable high productivity by enabling motion cycles at high speeds? And will the high-throughput/low-cost manufacturing of high-performance micro-/nanoscale products be possible? Development of the microfactory is essential for advancing the field of robotics. It is our vision that the "pilot microfactory" will do for nanotechnology what "molecular foundries" have done for biology.

25.3 A Pilot Microfactory for Nanomedicine Devices

A key component of the microfactory is two Kleindiek robotic manipulators with needle tip probes as shown inside an SEM in Fig. 25.17. The robotic tools can be moved in translation and rotation about two axes. Rotation about a third axis is possible using a rotational attachment. There is a coarse motion mode for slewing and a fine motion mode that can move with a few nanometer's precision. The robots can be controlled by a playstation joystick or by computer inputs. Since the SEM has a limited field of view, and limited depth perception, moving the robots is slow and tedious. Probes with needle tips are often used on the robots. The tips are used to measure the electrical properties of small bundles of nanotubes. Nanoparticles such as nanotubes tend to stick to the needles and releasing the particles is sometimes difficult. Often, nanotubes stick together to form small bundles of 10 or so nanotubes. Tweezers are one of the tools used to pick up nanotubes, but they are too coarse to pick up individual nanotubes that might be 10 nm in diameter. Also, the tweezers bend and break easily if they are accidentally driven into the sample holder or if the two robots crash, which happens occasionally. A force gage tool is used to measure the strength of small bundles of nanotubes. Special glue that cures when irradiated by the electron beam is used. The glue makes the nanotubes brittle at the glue junctions. Many small problems as mentioned above must be solved to make handling and building devices with nanotubes simpler and faster. In general, the nanorobots being built will use nanotube thread that is close to micro diameter, which is easier to manipulate than nanotubes. Building micron-sized robots is much simpler than trying to build nanosize robots. Students enjoy using the Kleindiek robots but every task requires a large amount of time. An important factor in making nanorobots and tiny machines practical will be to develop better manufacturing

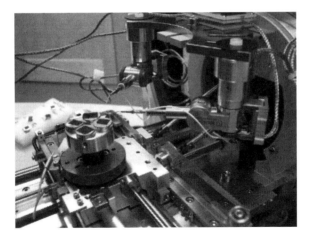

FIGURE 25.17

Two Kleindiek robots inside an SEM. A platform with four sample holder pins can be rotated to position the sample in the working position. (For color version of this figure, the reader is referred to the online version of this book.)

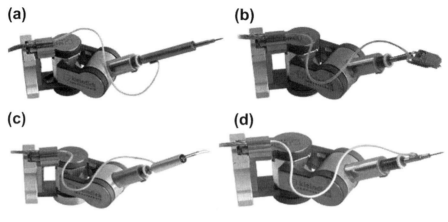

FIGURE 25.18

Kleindiek robot tools attachments: (a) rotational tip, (b) force gage, (c) tweezers, and (d) picoamp probe. (For color version of this figure, the reader is referred to the online version of this book.)

methods to handle microsize devices. The Kleindiek manipulators are extraordinary tools. A factor limiting their application is their high cost.

Kleindiek manipulators with different tools are shown in Fig. 25.18. The tools are used for characterizing nanotubes and making test samples of nanotube thread that have submicron diameter.

The tweezers are used to pick up a bundle of nanotubes in Fig. 25.19. Manipulating the nanotubes with the tweezers gives an indication of the quality of the nanotubes. Nanotubes with defects will break easily, whereas high-quality nanotubes may slip out of the grips.

25.4 TINY MACHINES CONCEPTS AND PROTOTYPE FABRICATION

Carbon materials and electric motor design are discussed to provide rationale for developing the first tiny machines. It is envisaged that most of the nanorobot will be manufactured using nanoscale materials. The housing could be cast using an elastomer or polymer and the carbon wire coil will be lighter than metal. The heat conduction of CNTs is very good and they exhibit a large maximum current density with the capacity for conducting large current, in relative terms, to generate strong magnetic fields. Electrical power conduction using lightweight nanotube material is limited by the temperature range of a given application. Thus the thermal design and cooling strategies for nanorobots will be critical design considerations. The resistivity of nanotube yarn decreases with increasing temperature, which is advantageous in comparison to copper. Reducing the electrical resistance of nanotubes and yarn

25.4 Tiny Machines Concepts and Prototype Fabrication

FIGURE 25.19

Tweezers gripping a bundle of nanotubes.

will be an important goal toward increasing the power and improving the efficiency of nanorobots [4]. The electrical conductivity of long nanotubes proposed to be manufactured will exceed that of copper on a per weight basis, and potentially approach it on a per area basis contingent on the doping, functionalization, and postprocess annealing of the nanotubes. Improvements are anticipated that will enable the conductivity of CNTs to be equivalent to that of copper (e.g. the electrical conductivity of short perfect armchair nanotubes is better than that of copper). Currents within the nanorobot windings will produce large magnetic fields [32—34,37].

25.4.1 Electromagnetic devices using CNT yarn

CNT can be used to form lightweight electromagnetic devices. Fig. 25.20(a) demonstrates the whirling of CNT yarn, and Fig. 25.20(b), the high-temperature capability of CNT thread that is used in the construction of nanomaterial robots and microdevices.

Reference [38] illustrates the comparison of a coil that is built using copper wire and CNT thread. The equation $\frac{B'}{B} = \frac{g'r}{gr'}$ shows that the magnetic flux density B' of the nanotube thread-based coil will be larger than the magnetic flux density of the copper coil B because the maximum current density g' of the nanotube thread is greater than the maximum current density g of the copper wire and the radius of the nanotube coil r' is smaller than the radius r of the copper coil. The CNT coil is smaller and has a larger flux density. This indicates that microdevices can have better specific

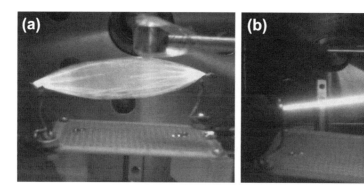

FIGURE 25.20

Carbon electromagnetics. (a) Whirling CNT yarn, demonstrating the principle of a carbon electric motor and (b) high current density of CNT yarn. (For color version of this figure, the reader is referred to the online version of this book.)

performance than macroscale electromagnetic devices, assuming that the duty cycle is low enough to ensure that the cooling of the device is satisfactory. By using long CNT; high-voltage, high-frequency, parallel CNT windings; and a superparamagnetic nanoparticle core, the nanomaterials-based motor will be smaller in radius r, lighter, and have higher flux density B' for an equivalent current i than existing motors.

The first devices to be developed in the microfactory are expected to include (1) a smart needle biosensor to diagnose lung tumors and prostate cancer, (2) a CNT thread nerve scaffold, (3) a sensor to monitor the subarachnoid space, (4) sensors for biodegradable metal implants, and (5) a longer term project to develop a sensor-cyte (sensor artificial blood cell) to screen for cancer. The new devices will have greater operational and computational capabilities than biological materials and will provide new understanding of disease and new possibilities for therapies.

The development of nanomaterial robots would revolutionize microscale engineering by making the construction of tiny machines feasible [14,24,39–51]. Especially robots that can function inside the human body have been a dream of engineers for decades [45,48]. A robot that is kinematically redundant and may be too complex to build at the microscale is shown in Minimally Invasive Therapy and Allied Technologies http://informahealthcare.com/toc/mit/15/3. This robot has kinematically redundant motion and may be too difficult to build at the microscale.

A simpler concept robot is shown in Fig. 25.21.

A simpler version of this robot may be feasible. A single-degree-of-motion robot is shown in Fig. 25.22. This robot is actuated using a solenoid with CNT wire and a nickel core. The diameter of this tweezer robot is designed to submillimeter size.

The device is being made using CNT thread [43] for coils and a strong solenoid to extend and retract the nickel nanowire core or shaft. The size can be submillimeter or larger. Modeling of the solenoid is given in Ref. [3]. The force versus displacement of the solenoid is shown in Fig. 25.23.

25.4 Tiny Machines Concepts and Prototype Fabrication

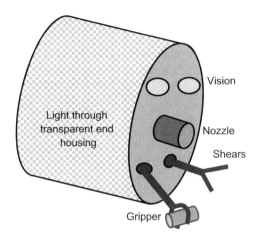

FIGURE 25.21

Concept robot. This robot has all the features needed including lighting, vision, probes, sampling, and drug delivery. (For color version of this figure, the reader is referred to the online version of this book.)

The tiny robot may be used for surgery and other in-body manipulation and may allow arterial atherosclerosis sensing and treatment. Clinicians would like to know if the plaque can flake off the wall. Arteries are 0.5–8.5 mm, veins are 0.5–2 mm, and capillaries are 0.0062 mm. The manipulator should be 0.1 mm in diameter to operate in arteries and veins. The micromanipulator may also be useful for some cases of "inoperable cancer". Surgery using current technology may not be appropriate for several reasons. Some cancers tend to spread early, and surgery is usually suggested only for small tumors; but 200 diseases are collectively called cancer. Surgery is usually considered for stages 1, 2 and 3A non-small cell lung cancer. Stages 3B and 4 lung cancers are most often treated with nonsurgical methods, such as chemotherapy and radiation therapy. If a tumor is located near vital structures, such as the

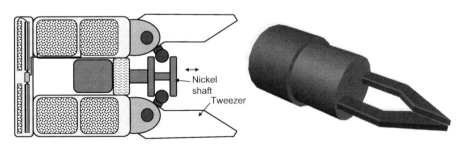

FIGURE 25.22

Biomedical micromanipulator platform: work in progress. (For color version of this figure, the reader is referred to the online version of this book.)

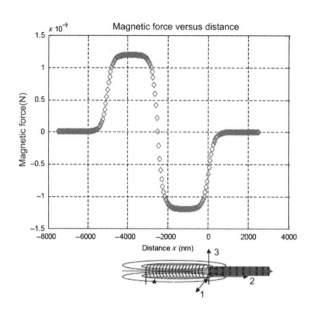

FIGURE 25.23

Magnetic force generated by a CNT solenoid on a Fe nanowire core changing with distance. The distance means the separation between the right side of the solenoid and the left side of the core. Iron nanowire Core is $D = 600$ nm in diameter and $L = 2.5\,\mu$m in length. There are 400 turns of CNT wiring on solenoid. There will be 8 layers. For each layer $N = 50$. When applying a 20V voltage across the solenoid, the current is about $70\,\mu$A.

Source: From [3].

heart, or the tumor is entangled in arteries and buried, surgery may be too difficult. Portions of the tumor often outgrow the immediate blood supply and die or "undergo necrosis". In contrast, peripheral regions of the tumor readily recruit the growth of new blood vessels (angiogenesis), enabling continued rapid growth of the tumor. The goal of surgery is to remove as much of the visible tumor as possible without damaging normal tissue. The invasive and infiltrating nature of some tumors make this a very challenging task. An array of new technologies, such as operating microscopes, microdissection techniques, intraoperative computerized image guidance, intraoperative ultrasound, intraoperative brain mapping, and most recently, real-time magnetic resonance imaging, makes surgical resection safer than ever. Due to the risks related to surgery and general anesthesia, some medical conditions could make surgery too dangerous. A much smaller surgical tool may be applied in some of the cases that are too risky using present bigger instrumentation. A sensor to identify cancer tissue integrated into the tweezer would be a great aid to the physician.

25.4.2 Failure mechanisms for in vivo devices

Biofouling coating with a protein is a problem for medical devices, as illustrated in Fig. 25.24(a). Heating the surrounding tissue may be a problem with in vivo medical

25.4 Tiny Machines Concepts and Prototype Fabrication

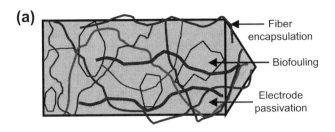

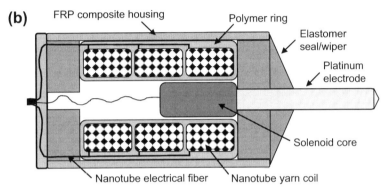

FIGURE 25.24

New sensor electrode using nanorobot technology. (a) Typical sensors face the problem of biofouling. (b) A self-cleaning telescoping biosensor electrode is proposed for in vivo impedance sensing. Two electrodes of the same or different materials may be used. Applications include tissue diagnostics for cancer screening, monitoring the integration of biodegradable metal implants, and sensing chemicals using electrodes made of different materials, e.g. sensing dissolved hydrogen using platinum and silver electrodes. The self-cleaning electrode uses the same basic solenoid design as in the micromanipulator.

devices, depending on the duty cycle. Low electrical resistance of CNT thread is desirable to reduce heating. Some devices may need to be cooled. Higher voltage might be used to reduce current and heating in conductors. A low duty cycle of the manipulator is desirable. Approximate power loss (neglects magnetic loss) due to heat generation is $P_{loss} = i^2 R = V^2 R$. A concept implantable biosensor with a retractable self-cleaning coaxial electrode is shown in Fig. 25.24(b). This is an antibiofouling design for a biosensor. The sensor is for in vivo use to measure the electrochemical impedance of tissue to detect cancer or measure chemicals.

A list of components and methods for manufacturing components of the manipulators are shown in Table 25.1. Conventional manufacturing such as using fasteners and machining holes is not practical when fabricating nano- and microdevices.

25.4.3 Nanomaterial electric motor

At small sizes, it is difficult to build conventional types of motors. Often, electrostatic motors are used at the microscale. A goal is to build a conventional type of

CHAPTER 25 Tiny Medicine

Table 25.1 Manufacturing Components of the Manipulators

Part Number	Component Name	Number Required	Materials	Manufacturing Method
BM-1	Tweezer	2	Nickel	Mask and photolithography
BM-2	Coil	2	CNT thread & epoxy	Winding, coating, encapsulating
BM-3	Mount	2	Nickel	Mask and photolithography
BM-4	Cushion/Seal	2	Elastomer	Casting in to mold
BM-5	Housing	1	Epoxy or bakelite	Cast around coils, end and center surfaces polished
BM-6	Shaft	1	Nickel	Etched from nickel wire
BM-7	Electronics	1	Au on Si	Microelectronics fabrication
BM-8	Electrode	1	Au on oxidized Ni	Etched and coating
BM-9	Scraper	1	Nickel	Etched

FIGURE 25.25

Motor designed with superparamagnetic nanoparticle composite and nanotube thread; the halbach motor could be used on the microscale. (For color version of this figure, the reader is referred to the online version of this book.)

motor at the microscale but to eliminate copper, iron, and rare earth magnets in electric motors. Long CNT; high-voltage, high-frequency, parallel windings; and a CNT superparamagnetic particle core are proposed. The nanomaterial motor is expected to be 50% lighter than conventional motors at the macro- or microscale. The concept motor is shown in Fig. 25.25.

In motor design, dense CNT thread, high-temperature insulation, cooling for high-power applications, and future chirality control are goals to improve the multifunctionality of CNT materials. An approach for operating CNT thread at high temperature is illustrated in Fig. 25.26.

25.4.4 Thermal analysis of nanomaterial components

A key consideration in the design of electromagnetic devices is the dissipation of heat. Metals are good thermal conductors, whereas CNTs are excellent thermal

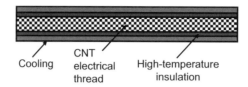

FIGURE 25.26

Concept for insulating and cooling CNT thread for high current density applications. (For color version of this figure, the reader is referred to the online version of this book.)

conductors in their axial direction and are thermal insulators in the transverse direction, and CNTs have low thermal mass. When a device is cycled on/off, the temperature of the CNT rises and then drops to ambient temperature almost immediately. In contrast, a copper wire requires several seconds to return to ambient temperature. CNT thread would allow the heat from the device to be dissipated more rapidly. It could thus allow the motor or solenoid to run at very high loads or the nanomaterial robot to charge quickly for short periods without building up excessive heat in the system. The thermal conductivity of perfect CNTs is greater than that of copper and the specific heat of CNT is lower than that of copper. If the resistivity of CNT is reduced to approach the level of copper, then CNT will be a superior conductor for electromagnetic devices in reducing the buildup of heat in the system. High-temperature dielectric materials such as boron nitride, aluminum oxide and glass may be needed to coat the CNT thread. Heat transfer in the nanorobot device must be modeled including the thermal properties of the nanoparticle core material and the CNT thread. Magnetic and eddy current losses are expected to be greatly reduced through the use of superparamagnetic materials, but high-frequency operation will generate more heat. A thermal analysis must be performed in the future to facilitate the production of a prototype nanorobot design. Important considerations when designing small components are that the surface area to volume ratio becomes large, which will increase heat transfer, and that when materials become thin the melting temperature may decrease significantly (e.g. the melting temperature of thin-film iron is about 800 °C and that of bulk iron is 1100 °C).

25.4.5 Composite electromagnetic material

Magnetic particles in a polymer may be used to form a magnetic composite material. The CNT thread is used as electrical wire to wind a coil around the magnetic composite. This coil and core form the secondary winding of a transformer and power the robot. Construction and initial bench testing of a large-size prototype core and coil have been performed.

The core material is a composite prepared by mixing particles in a polymer. This prototype does not have any other electronics beyond the coil and core. The final design can have diodes, capacitor, and antenna and the provision to release or sample

25.4 Tiny Machines Concepts and Prototype Fabrication

fluid. The wire used to wrap the flexible magnetic core is a thread spun from a vertically aligned CNT array synthesized using water-assisted chemical vapor deposition. A dry spinning method was used so that the thread is composed of highly aligned pristine CNT. The thread is then thermally treated and doped to increase its conductivity. The CNT thread must have an insulating coating to be used as a wire and wrapped into a coil in an electromagnetic device. This was accomplished by first coating the bare thread with a thin layer using an automatic coating system. The coating adheres well to the CNT thread and the coating system provides a very uniform coating. The thin layer is a base coat for a thicker layer of a second polymer, which is highly insulating and flexible. The second coating is the main insulating layer and bonds well to the thin layer.

It may be difficult to transfer power wirelessly to nanorobots at any position in the body. An option is to power the sensorcyte at one point when it passes near a coil. The coil might be worn as a wrist strap. A Helmholtz coil configuration is considered to power the nanorobot. This is one possible configuration of the primary coil. By using two identical coils of wire placed approximately one radius apart, a nearly uniform magnetic field can be generated at the center of the two coils. By alternating the current in the Helmholtz coils, an alternating magnetic field can be generated in the coil shown in Fig. 25.27(a). The flexible core increases the magnetic flux density and provides the electromotive force to power the nanorobot. The polymer core with the CNT wire wound onto it is shown in Fig. 25.27(b). Using this configuration, 18 mV peak voltage was generated in the prototype coil as pictured in Fig. 25.27(c) to demonstrate the concept. A larger number of turns can increase the power that can be transferred to the device. Future work will be to reduce the diameter of the core and coil for use in the body. Building the small-sized coil and core will require developing a miniaturized winding machine. Thus wireless powering of nanorobots is possible.

Future design may be a toroid transformer. This compact design has the secondary coil wound onto the primary coil. The secondary coil could power device components such as an antenna, sensor, drug delivery system, or fluid sampling system.

An elastomer nanocomposite material was developed using an elastomer and superparamagnetic nanoparticles, Fig. 25.28. The material mimics biological material for in vivo use. Fig. 25.29 shows magnetically trapping the magnetic nanocomposite material in a flow system.

Current clinically approved nanotechnology particles are relatively simple and generally lack active targeting or triggered drug release. *While all organisms on earth are composed of cells, no one has developed an artificial electromechanical cell.* A concept bionic artificial cell—electromagnetic device built using nanoscale materials—is shown in Fig. 25.30.

25.4.6 Biodegradability

In some cases, it may be desirable to include biodegradability as an inherent feature of nanorobot design. The materials used to construct the nanorobot should be

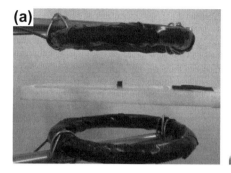

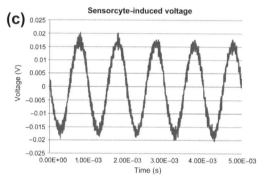

FIGURE 25.27

Testing the large-size prototype CNT and magnetic composite coil. (a) The large upper and lower coils are the circular coil primary windings of a Helmholtz coil transformer (number of turns is 25, coil diameter is 3 inches, coil spacing is 1.5 inches), no core is used for the primary coils, the small CNT coil and composite core are shown in the center. (b) Close-up of the hand-wound CNT coil and composite core transformer; diameter is 0.1 inch and height is ~0.15 inches. (c) Voltage generated by the CNT coil and composite core. Voltage is 18 mV at 5 kHz. (For color version of this figure, the reader is referred to the online version of this book.)

biodegradable and nontoxic, as are the superparamagnetic nanoparticle cores. Biodegradable polymers such as poly lactic-co-glycolic acid (PLGA), or an elastomer are available. Depending on their type and geometry, CNTs can potentially be toxic, hence there have been investigations into the potential biodegradability of CNTs [35]. Investigations into the biodegradability of each element utilized in the fabrication of nanorobots would have to be conducted.

25.4.7 Carbon nanotube wire

A carbon wire that is light and carries a large current with low impedance is possible using long CNTs that are postprocessed and operate at high frequency, where the impedance is low. Annealing, templating, coating, and doping techniques are being developed to lower the resistivity and electrical impedance of CNT yarn [52].

25.4 Tiny Machines Concepts and Prototype Fabrication

FIGURE 25.28

Testing a superparamagnetic nanocomposite material in water. (For color version of this figure, the reader is referred to the online version of this book.)

Presently, in the literature, the resistivity of CNT thread on a per weight basis is about equal to that of copper. At high frequencies, the advantage may be greater. Carbon wire is designed by drawing fine CNT strands, coating each strand with a dielectric, twisting the strands to form threads, and twisting the threads to form yarn. The yarn will essentially be a nanolitz wire that eliminates the skin effect. At high frequency, the impedance of CNT thread is lower than that of copper due to the capacitance of the thread and absence of the skin effect. As the CNT resistance is reduced, the cross-over frequency is moved to the left. It may be possible to power nanorobots at high frequency, which can reduce the impedance of the CNT [53–58]. A cylindrical magnesium pellet with attached Cu and then nanotube wire for implants is shown in Fig. 25.31.

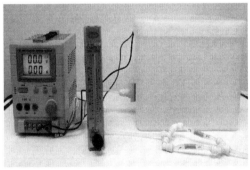

FIGURE 25.29

Trapping the magnetic nanocomposite material in a flow system. (For color version of this figure, the reader is referred to the online version of this book.)

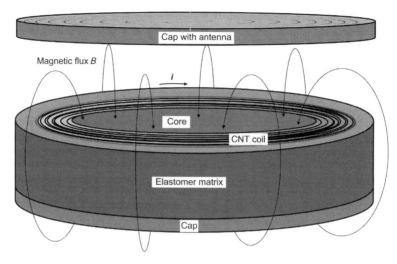

FIGURE 25.30

Concept bionic artificial cell for diagnostics and therapy. (For color version of this figure, the reader is referred to the online version of this book.)

25.4.8 Telescoping nanotubes

Telescoping nanotube arrays (TNAs) are being developed using NIL as envisioned in Fig. 25.32. The arrays have not been built yet. But NIL may provide the control needed to synthesize arrays that can be telescoped. TNAs will have many applications including electromagnetics, biomedical microdevices [38–40], and aerospace [41,42].

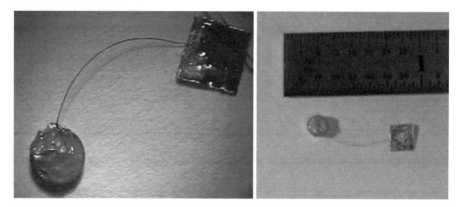

FIGURE 25.31

CNT thread used as a wire to cross the skin. Images are at different magnification. (For color version of this figure, the reader is referred to the online version of this book.)

25.5 Summary and Conclusions

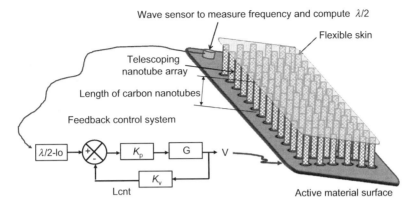

FIGURE 25.32
Telescoping array envisioned to be built using NIL. (For color version of this figure, the reader is referred to the online version of this book.)

FIGURE 25.33
Schematic of three-wall CNT telescoping. (For color version of this figure, the reader is referred to the online version of this book.)

This work to develop TNAs builds on the work of John Cummings and Alex Zettl who showed telescoping of individual nanotubes [40], Fig. 25.33.

25.4.9 Communication with tiny machines

Radio-frequency identification (RFID) chips in items should be tracked for safety or efficiency reasons, e.g. track guns to prevent crime, food for freshness, consumer products, sensors in the body, sensors for the environment, and many others (Fig. 25.34).

25.5 SUMMARY AND CONCLUSIONS

This chapter describes how devices that are microscale in size might be built using nanoscale materials. The devices may be called tiny machines or nanorobots as they are built using nanomaterials such as long CNTs, nanowires, and magnetic nanoparticles. The work in progress and the beginning of development of a pilot microfactory are significant steps toward fabricating the first nanomaterial robot

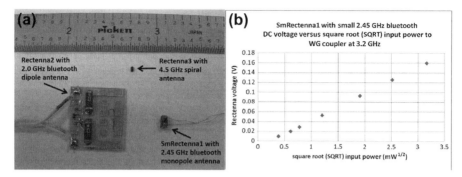

FIGURE 25.34

Antenna design for tiny machines. (a) Three prototype Rectennas for use with Sensorcyte. (b) Operation of SmRectenna1 at 3.2 GHz showing the correct linear dependence of output voltage on the square root of the low incident microwave power. Sm mean small; Rectenna means Rectifying Antenna; SQRT means square root. (For color version of this figure, the reader is referred to the online version of this book.)

devices. It is expected that nanorobot technological devices will serve as crucial elements in driving the future of microscale engineering and medicine. The capability to produce tiny machines will change engineering design at the microscale.

Acknowledgments

This research was supported by the NSF SNM GOALI: Carbon Nanotube Superfiber to Revolutionize Engineering Designs, the NSF ERC for Revolutionizing Metallic Biomaterials, the NSF MRI: Development of a Pilot Microfactory for Nanomedicine Devices, A Nano Antenna for Army Applications, and the ONR DURIP Mold Tool Set to Produce Breakthrough Carbon Nanotube Arrays. The opinions, findings, and conclusions or recommendations expressed in this chapter are those of the authors and do not necessarily reflect the views of the sponsors.

References

[1] http://en.wikipedia.org/wiki/Richard_Feynman#CITEREFFeynman1985.
[2] Transcript of the groundbreaking talk that Richard Feynman gave on 29 December 1959 at the annual meeting of the American Physical Society at the California Institute of Technology (Caltech), http://www.zyvex.com/nanotech/feynman.html.
[3] M.J. Schulz, V.N. Shanov, Y. Yun, Nanomedicine Design of Particles, Sensors, Motors, Implants, Robots, and Devices, with a Supplementary Materials and Solutions Manual, Artech House Publishers, 2009.
[4] R.A. Freitas Jr., Nanomedicine, In: Basic Capabilities, vol. I, Landes Bioscience, Georgetown, TX, 1999. http://www.nanomedicine.com/NMI.htm.
[5] R.A. Freitas Jr., Progress in nanomedicine and medical nanorobotics (Chapter 13), in: M. Rieth, W. Schommers (Eds.), Handbook of Theoretical and Computational

Nanotechnology, Bioinformatics, Nanomedicine, and Drug Design, vol. 6, American Scientific Publishers, Stevenson Ranch, CA, 2006, pp. 619–672, http://www.nanomedicine.com/Papers/ProgressNM06.pdf.
[6] http://www.foresight.org/, Foresight Institute.
[7] R.A. Freitas Jr., Nanotechnology and radically extended life span, Life Extension Magazine (January 2009). http://www.lef.org/.
[8] http://www.aananomed.org/, American Academy of Nanomedicine; http://www.amsocnanomed.org/, American Society for Nanomedicine.
[9] Nanoworld Laboratory, University of Cincinnati, http://www.min.uc.edu/nanoworldsmart.
[10] http://www.nanofactory.com/, Design of a Primitive Nanofactory.
[11] http://www.crnano.org/bootstrap.htm, Personal Nanofactory.
[12] Ethicon Endosurgery, http://www.ees.com/clinicians.
[13] University of Cincinnati College of Medicine, http://med.uc.edu/Home.aspx.
[14] Air Force Research Laboratory, Materials and Manufacturing Directorate, Thermal Materials and Sciences Branch, 2941 Hobson Way, Room 136, Wright-Patterson Air Force Base, OH 45433.
[15] H. Yoon, D.C. Deshpande, V. Ramachandran, V.K. Varadan, Aligned nanowire growth using lithography-assisted bonding of a polycarbonate template for neural probe electrodes, Nanotechnology 19 (2) (16 January 2008) 025304.
[16] L. Chen, J. Xie, M. Srivatsan, V.K. Varadan, Magnetic nanotubes and their potential use in neuroscience applications, Proceedings of the SPIE: The International Society for Optical Engineering, vol. 6172, 2006, pp. 61720J–61721J-8.
[17] V.K. Varadan, R.D. Hollinger, V.V. Varadan, J. Xie, P.K. Sharma, Development and characterization of micro-coil carbon fibers by a microwave CVD system, Smart Materials and Structures 9 (2000) 413–420.
[18] J.K. Abraham, H. Yoon, R. Chintakuntla, M. Kavdia, V.K. Varadan, Nanoelectronic interface for lab-on-a-chip devices, IET Nanobiotechnology 2 (3) (September 2008) 55–61.
[19] Q. Wang, K.M. Liew, V.K. Varadan, Molecular dynamics simulations of the torsional instability of carbon nanotubes filled with hydrogen or silicon atoms, Applied Physics Letters 92 (4) (28 January 2008) 043120–043121-3.
[20] J. Xie, L. Chen, V. Varadan, J. Yancey, M. Srivatsan, The effects of functional magnetic nanotubes with incorporated nerve growth factor in neuronal differentiation of PC12 cells, Nanotechnology 19 (10) (12 March 2008).
[21] V.K. Varadan, L. Chen, J. Xie, Nanomedicine: Design and Applications of Magnetic Nanomaterials, Nanosensors and Nanosystems, John Wiley & Sons, 2008.
[22] http://www.imego.com/Expertise/Electromagnetic-sensors/Magnetic-nanoparticles/index.aspx.
[23] D. Hasegawa, H. Yang, T. Ogawad, M. Takahashia, Challenge of ultra high frequency limit of permeability for magnetic nanoparticle assembly with organic polymer—application of superparamagnetism, Journal of Magnetism and Magnetic Materials 321 (7) (April 2009) 746–749. (Proceedings of the Forth Moscow International Symposium on Magnetism).
[24] A.-H. Lu, E.-L. Salabas, F. Schuth, Magnetic nanoparticles: synthesis, protection, functionalization, and application, Angewandte Chemie International Edition 46 (2007) 1222–1244.
[25] G. Reiss, A. Hütten, Magnetic nanoparticles, applications beyond data storage, Nature Materials 4 (October 2005). www.nature.com/naturematerials.

[26] P.M. Raj, et al., Novel Nanomagnetic Materials for High-Frequency RF Applications. Electronic Components and Technology Conference, 31 May–3 June, 2011, IEEE, Lake Buena Vista, FL, 2011, 1244–1249.
[27] http://www.inframat.com/magnetic.htm, Magnetic Nanocomposite Research.
[28] T. Dey, Polymer-coated magnetic nanoparticles: surface modification and end-functionalization, Journal of Nanoscience and Nanotechnology 6 (8) (August 2006) 2479–2483.
[29] http://www.popsci.com/science/article/2011-01/new-nanocomposite-magnets-could-reduce-demand-rare-earth-elements.
[30] J.W. Park, et al., Magnetic moment measurement of magnetic nanoparticles using atomic force microscopy, Measurement Science and Technology 19 (1) (2008).
[31] S.M. Moghimi, A.C. Hunter, J.C. Murray, Nanomedicine: current status and future prospects, The FASEB Journal 19 (3) (1 March 2005) 311–330.
[32] http://en.wikipedia.org/wiki/Superparamagnetism.
[33] http://en.wikipedia.org/wiki/Skin_effect.
[34] http://en.wikipedia.org/wiki/Permeability_(electromagnetism).
[35] A. Bianco, K. Kostarelos, M. Prato, Making carbon nanotubes biocompatible and biodegradable, Chemical Communications 47 (2011) 10182–10188.
[36] R. Venkatasubramanian, Composite Nanoparticle Materials for Electromagnetics, MS Thesis, University of Cincinnati, August 2012.
[37] http://en.wikipedia.org/wiki/Electromagnetism.
[38] M.J. Schulz, W. Li, B. Ruff, R. Venkatasubramanian, Y. Song, B. Suberu, W. Cho, P. Salunke, A. Sowani, J. Yin, D. Mast, V. Shanov, Z. Dong, S. Pixley, J. Hu, C. Muratore, Sensorcyte artificial cells for human diagnostics and analytics, in: F. Boehm (Ed.), Nanomedical Device and Systems Design: Challenges, Possibilities, Visions, CRC Press, 2012.
[39] Y. Yun, V.N. Shanov, S. Balaji, Y. Tu, S. Yarmolenko, S. Neralla, J. Sankar, S. Mall, J. Lee, L.W. Burggraf, L. Guangming, V.P. Sabelkin, M.J. Schulz, Developing a sensor, actuator, and nanoskin based on carbon nanotube arrays, SPIE Smart Structures and NDE Conference, March 2006, San Diego, CA.
[40] Telescoped Multiwall Nanotube and Manufacture Thereof, J P. Cumings (Oakland, CA), A.K. Zettl (Kensington, CA), S.G. Louie (Berkeley, CA), M.L. Cohen (Piedmont, CA), US Patent 6,874,668; Cumings, et al. 5 April 2005.
[41] F. Pinto, Nanopropulsion from high-energy particle beams via dispersion forces in nanotubes, 48th AIAA/ASME/SAE/ASEE Joint Propulsion Conference & Exhibit and 10th International Energy Conversion Engineering Conference, 29 July–1 August, 2012.
[42] InterStellar Technologies Corporation, 115 North Fifth Avenue, Monrovia, California, 91016 USA, http://www.interstellartechcorp.com/.
[43] V. Shanov, Y.-H. Yun, M.J. Schulz, Synthesis and characterization of carbon nanotube materials (review), Journal of the University of Chemical Technology and Metallurgy 41 (4) (2006) 377–390.
[44] P. Ball, Made to Measure, New Materials for the 21st Century, Princeton University Press, New Jersey, 1999.
[45] C. Montemagno, Integrative technology engineering emergent behavior into materials and systems, ICMENS, pp.2, International Conference on MEMS, NANO and Smart Systems (ICMENS'04), 2004.

[46] S. Hede, N. Huilgol, "Nano": the new nemesis of cancer (Review Article), Journal of Cancer Research and Therapeutics 2 (issue 4) (2006) 186–195. Available from:, http://www.cancerjournal.net/text.asp?2006/2/4/186/29829.

[47] http://www.ece.rice.edu/~halas/; http://bioe.rice.edu/FacultyDetail.cfm?RiceID=495; http://www.nanospectra.com/.

[48] T. Hogg, P. Kuekes, Mobile microscopic sensors for high-resolution in vivo diagnostics, Nanomedicine: Nanotechnology, Biology, and Medicine 2 (2006) 239.

[49] Kleindiek manipulators, http://www.nanotechnik.com/nanoindentation.html.

[50] http://www.journalamme.org/papers_amme06/192.pdf, Nanocrystalline Iron Based Powder Cores for High Frequency Applications.

[51] M. Alric, F. Chapelle, J.-J. Lemaire, G. Gogu, Potential applications of medical and non-medical robots for neurosurgical applications, 18 (4) (2009) 193–216.

[52] M. Schulz, V. Shanov, D. Qian, M. Sundaram, K. Vemaganti, Y. Liu, W. Li, N. Alvarez, B. Ruff, A. Krishnaswamy, J. Sullivan, G. Li, W. Jiang, R. Parlapalli, A. Johnson, A. Wang, M. Haase, R. Malik, A. Sowani, R. Venkatasubramanian, Carbon nanotube superfiber development, 5th WSEAS Int. Conf. on Nanotechnology, Cambridge, UK, 20–22 February 2013.

[53] Centers for Disease Control & Prevention, http://kidney.niddk.nih.gov/kudiseases/pubs/kustats/.

[54] National Kidney and Urologic Diseases Information Clearinghouse (NKUDIC).

[55] J. Ravi Kumar Reddy, et al., Nanomedicine and drug delivery—revolution in health system, Journal of Global Trends in Pharmaceutical Sciences, 2 (1) 2–11 March.

[56] O.C. Farokhzad, R. Langer, Impact of nanotechnology on drug delivery, ACS Nano 3 (1) (2009).

[57] P. Burke, C. Rutherglen, Towards a single-chip, implantable RFID system: is a single-cell radio possible? Biomedical Microdevices (24 January 2009).

[58] C.T. Phus, G. LIssorgues, Non-invasive measurement of blood flow using magnetic disturbance method, IEEE (2009). http://www.nve.com.

[59] Q. Wang, B. Geng, S. Wang, Y. Ye, B. Tao, Modified Kirkendall effect for fabrication of magnetic nanotubes, Chem. Commun. 46 (2010) 1899–1901.

CHAPTER

Carbon Nanotube Yarn and Sheet Antennas

26

Steven D. Keller, Amir I. Zaghloul
U.S. Army Research Laboratory

CHAPTER OUTLINE

- 26.1 Introduction 749
- 26.2 Carbon Nanotube Thread Antennas 751
  - 26.2.1 Electromagnetic theory of carbon nanotube thread dipole antenna ... 752
  - 26.2.2 Carbon nanotube thread conductivity 755
  - 26.2.3 Electromagnetic simulation of carbon nanotube thread dipole antenna current distribution 759
- 26.3 Carbon Nanotube Sheet Antennas 765
  - 26.3.1 Carbon nanotube sheet fabrication 765
  - 26.3.2 Carbon nanotube sheet patch antenna 765
- 26.4 Multifunctional Carbon Nanotube Antenna/Gas Sensor 773
  - 26.4.1 Meshed carbon nanotube thread patch antenna/gas sensor 773
  - 26.4.2 Effect of carbon nanotube thread spacing on antenna performance .. 778
  - 26.4.3 Effect of meshed carbon nanotube thread ground plane and feedline layers on antenna performance 781
- 26.5 Summary 785
- References 785

26.1 INTRODUCTION

Many functional challenges exist when wireless systems are integrated into textile-based platforms such as the clothing, helmet or personal gear of an individual such as a soldier or emergency first responder. Typical antennas for these systems are visually compromising and/or cumbersome to the user. Antennas fabricated from standard conductive materials such as copper or conductive ink fail to stand up to significant "wear-and-tear" due to lack of durability and are thus limited in their placement to areas on the platform that see minimal flexing and bending. The development of alternative materials that may be easily integrated into textile materials and that offer enhanced durability will be important for solving these challenges.

Carbon nanotube (CNT) structures have leapt to the forefront of materials research interest over the 2001–2010 timeframe because of their demonstration of a variety of attractive physical characteristics such as high durability (tensile strength exceeding 1–2 GPa) and extremely light weight in comparison with standard metallic materials such as copper and steel. In addition to possessing such valuable physical characteristics, it has been shown that CNTs may exhibit similar and in some cases enhanced electrical characteristics when compared with standard conductive materials such as copper [1]. The conductivity and power efficiency of an individual CNT may be exceptionally high since its quantum resistance does not inversely scale with the square of its radius like traditional metal structures and since electrons only flow along its carbon atom shell in two parallel propagation channels, referred to as π-bands, resulting in a negligible skin effect. The application of an individual single-walled CNT (SWNT) as an antenna has been analytically shown to yield very low radiation efficiency at microwave frequencies [2,3], mainly due to the large reactance resulting from classical and quantum effects of its nanometer radius. These include quantum capacitance, $C_Q = 100$ aF/μm, due to the finite density of states at the Fermi energy and kinetic inductance, $L_K = 16$ nH/μm, due to charge-carrier inertia of the electrons flowing along the nanotube [2]. While overall radiation efficiency of a single CNT structure may be lower than that of a traditional metal structure (since the resistance of CNT structures has been predicted to be quite high at the nanometer scale), the counterbalancing attractive physical and electrical characteristics make its use in RF and antenna applications worthy of investigation.

The electrical properties of CNTs have been applied to antenna designs by using the individual CNTs as conductors in an array format and also by harvesting individual CNTs and mixing them into a liquid to form a conductive solution. Specific examples of these applications include the fabrication of a flexible patch antenna structure from compressed CNT arrays incorporated into a polymer substrate [4] and a patch antenna printed onto a flexible substrate using CNT-enhanced conductive ink [5].

It has recently been shown through analysis [6,7] and measurement [8] that the dominant kinetic inductance and resistance of a CNT can be significantly reduced by bringing together a large number of CNTs into a bundle structure. It has also been predicted through simulations that dipole antennas constructed from CNT bundle structures may exhibit radiation efficiency orders of magnitude higher than that of individual CNTs [9] and that this radiation efficiency increases as the nanotube density with the bundle is increased [10]. Emerging fabrication techniques have made realizable the synthesis of large-scale CNT bundle structures such as threads, ribbons and sheets. By using these CNT threads to produce wireframe antennas (dipole, loop, etc.) and CNT sheets/ribbons to produce planar and waveguide antennas (patch, horn, etc.), it is possible to fabricate antennas with significantly reduced weight and enhanced flexibility/durability and power handling capabilities when compared with antenna structures fabricated out of traditional bulk conductive materials.

This chapter will discuss the application of bulk CNT materials to a variety of RF antenna designs through simulation and measurement. First, a technique for simulating the current distribution and radiation efficiency of CNT thread and rope dipole antennas will be presented and the results of example simulations will be compared with those of an antenna constructed from a traditional bulk conductive material (e.g. copper). Next, the application of CNT sheet material to antenna design will be discussed through the design, simulation and measurement of an aperture-coupled CNT sheet patch antenna. Finally, a variation on this patch antenna design constructed from meshed CNT threads will be presented and developed as a concept for a multifunctional communications antenna and reactive gas sensor. By understanding the benefits and tradeoffs associated with constructing antennas from bulk CNT materials, the groundwork is laid for design and development of complex antenna structures constructed from these materials, including textile-embedded CNT thread antennas and conformal CNT sheet antennas. As a point of reference, the experimental results presented in this chapter use as-grown short nanotubes to form sheet and yarn, and the results obtained serve as a useful baseline for nanotube antenna performance. However, the use of longer nanotubes along with high-temperature annealing and doping postprocessing methods under development will substantially improve the electrical properties of sheet and yarn. Thus, the performance of antennas described herein should likewise substantially improve.

26.2 CARBON NANOTUBE THREAD ANTENNAS

A conceptual diagram of a half-wavelength ($\lambda/2$) CNT thread dipole antenna, composed of many thousands of individual double-walled CNTs, is shown in Fig. 26.1.

The legs of the CNT thread dipole antenna, each $\sim \lambda/4$ in length, are connected to the output of a balun to match the balanced dipole load to an unbalanced coaxial feed connector and provide a 180° phase difference between the signal excitation in each leg of the dipole.

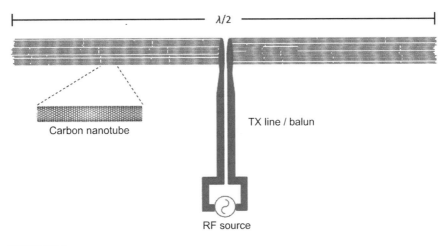

FIGURE 26.1

Carbon nanotube thread dipole antenna concept. (For color version of this figure, the reader is referred to the online version of this book.)

A diagram of the process for fabricating the CNT thread/rope that is used in this design is shown in Fig. 26.2. In order to fabricate bulk CNT thread material, a dense array of vertically aligned multiwall nanotubes (MWNTs), typically 10–12-nm-diameter double-walled CNTs, is grown using a chemical vapor deposition process on a 4″ silicon (Si) wafer with an array density of approximately 0.03 gm/cm^3 and an average tube length of 500 μm. The MWNTs are collected and spun into a thread-like structure, with the individual nanotubes within the thread held together by van der Waals forces. The final thread, referred to as 1-ply or single-ply thread, is typically 20–25 μm in diameter. By adding a weaving process into the overall spinning process with multiple spools of 1-ply thread, a variety of larger ply CNT rope may be fabricated, including 3-ply CNT rope and a 3 × 3-ply CNT rope (three 3-ply CNT ropes spun together).

An additional postprocessing technique of applying dimethyl sulfoxide (DMSO) to the CNT thread/rope samples may be employed to enhance the thread conductivity and tensile strength. In this process, the capillary forces caused by the evaporation of the applied DMSO solvent draw the neighboring CNTs within the threads closer together and yield a denser thread structure. This process both reduces the CNT thread/rope diameter and enhances the ability to work with and handle the samples. Environmental electron scanning microscope images of CNT thread/rope material are shown in Fig. 26.3.

26.2.1 Electromagnetic theory of carbon nanotube thread dipole antenna

In order to predict the radiation performance of a dipole antenna constructed from CNT thread, a Method of Moments simulation technique is applied to Hallén's

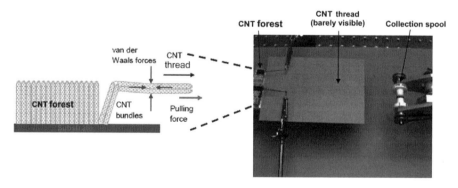

FIGURE 26.2

Carbon nanotube thread fabrication technique. (For color version of this figure, the reader is referred to the online version of this book.)

Source: Adapted from Ref. [11].

26.2 Carbon Nanotube Thread Antennas

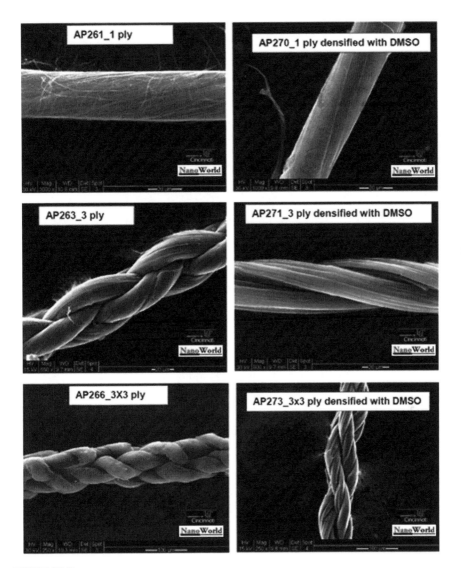

FIGURE 26.3

SEM images of standard and DMSO densified 1-ply, 3-ply and 3 × 3-ply carbon nanotube thread/rope. (For color version of this figure, the reader is referred to the online version of this book.)

integral equation for a thin wire [12]. The solution to the Helmholtz equation using the electric field, E, and the magnetic wave potential, A, serves as the starting point of this derivation. With this, the electric field may be represented in terms of the current density, J,

$$E = -j\omega\mu A + \frac{1}{j\omega\varepsilon}\nabla(\nabla \cdot A) \qquad (26.1)$$

where

$$A = \iiint \frac{Je^{-jk|r-r'|}}{4\pi|r-r'|}dr' \qquad (26.2)$$

In the above equation, k is the wave vector, ω is the angular frequency, ε is the permittivity, μ is the permeability, and $|r-r'|$ is the distance between a point source and a point in space. The current is assumed to travel axially along a wire with conductivity, σ. The surface current density, J_z, may be solved for with the general version of Ohm's law:

$$J_z = \sigma E_z \qquad (26.3)$$

Each of the cases that will be examined here will satisfy the criteria for a thin wire of half-length, L, and radius, a, with $ka \ll 1$ and $a \ll L$. As such, Eqn (26.1) may be rewritten with the standard thin wire kernel, $K(z-z')$,

$$E_z = \frac{1}{4\pi\omega\varepsilon}\left(k^2 + \frac{\partial^2}{\partial z^2}\right)\int_{-L}^{L} K(z-z')\,I(z')dz' \qquad (26.4)$$

where

$$K(z-z') = \frac{e^{-jk\sqrt{(z-z')^2+a^2}}}{\sqrt{(z-z')^2+a^2}} \qquad (26.5)$$

and

$$I(z) = 2\pi a J_z(z) \qquad (26.6)$$

In order to solve for the total electric field, $E^t = E^i + E^s$, from a radiating object with incident electric field, E^i, and scattered electric field, E_s, Eqn (26.3) becomes

$$J_z = \frac{I(z)}{2\pi a} = \sigma\left(E_z^i + E_z^s\right) \qquad (26.7)$$

By applying Eqn (26.4) as the scattered field, $E^s(z)$, in Eqn (26.7), and assuming a slice-gap voltage source for $E^i(z)$, a Hallén's integral equation is fully derived for the current distribution, $I(z)$, along a single thin wire of half-length, L, radius, a, and conductivity, σ, in Ref. [3] and results in

$$\int_{-L}^{L} \left(K(z-z') + \frac{\omega\varepsilon}{a\sigma}\frac{e^{-jk|z-z'|}}{k} \right) I(z') \mathrm{d}z' = c_1 \sin kz + c_2 \cos kz$$

$$- \frac{j2\pi\omega\varepsilon}{k} \sin k|z - z_0| \qquad (26.8)$$

Similarly, the current distribution for a bundle of thin wires composed of N wires of half-length, L, radius, a, and conductivity, σ, in Ref. [9] and results in

$$\int_{-L}^{L} \left(K_{\text{Bundle}}(z-z') + \frac{2\omega\varepsilon}{N\pi a\sigma}\frac{e^{-jk|z-z'|}}{k} \right) I(z') \mathrm{d}z'$$

$$= c_1 \sin kz + c_2 \cos kz - \frac{j2\pi\omega\varepsilon}{k} \sin k|z - z_0| \qquad (26.9)$$

For the bundle, the standard thin wire kernel, $K(z-z')$, is stated using a bundle radius, R, which also satisfies the thin wire criteria, $kR \ll 1$ and $R \ll L$,

$$K_{\text{Bundle}}(z-z') = \frac{e^{-jk\sqrt{(z')^2 + R^2}}}{\sqrt{(z')^2 + R^2}} \qquad (26.10)$$

According to Ref. [9], only the current along the wires at the outermost region of the bundle will contribute to far-field radiation while radiation from the inner wires will effectively be shielded. Thus, the number of wires, N, in Eqn (26.9) should be limited to the outermost tubes in the bundle, N_{outer}.

With the current distribution along the thin wire dipole, I_z, solved for in terms of frequency, geometry and material conductivity, a point matching Method of Moments solution may be applied to Eqns (26.8) and (26.9) as detailed in Ref. [3]. This solution is then calculated to compare the conductivity and current distribution of a copper wire, double-walled CNT, and a variety of multiwall CNT bundle (thread) dipole antennas.

26.2.2 Carbon nanotube thread conductivity

The dipole current distribution simulation described above assumes a thin wire with a surface current density. Thus, the conductivity for the dipole material should first be determined for this particular case. This has been derived for a general CNT geometry in Ref. [13] and for an infinitely thin SWNT and metal conductor in Ref. [3]. For the case of a thin metal cylinder,

$$\sigma_{\text{metal}} = -j\frac{e^2 N_e}{m_e(\omega - jv)} \qquad (26.11)$$

where e is the elementary electron charge (1.602×10^{-19} C), m_e is the electron mass (9.11×10^{-31} kg), N_e is the electron density, and v is the relaxation frequency. The value of N_e changes depends on whether the metal cylinder is modeled with a geometry similar to an SWNT with an infinitely thin wall (N_e^{2d}) or with a finite wall thickness (N_e^{3d}). In order to extend the comparison of this simulation to metal wire dipole antennas with realistic diameter sizes, the three-dimensional electron density, N_e^{3d}, will be used instead of the two-dimensional case, $N_e^{2d} = (N_e^{2d})^{2/3}$. For bulk copper, $N_e^{3d} \approx 8.46 \times 10^{28}$ electrons/m^3 and $v \approx (2.47 \times 10^{-14})^{-1}$.

For an infinitely thin SWNT with radius, a, the quantum conductivity is approximated as,

$$\sigma_{SWNT} = -j \frac{2e^2 v_F}{\pi^2 \hbar a (\omega - jv)} \qquad (26.12)$$

where v_F is the Fermi velocity for a CNT (approximately 9.71×10^5 m/s) and $\hbar$ is the Plank's constant (1.0546×10^{-34} m^{2}kg/s). The relaxation frequency of the SWNT is estimated as $v \approx (3 \times 10^{-12})^{-1}$.

As noted earlier, the CNT thread shown in Fig. 26.3 is composed of bundled MWNT structures. The conductivity must be adjusted to account for the multiple conductive paths offered by the multishell structures. According to Ref. [9], MWNTs may be thought of as a set of N coaxial cylindrical SWNTs with intershell spacing of approximately 3.4 Å. Assuming that electrons are confined to flow axially along each shell with no cross-shell hopping, it has been estimated that the MWNT may be treated as an SWNT with an effective radius of $R^{eff} = (R_1 + R_2 + \ldots + R_N)/N$, with R_N being the radius of the outermost shell, and with an effective conductivity of

$$\sigma_{MWNT} = \sum_{q=1}^{N} \sigma_{SWNT}^{(q)} \qquad (26.13)$$

Specifically, the MWNT structures in the CNT thread are double-walled ($N = 2$), and thus the conductivity of each CNT in the CNT bundle simulation will be approximated as

$$\sigma_{DWNT} \approx 2 \, \sigma_{SWNT} \qquad (26.14)$$

It has been experimentally shown that a large tunneling resistance, on the order of MΩ, exists between neighboring nanotubes within a CNT bundle [14]. Thus, each MWNT in this CNT thread simulation is assumed to be a parallel channel for electrons to flow with no cross-tube tunneling allowed. The CNT thread conductivity will be influenced by the number of MWNT conductive paths contained within the bundle. For this simulation, the CNT thread of radius, R, will be approximated as a square bundle of MWNTs of radius, a, with its cross-sectional length and width equal to the actual thread diameter, $2R$. For a densely packed bundle with uniform nanotube distribution, the total number of CNTs may be approximated as [15]

$$N_W = \frac{2(R-a)}{x} \quad ; \quad N_L = \frac{2(R-a)}{(\sqrt{3}/2)x} + 1 \quad (26.15)$$

$$N_{total} = N_W \cdot N_L - \frac{N_L}{2} \quad ; \quad N_{outer} = 2 \cdot (N_W + N_L) \quad (26.16)$$

$$\sigma_{Bundle} \approx N_{total} \cdot \sigma_{DWNT} \quad (26.17)$$

where x is the intertube spacing from center to center. Assuming a van der Waals spacing of 3.4 Å between nanotubes, the intertube spacing becomes $x = 2a + 3.4$ Å. Since the approximation of the CNT thread as a square bundle will yield slightly larger values of N_{total} and N_{outer}, the simulated CNT thread dipole conductivity and radiated power will be slightly higher than the measured values of the actual prototypes. In order to bring the simulated results closer to that of a circular bundle, N_{total} and N_{outer} will be scaled by the difference between the area of a circle inscribed within a square of side half-length, R,

$$\frac{A_{circle}}{A_{square}} = \frac{\pi R^2}{2R \cdot 2R} = \frac{\pi}{4} \quad ; \quad N_{total} \approx \frac{\pi}{4} N_{total} \quad ; \quad N_{outer} \approx \frac{\pi}{4} N_{outer} \quad (26.18)$$

The simulated conductivity of 30-gauge copper wire, a double-walled CNT, and CNT thread and rope (multiply thread) is shown in Fig. 26.4 (real part, Re[σ]) and

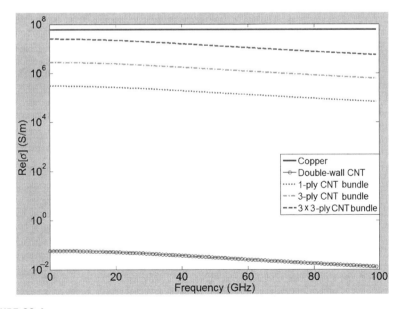

FIGURE 26.4

Real part of conductivity (log scale) vs frequency for carbon nanotube thread and copper wire. (For color version of this figure, the reader is referred to the online version of this book.)

Fig. 26.5 (imaginary part, Imag[σ]). The copper wire radius is 127.5 µm, the single CNT radius is 5 nm, the 1-ply CNT thread radius is 12.5 µm, the 3-ply CNT rope radius is 37.5 µm, and the 3 × 3-ply (9-ply) CNT rope radius is 112.5 µm. It should be noted that while the conductivity for the copper wire and CNT thread cases is calculated in siemens per meter, the accurate unit of conductivity for the individual double-walled CNT case is S since the SWNT is modeled as an infinitely thin tube.

The simulation data in Figs 26.4 and 26.5 indicate that by increasing the diameter of the CNT thread (larger ply rope), and thus increasing the number of conducting channels within the structure, Re[σ] should improve within one order of magnitude of bulk copper.

Fundamental differences between the conductivity of the CNT threads and 30 gauge copper wire become apparent when Imag[σ] is examined for each case. For copper, Imag[σ] remains 2–3 orders of magnitude lower than Re[σ] over the entire 0.1–100 GHz frequency range for copper, indicating that Re[σ] dominates and that the reactance losses in the structure attributed to Imag[σ] should be negligible. For each of the CNT threads/ropes, however, Imag[σ] is predicted to be 0–1 order of magnitude less than Re[σ], with Imag[σ] approaching Re[σ] as the thread ply is increased. For the 3 × 3-ply CNT rope, Imag[σ] is predicted to be on the same order of magnitude as Re[σ]. For the 1-ply CNT thread, Re[σ] is predicted to be ~2–3e5 S/m, just two orders of magnitude less than copper, while Imag[σ]

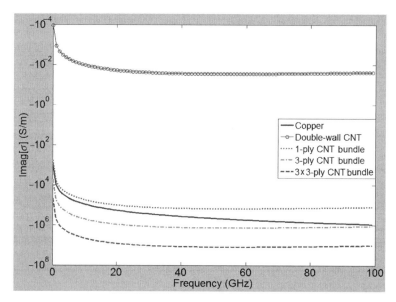

FIGURE 26.5

Imaginary part of conductivity (log scale) vs frequency for carbon nanotube thread and copper wire. (For color version of this figure, the reader is referred to the online version of this book.)

remains 1 order of magnitude lower than Re[σ] from 0.1–20 GHz. These results indicate that significant reactance effects will arise as the diameter of the CNT rope is increased. These reactance effects will increase the electrical length of the dipole antenna and will thus produce a resonant frequency shifted slightly lower than that predicted by its physical length. The quantum effects (quantum capacitance and kinetic inductance) that occur due to the nanometer radius of the individual CNTs within the threads are the likely mechanisms behind this predicted increased reactance. Thus, while overall conductivity may improve as the CNT ply is increased, efforts must be taken to mitigate the influence that increased reactance will have on the resonant length (and consequently the resonant frequency) of the antenna.

26.2.3 Electromagnetic simulation of carbon nanotube thread dipole antenna current distribution

With the conductivity of the copper wire, double-walled CNT and CNT threads accurately predicted, these values are then applied to the Method of Moments solution developed in Section 26.2.1 to simulate and compare the current distribution across a $\lambda/2$ dipole antenna constructed from each of these materials at a variety of frequencies. The current distribution for a 4″-long dipole antenna, with resonant frequency, f_0, of 1.475 GHz is shown in Fig. 26.6.

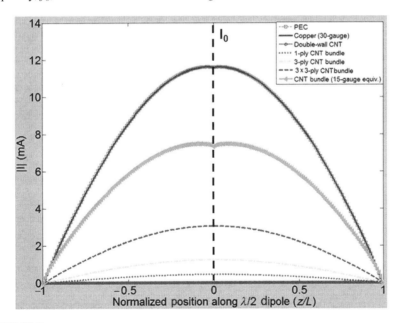

FIGURE 26.6

Comparison of current distribution (total magnitude) along $\lambda/2$ dipole antenna composed of different materials ($f_0 = 1.475$ GHz and $L = 4″$). (For color version of this figure, the reader is referred to the online version of this book.)

The simulation accurately predicts the peak current maximum, I_0, for the 4″ perfect electric conductor (PEC) and copper dipole antenna to be ~ 11.8 mA based on Ohm's law using a 1 V applied signal and the known feedpoint impedance of a $\lambda/2$ dipole antenna, $73 + j42.5$ Ω,

$$\left|\frac{V_0}{Z_{\text{feedpoint}}}\right| = |I_0| = \left|\frac{1 \text{ V}}{(73 + j42.5) \text{ }\Omega}\right| = 11.8 \text{ mA} \qquad (26.19)$$

The results indicate that each of the CNT thread dipole antennas will exhibit the traditional current distribution form of a $\lambda/2$ dipole antenna, with I_0 occurring at the center along the dipole structure (at $z/L = 0$), and that the value of I_0 should increase as the radius of the CNT thread increases. Since power delivered is proportional to I_0^2, it is possible to compare the approximate radiation efficiency of each CNT thread dipole with its copper dipole equivalent from

$$\text{Radiation Efficiency Difference} \propto 10 \cdot \log_{10}\left(\left(\frac{I_0^{\text{CNT}}}{I_0^{\text{Copper}}}\right)^2\right) \qquad (26.20)$$

A hypothetical CNT rope with a radius equivalent to 15-gauge ($r = 725$ μm) copper wire is included in this simulation to compare the performance of an extremely large radius CNT rope with that of a typical copper wire. Note that at the present time, it is not possible to fabricate a CNT rope with radius equivalent to a 15-gauge wire. The results indicate that I_0 of such a CNT rope is $\sim 65\%$ that of a dipole fabricated with 30-gauge copper wire at 1.475 GHz. Applying this ratio to Eqn (26.20) indicates that the radiation efficiency of the 15-gauge radius equivalent CNT rope dipole antenna may be ~ 3.8 dB less than that of the 30-gauge copper dipole antenna. I_0 for the 3 × 3-ply CNT rope dipole, which is approximately the same diameter as the 30-gauge copper wire ($r = 127.5$ μm), is $\sim 26\%$ that of a 30-gauge copper wire dipole. In turn, the radiation efficiency of the 3 × 3-ply CNT rope dipole is predicted to be ~ 11.6 dB less than that of the 30-gauge copper dipole. Evaluating the 3-ply and 1-ply CNT thread/rope dipoles in the same manner results in a predicted 19.4 dB lower radiation efficiency for the 3-ply rope and 28 dB lower radiation efficiency for the 1-ply thread. Thus, despite a predicted increase in reactance losses as CNT thread diameter is increased, the radiation efficiency of a dipole antenna constructed from CNT thread is predicted to improve as rope diameter is increased. However, even if the CNT rope radius is increased to over five times the radius of the 30-gauge copper wire, the dipole antenna radiation efficiency is still predicted to be ~ 3.8 dB less for an operational frequency of 1.475 GHz.

Figure 26.7 shows the current distribution for a 3-mm-long CNT thread/rope dipole antenna with f_0 in a higher frequency region (~ 50 GHz). The CNT rope dipole with a radius equivalent to 15-gauge ($r = 725$ μm) copper wire has been excluded from this simulation since the dipole radius approaches the $\lambda/2$ dipole length at 50 GHz.

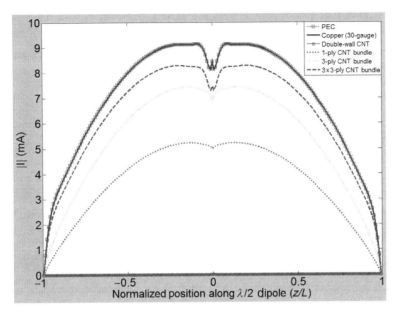

FIGURE 26.7

Comparison of current distribution (total magnitude) along $\lambda/2$ dipole antenna composed of different materials ($f_0 = 50$ GHz and $L = 3$ mm). (For color version of this figure, the reader is referred to the online version of this book.)

The results confirm that each of the CNT thread/rope dipole antennas will exhibit the traditional current distribution form of a $\lambda/2$ dipole antenna, with I_0 occurring at the center along the dipole structure. Note that I_0 does not reach the ideal maximum of 11.8 mA for the PEC and copper cases since the dipole length, L, is exactly $\lambda/2$ and yields a true $\lambda/2$ dipole antenna, while in practice the optimum input impedance match occurs when L is approximately 95% of $\lambda/2$ due to the fringing field end effects of the dipole structure. Also note that at this frequency, the effect of the nonzero dipole thread/wire radius causes a dip at the feedpoint ($z/L = 0$). The simulation again indicates that the value of I_0 should increase as the radius of the CNT thread increases.

Equation (26.20) can be applied to these results to provide a radiation efficiency comparison for a $\lambda/2$ dipole at this frequency. The results show a marked performance improvement for the CNT thread at this frequency compared with its performance at 1.475 GHz, with the 1-ply, 3-ply and 3×3-ply CNT thread/rope dipole antenna radiation efficiency predicted to be ~ 4.9, 1.8, and 0.9 dB lower than that of the 30-gauge copper wire dipole antenna, respectively. Thus, at higher frequencies, the RF performance tradeoff for using a CNT thread/rope dipole antenna in place of a copper wire antenna may be relatively low.

In Figs 26.6 and 26.7, the individual double-walled CNT current distribution is many orders of magnitude lower than the other data sets and appears to hug the bottom of the plot as a straight line. The magnified version of this data set is shown in Fig. 26.8, along with the simulated current distribution at $f_0 = 50$ GHz of a copper wire with comparable nanoscale radius, $r = 5$ nm.

The magnitude of the current distribution for these nanoscale dipole antennas is predicted to be approximately three orders of magnitude lower than that of the CNT thread and 30-gauge copper wire dipole antennas. A triangular current distribution form is also displayed, similar to that of an electrically short (Hertzian) dipole antenna. By applying Eqn (26.20) to these results, the radiation efficiency is predicted to be $\sim 60-70$ dB lower than that of the 30-gauge copper wire for both the double-walled CNT and the nanoscale radius copper wire. The results also reinforce that an individual CNT will have higher radiation efficiency than a metallic nanowire of comparable radius and will thus serve as a better radiator.

In order to compare the radiation efficiency of the CNT thread/rope dipole antennas with the 30-gauge copper dipole antenna over a wide range of frequencies, the Method of Moments simulation can be carried out at discrete frequencies and the peak current, I_0, located at $L = 0$ along the dipole, may be stored for each antenna

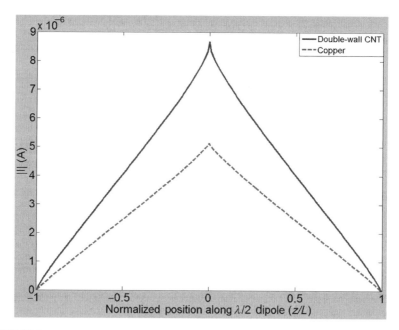

FIGURE 26.8

Comparison of current distribution (total magnitude) along $\lambda/2$ dipole antenna composed of nanoscale radius materials ($f_0 = 50$ GHz and $L = 3$ mm). (For color version of this figure, the reader is referred to the online version of this book.)

26.2 Carbon Nanotube Thread Antennas

type. For this simulation, the hypothetical CNT rope with a radius equivalent to 15-gauge ($r = 725$ μm) copper wire that was included in the 4″ dipole antenna simulation in Section 26.2.3 is included for all simulations from 0.1 to 1 GHz and is then excluded for 1–100 GHz simulations. In this way, we may avoid the points at higher frequencies, at which the dipole radius approaches its length.

As seen in Fig. 26.9, the radiation efficiency of larger ply CNT rope antennas (3 × 3-ply and rope with radius equivalent to 15-gauge wire) is predicted to increase more rapidly than the 1-ply and 3-ply thread/rope antennas from 100 MHz to 1 GHz. Below 500 MHz, radiation efficiency of the CNT thread/rope antennas is predicted to increase and diverge with that of the copper dipole antenna, regardless of rope radius. Above 500 MHz, the radiation efficiency of the large ply CNT rope antennas increasingly becomes more competitive with that of the copper dipole antenna, albeit with a minimum power loss of 4–5 dB.

The predicted dipole current maximum from 1 GHz to 100 GHz is shown in Fig. 26.10. For these higher frequencies, the radiation efficiency of the CNT thread/rope dipole antennas is predicted to become much more competitive with that of the 30-gauge copper wire above 10 GHz. Above 10 GHz, the radiation efficiency comparison relationship established by the previous 3 mm dipole antenna simulation remains relatively consistent up to 100 GHz. The

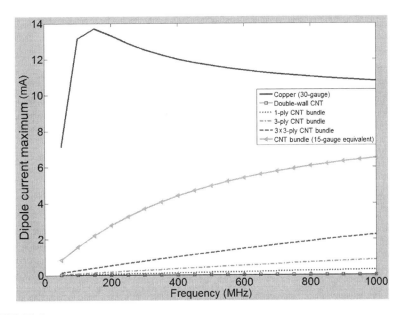

FIGURE 26.9

Current maximum (I_0) vs frequency for $\lambda/2$ dipole antenna composed of different materials in MHz frequency range ($L = \lambda/2$). (For color version of this figure, the reader is referred to the online version of this book.)

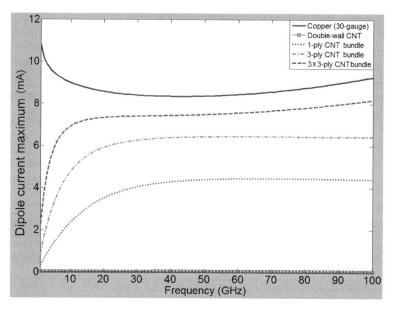

FIGURE 26.10

Current maximum (I_0) vs frequency for $\lambda/2$ dipole antenna composed of different materials in GHz frequency range ($L = \lambda/2$). (For color version of this figure, the reader is referred to the online version of this book.)

radiation efficiency of the 1-ply, 3-ply and 3×3-ply CNT thread/rope antennas is predicted to be $\sim$5.8, 2.5 and 1 dB lower than that of the 30-gauge copper wire, respectively.

It should be noted that while these current distribution simulations include complex nanoscale impedance effects such as the quantum capacitance and kinetic inductance (through the quantum conductance model detailed in Ref. [2]), certain factors have been approximated and/or unaccounted for and may limit the accuracy of the predicted results. One limitation to the applied CNT thread/rope simulation is that it does not account for defects in the individual CNTs which may disrupt electron flow and act as a resistive barrier. Another approximation is that the model assumes that each CNT in the thread of radius, R, is the full length of the dipole, $L = \lambda/2$. However, an actual CNT thread dipole prototype is composed of individual MWNTs with radius, $a = 10$ nm, and average length of 500 μm bundled together by van der Waals forces and aligned axially. The gaps and junctions between neighboring (lengthwise) CNTs may impact electrical conduction and cause electrons to hop laterally between nanotubes. Each of these approximations will likely increase the total resistance and reactance of the CNT thread, equal to both the longitudinal and lateral resistance of the nanotube bundle, and will consequently impact the thread conductivity and radiation efficiency.

26.3 CARBON NANOTUBE SHEET ANTENNAS
26.3.1 Carbon nanotube sheet fabrication

By modifying the CNT thread fabrication technique described in Section 26.2, it is possible to produce CNT sheet/ribbon materials for planar antenna designs. When the MWNTs are drawn away from their initial Si substrate, a twisting motion is applied to produce CNT thread, with the van der Waals forces between neighboring CNTs serving as the main binding mechanism. If the twisting motion is removed, it is possible to produce a CNT sheet that is initially the same width as the CNT forest from which the MWNTs are drawn. The width and thickness of this CNT sheet may be adjusted by collecting the sheet on a revolving substrate conveyor belt (e.g. Teflon belt) as shown in Fig. 26.11 [16]. SEM images of fully fabricated CNT sheet material are shown in Fig. 26.12. When the desired width and thickness is achieved, the finalized CNT sheet may then be removed from the initial substrate laminated and/or transferred to a different, preferred substrate. Average thickness for one layer of CNT sheet material is ~50 nm. Practical CNT sheets are at least 10 times this thickness, ranging from 0.5 to 5 μm.

26.3.2 Carbon nanotube sheet patch antenna

One antenna type that may be fabricated from CNT sheet material is a patch antenna. This antenna geometry has proven to be quite effective for a variety of applications,

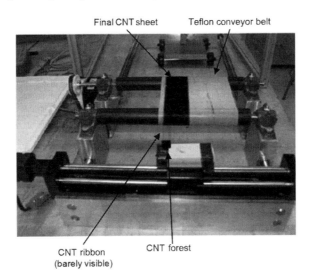

FIGURE 26.11

CNT sheet fabrication setup. (For color version of this figure, the reader is referred to the online version of this book.)

Source: Adapted from Ref. [16].

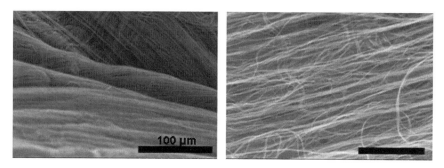

FIGURE 26.12

SEM images of fully fabricated CNT sheet.

including terrestrial and satellite communications systems and various radar electronic scanning arrays due to its low-profile, planar structure, reasonable bandwidth of typically 5–20%, and excellent gain of typically 7 dBi. The aperture-coupled patch antenna, shown in Fig. 26.13, provides an indirect feeding mechanism by which a microstrip feedline on the bottom surface of a lower substrate couples an applied signal up through an aperture located in a ground plane on the bottom surface of an upper substrate to a radiating patch located on the top surface.

In order to construct this design, CNT sheet material is fabricated and transferred to a Kapton tape substrate, which provides both an adhesive bonding mechanism on one side and a lamination protective mechanism on the other. This Kapton-laminated

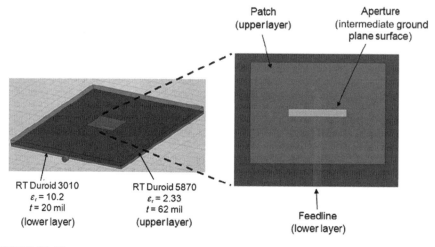

FIGURE 26.13

X-band aperture-coupled patch antenna design. (For color version of this figure, the reader is referred to the online version of this book.)

26.3 Carbon Nanotube Sheet Antennas

CNT sheet is then affixed to the top surface of a dielectric board (62-mil-thick RT/Duroid 5870 in this example) and cut to the precise patch antenna dimensions using a circuit board router to form the patch layer. This layer is then combined with the lower substrate (20-mil-thick RT/Duroid 3010 in this example) which contains the microstrip feedline and ground-plane layers. The patch is fabricated from copper for one prototype and from varying thicknesses of CNT sheet material for additional prototypes while the microstrip feedline and ground-plane layers are fabricated from copper for all prototypes. This facilitates the evaluation and comparison of the radiating properties of CNT sheet material vs that of the standard copper cladding. Examples of fully fabricated copper and CNT sheet patch antenna prototypes are shown in Fig. 26.14. The estimated thickness for the 0.5 oz copper cladding used on the RT/Duroid substrates in these prototypes is ∼17 μm.

Electromagnetic simulations using a full-wave solver may be conducted to predict the antenna performance. A simulation model of this design is shown in Fig. 26.13. The performance of a 0.5-μm-thick CNT sheet (∼10 layers of CNT sheet material) and a 5-μm-thick CNT sheet (∼100 layers of CNT sheet material) is evaluated through simulation. The CNT sheet patch is approximated as a finite conductivity boundary with the conductivity varied between simulations to evaluate its effect on the antenna performance. A starting value of ∼3e4 S/m has been applied based on reported conductivity measurements of CNT thread material. The actual conductivity of the CNT sheet material may be significantly higher than that of the CNT thread and can likely be improved to at least 1e5–1e6 S/m with postprocessing techniques. The finite conductivity boundary has been placed in between two layers of 5-mil-thick Kapton film in order to represent the Kapton laminate/bonding layers on the actual prototype. This film has an estimated permittivity of $\varepsilon_r = 3.5$. The predicted effects of CNT sheet thickness and conductivity on patch antenna radiation performance are shown in Fig. 26.15.

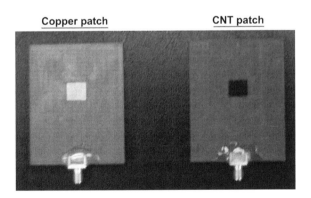

FIGURE 26.14

Copper and CNT sheet aperture-coupled patch antenna prototypes. (For color version of this figure, the reader is referred to the online version of this book.)

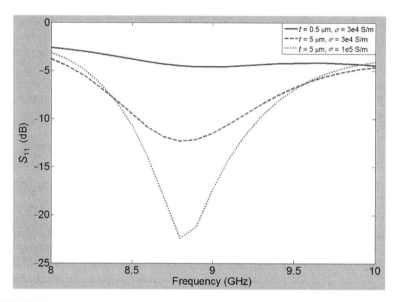

FIGURE 26.15

Effect of CNT sheet thickness and conductivity on reflection coefficient for CNT sheet patch antennas. (For color version of this figure, the reader is referred to the online version of this book.)

The high predicted reflection coefficient (and subsequently poor radiation performance) for the 0.5-μm-thick CNT sheet patch antenna is likely due to the fact that the thickness was much smaller than the skin depth of the material at 9 GHz. The skin depth represents the distance that an electromagnetic wave penetrates beneath the exterior surface of a conductor, since that is where the current density is largest, and is calculated as

$$\delta_s = \sqrt{\frac{2\rho}{\omega\mu}} \qquad (26.21)$$

where ρ is the material resistivity, ω is the angular frequency, and μ is the material permeability. For an estimated conductivity of $\sim$3e4 S/m, frequency of 9 GHz and estimated permeability of 4πe-7 H/m, the skin depth for the CNT sheet material is estimated to be $\delta_s \approx 30$ μm. This is higher than the 0.7 μm skin depth estimated for a copper sheet, because of its conductivity of $\sim$5.9e7 S/m. A significant improvement in the reflection coefficient occurs when the CNT sheet thickness is increased from 0.5 to 5 μm. An efficient patch antenna with $S_{11} < -10$ dB (<10% power reflected back to the RF source) is predicted to be attainable with a 5 μm CNT sheet with conductivity as low as 3e4 S/m. As the conductivity is improved to 1e5 S/m (which is expected to be surpassed with postprocessing techniques), a reflection coefficient and bandwidth that rivals a traditional copper patch antenna is predicted. Thus, with reasonable conductivity and by increasing the CNT sheet thickness closer to its estimated skin depth, a significant increase in radiation performance may be obtained. This has been verified by prototypes reflection coefficient measurements, as shown in Fig. 26.16.

While the 0.5 μm CNT sheet patch antenna does resonate at ~9 GHz, its reflection coefficient (−6 dB, or 25% power reflected back to RF source) is much higher than that of the standard copper antenna (−21 dB, or 0.8% power reflected back to RF source). The measured input impedance at 9 GHz is ~$44 + j8.8$ Ω for the copper patch antenna and is ~$40.5 + j44$ Ω for the 0.5 μm CNT sheet patch antenna. For these measurements, the real part represents the input resistance and the imaginary part represents the input reactance, with a positive value being inductive reactance and a negative value being capacitive reactance. A perfect match to the RF source would be a measurement of $50 + j0$ Ω. Both resistance values are close to the desired 50 Ω, though the inductive reactance value for the CNT sheet patch is ~5 times higher than that of the copper patch.

A significant performance improvement is observed for the 5 μm CNT sheet patch antenna, with a −10.5 dB reflection coefficient at resonance. A minor resonant frequency shift of ~100 MHz (~1.5%) from 9 to 8.85 GHz can be seen, potentially due to increased reactance from the CNT sheet layers and/or from fabrication tolerance errors. The measured input impedance at resonance is ~$75 + j28$ Ω. Thus,

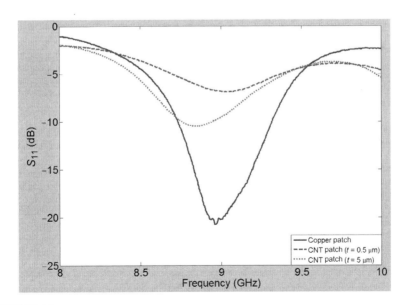

FIGURE 26.16

Measured reflection coefficient for CNT sheet and copper patch antenna prototypes. (For color version of this figure, the reader is referred to the online version of this book.)

while increasing the CNT sheet thickness does lower the reactance component of the input impedance, the resistance component increases and the overall reflection coefficient remains well above that of the copper patch antenna. It should be noted that a 5 μm sheet thickness is still well below the skin depth of ~30 μm for a 3e4 S/m conductor. A CNT sheet composed of ~600 layers of CNT sheet material would be needed in order to surpass the skin depth. With improved CNT sheet conductivity, the skin depth will decrease and the required number of CNT sheet layers will also decrease. While the example simulation results in Fig. 26.15 indicate a ~6.5–7% bandwidth (~600 MHz) of $S_{11} < -10$ dB for this antenna, the measured prototype exhibited only a 2.3% bandwidth (~200 MHz) due to the reflection coefficient just barely passing below -10 dB. As indicated by the simulation results, this bandwidth will increase as the CNT sheet conductivity is improved.

When the CNT sheet material is fabricated by pulling CNTs from an MWNT forest along a Teflon belt, the CNTs within the sheet bind together and are oriented in approximate alignment with the direction that the CNTs are pulled along the belt. Since electrons travel axially along the CNTs and rarely tunnel between neighboring nanotubes unless there is significant overlap, the RF performance of a CNT sheet patch antenna may be significantly impacted by the orientation of the CNTs within the sheet. For the prototype measured in Fig. 26.16, the CNTs within the sheet material are oriented parallel to the E-plane for the patch antenna (along the length of the patch, parallel to the microstrip feedline). It is possible to fabricate a CNT patch antenna with the CNTs oriented orthogonal to the antenna E-plane. This distinction is illustrated in Fig. 26.17.

The effect of CNT orientation on the patch antenna reflection coefficient is shown in Fig. 26.18. The reflection coefficient indicates that the orthogonal orientation of the CNT sheet to the E-plane of the antenna results in significant detuning at the expected resonance of ~8.8–9 GHz and a potential resonant frequency shift

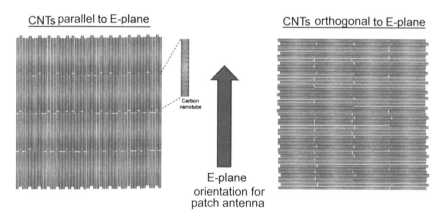

FIGURE 26.17

Variation in CNT sheet orientation for CNT sheet patch antenna prototypes. (For color version of this figure, the reader is referred to the online version of this book.)

26.3 Carbon Nanotube Sheet Antennas

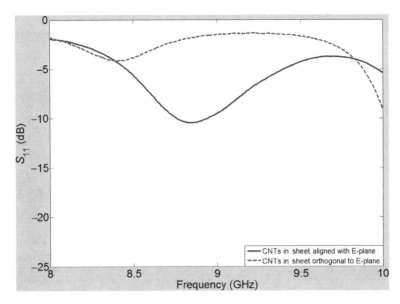

FIGURE 26.18

Effect of CNT sheet orientation on measured reflection coefficient for CNT sheet patch antennas. (For color version of this figure, the reader is referred to the online version of this book.)

down to ~8.4 GHz (4–5% shift). The measured input impedance at this shifted resonance is ~$68 + j91\ \Omega$. While the resistance component becomes closer to the desired 50 Ω, the inductive reactance component significantly increases when compared to the CNT sheet aligned with the E-plane.

The measured radiation pattern and gain data for each of the prototypes for this example design is shown in Figs 26.19–26.21. The baseline copper patch antenna and the CNT patch antennas with the CNTs aligned with the E-plane of the patch all display a typical patch antenna radiation pattern, with a half-power beamwidth (HPBW) of ~75–80° for both the E-plane and H-plane. Minor perturbations in some of the patterns are due to measurement error and/or radiation from the feedline and SubMiniature version A (SMA) connector feedpoint. The "orthogonal" CNT sheet patch antenna radiation pattern was not consistent between the E and H-planes and displays poor performance, as was expected from its poor measured reflection coefficient data.

The copper patch antenna displays a realized gain of ~5.6 dBi, just a little under the theoretical 7 dBi of an ideal patch antenna. The 5 μm "aligned" CNT sheet patch antenna exhibits a realized gain of 2.05 dBi, indicating an ~3.5 dB tradeoff in radiated power for using 5 μm CNT sheet material in place of standard copper for the patch. A -3.5 dBi realized gain is observed when the CNT sheet thickness is reduced to 0.5 μm, indicating that the reduction in thickness results in a gain reduction of ~9.1 dB from the standard copper patch and ~5.5 dB from the 5 μm CNT

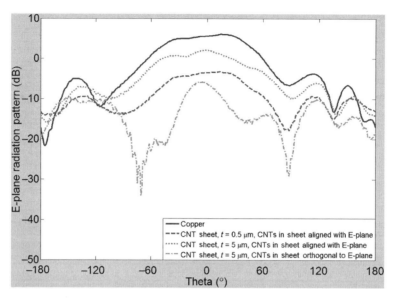

FIGURE 26.19

Measured E-plane radiation pattern for CNT sheet and copper patch antenna prototypes. (For color version of this figure, the reader is referred to the online version of this book.)

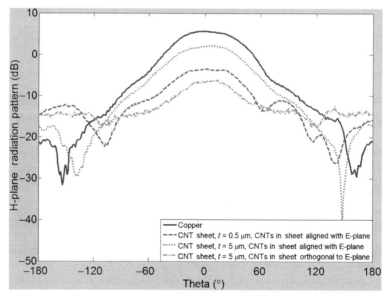

FIGURE 26.20

Measured H-plane radiation pattern for CNT sheet and copper patch antenna prototypes. (For color version of this figure, the reader is referred to the online version of this book.)

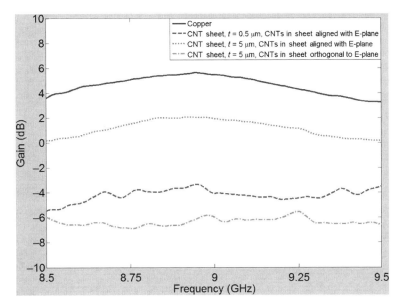

FIGURE 26.21

Measured realized gain for CNT sheet and copper patch antenna prototypes. (For color version of this figure, the reader is referred to the online version of this book.)

sheet patch antenna. The impact that the CNT sheet orientation has on the radiation performance of the patch antenna is clearly seen with the 5 μm "orthogonal" CNT sheet patch antenna, which exhibits a realized gain of −6.1 dBi. This is an ~8.2 dB difference in realized gain between prototypes constructed with the exact same design and from the exact same batch of CNT sheet material, but with varied CNT orientation. This indicates significant polarization selectivity for the CNT sheet patch antennas depending on whether the CNTs within the sheet material were generally aligned with the E-plane of the antenna. This behavior may have significant applications to polarization-specific antennas and may be useful for improving isolation between neighboring antennas on a shared platform.

26.4 MULTIFUNCTIONAL CARBON NANOTUBE ANTENNA/GAS SENSOR

26.4.1 Meshed carbon nanotube thread patch antenna/gas sensor

Recent research has shown that the permittivity, ε_r, and conductivity, σ, of a CNT is temporarily altered when it is subjected to certain gases, especially oxidizing/reducing gases such as NH_3 (ammonia) and O_2 (oxygen) [17]. This reaction has been exploited as a gas sensor by incorporating a thin layer of either randomly scattered SWNTs [18−21] or aligned MWNTs [22] as part of a microwave resonator. The resonator center frequency shifts in direct response to the change in the permittivity of the CNT layer that occurs due to the presence of a reacting gas. While the mechanism by which this permittivity shift occurs has not been conclusively determined, it is generally accepted that this shift arises from charge transfer between the reacting gas molecules and the nanotubes, with the molecules acting as either electron donors or acceptors [23]. All these unique properties may be exploited to design a multifunctional communications/gas sensor system that can be fully integrated into a textile substrate.

A variation on the patch antenna geometry discussed in Section 26.3.2 may be employed to demonstrate this concept when the patch structure is constructed from meshed CNT thread material. Previous research has shown that substituting a meshed conductive structure for the radiating patch and/or ground plane in place of a traditionally used solid conductive structure (e.g. copper and CNT sheet) yields no significant change to the antenna radiation pattern and moderate losses in gain and bandwidth depending on the density of the mesh [24]. By applying interwoven CNT thread/rope to such a design, it is possible to construct a patch antenna capable of being easily integrated into a textile material for applications in which weight, flexibility and durability are major concerns.

A model of the meshed CNT thread patch antenna is shown in Fig. 26.22. The antenna consists of a Ka-band patch constructed from 25-μm-diameter meshed

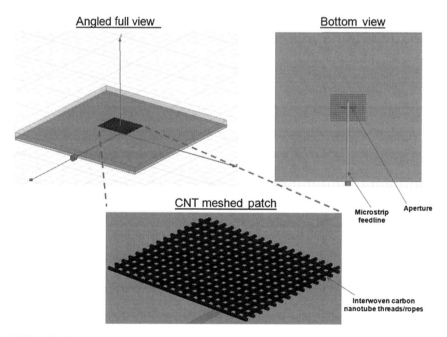

FIGURE 26.22

Model of meshed CNT thread patch antenna. (For color version of this figure, the reader is referred to the online version of this book.)

26.4 Multifunctional Carbon Nanotube Antenna/Gas Sensor

CNT threads residing on a dielectric substrate and fed with aperture coupling by a 50 Ω microstrip feedline. The substrates employed for this example are RT/Duroid 6010 ($\varepsilon_r = 10.2$ and $t = 10$ mil) for the feedline layer and RT/Duroid 5870 ($\varepsilon_r = 2.33$ and $t = 20$ mil) for the patch layer, with a metallic ground plane and aperture separating the two layers.

In order to achieve multifunctionality, the CNT threads that comprise the meshed patch are alternated between fully conducting thread fabricated from MWNTs and semiconducting threads fabricated using the same technique but with a significant number of defects added, as shown in Fig. 26.23. The additional defects introduced to the semiconducting threads ensure that the threads exhibit lower conductivity than their conductive thread neighbors (closer to a dielectric buffer) and also provide more locations for reactive gas molecules to donate or accept electrons, thus increasing the likelihood that a reactive gas will cause a noticeable change in the thread permittivity. The threads are spaced $\sim \lambda/60$ apart. With the conductive CNT threads serving as the meshed patch antenna structure and the semiconducting

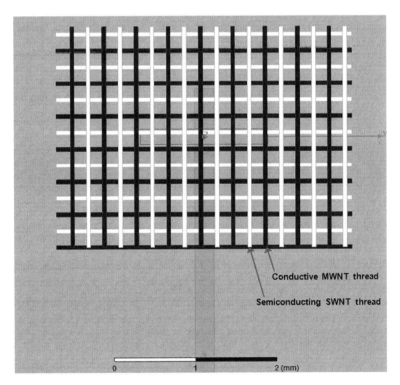

FIGURE 26.23

Model of patch antenna/gas sensor fabricated from meshed conductive and semiconducting MWNT thread. (For color version of this figure, the reader is referred to the online version of this book.)

CNT threads serving as dielectric spacer material with variable permittivity based on the presence of a reacting gas, this meshed CNT thread patch may simultaneously serve as both the radiating antenna for a communications system and the dielectric loaded resonator for a gas sensor system.

Based on the measured data, the conductive threads may be modeled in a full-wave electromagnetic solver as rectangular tubes with $\sigma = 3e4$ S/m [25] while the semiconducting threads are modeled as material with estimated $\varepsilon_r = 5$ [22]. It should be noted that this simulation does not account for defects in the CNT walls that may disrupt electron flow and act as a resistive barrier. Also unaccounted for are quantum level effects such as the quantum capacitance and kinetic inductance for the individual CNTs that comprise the threads. Each of these factors may increase the radiation losses of the antenna. The antenna exhibits a center frequency, $f_0 = 27.85$ GHz, and -10 dB bandwidth of 2.1 GHz. This is a center frequency shift of ~ 2 GHz (7%) and bandwidth reduction of ~ 400 MHz (16%) from the center frequency of 29.6 GHz and -10 dB bandwidth of 2.5 GHz predicted for a baseline unmeshed patch antenna constructed from copper cladding, as shown in Fig. 26.24. The simulated realized gain for the meshed CNT thread patch antenna is shown in Fig. 26.25. The peak total realized gain ($\phi = 0°$ and $\theta = 90°$) decreases ~ 0.46 dB (7%) from 6.79 to 6.33 dB when meshed CNT thread is used in place of unmeshed copper for the patch layer.

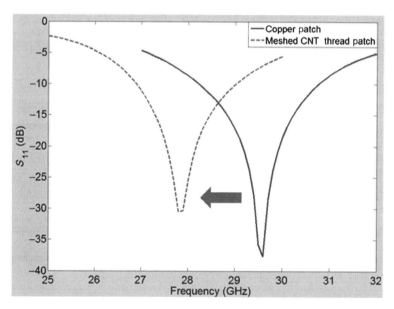

FIGURE 26.24

Simulated reflection coefficient for meshed CNT thread patch antenna. (For color version of this figure, the reader is referred to the online version of this book.)

26.4 Multifunctional Carbon Nanotube Antenna/Gas Sensor

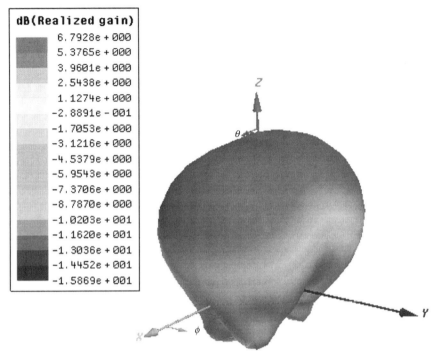

FIGURE 26.25

Simulated realized gain (3-D polar plot) for meshed CNT thread patch antenna. (For color version of this figure, the reader is referred to the online version of this book.)

Previous research has shown through measurement that ε_r for a thin layer of semiconducting SWNTs increases linearly when in the presence of increasing concentrations of NH_3 [18]. The estimated changes in ε_r from these measurements for NH_3 concentrations are

$$\begin{aligned} \varepsilon_r &\approx 5 + 0.00015x \; ; \; x \leq 1000 \text{ ppm} \\ \varepsilon_r &\approx 5.15 + 0.00005x \; ; \; x > 1000 \text{ ppm} \end{aligned} \quad (26.22)$$

By applying these discrete permittivity values to the semiconducting CNT threads in the simulation model of the meshed CNT thread patch antenna, it is possible to predict the change in f_0 expected for the meshed CNT thread patch antenna in the presence of a reacting gas. The resulting shifted f_0 for this example is shown in Fig. 26.26. A small, but measurable resonant frequency shift of -60 MHz ($\sim 0.25\%$) is predicted to occur as the concentration of NH_3 is increased around the meshed patch antenna. This shift in f_0 is large enough to measure with the appropriate complementing gas sensor circuitry and small enough to guarantee continuous bandwidth for communications functionality.

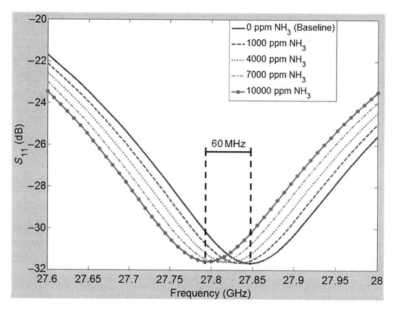

FIGURE 26.26

Simulated change in meshed CNT thread patch antenna resonant frequency in presence of varying concentrations of NH₃ (ammonia) gas. (For color version of this figure, the reader is referred to the online version of this book.)

26.4.2 Effect of carbon nanotube thread spacing on antenna performance

A number of degrees of freedom exist in the design of the meshed CNT thread patch antenna, including the CNT thread spacing and which layers are fabricated from meshed CNT thread vs standard copper cladding. Using the model developed in Section 26.4.1, it is possible to predict the effects that these design variations will have on important antenna characteristics such as resonant frequency, bandwidth, radiation pattern and realized gain. For this study, the CNT thread spacing will be varied from $\lambda/60$ and $\lambda/15$, as shown in Fig. 26.27. In order to focus solely on the effects of thread spacing and to minimize simulation complexity, the feedline and ground plane layers are modeled as ideal PEC surfaces.

The borders of the meshed thread patch have been kept constant and centered over the aperture/feedline as the spacing between neighboring CNT threads is adjusted from $\lambda/60$ to $\lambda/15$. Since the patch dimensions remain fixed, the density of the meshed CNT thread within the patch antenna area decreases as the thread spacing is increased. The simulated reflection coefficient, realized gain and bandwidth of this antenna as a function of thread spacing are shown in Figs 26.28–26.30.

26.4 Multifunctional Carbon Nanotube Antenna/Gas Sensor

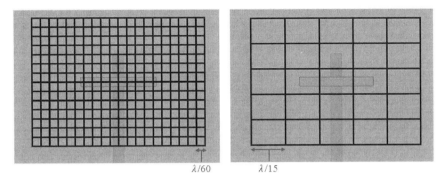

FIGURE 26.27

Meshed CNT thread spacing variation. (For color version of this figure, the reader is referred to the online version of this book.)

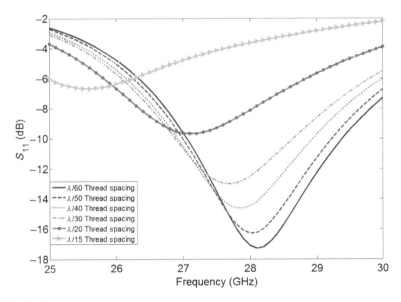

FIGURE 26.28

Effect of thread spacing on reflection coefficient for meshed CNT thread patch antennas. (For color version of this figure, the reader is referred to the online version of this book.)

As the thread spacing is increased, the patch antenna experiences a center frequency shift from ~28.1 GHz with $\lambda/60$ spacing to ~25.6 GHz with $\lambda/15$ spacing (8.9% shift). The $S_{11} < -10$ dB bandwidth decreases as this frequency shift occurs, from ~2.3 GHz at $\lambda/60$ spacing to 0.71 GHz at $\lambda/30$ spacing (70% reduction). As the spacing is increased beyond $\lambda/30$, the bandwidth with $S_{11} < -10$ dB disappears completely, indicating a severe reduction in antenna radiation efficiency.

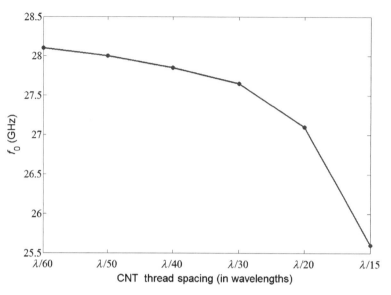

FIGURE 26.29

Effect of thread spacing on center frequency for meshed CNT thread patch antennas. (For color version of this figure, the reader is referred to the online version of this book.)

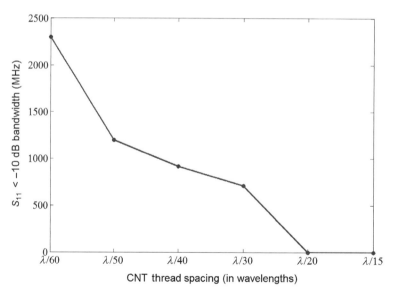

FIGURE 26.30

Effect of thread spacing on bandwidth for meshed CNT thread patch antennas. (For color version of this figure, the reader is referred to the online version of this book.)

From the E-plane radiation pattern plot shown in Fig. 26.31, the HPBW is shown to remain relatively steady at $\sim 120°$ for $\lambda/60$ through $\lambda/30$ spacing and then sharply declines as the spacing is increased further, being $\sim 100°$ at $\lambda/20$ spacing and $80°$ at $\lambda/15$ spacing. The realized gain also drops from ~ 5.2 dBi with $\lambda/60$ spacing to ~ 2 dBi with $\lambda/15$ spacing, as shown in Fig. 26.32. This is likely due to the impedance mismatch that is seen in Fig. 26.28, which shows that the S_{11} value at resonance increases as the thread spacing increases, from a minimum of -17.3 dB with $\lambda/60$ spacing to a minimum of -6.7 dB with $\lambda/15$ spacing.

The data from these simulations indicate that a thread spacing of $d \leq \lambda/30$ will ensure sufficient antenna radiation efficiency and usable bandwidth, with the caveat that antenna bandwidth will shrink as the thread spacing is increased. The minor center frequency shift and gain reduction as thread spacing is increased should also be accounted for in the design process.

26.4.3 Effect of meshed carbon nanotube thread ground plane and feedline layers on antenna performance

Using the model from Section 26.4.1, it is also possible to investigate the viability of constructing the antenna feedline and ground plane from the meshed CNT thread material. For each of these cases, the CNT threads are generally spaced $\lambda/60$ apart. An electromagnetic model of the patch antenna with all layers (patch, ground plane

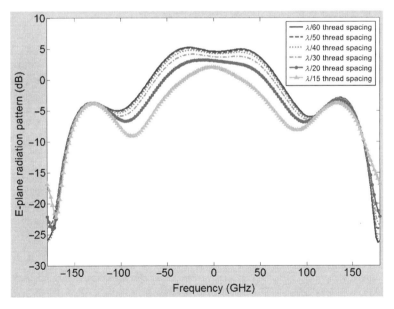

FIGURE 26.31

Effect of thread spacing on radiation pattern for meshed CNT thread patch antennas. (For color version of this figure, the reader is referred to the online version of this book.)

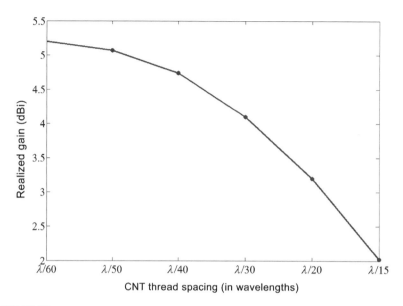

FIGURE 26.32

Effect of thread spacing on realized gain for meshed CNT thread patch antennas. (For color version of this figure, the reader is referred to the online version of this book.)

and feedline) composed of meshed CNT thread is shown in Figs 26.33 and 26.34. In order to accurately realize the dimensions and location of the aperture for the antenna, the spacing density of the meshed CNT threads that compose the ground plane layer are increased for the area near the aperture, with a thread spacing of $\lambda/120$. This spacing is also used for the threads composing the feedline layer in the E-plane direction in order to accurately realize the feedline width.

As seen in Fig. 26.35, the reflection coefficient improves as the feedline and ground plane are constructed from meshed CNT thread. The $S_{11} < -10$ dB bandwidth also increases well beyond that of the baseline case, indicating an improved and wider bandwidth input impedance match for these designs.

While these characteristics look promising, the E-plane radiation pattern data in Fig. 26.36 shows the cost of replacing high-conductivity copper material with moderate conductivity meshed CNT thread material for the feedline and ground (GND) plane. When the antenna feedline is fabricated from meshed CNT thread, the realized gain of the antenna drops from 5.2 to 0.9 dBi—a reduction to under 40% of the radiated power level for the baseline case. When the GND plane is also fabricated from meshed CNT thread, the realized gain is further reduced to less than -1.8 dBi—a further reduction to <20% of the baseline meshed patch antenna radiated power level. The drop in realized gain is likely due to increased conductor losses in the meshed feedline and ground plane that overshadow the reduction in impedance mismatch losses. Thus, while the impedance match and

26.4 Multifunctional Carbon Nanotube Antenna/Gas Sensor

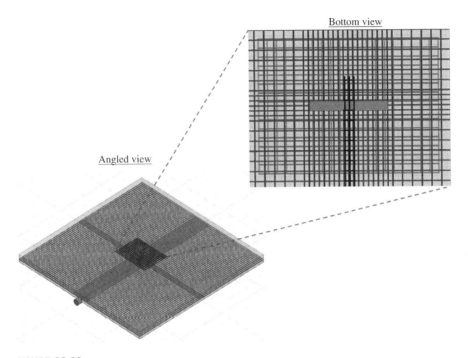

FIGURE 26.33

Meshed CNT thread patch antenna with fully-meshed feedline and ground plane. (For color version of this figure, the reader is referred to the online version of this book.)

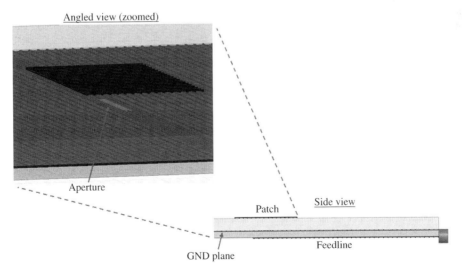

FIGURE 26.34

Cross-sectional view of CNT thread patch antenna with fully meshed feedline and ground plane. (For color version of this figure, the reader is referred to the online version of this book.)

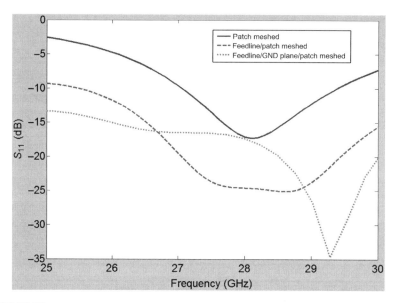

FIGURE 26.35

Effect of meshed feedline/ground plane on reflection coefficient for meshed CNT thread patch antennas. (For color version of this figure, the reader is referred to the online version of this book.)

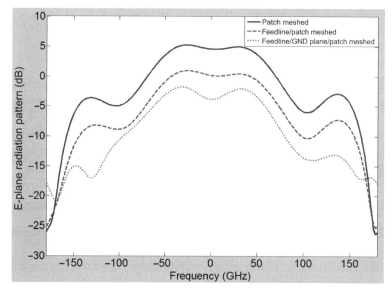

FIGURE 26.36

Effect of meshed feedline/ground plane on radiation pattern for meshed CNT thread patch antennas. (For color version of this figure, the reader is referred to the online version of this book.)

bandwidth may improve by fabricating the entire antenna from the meshed CNT thread material, the reduction in realized gain will have a severe impact on the radiation performance of the antenna. It is expected that by improving the conductivity of the CNT thread material, the realized gain may be improved.

It is important to note that RT/Duroid materials were used as the antenna substrate for this example mainly due to their low-loss nature and their well-defined permittivity values since the focus of the example was on the meshed thread spacing and the viability of its use as the feedline and/or ground plane. This specific design is not flexible and would not be a practical application of the meshed CNT thread materials. A practical embodiment of this concept should employ one or more fabric materials as the antenna substrates in order to facilitate the integration of the meshed CNT thread patch antenna/gas sensor into a high-durability, lightweight body-wearable system.

26.5 SUMMARY

The application of CNT superfiber materials to a variety of RF antenna designs has been explored through theory, simulation and measurement. These materials function as viable conductors for antenna designs and offer significant physical benefits while yielding minor to significant RF performance reduction, depending on the design. The electrical tradeoffs of employing CNT superfiber materials as RF radiators, including frequency shift and realized gain reduction, have been explored with the fabrication and measurement of a variety of CNT thread and sheet antenna designs. The minimization of this reduction in RF performance is dependent upon the state-of-the-art for the CNT material conductivity, which is being continually improved through fabrication technique research, and on the fidelity of the physical connection made between the RF connector/feedline and the CNT material. The development of efficient and consistently successful fabrication techniques for establishing an electrical connection between an RF cable/connector and CNT superfiber material will be required in order for these materials to have practical application to modern antenna designs. Improvements to the physical and electrical characteristics of the CNT thread and sheet materials will lead to increased performance and lower fabrication costs. Future work will also place specific emphasis on textile-integrated CNT thread designs, particularly on the fabrication of CNT thread designs with the use of a sewing machine, in order to produce lightweight, durable clothing/backpack/tent-integrated antennas for Army systems.

References

[1] I. Kang, et al., Introduction to carbon nanotube and nanofiber smart materials, Composites Part B: Engineering 37 (6) (2006) 382–394.

[2] P.J. Burke, S. Li, Z. Yu, Quantitative theory of nanowire and nanotube antenna performance, IEEE Transactions on Nanotechnology 5 (4) (July 2006) 314–334.

[3] G.W. Hanson, Radiation efficiency of nano-radius dipole antennas in the microwave and far-infrared regimes, IEEE Antennas and Propagation Magazine 50 (3) (June 2008) 66–77.

[4] Y. Zhou, Y. Bayram, F. Du, L. Dai, J.L. Volakis, Polymer-carbon nanotube sheets for conformal load bearing antennas, IEEE Transactions on Antennas and Propagation 58 (7) (July 2010) 2169–2175.

[5] L. Yang, R. Zhang, D. Staiculescu, C.P. Wong, M.M. Tentzeris, A novel conformal RFID-enabled module utilizing inkjet-printed antennas and carbon nanotubes for gas-detection applications, IEEE Antennas and Wireless Propagation Letters 8 (2009) 653–656.

[6] S. Salahuddin, M. Lundstrom, S. Datta, Transport effects on signal propagation in quantum wires, IEEE Transactions on Electron Devices 52 (8) (August 2005) 1734–1742.

[7] A. Raychowdhury, K. Roy, Modeling of metallic carbon-nanotube interconnects for circuit simulations and a comparison with cu interconnects for scaled technologies, IEEE Transactions on Computer-Aided Design 25 (1) (January 2006) 58–65.

[8] J.J. Plombon, K.P. O'Brien, F. Gstrein, V.M. Dubin, High-frequency electrical properties of individual and bundled carbon nanotubes, Applied Physics Letters 90 (2007).

[9] Y. Huang, W.-Y. Yin, Q.H. Liu, Performance prediction of carbon nanotube bundle dipole antennas, IEEE Transactions on Nanotechnology 7 (3) (May 2008) 331–337.

[10] S. Choi, K. Sarabandi, Performance assessment of bundled carbon nanotube for antenna applications at terahertz frequencies and higher, IEEE Transactions on Antennas and Propagation 59 (3) (March 2011) 802–809.

[11] D. Mast, The Future of Carbon Nanotubes in Wireless Applications, Antenna Systems/Short-Range Wireless Conference, September 2009.

[12] G. Fikioris, On the application of numerical methods to Hallén's equation, IEEE Transactions on Antennas and Propagation 49 (3) (March 2001) 383–392.

[13] S.A. Maksimenko, G.Y. Slepyan, A. Lakhtakia, O. Yevtushenko, A.V. Gusakov, Electrodynamics of carbon nanotubes: dynamic conductivity, impedance boundary conditions, and surface wave propagation, Physical Review B 60 (December 1999) 17136–17149.

[14] H. Stahl, J. Appenzeller, R. Martel, Ph. Avouris, B. Lengler, Intertube coupling in ropes of single-wall carbon nanotubes, Physical Review Letters 85 (24) (December 2000) 5186–5189.

[15] N. Srivastava, R.V. Joshi, F. Banerjee, Carbon Nanotube Interconnects: Implications for Performance, Power Dissipation and Thermal Management, Electron Devices Meeting, 2005, IEDM Technical Digest, pp. 249–252, December 2005.

[16] J.-H. Pohls, M.B. Johnson, M.A. White, R. Malik, B. Ruff, C. Jayasinghe, M.J. Schulz, V. Shanov, Physical properties of carbon nanotube sheets drawn from nanotube arrays, Carbon 50 (11) (September 2012) 4175–4183.

[17] J. Kong, N.R. Franklin, C. Zhou, M.G. Chapline, S. Peng, K. Cho, H. Dai, Nanotube molecular wires as chemical sensors, Science 287 (January 2000) 622–625.

[18] K.G. Ong, K. Zeng, C.A. Grimes, A wireless, passive carbon nanotube-based gas sensor, IEEE Sensors Journal 2 (2) (April 2002) 82–88.

[19] S. Chopra, A. Pham, J. Gaillard, A. Parker, A.M. Rao, Carbon-nanotube-based resonant-circuit sensor for ammonia, Applied Physics Letters 80 (24) (June 2002) 4632–4634.

[20] S. Chopra, A. Pham, J. Gaillard, A.M. Rao, Development of RF Carbon Nanotube Resonant Circuit Sensors for Gas Remote Sensing Applications, 2002 IEEE MTT-S Digest, 2002, pp. 639–642.

[21] M. Mcgrath, A. Pham, Carbon nanotube based microwave resonator gas sensors, International Journal of High Speed Electronics and Systems 16 (4) (December 2006) 913–935.

[22] M.P. McGrath, A. Pham, Microwave Vertically Aligned Carbon Nanotube Array Sensors for Ammonia Detection, 2005 IEEE Sensors Conference Proceedings, 2005, pp. 837–840.

[23] Y. Wang, J.T.W. Yeow, A review of carbon nanotubes-based gas sensors, Journal of Sensors 2009 (2009) 1–24.

[24] G. Clasen, R. Langley, Meshed patch antennas, IEEE Transactions on Antennas and Propagation 52 (6) (June 2004) 1412–1416.

[25] N. Mallik, et al., Study on carbon nano-tube spun thread as piezoresistive sensor element, Advanced Materials Research 67 (2009) 155–160.

CHAPTER

Energy Storage from Dispersion Forces in Nanotubes

27

Fabrizio Pinto
InterStellar Technologies Corporation, Monrovia, CA, USA[1]

CHAPTER OUTLINE

27.1 Introduction	789
27.2 Idealized Parallel-Plate System	791
27.3 Orders of Magnitude	792
27.4 Performance Simulations	797
27.5 Conclusions	800
Acknowledgments	802
References	803

27.1 INTRODUCTION

Interest in market-disruptive technologies capable to deliver order-of-magnitude improvements over present-day energy storage performance has been exponentially growing in recent years, ultimately driven by anticipated applications of electrical energy on all scales, from powering next-generation microdevices within the human body, to transportation solutions capable to compete with hydrocarbon-powered engines, to stabilization of the national grids against the severe fluctuations typical of such "clean" alternatives as solar and wind energy. However, such a formidable challenge, critically important to our civilization for environmental, economic, and geopolitical reasons, has so far been addressed by employing approaches rooted, in some cases, in storage mechanisms known since the nineteenth century, including batteries, capacitors, and fuel cells [1,2].

Appreciable progress has certainly been made in all areas of energy storage based on such traditional technologies but there exist unavoidable, quite restrictive limitations set by fundamental physical principles to any foreseeable future improvement. Although we are periodically reminded of such impenetrable barriers [3], warnings are largely ignored with the result that, for instance, an expert recently reported "a false perception among the public and policymakers

---
[1]Now at: Physics Department, Faculty of Science, Jazan University, Jazan 22822, P.O. Box 114, Kingdom of Saudi Arabia.

Nanotube Superfiber Materials. http://dx.doi.org/10.1016/B978-1-4557-7863-8.00027-X
Copyright © 2014 Elsevier Inc. All rights reserved.

that present battery performance is adequate for widespread acceptance of battery-electric vehicles" [4].

Among most recent strategies in the long history of incremental improvements in traditional storage systems has been the adoption of nanotubes in a wide variety of performance-enhancing roles [5–10] although applications of nanotubes and other nanostructures for energy storage by means of completely different mechanisms have also been proposed [11,12].

The author has recently highlighted a novel approach to the acceleration and high-speed ejection of nanocores sliding within the outer walls of multiwalled nanotubes (MWNTs) by phased dispersion force manipulation [13]. The first critical enabling factor introduced by the inclusion of dispersion forces in such a nanoaccelerator is the rapid divergence of their magnitude as a function of interboundary separation, which makes dispersion forces of pervasive importance in the synthesis and selection, structure, stability, and mutual interaction of structures on the nanoscale [14–16]. For instance, the intershell van der Waals interaction between the inner core of a double-walled nanotube and the outer wall is exclusively responsible for the needed restoring force in proposed low-friction bearings [17], ultra-high-frequency oscillators [18], and nonvolatile memory devices [19].

The second element characteristic of the suite of designs developed by the author is "dispersion force modulation", that is, a set of available strategies to manipulate dispersion forces in space and time by exchanging energy with the system to alter the optical properties of the interacting boundaries. As a historically important example, carrier density photomodulation in bulk semiconductors represented the first system in which the van der Waals force between two surfaces was experimentally observed to depend on time-dependent illumination [20]. Generally, the possibility to modulate dispersion forces by acting on their optical properties is suggested from their full expression as provided by the Lifshitz theory [21], which displays an explicit dependence on the dielectric function of the bulk materials. A possible strategy to extend this approach to the core outer wall van der Waals forces in nanotubes by reflectance modulation due to photogenerated free carrier screening of excitonic spectra was also discussed [22].

It can be shown that dispersion forces can be employed to do mechanical work on the plates of an idealized biased capacitive converter thus transforming part of the van der Waals potential energy into electrostatic energy and heat due to an electric current flowing across a load [23]. A detailed analysis of engine cycles resulting from dispersion force modulation in such systems shows the existence of nontrivial issues of compatibility with the laws of thermodynamics [24] that may even require consideration of departures from standard Lifshitz theory in the case of excited states of matter [25].

In this chapter, we consider the potential technological advantages, and some of the challenges, of storing energy in the van der Waals field of interacting nanostructures for later retrieval as an ordinary electric current. We shall not consider nanoscale engine cycles to convert electromagnetic or thermal energy into mechanical or electrical energy via dispersion force modulation as is the case for the nanoaccelerator mentioned above. In other words, the ultimate origin of the energy assumed to have "charged" the system will not concern us. However, as we shall see, both energy density and power density available from storage systems of the type described herein can, in principle, be quite competitive with the state of the art of traditional technologies.

The fascinating contemporary role of dispersion interactions in nanotechnology exemplified by this particular application is perhaps best illustrated by the fact that the Casimir force, once described by Schwinger et al. as "one of the least intuitive consequences of quantum electrodynamics", [26] is now routinely employed[2] to enhance device performance or to enable novel technological capabilities [27–29].

27.2 IDEALIZED PARALLEL-PLATE SYSTEM

In order to introduce the fundamental elements of an energy storage approach based on dispersion force interactions, let us consider the system shown in Fig. 27.1. A perfectly conducting cavity, whose size is assumed to be far larger than that of the system, contains a capacitor, represented by a middle, movable plate (piston), maintained by an external voltage source at a potential difference $V(t)$ with respect to an upper, fixed (red) plate in general of different radius and thickness, located at a variable distance δ. At a distance s from the piston, and at its same potential, is an

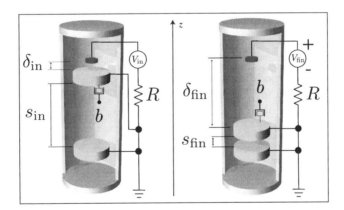

FIGURE 27.1

Cutout of an idealized thermodynamical engine storing energy in the cavity zero-point-field (see text). The movable capacitor plate is shown as the middle yellow plate. (For interpretation of the references to color in this figure legend, the reader is referred to the online version of this book.)

---

[2]See also US Patents Nos. 8,299,761; 8,174,706; 8,149,422; 6,920,032; 6,842,326; 6,665,167; 6,661,576; 6,650,527; 6,593,566; and 6,477,028 issued to InterStellar Technologies Corporation (1999–2012).

identical plate (lowest yellow plate), also fixed. The net force acting on the piston as a function of its coordinate z is therefore given by the electrostatic and the dispersion forces due to the capacitor plate, to the dispersion force due to the bottom plate, and to other possible terms such as frictional forces which, in the present model, are indicated by the constant b.

In what follows, we shall neglect fringe effects and adopt, for the calculation of dispersion forces, the highly idealized geometry of two parallel-plane, perfect conductors, also assuming a temperature $T = 0$ K, which immediately leads to the famous expression for the attractive force between two neutral plates obtained by Casimir [26], $F_{Cas} = \pi^2 \hbar c A/(240 s^4)$, where A is the area of the piston. Let us consider a transformation corresponding to the displacement of the piston from the initial position, $s_{in} \to s_{fin}$ (with $s_{in} \simeq 3.33$ μm and $s_{fin} = 0.5$ μm).

By assuming that $\delta_{fin} \gg s_{fin}$, it is immediate to verify that the van der Waals energy available to be converted, in part, into mechanical work and electrostatic energy is only $\Delta U_{Cas} \simeq -1.08 \times 10^{-14}$ J. An order of magnitude of the maximum available energy per unit mass (specific energy) is given by assuming $s_{in} \to \infty$ and by writing $\varepsilon_{Cas}(s_{fin}) = -\pi^2 \hbar c A/(720 s_{fin}^3)(1/(2AD\rho))$, where ρ is the mass density and D is the thickness of the piston. For instance, if $D = 10$ μm, $s_{fin} = 1$ μm, and $\rho = 18 \times 10^3$ kg/m^3 (gold), the maximum specific energy available is $\varepsilon_{Cas}(s) \simeq 1.2 \times 10^{-9}$ J/kg $\simeq 3.3 \times 10^{-13}$ W·hr/kg, which compares quite unfavorably with specific energy values $\varepsilon \sim 10^0$–10^3 W h/kg typical of batteries, supercapacitors, and fuel cells [30]. This extremely unpromising result, which pertains to the simplistic parallel-plate geometry, explains the little consideration so far given to the practicality of storing energy in the dispersion force field of interacting surfaces.

It is historically appropriate to mention at the end of this section that, 30 years ago, a "vacuum-fluctuation battery", "… along the lines of a Slinky^{TM} toy model" was proposed as a possible actual experiment to demonstrate for the first time that energy stored in the van der Waals field of interacting boundaries can be converted into an electric current [31]. Initially, estimates of the energy density available from this curious design were incorrectly reported as being "comparable to that of a good chemical battery" [31], apparently due to an undetected trivial error in the computation of the Casimir energy per unit area of two perfectly conducting plates (see p. A-12 of Ref. [31], where the energy of two such planes with $s = 20$ nm is given as 5.4 J/m^2 instead of 5.4×10^{-5} J/m^2). However, when the same picturesque device was finally described in the referred literature, corrected energy density estimates were not reported [32].

27.3 ORDERS OF MAGNITUDE

In order to address the severe quantitative limitation outlined in the previous section, let us turn to an alternative strategy for energy storage based on the van der Waals field of interacting nanostructures (Fig. 27.2). The idealized model consists of two

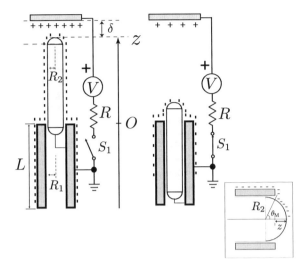

FIGURE 27.2

Geometry of a polarized telescoping nanotube as a dispersion force field storage system. As the interwall dispersion force attracts the inner core back inside the outer wall while switch S_1 is closed, a current appears across the load R, which includes the internal resistance of the nanotube itself. Here, the interboundary gap is $s = R_1 - R_2$. In the inset, the geometry adopted to model the electrostatic repulsion between the outer and inner walls while the cap is partially extruded.

perfectly conducting, telescoping, concentric cylinders of length L and radii R_1 and R_2, respectively, held at the same electrical potential. We expect the assumption of perfect conductivity to yield very high upper bounds on the performance of this device although detailed analyses have shown that the dielectric function idealization at the basis of the Lifshitz theory of dispersion forces [26,33] is already realized in multiwalled CNTs (MWCNTs) containing just two or three walls [34,35]. The outer base, or cap, of the inner cylinder interacts electrostatically with a facing electrode held by an external voltage source at a potential difference V with respect to the two telescoping tubes.

From the standpoint of dispersion force computation, this cylindrical geometry is equivalent to that of Fig. 27.1 in the limit for $s/R_{1,2} \ll 1$ [35,36] although the electrostatic interactions are here modeled by means of far-more realistic, non-parallel-plane conductors. In this configuration, the van der Waals energy field manifests itself as a restoring force on the inner cylinder, which is attracted back within the outer shell by a force that is approximately constant if $R_{1,2} \ll L$ [14]. Since the inner tube interacts capacitively with the fixed electrode, any change in their relative position will result in an electric current if the circuit is closed on a load, as previously illustrated in the archetypal system of Fig. 27.1. When the inner tube is extruded, van der Waals energy is stored in the nanodevice; as the inner tube is adiabatically retracted under the action of dispersion forces in *quasi*-equilibrium (equivalently to the aforementioned piston downstroke), this initial energy is converted, in part, into electrical energy, the balance being owed to losses, such as friction, Joule heating on the load and other losses associated to transport processes on the nanoscale [37]. Therefore, thermodynamically, such a polarized telescoping nanotube (PTNT) system is a rechargeable, nonelectrochemical "battery" which stores energy in the dispersion force field [26] instead of the ordinary materials typical of electrochemical batteries and supercapacitors (Fig. 27.3).

The notable improvement offered by the dispersion force field approach, even under conservative assumptions, is demonstrated by considering a dense assembly of such aligned PTNTs whose mass fabrication has been already demonstrated and is steadily improving [38–40]. The specific energy for a storage system with an effective area per unit mass S_{eff} is $\varepsilon_{Cas} = E_{Cas} S_{eff}$, where E_{Cas} is the customary Casimir energy per unit area. The expression for the Casimir energy of two concentric, perfectly conducting cylinders separated by a gap $s \equiv R_1 - R_2 \ll R_1, R_2$ is

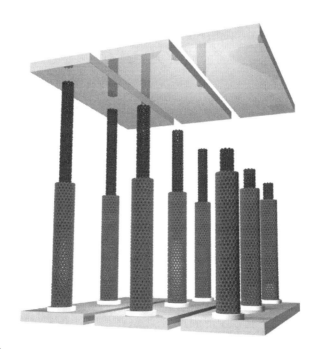

FIGURE 27.3

Schematic view of the macroscopic architecture of an energy storage system based on PTNTs. In this scalable configuration, nine nanotubes are organized in three parallel subdomains (from right to left) addressed by three independent electrodes. Three PTNTs on the far left are shown "fully charged", whereas on the far right, the nanotube cores have almost completely been retracted into the outer walls at the end of the energy release process. (For color version of this figure, the reader is referred to the online version of this book.)

$E_{\text{Cas}}(z) \simeq -(L-z)(\hbar c \pi^3/360)R_2/s^3$, where $(L-z)$ is the length of the overlapping region [36] (Fig. 27.2). By writing $E_{\text{Cas}}(z)$ for $z=0$ in terms of the total lateral area, $A \simeq 2\pi R_2 L$, we arrive at the following expression for the specific dispersion field energy of the concentric, unextruded cylinders:

$$\varepsilon_{\text{Cas}} \simeq -\frac{\hbar c \pi^2}{720}\frac{S_{\text{eff}}}{s^3} \simeq -1.2 \times 10^{-4} \frac{S_{\text{eff}}}{s_{\text{nm}}^3} \frac{\text{W} \cdot \text{hr}}{\text{kg}}, \qquad (27.1)$$

where s_{nm} is the interboundary gap in nanometers. In the case of double-walled carbon nanotubes (DWCNTs), values of $S_{\text{eff}} \approx 2 \times 10^6$ m²/kg are observed, which yields $\varepsilon_{\text{Cas}} \approx -240/s_{\text{nm}}^3$ W·hr/kg. For $s=0.34$ nm, equal to that of typical DWCNTs [14], we obtain the upper limit $|\varepsilon_{\text{Cas}}| \approx 6 \times 10^3$ W·hr/kg for idealized, perfectly conducting cylinders. However, by repeating this calculation upon replacing E_{Cas} by the van der Waals energy per unit area of realistic DWCNTs, $E_{vdW}(x) \simeq -\gamma C x/(2\pi R_2 L)$ [14,17], where the interlayer cohesive energy density $\gamma \simeq 0.16$ J/m² and $C = 2\pi R_2$, we find, for $x=L$, $|\varepsilon_{vdW}| = \gamma S_{\text{eff}} \approx 88$ W·hr/kg, typical of practical batteries [30] and an improvement over supercapacitors [7]. An immediate demonstration that these initial estimates can be improved by further research is given by considering other nanotube species, which are already known to yield higher retracting van der Waals forces. For instance, molecular dynamics computations on boron nitride nanotube oscillators have been reported to yield retracting van der Waals forces approximately three times as high as those in the corresponding carbon structures, consistent with $|\varepsilon_{vdW}| \approx 300$ W·hr/kg [41].

An analysis of the forces which determine the electromechanical behavior of the elemental PTNT yields estimates of the theoretical capacity, power density, and discharge characteristics of a macroscopic storage device based on this approach. Let us model the extruded inner tube cap and the external electrode as a perfectly conducting, infinite plane-sphere capacitor, so that their mutual electrostatic force in the reference frame at Fig. 27.2 is $F_{\text{el,cap}} = +\pi \varepsilon_0 R_2^2/[\delta(\delta + R_2)]V^2$, where δ is the gap between the cap center and the electrode [19,42–44].

An additional electrostatic interaction is the repulsive interwall force between the outer tube and the extruded part of the inner tube, which are at the same potential, due to charges confined to the PTNT surface by the Faraday cage effect [19,45]. This force can be described as the repulsion of two equipotential spherical caps composing a conducting sphere of total charge Q_{NT}, radius equal to the inner tube radius, and separated by an infinitesimally thin slit (Fig. 27.2, inset). The force on a partially extruded cap ($0 \leq z \leq R_2$) is obtained by the customary calculation of the repulsive force between two equipotential hemispheres but integrating the surface force only over the range $\theta \in [0, \theta_M]$ instead of $\theta \in [0, \pi/2]$ as usual; this yields $F_{\text{rep}} = Q_{\text{NT}}^2/(16\pi \varepsilon_0 R_2^2)[1 - (1 - z/R_2)^2]$. If the cap is fully extruded ($z \geq R_2$), the force is assumed to be constant as a function of z and equal to the well-known result [46], $F_{\text{rep}} = Q_{\text{NT}}^2/(16\pi \varepsilon_0 R_2^2)$. In this case, if $R_2 \ll b$, the mutual capacitance becomes $C_{\text{cap,el}} \simeq 4\pi \varepsilon_0 R_2$; thus, by replacing $Q_{\text{NT}} = C_{\text{cap,el}} V$, we obtain $F_{\text{rep}} = +\pi \varepsilon_0 V^2$ [19].

The dispersion force between the spherical cap and the flat electrode, both assumed perfectly conducting in these first estimates, can be obtained from the Derjaguin approximation [26,33,47–51], that is, the proximity force theorem applied to the familiar Casimir force between two flat planes, which yields, in our coordinate system, $F_{\text{cap,el}} = +2\pi R_2 |U_{\text{Cas}}|$, where $U_{\text{Cas}} = -\pi^2 \hbar c/(720\delta^3)$ is the Casimir energy per unit area and $\delta = b - R_2$. Finally, the only force attracting the extruded core back within the outer tube is obtained from the Casimir energy given above as [14] $F_{\text{Cas}} = -\partial E_{\text{Cas}}(z)/\partial z = -(\hbar c \pi^3/360) R_2/s^3$.

The requirement that the system be in equilibrium at rest in its initial state yields a condition on the charge on the nanotube, Q_{NT}, or, equivalently, on the external voltage V required to maintain the tube, assumed fully extruded, in equilibrium at a generic position, z:

$$Q_{\text{NT}}^2 = \frac{2}{50} \pi^4 \varepsilon_0 \hbar c \, R_2^3 \left[1 + \frac{R_2^2}{\delta(\delta + R_2)}\right]^{-1} \left(\frac{1}{s^3} - \frac{1}{\delta^3}\right). \quad (27.2)$$

Since $Q_{\text{NT}}^2 \geq 0$, within the assumptions of this model, a nonvanishing charge makes an initial state of equilibrium possible only if $\delta > s$, whereas no charge is needed if $\delta = s$ and no equilibrium is possible if $\delta < s$ (the characterization of equilibrium positions, if any exist, is dealt with in the following section).

Therefore, we can envision a discharge cycle commencing at $\delta \geq s$ so that the charge on the nanotube is very small; if the potential is maintained at a level infinitesimally smaller than the value needed for equilibrium, the core is sucked back into the outer tube over a sequence of near-equilibrium positions. The interwall dispersion force does work by causing charge to move from the voltage source to the nanotube–electrode system [27,52]. The upper bound on the charge is achieved in the limits (δ/R_2) and $(\delta/s) \to \infty$, for a fully extruded cap, where it converges to $Q_{\text{NT,max}} \to \sqrt{\frac{2}{50} \pi^4 \varepsilon_0 \hbar c (R_2/s)^3}$. By writing the charge per unit area as $Q_{\text{NT,max}}/(2\pi R_2 L)$, we have that the theoretical capacity of such a PTNT assembly is $Q_{\text{vdW,max}} \simeq S_{\text{eff}} Q_{\text{NT,max}}/(2\pi R_2 L)$, or

$$Q_{\text{vdW,max}} \simeq 4.61 \times 10^{-5} \frac{1}{L_{\text{nm}}} \sqrt{\frac{R_{2,\text{nm}}}{s_{\text{nm}}^3}} \, S_{\text{eff}} \frac{\text{A} \cdot \text{hr}}{\text{kg}}, \quad (27.3)$$

where L_{nm}, $R_{2,\text{nm}}$, and s_{nm} are expressed in nanometers. For instance, with $s = 1.4$ nm, $R_2 = 5$ nm, and $L = 12$ nm, we find $Q_{\text{NT,max}} \simeq 44e$ and, finally, $V \simeq 16.0$ V, consistent with values typical of proposed MWCNT-based memory devices [19]. The system capacity is found to be $Q_{\text{vdW,max}} \simeq 10.4$ A·h/kg, which compares favorably with that of practical batteries [30] (the relevance of the value $s = 1.4$ nm is that it corresponds to an interwall distance such that the specific energy predictions from our idealized model and those from the van der Waals calculations coincide, that is, $240/s_{\text{nm}}^3$ W·hr/kg $= 88$ W·hr/kg). For $s = 0.34$ nm, all else being the same, the idealized, perfect conductor model yields an upper bound $Q_{Cas,max} \approx 86.8$ A·h/kg and $V = 106$ V.

A key feature of the energy storage strategy explored in this chapter is that the conversion of van der Waals energy to a current is not limited by fundamental electrochemical processes but it is regulated by an external voltage source, V [23]. The ability to directly control the retraction of the core back into the outer nanotube wall presents the user with the opportunity to impose the desired discharge characteristics by programming a specific voltage time profile. An important characteristic time, τ_{Cas}, can be found by assuming that $\delta \gg s$ and that the actual charge on the PTNT be $Q_{act} \ll Q_{NT, max}$. In this limit, the dynamics of the inner core is determined by the interwall dispersion force, F_{Cas}, and $\tau_{Cas} = \sqrt{2m_{core}L/F_{Cas}}$, where m_{core} is the mass of the inner tube. If the core is described as a hollow cylinder with thickness $\Delta R = 1$ nm and mass density $\rho \approx 10^3$ kg/m^3, and for $s = 1$ nm, we find $\tau_{Cas} \approx 2.6 \times 10^{-11}$ s, consistent with the upper limits for the frequency of MWCNT-based oscillators ($\nu \geq 10^0 - 10^2$ GHz) [14,53,54].

In general, the characteristic discharge time is $\tau_{dis} = \sqrt{2m_{core}L/(fF_{Cas})} = \tau_{Cas}/\sqrt{f}$, where $f \in [0, 1]$ is the fraction of the interwall dispersion force not balanced by electrostatic effects in this limit ($f = 1$ corresponds to the case above and $f = 0$ to the equilibrium). Since the specific power is $W_{Cas} = \varepsilon_{Cas}/\tau_{dis}$, with τ_{dis} expressed in hours, the possibility to program extremely high, specific power densities is predicted by our model. For instance, in the previous realistic case of van der Waal force interaction between the walls ($|\varepsilon_{Cas}| \approx 88$ W·hr/kg), achieving $W_{Cas} = 10^4$ W/kg would require $\tau_{dis} \approx 32$ s, or $f \approx 6 \times 10^{-25}$ during discharge, or conditions of near-equilibrium. In this regime, ultrahigh power might be obtained from subdomains of the macroscopic nanotube ensemble by sequentially addressing an array of independent electrodes, so as to cause the corresponding cores to be retracted in "nearly free-fall" followed by damped oscillatory motion (Fig. 27.7). In the $\delta \approx s$ regime, a smaller range of motion of the nanotube cores would yield lower power densities.

An ultrafast recharge can be achieved by reversing the cycle described above. Assigning the specific volume or, importantly, the surface of the storage system actively delivering power and the process of release on the nanoscale could provide an attractive answer in nanorobotics medical applications and for a more efficient integration of power generation, storage, and delivery in solar panels.

27.4 PERFORMANCE SIMULATIONS

In the previous sections, we have analyzed the potential advantages of dispersion force-based storage approach largely from an energetic standpoint, that is, without consideration for the dynamics of the system. An analysis of the various forces acting on an individual nanotube core as a function of the coordinate z of its center of mass reveals that the dynamical implementation of the proposed storage scheme is far from trivial. As a specific example, let us consider an MWCNT whose outer

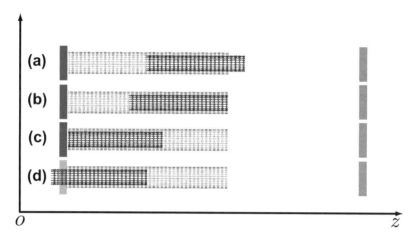

FIGURE 27.4

Showing positions of the core (black) in relation to the electrode at the far right (gold), and the two segments of the nanotube outer wall where the van der Waals energy is larger (blue) and smaller (pink) in magnitude, respectively. In this case, one or two positions of equilibrium appear in the range of z-coordinate between states (a) and (b). The electrostatic interaction is completely shielded by the Faraday cage effect in state (c). State (d) is only shown for completeness in the assumption the connecting pad is removed, which corresponds to the dynamics of MWCNT oscillators. (For interpretation of the references to color in this figure legend, the reader is referred to the online version of this book.)

wall is of length $L_{OW} = 2$ µm, radius $R_{OW} = 10$ nm, with a core consisting of four shells and of length $L_{core} = 1.2$ µm. Let us also assume that the biasing electrode be located at a distance of 1.5 µm from the outer wall edge (Fig. 27.4).

For the greatest clarity at this early stage, we shall adopt the well-known equation derived from a simplified treatment of the dielectric function in the unretarded limit of Lifshitz theory [55,56], which yields the following expression for the retracting van der Waals force on a partially extruded core [22]:

$$F_{vdW,z} = -\frac{1}{32\sqrt{\pi}}\frac{\bar{R}}{s^2}\frac{\left(\hbar\bar{\varepsilon}_{I,G}\right)^2}{\left(\pi\hbar\omega_0 + \hbar\bar{\varepsilon}_{I,G}\right)^{3/2}}\sqrt{\hbar\omega_0}. \quad (27.4)$$

For graphite, we choose $\hbar\omega_0 \approx 10$ eV, $\hbar\bar{\varepsilon}_{I,G} \approx 75$ eV, $\bar{R} = 10$ nm and a gap width equal to the graphite interlayer separation, $s = 0.34$ nm. This yields $\hbar I_{Lif} \approx 7.2$ eV and $|F_{vdW,z}| \approx 4$ nN on a partially extruded core, consistent with the values estimated from direct measurements in MWCNTs [17,18,57]. Importantly, this force is constant as a function of the coordinate z of the core while the core is fully extruded (see for instance Fig. 27.7(c)).

A similar behavior is observed in the electrostatic self-repulsion between the cap and the nanotube, which is also found to not depend on the core coordinate.

Therefore, neglecting for the moment as much smaller both the dispersion force and the electrostatic attraction between the cap and the electrode, we find that the electrostatic self-repulsion and the van der Waals retracting force are approximately equal if the biasing voltage is $V \simeq 12$ V, and that for larger values of the potential, the core will be expelled at high speed from the outer wall, whereas for smaller values, it will be retracted back. This two-state dynamics, which is actually exploited as the foundation of proposed memory systems based on nanotube arrays [19,58,59], does not allow for any position of stable equilibrium where the core might be placed on the user's command to execute long-term energy storage. The situation only marginally improves if the other two forces are included into consideration with the appearance of the position of equilibrium—always unstable—corresponding to the analysis carried out in the previous section (Fig. 27.5(a)). The immediate conclusion one must draw from this analysis is that, in this scenario, long-term energy storage cannot be achieved because the core will rapidly move away from any initial position on a timescale ~ 1 ns.

However, this unpromising situation radically changes if the dispersion force between the core and the outer wall is appropriately engineered. For instance, if the outer wall is divided into two segments of equal length, and the van der Waals interaction energy between the core and the outer wall segment closer to the electrode is made very slightly stronger than that between the segment in contact with the substratum and the core, one or even two positions of stable equilibrium can be created (Fig. 27.5(b) and (c)).

The complex details of the dynamics under such conditions will be presented elsewhere but here we show the main features of oscillatory motion near the two positions of equilibrium corresponding to Fig. 27.5(b). In this particular example,

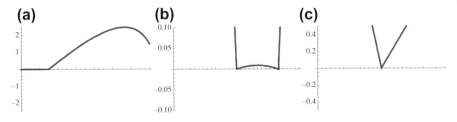

FIGURE 27.5

Topology of the effective potential corresponding to all four forces acting on the core (see text), in arbitrary units. Case (a) shows the full range of the effective potential from state (c) in Fig. 27.4 to the neighborhood of the electrode position, including a position of unstable equilibrium. Case (b) shows the potential for core coordinates between states (a) and (b) of Fig. 27.4 for a van der Waals energy larger in the exit segment by a relative amount of $+0.001$ with respect to the nanotube segment anchored to the connecting pad, and with a bias voltage $V = 11.93$ V. Case (c) shows the disappearance of one of the two positions of equilibrium under the same conditions as Case (b) but with $V = 11.929$ V. (For color version of this figure, the reader is referred to the online version of this book.)

the bias potential is caused to adiabatically change over a period of time corresponding to many cycles near the positions of equilibrium of the core. As typical of systems of this type studied by this author, the dynamics is sensitive to the interplay of small changes in the time-dependent potential and of the time-independent "contrast" between the van der Waals energy in the two nanotube segments. By managing the behavior of the bias potential, it is possible to control the average position of the core without the need for more sophisticated feedback control. In this example, the core draws closer to the outer wall—corresponding to a "discharge"—by ≈ 0.2 μm as the bias potential decreases by V.

27.5 CONCLUSIONS

In this contribution, we proposed an entirely novel method to achieve high-density energy storage by exploiting the van der Waals field of interacting nanostructures, and by focusing, in particular, on cores strongly interacting with their parent nanotube outer walls. In the first part of this work, we showed that dense arrays of parallel, telescoping, MWNTs can deliver energy densities exceeding that of practical electrochemical batteries. In addition, consideration of nanotubes of different species showed that energy densities as large as three times that of ordinary batteries can be certainly achieved. Since energy is being stored in the van der Waals field, which is ultimately described as a gas of virtual photons in the quantum vacuum, the limitations on this approach are extremely high compared to traditional electrochemical technologies and maximum achievable energy densities possibly approach those of hydrocarbons in the idealized case of perfectly conducting boundaries.

In the second part of this work, we analyzed the dynamical challenge of managing nanotube cores under the combined action of van der Waals, electrostatic, and frictional forces. Since the electrostatic self-repulsion and the van der Waals retracting force are both independent from the translational coordinate to the first approximation, no positions of equilibrium can exist after the core extrudes from the outer walls. In practice, this conclusion does not change even when electrode-core electrostatic and dispersion forces are considered. The solution to this apparently crippling challenge represents yet another reminder of the powerful opportunity represented by dispersion force engineering. In this case, we showed that by introducing a very small "contrast" in the van der Waals energy density across two subsegments of the outer walls, one or even two positions of equilibrium can appear if the bias voltage is appropriate. We also demonstrated that adiabatically changing the bias potential over time causes the properties of the system in phase space to evolve in such a way that managing the position of the core becomes feasible (Fig. 27.6).

This solution introduces an additional complication into the energy storage system proposed herein since the bias voltage must be "intelligently" managed in order to store and release the energy as needed (Fig. 27.7). However, this additional requirement is only an apparent complication, which should actually be considered

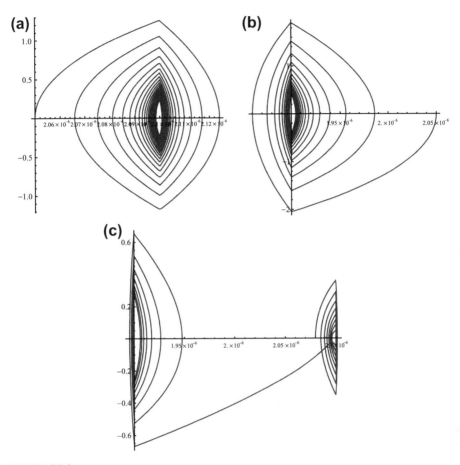

FIGURE 27.6

Cases (a) and (b) correspond to the oscillations in the neighborhood of the two points of equilibrium shown at Case (b) of Fig. 27.5 in $(z, \dot{z})$ phase space and arbitrary units. In Case (c), the bias voltage slowly evolves over time until one position of equilibrium disappears (see Case (c) at Fig. 27.5) and the core, which was near position (a) in this figure, "migrates" from that position to orbiting the second position of equilibrium shown at (b). Dynamical frictional forces were modeled as $F_{f,z}^{dyn}(t) = -\Gamma_1 |\dot{z}(t)| \text{sgn} \dot{z}(t)$, where, typically, the friction coefficient was $\Gamma_1 = 10^{-13}$ kg/s [13]. (For color version of this figure, the reader is referred to the online version of this book.)

as a unique technological opportunity. In fact, it appears likely that dispersion force engineering, along with the inherent energetic advantages presented by telescoping nanotubes, will enable the demonstration of extremely high-density storage systems, in which the process of charge/discharge is not determined by inaccessible electrochemical processes but can be user-assigned. In the future, it may be possible to

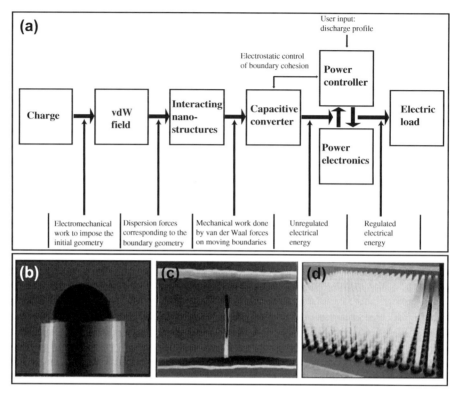

FIGURE 27.7

(a) Block diagram of a high-density energy storage system based on the electrostatic conversion of van der Waals energy. Artist's 3D graphics illustration of simulated TEM images of (b) nanotube detail; (c) individual extruded nanotube core; and (d) nanotube array.

fabricate dispersion force-enabled energy storage systems managed via computer chips tightly integrated on the nanoscale so that the device, or its addressable subdomains, can be instructed to carry out the duties presently typical of such distinct traditional technologies as batteries or supercapacitors, switching from storage to hyperfast recharge–discharge on command.

Acknowledgments

My exploration into the multidimensional connections among the dispersion force research described herein, product and patent design, and the complex interface to the energy storage marketplace greatly benefited from many lively, technical discussions with Wayne S. Breyer, Stephen G. Eichenlaub and Joyce Chung. I am grateful to Mark J. Schulz and Vesselin Shanov for sharing their unique know-how on nanotube fabrication and applications, and for continued encouragement throughout this project. I dedicate this chapter to my father, Italo Pinto, whose unwavering faith in my abilities always challenges me to dare.

References

[1] Martin Winter, Ralph J. Brodd, What are batteries, fuel cells, and supercapacitors? Chemical Reviews 104 (10) (October 2004) 4245−4269.

[2] B. John Goodenough. Basic Research Needs for Electrical Energy Storage—Report of the Basic Energy Sciences Workshop on Electrical Energy Storage, 2−4 April 2007. Technical report, Office of Science, U.S. Department of Energy, 2007.

[3] Kurt Zenz House, The limits of energy storage technology, Bulletin of the Atomic Scientists (2−3 January 2009).

[4] Fred Schlachter, Has the battery bubble burst? APS News—The Back Page 21 (8) (2012) 1−3.

[5] Elzbieta Frackowiak, Francois Beguin, Electrochemical storage of energy in carbon nanotubes and nanostructured carbons, Carbon 40 (2002) 1775−1787.

[6] Antonino Salvatore Arico', Peter Bruce, Bruno Scrosati, Jean-Marie Tarascon, Walter van Schalkwijk, Nanostructured materials for advanced energy conversion and storage devices, Nature Materials 4 (May 2005) 366−377.

[7] Yong Jung Kim, Yoong Ahm Kim, Teruaki Chino, Hiroaki Suezaki, Morinobu Endo, Mildred S. Dresselhaus, Chemically modified multiwalled carbon nanotubes as an additive for supercapacitors, Small 2 (3) (2006) 339−345.

[8] A. Kiebele, G. Gruner, Carbon nanotube based battery architecture, Applied Physics Letters 91 (2007) 144104.

[9] Ray H. Baughman, Anvar A. Zakhidov, Walt A. de Heer, Carbon nanotubes—the route toward applications, Science 297 (2010) 787−792.

[10] Chang Liu, Feng Li, Lai-peng Ma, Hui-ming Cheng, Advanced materials for energy storage, Advanced Energy Materials 22 (2010) 28−62.

[11] F. a Hill, T.F. Havel, C. Livermore, Modeling mechanical energy storage in springs based on carbon nanotubes, Nanotechnology 20 (25) (June 2009) 255704.

[12] Alfred W. Hubler, Onyeama Osuagwu, Digital quantum batteries: energy and information storage in nano vacuum tube arrays, Complexity 15 (5) (2009) 48−55.

[13] Fabrizio Pinto, Nanopropulsion from high-energy particle beams via dispersion forces in nanotubes, in: 48th AIAA/ASME/SAE/ASEE Joint Propulsion Conference (JPC), AIAA, Atlanta, July 2012, pp. 1−31.

[14] Andras Kis, A. Zettl, Nanomechanics of carbon nanotubes, Philosophical Transactions of the Royal Society of London. Series A: Mathematical and Physical Sciences 366 (2008) 1591−1611.

[15] Thomas Rueckes, Kyoungha Kim, Ernesto Joselevich, Greg Y. Tseng, Chin-Li Cheung, Charles M. Lieber, Carbon nanotube-based nonvolatile random access memory for molecular computing, Science 289 (2000) 94−97.

[16] K. Tatur, L.M. Woods, I.V. Bondarev, Zero-point energy of a cylindrical layer of finite thickness, Physical Review A: Atomic, Molecular, and Optical Physics 78 (012110) (2008).

[17] John Cumings, A. Zettl, Low-friction nanoscale linear bearing realized from multiwall carbon nanotubes, Science 289 (28 July 2000) 602−604.

[18] Quanshui Zheng, Jefferson Z. Liu, Qing Jiang, Excess van der Waals interaction energy of a multiwalled carbon nanotube with an extruded core and the induced core oscillation, Physical Review B: Condensed Matter and Materials Physics 65 (2002) 245409.

[19] Jeong Won Kang, Qing Jiang, Electrostatically telescoping nanotube nonvolatile memory device, Nanotechnology 18 (095705) (2007).
[20] W. Arnold, S. Hunklinger, K. Dransfeld, Influence of optical absorption on the van der Waals interaction between solids, Physical Review B: Condensed Matter and Materials Physics 19 (12) (1979) 6049–6056.
[21] E.M. Lifshitz, The theory of molecular attractive forces between solids, Soviet Physics JETP 2 (1) (1956) 73–83.
[22] Fabrizio Pinto, Reflectance modulation by free-carrier exciton screening in semiconducting nanotubes. Journal of Applied Physics, in press, http://dx.doi.org/10.1063/1.4812495.
[23] Fabrizio Pinto, Engine cycle of an optically controlled vacuum energy transducer, Physical Review B: Condensed Matter and Materials Physics 60 (21) (December 1999) 14740–14755.
[24] M. Bordag, G.L. Klimchitskaya, U. Mohideen, V.M. Mostepanenko, Advances in the Casimir Effect, Oxford University Press, Oxford, 2009.
[25] Yury B. Sherkunov, van der Waals interaction of excited media, Physical Review A: Atomic, Molecular, and Optical Physics 72 (052703) (2005).
[26] Peter W. Milonni, The Quantum Vacuum, Academic Press, San Diego, 1994.
[27] Fabrizio Pinto, Membrane actuation by Casimir force manipulation, Journal of Physics A 41 (2008) 164033.
[28] Fabrizio Pinto, Adaptive optics actuation by means of van der Waals forces: a novel nanotechnology strategy to steer light by light, in: Yukitoshi Otani, et al. (Eds.), Optomechatronic Technologies 2008, vol. 7266, SPIE, San Diego, 2008, p. 726616.
[29] Fabrizio Pinto, The economics of van der Waals force engineering, in: Mohamed S. El-Genk (Ed.), AIP Conf. Proc., Number 626, AIP, 2008, pp. 959–968.
[30] D. Linden, Handbook of Batteries, third ed., McGraw-Hill Book Company, New York, 2002 (chapter 1)1.3–1.18.
[31] Robert L. Forward. March 1983 to 21 September 1983, AFRPL TR-83–067. Technical report, Alternate Propulsion Energy Sources, Final Report for the Period, vol. 3, Edwards Air Force Base: Air Force Rocket Propulsion Laboratory, 1983.
[32] Robert L. Forward, Extracting electrical energy from the vacuum by cohesion of charged foliated conductors, Physical Review B: Condensed Matter and Materials Physics 30 (4) (1984) 1700–1702.
[33] V. Adrian Parsegian, Van der Waals Forces, Cambridge University Press, Cambridge, 2006.
[34] E.V. Blagov, G.L. Klimchitskaya, V.M. Mostepanenko, van der Waals interaction between a microparticle and a single-walled carbon nanotube, Physical Review B: Condensed Matter and Materials Physics 75 (2007) 235413.
[35] K. Tatur, L.M. Woods, Zero-point energy of N perfectly conducting cylindrical shells, Physics Letters A 372 (2008) 6705–6710.
[36] Francisco D. Mazzitelli, Maria J. Sanchez, Norberto N. Scoccola, Javier von Strecher, Casimir interaction between two concentric cylinders: exact versus semiclassical results, Physical Review A: Atomic, Molecular, and Optical Physics 67 (013807) (2003).
[37] John Cumings, A. Zettl. Resistance of telescoping nanotubes. In Stuttgart Hans Kuzmany (Universität Wien), Jörg Fink (Institut Für Festkörper-und Werkstoff-Forschung, Dresden), Michael Mehring (Universität Stuttgart), Siegmar Roth Max-

Planck-Institut Für Festkörperforschung, Ed., *Structural and Electronic Properties of Molecular Nanostructures*: XVI International Winterschool on Electronic Properties of Novel Materials AIP Conference Proceedings Volume 633, vol. 633, Kirchberg, Tirol, Austria, 2–9 March 2002, pp. 227–230.

[38] S.H. Jo, Y. Tu, Z.P. Huang, D.L. Carnahan, D.Z. Wang, Z.F. Ren, Effect of length and spacing of vertically aligned carbon nanotubes on field emission properties, Applied Physics Letters 82 (20) (2003) 3520–3523.

[39] YeoHeung Yun, Vesselin N. Shanov, Yi Tu, Srinivas Subramaniam, Mark J. Schulz, Growth mechanism of long aligned multiwall carbon nanotube arrays by water-assisted chemical vapor deposition, Journal of Physical Chemistry B 110 (2006) 23920–23925.

[40] Yeoheung Yun, Vesselin N. Shanov, Swathi Balaji, Yi Tu, Sergey Yarmolenko, Sudhir Neralla, Developing a sensor, actuator, and nanoskin based on carbon nanotube arrays, in: Masayoshi Tomizuka (Ed.), Smart Structures and Materials 2006 Proc. of SPIE Volume 6174, vol. 6174, 2006, pp. 61743Z–61751Z.

[41] Jun Ha Lee, A study on a boron–nitride nanotube as a gigahertz oscillator, Journal of the Korean Physical Society 49 (1) (2006) 172–176.

[42] William R. Smythe, Static and Dynamic Electricity, McGraw-Hill Book Company, New York, 1950.

[43] E. Durand, Électrostatique, Masson, Paris, 1966.

[44] S. Hudlet, M. Saint Jean, C. Guthmann, J. Berger, Evaluation of the capacitive force between an atomic force microscopy tip and a metallic surface, European Physical Journal B 2 (1998) 5–10.

[45] László Forró, Nanotechnology: beyond gedanken experiments, Science 289 (5479) (2000) 560–561.

[46] D. Landau, E.M. Lifshitz, L.P. Pitaevskii, Electrodynamics of Continuous Media, Butterworth-Heinemann, Oxford, 1998.

[47] B.V. Derjaguin, Untersuchungen uber die reibung und adhesion IV, Kolloid-Zeitschrift 69 (1934) 155–164.

[48] B.V. Derjaguin, I.I. Abrikosova, E.M. Lifshitz, Direct measurement of molecular attraction between solids separated by a narrow gap, Quarterly Reviews (London) 10 (1956) 295–329.

[49] B.V. Derjaguin, I.I. Abrikosova, Direct measurement of the molecular attraction of solid bodies. I. Statement of the problem and method of measuring forces by using negative feedback, Soviet Physics JETP 3 (6) (1957) 819–829.

[50] J. Blocki, J. Randrup, W.J. Swiatecki, C.F. Tsang, Proximity forces, Annals of Physics 105 (1977) 427–462.

[51] B.V. Derjaguin, Y.I. Rabinovich, N.V. Churaev, Direct measurement of molecular forces, Nature 272 (1978) 313–318.

[52] S. Meninger, J.O. Mur-Miranda, R. Amirtharajah, A. Chandrakasan, J.H. Lang, Vibration-to-electric energy conversion, IEEE Transactions on Very Large Scale Integration (VLSI) Systems 9 (1) (2001) 64–76.

[53] Wanlin Guo, Yufeng Guo, Huajian Gao, Quanshui Zheng, Wenyu Zhong, Energy dissipation in gigahertz oscillators from multiwalled carbon nanotubes, Physical Review Letters 91 (12) (2003) 125501.

[54] S.B. Legoas, V. R Coluci, S. F Braga, P. Z Coura, S. O Dantas, D.S. Galvao, Gigahertz nanomechanical oscillators based on carbon nanotubes, Nanotechnology 15 (2004) S184–S189.
[55] H. Krupp, G. Sandstede, K.H. Schramm, Beitrage zur entwicklung des chemischen apparatewesens, Dechema Monograph. 38 (1960) 115–147.
[56] H. Krupp, Particle adhesion: theory and experiment, Advances in Colloid and Interface Science 1 (1967) 111–239.
[57] Seiji Akita, Yoshikazu Nakayama, Interlayer sliding force of individual multiwall carbon nanotubes, Japanese Journal of Applied Physics 42 (7B) (2003) 4830–4833.
[58] Jeong Won Kang, Jun Ha Lee, Ho Jung Hwang, Schematics and simulations of a nanotube shuttle nonvolatile memory device, Journal of the Korean Physical Society 50 (3) (March 2007) 733–738.
[59] Leonid Maslov, Concept of nonvolatile memory based on multiwall carbon nanotubes, Nanotechnology 17 (2006) 2475–2482.

Index

Note: Page numbers followed by "f" and "t" indicate figures and tables respectively.

A

AAO templates. *See* Anodized aluminum oxide templates
AC motors. *See* Alternating current motors
2-acrylamido-2-methyl-1propane sulfonic acid (AMPSA), 196
Active optoelectronic materials
   metallic nanotubes and graphene, 554–555
   semiconducting SWNTs, 553–554, 554f
AFM. *See* Atomic force microscopy
Alginate fibers, 200
Aligned carbon nanotube composite prepregs
   CNT alignment and straightness effect
      nanotube orientation direction, 660–661
      polarized Raman spectroscopy, 661–662
      pristine CNT sheet, 661f
   CNT
      composite effects, 662–663
      fibers, 651–652
      sheets, 652
      volume fraction effect, 662
   composite fabrication process, 664
   dispersion, 650
   fabrication, 652–659
   high-strength CNT composites, 664
   matrix type CNT effect
      high-toughness composites, 663
      stress-transfer efficiency, 663
   mechanical and physical properties, 660
   paper-like CNT films, 650–651
Aligned composite prepreg fabrication
   from aligned CNT arrays, 658f, 660f
      advantages, 659
      angle of pressing, 659
      composite fabrication, 657–659
      resin-infused prepregs, 659
   CNT/polymer composites, 652
   from drawable superaligned CNT sheets, 652–653
      CFRP composites, 654–655
      CNT waviness adverse effect, 654
      polymer matrix, 653–654
      prepregs postfabrication stretching, 655–657
      spraywinding process, 653
Aligning and tensioning system (ATS), 219
α-SiC. *See* hexagonal SiC
Alternating current motors (AC motors), 601

Ammonia, 150–151
Amphiphilic agents, 170–171
AMPSA. *See* 2-acrylamido-2-methyl-1propane sulfonic acid
Analysis of variance (ANOVA), 686–688
   for CNT thread gauge factor optimization, 689f
Anodized aluminum oxide templates (AAO templates), 469–470
ANOVA. *See* Analysis of variance
Applied Sciences, Inc. (ASI), 315–316
Arc discharge method, 249, 250f
Armchair carbon nanotube, 4–6, 5f, 248
As-spun yarns, 19
   CNT length dependence, 405
   postspin twisting effect
      CNT yarn, 402
      cross-sectional observations, 403, 404f
      tensile strength, 402, 403t
      twisting density dependence, 403, 404f
      Young's modulus, 403t
   stress–strain curves, 402f
   yarns' diameter dependence, 405–406
      MWCNT length dependence, 405f
      for single-spun yarns, 406
      Yarn diameter dependence, 406, 406f
ASI. *See* Applied Sciences, Inc.
Atmospheric pressure plasma, 358
Atmospheric pressure plasma functionalization
   pristine CNT sheets, 362–363
   using Raman spectroscopy, 363
   Surfx Atomflo 400-D plasma controller, 363f
Atomic force microscopy (AFM), 50, 542, 542f
ATS. *See* Aligning and tensioning system
Au-coated yarns, 156
Au-decorated yarns, 156–158
Au–CNT yarn, 231, 231f

B

Ball milling, 249–250, 251f
Bandwidth
   calibration measurement, 695
   CNT thread bandwidth, 697–699
   custom double-cantilever calibration rig, 695–696, 695f
   epoxy mounting points, 696, 696f
   FRF, 697
   sine wave of CNT thread, 698f–699f

Bandwidth (*Continued*)
  stiffer epoxy, 697
  voltage output measurement, 696–697
  wave excitation, 697
BCPs. *See* Block copolymers
BEM. *See* Boundary element methods
β-SiC. *See* cubic SiC
BIE. *See* Boundary integral equation
Bingham Canyon Copper Mine, 36f
Bingham Canyon Mine, 35–36
Biodegradability, 739–740
Biofouling coating, 734–735
Bismaleimide (BMI), 650–651
Block copolymers (BCPs), 535–536
Blood cells. *See* Hematocytes
BMI. *See* Bismaleimide
BN. *See* Boron nitride
BNNT. *See* Boron nitride nanotube
Boron nitride (BN), 2–3
Boron nitride nanotube (BNNT), 246, 248f, 268, 449, 727
  see also Carbon nanotube (CNT); Silicon carbide nanotube (SiCNT)
  applications
    electric motors, 282–283
    extreme environment application for, 281–282
    nanotube-reinforced aluminum, 283–284
    NASA-GRC, 281
    NIOSH regulations, 285
    potential NASA application, 283t, 284f
    radiation shielding, 283
    spur commercialization, 284
    superior thermal gradient cyclic stress resistance, 282f
  comparison
    elastic modulus, 274
    mechanical properties and thermal conductivity, 263
    metallic character, 274–275
    properties, 276t
    structural similarities, 261–262
    TGA, 275f
    thermal conductivity with CNT, 274
    thermogravimetric analysis data, 262f
    water, 262f
    wide-bandgap semiconductors, 262
  composites
    in barium calcium aluminosilicate glass composite, 277f
    BNNTsurvived hot pressing, 280f
    fuzzy fiber concept, 278, 278f
    growth on SiC tows and woven fabric, 278–281, 279f
    NASA-GRC, 277
    polyaniline and polystyrene, 276–277
    SiC_f/SiC_m composites, 278, 279f
    tensile strength, 280f
  morphologies and infiltrations, 271f
  properties, 259
    current–voltage characteristics, 260f
    electrical properties, 259
    mechanical properties, 259–260
    optical properties, 260
    thermal conductivity values, 261f
    thermal properties and thermal stability, 260–261
    wetting behavior, 260–261
  structure
    atomic, 247f
    cubic and hexagonal polytypes, 268
    h-BN, 268–269
    hexagonal sheets, 269f
  synthesis, 249, 270
    arc discharge, 249, 250f
    ball milling, 249–250, 251f
    carbothermal synthesis, 250–251
    chemical vapor deposition, 251–258
    laser heating and ablation, 258
    SWCNT bundles and B_xC_{1-x} bundles, 252f
Boundary element methods (BEM), 572
  BIE, 578–579
  cohesive interface model, 579
  equilibrium of forces, 580
  FMM
    conventional BEM, 581f
    direct or iterative solvers, 580
    fast multipole BEM, 581f
    FMBEM, 580–581
  microscale simulations, 578
  rigid-fiber model, 579–580
  stresses and strains, 580
  3-D elastic domain V, 578, 578f
Boundary integral equation (BIE), 574
Bridging coagulation, 186
Bridging flocculation, 173–174
Brittle fracture
  atomistic simulations *vs.* QFM predictions, 500t
  cable strength, 502
  LEFM, 499–502
  QFM, 499
  taper ratio, 503
Buckypaper and thin films
  CNT film structure, 440–441, 440f

Index

in cross-plane direction, 440
macroscopic materials, 437–439
SWCNT buckypaper, 437f
Buckypapers. *See* Paper-like CNT films
Bulk boron nitride, 268
Bulk staple fibers, 17

C

Capstan effect rod system (CERS), 219
Carbon
  carbon-based fibers, 61
  sources, 102, 108
  tubing, 36
Carbon fiber-reinforced polymer composites
    (CFRP composites), 654–655, 656f
Carbon nanofiber (CNF), 1–2, 313
  ASI, 314f, 315–316
  cylindrical graphitic structures, 10
  ILSS and resin, 314
  stacked-cup morphology, 316–317
  stiffness or tensile modulus, 11
  wider edges of cones, 10–11
Carbon nanofiber mat (CNF mat), 315–317
  using Cytec 5240-4 BMI resin film, 320–321
  Cytec's Anaheim facility, 319
  80-ft-long roll of nanofiber mat, 319f
  fabrication mat-reinforced composites, 321–322
  fully cured laminate, 320f
  nanomaterials, 319
  processing on industrial-scale prepreg line, 321f
  production
    via ASI's proprietary process, 318f
    via filtration, 318f
  PR-19 carbon nanofiber, 317f
  Pyrograf III material, 316f
  reinforced composites
    composite configurations, 321–322
    fabricated laminated composites, 323f
    fabrication, 321
    test coupons, 321
  Renegade Materials, 321
  short beam shear testing, 322–328
  small-scale impregnation trials, 319
  using traditional wet-laid papermaking process, 318–319
  on 12" pilot line at Cytec, 320f
Carbon nanomaterial (CNM), 313
Carbon nanotube (CNT), 1–2, 87–88, 649–650
  see also Boron nitride nanotube (BNNT);
    Carbon nanotube yarns; Ultralong carbon
    nanotube (Ultralong CNT)
  aerogel, 213

aligning and tensioning system, 219, 660–661
assembly in macroscopic structures, 213
  interrelations, 214–215
  potential energy function, 215–216
benefits in nanocomposites, 313
biscrolling process particles, 230, 230f
bundles
  and fibers, 290
  measurements on individual CNT, 432
  SWCNT, 432–435
charge transfer kinetics, 138–139
CNF, 4
CNT-based carbon fibers, 62f
coagulation
  in bad solvents, 170–172
  fibers spun from CNT dispersions, 171–172
comparison
  mechanical properties and thermal
    conductivity, 263
  structural similarities, 261–262
  thermogravimetric analysis data, 262f
  water, 262f
  wide-bandgap semiconductors, 262
composites, 435, 436f
  CNT-enabled materials, 435
  CNT–CNT junctions, 435–436
  EMT models, 436–437
CVD process, 4
development for use in aggressive environments,
    267–268
diameter and growth process, 8–9
discrete fibrous elements, 63f
dispersion approach, 350–351
doping, 290
dry spinning method, 216f
electrical properties, 40t
experimental challenges in characterization
  as-spun, densification and doping, 305f
  CNT thread and Ag paste, 304f
  diameter measurements, 304–305
  electrical contact, 303–304
fibers, 334, 651–652
  dilatation, 222–223
  dilatational separation, 223f
forest/matrix, 217
freestanding, 3
  film preparation, 351–352
gas sensors, 152
graphene, 4
growth via CVD, 352
material design, 72
  analytical models, 72–74

Carbon nanotube (CNT) (*Continued*)
  atomic-scale and nanoscale defects, 76–77
  CNT–CNT interactions, 74
  CNT maximum current density measurement, 720f
  CNT thread properties, 719–720, 720f
  degree of CNT alignment, 77
  in elastic regime, 75
  electrical conduction, 721
  electrical properties, 720
  using factors, 76
  fiber diameter and large-scale defects, 78–79
  forms, 719, 719f
  high-temperature annealing, 720–721, 721f
  macroscale helical structure, 78
  material properties, 74
  multiscale mechanical factors, 79
  nanoscale and macroscale geometry, 73f
  NIL, 721, 722f
  number of walls, 722f
  optimal overlap length, 74–76
  in strength, 335f
  tube diameter, 76
  twist-spun fiber from MWNT forest, 78f
  2D unit cell using in analytical model, 75f
  material development and characterization, 601
    CNT yarn and wire forming, 604–606
    coating CNT thread and yarn, 604
    nanocomposite material fabrication, 606, 607f
    postprocessing yarn, 603
    synthesis research efforts, 602
  mechanical and physical properties, 22t, 62–63
  metal nanoparticle deposition
    on CNT yarns, 143–149
    on macrostructures, 142–149
  metals, 138
  nanodevice, 42–43
  natural composites, 64f
  optical characterization
    CNT sheet sample variation, 373
    D. I. water droplet on pristine, 371f
    IR transparency, 371
    orientation, 371, 373f
    percent transmittance *vs.* wavelength, 374f
    UV–visible region, 371, 372f
  packing fibers
    CNT fibers in yarns, 220f
    CNT web/sliver, 219f
    for dry spinning process, 220–221
    loss of strength, 220
    modified spinning process, 221–222
    snarl of yarn, 221f
    tensile strength, 220
    in traditional textile, 219
  perturbations
    C–C bonding symmetry and energy, 11
    external hydrostatic pressure effect, 11–12
    single-CNT shell, 11
    strength or weakness, 12
    SWCNT on flat surface, 12
    torsion, axial compression and bending, 12
  polygonization, 9
  polymer-induced coagulation, 173
    composition, structure and properties, 177–185
    fiber formation and physicochemical mechanisms, 173–177
  properties, 276t, 350t, 425–426
  public and scientific attention, 3–4
  reactor design, 44–45
    CVD, 44–45
    methods, 45
    synthesis, 45
  shear interactions, 61–62
  significance of spacing, 9–10
  structures, 5f, 749–750
    angstroms calculation, 6
    armchair type and diameters, 8t
    atomic, 212, 247f
    and behavior, 8
    carbon atom pattern, 7
    chiral angle calculation, 6
    CNF section with tilt angle, 10f
    electrical properties, 750
    fabrication techniques, 750
    MWCNT, 7
    RF antenna designs, 750–751
    sp^2-bonded carbon, 4–6
    SWNT, 7
    tube indices, 6
  synthesis
    catalyst or precursor, 13
    catalyst particles, 13–14
    evaporation of carbon, 14
    growth processes, 13
    in laminar gas stream, 14
    requirements for, 12–13
  techniques for sheet production
    CNT buckypapers with aligned nanotubes, 353
    domino pushing method, 354f
    iron particles with diameters, 353
    long roll, 353f
  TEM images, 139–140, 139f
  textile spinning technology, 216

dry spinning method, 216, 216f
packing of fibers, 217
thermal annealing, 46f
  in healing, 47
  high-temperature annealing, 45–47
  postannealing Raman spectroscopy, 47
threads, 34
waviness, 660–661
web model
  dual slip-link-based, 218–219
  formation, 217
  structure, 218f
web separation
  fibers dilatation, 222–223
  silver/web dilatation, 223–224
  two-step spinning process model, 222f
wire
  with low impedance, 740–741
  thread, 742f
yarn fabrication, 138f
yarns, 212
Carbon nanotube field-effect transistor (CNTFET), 152–153
Carbon nanotube sheet manufacturing process, 361f
  accumulation on Teflon substrate, 362f
  applications, 373
    composites, 377–380
    electromagnetic interference shielding, 373–377
  atmospheric pressure plasma functionalization
    pristine CNT sheets, 362–363
    using Raman spectroscopy, 363
    Surfx Atomflo 400-D plasma controller, 363f
  characterization
    contact angle testing, 371
    FTIR analysis, 369–370
    optical characterization, 371–373
    Raman spectroscopy, 363–365
  solvent densification effect, 362t
  at UC Nanoworld laboratories, 361
Carbon nanotube sheets, 350
  applications, 373
    composites, 377–380
    EM interference shielding, 373–377
  fabrication, 765
    SEM images, 766f
    setup, 765f
  functionalization
    requirement for, 354–355
    techniques for, 355–361
  mechanical properties, 379t, 381t

potential applications, 354
PVA/CNT prepregs, 378f, 380f
Carbon nanotube yarn, 333–334
see also Multiply twisted carbon nanotube yarns
  advantage, 410
  using Ag epoxy, 144–145
  Au and Cu deposition, 147
  Au-CNT and Cu-CNT yarns, 148–149
  CNT web, 401–402, 401f
  coating with, 145f
  comparison, 409f, 409t
  doped and coated macrostructures, 149t
  electrical conductivity, 140, 141t, 145, 149f
    approaches, 140–141
    in DWNT, 141
    MWCNT-based macrostructures, 140
    postsynthesis treatment, 142
    SWCNT, 142
  fabrication by dry spinning method, 399–400, 400f
  mechanical properties, 401–402
  metal-carbon nanoyarn hybrid yarns, 148f
  MNP deposition on, 146–147
  optical images, 146f
  parameters, 144t
  porous barriers, 143–144
  postspin twisting effect, 402–403
    CNT length dependence, 405
    yarns' diameter dependence, 405–406
  pressure, 409
  Pt-CNT hybrid yarns, 147f
  SFED deposition, 143, 144f
  specimen for tensile strength measurement, 401f
  temperature dependence, 150f
  XRD spectroscopy, 146t
  yarn evaluation methods, 400–401
Carbothermal synthesis
  reaction temperature, 250
  for synthesizing BNNT, 250–251
  variations, 251
Carrier mobility, 527, 528t
Catalyst activity probability, 121
Catalyst co-injection (CCI), 13
Catalyst deactivation probability, 121
Catalyst predeposition (CPD), 13
CCI. *See* Catalyst co-injection
CERS. *See* Capstan effect rod system
CFRP composites. *See* Carbon fiber-reinforced polymer composites
CGR. *See* Cosmic galactic radiation

Chemical vapor deposition (CVD), 44–45, 87–88, 251–252
  arc discharge evaporation method, 350
  BNNT synthesis
    β-rhombohedral boron and h-BN, 258
    RF induction heating, 258
  CNT synthesis by, 294
  plasma-based CVD synthesis
    MPECVD, 252–256
    PE-PLD, 256–257
  in situ synthesis
    $B_4N_3O_2H$, 257f
    BN precursors, 257
  synthesis, 629, 629f
  ultralong CNT synthesis, 101
    carbon sources, 102
    catalysts and CVD process, 101–102
    methods for, 103t
    substrates, 102
Chiral carbon nanotube, 4–6, 5f
Chiral nanotubes, 248
Chiral vector, 247–248
Chito (1 → 4)-2-amino-2-deoxy-b-D-glucose, 200
Chitosan/CNT composite fibers, 200
CNF. *See* Carbon nanofiber
CNF mat. *See* Carbon nanofiber mat
CNM. *See* Carbon nanomaterial
CNT. *See* Carbon nanotube
CNT network synthesis, 468–469
  carbon nanotube sponges, 479f
    amount of boron, 478–479
    B-doped MWNT sponges, 478–479
    3D CNT sponge, 477–478
  electron beam irradiation
    at high temperatures, 472–473
    SWNT welding technique, 473
    X, Y and T junctions, 474f
  noncovalent synthesis
    binary digit representation, 471f
    CNT aerogels, 472, 473f
    electronic applications, 470
    flat substrates, 472
    number of substrates, 470–472
    orthogonal CNT networks, 472
  role of sulfur, 476f
    catalytic processes, 475
    energy-dispersive X-ray spectroscopy, 477
    sea urchin nanoparticles, 476–477
    sulfur/carbon ratios, 478f
  secondary growth approaches
    carbon nanostructures, 475f
    substrates and catalytic particles, 473–474
    undoped and nitrogen-doped MWNTs, 474–475
  template approaches, 469–470, 470f
CNT-based composite yarns, 227
  Au–CNT and Cu–CNT yarns, 231, 231f
  biscrolling CNT web and particles, 230, 230f
  CNT-based wet-spun fibers, 170
  CNT–metal composite yarns, 230
  CNT–polymer composite yarns
    composite manufacturing processes, 227–228
    distribution of polymer, 228
    multiweb spinning from multiwafers, 229–230, 229f
    polymer/CNT composite manufacturing process, 228f
    PU/CNT composite sliver, 228–229, 228f
    pure CNT and PU/CNT composite yarns, 229f
CNT-fiber-electrode-based dye-sensitized photovoltaic wire, 344
CNT–conductive polymer composite fibers, 195–196
  electrical properties, 199f
  internanotube contacts, 196–197
  mechanical properties, 199f
  opportunities to develop electrodes, 197
  PANi–CNT composite fibers, 196
  PPy-alginate fibers, 197
  PPy/alginate/SWNT composite fibers, 197
  SWNT and PEDOT, 197
  wet-spinning and polymerization line, 198f
CNTFET. *See* Carbon nanotube field-effect transistor
CNT–natural polymer composite fibers
  alginate fibers, 200
  biopolymers, 200–201
  chitosan/CNT composite fibers, 200
  formulations, 199
  wet-spinning technologies, 198–199
Cohesive interface model development
  axial cohesive forces, 577–578, 577f
  CNT pull-out simulations, 575
  compliance matrix, 575–576
  cylindrical coordinates, 576f
  PE system models, 576–577
  Tersoff–Brenner potential, 576–577
Composite electromagnetic material
  see also Elastomer nanocomposite material
  bionic artificial cell, 739, 742f
  dry spinning method, 738–739
  Helmholtz coil configuration, 739, 740f
  in polymer, 738

toroid transformer, 739
Constituent fibers, 16
Contact angle testing, 371
Continuous wave (CW), 258
Continuum mechanics, 571–572
Copper wire, 37
Cosmic galactic radiation (CGR), 283
Covalent CNT junctions, 460–461
   see also Noncovalent CNT junctions
Covalent functionalization, 170–171
CPD. See Catalyst predeposition
Crystal lattice-directed growth, 113–114
cubic SiC (β-SiC), 268
Cu–CNT yarn, 231, 231f
Current–voltage measurement (I–V measurement), 259
CVD. See Chemical vapor deposition
CW. See Continuous wave

D

DAC. See Diamond anvil cell
Data acquisition (DAQ), 683–684
DBD. See Dielectric barrier discharge
DC. See Direct current
DCAA. See Dichlororacetic acid
Density functional theory (DFT), 462–466
Deoxyribonucleic acid (DNA), 200–201, 481–483
DFT. See Density functional theory
Diamond anvil cell (DAC), 258
Dichlororacetic acid (DCAA), 196
Dielectric barrier discharge (DBD), 360–361
Dimethyl methylphosphonate (DMMP), 484
Dimethyl sulfoxide (DMSO), 174, 752
Dip-pen nanolithography (DPN), 51–52
   catalyst particle deposition, 52f
   silicon surface, 52
Direct current (DC), 37, 223–224, 612–614
Direct spinning, 213
   see also Dry spinning
Direct synthesis technique
   in continuous spinning process, 337
      CVD, 337
         spinning apparatus, 338
      direct spinning, 338f
      mechanical properties, 336–337
      production, 337
      salient feature, 336
      shrinking process, 337–338
      SWNT, 335–336
Dispersion forces, 790
DMMP. See Dimethyl methylphosphonate

DMSO. See Dimethyl sulfoxide
DNA. See Deoxyribonucleic acid
Domino pushing method, 354f, 657–659
Doped carbon nanotube characterization
   Au-doped CNT fibers, 301f
   electrical conductivity, 300
   highly sensitive to π-electronic structures, 302
   Raman spectroscopy, 301–302, 302f
   ^{13}C-NMR, 302–303
   XPS, 302
Doping, 290, 603–604
   CNT doping types, 291
   endohedral, 291–292
   exohedral
      electronic density, 292
      p-type dopants, 292–293
      SWCNT, 293
   fullerene encapsulation, 292f
   functionalization, 296
   SEM, 603f
   substitutional doping, 293–294
      B-N codoping, 295
      B-or N-containing precursors, 294
      carbon replacement, 295
      endo-and exo-substitutional doping, 295f
      hexagonal atom configuration, 295
      pyridine-like configuration, 294
Double-walled carbon nanotube (DWCNT), 7, 65–66
   see also Multiwall carbon nanotube (MWCNT); Single-walled carbon nanotube (SWCNT)
   CNT fibers, 441
   electrical conductivity in, 141
   van der Waals forces, 794–795
   wet/solution functionalization, 355–356
DPN. See Dip-pen nanolithography
Drawable arrays, 652–653
Drawable superaligned CNT arrays, 652–653
Dry spinning, 213
   see also Wet spinning
   CNT, 216, 216f
   method, 738–739
   modified schema, 217f
   process, 217
   thread, 55, 55f
Dry-jet wet spinning, 169
Dual slip-link-based CNT web model
   CNT fibers, 218–219
   CNT web formation model, 218f
   SEM analysis, 218

DWCNT. *See* Double-walled carbon nanotube
DWNT. *See* Double-walled carbon nanotube (DWCNT)

E

EB. *See* Emeraldine base
EBL. *See* Electron-beam lithography
EELS. *See* Electron energy loss spectroscopy
E_F. *See* Fermi energy level
Effective medium theory (EMT), 435–436
Elastic-plasticity, 503
Elasticity
   elastic degradation, 506
   single-defect typology, 505
   transversal crack, 505
Elastomer nanocomposite material, 739
   see also Composite electromagnetic material
   testing, 741f
   trapping, 741f
Electric field-directed growth, 115–116
Electric motor development
   carbon materials, 598
   material preparation and technical design, 602f
      CNT material characterization, 601
      electromagnetic materials development, 606
      motor components design and testing, 612
   using nanoscale materials
      CNT thread resistivity, 600
      energy recovery, 599–601
      heat conduction, 598–599
      magnetization curve, 600
Electromagnetic devices, 35
   carbon motors and actuators, 596, 596f
   CNT wire, 595–596
   Faraday's law of induction, 597
   nanomaterials, 598
   nex-gen motor, 597–598
Electromagnetic Halbach array fabrication
   component testing, 619f
   magnetic flux density, 619–620
   motor torque, 619
Electromagnetic interference shielding (EMI shielding), 350
   electric and magnetic fields, 373
   electric field on surface, 373–374
   EM radiation, 375
   frequency effect, 376, 376f
   incident, 374f
   nanotube orientation effect, 375f
   shielding measurements, 375–376
Electromagnetic material development
   carbon wire design and testing, 611
   and characterization, 606
   magnetic nanotubes and nanowires
      magnetic nanoscale materials, 610f
      nanomaterials, 609–610
      superparamagnetic materials, 610–611
   superparamagnetic particle core design, 607
      eddy currents and magnetic losses, 608f
      exchange length, 607
      iron replacing, 609f
      magnetization, 607–608
   thermal testing, 611, 612f
Electromagnetic simulation, thread antennas
   comparison of current distribution, 759f
   current distribution, 760, 761f
   double-walled CNT current distribution, 762
   large ply CNT rope antennas, 763, 763f
   maximum dipole current, 763–764, 764f
   moments solution method, 759
   nanoscale dipole antennas, 762, 762f
   nanoscale impedance effects, 764
   nonzero dipole thread effect, 761
   Ohm's law, 760
   radiation efficiency comparison, 760, 762–763
   RF performance tradeoff, 761
electromotive force (emf), 615
Electron beam evaporation, 253
Electron beam irradiation
   at high temperatures, 472–473
   SWNT welding technique, 473
   X, Y and T junctions, 474f
Electron diffraction analysis, 253–254
Electron energy loss spectroscopy (EELS), 556
Electron-beam lithography (EBL), 49
Electronic band structures
   GNR
      edge morphology, 525–526, 526f
      FET, 525f
      2D grapheme sheets, 524–525, 525f
   monolayer graphene
      carbon atoms, 520–521
      conduction bands, 522
      energy dispersion, 522
      first Brillouin zone, 521f
      lattice structure, 521f
   SWNTs derived from graphenes, 523f
      carbon atoms, 522–523
      Fermi level, 523–524
      2D graphene sheets, 523
Electrostatic interaction, 795
Electrostatic self-repulsion, 798–799
Embedded carbon nanotube sensor thread
   carbon nanotube thread SHM architectures

cross-weave or grid design, 699—702
damage modes, 699
F-22 raptor, 703—704
micromechanical failure modes in laminated composites, 703f
multiplexing, 702—703
for retrofit to aircraft, 705f
sensors array configuration, 703f
SHM sensor development, 699
SHM system communication, 704f
CNT materials, 675f
CNT sensor thread performance
bandwidth, 695—699
consistency, 691—692
hysteresis, 688—691
polymeric material, 682
sensing strain and damage, 680—681
sensitivity, 682
stability, 692—694
CNT thread, 674
composite materials, 671—672
embedded sensing proof
CNT thread, 676
epoxy beam, 676f
epoxy brick with beam loading conditions, 677f
four-point bending test, 676
gauge factor, 677—678, 678f
mode I delamination, 678—679, 680f
mode II delamination, 678, 679f
piezoresistive measurements, 680
strain resistance response, 677, 677f
F-22 raptor wings, 673—674, 673f
high volume fraction of CNT Thread, 676
ply delamination, 672—673, 672f
polymeric composite materials, 674—676
strong multifunctional potential areas
CNT thread, 704
energy dissipation, 706f
epoxy resin, 707
hysteresis, 704—705
loss factor comparison, 707f
specific damping capacity, 705—707
toughness or damage limiting, 707—708
ultrasonic inspection, 673
Embedded sensing proof
CNT thread, 676
delamination
mode I, 678—679, 680f
mode II, 678, 679f
epoxy beam, 676f
epoxy brick with beam loading conditions, 677f
four-point bending test, 676
gauge factor, 677—678, 678f
piezoresistive measurements, 680
strain resistance response, 677, 677f
Emeraldine base (EB), 196
emf. *See* electromotive force
EMI shielding. *See* Electromagnetic interference shielding
EMT. *See* Effective medium theory
Endohedral doping, 291—292, 291f
Energy density, 792
Energy dispersion, 522
Energy spectrum. *See* Energy dispersion
Energy storage from dispersion forces, 789
dispersion forces, 790
MWNT, 790
nanoscale engine cycles, 790—791
nanotubes adoption, 790
orders of magnitude
Derjaguin approximation, 796
equilibrium position, 796
external voltage source, 797
interwall dispersion force, 796—797
mass fabrication, 794—795
nanostructures interaction, 792—793
polarized telescoping nanotube geometry, 793f
power densities, 797
quasi-equilibrium, 793—794
parallel-plate system idealization
dispersion force interactions, 791—792
energy density, 792
parallel-plate geometry, 792
thermodynamical engine storing energy, 791f
transformation, 792
performance simulations
biasing electrode, 798f
dynamical implementation, 797—798
electrostatic self-repulsion, 798—799
graphite interlayer separation, 798
Lifshitz theory, 798
oscillatory motion features, 799—800
van der Waals interaction energy, 799, 799f
Erythrocytes. *See* Red blood cells
Exohedral doping, 291f, 292
electronic density, 292
p-type dopants, 292—293
SWCNT, 293

F

Fast multipole BEM (FMBEM), 573
Fast multipole method (FMM), 578
conventional BEM and fast multipole BEM, 581f

Fast multipole method (FMM) (*Continued*)
   direct or iterative solvers, 580
   FMBEM, 580–581
Fatigue
   failure reduction, 504
   finite width role, 505
   mean-field approach, 505
   Paris' law, 504
FE. *See* Field emitter
FEM. *See* Finite element methods
Fermi energy level (E_F), 549
FET. *See* Field effect transistor
FIB lithography. *See* Focused ion-beam lithography
Fibers, 1–2
   see also Nanofibers
   carbon, 16
   chiral angle, 6
   spinning, 187
   staple, 2
   twisting, 2
Field effect transistor (FET), 524–525
Field emitter (FE), 259
Finite element methods (FEM), 572
FMBEM. *See* Fast multipole BEM
FMM. *See* Fast multipole method
Focused ion-beam lithography (FIB lithography), 49
Four-point bending test, 676
Fourier transform infrared spectroscopy (FTIR spectroscopy), 356, 369–370, 370f
Frequency response function (FRF), 696–697
   calibration rig's base acceleration and strain, 700f
   CNT thread and strain gauge, 701f
   10-to100-Hz FRF, 702f
FTIR spectroscopy. *See* Fourier transform infrared spectroscopy
Full-wave solver, 767
Functionalization
   CNT, 296, 355
   covalent, 355
   plasma functionalization, 356
      atmospheric pressure plasma, 358
      comparison, 358t
      DBD plasma, 360–361
      FTIR spectroscopy, 356, 358f
      low-pressure RF plasma, 356
      sedimentation behavior, 357, 359f
      SEM micrograph, 360f
      using TPD technique, 359–360
      X-ray photoelectron spectra, 357, 359f
      XPS analysis, 356
      XPS survey scan and C1s spectra, 357f
   wet/solution functionalization, 355–356
Furnace-moving method, 121
Fuzzy fiber concept, 278, 278f

G

Gas flow-directed growth, 95–97
Gas sensing applications
   biological and chemical species, 150–151
   CNT gas sensors, 152
   using CNT macrostructures
      adsorption–desorption response, 156f, 157f
      Au-coated yarns, 156
      Au-decorated yarns, 156–158
      change in electrical conductance, 154–155
      CNT rope, 155f
      electrical resistance in MWCNT yarns, 157f
      hydrogen detection, 160–161, 160f
      MWCNT yarns, 155–156
      Pd-decorated yarn, 159
      plasma-treated samples, 158
      response cycles, 159f
      SEM diagrams, 158f
   gases and organic vapors, 151–152
   metal-CNT heterostructures, 152
   rapid recovery of conductance, 151
   sensing mechanisms
      Au–CNT contact regions, 153
      conductance, 152–153
      MNPCNT devices, 154
      using optical spectroscopic technique, 154
   simple and straightforward methods, 150
Gas sensor. *See* Multifunctional carbon nanotube antenna
Gas-phase precursors, 253
Gauge factor, 677–678, 678f, 682
Gelatin (GE), 187
Glenn Research Center (GRC), 267–268
GNM. *See* Graphene nanomeshe
GNR. *See* Graphene nanoribbon
GO films. *See* Graphene oxide films
Graphene and CNT comparison
   applications in bendable electronics
      roll-to-roll method, 544
      2D honey-comb lattice structure, 543–544
   bandgap engineering
      BCP, 535–536
      chemical unzipping, 536f
      dual-gated bilayer graphene, 534f
      GNRs fabrication, 533
      plasma-free methods, 534–535
      semiconductors GNR, 535f

single-layer graphene sheets, 532–533
ultrasmooth GNR, 535
elastic properties
  AFM, 542
  anisotropic indentation, 542–543
  elastic buckling, 544f
  elastic stress response, 541
  MWCNT, 545f
  stable molecular structures, 541
  stress–strain curves, 543f
  tensile-loading setup, 542
electrical properties, 527–541
high-frequency electronic applications
  drift velocity, 530–531, 530f
  high-frequency FET, 529f
  intrinsic mobility and field-effect mobility, 532f
  metallic SWNT, 530
  multichannel FET, 533f
  Schottky contact barriers, 531
mobility characteristics
  hole mobility, 527
  linear energy dispersion, 527
  mobility, 529
  RF applications, 528–529
  substrate-supported geometry, 527–528
ON/OFF switching characteristics
  CNT-based FET, 539–541
  gate-dependent transport characteristics, 539f
  multi-CNT FET, 540f
  multi-GNR FET, 536–538, 538f
  1D semiconducting SWNT, 538
  stepwise evolution, 541f
physical properties, 526–527
PV cells
  graphene-based cell, 544–546, 547f
  semiconducting CNT, 546–547
Graphene nanomeshe (GNM), 535–536, 537f
Graphene nanoribbon (GNR), 524–525
Graphene oxide films (GO films), 66–67
GRC. *See* Glenn Research Center

H

h-BN. *See* hexagonal boron nitride
HA acid. *See* Hyaluronic acid
Halbach array formation, 37–38
Half-power beamwidth (HPBW), 771
Hallén's integral equation, 754–755
Helical nanotubes. *See* Chiral nanotubes
Hematocytes, 716
Hemoglobin, 716
hexagonal boron nitride (h-BN), 246

bulk, 268
bulk properties, 268–269
graphene devices, 527–528
hexagonal SiC (α-SiC), 268
HF acid. *See* Hydrofluoric acid
High Volume Fraction of CNT Thread, 676
High-energy ball milling, 249
High-performance composites
  development, 314–315
  filtration of nanoparticles, 315
  nanomaterials, 315
High-performance synthetic polymers, 186
  PAN–CNT fibers, 190
    DMAc or DMF, 191
    using dry-jet wet-spinning technique, 190–191
    fiber spinning conditions, 192
    PAN–SWNT fibers, 191–192
    section of oxidized fibers, 192f
    stabilization mechanisms, 193
  PBO–CNT fibers, 193
    mechanical properties, 193–194
    neat PBO fibers, 193
    stress *vs.* strain curves, 194f
  PVA–CNT fibers
    at high weight fractions, 188–190
    at low weight fractions, 186–188
    using 250-mm-diameter spinneret, 193
  UHMW PE–CNT fibers, 194
    using decalin solvent, 195
    nanotubes, 194–195
High-quality carbon nanotube
  atomic-resolution imaging, 340–341
  growth
    approaches, 340
    atomic configurations, 340
    atomic structures, 340
    mechanism and chirality control, 341f
High-strength polymer composites, 377
High-temperature annealing, 45–47
High-temperature nanotubes
  applications
    electric motors, 282–283
    extreme environment application for, 281–282
    nanotube-reinforced aluminum, 283–284
    NASA-GRC, 281
    NIOSH regulations, 285
    potential NASA application, 283t, 284f
    radiation shielding, 283
    spur commercialization, 284
    superior thermal gradient cyclic stress resistance, 282f
  BNNT composites

High-temperature nanotubes (*Continued*)
    in barium calcium aluminosilicate glass composite, 277f
    BNNT survived hot pressing, 280f
    fuzzy fiber concept, 278, 278f
    growth on SiC tows and woven fabric, 278−281, 279f
    NASA-GRC, 277
    polyaniline and polystyrene, 276−277
    SiC_f/SiC_m composites, 278, 279f
    tensile strength, 280f
    SiCNT composites, 281
Highly spinnable multiwalled carbon nanotube arrays
    characterization, 392−393
    dry spinning, 393−394
    millimeter-long spinnable, 391−392
    two-millimeter-long spinnable, 391f
    web initiation using tweezers, 394f
High−resolution transmission electron microscopy (HRTEM), 340, 466−467, 534−535
Hole mobility, 527
HPBW. *See* Half-power beamwidth
HRTEM. *See* High−resolution transmission electron microscopy
Hyaluronic acid (HA acid), 200−201
Hydrofluoric acid (HF acid), 469−470
Hysteresis
    behavior with statistics, 691f
    CNT thread analysis, 690f
    loading and unloading integrations, 688−691
    single resistance−strain cycle, 688
    stress−strain curve, 688

I

IL. *See* Ionic liquid
IL-MWNT. *See* Ionic liquid-multiwalled carbon nanotube
ILSS. *See* Improve interlaminar shear strength
IM7-CNT array layup consolidation
    ply orientation, 630
    quasi-static tests, 630
    in vacuum bag, 630f
Improve interlaminar shear strength (ILSS), 314
In situ precursor synthesis approach, 257
    $B_4N_3O_2H$, 257f
    BN precursors, 257
Indium tin oxide (ITO), 480, 544, 546f
Inert buffer gases, 151
Infrared (IR), 371
    damage sensor, 680
    remote damage detection, 681f
Interfacial shear properties, 68
Ionic liquid (IL), 199
Ionic liquid-multiwalled carbon nanotube (IL-MWNT), 354−355
Iosipescu interlaminar shear test, 633−634, 633f, 642f, 643t
    CNT array-reinforced multiscale composite, 641f, 642f
    CNT reinforcement, 642−644
    crack bridging effect, 644, 644f
    interlaminar composite failure, 633−634
    load−displacement results, 641−642
    strain gage readings, 634
IR. *See* Infrared
Irradiation cross-linking, 64−65
    see also Carbon nanotube (CNT)
    bridging mechanism, 65
    cross-linked CNT material simulations, 66f
    cross-linking bridges, 65−66
    electron irradiation strengthening and stiffening, 67f
    mechanical property, 66−67
    particle irradiation, 67−68
ITO. *See* Indium tin oxide
I−V measurement. *See* Current−voltage measurement

K

Kite mechanism, 91

L

Laser ablation, 258−259
Laser heating, 258−259
Lattice unit vectors, 246
LDOS. *See* Local density of states
LEFM. *See* Linear elastic fracture mechanics
Lennard Jones potential (LJP), 215
Leukocytes. *See* White blood cells
Lightweight sustainable electric motors development
    electric motor development, 598−620
    electromagnetic devices
        carbon motors and actuators, 596, 596f
        CNT wire, 595−596
        Faraday's law of induction, 597
        nanomaterials, 598
        nex-gen motor, 597−598
Linear elastic fracture mechanics (LEFM), 499−502
Lithography techniques, 53
LJP. *See* Lennard Jones potential

Load-bearing fibers, 64–65
Local density of states (LDOS), 458–459
Long carbon nanotube yarns
  development in laboratory facilities, 338–340
  direct synthesis technique
    in continuous spinning process, 337
    CVD, 337–338
    direct spinning, 338f
    mechanical properties, 336–337
    production, 337
    salient feature, 336
    shrinking process, 337–338
    SWNT, 335–336
  parameters to synthesize CNT fibers, 339t
Low-pressure radio frequency plasma, 356

M

Macroscopic structures
  CNT interrelations, 214f
    diameter and length, 215
    during synthesis, 214
  CNT spinning process, 214f
  hierarchical structure, 213
  potential energy function
    CNT interaction force, 215
    fibers/yarns, 215–216
    interaction energies, 216
    potential interaction calculation, 215
    structure and forest matrix, 216f
    universal potential function parameters, 215t
Magnetic core materials, 724–725
Magnetic flux density, 731–732
Magnetic nanoparticle materials
  magnetic core materials, 724–725
  nanometer-diameter spherical particles, 724
Magnetic nanotube materials, 721–724, 723f
Magnetic resonance imaging, 243–244
Magnetization, 726
Material science research, 243–244
  electron energy levels, 244
  material property parameters, 245–246
  nanoscale materials, 244
  nanostructured materials, 244
  physical laws, 244
  physical properties, 246
  plot of Si tetrahedral nanoparticle melting point, 245f
MD. See Molecular dynamics
Meshed CNT thread effect
  antenna feedline and ground plane, 781–782, 783f
  E-plane radiation pattern, 782–785
  on radiation pattern, 784f
  reflection coefficient, 782, 784f
  RT/Duroid materials, 785
Meshed CNT thread patch antenna, 774f
  conductive threads, 776
  gain realization simulation, 777f
  oxidizing/reducing gases, 773–774
  reflection coefficient simulation, 776f
  resonant frequency, 778f
  semiconducting
    CNT threads, 777
    threads fabrication, 775–776, 775f
  substrates, 774–775
  SWNTs, 777
  variation on patch antenna geometry, 774
Metal nanoparticle deposition
  on CNT yarns
    using Ag epoxy, 144–145
    Au and Cu deposition, 147
    Au-CNT and Cu-CNT yarns, 148–149
    coating with, 145f
    doped and coated macrostructures, 149t
    electrical conductivity, 145, 149f
    metal-carbon nanoyarn hybrid yarns, 148f
    MNP deposition on, 146–147
    optical images, 146f
    parameters, 144t
    porous barriers, 143–144
    Pt-CNT hybrid yarns, 147f
    SFED deposition, 143, 144f
    temperature dependence, 150f
    XRD spectroscopy, 146t
  on macrostructures
    chemical reduction process, 142
    electrochemical and electroless, 142
    SFED techniques, 143
Metal nanoparticle-decorated carbon nanotube (MNPCNT), 154
Metal nanoparticles (MNP), 137–138
Metal-carbon nanoyarn hybrid yarns, 148f
MFSL. See Multifractal scaling law
Microcontact printing (μCP), 49
Microwave plasma-enhanced chemical vapor deposition (MPECVD), 252–253
  BNNT bundles, 253
  boron nanostructures, 256
  catalyst layer, 253
  electron diffraction analysis, 253–254
  gas-phase precursors, 253–254
  individual BNNT, 254f
  Ni catalyst and Co catalyst, 254f
  OES and QMS data, 255

Microwave plasma-enhanced chemical vapor deposition (MPECVD) (*Continued*)
  patterned catalyst, 255–256, 256f
  reaction chamber, 253
  two aligned BNNT, 255f
  VLS mechanism, 255
Millimeter-long spinnable multiwalled carbon nanotube arrays, 392
MM. *See* Molecular mechanics
MNP. *See* Metal nanoparticles
MNPCNT. *See* Metal nanoparticle-decorated carbon nanotube
Modified Tersoff–Brenner potential of second generation (MTB-G2), 506
Moisture sensing, 674
  see also Oxidation sensing
Molecular dynamics (MD), 542–543, 570
  CNT–polymer composites, 571
  nanoscale simulations
    atomic systems simulation, 574–575
    cohesive interface model development, 575–578
  simulation method, 91
Molecular mechanics (MM), 506
Moment simulation technique, 752–754
Monolayer grapheme electronic band structure
  carbon atoms, 520–521
  conduction bands, 522
  energy dispersion, 522
  first Brillouin zone, 521f
  lattice structure, 521f
Morse's formulation, 581–582
Motor components, 612
  composite motor structure design, 620
  design and analysis
    CNT materials, 612
    conventional materials electrical properties, 613t
    MATLAB model, 615
    noncompensated AC motor, 614–615
    power-to-weight ratio, 616
    series-wound motor, 612–614
    steady-state torque–speed relationship, 615
    universal motor, 614
  electromagnetic Halbach array fabrication
    component testing, 619f
    magnetic flux density, 619–620
    motor torque, 619
  universal motor prototype, 617t
    AC commutated carbon motor, 618f
    circular power rails, 616–619
    high-temperature dielectric coatings, 619

MPECVD. *See* Microwave plasma-enhanced chemical vapor deposition
MTB-G2. *See* Modified Tersoff–Brenner potential of second generation
μCP. *See* Microcontact printing
Multifractal scaling law (MFSL), 497–499
Multifunctional carbon nanotube antenna
  CNT thread spacing effect
    on antenna performance, 778
    antenna radiation efficiency, 781
    on bandwidth, 780f
    on center frequency, 780f
    E-plane radiation pattern, 781
    meshed CNT thread spacing variation, 779f
    patch dimensions, 778
    on radiation pattern, 781f
    on realized gain, 782f
    on reflection coefficient, 779f
  meshed CNT thread effect
    antenna feedline and ground plane, 781–782, 783f
    E-plane radiation pattern, 782–785
    on radiation pattern, 784f
    reflection coefficient, 782, 784f
    RT/Duroid materials, 785
  meshed CNT thread patch antenna, 774f
    conductive threads, 776
    gain realization simulation, 777f
    oxidizing/reducing gases, 773–774
    reflection coefficient simulation, 776f
    resonant frequency, 778f
    semiconducting CNT threads, 777
    semiconducting threads fabrication, 775–776, 775f
    SWNT, 777
    variation on patch antenna geometry, 774
Multifunctional structural systems, 313
Multiply twisted carbon nanotube yarns, 407f
  multiply twisting
    as-spun CNT yarns, 407
    measurement errors, 407–408
    two-ply twisting process, 407
  twisting tension dependence, 408
    higher twisting tension, 408–409
    tensile strength, 408f
    by two-ply twisting, 409
    two-ply-twisted yarns, 408t
Multiply twisted yarns, 227, 227f
  as-spun CNT yarns, 407
  measurement errors, 407–408
  two-ply twisting process, 407
Multiscale laminated composite materials

Index   **821**

aluminum, 627—628
MWCNT array-reinforced laminated composites, 628—634
Multiscale modeling
  atomistic models, 571
  CNT fibers, 572—573
  continuum mechanics approaches, 571—572
  FEM model, 572
  FMBEM, 573
  hierarchical approach
    BIE, 574
    fillers in nanocomposites, 573
    FMBEM, 573—574
    using MD simulations, 573
    microscale model, 574f
  interface models, 570—571
  MD, 570
  microscale simulations, 578—581
  nanocomposites, 570, 570f
  nanoscale simulations, 574—578
Multisliver yarn, 225—226, 226f
Multistep spinning process, 225f
  CNT web separation
    CNT fibers dilatation, 222—223
    CNT silver/web dilatation, 223—224
  using drafting system, 221—222
  multistep CNT dry spinning process
    alignment, 224
    ATS, 224
    CNT fibers' elongation, 224f
    experimental studies, 225
    role of electrostatic field, 224
    for spinning ultrafine yarns, 225
Multiwall boron nitride nanotube (MWBNNT), 248
Multiwall carbon nanotube (MWCNT), 7, 61—62
  see also Single-walled carbon nanotube (SWCNT); Double-walled carbon nanotube (DWCNT)
  arrays, 391
  bulk CNT thread material, 752
  characterization, 392—393
  CNM, 314
  using CVD methods, 137—138, 390
  data set on, 507
  doping, 299—300
  dry spinning, 390—391
  pentagon—heptagon pair, 458—459
  by phased dispersion force manipulation, 790
  polarized Raman spectra, 393f
  SiCNT product, 271—274
  TEM image, 392f
  thermal conductivity, 430—431
  web initiation using tweezers, 394f
Multiwalled nanotubes array-reinforced laminated composites
  CNT array synthesis, 629
  CNT array transfer onto prepreg laminae
    using hot iron and dry ice, 629—630
    resin infiltration, 629
  fabrication and characterization, 628—629
  IM7-CNT array layup consolidation
    ply orientation, 630
    quasi-static tests, 630
    in vacuum bag, 630f
  Iosipescu interlaminar shear test, 633—634, 633f
    interlaminar composite failure, 633—634
    strain gage readings, 634
  SBS test, 631f
    fiber direction, 631—632
    interlaminar shear stress, 631
    span-to-thickness ratio, 631
  three-point bending test, 633f
    ASTM specification, 632
    maximum flexural strength, 632
    span length, 632
MWBNNT. See Multiwall boron nitride nanotube
MWCNT. See Multiwall carbon nanotube
MWNT. See Multiwall carbon nanotube (MWCNT)

N

N-MWNT. See Nitrogen-doped multi-walled carbon nanotubes
Nanocomp technologies, 68
Nanoelectric motor design
  advantage, 39
  application concepts, 39f
  CNT fiber, 37
  electrical properties, 40t
  Halbach array formation, 37—38
  lightweight and high strength, 38
  nanoelectric motor concepts, 37f
  thermal management, 39
  well-established motivation, 38
Nanofibers
  see also Carbon nanotube yarn fiber fabrication
  CNT, 2—3
  natural fibers, 2
Nanographene platelets (NGP), 314
Nanoimprint lithography (NIL), 52
  using silicon stamps, 53
  tiny machines, 714
Nanoindentation process, 50

Nanomaterial electric motor
   CNT thread operation, 737, 738f
   electrostatic motors, 735—737
   superparamagnetic nanoparticle composite, 737f
Nanomaterial robotic devices, 714
Nanorobots, 40—42, 714—715
Nanoscale materials, 39—40, 244, 718
   carbon nanotube materials
      CNT maximum current density measurement, 720f
      CNT thread properties, 719—720, 720f
      electrical conduction, 721
      electrical properties, 720
      high-temperature annealing, 720—721, 721f
      material forms, 719, 719f
      NIL, 721, 722f
      number of walls, 722f
   magnetic nanoparticle materials
      magnetic core materials, 724—725
      nanometer-diameter spherical particles, 724
   magnetic nanotube and nanowire materials, 721—724
   nanorobot devices, 718—719
   superparamagnetic core design and testing
      eddy currents and magnetic losses, 725f
      iron oxide nanoparticles, 727
      magnetic nanoparticle composite, 726, 726f
      magnetization, 726
      powder synthesis, 725
Nanoscale simulations using molecular dynamics
   atomic systems simulation, 574—575
   cohesive interface model development
      axial cohesive forces, 577—578, 577f
      CNT pull-out simulations, 575
      compliance matrix, 575
      cylindrical coordinates, 576f
      Tersoff—Brenner potential, 576—577
Nanostructured materials, 244
Nanotechnology enthusiasm, 33—34
Nanotensile tests
   crack length, 508
   QFM, 507
   Weibull statistics, 507, 508f
Nanotube superfiber material development, 33—34
   applications
      biomedical applications, 43
      electromagnetics applications, 35—36
      nanodevices made from one long CNT, 42—43
      nanoelectric motor design, 37—39
      nanovivo robots to change interventional medicine, 39—42
      CNT threads, 34
      elevator cable to orbit, 34
   techniques, 43—44
      carbon nanotubes thermal annealing, 45—47
      reactor design for growing carbon nanotubes, 44—45
      spinning carbon nanotube thread, 53—56
      substrate patterning, 47—53
   networks, 459, 460f
      covalent CNT junctions, 460—461
      noncovalent CNT junctions, 461
Nanotube—polymer composites, 342—343
Nanotube(s), 246
   chiral nanotubes, 248
   chiral vector, 247—248
   dispersions, 174
   lattice unit vectors, 246, 247f
   materials for electromagnetic applications, 35
      carbon tubing, 36
      copper manufacturing, 36f
      impact in electric motors, 35—36
   superfiber, 39—40
   thread
      biomedical applications, 43
      CNT thread for, 43f
   2-D hexagonal arrangement, 246
   2-D layered compounds, 246
Nanovivo robots, 39—40, 41f
Nanowire materials, 721—724
NASA-GRC, 270
National Aeronautics and Space Administration (NASA), 267—268
Natural fibers, 2
Natural materials, 70, 168
NGP. *See* Nanographene platelets
NHE. *See* Normal hydrogen electrode
NIL. *See* Nanoimprint lithography
Nitrogen-doped multi-walled carbon nanotubes (N-MWNT), 171—172
Nitrosyl tetrafluoroborate ($NOBF_4$), 298
NMR. *See* Nuclear magnetic resonance
$NOBF_4$. *See* Nitrosyl tetrafluoroborate
Noncovalent CNT junctions, 461
   see also Covalent CNT junctions
Noncovalent functionalization. *See* Exohedral doping carbon nanotube doping
Normal hydrogen electrode (NHE), 299
Nuclear magnetic resonance (NMR), 69

O

OES. *See* Optical emission spectroscopy
Oil spill absorption, 486—487

Index

ON/OFF switching characteristics
    BCP lithography techniques, 535–536
    CNT-based FET, 539–541
    dual-gated bilayer graphene, 534f
    fabrication of GNR, 533
    gate-dependent transport characteristics, 539f
    HR-TEM images, 534–535
    multi-CNT FET, 536–538, 538f, 540f
    1D semiconducting SWNT, 538
    ratio in graphene-based FET, 532–533
    stepwise evolution, 541f
    transistors and electronic devices, 535
One-step spinning method. *See* Direct spinning
Optical emission spectroscopy (OES), 253
Optical properties, graphene
    active optoelectronic materials, 553–555
    electromagnetic excitations, 556
    linear optical response
        gate voltage, 549
        infrared spectroscopy, 549
        SWNT, 551, 552f
        SWNT-based polarizers, 551–552, 553f
        in terahertz and infrared spectral ranges, 547–549, 551f
        total optical conductivity, 549, 550f
    plasmon
        graphene microribbon arrays, 556f
        light-plasmon coupling, 555–556
        2D, 555
Optimal nanoscale systems, 77
Orders of magnitude
    Derjaguin approximation, 796
    equilibrium position, 796
    external voltage source, 797
    interwall dispersion force, 796–797
    mass fabrication, 794–795
    nanostructures interaction, 792–793
    polarized telescoping nanotube geometry, 793f
    power densities, 797
    quasi-equilibrium, 793–794
Organic vapors, 151–152
Oxidation sensing, 674–675

P

p-type dopants, 292–293
PAN. *See* Polyacrylonitrile
PAN–CNT fibers, 190
    DMAc or DMF, 191
    using dry-jet wet-spinning technique, 190–191
    fiber spinning conditions, 192
    PAN–SWNT fibers, 191–192
    section of oxidized fibers, 192f
    stabilization mechanisms, 193
Paper-like CNT films, 650–651
Parallel thermal conductance (PTC), 427–428, 429f
Parallel-plate system idealization
    dispersion force interactions, 791–792
    energy density, 792
    parallel-plate geometry, 792
    thermodynamical engine storing energy, 791f
    transformation, 792
Paris' law, 504
Particle coagulation spinning process, 173–174
parts per billion (ppb), 150–151
PBO, 6-benzobisoxazole). *See* Poly(p-phenylene-2
PBO–CNT fibers, 193
    using 250-mm-diameter spinneret, 193
    mechanical properties, 193–194
    neat PBO fibers, 193
    stress *vs.* strain curves, 194f
PE systems. *See* Polyethylene systems
PE-PLD. *See* Plasma-enhanced pulsed laser deposition
PEI. *See* Polyethyleneimine
PET. *See* Polyethylene terephthalate
Phonon thermal conductivity, 427
Photolithography, 48, 48f
Photoluminescence (PL), 260
Photovoltaic cells (PV cells), 544–546
    graphene-based cell, 544–546, 547f
    semiconducting CNT, 546–547
Physical laws, 244
Pilot microfactory
    commercial Kleindiek robots, 728
    Kleindiek robotic manipulators, 729–730, 729f
    tools, 730, 730f
    tweezers, 730, 731f
    microfabrication technology, 727–728
    transitioning engineering breakthroughs, 727
π-stacking. *See* Exohedral doping
PL. *See* Photoluminescence
Plasma, 716
    functionalization, 356
        atmospheric pressure plasma, 358
        comparison, 358t
        DBD plasma, 360–361
        FTIR spectroscopy, 356, 358f
        low-pressure RF plasma, 356
        sedimentation behavior, 357, 359f
        SEM micrograph, 360f
        using TPD technique, 359–360
        X-ray photoelectron spectra, 357, 359f
        XPS analysis, 356
        XPS survey scan and C1s spectra, 357f

Plasma-based chemical vapor deposition
    synthesis, 252
  MPECVD, 252−253
    aligned BNNT, 255f
    BNNT bundles, 253
    boron nanostructures, 256
    catalyst layer, 253
    electron diffraction analysis, 253−254
    gas-phase precursors, 253−254
    individual BNNT, 254f
    Ni catalyst and Co catalyst, 254f
    OES and QMS data, 255
    patterned catalyst, 255−256, 256f
    reaction chamber, 253
    VLS mechanism, 255
  PE-PLD, 256−257
Plasma-based synthesis methods, 252
Plasma-enhanced pulsed laser deposition
    (PE-PLD), 256−257
Platelets, 716
PLD. See Pulsed laser deposition
PMC. See Polymer matrix composite
PMMA. See Poly(methyl methacrylate)
Point matching method, 755
Polarized telescoping nanotube (PTNT),
    793−794
  electromechanical behavior, 795
  macroscopic architecture, 794f
  mass fabrication, 794−795
Poly(methyl methacrylate) (PMMA), 69, 153
Poly(p-phenylene-2, 6-benzobisoxazole) (PBO),
    186
  CNT fibers, 193
  fiber, 193
  using 250-mm-diameter spinneret, 193
    mechanical properties, 193−194
    neat PBO fibers, 193
    stress vs. strain curves, 194f
Poly(vinyl alcohol) (PVA), 70−72
Polyacrylonitrile (PAN), 168−169
Polyethylene systems (PE systems), 576−577
Polyethylene terephthalate (PET), 544
Polyethyleneimine (PEI), 174−175
Polygonization, 9
Polymer matrix composite (PMC), 313
Polymer-induced coagulation, 173
  from composite to neat CNT fibers, 177
  electrochemical and electromechanical properties,
    183−184
  fiber formation mechanisms, 173−174
    continuous spinning system, 176f
    fibers in coflowing stream, 175−176, 175f
    flow-induced alignment, 174
    laboratory-scale experimental setup, 173f
    nanotube dispersions, 174
    robustness, efficiency and production rate,
      176−177
    wet-spun CNT fiber, 174−175
  mechanical properties, 178−180
  shape memory effects, 180−183
  structure and alignment, 177−178
  thermomechanical properties, 180−183
Polymer−CNT fibers, 169−170, 185
  see also High-performance synthetic polymers
  CNT−conductive polymer composite fibers,
    195−196
    continuous wet-spinning and polymerization
      line, 198f
    electrical properties, 199f
    internanotube contacts, 196−197
    mechanical properties, 199f
    opportunities to develop electrodes, 197
    PANi−CNT composite fibers, 196
    PPy-alginate fibers, 197
    PPy/alginate/SWNT composite fibers,
      197
    SWNT and PEDOT, 197
  CNT−natural polymer composite fibers
    alginate fibers, 200
    biopolymers, 200−201
    chitosan/CNT composite fibers, 200
    formulations, 199
    wet-spinning technologies, 198−199
  developments, 185−186
Polymeric materials, 168−169
  see also Natural materials
Polymers
  conducting, 195−196
  conductive, 168−169
  matrix, 653−654
Polystyrene (PS), 355−356
Polyvinyl alcohol (PVA), 168−169, 653
Porous inorganic materials, 106
Postfabrication stretching, 655−657
Postspin twisting effect
  CNT yarn, 402
  cross-sectional observations, 403, 404f
  tensile strength, 402, 403t
  twisting density dependence, 403, 404f
  Young's modulus, 403t
Postsynthesis treatment, 174
ppb. See parts per billion
PPy fiber, 197
Prototype fabrication, 730−731

biodegradability, 739–740
carbon nanotube wire, 740–741
communication with tiny machines, 743, 744f
composite electromagnetic material
  bionic artificial cell, 739, 742f
  dry spinning method, 738–739
  elastomer nanocomposite material, 739
  Helmholtz coil configuration, 739, 740f
  in polymer, 738
  toroid transformer, 739
electromagnetic devices
  biosensor, 732
  carbon electromagnetics, 732f
  high-temperature capability, 731
  magnetic flux density, 731–732
  nanomaterial robots development, 732
  robot, 732, 733f
  single-degree-of-motion robot, 732, 733f
  solenoid, 732, 734f
  tiny robot, 733–734
failure mechanisms
  biofouling coating, 734–735
  implantable biosensor, 735f
  manipulators manufacturing components, 735, 736t
nanomaterial components thermal analysis, 737–738
nanomaterial electric motor
  CNT thread operation, 737, 738f
  electrostatic motors, 735–737
  superparamagnetic nanoparticle composite, 737f
TNA, 742–743
PS. See Polystyrene
PTC. See Parallel thermal conductance
PTNT. See Polarized telescoping nanotube
Pull-out test, CNT, 575
  with amorphous polymer chains, 576f
  CNT/polymer composite fracture analysis, 584, 584f
  cohesive interface effect, 589f
    axial component, 587
    CNT volume fractions, 587–588
    CNT/polymer composites Young's moduli, 586
    interface properties cases, 587
  with crystalline polymer chains, 577f
  large-scale BEM models
    bonding interface model by BEM, 585–586, 588f
    CNT composites, 584–585
    RVE, 585, 585f, 586f, 587f
  using MD

    C−C interactions, 582
    cosine potential parameters, 582t
    harmonic cosine potential parameters, 582t
    Morse potential parameters, 582t
    Morse's formulation, 581–582
    nanotube pulling out process images, 582–584, 583f
    polystyrene matrix, 581
    shear stress, 583f
Pulsed laser deposition (PLD), 256–257
PV cells. See Photovoltaic cells
PVA. See Poly(vinyl alcohol); Polyvinyl alcohol
PVA-based composite fibers, 187
PVA−CNT fibers
  alignment, 178
  electrochemical and electromechanical properties, 183–185
  hierarchical structure, 177–178
  at high weight fractions
    mechanical properties, 190t
    properties and requirements, 188–190
    stress vs. strain, 189f
  at low weight fractions
    bridging coagulation, 186
    DMSO-based dispersions, 187–188
    gel-spinning, 186–187
    PVA-based composite fibers, 187
  mechanical properties, 178–180
  peak recovery stress, 181–183
  polymer-induced coagulation, 177
  qualitative evidence, 181f
  SMP, 180–181
  stress and energy absorption vs. strain, 179f
  stress generation by nanocomposite fiber, 182f
  stress vs. strain curves, 181f
Pyrograf III carbon nanofiber, 315–316
Pyrograf III material, 316f

Q

Quadrupole mass spectroscopy (QMS), 253
Quantized fracture mechanics (QFM), 496
Quantum mechanics (QM), 506

R

Radial breathing mode (RBM), 296–297
Radio frequency (RF), 528–529
Raman spectroscopy, 301–302
  see also X-ray photoelectron spectroscopy (XPS)
  analysis, 363
  I_D/I_G ratio, 364, 366f, 366t
  medium and low power ranges, 364

Raman spectroscopy (*Continued*)
    modification by oxygen plasma, 363, 365f
    plasma power effect, 363, 366f
    plasma treatment effect, 365, 367t
    for untreated and treated CNT sheet samples, 364f
Rayon. *See* Silk, artificial
RBM. *See* Radial breathing mode
Reciprocyte. *See* Artificial red blood cell
Red blood cells, 716
    artificial, 717
Reformable bonding
    see also Carbon nanotube (CNT)
    chemical methods, 68
    combination of oligomeric chains, 69
    CVD fabrication, 72
    DWNT bundles, 69
    hydrogen-bond network, 70
    interfacial shear properties, 68
    molecular unfolding mechanism, 71f
    Nanocomp Technologies, 68
    natural materials, 70
    PMMA, 69
    PVA, 70–72
    spider silk, 70
    wet and dry treatments, 68–69
Regular solid, 244
Renegade Materials, 321
Representative volume element (RVE), 570
RF. *See* Radio frequency
Ring motors, 38

S

SBS testing. *See* Short beam shear testing
Scanning electron microscope (SEM), 77, 122–125
    analysis, 218
    BNNT, 253
    branched MWNT, 441–444, 469–470
    CNT array metalized with gold, 603f
    local anodic oxidation, 50–51
    methods, 51f
    tiny machines, 714
Scanning tunneling microscope (STM), 555–556
Schulz–Flory distribution (SF distribution), 119–121
    CNT number density, 121
    furnace-moving method, 121
    one-dimensional carbon molecules, 121
    relationship of catalyst activity probability, 122f
SDS. *See* Sodium dodecyl sulfate
Sea urchin, 476–477
Self-fueled electro deposition (SFED), 143, 231

SEM. *See* Scanning electron microscope
Sensitivity. *See* Gauge factor
Sensor thread performance
    bandwidth
        calibration measurement, 695
        CNT thread bandwidth, 697–699
        custom double-cantilever calibration rig, 695–696, 695f
        epoxy mounting points, 696, 696f
        FRF, 697
        sine wave of CNT thread, 698f–699f
        stiffer epoxy, 697
        voltage output measurement, 696–697
        wave excitation, 697
    consistency
        CNT thread strain sensor, 691–692
        gauge factor, 692t
    hysteresis
        analysis of CNT thread, 690f
        behavior with statistics, 691f
        loading and unloading integrations, 688–691
        single resistance—strain cycle, 688
        stress—strain curve, 688
    polymeric material, 682
    sensing strain and damage, 680–681
    sensitivity, 682f
        as-spun CNT thread, 674
        custom-designed coupon, 682–683
        piezoresistive coefficient, 686
        resistance response, 685f, 687f
        strain measurement, 682
        tensile test coupon, 684
        theoretical response, 686f
        thread's geometry, 686–688
        Wheatstone bridge circuit, 683–684, 683f
    stability
        CNT thread resistance measurement, 692, 693f
        reference thread, 693–694
        resistance stability, 694f
        temperature variations, 692–693
Sensorcytes, 716–717, 717f
Series-wound motor, 612–614
SF distribution. *See* Schulz–Flory distribution
SFED. *See* Self-fueled electro deposition
SG. *See* Specific gravity
Shape memory polymer (SMP), 180
Sheet antennas, CNT
    see also Thread antennas, CNT
    sheet fabrication, 765, 765f
    sheet patch antenna
        antenna geometry, 765–766

aperture-coupled patch antenna prototypes, 767f
CNT orientation effect, 770–771, 771f
CNT sheet thickness and conductivity, 768f
E-plane radiation pattern measurement, 772f
electromagnetic wave, 768–769
full-wave solver, 767
gain realization measurement, 773f
H-plane radiation pattern measurement, 772f
HPBW, 771
inductive and capacitive reactance, 769
input impedance, 769–770
Kapton tape substrate, 766–767
reflection coefficient for prototypes, 769f
variation in CNT sheet orientation, 770, 770f
X-band aperture-coupled patch antenna design, 766f
SHM. *See* Structural health monitoring
Short beam shear testing (SBS testing), 322, 631, 631f, 636t
  BL composite, 326
  CFRP composites, 326–328
  cylindrical supports, 322
  deflection test curves *vs.*, 635f
  using equation, 322–323
  failure behavior changes, 325–326
  fiber direction, 631–632
  fiber/matrix debonding, 637
  at high temperatures, 328
  ILSS, 323–325, 326f, 327f
    composite configurations testing, 327t
    strength testing, 327t
  interlaminar shear stress, 631
  for lightweight structural composites, 325
  loading configuration, 324f
  load–displacement curves
    composites tested at 200 °C, 325f
    composites tested at 23 °C, 324f
  shear strength–deflection plots, 634–635
  span-to-depth ratio, 634
  span-to-thickness ratio, 631
  20-μm-long CNT array, 635–637, 637f
SiC_f/SiC_m composites, 278, 279f
SiCNT. *See* Silicon carbide nanotube
Silicon carbide nanotube (SiCNT), 268
  see also Boron nitride nanotube (BNNT)
  applications
    electric motors, 282–283
    extreme environment application for, 281–282
    nanotube-reinforced aluminum, 283–284
    NASA-GRC, 281
    NIOSH regulations, 285
    potential NASA application, 283t, 284f
    radiation shielding, 283
    spur commercialization, 284
    superior thermal gradient cyclic stress resistance, 282f
  composites, 281
  property comparison
    band gap semiconductors, 275–276
    elastic modulus, 274
    properties, 276t
    thermal stability, 276
  structure
    cubic and hexagonal polytypes, 268
    h-BN, 268–269
    hexagonal sheets, 269f
    two-dimensional graphene structure, 269–270
  synthesis, 271
    using chemical templating approach, 271–274
    CNT to SiCNT conversion, 271f
    EDS and X-ray data, 272f
    isolated SiC grains, 273f
    SiCNT and CNT wall structures, 273f
Silk, artificial, 168
Silver epoxy, 143–144
Single-CNT shell, 11
Single-crystal quartz, 102
Single-crystal silicon wafers, 101–102
Single-walled boron nitride nanotube (SWBNNT), 248
Single-walled carbon nanotube (SWCNT), 7, 65–66, 415–416
  see also Double-walled carbon nanotube (DWCNT); Multiwall carbon nanotube (MWCNT)
  application, 749–750
  conduction mechanisms, 420
  cutting and wrapping, 520
  using dispersions, 172
  endohedral doping, 291
  ES-VRH mechanism, 418–420
  during growth, 350
  high-purity samples, 416
  magnetoresistance in nanotube networks, 421f
  MS ratio and conductivity, 416
  multifunctional additives, 314
  normalized resistance, 419f
  optical absorption spectra, 417, 417f
  parallel and serial circuits, 458
  phase diagram, 422f
  room temperature thermal conductivities, 649–650
  strength, 507

Single-walled carbon nanotube (SWCNT) (*Continued*)
  using TEM metals, 138–139
  temperature dependence, 417, 418f
  tensile strength and critical strain, 90
  during tensile tests, 495–496
  two-dimensional WL theory equation, 421
  using VRH models, 418
Single-walled carbon nanotube doping (SWCNT doping), 296–297
  redox potential, 297
    fitting parameters, 297–298
    function of diameter, 297f
    photoluminescence spectroscopy, 298
  work functions
    doping concentrations, 299
    individual metallic and semiconducting, 299f
    physical parameters, 298–299
    relationship, 299
    sheets and electrical resistance, 298f
Sirospun single yarn, 225–226
Sliding failure
  defect size distribution, 514
  dissipative mechanisms, 515
  energy balance, 514–515
  fracture mechanics approach, 513
  longitudinal delamination, 513–514
  maximum achievable strength, 515
  overlapping length, 514
Smart stent, 715
SMP. *See* Shape memory polymer
Sodium dodecyl sulfate (SDS), 172
Space elevator, 495
Specific gravity (SG), 430
Spider silk, 70, 168
Spinning carbon nanotube thread
  dispersing long CNT, 56
  dry spinning thread, 55, 55f
  long-range goal, 53–54
  wet spinning thread, 55–56, 56f
Spraywinding process, 653, 654f
Stacked-cup carbon nanotubes. *See* Carbon nanofiber (CNF)
Staple fiber, 2
STM. *See* Scanning tunneling microscope
Strain sensing, 674
  see also Thermal sensing
Strain sensor calibration sheet, 680–681
  bandwidth, 681
  consistency, 681
  hysteresis, 681
  sensitivity, 681
  stability, 681
Stretch-winding process, 654
  CNT/BMI composites, 654–655, 657f
  postfabrication stretching of prepregs, 655–657
  straightening CNTs, 655f
Strong/tough CNT bundles, 68–72
Structural health monitoring (SHM), 674
Substitutional doping, 291f, 293–294
  B-N codoping, 295
  B-or N-containing precursors, 294
  carbon replacement, 295
  endo-and exo-substitutional doping, 295f
  hexagonal atom configuration, 295
  pyridine-like configuration, 294
Substrate patterning, 47–48
  see also Carbon nanotube (CNT)
  applications, 48–49
  DPN, 51–52
  EBL, 49
  FIB lithography, 49
  lithography techniques, 53
  local anodic oxidation, 50–51
  metal catalyst, 48, 52–53
  μCP, 49
  nanoimprint lithography results, 54f
  nanoindentation process, 50
  NIL, 52
  photolithography, 48, 48f
  precise control, 47–48
  using silicon stamps, 53
  SPM, 49–50, 51f
  surface patterns by tip-based nanofabrication, 50f
Substrate surface force, 101
Substrate surface-directed growth
  crystal lattice-directed growth, 113–114
  crystal-edge orientation, 115
  electric field-directed growth, 115–116
  gas flow tuning, 113
  R-plane and A-plane surfaces, 115
  wake-growth mechanism, 114–115
  on Y-cut quartz substrates, 115
Substrates, 102
Sulfur, 475
Superacids, 172
Superfiber materials, CNT
  CNT arrays
    anisotropy, 441–444
    electrical conductivity, 444–445
    MWCNT arrays, 444f
    MWCNT quality effect, 446f
    thermal interface materials, 441
  CNT sheets and yarns

acid spinning, 447
MWCNT arrays, 445–447, 445f
thermal conductivity comparison, 446f
Superparamagnetic composites
eddy currents and magnetic losses, 725f
iron oxide nanoparticles, 727
magnetic nanoparticles, 726, 726f
magnetization, 726
powder synthesis, 725
Superstrong CNT and graphene fibers
atomistic simulations
elasticity, 506–507
nanocracks and circular nanoholes strength, 506
stress–strain curves, 506
brittle fracture
atomistic simulations vs. QFM predictions, 500t
cable strength, 502
LEFM, 499–502
QFM, 499
taper ratio, 503
crack roughness, 503
elastic-plasticity, 503
elasticity
elastic degradation, 506
single-defect typology, 505
transversal crack, 505
fatigue
failure reduction, 504
finite width role, 505
mean-field approach, 505
Paris' law, 504
finite width, 504
hierarchical simulations and size effects
multiscale simulation procedure, 498f
and scaling law comparison, 498f
scaling laws, 497–499
SE^3 algorithm, 497
in silico stress–strain experiments, 497
strains, 497
nanotensile tests
crack length, 508
QFM, 507, 509t
strength and elasticity, 512t
Weibull statistics, 507, 508f
QFM, 496
sliding failure
defect size distribution, 514
dissipative mechanisms, 515
energy balance, 514–515
fracture mechanics approach, 513
longitudinal delamination, 513–514
maximum achievable strength, 515
overlapping length, 514
SWCNT, 495–496
thermodynamic limit
maximum achievable strength, 513
nanocrack, 512–513
vacancy fraction, 508
uniform tensile stress, 496
Surfx Atomflo 400-D plasma controller, 363f
SWBNNT. See Single-walled boron nitride nanotube
SWCNT. See Single-walled carbon nanotube
SWCNT doping. See Single-walled carbon nanotube doping
SWNT. See Single-walled carbon nanotube (SWCNT)

T

TBMD. See Tight-binding molecular dynamics
TD. See Topological defect
TE. See Transparent electrode
Telescoping nanotube array (TNA), 742
using NIL, 742, 743f
three-wall CNT telescoping, 743, 743f
TEM. See Transmission electron microscopy
Temperature programmed desorption (TPD), 359–360
TFT. See Thin-film transistor
TGA. See Thermogravimetric analysis
Thermal annealing
carbon nanotubes, 46f
in healing, 47
high-temperature annealing, 45–47
postannealing Raman spectroscopy, 47
Thermal conductivity
boron nitride nanotubes, 449
carbon nanotube bundles
measurements on individual CNT, 432
MWCNT bundles, 434f
SWCNT, 432–435
carbon nanotube composites, 435, 436f
CNT-enabled materials, 435
CNT–CNT junctions, 435–436
EMT models, 436–437
CNT buckypaper and thin films
CNT film structure, 440–441, 440f
in cross-plane direction, 440
macroscopic materials, 437–439
SWCNT buckypaper, 437f
CNT superfiber materials
CNT arrays, 441–445

Thermal conductivity (*Continued*)
    CNT sheets and yarns, 445–447
    individual carbon nanotubes, 430–431
        CVD CNTs, 431–432
        self-heating method, 431
        SWCNT thermal conductivity, 431f
    measurement issues, 426, 428f
        CNT fibers, yarns, and array-drawn sheets, 448t
        CNT mats and bundles, 433t
        Fourier's law, 426
        heterogeneous materials, 429–430
        in-plane oriented CNT and buckypaper, 438t
        phonon thermal conductivity, 427
        steady-state temperature difference, 427–428
        temperature response, 428–429
        vertically grown CNT arrays, 442t
        Wiedemann–Franz law, 426–427
Thermal interface material (TIM), 282–283
Thermal management, 39
Thermal sensing, 674
    see also Strain sensing
Thermionic cathode, 345
Thermodynamic limit
    maximum achievable strength, 513
    nanocrack, 512–513
    vacancy fraction, 508
Thermogravimetric analysis (TGA), 69, 271–274
Thin-film transistor (TFT), 480, 539–541, 546–547, 548f
^{13}C-nuclear magnetic resonance (^{13}C-NMR), 302–303
Thread antennas, CNT
    CNT fabrication, 752, 752f
    CNT thread conductivity
        CNT bundle simulation, 756
        conductivity simulation, 757–758
        dipole current distribution simulation, 755–756
        imaginary part conductivity *vs.* frequency for carbon nanotube, 758f
        MWNTs, 756
        order of magnitude, 758–759
        quantum conductivity, 756
        real part conductivity *vs.* frequency for carbon nanotube, 757f
        uniform nanotube distribution, 756–757
    DMSO, 752, 753f
    electromagnetic simulation
        current distribution, 760, 761f
        double-walled CNT current distribution, 762
        large ply CNT rope antennas, 763, 763f
        maximum dipole current, 763–764, 764f
        moments solution method, 759
        nanoscale dipole antennas, 762, 762f
        nanoscale impedance effects, 764
        nonzero dipole thread effect, 761
        Ohm's law, 760
        radiation efficiency comparison, 760, 762–763
        RF performance tradeoff, 761
    electromagnetic theory
        electric field, 754
        far-field radiation, 755
        Hallén's integral equation, 754–755
        Method of Moments simulation technique, 752–754
        Ohm's law, 754
        point matching Method of Moments solution, 755
    signal excitation, 751
    thread dipole antenna, 751, 751f
Thread SHM architectures, 699
    cross-weave or grid design, 699–702
    damage modes, 699
    F-22 Raptor, 703–704
    micromechanical failure modes, 703f
    multiplexing, 702–703
    sensors array configuration, 703f
    SHM system communication, 704f
Thread spacing effect, CNT
    on antenna performance, 778
    antenna radiation efficiency, 781
    on bandwidth, 780f
    on center frequency, 780f
    E-plane radiation pattern, 781
    meshed CNT thread spacing variation, 779f
    patch dimensions, 778
    on reflection coefficient, 779f
Three-dimensional nanotube networks (3D nanotube networks), 458
    bioapplications, 482f–483f
        biosensing characteristics, 483
        carbon nanofoams, 484
        long aerogel sheets, 484
        nerve agents detection, 484
        SWNT–cotton yarns, 484
    electronic applications, 483
        memory devices, 479–480, 481f
        orthogonal van der Waals networks, 480
        source and drain electrodes, 481–483
        TFTs, 480
    junctions and networks
        aspects, 462–466
        electronic density of states, 469f

LDOS, 467−468, 468f
TBMD calculations, 466−467, 467f
2D SWNT networks atomic models, 466f
mechanical applications
  conductive percolating CNT networks, 485−486
  resistance-type strain sensors, 484−485
  strain gauge, 485f
nanotube junction, 461
nanotube network types, 459, 460f
  covalent CNT junctions, 460−461
  noncovalent CNT junctions, 461
networks synthesis, 468−479
oil absorption applications, 486f
  CNT sponges, 486
  hydrogen storage capacity, 487
  oil spill absorption, 486−487
  superhydrophobicity, 486
pentagon−heptagon pair defects, 458−459
schwarzites, 461−462
  3D nanotube networks, 463f
  3D superdiamond network, 465f
  zigzag SWNT, 464f
TEM image, 459f
3-D structured materials, 246
Three-point bending test, 632, 633f, 639t
  ASTM specification, 632
  baseline laminated composite, 637−640, 638f, 640f
  CNT array reinforcement, 640−641
  IM7 20-ply CNT array-reinforced multiscale composites, 638f
  maximum flexural strength, 632
  span length, 632
  stress *vs.* deflection curves, 637
Thrombocytes. *See* Platelets
Tight-binding molecular dynamics (TBMD), 466−467
TIM. *See* Thermal interface material
Tiny machines, 714
  artificial red blood cell, 717
  circulatory system, 716
  da Vinci robot, 718, 719f
  macrosize devices, 717, 718f
  microsize nanomedicine device, 716f
  nanorobot devices, 714−715, 715f
  nanoscale materials, 718
    carbon nanotube materials, 719−721
    magnetic nanoparticle materials, 724−725
    magnetic nanotube and nanowire materials, 721−724
    nanorobot devices, 718−719
    superparamagnetic core design and testing, 725−727
  nanosize device, 716−717
  nanotube artificial vein, 718f
  pilot microfactory
    commercial Kleindiek robots, 728
    Kleindiek robotic manipulators, 729−730, 729f
    microfabrication technology, 727−728
    transitioning engineering breakthroughs, 727
  prototype fabrication, 730−731
    biodegradability, 739−740
    carbon nanotube wire, 740−741
    communication with tiny machines, 743
    composite electromagnetic material, 738−739
    electromagnetic devices using CNT yarn, 731−734
    failure mechanisms for in vivo devices, 734−735
    nanomaterial components thermal analysis, 737−738
    nanomaterial electric motor, 735−737
    TNA, 742−743
  Richard Feynman and Robert Freitas, Jr, 714f
  smart stent, 715
  used in medicine, 714
Tip-growth mode, 95−97
TNA. *See* Telescoping nanotube array
Topological defect (TD), 87−88
Torsion sensor, 345
Tow, 16
TPD. *See* Temperature programmed desorption
TPMS. *See* Triply periodic minimal surfaces
Transfer-printing approach, 650−651
Transition metal catalyst, 101
Transmission electron microscopy (TEM), 3, 61−62
  BNNT, 253
  diffraction patterns, 77
  MWNT, 458−459
  TWCNT, 108
Transparent electrode (TE), 472
Triple-walled carbon nanotube (TWCNT), 7, 95−97
Triply periodic minimal surfaces (TPMS), 461−462
Tube diameter, 76
Tuning areal density, 119
  average space calculation, 119
  CNT arrays, 119
  high-density and perfectly aligned arrays, 120f
  MWCNT, 119
  theoretical number distribution, 120f

Tuning morphology, ultralong CNTs, 113
　orientation and arrangement
　　gas flow-directed growth, 113
　　high on/off ratio FETs fabrication, 112f
　　sample rotation for fabrication CNT, 114f
　　substrate surface-directed growth, 113−115
　　ultralong CNT arrays, 113
　SF distribution-controlled growth, 119−121
　　CNT number density, 121
　　furnace-moving method, 121
　　one-dimensional carbon molecules, 121
　　relationship of catalyst activity probability, 122f
　tuning areal density, 119
　　average space calculation, 119
　　CNT arrays, 119
　　high-density and perfectly aligned arrays, 120f
　　MWCNT, 119
　　theoretical number distribution, 120f
　tuning length, 116
　　length of furnace heating zone, 117−119
　　lifetime of catalysts and effects, 116
　　long CNTs grown on Si substrates, 118f
　　time-dependent characteristic, 116
　　water concentration effect on growth rate, 117f
　　water effect on improvement, 116
TWCNT. See Triple-walled carbon nanotube
2D electron gases (2DEG), 555
2-D layered compounds, 246
Two-dimensional woven laminated composite, 322
Two-millimeter-long spinnable multiwalled carbon nanotube array, 391f
Two-ply twisting process, 407

U

UC. See University of Cincinnati
UHMW PE. See Ultrahigh molecular weight polyethylene
UHMW PE−CNT fibers, 194
　using decalin solvent, 195
　nanotubes, 194−195
Ultrahigh molecular weight polyethylene (UHMW PE), 168−169, 194
Ultralong carbon nanotube (Ultralong CNT), 88
　growth mechanism
　　axial screw dislocation, 99f
　　base-growth mode, 97
　　chirality distribution, 100t
　　chirality selectivity, 97−99
　　CNT−catalyst interface, 95
　　defect healing, 94f−95f
　　dissolved carbon content, 92f
　　kite mechanism, 91
　　metal NP and interaction, 92−95
　　morphology formation, 98f
　　SCWNT diameter dependence, 93f
　　SWCNT growth at different temperatures, 92f
　　SWCNT growth scenarios, 93f
　　temperature, 91
　　tip-growth mode, 95−97, 96f
　　unsaturation stage, 91
　　VLS mechanism, 90−91
　mechanical properties
　　gas-flow-blowing system, 127f
　　high-aspect-ratio structure, 122
　　MWCNT, 122−125, 124f
　　plot of stress vs. strain curves, 124f
　　Raman spectrum, 128f
　　scaled Young's modulus, 125, 126f
　　synthesis, 123f
　potential applications, 127−129, 129f
　structure
　　atomic structures, 89−90
　　carbon−carbon bond, 90
　　chiral index, 89
　　graphene sheet into CNT shell with chiral index, 89f
　　relationship, 89
　　single or multiple graphene sheets, 88−89
　　SWCNT, 90
　synthesis by CVD, 101
　　carbon sources, 102
　　catalysts and CVD process, 101−102
　　methods for, 103t
　　substrates, 102
　tuning chiral consistency
　　as-grown ultralong SWCNT, 107f
　　carbon source, 108
　　for centimeters-long CNT, 106−107
　　chiral angle distribution, 110f
　　chiral angle for long TWCNT, 108
　　long-range homogeneous atomic structure, 109f
　　PLE intensity map, 111f
　　semiconducting and metallic shells, 110f
　　SWCNT on Si/SiO$_2$ substrates, 107
　tuning electrical properties, 108−112
　tuning morphology, 113
　　Schulz−Flory distribution-controlled growth, 119−121
　　tuning areal density, 119

tuning length, 113–119
tuning number of walls and diameters, 104
  catalyst impact on growth, 106
  structure distribution, 106f
  temperature effect on diameter, 105f
  thermodynamic and kinetic behavior, 104–105
Ultralong CNT. See Ultralong carbon nanotube
Ultraviolet (UV), 143–144, 371, 556
Unidirectionally aligned carbon nanotube sheet
  cross-sectional area, 398–399, 399f
  current–voltage characteristics, 396f, 397f
  diffusivity and conductivity, 397t
  electrical properties, 395
  fabrication, 394–395, 395f
  interfaces and surfaces, 398
  mechanical properties, 396–397
  sheet resistance and resistivity, 396t
  thermal properties, 397–398
Universal motor block, 615, 615f
Universal motor prototype, 617t
  AC commutated carbon motor, 618f
  circular power rails, 616–619
  high-temperature dielectric coatings, 619
University of Cincinnati (UC), 352, 673–674
Unsaturation stage, 91
UV. See Ultraviolet

V

VACNT. See Vertically aligned CNT
van der Waals interaction (vdW interaction), 114–115
Vapor–liquid–solid mechanism (VLS mechanism), 90–91, 255
Variable range hopping model (VRH model), 415–416
vdW interaction. See van der Waals interaction
Vertically aligned CNT (VACNT), 628–629
Viscose
  See Artificial silk
VLS mechanism. See Vapor–liquid–solid mechanism
VRH model. See Variable range hopping model

W

w-BN. See wurtzite BN
Wake-growth mechanism, 114–115
Weak localization (WL), 415–416
Weibull statistics, 507
Wet spinning, 167–168, 212–213
  see also Direct spinning
  CNT-based wet-spun fibers, 170
  development, 168

dry-jet wet spinning, 169
homogeneous fluids, 170–171
natural materials, 168
for polymer fiber production, 168f
polymer–CNT fibers, 169–170
polymeric materials, 168–169
technologies, 169
thread, 55–56, 56f
Wet-spinning technologies, 198–199
Wet/solution functionalization, 355–356
Wheatstone bridge circuit, 683–684, 683f
Wheel motors, 38
White blood cells, 716
Wiedemann–Franz law, 426–427
WL. See Weak localization
wurtzite BN (w-BN), 268

X

X-ray photoelectron spectroscopy (XPS), 146–147, 302
  see also Raman spectroscopy
  for CNT sheets, 368f
  elemental composition results, 367t
  high-resolution, 367–368
    plasma-functionalities, 369f
    plasma-treated sample, 368
  Mg K-alpha X-rays, 367
  N_2 plasma-treated MWNT buckypapers, 356
  oxygen incorporation, 367

Y

Yarn fiber production, CNT, 17
  challenge, 15–16
  CNT to CNT yarns/fibers, 334f
    applications, 341–345
    in composite fibers, 342f
    in photovoltaics, 343f, 344
    in sensors, 344f, 345
  by direct spinning, 18–19
    as-spun yarns, 19
    cohesive bundles, 19
    poor tensile performance, 19
  by direct synthesis, 19–20
    high-temperature process, 20
    SWCNT and DWCNT, 20
  by dispersion
    dispersion rheology, 18
    high-load or pure CNT yarn formation, 18
    routes, 17–18
  diversion into conventional fibers
    conventional bulk staple fibers, 17
    frictional interaction, 16

Yarn fiber production, CNT (*Continued*)
    tow of monofilaments, 16
    un-or very slightly twisted bundle, 16
  mechanical and physical properties, 22t
  on physical properties
    length and strength, 21
    Lucent patent, 21
    nanoscale properties, 21
    quantity and quality, 20–21
    twist-induced radial compression, 24–25
    twisting, 25
    yarn-like qualities, 21–24
  properties and manifestation, 14–15
  vision for space elevator, 15
Yarns, 212
  applications, 231–232
    for actuators, 232
    in bioengineering, 233
    CNT-based high-performance yarns, 234
    sensors, 232–233
    textile electrodes and supercapacitors, 232
    three-dimensional knitted scaffold, 233f
  Au-coated, 156
  Au-decorated, 156–158
  Au–CNT, 231, 231f
  CNT-based high-performance, 234
  diameter dependence, 405–406
    MWCNT length dependence, 405f
    for single-spun yarns, 406
    Yarn diameter dependence, 406, 406f
  multisliver, 225–226, 226f
  packing fibers
    CNT fibers in yarns, 220f
    CNT web/sliver, 219f
    for dry spinning process, 220–221
    loss of strength, 220
    modified spinning process, 221–222
    snarl of yarn, 221f
    tensile strength, 220
    in traditional textile, 219
  treatments for improvement, 225
    compaction, 225–226
    multiply twisted yarns, 227
    multisliver yarn based on Sirospun principle, 226f
    strength improvement, 226–227
    web heat treatment, 226
  and wire forming
    spinning multiple threads, 605–606
    spinning process, 604, 604f
    strong and conductive yarn development, 605f
    thread, 604–605

Z

Zigzag carbon nanotube, 4–6, 5f, 98–99
Zigzag nanotubes, 248
Zone-folding scheme, 523, 524f

Edwards Brothers Malloy
Ann Arbor MI. USA
October 31, 2013